Analysis and Thermomechanics

WALTER NOLL

Analysis and Thermomechanics

A Collection of Papers Dedicated to W. Noll on His Sixtieth Birthday

Invited by

B. D. Coleman, M. Feinberg, and J. Serrin

Reprinted from
Archive for Rational Mechanics and Analysis
edited by C. Truesdell and J. Serrin

With 24 Figures

Springer-Verlag
Berlin Heidelberg New York
London Paris Tokyo

Professor Dr. Bernard D. Coleman

Department of Mathematics, Carnegie-Mellon University
Pittsburgh, PA 15213, USA

Professor Dr. Martin Feinberg

Department of Chemical Engineering, The University of Rochester
Rochester, NY 14627, USA

Professor Dr. James Serrin

School of Mathematics, University of Minnesota
Minneapolis, MN 55455, USA

ISBN-13: 978-3-540-18125-5 e-ISBN-13: 978-3-642-61598-6
DOI: 10.1007/978-3-642-61598-6

Library of Congress Cataloging-in-Publication Data. Analysis and thermomechanics. "Reprinted from Archive for rational mechanics and analysis." 1. Mechanics, Analytic. 2. Mathematical physics. 3. Noll, W. (Walter), 1925-. I. Noll, W. (Walter), 1925-. II. Coleman, Bernard D. (Bernard David), 1930-. III. Feinberg, M. (Martin), 1942-. IV. Serrin, J. (James), 1926-. V. Truesdell, C. (Clifford), 1919-. VI. Archive for rational mechanics and analysis. QA807.5.A53 1987 531 87-23243

2153/3150-543210

Preface

This volume collects papers dedicated to *Walter Noll* on his sixtieth birthday, January 7, 1985. They first appeared in Volumes 86–97 (1984–1987) of the Archive for Rational Mechanics and Analysis. At the request of the Editors the list of authors to be invited was drawn up by B. D. Coleman, M. Feinberg, and J. Serrin.

Walter Noll's influence upon research into the foundations of mechanics and thermodynamics is plain, everywhere acknowledged. Less obvious is the wide effect his writings have exerted upon those who apply mechanics to special problems, but it is witnessed by the now frequent use of terms, concepts, and styles of argument he introduced, use sometimes by young engineers who have learnt them in some recent textbook and hence take them for granted, often with no idea whence they come. Examples are "objectivity", "material frame-indifference", "constitutive equation", "reduced form" of the last-named, "simple material", "simple solid", "simple fluid", "isotropy group", and the associated notations and lines of reasoning.

Noll is most famous for his penetration to the mathematical essence of physical principles and circumstances. On the one hand, he has perceived and illustrated the clarity that a successful axiomatic treatment can bring to an aspect of natural science. In Hilbert's words,

> [While] the creative power of pure thought is at work, the outside world asserts itself again; through the real phenomena it forces new questions upon us; it opens up new fields of mathematical science; and while we try to gain these new fields of science for the realm of pure thought, we often find the answers to old unsolved problems and so at the same time best further the old theories....
>
> Besides, it is wrong to think that rigor in proof is the enemy of simplicity. Numerous examples establish the opposite, that the rigorous method is also the simpler and the easier to grasp. The pursuit of rigor compels us to discover simpler arguments; also, often it clears the path to methods susceptible of more development than were the old, less rigorous ones....
>
> While I insist upon rigor in proofs as a requirement for a perfect solution of a problem, I should like, on the other hand, to oppose the opinion that only the concepts of analysis, or even those of arithmetic alone, are susceptible of a fully rigorous treatment. This opinion, occasionally advocated by eminent men, I consider entirely mistaken. Such a one-sided interpretation of the requirement of rigor would soon lead us to ignore all concepts that derive from geometry, mechanics, and physics, to shut off the flow of new material from the outside world, and finally, indeed, as a last consequence to reject the concepts of the continuum and of the irrational number. What an important, vital nerve would be cut, were we to root out geometry and mathematical physics! On the contrary, I think that wherever mathematical ideas

> come up, whether from the theory of knowledge or in geometry, or from the theories of natural science, the task is set for mathematics to investigate the principles underlying these ideas and establish them upon a simple and complete system of axioms in such a way that in exactness and in application to proof the new ideas shall be no whit inferior to the old arithmetical concepts.

Part of *Noll*'s work contributes directly to solution of Hilbert's Sixth Problem:

> *Mathematical Treatment of the Axioms of Physics*
>
> The researches into the foundations of geometry strongly suggest we undertake, *following that model, to treat axiomatically those disciplines of physics in which mathematics already plays a prominent role: these today are, first of all, probability theory and mechanics.*

Noll's studies respond also to Hilbert's more specific advice:

> If we do not succeed in solving a mathematical problem, it is often because we have failed to recognize the more general standpoint from which the problem before us appears only as a single link in a chain of related problems.... This way to find general methods is certainly the most practicable and the surest, for he who seeks for methods without having a definite problem in mind mainly seeks in vain.

Noll's training as an engineer has helped him heed another dictum of Hilbert:

> Indeed, a satisfactory treatment of the foundations of a science prerequires deep understanding of its special theories; only that architect is qualified to dispose securely the foundations of a structure who himself knows in detail all the functions it is to discharge.

For example, *Noll* was one of the first to see in Cauchy's proof of the existence and nature of the stress tensor a basic theorem of analysis, applicable whenever a statement of balance or inequality relates a volume integral to a surface integral. Among the responses to his researches on such matters are the papers in this volume by Šilhavý and by Gurtin, Williams & Ziemer.

Noll has shown equal zeal and capacity in dealing with old problems not yet solved, especially if imprecision of statement had left them vague. An outstanding example is his basic theorem regarding the phase rule; for a long time this achievement of his aroused little notice. The paper by Fosdick & Patino in this volume is one of several recent works which speak to that matter.

Another major part of *Noll*'s publication is seen in the papers Coleman & he wrote jointly in 1959–1964. Those did much to bring continuum mechanics into man's estate in our time, estate comparable to that which the work of Cauchy and others had given it for theirs, a century and more earlier.

Not only that, *Coleman & Noll*'s joint work and later researches by Coleman and others took up the thermomechanics of continua where Duhem had left it and made of it a mathematical science. The paper by Man, giving a particular dynamical status to one of Gibbs's major thermostatic tenets, is much in *Coleman & Noll*'s spirit.

The spread of topics treated by the works reprinted below – in pure mathematics, in analysis designed for application, in rational thermomechanics, in physics, in engineering – witnesses to the breadth of *Noll*'s influence and interests.

C. Truesdell

Contents

On the Failure of the Maximum Principle in Coupled Thermoelasticity

W. A. DAY

Dedicated to WALTER NOLL

Introduction

Our concern is with the conduction of heat in a deformable solid slab which is both homogeneous and isotropic. We take it that the slab occupies the region $0 \leqq x \leqq 1$, where x is a cartesian coordinate; thus the slab has unit (scaled) thickness and its faces are the planes $x = 0$ and $x = 1$. If the displacement vector remains parallel to the x-axis, and if the temperature $\theta(x, t)$ and the displacement $u(x, t)$ depend upon x and the time t only, the linear theory of thermoelasticity implies that $\theta(x, t)$ and $u(x, t)$ satisfy the energy equation

$$\frac{\partial^2 \theta}{\partial x^2} = \frac{\partial \theta}{\partial t} + \sqrt{a}\,\frac{\partial^2 u}{\partial x\,\partial t},$$

and the equation of motion

$$\frac{\partial^2 u}{\partial x^2} = \sqrt{a}\,\frac{\partial \theta}{\partial x} + b\,\frac{\partial^2 u}{\partial t^2}.$$

The derivation of these equations from the equations of thermoelasticity, for an account of which we refer to the *Handbuch* article of CARLSON [1], involves scaling the variables in appropriate ways; we shall forgo the details which are to be found in [2].

The constants a and b measure, respectively, the coupling between thermal and mechanical effects and the effect of inertia.

If we were to ignore coupling and set $a = 0$, and if we were to suppose the slab rigid, in the sense that $u(x, t) \equiv 0$, we should return to the classical theory of heat conduction, according to which the temperature is a solution of the heat equation

$$\frac{\partial^2 \theta}{\partial x^2} = \frac{\partial \theta}{\partial t}.$$

It scarcely need be said that the classical theory is highly successful and provides an account of heat conduction which is adequate for a great many practical purposes. Since this is so it might be thought unnecessary or even perverse, to study the implications for heat conduction of the more difficult thermoelastic theory. It seems to me, though, that that view is mistaken, for the study of heat conduction within the setting of linear thermoelasticity offers insight into the nature of thermomechanical coupling—a topic which is understood but poorly at the present time.

In what follows we shall make the quasi-static approximation, that is to say we shall set $b = 0$, thereby ignoring the effect of inertia and reducing the equation of motion to

$$\frac{\partial^2 u}{\partial x^2} = \sqrt{a}\,\frac{\partial \theta}{\partial x}.$$

For the sake of comparison with the classical theory, we shall suppose the faces of the slab to be clamped in an attempt to hold the lab as nearly rigid as can be, that is to say

$$u(0, t) = u(1, t) = 0.$$

If we integrate the reduced equation of motion with respect to x and take into account these conditions on the displacement we find that the strain is

$$\frac{\partial u}{\partial x}(x, t) = \sqrt{a}\,\theta(x, t) - \sqrt{a}\int_0^1 \theta(y, t)\,dy.$$

Upon differentiating with respect to t and substituting for $\partial^2 u/\partial x\,\partial t$ in the energy equation we conclude that the temperature is a solution not of the heat equation but of the integro-partial differential equation

$$\frac{\partial^2 \theta}{\partial x^2}(x, t) = (1 + a)\frac{\partial \theta}{\partial t}(x, t) - a\int_0^1 \frac{\partial \theta}{\partial t}(y, t)\,dy.$$

This equation reduces to the heat equation when $a = 0$; we shall suppose that the constant a is positive but we shall place no restriction upon how small it may be.

It is known that in certain circumstances the integro-partial differential equation provides better approximations to the temperature than does the heat equation [2]. A similar equation involving three spatial variables arises in the course of approximating to the temperature in a thermoelastic fluid [3].

The familiar Maximum Principle embodies a qualitative feature of the heat equation which is extremely important and has had ramifications in many directions. In their treatise on Maximum Principles in Differential Equations [7] PROTTER & WEINBERGER attribute the Principle to E. E. LEVI [5, 6].

One version of the Principle asserts that, if the faces of the slab are maintained at zero temperature and the initial temperature is nonnegative, the temperature is nonnegative everywhere at every subsequent instant. In more formal terms:

If $\theta(x, t)$ is a solution of the heat equation which is continuous in $0 \leqq x \leqq 1$, $t \geqq 0$ and satisfies the boundary conditions

$$\theta(0, t) = \theta(1, t) = 0, \quad t \geqq 0,$$

and the initial condition

$$\theta(x, 0) = f(x), \quad 0 \leqq x \leqq 1,$$

where $f(0) = f(1) = 0$ and $f(x) \geqq 0$ in $0 < x < 1$ then $\theta(x, t) \geqq 0$ throughout $0 \leqq x \leqq 1, \ t \geqq 0$.

In this instance the temperature can be represented by the integral

$$\theta(x, t) = \int_0^1 G(x, y, t) f(y)\, dy,$$

in which the kernel

$$G(x, y, t) = 2 \sum_{n=1}^{\infty} \sin n\pi x \cdot \sin n\pi y \, e^{-n^2\pi^2 t}.$$

Jacobi's imaginary transformation in the theory of theta functions, or alternatively an appeal to Poisson's summation formula, enables us to express the kernel in the alternative form

$$\frac{1}{2\sqrt{\pi t}} \sum_{n=-\infty}^{\infty} \left(e^{-(x-y+2n)^2/4t} - e^{-(x+y+2n)^2/4t}\right).$$

Thus $G(x, y, t) > 0$ if $0 < x, y < 1, \ t > 0$ and this fact makes it plain that $\theta(x, t) \geqq 0$ if $f(x) \geqq 0$.

If we turn to the coupled theory, in which the temperature is a solution of the integro-partial differential equation and the constant a is a small positive number, we expect that the presence of coupling will make only small quantitative changes to the temperature. We may ask, however, if the qualitative features, namely the nonnegative properties of the temperature and of the associated kernel, are preserved. My purpose is to point out that they are not preserved and are peculiar to the classical theory; any amount of coupling, no matter how small, is sufficient to obliterate them.

As we shall see, and this is the content of our Theorem 1, the nonnegative property of the temperature fails in quite a striking way for it is possible to have $f(x) > 0$ throughout the interior $0 < x < 1$ and, at the same time, to have $\theta(x, t) < 0$ throughout $0 < x < 1, \ t > t_0$, where t_0 is some appropriately large positive number. An example of such an initial temperature $f(x)$ is provided by a function which is approximately a delta function and whose graph has a high but thin positive peak at a point $x = \xi$ just inside one of the boundaries $x = 0$ or $x = 1$.

This result leads us to conjecture that the kernel $K(x, y, t)$ which corresponds, when $a > 0$, to the classical $G(x, y, t)$ must be negative at some points of the unit square $0 \leqq x, y \leqq 1$, at least if t is sufficiently large. Theorem 2 tells us that the conjecture is correct but underestimates the true state of affairs for the kernel must have changes of sign in the unit square for each $t > 0$ whether large or not.

Once we have established that the kernel changes sign we shall be in position to prove, as we do in Theorem 3, that it is not necessary to wait a long time in order to violate the conclusion of the Maximum Principle. For, given any $t_0 > 0$ no matter how small we can choose an initial temperature $f(x)$, which is positive in $0 < x < 1$, in such a way that the mean temperature

$$\int_0^1 \theta(x, t_0)\, dx$$

is negative. Here the required initial distribution of temperature is similar to that in Theorem 1, namely one which approximates a delta function with its peak just inside one face.

When we come to prove Theorems 1 and 3 we shall appeal to a Lemma depending upon arguments, which we omit, of a kind which occur everywhere in the theory of generalized functions.

Lemma. *Let $g(x)$ be continuous in $0 \leqq x \leqq 1$ and let ξ be any number in $0 < \xi < 1$. Then there is a sequence of functions $\varphi_n(x)$ $(n = 1, 2, 3, \ldots)$ with the properties*

(i) *each $\varphi_n(x)$ has derivatives of all orders in $0 \leqq x \leqq 1$;*

(ii) *$\varphi_n(0) = \varphi_n(1) = 0$ and $\varphi_n(x) > 0$ in $0 < x < 1$;*

(iii) $\int_0^1 \varphi_n\, dx = 1;$

(iv) $\int_0^1 \varphi_n g\, dx \to g(\xi)$ as $n \to \infty$.

A temperature eventually negative everywhere

We begin by establishing

Theorem 1. *There is a solution $\theta(x, t)$ of the integro-partial differential equation which is continuous in $0 \leqq x \leqq 1$, $t \geqq 0$ and satisfies the boundary conditions*

$$\theta(0, t) = \theta(1, t) = 0, \quad t \geqq 0,$$

and the initial condition

$$\theta(x, 0) = f(x), \quad 0 \leqq x \leqq 1,$$

whese $f(0) = f(1) = 0$ and $f(x) > 0$ in $0 < x < 1$, but $\theta(x, t) < 0$ throughout $0 < x < 1$, $t > t_0$ for some $t_0 > 0$.

We shall construct $\theta(x, t)$ explicitly by using separation of variables and generalized Fourier analysis. To this end we introduce the inner product

$$(g, h) = \int_0^1 gh\, dx - \frac{a}{1+a} \int_0^1 g\, dx \cdot \int_0^1 h\, dx$$

of functions $g(x)$, $h(x)$, ... which are continuous in the closed unit interval. Since, as the Schwarz inequality shows,

$$\frac{1}{1+a}\int_0^1 g^2\,dx \leqq (g,g) \leqq \int_0^1 g^2\,dx$$

the corresponding norm $\|g\| = \sqrt{(g,g)}$ is equivalent to the norm in the Lebesgue space $L^2[0,1]$.

The integro-partial differential equation has separable solutions

$$s_n(x)\,e^{-\lambda_n^2 t/(1+a)}, \qquad n = 1, 2, 3, \ldots,$$

in which the functions $s_n(x)$ and the numbers λ_n are determined by the eigenvalue problem

$$s_n''(x) = -\lambda_n^2\left(s_n(x) - \frac{a}{1+a}\int_0^1 s_n(y)\,dy\right), \qquad s_n(0) = s_n(1) = 0.$$

We suppose the eigenfunctions to be normalized by the requirement $\|s_n\| = 1$.

It is easily checked that the numbers λ_n, arranged in order of increasing magnitude, are the positive roots of the transcendental equation

$$\lambda \sin\lambda + 2a(1 - \cos\lambda) = 0.$$

Since a is a small positive constant we see that if n is odd λ_n is close to, and slightly greater than, $n\pi$, while if n is even $\lambda_n = n\pi$ *exactly*.

The corresponding eigenfunctions are, if n is odd,

$$s_n(x) = (\sin\lambda_n x + \sin\lambda_n(1-x) - \sin\lambda_n)/D_n,$$

where the denominator is

$$D_n = (1 - \cos\lambda_n)^{\frac{1}{2}}\left(1 - \frac{1}{\lambda_n}\sin\lambda_n\right)^{\frac{1}{2}},$$

and

$$s_n(x) = \sqrt{2}\sin n\pi x$$

if n is even.

The eigenfunctions are mutually orthogonal, in the sense that $(s_m, s_n) = 0$ if $m \neq n$, and they can be given the usual variational characterizations. Thus $s_1(x)$ minimizes the functional $\int_0^1 g'^2\,dx$ among all functions $g(x)$ which are continuous and piecewise continuously differentiable in the closed unit interval and satisfy $g(0) = g(1) = 0$ and $\|g\| = 1$. Each subsequent $s_n(x)$ minimizes the same functional among all $g(x)$ which satisfy the orthogonality requirements $(g, s_1) = \ldots = (g, s_{n-1}) = 0$ in addition.

With the aid of standard considerations of the Calculus of Variations* we can conclude that the initial temperature $f(x)$ can be developed as a Fourier series of

* It is necessary to make only minor modifications to the arguments of H. Lewy's lectures [4, Chapter II, Sections 5–10].

eigenfunctions provided we can be assured first that the eigenfunctions are uniformly bounded and second that the series $\sum_{n=1}^{\infty} \lambda_n^{-2}$ converges.

The first is true because $\sin \lambda_n < 0$ and $\cos \lambda_n < 0$ when n is odd and, therefore, the denominator $D_n > 1$. Thus $|s_n(x)| \leqq 3$ whether n is odd or even. The second is true because $\lambda_n \geqq n\pi$ for every n. Accordingly, we can prove, for example, that if $f(x)$ is twice continuously differentiable in $0 \leqq x \leqq 1$ and $f(0) = f(1) = 0$ then

$$f(x) = \sum_{n=1}^{\infty} (f, s_n)\, s_n(x), \quad 0 \leqq x \leqq 1,$$

the series converging uniformly and absolutely.

For such a function the Fourier coefficients satisfy an estimate $|(f, s_n)| \leqq A n^{-2}$, where A depends upon $f(x)$ and a only. To prove this we note that

$$\begin{aligned}(f, s_n) &= \int_0^1 f s_n \, dx - \frac{a}{1+a} \int_0^1 f \, dx \cdot \int_0^1 s_n \, dx \\ &= \int_0^1 f(x) \left(s_n(x) - \frac{a}{1+a} \int_0^1 s_n(y) \, dy \right) dx \\ &= -\frac{1}{\lambda_n^2} \int_0^1 f s_n'' \, dx\end{aligned}$$

and, when we integrate by parts twice, we deduce that

$$(f, s_n) = -\frac{1}{\lambda_n^2} \int_0^1 f'' s_n \, dx.$$

On appealing to the Schwarz inequality and to the inequalities $\lambda_n \geqq n\pi$ and

$$\int_0^1 s_n^2 \, dx \leqq (1 + a) \, \|s_n\|^2 = 1 + a$$

we obtain

$$|(f, s_n)| \leqq (n\pi)^{-2} \sqrt{(1 + a) \int_0^1 f''^2 \, dx},$$

as required.

Next we set

$$\theta(x, t) = \sum_{n=1}^{\infty} (f, s_n)\, s_n(x)\, e^{-\lambda_n^2 t/(1+a)}.$$

In view of the estimates on the Fourier coefficients and the fact that the eigenfunctions are uniformly bounded, this series converges uniformly and absolutely in $0 \leqq x \leqq 1$, $t \geqq 0$ and defines a continuous function $\theta(x, t)$ which has derivatives of all orders in $0 \leqq x \leqq 1$, $t > 0$ and satisfies the integro-partial differential equation, the boundary conditions

$$\theta(0, t) = \theta(1, t) = 0, \quad t \geqq 0,$$

and the initial condition

$$\theta(x, 0) = f(x), \quad 0 \leqq x \leqq 1.$$

In order to arrive at our conclusion that $\theta(x, t)$ can become negative even if $f(x)$ is positive in $0 < x < 1$ we need to examine the eigenfunctions more closely.

We turn first to $s_1(x)$, which we can write as

$$\frac{4}{D_1} \sin \tfrac{1}{2}\lambda_1 \cdot \sin \tfrac{1}{2}\lambda_1 x \cdot \sin \tfrac{1}{2}\lambda_1 (1 - x).$$

Because $\sin \frac{1}{2}\lambda_1 x$ is a concave function of x in $0 \leqq x \leqq 1$, we have

$$\sin \tfrac{1}{2}\lambda_1 x \geqq (\sin \tfrac{1}{2}\lambda_1)\, x, \quad \sin \tfrac{1}{2}\lambda_1 (1 - x) \geqq (\sin \tfrac{1}{2}\lambda_1)(1 - x)$$

and, therefore,

$$s_1(x) \geqq \frac{4}{D_1} (\sin \tfrac{1}{2}\lambda_1)^3 \cdot x(1 - x).$$

This inequality confirms incidentally that the first eigenfunction vanishes at $x = 0$ and $x = 1$ only, as we know to be the case from general considerations.

Next we establish that all subsequent eigenfunctions satisfy

$$|s_n(x)| \leqq B(n + 1)\, s_1(x),$$

where B is a constant depending upon a only. To prove this when n is odd we express $s_n(x)$ in the form

$$\frac{\lambda_n}{D_n} (1 - x) \int_0^x (\cos \lambda_n y - \cos \lambda_n (1 - y))\, dy - \frac{\lambda_n}{D_n} x \int_x^1 (\cos \lambda_n y - \cos \lambda_n (1 - y))\, dy.$$

Since $D_n > 1$ and $\lambda_n < (n + 1)\pi$ we have

$$|s_n(x)| \leqq 4(n + 1)\pi x(1 - x) \leqq \frac{\pi D_1 (n + 1)}{(\sin \frac{1}{2}\lambda_1)^3} s_1(x).$$

On the other hand,

$$s_n(x) = \sqrt{2}\, n\pi (1 - x) \int_0^x \cos n\pi y\, dy - \sqrt{2}\, n\pi x \int_x^1 \cos n\pi y\, dy$$

when n is even and, therefore,

$$|s_n(x)| \leqq 2^{3/2} n\pi x(1 - x) \leqq \frac{\pi D_1 n}{\sqrt{2}\, (\sin \frac{1}{2}\lambda_1)^3} s_1(x).$$

Thus every subsequent eigenfunction satisfies the required estimate, with $B = \pi D_1 (\sin \frac{1}{2}\lambda_1)^{-3}$.

When t is large it is the leading term which dominates the series expansion for $\theta(x, t)$. We can estimate the sum of the remaining terms as

$$\left| \sum_{n=2}^{\infty} (f, s_n)\, s_n(x)\, e^{-\lambda_n^2 t/(1+a)} \right| \leq \sum_{n=2}^{\infty} \frac{A}{n^2} \cdot B(n+1)\, s_1(x)\, e^{-\lambda_n^2 t/(1+a)}$$

$$\leq ABs_1(x)\, e^{-\lambda_1^2 t/(1+a)} \sum_{n=2}^{\infty} e^{(\lambda_1^2 - \lambda_n^2)t/(1+a)},$$

where

$$\sum_{n=2}^{\infty} e^{(\lambda_1^2 - \lambda_n^2)t/(1+a)} \to 0 \quad \text{as} \quad t \to \infty,$$

and, therefore,

$$\theta(x, t) = [(f, s_1) + o(1)]\, s_1(x)\, e^{-\lambda_1^2 t/(1+a)},$$

the term $o(1)$ being uniform with respect to x in the unit interval.

Our assertion about the qualitative behaviour of $\theta(x, t)$ rests upon the fact, which is curious at first sight, that even though $s_1(x)$ is positive in $0 < x < 1$ we can choose $f(x)$ in such a way that it too is positive in $0 < x < 1$ but the Fourier coefficient (f, s_1) is nonetheless negative.

For, if we confine our attention to functions $f(x)$ with $\int_0^1 f\,dx = 1$, the coefficient is

$$(f, s_1) = \int_0^1 fs_1\,dx - \frac{a}{1+a} \int_0^1 f\,dx \cdot \int_0^1 s_1\,dx$$

$$= \int_0^1 fs_1\,dx - \frac{a}{1+a} \int_0^1 s_1\,dx,$$

where, as direct calculation shows,

$$\frac{a}{1+a} \int_0^1 s_1\,dx = \frac{|\sin \lambda_1|}{D_1} > 0.$$

However, $s_1(x)$ vanishes at $x = 0$ and, therefore, there is ξ in $0 < \xi < 1$ such that

$$s_1(\xi) < \frac{|\sin \lambda_1|}{D_1}.$$

If now we appeal to the Lemma, with $g(x) = s_1(x)$, we see that the sequence of functions $\varphi_n(x)$, whose existence is asserted by the Lemma, has the property that

$$\int_0^1 \varphi_n s_1\,dx \to s_1(\xi)$$

and, therefore,

$$(\varphi_n, s_1) \to s_1(\xi) - \frac{|\sin \lambda_1|}{D_1} < 0 \quad \text{as} \quad n \to \infty.$$

Thus if we choose $f(x) = \varphi_n(x)$ for some large n we shall have $(f, s_1) < 0$. With $f(x)$ chosen in this way it follows that $\theta(x, t) < 0$ in $0 < x < 1$, $t > t_0$ if t_0 is sufficiently large, and the proof is complete.

Behaviour of the kernel

Let us define the kernel $K(x, y, t)$ to be

$$\sum_{n\,\mathrm{odd}} \frac{1}{D_n^2} (\sin \lambda_n x + \sin \lambda_n(1-x) - \sin \lambda_n)(\sin \lambda_n y + \sin \lambda_n(1-y))\, e^{-\lambda_n^2 t/(1+a)}$$

$$+ 2 \sum_{n\,\mathrm{even}} \sin n\pi x \cdot \sin n\pi y\, e^{-n^2\pi^2 t/(1+a)},$$

where the first sum is taken over all odd positive integers and the second over all even positive integers. The series converge uniformly and absolutely in $0 \leqq x, y \leqq 1$, $t \geqq \delta$ for each $\delta > 0$ and define a kernel which is continuous in $0 \leqq x, y \leqq 1$, $t > 0$. The kernel reduces to the classical $G(x, y, t)$ when $a = 0$.

Since, as we can readily verify,

$$\frac{a}{1+a} \int_0^1 s_n\, dx = -\frac{\sin \lambda_n}{D_n}$$

if n is odd, and

$$\int_0^1 s_n\, dx = 0$$

if n is even, the kernel equals

$$\sum_{n=1}^{\infty} s_n(x) \left(s_n(y) - \frac{a}{1+a} \int_0^1 s_n(z)\, dz \right) e^{-\lambda_n^2 t/(1+a)}.$$

Now let $f(x)$ be twice continuously differentiable in $0 \leqq x \leqq 1$, let $f(0) = f(1) = 0$ and let $\theta(x, t)$ be defined as before, that is as

$$\sum_1^{\infty} (f, s_n)\, s_n(x)\, e^{-\lambda_n^2 t/(1+a)}.$$

On multiplying the expression just derived for the kernel by $f(y)$ and integrating term-by-term with respect to y, as the uniform nature of the convergence permits, we obtain the representation

$$\theta(x, t) = \int_0^1 K(x, y, t) f(y)\, dy$$

for the temperature.

In order to prove that the kernel must change sign we introduce the sum of positive terms

$$k(t) = 4(1+a) \sum_{n\,\mathrm{odd}} \left(\frac{1 - \cos \lambda_n}{D_n \lambda_n} \right)^2 e^{-\lambda_n^2 t/(1+a)}, \qquad t \geqq 0.$$

Theorem 2. *Let $t > 0$ be arbitrary. Then there are points of the unit square $0 \leqq x, y \leqq 1$ at which $K(x, y, t) \geqq k(t)$ and there are points at which $K(x, y, t) \leqq -ak(t)$.*

The desired conclusions follow from the equation

$$\int_0^1 \int_0^1 K(x, y, t)\, dx\, dy = \sum_{n \text{ odd}} \frac{1}{D_n^2}\left(\frac{2}{\lambda_n}(1 - \cos \lambda_n) - \sin \lambda_n\right) \times \frac{2}{\lambda_n}(1 - \cos \lambda_n)\, e^{-\lambda_n^2 t/(1+a)}$$

and the equation

$$\int_0^1 K(x, 0, t)\, dx = \sum_{n \text{ odd}} \frac{1}{D_n^2}\left(\frac{2}{\lambda_n}(1 - \cos \lambda_n) - \sin \lambda_n\right) \sin \lambda_n\, e^{-\lambda_n^2 t/(1+a)},$$

for, on replacing $\sin \lambda_n$ by $-2a(1 - \cos \lambda_n)/\lambda_n$, we arrive at the formulae

$$\int_0^1 \int_0^1 K(x, y, t)\, dx\, dy = k(t),$$

$$\int_0^1 K(x, 0, t)\, dx = -ak(t),$$

and thus the Theorem is correct.

It should be noted that the points of the unit square whose existence has just been established may well be different for different values of t.

It is easily checked that

$$K(x, 0, t) = -a \int_0^1 K(x, y, t)\, dy.$$

When a is small the right-hand side is approximately

$$-a \int_0^1 G(x, y, t)\, dy$$

which is negative whenever $0 < x < 1$ and $t > 0$. It is almost certainly the case that $K(x, 0, t) < 0$ for $0 < x < 1$, $t > 0$ and every $a > 0$ but I have not been able to establish this.

Negative mean temperature at an arbitrary time

We show finally that, given any $t_0 > 0$ no matter how small, we can choose an initial temperature $f(x)$, which is positive in $0 < x < 1$, in such a way that the mean temperature at the time t_0 is negative; the choice of the initial temperature will vary as t_0 varies.

Theorem 3. *Let* $t_0 > 0$ *be arbitrary and let* ε *be any number in* $0 < \varepsilon < k(t_0)$. *Then there is a solution* $\theta(x, t)$ *of the integro-partial differential equation which is continuous in* $0 \leqq x \leqq 1$, $t \geqq 0$ *and satisfies the boundary conditions*

$$\theta(0, t) = \theta(1, t) = 0, \quad t \geqq 0,$$

and the initial condition

$$\theta(x, 0) = f(x), \quad 0 \leqq x \leqq 1,$$

where $f(0) = f(1) = 0$ *and* $f(x) > 0$ *in* $0 < x < 1$, *but*

$$\int_0^1 \theta(x, t_0)\, dx < -a\varepsilon \int_0^1 f(x)\, dx.$$

Consider

$$H(y, t) = \int_0^1 K(x, y, t)\, dx,$$

which is continuous in $0 \leqq y \leqq 1$, $t > 0$. If we integrate with respect to x each side of the equation

$$\theta(x, t) = \int_0^1 K(x, y, t) f(y)\, dy,$$

in which $f(x)$ has yet to be selected, and interchange the orders of integration on the right-hand side we find the mean temperature to be

$$\int_0^1 \theta(x, t)\, dx = \int_0^1 H(y, t) f(y)\, dy.$$

We discovered in the course of proving Theorem 2 that $H(0, t) = -ak(t) < 0$. Thus, given any $t_0 > 0$, we can choose a number ξ in $0 < \xi < 1$, depending upon t_0 and ε, in such a way that $H(\xi, t_0) < -a\varepsilon$. On appealing to the Lemma, with $g(x) = H(x, t_0)$, we deduce that

$$\int_0^1 H(y, t_0)\, \varphi_n(y)\, dy \to H(\xi, t_0)$$

as $n \to \infty$ and, therefore, by choosing $f(x) = \varphi_n(x)$ for some large n, we can arrange that

$$\int_0^1 H(y, t_0) f(y)\, dy < -a\varepsilon = -a\varepsilon \int_0^1 f(x)\, dx,$$

that is to say

$$\int_0^1 \theta(x, t_0)\, dx < -a\varepsilon \int_0^1 f(x)\, dx.$$

References

1. CARLSON, D. E., Linear Thermoelasticity. Handbuch der Physik. Bd. VI a/2. Springer-Verlag, Berlin, Heidelberg, New York (1972).
2. DAY, W. A., A comment on approximations to the temperature in dynamic linear thermoelasticity. Arch. Rational Mech. Anal. To appear.
3. DAY, W. A., Approximations to the temperature in a heated thermoelastic fluid. Mathematika. To appear.

4. Green, J. W., Aspects of the Calculus of Variations. Notes after lectures by Hans Lewy. University of California Press. Berkeley (1939).
5. Levi, E. E., Sull'equazione del calore. Reale Accad. dei Lincei, Roma, Rendiconti (5) **162**, 450–456 (1907).
6. Levi, E. E., Sull'equazione del calore. Annali di Mat. Pura ed. Appl. **14**, 187–264 (1908).
7. Protter, M. H., & H. F. Weinberger, Maximum Principles in Differential Equations, Prentice-Hall, Englewood Cliffs, N. J. (1967).

Hertford College,
Oxford

(Received December 28, 1983)

Decay Estimates for a Class of Second-Order Quasilinear Equations in Three Dimensions

C. O. HORGAN & L. E. PAYNE

Dedicated to Walter Noll on the occasion of his 60th birthday

Introduction

In a recent paper [1] the authors made use of differential inequality techniques to derive exponential decay estimates for a class of second-order quasilinear equations (not necessarily elliptic) in the plane. Specifically they were concerned with solutions of equations in divergence form defined in a semi-infinite strip, with non-zero boundary data on one end only. Decay estimates in various norms were derived.

The methods of [1] do not, however, carry over to equations in more than two dimensions. In the present paper we derive analogous results in $\mathbb{R}^3$, but in this case the assumptions on the form of the nonlinearity are slightly more restrictive and in some cases the decay rate is somewhat less sharp. We compute explicit decay bounds in both the energy and L_2 norms.

The spatial decay estimates of concern in [1] and in the present paper are of interest in connection with studies on Saint-Venant's principle in elasticity theory. (See [2] for a review of recent results on Saint-Venant's principle.) The results may also be viewed as giving rise to theorems of Phragmén-Lindelöf type (see *e.g.* the references cited in [1]).

We are specifically concerned with classical solutions of the second-order quasilinear equation in three independent variables

$$[\varrho(\boldsymbol{x}, u, \nabla u)\, u_{,j}]_{,j} = 0, \qquad j = 1, 2, 3, \tag{1.1}$$

where the usual summation convention is employed with subscripts preceded by a comma denoting partial differentiation with respect to the corresponding cartesian coordinate. Our attention is focused on a Dirichlet problem for (1.1) on the semi-infinite cylinder $R = \{(x_1, x_2, x_3) \mid (x_1, x_2) \in S, x_3 > 0\}$, where S denotes the open simply-connected cross-section of R. We assume the existence of solutions $u \in C^2(R) \cap C^1(\bar{R})$ satisfying (1.1) in R subject to the boundary conditions

$$u(x_1, x_2, x_3) = 0, \quad (x_1, x_2) \in \partial S, \quad x_3 > 0, \tag{1.2}$$

$$u,\ \partial u/\partial x_3 \to 0, \quad \text{(uniformly in } x_1, x_2) \text{ as } x_3 \to \infty, \tag{1.3}$$

$$u(x_1, x_2, 0) = f(x_1, x_2), \quad (x_1, x_2) \in S, \tag{1.4}$$

where the prescribed function f is assumed to be sufficiently smooth and $f = 0$ for $(x_1, x_2) \in \partial S$.

As in the previous paper [1] it would be possible to obtain the estimates of this paper without making the strong *a priori* assumption (1.3) and without insisting on strong solutions, but in the interests of simplicity we avoid this generalization.

The function ϱ which appears in (1.1) is assumed to satisfy, for all solutions of (1.1)–(1.4), one or the other of the following assumptions for positive constants m_i, M_i and non-negative constants K_i:

$$\textit{Case } 1\colon \quad 0 < m_1 \leqq \varrho \leqq M_1 + K_1[q^2\varrho]^{\frac{1}{p}} \text{ for some } p > 1, \tag{1.5}$$

or

$$\textit{Case } 2\colon \quad 0 < m_2 \leqq \varrho^{-1} \leqq M_2 + K_2 q^2 \varrho. \tag{1.6}$$

Here we have introduced the notation $q^2 = |\nabla u|^2$. In [1], Case 2 had the same form as (1.6) while Case 1 was assumed to hold for $p = 1$. As we pointed out in [1], the hypotheses (1.5) or (1.6) provide some restrictions on the behavior of ϱ for small and large values of q^2, without which one would not expect exponential decay of solutions of (1.1)–(1.4). An example of a function ϱ for which (1.5) holds is $\varrho = 1 + q^2$. In this case, we may take $m_1 = 1$, $M_1 = 1$, $K_1 = 1$ and $p = 2$. (Of course, smaller values of p may also be chosen.) As an example of a function ϱ satisfying (1.6), consider $\varrho = (1 + q^2)^{-\frac{1}{2}}$, in which case (1.1) is the minimal surface equation

$$(1 + u_{,i}u_{,i})\, u_{,kk} - u_{,i}u_{,j}u_{,ij} = 0. \tag{1.7}$$

In this case, (1.6) is satisfied with $m_2 = 1$, $M_2 = 1$ and $K_2 = 1$. As was observed in [1], we note that neither (1.5) nor (1.6) implies the ellipticity of (1.1).

For functions ϱ satisfying (1.5) or (1.6) we first show that the s-energy

$$E_s(z) = \int_{R_z} |u|^s \varrho q^2 \, dV, \quad s \geqq 0, \tag{1.8}$$

contained in the subdomain $R_z = \{(x_1, x_2, x_3) \mid (x_1, x_2) \in S, 0 \leqq z < x_3 < \infty\}$ has exponential decay in z. From this we then obtain explicit exponential decay estimates for L_2-norms of u. In Case 1, where (1.5) is satisfied, it turns out that the latter results follow from a basic decay estimate for the usual energy $E_0(z) \equiv E(z)$ while in Case 2, where (1.6) holds, we require estimates for both $E(z)$ and $E_2(z)$.

The energy decay estimate established in Section 2 for Case 1 (Theorem 1) leads to a result of the form

$$E(z) \leqq Q_1 e^{-2k_1 z\{1-\varepsilon(z)\}}, \quad z > 0, \tag{1.9}$$

where the constant Q_1 depends on the total energy $E(0)$ contained in R, the constants m_1, M_1, K_1 and p of (1.5) and on the properties of the cross-section S. The estimated decay rate k_1 in (1.9) is given by

$$k_1 = [m_1/M_1]^{\frac{1}{2}} \nu_2, \tag{1.10}$$

and $\varepsilon(z)$ tends to zero like $z^{-1/p}$ as z tends to infinity. Here ν_2 denotes the smallest fixed membrane eigenvalue for S, that is,

$$\phi_{,\alpha\alpha} + \nu_2^2 \phi = 0 \quad \text{on } S, \quad (\alpha = 1, 2), \tag{1.11}$$

$$\phi = 0 \quad \text{on } \partial S, \tag{1.12}$$

or equivalently,

$$\nu_2^2 = \inf_{v(x_\alpha)\in H_1^0(S)} \frac{\int_S v_{,\alpha} v_{,\alpha} dA}{\int_S v^2 \, dA}. \tag{1.13}$$

For the special case in which $\varrho \equiv$ constant in R, (1.1) reduces to Laplace's equation and we may choose $m_1 = M_1$, $K_1 = 0$ in (1.5). The estimate (2.1) of Theorem 1 (and the result (1.9)) now read

$$E(z) \leqq E(0)\, e^{-2\nu_2 z}, \quad z \geqq 0, \tag{1.14}$$

which is a well-known result for Laplace's equation [2]. The decay rate is optimal in this case [2]. Since from (1.5) we have $m_1 \leqq M_1$, the estimated decay rate k_1 given by (1.10) in Case 1 for the general quasilinear equation (1.1) is never faster than that for Laplace's equation. Whenever $m_1 = M_1$, the estimated decay rate is at least as fast as that for Laplace's equation. (Note that the estimated decay rate in (1.9) provides a *lower bound* for an actual rate of decay.)

In Section 3, the decay estimates are established in Case 2. For this case, somewhat less sharp results are obtained.

2. Decay Estimates in Case 1

2.1. An Energy Estimate. In this sub-section, we first establish the following theorem:

Theorem 1. *Let* $u \in C^2(R) \cap C^1(\bar{R})$ *be a solution of* (1.1)–(1.4) *where* ϱ *satisfies* (1.5). *Then for* $z > 0$,

$$E(z) \leqq E(0) \exp\left\{-2\nu_2 \left(\frac{m_1}{M_1}\right)^{\frac{1}{2}} z \left[1 - \frac{\nu_2^2 K_1}{2\nu_\sigma^2 M_1} \left(\frac{E(0)}{z}\right)^{\frac{1}{p}}\right]\right\}. \tag{2.1}$$

where ν_2 *is defined in* (1.13) *and* $\sigma = 2p/(p-1)$. *Here* ν_σ *denotes the Sobolev constant*

$$\nu_\sigma = \inf_{v \in \overset{\circ}{H}_1(S)} \frac{\left[\int_S v_{,\alpha} v_{,\alpha}\, dA\right]^{\frac{1}{2}}}{\left[\int_S |v|^\sigma\, dA\right]^{\frac{1}{\sigma}}}, \qquad \sigma > 1. \tag{2.2}$$

Proof. The result (2.1) is established conveniently in two stages. First, we derive the differential inequality

$$E'(z) + k_1(z,p)\, E(z) \leqq 0, \qquad z > 0, \tag{2.3}$$ [1]

with

$$k_1(z,p) = 2(m_1)^{\frac{1}{2}}\left[M_1 \nu_2^{-2} + K_1\left(\int_{S_z} \varrho q^2\, dA\right)^{\frac{1}{p}} \nu_\sigma^{-2}\right]^{-\frac{1}{2}}, \tag{2.4}$$

where $\sigma = 2p/(p-1)$. Here S_z denotes the interior of the cross-section $(x_1, x_2) \in S$ at $x_3 = z$. The inequality (2.3) may be immediately integrated to yield

$$E(z) \leqq E(0) \exp\left\{-\int_0^z k_1(\eta, p)\, d\eta\right\}. \tag{2.5}$$

The second stage of the proof consists of using the hypothesis (1.5) to bound the exponential in (2.5) leading to the result (2.1).

The derivation of (2.3)–(2.4) proceeds as follows: Use of the divergence theorem and (1.2), (1.3) shows that

$$E(z) = -\int_{S_z} \varrho u u_{,3}\, dA. \tag{2.6}$$

Direct differentiation of (1.8) (with $s = 0$) gives

$$E'(z) = -\int_{S_z} \varrho\, |\nabla u|^2\, dA, \tag{2.7}$$

and so we have, for any function $k_1(z,p)$,

$$E'(z) + k_1(z,p)\, E(z) = -\int_{S_z} [\varrho\, |\nabla u|^2 + k_1(z,p)\, \varrho u u_{,3}]\, dA. \tag{2.8}$$

Our object now is to show that the right hand side of (2.8) is negative for the choice of the (positive) function $k_1(z,p)$ given in (2.4). We now investigate the last integral in (2.8) (noting that $k_1(z,p)$ does not depend on x_1, x_2) and using Schwarz's inequality obtain

$$\int_{S_z} \varrho u u_{,3}\, dA \leqq \left[\int_{S_z} \varrho u^2\, dA \int_{S_z} \varrho u_{,3}^2\, dA\right]^{\frac{1}{2}}. \tag{2.9}$$

But from (1.5) it follows that

$$\int_{S_z} \varrho u^2\, dA \leqq M_1 \int_{S_z} u^2\, dA + K_1 \int_{S_z} (\varrho q^2)^{\frac{1}{p}} u^2\, dA. \tag{2.10}$$

---

[1] The prime denotes differentiation with respect to z.

Applying (1.13), (2.2) with $\sigma = 2p/(p-1)$, and Hölder's inequality (recall from (1.5) that $p > 1$) we are led to

$$\int_{S_z} \varrho u^2 \, dA \leqq \frac{M_1}{\nu_2^2} \int_{S_z} u_{,\alpha} u_{,\alpha} \, dA + K_1 \left[\int_{S_z} \varrho q^2 \, dA \right]^{\frac{1}{p}} \left[\int_{S_z} |u|^\sigma \, dA \right]^{\frac{2}{\sigma}}$$

$$\leqq \left\{ \frac{M_1}{\nu_2^2} + \frac{K_1}{\nu_\sigma^2} \left[\int_{S_z} \varrho q^2 \, dA \right]^{\frac{1}{p}} \right\} \int_{S_z} u_{,\alpha} u_{,\alpha} \, dA \tag{2.11}$$

$$\leqq 4 \, [k_1(z, p)]^{-2} \int_{S_z} \varrho u_{,\alpha} u_{,\alpha} \, dA,$$

where the left hand inequality in (1.5) has been used in the last step in (2.11) and $k_1(z, p)$ is given in (2.4). The insertion of (2.11) into (2.9) and the resulting expression into (2.8) yields

$$E'(z) + k_1(z, p) \, E(z) \leqq - \int_{S_z} \varrho \, |\nabla u|^2 \, dA + 2 \left\{ \int_{S_z} \varrho u_{,\alpha} u_{,\alpha} \, dA \int_{S_z} \varrho u_{,3}^2 \, dA \right\}^{\frac{1}{2}}, \tag{2.12}$$

which leads to directly (2.3), and hence to (2.5).

We now proceed to obtain an upper bound for the exponential term in (2.5). From (2.4) it follows that $k_1(z, p)$ may be written as

$$k_1(z, p) = 2\nu_2 \left(\frac{m_1}{M_1} \right)^{\frac{1}{2}} \left[1 + \frac{\nu_2^2 K_1}{\nu_\sigma^2 M_1} \left(\int_{S_z} \varrho q^2 \, dA \right)^{\frac{1}{p}} \right]^{-\frac{1}{2}}$$

$$\equiv 2\nu_2 \left(\frac{m_1}{M_1} \right)^{\frac{1}{2}} [h(z, p)]^{-\frac{1}{2}}. \tag{2.13}$$

On using Schwarz's inequality in the form

$$\int_0^z h^{-\frac{1}{2}}(\eta, p) \, d\eta \geqq z^2 \left[\int_0^z h^{\frac{1}{2}}(\eta, p) \, d\eta \right]^{-1} \tag{2.14}$$

and the inequality $(1 + a)^{\frac{1}{2}} \leqq 1 + a/2, \ a \geqq 0,$ we see from (2.13) that

$$\int_0^z k_1(\eta, p) \, d\eta \geqq 2\nu_2 \left(\frac{m_1}{M_1} \right)^{\frac{1}{2}} z^2 \left[\int_0^z \left\{ 1 + \frac{\nu_2^2 K_1}{2\nu_\sigma^2 M_1} \left(\int_{S_\eta} \varrho q^2 \, dA \right)^{\frac{1}{p}} \right\} d\eta \right]^{-1}. \tag{2.15}$$

Use of Hölder's inequality and the obvious inequality $E(z) \leqq E(0)$ in the last term of the denominator in (2.15) leads to

$$\int_0^z k_1(\eta, p) \, d\eta \geqq 2\nu_2 \left(\frac{m_1}{M_1} \right)^{\frac{1}{2}} z \left[1 + \frac{\nu_2^2 K_1}{2\nu_\sigma^2 M_1} \left(\frac{E(0)}{z} \right)^{\frac{1}{p}} \right]^{-1}$$

$$\geqq 2\nu_2 \left(\frac{m_1}{M_1} \right)^{\frac{1}{2}} z \left[1 - \frac{\nu_2^2 K_1}{2\nu_\sigma^2 M_1} \left(\frac{E(0)}{z} \right)^{\frac{1}{p}} \right], \tag{2.16}$$

the last step following from the inequality $(1 + a)^{-1} \geqq 1 - a,\ a \geqq 0$. Substitution of (2.16) into (2.5) yields the result (2.1). This completes the proof of Theorem 1.

2.2. L_2 Estimates. The energy decay estimate (2.1) may be regarded as providing a *weighted mean-square* estimate on R_z for first derivatives of solutions of the Dirichlet problem (1.1)–(1.4). Other mean-square bounds, as well as pointwise estimates, may be obtained from this basic result (*cf.* [1, 2]). For example, on using (1.5), it follows immediately that

$$\int_{R_z} |\nabla u|^2\, dV \leqq E(z)/m_1\,. \tag{2.17}$$

Similarly, on using (1.13), it follows that

$$\int_{R_z} u^2\, dV \leqq \frac{1}{\nu_2^2} \int_{R_z} u_{,\alpha} u_{,\alpha}\, dV \leqq E(z)/m_1\nu_2^2. \tag{2.18}$$

An estimate analogous to (2.18) on the cross-section S_z is readily obtained. Thus we have

$$\begin{aligned}\int_{S_z} u^2\, dA &= -\int_{R_z} (u^2)_{,3}\, dV = -2 \int_{R_z} uu_{,3}\, dV \\ &\leqq 2\left(\int_{R_z} u^2\, dV\right)^{\frac{1}{2}} \left(\int_{R_z} u_{,3}^2\, dV\right)^{\frac{1}{2}}.\end{aligned} \tag{2.19}$$

On using (1.13) and (1.5), it follows that

$$\begin{aligned}\int_{S_z} u^2\, dA &\leqq \frac{2}{\nu_2 m_1} \left(\int_{R_z} \varrho u_{,\alpha} u_{,\alpha}\, dV\right)^{\frac{1}{2}} \left(\int_{R_z} \varrho u_{,3}^2\, dV\right)^{\frac{1}{2}} \\ &\leqq E(z)/m_1\nu_2\,.\end{aligned} \tag{2.20}$$

Other norm estimates can be obtained (*cf.* [1]) but we will not pursue such questions here.

2.3. Discussion. In order to render the basic decay estimate (2.1) fully explicit, one would require an upper bound for the total energy $E(0)$ in terms of boundary data. Such a bound may be obtained by use of the methods sketched in Appendix A of [1]. These methods do not depend on dimension. Of course, it would be also desirable to have estimates for the constants ν_2 and ν_σ $(\sigma = 2p/(p-1))$ in terms of the geometry of the cross-section S. Such results are well known for the fixed membrane eigenvalue ν_2. Some discussion regarding the Sobolev constant ν_σ and its characterization as an eigenvalue may be found in [3].

Finally here we note that the basic estimate (2.1) may be readily shown to yield a result of the form (1.9), described in the Introduction. As was discussed there, the resulting estimated decay rate (1.10) is never greater than that for Laplace's equation. For the special case of Laplace's equation, the estimate (2.1) reduces to the (best possible) result (1.14).

3. Decay Estimates in Case 2

3.1. Energy Estimates. In Case 2, we shall require decay estimates for both $E(z)$ and $E_2(z)$. Thus we establish the following theorem for $E_s(z)$, where s is an even non-negative integer.

Theorem 2. *Let* $u \in C^2(R) \cap C^1(\bar{R})$ *be a solution of* (1.1)–(1.4) *where* ϱ *satisfies* (1.6). *Then for* $z \geqq 0$,

$$E_s(z) \leqq E_s(0) \exp\left\{\frac{4(s+1)}{s+2}\left(\frac{2m_2}{M_2^3 A^3}\right)^{\frac{1}{2}} K_2 E(0)\right\} e^{-2k_2(s)z}, \tag{3.1}$$

where

$$k_2(s) = \frac{4(s+1)}{s+2}\left(\frac{2m_2}{M_2 A}\right)^{\frac{1}{2}}. \tag{3.2}$$

Here $s \geqq 0$ *is an even integer and* A *denotes the cross-sectional area of* S.

Proof. As in the proof of Theorem 1, we proceed in two stages. First we show that

$$E_s'(z) + k_2(z, s)\, E_s(z) \leqq 0, \quad z \geqq 0, \tag{3.3}$$

where

$$k_2(z, s) = \frac{8(s+1)}{s+2}(2m_2)^{\frac{1}{2}}\left[\int_{S_z} \varrho^{-1}\, dA\right]^{-\frac{1}{2}}. \tag{3.4}$$

Thus, on integration, (3.3) yields

$$E_s(z) \leqq E_s(0) \exp\left\{-\int_0^z k_2(\eta, s)\, d\eta\right\}. \tag{3.5}$$

Then we use the hypothesis (1.6) to bound the exponential in (3.5) leading to the result (3.1), (3.2).

Again, it readily follows from its definition in (1.8) that $E_s(z)$ satisfies

$$E_s'(z) + k_2(z, s)\, E_s(z) = -\int_{S_z}\left[u^s \varrho q^2 + \frac{k_2(z, s)}{s+1} u^{s+1} \varrho u_{,3}\right] dA, \tag{3.6}$$

for any function $k_2(z, s)$. By Schwarz's inequality and use of the left hand inequality in (1.6) we obtain

$$\left[-\int_{S_z} u^{s+1} \varrho u_{,3}\, dA\right]^2 \leqq \frac{1}{m_2}\int_{S_z} u^{s+2}\, dA \int_{S_z} \varrho u^s u_{,3}^2\, dA. \tag{3.7}$$

The first integral on the right of (3.7) is now to be bounded. Extending u as zero outside S_z we have for fixed z

$$\begin{aligned}[u(x_1, x_2, z)]^{\frac{s+2}{2}} &= -\left(\frac{s+2}{2}\right)\int_{x_1}^{\infty} [u(\xi, x_2, z)]^{\frac{s}{2}} \frac{\partial u}{\partial \xi}(\xi, x_2, z)\, d\xi \\ &= \left(\frac{s+2}{2}\right)\int_{-\infty}^{x_1} [u(\xi, x_2, z)]^{\frac{s}{2}} \frac{\partial u}{\partial \xi}(\xi, x_2, z)\, d\xi\end{aligned} \tag{3.8}$$

from which it follows that

$$[u(x_1, x_2, z)]^{\frac{s+2}{2}} \leqq \frac{s+2}{4} \int_{-\infty}^{\infty} |u(\xi, x_2, z)|^{\frac{s}{2}} \left| \frac{\partial u}{\partial \xi}(\xi, x_2, z) \right| d\xi. \tag{3.9}$$

Similarly

$$[u(x_1, x_2, z)]^{\frac{s+2}{2}} \leqq \frac{s+2}{4} \int_{-\infty}^{\infty} |u(x_1, \eta, z)|^{\frac{s}{2}} \left| \frac{\partial u}{\partial \eta}(x_1, \eta, z) \right| d\eta. \tag{3.10}$$

Multiplying the inequalities (3.9) and (3.10) and integrating over S_z we find

$$\begin{aligned} \int_{S_z} u^{s+2}\, dA &\leqq \left(\frac{s+2}{4}\right)^2 \int_{S_z} |u|^{\frac{s}{2}} |u_{,1}|\, dA \int_{S_z} |u|^{\frac{s}{2}} |u_{,2}|\, dA \\ &\leqq \left(\frac{s+2}{4}\right)^2 \left\{ \int_{S_z} u^s \varrho (u_{,1})^2\, dA \int_{S_z} u^s \varrho (u_{,2})^2\, dA \right\}^{\frac{1}{2}} \int_{S_z} \varrho^{-1}\, dA. \end{aligned} \tag{3.11}$$

Substituting back into (3.7) we have, after use of the fact that for positive A, B, C

$$AB^{\frac{1}{2}}C^{\frac{1}{2}} \leqq (A + B + C)^2/8 \tag{3.12}$$

the inequality

$$\left[-\int_{S_z} \varrho u^{s+1} u_{,3}\, dA \right]^2 \leqq \frac{(s+2)^2}{128 m_2} \left[\int_{S_z} u^s \varrho q^2\, dA \right]^2 \int_{S_z} \varrho^{-1}\, dA. \tag{3.13}$$

If (3.13) is now introduced into the right side of (3.6) we easily conclude (3.3), (3.4) and thus (3.5).

The proof of Theorem 2 clearly goes through for any non-negative real constant s, but by restricting s to be an even integer we have avoided the necessity of using absolute value signs in some of the inequalities.

It remains to simplify the exponential term in (3.5). On using Schwarz's inequality in the form

$$\int_0^z \left[\int_{S_\eta} \varrho^{-1}\, dA \right]^{-\frac{1}{2}} d\eta \geqq z^2 \left[\int_0^z \left(\int_{S_\eta} \varrho^{-1}\, dA \right)^{\frac{1}{2}} d\eta \right]^{-1}, \tag{3.14}$$

and the right hand inequality in (1.6), from (3.4) we conclude that

$$\int_0^z k_2(\eta, s)\, d\eta \geqq \frac{8(s+1)}{s+2} (2m_2)^{\frac{1}{2}} z^2 \left[\int_0^z \left\{ \int_{S_\eta} (M_2 + K_2 \varrho q^2)\, dA \right\}^{\frac{1}{2}} d\eta \right]^{-1}. \tag{3.15}$$

Using the inequality $(1 + a)^{\frac{1}{2}} \leqq 1 + a/2$, $a \geqq 0$, and the obvious inequality $E(z) \leqq E(0)$, we have

$$\begin{aligned} \int_0^z \left\{ \int_{S_\eta} (M_2 + K_2 \varrho q^2)\, dA \right\}^{\frac{1}{2}} d\eta &= \int_0^z \left[M_2 A + K_2 \int_{S_\eta} \varrho q^2\, dA \right]^{\frac{1}{2}} d\eta \\ &\leqq \int_0^z (M_2 A)^{\frac{1}{2}} \left[1 + \frac{K_2}{2 M_2 A} \int_{S_\eta} \varrho q^2\, dA \right] d\eta \\ &\leqq (M_2 A)^{\frac{1}{2}} z \left[1 + \frac{K_2 E(0)}{2 M_2 A z} \right], \end{aligned} \tag{3.16}$$

and so from (3.15) we obtain

$$
\begin{aligned}
\int_0^z k_2(\eta, s)\, d\eta &\geqq \frac{8(s+1)}{s+2}\left(\frac{2m_2}{M_2 A}\right)^{\frac{1}{2}} z\left[1+\frac{K_2 E(0)}{2M_2 A z}\right]^{-1} \\
&\geqq \frac{8(s+1)}{s+2}\left(\frac{2m_2}{M_2 A}\right)^{\frac{1}{2}}\left[z-\frac{K_2 E(0)}{2M_2 A}\right],
\end{aligned}
\tag{3.17}
$$

the last step following from the inequality $(1+a)^{-1} \geqq 1-a,\ a \geqq 0$. Substitution of (3.17) into (3.5) yields the results (3.1), (3.2). This completes the proof of Theorem 2.

3.2. L_2 Estimates. To obtain an estimate for the L_2 integral of u over R_z, we start with the identity

$$
\int_{R_z} u^2\, dV = -\int_z^\infty \int_{S_\eta} x_\alpha u u_{,\alpha}\, dA\, d\eta, \tag{3.18}
$$

where the origin in S_η in the x_1, x_2 coordinates is taken at some suitable point in S_η. Thus

$$
J^2 \equiv \left[\int_{R_z} u^2\, dV\right]^2 \leqq d^2\left(\int_{R_z} \varrho^{-1} u^2\, dV\right)\left(\int_{R_z} \varrho q^2\, dV\right), \tag{3.19}
$$

where d is the maximum distance from the origin to the boundary of S. On using the right hand inequality in (1.6), this may be written as

$$
\begin{aligned}
J^2 &\leqq d^2 M_2 E(z)\, J + d^2 K_2 E_2(z)\, E(z) \\
&\leqq \tfrac{1}{2} d^4 M_2^2 E(z)^2 + \tfrac{1}{2} J^2 + d^2 K_2 E_2(z)\, E(z),
\end{aligned}
\tag{3.20}
$$

and so we obtain

$$
J \leqq M_2\, d^2\left[E^2(z) + \frac{2K_2}{M_2\, d^2} E(z)\, E_2(z)\right]^{\frac{1}{2}}. \tag{3.21}
$$

Since the estimated decay rate $k_2(2)$ of (3.2) associated with $E_2(z)$ is greater than $k_2(0)$ associated with $E(z)$, it follows from (3.1) and (3.21) that

$$
\int_{R_z} u^2\, dV \leqq B e^{-2k_2(0)z}, \quad z \geqq 0, \tag{3.22}
$$

for some computable positive constant B, where

$$
k_2(0) = 2(2m_2/M_2 A)^{\frac{1}{2}}. \tag{3.23}
$$

To obtain a cross-sectional estimate analogous to (3.22), we have

$$
\begin{aligned}
\int_{S_z} u^2\, dA &= -2\int_{R_z} u u_{,3}\, dV \\
&\leqq 2\left(\int_{R_z} \varrho^{-1} u^2\, dV\right)^{\frac{1}{2}}\left(\int_{R_z} \varrho q^2\, dV\right)^{\frac{1}{2}} \\
&\leqq 2\left[M_2 \int_{R_z} u^2\, dV + K_2 E_2(z)\right]^{\frac{1}{2}} [E(z)]^{\frac{1}{2}},
\end{aligned}
\tag{3.24}
$$

where the right hand inequality in (1.6) has been used in the last step of (3.24). On using (3.22), the fact that $E_2(z)$ has an estimated decay rate which is greater than that for $E(z)$ and (3.1), (3.2) allows us to again conclude that

$$\int_{S_z} u^2 \, dA \leqq C e^{-2k_2(0)z}, \quad z \geqq 0, \tag{3.25}$$

for a computable positive constant C, where $k_2(0)$ is given by (3.23).

3.3. Discussion. As in Case 1, in order to render the estimate (3.1) fully explicit, one would require an upper bound for the total energy $E(0)$ in terms of boundary data. Such a bound may again be obtained by use of the methods sketched in Appendix A of [1]. These methods may also be readily modified to yield an analogous upper bound for $E_2(0)$, which is required for the L_2 estimates (3.22) and (3.25).

In conclusion, we observe that the results established here in Case 2 for the estimated decay rate $k_2(0)$ of (3.23) occurring in (3.1) for $s = 0$ and in (3.22), (3.25) are not as sharp as those obtained in Case 1. For example, if we consider the special case of $\varrho \equiv$ constant in R, for which (1.1) becomes Laplace's equation, we may choose $m_2 = M_2$, $K_2 = 0$ in (1.6). In this case, (3.1) with $s = 0$ reads

$$E(z) \leqq E(0)\, e^{-4(2/A)^{\frac{1}{2}} z}, \quad z \geqq 0, \tag{3.26}$$

which falls short of the (optimal) result (1.14). In fact, if we make use of Faber-Krahn inequality (see *e.g.* [6])

$$\nu_2 \geqq 2.4048 \sqrt{\frac{\pi}{A}} = \frac{4.2624}{\sqrt{A}} \tag{3.27}$$

the result (1.14) for Laplace's equation implies the (weaker) result

$$E(z) \leqq E(0)\, e^{-\left(\frac{8.5248}{\sqrt{A}}\right) z} \tag{3.28}$$

while (3.26) reads

$$E(z) \leqq E(0)\, e^{-\left(\frac{5.5669}{\sqrt{A}}\right) z}. \tag{3.29}$$

Our results are also conservative for the special case of the minimal surface equation (1.7). For this equation (1.6) holds with $m_2 = M_2 = 1$, $K_2 = 1$ and so the estimated decay rate (3.23) is again the same as that of (3.26). It has been shown recently in [4], using comparison principle arguments, that solutions of the Dirichlet problem (1.1)–(1.4) for the minimal surface equation decay at least as fast as do harmonic functions[1], provided the cross-section S is *convex*. In fact such a result holds for equations of the form (1.1) with $\varrho = \varrho(q^2)$ provided ϱ' is non-positive and (1.1) is elliptic [4]. Thus solutions of (1.1)–(1.4), for any ϱ satisfying these hypotheses, will decay at least as fast as do harmonic functions provided S is convex. It remains to be seen whether the techniques of the present paper can be modified in Case 2 to yield such sharper estimates in these situations.

[1] This is a generalization to three dimensions of an earlier result of KNOWLES [5] for the two-dimensional case (see also [1]).

Acknowledgment. Our work, respectively, was supported by NSF Grants MEA-78-26071 and MCS-82035597.

References

1. HORGAN, C. O., & L. E. PAYNE, Decay estimates for second-order quasilinear partial differential equations. Advances in Applied Mathematics **5** (1984).
2. HORGAN, C. O., & J. K. KNOWLES, Recent developments concerning Saint-Venant's principle, Advances in Applied Mechanics, J. W. HUTCHINSON ed., Vol. 23, pp. 179–269. Academic Press, New York, 1983.
3. CROOKE, P. S., & R. P. SPERB, Isoperimetric inequalities in a class of nonlinear eigenvalue problems. SIAM J. Math. Anal. **9**, 671–681 (1978).
4. HORGAN, C. O., A note on the spatial decay of a three-dimensional minimal surface over a semi-infinite cylinder. J. Math. Anal. Appl. (in press).
5. KNOWLES, J. K., A note on the spatial decay of a minimal surface over a semi-infinite strip. J. Math. Anal. Appl. **59**, 29–32 (1977).
6. BANDLE, C., Isoperimetric Inequalities and Applications, Monographs and Studies in Mathematics # 7, Pitman Publishing Inc., Mansfield, Massachusetts (1980).

Michigan State University
East Lansing
and
Cornell University
Ithaca, New York

(Received March 1, 1984)

Hyperbolicity and Change of Type in the Flow of Viscoelastic Fluids

DANIEL D. JOSEPH, MICHAEL RENARDY & JEAN-CLAUDE SAUT

Dedicated to Walter Noll on the Occasion of his 60th Birthday

Abstract

The equations governing the flow of viscoelastic liquids are classified according to the symbol of their differential operators. Propagation of singularities is discussed and conditions for a change of type are investigated. The vorticity equation for steady flow can change type when a critical condition involving speed and stresses is satisfied. This leads to a partitioning of the field of flow into subcritical and supercritical regions, as in the problem of transonic flow.

Table of Contents

1. Introduction

The equations of steady gas dynamics change type when the speed of the fluid at some point exceeds the speed of sound. If this happens, then discontinuities can appear in the supersonic region. We are interested in the possibility that many strange effects in the flow of viscoelastic liquids, as well as difficulties in numerical simulation, are also associated with the appearance of real characteristics and a change of type, analogous to the sonic transition.

For a physical interpretation, it is necessary to identify the variables which may propagate and become discontinuous. In gas dynamics, there are compression waves and shock waves of compression. In the present paper, we deal with incompressible materials, so compression is impossible. Instead, we can exhibit cases where singular shear surfaces propagate along characteristics (Chapters 6, 7). In steady flow, the vorticity is the variable which is affected by a change of type and may become discontinuous (Chapters 8–11). The implications of hyperbolicity and change of type for the interpretation of experiments are not yet well understood.

The organization of our paper is shown in the Table of Contents. In § 2, we motivate our study by suggesting that one of the main unsolved practical problems of computation of viscoelastic flow may be partly due to the problem of change of type. We suggest that the solution of this problem is to be found in recently developed switching algorithms of the type used in transonic flow. In § 3 we define some basic concepts needed in our study, including elliptic, hyperbolic, characteristic, symbol of an operator and Hadamard instability. We also give some applications of these concepts which arise in modeling phase changes and may be relevant in analyzing some instabilities in the extrusion of polymers from capillary tubes. Chapter 4 discusses characteristics and classification of type for first order quasilinear systems.

In Chapter 5, we look at constitutive equations for viscoelastic fluids from the point of view of classification of type. For this, we have to maintain a distinction between fluids with and without Newtonian viscosity. In Oldroyd models, the term with Newtonian viscosity is the one associated with a retardation time. The addition of even small amounts of Newtonian viscosity can smooth discontinuities, replacing sharp fronts by thin layers and thus masking the underlying dynamics. To emphasize the effect of hyperbolicity, we confine our attention to models without Newtonian viscosity. In particular, we focus on a three-parameter family of nonlinear Oldroyd models containing the upper and lower convected and corotational Maxwell models. The occurrence of instabilities of the Hadamard type for these models is discussed. These models also form the basis for the discussion of steady flows in Chapters 9–11. We also discuss more general models of integral type. It is shown that the principal part of the linearization at any given motion has the form of a rate equation not involving integrals, provided that the integral kernels have sufficient smoothness. Thus the discussion of change of type does not necessarily require a special constitutive model.

Chapter 6 discusses the linear system of equations for motion perturbing rest. The wave speed along characteristics is given by $\sqrt{G(0)/\varrho}$, where $G(0)$ is the instantaneous value of the relaxation modulus $G(s)$ and ϱ is the density. We review recent results on the propagation of slip surfaces for velocity and displacement, which show in particular the crucial dependence on the nature of the kernel $G(s)$. In particular, consideration is given to the possibility that $G(0)$ or $G'(0)$ may be infinite. In Chapter 7, we discuss the formation and propagation of slip surfaces in nonlinear shearing problems treated by COLEMAN & GURTIN [6], [7] and SLEMROD [43], [44]. We discuss the application of their results to melt fracture.

In Chapter 8 we take up the analysis of change of type in steady problems. This is a natural question from a mathematical point of view, but the first studies of it in the theory of viscoelastic fluids seem to be in the work of RUTKEVICH [40, pp. 44–45], who analyzed the two dimensional equations for an upper convected Maxwell model. ULTMAN & DENN [49] and LUSKIN [27] classified the linearized equations perturbing uniform flow with velocity U of an upper convected Maxwell fluid. Our analysis in Chapter 8 generalizes the results of ULTMAN & DENN and LUSKIN to a wider class of constitutive laws. There is a change of type leading to real characteristics when the viscoelastic Mach number

$$M = U/c, \qquad c = \sqrt{G(0)/\varrho}$$

exceeds one. The vorticity is identified as the variable which can become discontinuous along these characteristics. We shall, somewhat loosely, say that "the vorticity changes type". In Chapter 9, we give a complete classification of the quasilinear system describing the upper convected Maxwell model in arbitrary steady two-dimensional motions. The streamlines are double characteristics. The vorticity changes its type when the speeds are great enough. In the supercritical (hyperbolic) case, there are two families of real characteristics for the vorticity, but the formula for the characteristics depends on the solution. There are also complex roots to the characteristic equation associated with the elliptic equation giving the vorticity as the Laplacian of the stream function. In Chapter 10, we discuss a number of specific flows for an upper convected Maxwell fluid. These flows include plane parallel shear flow, steady extensional flow, sink flow in the plane and shear flow outside a rotating cylinder. We discuss characteristics for motions perturbing those flows and characterize the regions of flow where the vorticity equation is hyperbolic. In Chapter 11, we extend our results to a three parameter family of Oldroyd models which contains the upper and lower convected and corotational Maxwell models as special cases. The vorticity is again identified as the variable which changes its type. We compute the characteristic directions for the nonlinear problem without approximation. We exhibit special cases which show that the partitioning of the flow into sub- and supercritical regions is model sensitive. It is therefore desirable to develop this type of theory on a high level of generality, suppressing models. We take some steps in that direction in Chapter 12, where we study fading memory fluids of Coleman-Noll type.

2. Numerical Simulation of Steady Flows of Changing Type

There are some unsolved problems of numerical simulation of the flow of viscoelastic fluids. One problem is that the equations cannot be integrated when the relaxation time is large. Though relaxation times appear expiicitly only in very special models, the concept of a relaxation time is a useful one which can be expressed mathematically in a general context (*e.g.*, C. TRUESDELL [52]). A large relaxation time means that the elastic response of the fluid is persistent; the fluid can be said to have a long memory. The simulation problem associated with highly elastic viscoelastic fluids is sometimes called "the high Weissenberg number problem". Different dimensionless ratios are called "Weissenberg numbers" by different authors. Some authors call the ratio of the first normal stress to the shear stress a "Weissenberg number". This definition leads to a dimensionless function of the rate of shear. Other authors define a different "Weissenberg number" as the ratio of the relaxation time of the fluid to an externally given time which is usually expressed as d/U, where U and d are a typical velocity and length in the flow.

The "high Weissenberg number problem" refers to the failure of numerical simulations when the second of the two "Weissenberg numbers" is large. The problem occurs with different constitutive models and different methods of numerical integration. Maybe there are some underlying mathematical reasons.

Our study here is not framed in terms of a "Weissenberg number". The quantities of interest in our study are values of velocity and stress which in steady flow may lead to a change of type as in the problem of transonic flow. The criterion for change of type may be framed in terms of a viscoelastic. Mach number defined as the pointwise ratio of some speed to a characteristic wave speed. If all other quantities are fixed, this "Mach number" increases with the relaxation time, but our "Mach number" and the "Weissenberg number" are in principle independent. It is probable however that some numerical problems at high "Weissenberg numbers" are actually associated with a change of type, like the transition from subsonic flow to supersonic flow, and that the solution of the problem is to be sought in various hyperbolic algorithms, especially those recently introduced for transonic flow.

To compute subsonic flow you use some central differences. To compute supersonic flow you use the method of characteristics. It would be a disaster to try to do supersonic flow by central differencing of the type used for Laplace's equation.

In the flow over a bump, say an airfoil with the free stream slightly less than $M = 1$,

$$.75 \leqq M < 1,$$

we get a supersonic bubble with unknown boundaries (see Fig. 1).

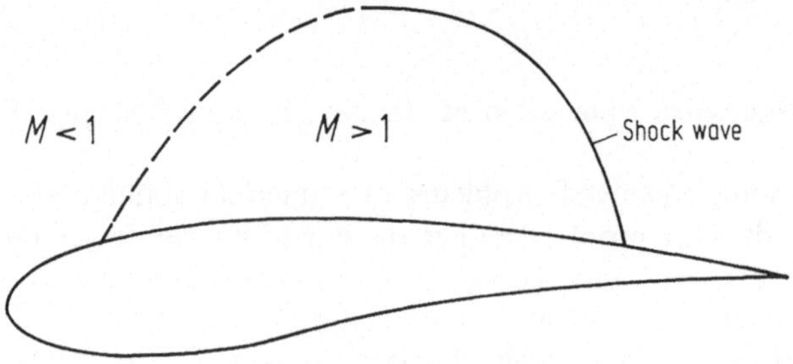

Fig. 1

To solve this problem you have to find the sonic line, the position and strength of the shock wave. This is a very hard free boundary problem. It wasn't solved until 1971 when MURMAN & COLE [32] realized that upwind differencing was necessary in the supersonic part of the flow. In central differencing the nodal point is at the center. In upwind differencing, the information at a nodal point is determined only by the flow upstream (see Fig. 2).

Fig. 2

MURMAN & COLE studied the small disturbance equations. They derived a switching scheme of numerical analysis which tells the computer to use central differencing if the flow is subsonic and upwind differencing if the flow is supersonic. Their upwind differencing equation can be interpreted as approximating a differential equation with an artificial viscosity proportional to the mesh size [31].

MURMAN & COLE'S method was the first success. But this method is too simple for the full nonlinear potential. This more complicated problem was successfully attacked by the artificial viscosity method of JAMESON [19], [20], whose work makes transonic computation possible in a practical sense.

People doing flow computations for viscoelastic fluids are also able to go to higher Weissenberg numbers when they have constitutive equations with more Newtonian viscosity. This procedure masks the problem of dealing with change of type instead of solving it.

3. Concepts and Some Applications of Change of Type

This paper deals with equations which undergo a change of type. To make our notions precise we shall need some classical definitions related to the type of a partial differential equation.

Consider the linear differential operator

$$P\left(\boldsymbol{x}, t, \frac{\partial}{\partial t}, \frac{\partial}{\partial x_1}, \ldots, \frac{\partial}{\partial x_n}\right), \tag{3.1}$$

where

$$\boldsymbol{x} = (x_1, x_2, \ldots, x_n) \text{ and } t$$

are space and time coordinates. We define the

$$\text{Symbol of } P = P(\boldsymbol{x}, t, i\xi_0, i\xi_1, \ldots, i\xi_n), \tag{3.2}$$

where $i = \sqrt{-1}$. To form the symbol we replace the arguments $\frac{\partial}{\partial t}, \frac{\partial}{\partial x_1}, \ldots, \frac{\partial}{\partial x_n}$ of P with the Fourier variables $i\xi_0, i\xi_1, \ldots, i\xi_n$. In this way we obtain a polynomial in the real variables ξ. The symbol of the Laplace operator $-\Delta$ is $\sum_{i=1}^n \xi_i^2$; the symbol of the wave operator $\frac{\partial^2}{\partial t^2} - \Delta$ is $-\xi_0^2 + \sum_{i=1}^n \xi_i^2$; the symbol of the heat operator $\frac{\partial}{\partial t} - \Delta$ is $i\xi_0 + \sum_{i=1}^n \xi_i^2$. The symbol for a system of equations is defined in a similar fashion and is a matrix with polynomial entries.

Characteristic curves are lines along which discontinuous data may propagate. In dimensions higher than two we may speak of characteristic surfaces. Let m be the highest order of the derivatives in P. Then

$$P = \sum_{|\alpha|=m} a_\alpha(\boldsymbol{x}, t)\, \partial^\alpha + \sum_{|\alpha|<m} a_\alpha(\boldsymbol{x}, t)\, \partial^\alpha,$$

where $\alpha = (\alpha_0, \alpha_1, \ldots, \alpha_n)$ is a multi-index, $|\alpha| = \Sigma \alpha_i$ and

$$\partial^\alpha = \frac{\partial^{|\alpha|}}{\partial t^{\alpha_0}\, \partial x_1^{\alpha_1} \ldots \partial x_n^{\alpha_n}}.$$

The equation

$$\sum_{|\alpha|=m} a_\alpha(\boldsymbol{x}, t)\, \sigma^\alpha = 0, \qquad \sigma = (\sigma_0, \ldots, \sigma_n), \tag{3.3}$$

$$\sigma^\alpha = \sigma_0^{\alpha_0} \ldots \sigma_n^{\alpha_n}$$

is called the characteristic equation for P. Only the principal part of P, the terms of highest order, appears in (3.3).

A surface S in $(\boldsymbol{x}, t)$ space is characteristic for P at a point $s \in S$ if the normal vector to S at s satisfies the characteristic equation. If $\sigma = (\sigma_0, \ldots, \sigma_n)$ is a unit normal vector at s, S is characteristic for P if and only if

$$\sum_0^n \sigma_k^2 = 1 \quad \text{and} \quad \sum_{|\alpha|=m} a_\alpha(\boldsymbol{x})\, \sigma^\alpha = 0. \tag{3.4}$$

The characteristic equation for Laplace's equation $\sum_{k=1}^n \partial^2 u/\partial x_k^2 = 0$ is $\sum_{k=1}^n \sigma_k^2 = 0$. There are no real characteristics because $(3.4)_1$ is not satisfied. More generally, the operators P for which, at every point $(\boldsymbol{x}, t)$ the equation $(3.4)_2$ has no nontrivial real zeroes are called *elliptic*. For systems, ellipticity means that the only real zeroes of the determinant of the matrix symbol $A(\boldsymbol{x}, \xi_1, \ldots, \xi_n)$ are $(\xi_1, \xi_1, \ldots, \xi_2) = (0, 0, \ldots, 0)$.

Elliptic problems have existence, uniqueness and continuous dependence on data (are well posed) as boundary value problems [1], [26].

The initial value problem, the Cauchy problem, is not well posed for elliptic equations. For example, a Cauchy problem for Laplace's equation in the domain

$D = \{x, y; x > 0, -\infty < y < \infty)$ is$\}$

$$\begin{aligned} \Delta u &= 0 \quad \text{in } D, \\ u(0, y) &= 0, \\ \frac{\partial u}{\partial x}(0, y) &= U(y), \end{aligned} \tag{3.5}$$

where

$$U(y) = \frac{1}{n^p} \sin ny, \qquad p > 0.$$

The solution of (3.5) is

$$u(x, y) = \frac{1}{n^{1+p}} \sin ny \sinh nx.$$

The mapping $\left(u, \frac{\partial u}{\partial x}\right)\Big|_{x=0} \to u$ for $x > 0$ is not continuous since $U(y)$ is small when n is large and $u(x, y)$ is very big. Small data at $x = 0$ lead to larger and larger oscillations for $x > 0$. This lack of continuous dependence is called *Hadamard instability*. It can be shown that (3.5) has no solution if $U(\cdot)$ is not analytic.

The initial value problem, or mixed initial-boundary value problems are well-posed for hyperbolic equations like the wave equations. For example, the characteristic equation $(3.4)_2$

$$\sigma_0^2 - c^2 \sum_1^n \sigma_k^2 = 0$$

for the n-dimensional wave equation

$$\frac{\partial^2 u}{\partial t^2} = c^2 \Delta u$$

satisfies the characteristic equation $(3.4)_1$ when $\sigma_0 = \pm c/\sqrt{c^2+1}$. Therefore a surface is characteristic for the wave equation if and only if its normal makes an angle β, $\cos\beta = c/\sqrt{c^2+1}$, with the t axis. For the one-dimensional wave equation $\Delta = \partial^2/\partial x^2$, this implies that the family of lines $x \pm ct = \text{const}$ are characteristic.

The operator P of (3.1) is called strictly hyperbolic if all the roots ξ_0 of the principal part of its symbol (3.2) are real and distinct for all $(\xi_1, \ldots, \xi_n) \in \mathbb{R}^n \setminus 0$. The Cauchy problem is well posed and the boundary value problem is ill posed for hyperbolic equations. The backward Cauchy problem where t is replaced with $-t$ is also well-posed for hyperbolic equations.

The Cauchy problem is well posed for parabolic problems but the backward Cauchy problem is ill posed. The classic example of a parabolic equation is the heat equation, $\partial u/\partial t = \Delta u$. The characteristic equation $(3.4)_2$ is

$$\sum_{l=1}^n \sigma_l^2 = 0.$$

Hence, from $(3.4)_1$, $\sigma_0^2 = 1$ and the characteristic surfaces are the hyperplanes $t = \text{const}$. The Cauchy problem is not well posed for the backward heat equation $\frac{\partial u}{\partial t} = -\Delta u$. Operators of the form $\frac{\partial u}{\partial t} + Lu$, where L, like $-\Delta$, is a positive definite elliptic operator, are parabolic. These operators are strongly dissipative and lead to diffusion rather than to propagation. Unlike hyperbolic operators, parabolic operators will smooth initially discontinuous Cauchy data.

Two homogeneous scalar operators are said to be of the same type, if up to a transformation of the independent variables, their symbols have the same asymptotic behavior at infinity. If the asymptotic behavior of the symbol changes, then we say that the equation changes type. For example, the Tricomi equation

$$y \frac{\partial^2 u}{\partial x^2} + \frac{\partial^2 u}{\partial y^2} = 0$$

is hyperbolic when $y < 0$ and elliptic when $y > 0$. Another example is the quasi-linear system

$$\begin{aligned} \frac{\partial u}{\partial t} &= \frac{\partial \sigma(v)}{\partial x}, \\ \frac{\partial v}{\partial t} &= \frac{\partial u}{\partial x}, \end{aligned} \qquad \left(\text{or} \quad \frac{\partial^2 v}{\partial t^2} = \frac{\partial}{\partial x}\left(\sigma'(v)\frac{\partial v}{\partial x}\right)\right) \tag{3.6}$$

which is hyperbolic for $\sigma'(v) > 0$ and elliptic for $\sigma'(v) < 0$. These problems all involve a change in the sign of the symbol and Hadamard instabilities, which occur if the solution of the Cauchy problem with initial data in the hyperbolic region enters the elliptic region.

Problems of the form (3.6) suggest models for theories of phase changes in solids and fluids. The van der Waals gas is a well-known classical example. In solid mechanics, ideas of this type were introduced by J. L. ERICKSEN [12] in his study of elastic bars. We may suppose that the graph of $\sigma(v)$ is as shown in Fig. 3.

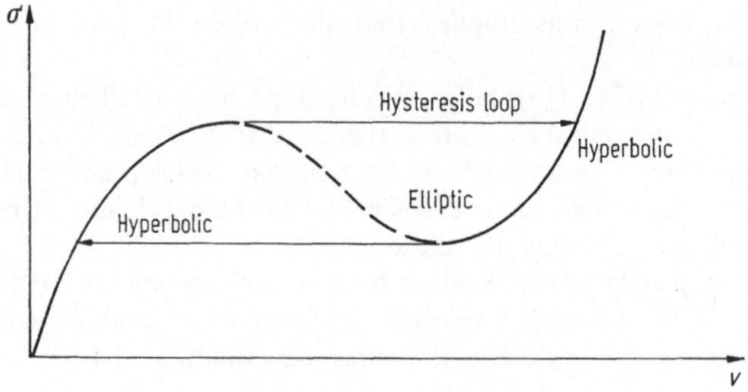

Fig. 3. The system (3.6) is hyperbolic when $\sigma'(v) > 0$. The elliptic branch is unstable in the sense of Hadamard

The solid lines, where $\sigma'(v) > 0$, lead to a hyperbolic equation and the dashed line leads to an elliptic equation. The elliptic portion is rejected because it will exhibit Hadamard instabilities; and actual solutions are required to operate only on the hyperbolic parts of the curve. This leads to spatially segregated solutions, separated by lines of diccontinuity, each part operating on a different hyperbolic branch of the curve. There is hysteresis and abrupt transitions in the response of such models. These features are all present in the recent study of HUNTER & SLEMROD [17], which attempts to explain some observations of TORDELLA [47] of a type of melt fracture called ripple. This phenomenon shows hysteresis loops, double-valued shear rates at certain stresses and spatially segregated flow regimes. Similar ideas have also been used to explain to phenomenon of necking occurring in cold drawing of polymers [8].

REGIRER & RUTKEVICH [36] have considered fluids of the Reiner-Rivlin type which exhibit change of type. Their constitutive law is

$$T = -p\mathbf{1} + \eta f(II)\,\boldsymbol{D},$$

where $\boldsymbol{D} = \frac{1}{2}(\nabla\boldsymbol{u} + (\nabla\boldsymbol{u})^T)$, $II = \operatorname{tr} \boldsymbol{D}^2$. Written in terms of a stream function $\boldsymbol{u} = (u, v) = (\psi_y, -\psi_x)$, the equation governing steady two-dimensional flows is as follows.

$$(3.7)\qquad L\psi \overset{\text{def}}{=} a_1\left[\frac{\partial^4\psi}{\partial x^4} + \frac{\partial^4\psi}{\partial y^4}\right] + 2a_2\frac{\partial^4\psi}{\partial x^2\,\partial y^2} + 4a_3\frac{\partial^2}{\partial x\,\partial y}\left(\frac{\partial^2\psi}{\partial y^2} - \frac{\partial^2\psi}{\partial x^2}\right) = H(\psi),$$

where $H(\psi)$ is a nonlinear third order operator and the coefficients a are nonlinear functions of the second derivatives of ψ. The characteristic curves $y(x)$ are solutions of

$$(3.8)\qquad a_1 y_x^4 + 4a_3 y_x^3 + 2a_2 y_x^2 - 4a_3 y_x + a_1 = 0.$$

There are three cases:

(i) $f + 2IIf' > 0$ (no real roots, elliptic),
(ii) $f + 2IIf' = 0$ (parabolic),
(iii) $f + 2IIf' < 0$ (four real roots, hyperbolic).

The hyperbolic regions are those where the stress decreases as a function of shear rate, and the elliptic regions are those where it increases. The unsteady problem corresponding to (3.7) is

$$\varrho\frac{\partial}{\partial t}(\Delta\psi) = L(\psi) - H(\psi).$$

When the right side is elliptic, this problem is parabolic and evolutionary (see GELFAND [14]). When the right-hand side changes type, the problem is neither parabolic nor evolutionary and Hadamard instability occurs. Changes of type and Hadamard instabilities can occur in rheological problems which are not one-dimensional and they need not be associated with non-monotone constitutive equations. An interesting case of this type arises in a stability analysis of plane

Couette flow by AKBAY, BECKER, KROZER & SPONAGEL [3]. In order to obtain a manageable equation, they introduce the "short memory approximation". This means that, in the memory integrals occurring in the equation for the disturbances, only terms of first order in the relaxation time of the fluid are kept. Proceeding thus, they find the following linearized equation for the stream function in two dimensions:

$$\varrho\left(\frac{\partial}{\partial t}+\varkappa x_2\frac{\partial}{\partial x_1}\right)\Delta\psi=\left(N_1'-\frac{N_1}{\varkappa}\right)\frac{\partial^2}{\partial x_1\,\partial x_2}L\psi+\tau' L^2\psi+\frac{4\tau}{\varkappa}\frac{\partial^4\psi}{\partial x_1^2\,\partial x_2^2},\tag{3.9}$$

$$\psi=\frac{\partial\psi}{\partial x_2}=0 \quad \text{at} \quad x_2=0,\ x_2=h.$$

Here L denotes the operator $\frac{\partial^2}{\partial x_2^2}-\frac{\partial^2}{\partial x_1^2}$. The problem is posed in the strip $-\infty<x_1<\infty$, $0<x_2<h$. $\varkappa$ is the shear rate of the basic Couette flow and $\tau(\varkappa)$, $N_1(\varkappa)$ are the shear stress and the first normal stress difference as functions of the rate of shear. AKBAY *et al.* find that (3.9) admits exponentially growing solutions if

$$\mathrm{We}=\frac{\left(\frac{N_1}{\varkappa}\right)'\varkappa}{\sqrt{\frac{\tau}{\varkappa}\tau'}}>4.\tag{3.10}$$

It was pointed out by AHRENS, JOSEPH, RENARDY & RENARDY [2] that this instability is associated with a change of type. If we consider the symbol of the differential operator, i.e. if we formally set $\frac{\partial}{\partial t}=\sigma$, $\frac{\partial}{\partial x_1}=i\alpha$, $\frac{\partial}{\partial x_2}=i\beta$, then the left-hand side of (3.9) becomes

$$\varrho(\sigma+\varkappa x_2 i\alpha)(-\alpha^2-\beta^2),\tag{3.11}$$

and the right-hand side becomes

$$-\left(N_1'-\frac{N_1}{\varkappa}\right)(\alpha^2-\beta^2)\,\alpha\beta+\tau'(\alpha^2-\beta^2)^2+\frac{4\tau}{\varkappa}\alpha^2\beta^2.\tag{3.12}$$

This homogeneous polynomials of is fourth degree positive definite for $\mathrm{We}<4$, but indefinite for $\mathrm{We}>4$. For $\mathrm{We}>4$, one thus expects short-wave instabilities of a catastrophic nature, *i.e.* $\mathrm{Re}\,\sigma$ becomes arbitrarily large as the wave length tends to zero. This type of instability seems to occur in some types of melt fracture (see [2] for a more complete discussion).

4. Quasilinear Systems

The analysis of the equations of viscoelastic flow will be framed in terms of systems of equations of first order. We consider linear systems and quasilinear systems. We write the quasilinear system as

(4.1) $$\sum_{l=0}^{n} A_l \frac{\partial \boldsymbol{u}}{\partial x_l} = f, \qquad \boldsymbol{x} = (t, x_1, x_2, \ldots, x_n),$$

where $\boldsymbol{u} = (u_1, u_2, \ldots, u_k)$ is a k vector and A_l are $k \times k$ matrices which like f may depend on x_l and on the components of $\boldsymbol{u}$. If A_l is independent of $\boldsymbol{u}$, and $f = \boldsymbol{B} \cdot \boldsymbol{u}$, then (3.7) is a linear system.

The following definitions apply to both linear and quasilinear systems. A surface S defined by the equation $\phi(t, x_1, \ldots, x_n) = 0$, is characteristic with respect to (4.1) at $\boldsymbol{x} = (t, x_1, \ldots, x_n)$ if

(4.2) $$\det \left(\sum_{l=0}^{n} A_l \frac{\partial \phi}{\partial x_l} \right) (\boldsymbol{x}) = 0.$$

If $\phi = x_n - f(x_0, \ldots, x_{n-1})$, then

(4.3) $$\det \left(A_n - \sum_{l=0}^{n-1} A_l \frac{\partial f}{\partial x_l} \right) = 0.$$

Any one of the $n + 1$ quantities $\partial\phi/\partial x_l$ in (4.2) may be regarded as an eigenvalue. We shall say (4.1) is hyperbolic if $\boldsymbol{A} = \boldsymbol{A}_\mu$ is non-singular and for any choice of the real parameters $(\lambda_l, l = 0, 1, \ldots, n; l \neq \mu)$, the roots α of

(4.4) $$\det \left(\alpha \boldsymbol{A} - \sum_{\substack{l=0 \\ l \neq \mu}}^{n} \lambda_l \boldsymbol{A}_l \right) = 0$$

are real and are associated with k linearly independent characteristic vectors $\boldsymbol{v}$:

(4.5) $$\alpha \boldsymbol{A} \boldsymbol{v} = \sum_{\substack{l=0 \\ l \neq \mu}}^{n} \lambda_l \boldsymbol{A}_l \boldsymbol{v}.$$

First-order systems can be of *mixed type* with real and complex eigenvalues, neither totally elliptic or totally hyperbolic.

We are interested in two-dimensional quasilinear problems of the form

(4.6) $$\boldsymbol{A} \frac{\partial \boldsymbol{u}}{\partial t} + \boldsymbol{B} \frac{\partial \boldsymbol{u}}{\partial x} + \boldsymbol{C} \frac{\partial \boldsymbol{u}}{\partial y} = \boldsymbol{f}.$$

We consider one-dimensional evolutionary problems in which $\boldsymbol{C} = 0$ and steady problems in which $\boldsymbol{A} = 0$. For evolutionary problems we suppose that $\boldsymbol{A}$ is not singular and write

(4.7) $$\frac{\partial \boldsymbol{u}}{\partial t} + \boldsymbol{B} \frac{\partial \boldsymbol{u}}{\partial x} = \boldsymbol{f}.$$

The characteristic surface is $\phi(x, t) = 0$ and (4.2) becomes

(4.8) $$\det \left(\frac{\partial \phi}{\partial t} \mathbf{1} + \boldsymbol{B} \frac{\partial \phi}{\partial x} \right) = 0.$$

On $\phi(x, t) = 0$, we have

$$d\phi = \frac{\partial \phi}{\partial x} dx + \frac{\partial \phi}{\partial t} dt = 0.$$

Hence, (4.8) may be written

$$\det\left(\boldsymbol{B} - \frac{dx}{dt}\boldsymbol{1}\right) = 0, \tag{4.9}$$

where dx/dt is the slope of the characteristic.

A linear system of the form (4.7) is said to be of *evolutionary* type if $\boldsymbol{B}$ has only real eigenvalues, *hyperbolic* if of evolutionary type and if $\boldsymbol{B}$ can be made diagonal, *strictly hyperbolic* if $\boldsymbol{B}$ has simple real eigenvalues.

If the Cauchy problem for (4.7) with $\boldsymbol{f} = 0$ is well-posed, then (4.7) must be of evolutionary type. Solutions of the form $\boldsymbol{u}(x, t) = \boldsymbol{B}e^{i(\lambda t + \mu x)}$ are bounded for large $|t|$ if and only if the eigenvalues $-\lambda/\mu$ of $\boldsymbol{B}$ are real (GELFAND [14]).

Suppose $\boldsymbol{u}(x, t)$ is given on a curve $\phi(x, t) = 0$. If this line is characteristic, then the equation

$$\frac{d\boldsymbol{u}}{ds} = \frac{\partial \boldsymbol{u}}{\partial t}\frac{dt}{ds} + \frac{\partial \boldsymbol{u}}{\partial x}\frac{dx}{ds},$$

where $t(s)$, $x(s)$ is a parametric representation for the curve $\phi = 0$, and the quasilinear equation (4.7) cannot be uniquely solved for the $2k$ derivatives $\partial\boldsymbol{u}/\partial t$ and $\partial\boldsymbol{u}/\partial x$. This special condition requires that the determinant of the coefficients of the derivatives vanish

$$\det\begin{bmatrix} \boldsymbol{1} & \boldsymbol{B} \\ \boldsymbol{1}\,dt & \boldsymbol{1}\,dx \end{bmatrix} = \det\left[\boldsymbol{1}\frac{dx}{ds} - \boldsymbol{B}\frac{dt}{ds}\right] = 0.$$

The same considerations apply for the quasilinear steady problem

$$\boldsymbol{B}\frac{\partial \boldsymbol{u}}{\partial x} + \boldsymbol{C}\frac{\partial \boldsymbol{u}}{\partial y} = \boldsymbol{f}.$$

Such problems are frequently associated with a change of type, like transonic flow, in which some regions of flow are subcritical and some supercritical. A typical example is Tricomi's equation. Other, more applicable examples are derived in §§ 8–11.

It is not always possible to assign a definite type to a system of quasilinear equations. There can be both real and complex eigenvalues. Nonlinear problems of mixed type have not been thoroughly studied by mathematicians. Some special results have been given by MOCK [30]. Here it is perhaps useful to give some simple examples from hydrodynamics.

Consider first the Euler equations for the flow of inviscid, incompressible fluids in two dimensions

$$\varrho(\boldsymbol{u} \cdot \nabla)\, \boldsymbol{u} + \nabla p = \boldsymbol{f}, \qquad \operatorname{div} \boldsymbol{u} = 0. \tag{4.10}$$

Let (u, v) be the components of $\boldsymbol{u}$ with respect to x and y. Then we can write (4.10) as

$$A_1 q_x + A_2 q_y = f, \tag{4.11}$$

where

$$q = (u, v, p), \quad q_x = \frac{\partial q}{\partial x}, \quad q_y = \frac{\partial q}{\partial y},$$

$$f = (f_1, f_2, 0),$$

$$A_1 = \begin{bmatrix} u & 0 & 1/\varrho \\ 0 & u & 0 \\ \varrho & 0 & 0 \end{bmatrix},$$

$$A_2 = \begin{bmatrix} v & 0 & 0 \\ 0 & v & 1/\varrho \\ 0 & \varrho & 0 \end{bmatrix}.$$

The characteristic equation for (4.11) is

$$\det\left[\frac{dy}{dx} A_1 - A_2\right] = \left(v - \frac{dy}{dx} u\right)\left(\left(\frac{dy}{dx}\right)^2 + 1\right) = 0.$$

Hence

$$\frac{dy}{dx} = \frac{v}{u}, \quad \text{streamlines are characteristic, and}$$

$$\frac{dy}{dx} = \pm i.$$

The presence of imaginary roots means that (4.11) is not hyperbolic. It is not elliptic beause the determinant of the matrix symbol of (4.11)

$$\det \begin{bmatrix} u\xi_1 + v\xi_2 & 0 & \frac{1}{\varrho}\xi_1 \\ 0 & u\xi_1 + v\xi_2 & \frac{1}{\varrho}\xi_2 \\ \varrho\xi_1 & \varrho\xi_2 & 0 \end{bmatrix} = -(u\xi_1 + v\xi_2)(\xi_1^2 + \xi_2^2)$$

vanishes for $u\xi_1 + v\xi_2 = 0$.

Another example is from the theory of irrotational water waves. In this case the velocity potential is elliptic but the height function is governed by a hyperbolic equation giving rise to water waves.

5. Constitutive Equations

A constitutive equation relates stress to deformation. The stress in viscoelastic fluids depends on the history of the deformation. Usually the history is defined on some strain measure. The stress in Newtonian fluids depends on the instantaneous value of the velocity gradient, not on the prior history of the deformation. Viscoelastic fluids have instantaneous elasticity. Elasticity is present also in inviscid compressible fluids. For unsteady problems elasticity is associated with hyperbolic, rather than parabolic response. It is necessary to be more precise about the difference between elastic and viscous responses.

Many constitutive models have been proposed. Each one leads to different answers for the same problem though some groups have similar qualitative properties. In problems of changing type the linearized part is of primary importance. The linear part may be of three types:

1) Constitutive equations with some viscosity. The viscosity which we have in mind is that which rheologists sometimes associate with a retardation time.

2) Constitutive equations without "viscosity". Constitutive equations of integral type with smooth kernels, and various types of rate equations in the class called Maxwell models are of this type. These kind of equations allow propagation of rather than smoothing of discontinuities. In some nonlinear models [16], [28], [43], [44], [51] discontinuities may arise, as do shocks in gas dynamics, from smooth data.

3) Constitutive equations of integral type with singular kernels. These are in a sense intermediate between 1) and 2). Depending on the type of the singularity, the wave speeds may be finite or infinite. However, even if they are finite, *i.e.* real characteristics exist, there is no propagation of discontinuities (see Chapter 6).

The stress in an incompressible fluid is given by

$$\boldsymbol{T} = -p\mathbf{1} + \boldsymbol{\tau}, \tag{5.1}$$

where $\boldsymbol{\tau}$, the determinate stress [50], p. 176 (sometimes called the extra stress [48]), may be related to the deformation, whilst p, the reaction pressure, is determined only through the equations of motion. An example of $\boldsymbol{\tau}$ in the class 1) of constitutive equation with some viscosity is the Jeffreys model with two time constants, a relaxation time λ_1 and a retardation time λ_2. This model may be written in rate form

$$\boldsymbol{\tau} + \lambda_1 \frac{\partial \boldsymbol{\tau}}{\partial t} = \eta \boldsymbol{A} + \eta\lambda_2 \frac{\partial \boldsymbol{A}}{\partial t}, \tag{5.2}$$

where η is a constant, called the zero shear-rate viscosity, $\boldsymbol{A} = \nabla \boldsymbol{u} + \nabla \boldsymbol{u}^T$, $\boldsymbol{u} = \boldsymbol{u}(\boldsymbol{x}, t)$, or in integral form

$$\boldsymbol{\tau} = \frac{\eta\lambda_2}{\lambda_1} A[\boldsymbol{u}(\boldsymbol{x}, t)] + \frac{\eta}{\lambda_1}\left(1 - \frac{\lambda_2}{\lambda_1}\right) \int_{-\infty}^{t} A[\boldsymbol{u}(\boldsymbol{x}, \tau)] \exp\left(\frac{-(t-\tau)}{\lambda_1}\right) d\tau. \tag{5.3}$$

The constant $\eta\lambda_2/\lambda_1$ is a second viscosity which is equal to the zero shear-rate viscosity when $\lambda_2 = \lambda_1$. It is this viscosity that we have in mind when we talk about "with" or "without" viscosity. In § 6 we show that (5.2) and (5.3) enjoy a certain general status when they are regarded as holding only in motions which perturb a state of rest.

The Maxwell model arises from (5.2) and (5.3) when λ_2 is put equal to zero. In § 6 we note that the Maxwell model permits propagation of *waves* with a finite velocity of propagation, but such propagation cannot occur for Jeffreys models; more precisely the viscosity term $\lambda_2\eta/\lambda_1$ smooths discontinuities in the same way that viscosity smooths the discontinuities of solutions of Euler's equations for an inviscid fluid.

Nonlinear models can be classified according to the type of their linearization. Popular models of the rate type include those due to OLDROYD [34], LEONOV [25], GIESEKUS [15]. These models generalize both Maxwell and Jeffreys type fluids, *i.e.* some have "viscosity", some do not. Popular models of integral type include K-BKZ single integral models [4], [22] which may be of type 2) or 3) depending on the nature of the kernel, and the model of CURTISS & BIRD [11], which contains a viscosity term.

We shall do some work with rate equations of Oldroyd type depending upon three constants:

$$\lambda \frac{D\boldsymbol{\tau}}{Dt} + \boldsymbol{\tau} = \eta \boldsymbol{A}. \tag{5.4}$$

Here D/Dt is an invariant time derivative

$$\frac{D\boldsymbol{\tau}}{Dt} = \frac{\partial \boldsymbol{\tau}}{\partial t} + (\boldsymbol{u}\cdot\nabla)\,\boldsymbol{\tau} + \boldsymbol{\tau}\boldsymbol{\Omega} - \boldsymbol{\Omega}\boldsymbol{\tau} - a(\boldsymbol{D}\boldsymbol{\tau} + \boldsymbol{\tau}\boldsymbol{D}), \tag{5.5}$$

where $-1 \leqq a \leqq 1$, $\boldsymbol{D} = \frac{1}{2}(\nabla\boldsymbol{u} + \nabla\boldsymbol{u}^T)$, $\boldsymbol{\Omega} = \frac{1}{2}(\nabla\boldsymbol{u} - \nabla\boldsymbol{u}^T)$. The upper convected Maxwell model has $a = 1$ and

$$\frac{D\boldsymbol{\tau}}{Dt} = \frac{\partial \boldsymbol{\tau}}{\partial t} + (\boldsymbol{u}\cdot\nabla)\,\boldsymbol{\tau} - \nabla\boldsymbol{u}\boldsymbol{\tau} - \boldsymbol{\tau}\,\nabla\boldsymbol{u}^T, \tag{5.6}$$

where $(\nabla\boldsymbol{u})_{ij} = \partial u_i/\partial x_j$. The lower convected Maxwell model has $a = -1$ and

$$\frac{D\boldsymbol{\tau}}{Dt} = \frac{\partial \boldsymbol{\tau}}{\partial t} + (\boldsymbol{u}\cdot\nabla)\,\boldsymbol{\tau} + \boldsymbol{\tau}\,\nabla\boldsymbol{u} + \nabla\boldsymbol{u}^T\boldsymbol{\tau}. \tag{5.7}$$

The corotational Maxwell model has $a = 0$. The integral model

$$\boldsymbol{\tau} = \frac{\eta}{\lambda^2} \int_{-\infty}^{t} \exp\,[-(t-\tau)/\lambda]\,[\boldsymbol{C}_t^{-1}(\tau) - \boldsymbol{1}]\,d\tau \tag{5.8}$$

is an alternative form of the upper convected Maxwell model ($a = 1$). The rate form (5.4, 6) may be obtained by differentiating (5.8) partially with respect to t, holding $\boldsymbol{x}$ fixed. The expression

$$\boldsymbol{\tau} = \frac{\eta}{\lambda^2} \int_{-\infty}^{t} \exp\,[-(t-\tau)/\lambda]\,[\boldsymbol{1} - \boldsymbol{C}_t(\tau)]\,d\tau \tag{5.9}$$

is equivalent to (5.4, 7) with $a = -1$, in the same way.

RUTKEVICH [40], [41] studies differential constitutive models (5.4) of Oldroyd type. He linearizes these equations and the equations of motion at a state of no motion ($\boldsymbol{u} = 0$) and constant stress. He finds that a change of type leading to imaginary wave speeds and a instability of Hadamard type occur if the principal values of $\boldsymbol{\tau}$ satisfy certain inequalities. If we denote these principal values by $\tau_1 \geqq \tau_2 \geqq \tau_3$, then RUTKEVICH'S instability criterion reads as follows:

$$\tau_3 < -\frac{\eta}{\lambda}, \qquad a = 1,$$

(5.10)
$$\tau_1 > \frac{\eta}{\lambda}, \qquad a = -1,$$

$$\tau_1 - \tau_3 > \frac{2\eta}{\lambda}, \qquad a = 0.$$

In order to assess the significance of these results, it is necessary to look also at the integral models (5.8) and (5.9) corresponding to (5.4). Since $\boldsymbol{C}_t^{-1}$ and $\boldsymbol{C}_t$ are positive definite matrices, we find that $\tau_3 > -\frac{\eta}{\lambda}$ for $a = 1$ and $\tau_1 < \frac{\eta}{\lambda}$ for a -1. Thus Rutkevich's instability criterion cannot be achieved on solutions and must be considered unphysical. This was also noticed by CROCHET [10] who proved directly from the differential equation that $\tau_3 > -\frac{\eta}{\lambda}$ or, respectively, $\tau_1 < \frac{\eta}{\lambda}$ for all time, if this is the case initially.

M. RENARDY [38, 39] did two independent studies, related in part to our joint work. He has shown that the upper and lower convected Maxwell models are always evolutionary. These two models are special cases of the Kaye-BKZ model [22], [4], which has the form

(5.11)
$$\boldsymbol{\tau} = \int_{-\infty}^{t} a(t-s)\left(\frac{\partial W}{\partial I_1}\boldsymbol{C}_t^{-1}(s) - \frac{\partial W}{\partial I_2}\boldsymbol{C}_t(s)\right) ds.$$

Here a is a positive kernel and the "strain energy" W is a scalar function of $I_1 = \operatorname{tr} \boldsymbol{C}_t^{-1}(s)$ and $I_2 = \operatorname{tr} \boldsymbol{C}_t(s)$. The initial value problem is always well posed, *i.e.* instability of Hadamard type cannot occur, if W satisfies a strong ellipticity condition of the same form as in nonlinear elasticity. This condition is satisfied if W is monotone in both arguments and a convex function of $\sqrt{I_1}$ and $\sqrt{I_2}$. This obviously includes the cases $W = I_1$, $W = I_2$, corresponding to (5.8), (5.9). No such objections can be raised in the corotational case, $a = 0$. In fact, we need only restrict attention to motions where $\boldsymbol{\Omega} = 0$, and $\boldsymbol{\tau}$ and $\boldsymbol{D}$ are spatially homogeneous, to see that (5.4) allows $\tau_1 - \tau_3$ to reach any value. The instability in this case is therefore genuine.

It is instructive to look at special flow geometries. For time-dependent simple shear-flow, we have [21]

$$\tau = \int_{-\infty}^{t} \frac{\eta}{\lambda} e^{-(t-s)/\lambda} \varkappa(s) \cos\left(\sqrt{1-a^2}\int_{s}^{t} \varkappa(t')\,dt'\right) ds.$$

Differentiating this with respect to t, we find

$$\tau = \varkappa(t)\left[\frac{\eta}{\lambda} - \int_{-\infty}^{t} \frac{\eta}{\lambda} e^{-(t-s)/\lambda} \varkappa(s) \sqrt{1-a^2} \sin\left(\sqrt{1-a^2} \int_{s}^{t} \varkappa(t')\, dt'\right) ds\right]$$

$$+ \text{ terms of lower order.}$$

A change of type occurs when the expression in brackets changes sign. For $|a| < 1$, it is possible to construct histories for the shear rate $\varkappa(s)$ such that this is the case. For steady shear flows, $\varkappa \equiv \varkappa_0$, the expression in brackets is positive and there is no change of type, even though the shear stress-shear rate law for $a \neq \pm 1$ is not monotone. The case of simple elongation was considered by RENARDY [38]. He finds that a change of type can occur for $-1 < a < \frac{1}{2}$. Although RUTKEVICH'S instability for $a = \pm 1$ is unphysical it is nevertheless relevant for numerical calculations (see CROCHET [10]). The reason is that at high Weissenberg numbers the eigenvalues of $\boldsymbol{\tau}$ can be arbitrarily close to the stability boundary $\pm\eta/\lambda$. Numerical errors can push them beyond this limit, with disastrous results.

Most of the constitutive equations which have been proposed, all the ones considered here, are simple fluids in which the stress is determined by the history of the relative gradient of the deformation

$$\boldsymbol{F}_t(\boldsymbol{x}, t) = \nabla \boldsymbol{\chi}_t(\boldsymbol{x}, t),$$

where

$$\boldsymbol{u}(\boldsymbol{\chi}_t(\boldsymbol{x}, \tau), \tau) = \partial \boldsymbol{\chi}_t / \partial \tau$$

is the velocity of the particle $\boldsymbol{x} = \boldsymbol{\chi}_t(\boldsymbol{x}, t)$ at times $\tau \leqq t$. The constitutive equation satisfying material frame-indifference may be expressed by a functional [48, p. 80]

$$\boldsymbol{\tau}(\boldsymbol{x}, t) = \underset{\tau=-\infty}{\overset{t}{F}} [\boldsymbol{C}_t(\boldsymbol{x}, \tau)] \tag{5.12}$$

on the history of the relative Cauchy strain

$$\boldsymbol{C}_t(\boldsymbol{x}, \tau) = \boldsymbol{F}_t^T \boldsymbol{F}_t.$$

In order to give a precise meaning to an expression such as (5.12), it is necessary to specify the set of arguments on which F is defined, called the domain of F or dom F. SAUT & JOSEPH [42], extending ideas of COLEMAN & NOLL [9], have proposed a classification of constitutive laws according to the choice of this domain. Roughly, this associates the nature of the stress-strain relation with the deformations which are allowed. In the linearized case the stresses which are allowed are functionals in the topological dual of dom F. The smaller dom F is, the larger is the dual, *i.e.* the more constitutive laws are allowed. To be more precise, let $\boldsymbol{C}_t^0(\tau)$ be some given history and $\boldsymbol{c}(\tau) = \boldsymbol{c}(x, t-\tau)$ be a small perturbation. The linearization of $\boldsymbol{F}$ relative to $\boldsymbol{C}_t^0(\tau)$ can formally be written as an integral

$$\boldsymbol{F}_1\left[\underset{\tau=-\infty}{\overset{t}{\boldsymbol{C}_t^0(\tau)}} \,\middle|\, \boldsymbol{c}(\tau)\right] = \int_{-\infty}^{t} \mathbf{K}\left(\underset{\tau'=-\infty}{\overset{t}{\boldsymbol{C}_t^0(\tau')}}, t-\tau\right) \boldsymbol{c}(\tau)\, d\tau, \tag{5.13}$$

where $\mathbf{K}$ is a function of $t-\tau$ and functional of the history of C_t^0 whose values are tensors of fourth order. The class of admissible kernels $\mathbf{K}$ depends on which deformations $c(\tau)$ are allowed. If, following COLEMAN & NOLL, c is restricted to a weighted L^2-space, then $\mathbf{K}$ must also be in a weighted L^2-space. If c is restricted to a Sobolev space, then Dirac measures and derivatives of Dirac measures can be included in $\mathbf{K}$, thus admitting Jeffreys type models.

To see the effects of hyperbolicity clearly we exclude these cases and adopt (5.13) with a smooth kernel as the basis for our study of change of type.

It will be useful in what follows to specify the quantities in (5.13) more precisely. Let

$$\chi_t(\boldsymbol{x}, \tau) = \xi^0(\boldsymbol{x}, \tau) + \xi(\boldsymbol{x}, \tau) \tag{5.14}$$

be the particle path for

$$\boldsymbol{x} = \chi(\boldsymbol{x}, t) = \xi^0(\boldsymbol{x}, t),$$

where ξ^0 is the position of $\boldsymbol{x}$ in some given motion and $\xi(\boldsymbol{x}, \tau)$ is a perturbation with $\xi(\boldsymbol{x}, t) = 0$. Quadratic and terms of higher-order are neglected. To this order of approximation

$$C_t(\boldsymbol{x}, \tau) = C_t^0(\boldsymbol{x}, \tau) + c(\boldsymbol{x}, \tau), \tag{5.15}$$

where

$$\begin{aligned} C_t^0(x, \tau) &= (\nabla \xi^0)^T \nabla \xi^0, \quad C_t^0(x, t) = \mathbf{1}, \\ c(x, \tau) &= (\nabla \xi^0)^T \nabla \xi + (\nabla \xi)^T \nabla \xi^0, \quad c(x, t) = 0. \end{aligned} \tag{5.16}$$

Let the perturbed extra stress be denoted by $\boldsymbol{\tau}$ and the unperturbed extra stress by $\boldsymbol{\tau}_0$. The total extra stress is $\boldsymbol{\tau}_0 + \boldsymbol{\tau}$, where

$$\begin{aligned} \boldsymbol{\tau}_0 &= \mathop{F}_{\tau=-\infty}^{t} [C_t^0(x, \tau)], \\ \boldsymbol{\tau} &= \int_{-\infty}^{t} \mathop{\mathbf{K}}_{\tau'=-\infty}^{t} [C_t^0(x, \tau'), t-\tau]\, c(x, \tau)\, d\tau. \end{aligned} \tag{5.17}$$

We assume that $\mathbf{K}$ is a smooth function of τ and find that

$$\left(\frac{\partial}{\partial t} + \boldsymbol{u}^0 \cdot \nabla\right) \boldsymbol{\tau} = \int_{-\infty}^{t} \mathbf{K} \left(\frac{\partial}{\partial t} + \boldsymbol{u}^0 \cdot \nabla\right) c(x, \tau)\, d\tau \tag{5.18}$$

$$+ \text{ terms of lower differential order},$$

in which $\boldsymbol{u}^0$ pertains to the unperturbed motion. The position at time τ of any given particle is independent of the reference time t.

$$\frac{d\chi_t}{dt} = \frac{\partial \chi_t(x, \tau)}{\partial t} + \boldsymbol{u}(x, t) \cdot \nabla_x \chi_t(x, \tau) = 0. \tag{5.19}$$

Hence

$$\frac{\partial \xi^0}{\partial t} + u^0(x,t)\cdot\nabla_x \xi^0 = 0, \tag{5.20}$$

$$\frac{\partial \xi}{\partial t} + u^0(x,t)\cdot\nabla_x \xi + v(x,t)\cdot\nabla_x \xi^0 = 0.$$

This yields

$$\begin{aligned}\left(\frac{\partial}{\partial t} + (u^0\cdot\nabla)\right) c &= \left(\frac{\partial}{\partial t} + (u^0\cdot\nabla)\right)(\nabla\xi^{0T}\nabla\xi + \nabla\xi^T\nabla\xi^0)\\ &= \nabla\xi^{0T}\nabla\left\{\left(\frac{\partial}{\partial t} + (u^0\cdot\nabla)\right)\xi\right\} + \left(\nabla\left\{\left(\frac{\partial}{\partial t} + u^0\cdot\nabla\right)\xi\right\}\right)^T\nabla\xi^0\\ &\quad + \text{terms of lower order}\\ &= -\nabla\xi^{0T}\cdot\nabla(v\cdot\nabla\xi^0) - (\nabla(v\cdot\nabla\xi^0))^T\nabla\xi^0\\ &\quad + \text{terms of lower order}\\ &= -(\nabla\xi^0)^T\nabla v\,\nabla\xi^0 + \text{transpose} + \text{terms of lower order}.\end{aligned}$$

By inserting this in (5.18), we find that the principal part of the constitutive equation is given by

$$\left[\frac{\partial}{\partial t} + u_0\cdot\nabla\right]\tau = \mathbf{M}(x,t)\,\nabla v(x,t), \tag{5.21}$$

where

$$M_{ijmn} = -\int_{-\infty}^{t} K_{ijkl}\frac{\partial C_m^0}{\partial x_k}\frac{\partial C_n^0}{\partial x_l}\,d\tau.$$

Equations like (5.21) can also be derived in general, without linearization (§ 12).

6. Slip Surface Propagation in Problems Perturbing Rest

In problems which perturb rest we have $\xi^0 = x$, $\tau^0 = 0$ and

$$\tau = \int_{-\infty}^{t} G(t-\tau)\,A_1[u(x,\tau)]\,d\tau. \tag{6.1}$$

A Newtonian part of the stress arises when

$$G(s) = \mu\,\delta(s) + g(s),$$

where $\delta(s)$ is a Dirac measure and $g(s)$ is a smooth kernel. Thus

$$\tau = \mu A_1([u(x,t)] + \int_{-\infty}^{t} g(t-\tau)\,A_1[u(x,\tau)]\,d\tau. \tag{6.2}$$

In the special case when $g(s)$ is in exponential form (6.2) reduces to the model (5.3) of Jeffreys with $\mu = \eta\lambda_2/\lambda_1$. We introduce μ here and elsewhere only to notice that if μ is small the underlying dynamics is close to $\mu = 0$ dynamics. When $\mu = 0$, the dynamics is governed by

$$\frac{\partial \boldsymbol{\tau}}{\partial t} = G(0)\, \boldsymbol{A}_1[\boldsymbol{u}(\boldsymbol{x}, t)] + \int_{-\infty}^{t} G'(t-\tau)\, \boldsymbol{A}_1[\boldsymbol{u}(\boldsymbol{x}, \tau)]\, d\tau, \tag{6.3}$$

and

$$\varrho \frac{\partial \boldsymbol{u}}{\partial t} + \nabla p - \operatorname{div} \boldsymbol{\tau} = 0, \tag{$6.3)_2$}$$

where

$$\operatorname{div} \boldsymbol{u} = 0. \tag{$6.3)_3$}$$

Equations (6.3) are a first-order system, linear in the derivatives of p and the components of $\boldsymbol{u}$ and $\boldsymbol{\tau}$. If $G(t-\tau) = \text{const} \exp\{-(t-\tau)/\lambda\}$, then the last term in $(6.3)_1$ reduces to $-\boldsymbol{\tau}/\lambda$. In the analysis of characteristics using the Maxwell model the term $\boldsymbol{\tau}/\lambda$ is of lower order and it does not enter into the analysis of characteristics. In general, for smooth G', in (6.3) the integral is of order -1 in t and $+1$ in x, hence of order 0 as an operator in x and t, and is thus also of lower order. This shows that we can analyze first-order systems for general kernels; *we do not need special models.*

The case of one-dimensional shearing motion can be used to discuss the effects of viscosity, wave speed, wave amplitude and classify the kernels. In this case (6.3) reduces to

$$\begin{aligned} \frac{\partial \tau}{\partial t} &= G(0)\frac{\partial u}{\partial x} + \int_{-\infty}^{t} G'(t-\tau)\frac{\partial u(x,\tau)}{\partial x}\, d\tau, \\ \varrho \frac{\partial u}{\partial t} &= \frac{\partial \tau}{\partial x}. \end{aligned} \tag{6.4}$$

We may write (6.4) as

$$\frac{\partial \boldsymbol{q}}{\partial t} + \boldsymbol{B}\frac{\partial \boldsymbol{q}}{\partial x} + \boldsymbol{f} = 0, \tag{6.5}$$

where

$$\boldsymbol{q} = \begin{bmatrix} u \\ \tau \end{bmatrix},$$

$$\boldsymbol{B} = -\begin{bmatrix} 0 & 1/\varrho \\ G(0) & 0 \end{bmatrix},$$

$$\boldsymbol{f} = \begin{pmatrix} 0 \\ \int_{-\infty}^{t} G'(t-\tau)\frac{\partial u(x,\tau)}{\partial x}\, d\tau \end{pmatrix}.$$

The eigenvalues

$$C = \pm\sqrt{G(0)/\varrho} \tag{6.6}$$

of $\boldsymbol{B}$ are the wave speeds along the characteristics

$$x \pm ct = \text{const.}$$

It is instructive to review the results of analysis of the following problem, known as Stokes' or Rayleigh's problem, for a viscoelastic fluid. A fluid occupying a half space is at rest for $t \leqq 0$. At times $t \geqq 0$, the boundary of the fluid at $x = 0$ is made to move forward with a constant speed. The problem may be described by the following equations

$$\varrho \frac{\partial u}{\partial t} = \mu \frac{\partial^2 u}{\partial x^2} + \int_{-\infty}^{t} g(t-\tau) \frac{\partial^2 u(x, \tau)}{\partial x^2} \, d\tau,$$

(6.7) $$u(x, \tau) = 0, \qquad \tau \leqq 0,$$

$u(x, t)$ is bounded for positive values of x and t,

$$u(0, t) = H(t) = \begin{cases} 0, & t < 0 \\ 1, & t > 0 \end{cases}.$$

This problem has been studied by NARAIN & JOSEPH [33] when $\mu \neq 0$ and $\mu = 0$ and by RENARDY [37] for $\mu = 0$ and singular kernels. Other authors, TANNER [46], STRAUB [45], BÖHME [5], KAZAKIA & RIVLIN [24] have studied the problem for special constitutive models. TANNER used an Oldroyd model B, which contains a contribution due to Newtonian viscosity. This model does not require linearization; it automatically leads to a linear problem. All authors solve the problem using a Laplace transform with respect to the time.

In the context of the present paper, the most interesting issue is the qualitative behavior of solutions in relation to properties of the kernel g. In particular, wave speeds and the presence or absence of discontinuities are of interest. These questions were addressed in [33], [37]. Not surprisingly, the crucial factor is the asymptotic behavior of the Laplace transform of g at infinity or in other words, the symbol of the operator in (6.7). The qualitative nature of solutions is thus determined by the type of the equation. With $\hat{u}(x, \omega)$ denoting the Laplace transform of $u(x, t)$, equation (6.7) becomes

(6.8) $$\omega \varrho^2 \hat{u} = \omega \hat{G}(\omega) \, \hat{u}_{xx}, \qquad \hat{u}(x = 0) = \frac{1}{\omega},$$

where $\hat{G}(\omega)$ is the Laplace transform of G:

$$\hat{G}(\omega) = \int_0^{\infty} G(t) \, e^{-\omega t} \, dt.$$

One finds that

$$\hat{G}(\omega) = \frac{G(0)}{\omega} + \frac{G'(0)}{\omega^2} + o\left(\frac{1}{\omega^2}\right),$$

if G'' is integrable. Since the character of solutions is determined by the symbol, we should look at the problem

$$\varrho \omega^2 \hat{u} = G(0) \, \hat{u}_{xx} + \frac{1}{\omega} G'(0) \, \hat{u}_{xx}.$$

Since

$$\frac{1}{G(0)+\frac{1}{\omega}G'(0)}=\frac{1}{G(0)}-\frac{1}{\omega}\frac{G'(0)}{G(0)^2},$$

an equivalent statement is

$$\frac{\varrho\omega^2}{G(0)}\hat{u}-\frac{\varrho\omega G'(0)}{G(0)^2}\hat{u}=\hat{u}_{xx}.$$

This corresponds to the equation

$$\frac{\varrho}{G(0)}u_{tt}-\frac{\varrho G'(0)}{G(0)^2}u_t=u_{xx}.$$

Hence the singularity may be expected to propagate at the wave speed $c=\sqrt{G(0)/\varrho}$ and to decay with an amplitude factor $\exp(tG'(0)/2G(0))$.

COLEMAN & GURTIN [6], [7] proved that if the acceleration was discontinuous, then its wave speed and amplitude had to be given by these expressions. In [33] and [37] it is shown that these expressions for the speed and amplitude could be derived for the problem (5.7) without assuming discontinuities of acceleration. The demonstration of propagation in [33] applies to propagating steps in displacement (propagating delta functions) as well as steps in velocity.

If a Newtonian term is included in the constitutive law, G contains a contribution $\mu\,\delta(0)$, and $\hat{G}(\omega)$ at ∞ behaves like $O(1)$. The equation becomes parabolic, and smooth (analytic) solutions are obtained. As $\mu\to 0$, a boundary layer forms around the shock front. This was shown numerically in [46]. An analysis of this boundary layer is given by NARAIN & JOSEPH [33].

RENARDY [37] studied the case where G does not contain a δ-function, but some weaker singularity. Specifically, he considers the kernels

$$-G'(t)=\sum_{n=1}^{\infty}e^{-n^{\alpha}t},\qquad \alpha>\tfrac{1}{2}. \tag{6.9}$$

For $\alpha>1$, $G(0)$ is finite, but $G'(0)$ is not. The asymptotic behavior of $\hat{G}(\omega)$ is given by

$$\hat{G}(\omega)=\frac{G(0)}{\omega}+O(\omega^{-2+1/\alpha}). \tag{6.10}$$

As one expects, there is a finite wave speed $c=\sqrt{G(0)/\varrho}$. Solutions are zero in front of the wave and not zero behind it. Across the wave, however, they are of class C^{∞}. If $\alpha<1$, $G(0)$ is infinite, and the asymptotic behaviour of $\hat{G}(\omega)$ is dominated by a term $O(\omega^{1/\alpha-2})$ or $O(\ln\omega/\omega)$ in the limiting case $\alpha=1$. The wave speed is infinite and the solution becomes analytic everywhere except at $t=0$.

Similar studies can be done for small perturbations of rigid motions; see the work of KAZAKIA [23].

7. Slip Surface Propagation in Nonlinear Shearing Problems

COLEMAN & GURTIN [6], [7] studied simple shear flows of a viscoelastic liquid, which involve a surface across which the acceleration is discontinuous. They showed that in a material with a smooth memory function (no Newtonian viscosity), acceleration discontinuities propagate at the speed $c = \sqrt{G(0)/\varrho}$. If the amplitude is smaller than a critical amplitude, it decays with a factor $\exp\,(tG'(0)/2G(0))$, if the amplitude is larger than the critical amplitude, it will reach infinity in a finite time ("blow-up"). This blow-up of acceleration waves might be interpreted as development of a slip surface for the velocity. No such result has been proven.

SLEMROD [44], MALEK-MADANI & NOHEL [28], GRIPENBERG [51] and HATTORI [16] have studied simple nonlinear models. They show by contradiction using Riemann invariants and the method of characteristics that, for suitably chosen initial data, a global C^1-solution for the equation of motion cannot exist. This may mean the formation of a slip surface for the velocity [28]. Numerical evidence for the development of such discontinuities was found by MARKOVICH & RENARDY [29].

SLEMROD [43], following a suggestion of COLEMAN & GURTIN [7], uses his result to explain some of the various instabilities of shear flows collectively called melt fracture. Melt fracture is an instability of flow of molten polymers or polymeric solutions down capillaries. In the experiments (TORDELLA [47]), the polymer is forced down the capillary by high pressure. Extrudates leaving the capillary which at lower shear rates are smooth and continuous, become rough (shark skin effect), irregular and ultimately disintegrate. There are different explanations of the different types of instability which can occur. These are reviewed in the paper of PETRIE & DENN [35]. None of the explanations can be regarded as established. The mechanism proposed by SLEMROD has some possibilities when fracture is associated with a stick-slip phenomenon. There is some controversy about the presence or absence of slip in experiments. If it does occur periodically, as it might in SLEMROD'S theory, it would be a candidate for the explanation of the wavy surfaces shown in the pictures of TORDELLA [47]. It should be noted that the theory of HUNTER & SLEMROD [17] requires an entirely different type of shear stress to explain the type of hysteretic melt fracture which TORDELLA calls ripple.

It would be interesting to have the conditions under which slip surfaces for the velocity might develop from weaker slip surfaces or from smooth data.

8. Classification of Equations for Flows Perturbing Uniform Motion

ULTMAN & DENN [49] consider the equations for two-dimensional steady flow of an upper convected Maxwell fluid. They linearize at a motion with uniform velocity and zero stress, and they show that these linearized equations change type when a viscoelastic "Mach" number

$$M = \frac{U}{c} \tag{8.1}$$

exceeds one. Here U is the velocity of the unperturbed uniform flow, and c is the wave speed for propagation of singularities as considered in § 6: $c = \sqrt{G(0)/\varrho} = \sqrt{\eta/(\varrho\lambda)}$.

ULTMAN & DENN attempted to correlate some experimental observations of D. F. JAMES [18] with this change of type. JAMES observes a sudden change in the slope of the heat transfer curve as a function of velocity. This happens at a critical velocity, which, for the polyox solution used by JAMES, was about 1 cm/sec. It is not clear from the graphs how abrupt this change in slope is, but there is a change of slope. ULTMAN & DENN also suggest that the transition from subcritical to supercritical flow might explain abrupt changes in the drag coefficient observed by A. FABULA [13]. Again, the idea is that the critical velocity at transition is the wave speed c. They make an estimate of c from a molecular theory and correlate this prediction with the data of JAMES. Of course any such estimate can be expected to give at best an order of magnitude, since the fluids used in experiments are not really Maxwell fluids.

M. LUSKIN [27] studies the equations of ULTMAN & DENN as a first order system. He reduces them to canonical form and investigates characteristics. The streamline is a double characteristic, two characteristics are always complex, and the remaining two are complex for $M < 1$ and real for $M > 1$. Each characteristic has associated with it a canonical variable, which is a linear combination of the two velocity components, the three components of the extra stress and the pressure. This means that two of these variables can be discontinuous across streamlines, and two others can be discontinuous only for $M > 1$.

We shall now show how results similar to those of ULTMAN & DENN apply to general constitutive models without "viscosity". In linearization at uniform motion, the extra stress is given by

$$\boldsymbol{\tau} = \int_{-\infty}^{t} G(t-\tau)\, A[\boldsymbol{u}(\zeta, \tau)]\, d\tau, \tag{8.2}$$

where

$$\zeta = \begin{pmatrix} x_1 - U(t-\tau) \\ x_2 \end{pmatrix}.$$

We assume G is smooth, positive and monotone decreasing. By differentiating (8.2) with respect to t, holding $\boldsymbol{x}$ fixed, we find

$$\frac{\partial \boldsymbol{\tau}}{\partial t} + U \frac{\partial \boldsymbol{\tau}}{\partial x_1} = G(0)\, A[\boldsymbol{u}(\boldsymbol{x}, t)] + \int_{-\infty}^{t} G'(t-\tau)\, A[\boldsymbol{u}(\zeta, \tau)]\, d\tau, \tag{8.3}$$

where the last term is of lower differential order. The leading term is the same as for the Maxwell model, if $G(0)$ is replaced by η/λ.

It can be shown that the change of type is primarily associated with the behavior of the vorticity. Since we study motions in the plane, the vorticity curl $\boldsymbol{u}$ has only one component, which we denote by α. We take the curl of the equation of motion, apply the operation curl div to (8.3) and combine the two. In steady flow this leads to

$$(M^2 - 1)\frac{\partial^2 \alpha}{\partial x_1^2} - \frac{\partial^2 \alpha}{\partial x_2^2} = \int_{-\infty}^{t} G'(t-\tau)\, \Delta\alpha(\zeta)\, d\tau. \tag{8.4}$$

The right side is again of lower order. The left hand side is elliptic when $M < 1$ and hyperbolic when $M > 1$. The elliptic roots found by LUSKIN [27] correspond to the elliptic equatlon $\alpha = -\Delta\psi$ expressing the vorticity in terms of the stream function.

9. Classification of the Quasilinear System of Equations Governing the Steady Flow of an Upper Convected Maxwell Fluid

The flow of an upper convected Maxwell fluid is governed by the following system of equations

$$\varrho \frac{d\boldsymbol{u}}{dt} + \nabla p - \operatorname{div} \boldsymbol{\tau} = 0, \qquad \operatorname{div} \boldsymbol{u} = 0,$$

(9.1)

$$\boldsymbol{\tau} + \lambda \left[\frac{d\boldsymbol{\tau}}{dt} - \nabla \boldsymbol{u} \boldsymbol{\tau} - \boldsymbol{\tau} \nabla \boldsymbol{u}^T \right] = \eta \boldsymbol{A} \stackrel{\text{def}}{=} \mu\lambda \boldsymbol{A}.$$

We consider two-dimensional flows. In view of the applications in Chapter 10, we want to write the equations in both Cartesian and polar coordinates. In cartesian coordinates, we set

(9.2)
$$\boldsymbol{u} = (u, v), \qquad \boldsymbol{\tau} = \begin{bmatrix} \sigma & \tau \\ \tau & \gamma \end{bmatrix}.$$

We then obtain the equations

(9.3)
$$\begin{gathered}
\sigma_t + u\sigma_x + v\sigma_y - 2\sigma u_x - 2\tau u_y - 2\mu u_x = -\sigma/\lambda, \\
\tau_t + u\tau_x + v\tau_y - \gamma u_y - \sigma v_x - \mu(u_y + v_x) = -\tau/\lambda, \\
\gamma_t + u\gamma_x + v\gamma_y - 2\tau v_x - 2\gamma v_y - 2\mu v_y = -\gamma/\lambda, \\
\varrho(u_t + uu_x + vu_y) + p_x - \sigma_x - \tau_y = 0, \\
\varrho(v_t + uv_x + vv_y) + p_y - \tau_x - \gamma_y = 0, \\
u_x + v_y = 0.
\end{gathered}$$

The subscripts denote differentiation. In polar coordinates, we denote by u and v the radial and azimuthal velocities and by σ, τ, γ the components of the stress in polar coordinates. We then obtain the following system

(9.4)
$$\sigma_t + u\sigma_r + \frac{v\sigma_\theta}{r} - 2\sigma u_r - 2\tau \frac{u_\theta}{r} - 2\mu u_r = -\frac{\sigma}{\lambda},$$

$$\tau_t + u\tau_r + \frac{v\tau_\theta}{r} - \frac{\gamma u_\theta}{r} - \sigma v_r - \mu\left(v_r + \frac{u_\theta}{r}\right) = -\frac{\tau}{\lambda} - \frac{\sigma v}{r} - \frac{\mu v}{r},$$

$$\gamma_t + u\gamma_r + \frac{v\gamma_\theta}{r} - 2\tau v_r - 2\gamma \frac{v_\theta}{r} - 2\mu \frac{v_\theta}{r} = -\frac{\gamma}{\lambda} - \frac{2v\tau}{r} + \frac{2u}{r}(\gamma + \mu),$$

$$\varrho\left(u_t + uu_r + v\frac{u_\theta}{r}\right) + p_r - \sigma_r - \frac{\tau_\theta}{r} = \varrho\frac{v^2}{r} + \frac{\sigma - \gamma}{r},$$

$$\varrho\left(v_t + uv_r + \frac{vv_\theta}{r}\right) + \frac{p_\theta}{r} - \tau_r - \frac{\gamma_\theta}{r} = -\frac{\varrho uv}{r} + \frac{2\tau}{r},$$

$$u_r + \frac{v_\theta}{r} = -\frac{u}{r}.$$

The terms on the right of (9.3) and (9.4) are of lower order and do not enter into the analysis of characteristics. The systems on the left are identical if we make the identifications $\frac{\partial}{\partial x} \sim \frac{\partial}{\partial r}$ and $\frac{\partial}{\partial y} \sim \frac{1}{r}\frac{\partial}{\partial \theta}$.

We consider steady solutions of (9.3) and (9.4). Thus we put the time derivatives equal to zero and introduce $\boldsymbol{q} = (u, v, \sigma, \tau, \gamma, p)$. Equations (9.3) and (9.4) are of the respective forms

(9.5) $$\boldsymbol{A}\boldsymbol{q}_x + \boldsymbol{B}\boldsymbol{q}_y = \boldsymbol{f},$$

and

(9.5) $$\hat{\boldsymbol{A}}\boldsymbol{q}_r + \hat{\boldsymbol{B}}\frac{\boldsymbol{q}_\theta}{r} = \hat{\boldsymbol{f}}.$$

Characteristics of (9.5) are determined by the equations

(9.7) $$\det\left(\frac{dy}{dx}\boldsymbol{A}(\boldsymbol{q}) - \boldsymbol{B}(\boldsymbol{q})\right) = 0.$$

With $\alpha = -\frac{dy}{dx}$, this leads to

(9.8) $$(\alpha u + v)^2 (1 + \alpha^2) [2\alpha\tau - \varrho(\alpha u + v)^2 + \alpha^2\sigma + \gamma + (1 + \alpha^2)\mu] = 0.$$

Hence the streamlines

(9.9) $$\frac{dy}{dx}u - v = 0$$

are double characteristics. There are two imaginary roots $\alpha = \pm i$. Finally, the last bracket yields

(9.10)

$$-\frac{dy}{dx} = \frac{\varrho uv - \tau}{\mu + \sigma - \varrho u^2} \pm \left\{\frac{\tau^2 - 2\varrho\tau uv - (\mu + \gamma)(\mu + \sigma) + \varrho v^2(\mu + \sigma) + \varrho u^2(\mu + \gamma)}{(\mu + \sigma - \varrho u^2)^2}\right\}^{\frac{1}{2}}.$$

These characteristics are real or complex depending on whether the argument of the square root is positive or negative. This argument of course depends on the solution. It should be kept in mind here that the integral form of the Maxwell model imposes certain restrictions on the stresses (see Chapter 5). These roots are the ones which exhibit a change of type.

The imaginary roots $\frac{dy}{dx} = \pm i$ arise from the identity

$$\zeta = -\nabla^2\psi, \tag{9.11}$$

where ζ is the vorticity and ψ is the stream function. It is easy to show that

(9.12)
$$(\mu + \sigma - \varrho u^2)\frac{\partial^2\zeta}{\partial x^2} + (\mu + \gamma - \varrho v^2)\frac{\partial^2\zeta}{\partial y^2} + 2(\tau - \varrho uv)\frac{\partial^2\zeta}{\partial x\,\partial y} = \text{lower order}.$$

This shows that the interesting characteristic roots (9.10) which can change type are associated with the vorticity. Using (9.11), we find that the quantity on left of (9.12) with ψ replacing ζ, is harmonic to within terms of lower order. In polar coordinates, the equation for characteristics becomes

$$-r\frac{d\theta}{dr} = \text{the right-hand side of (9.10)}, \tag{9.13}$$

with u, v, τ, γ, σ interpreted as the physical components of velocity and stress in polar coordinates.

10. Change of Type in Shear Flow, Extensional Flow, Sink Flow and Circular Couette Flow of an Upper Convected Maxwell Fluid

In this section we consider the problem of hyperbolicity of the linear equations perturbing some special solutions of (9.1). The characteristics equations (9.10) and (9.13) are useful for this. To compute the characteristics of the perturbation we need only to replace the u, v, τ, σ, γ in these formulas with the special values that (9.1) requires for special flows.

(i) First, we again conider the uniform flows discussed in § 8; $u = U$, $v = \tau = \gamma = \sigma = 0$. For these we find from (9.10) that

$$\frac{dy}{dx} = \pm\sqrt{\frac{1}{M^2 - 1}}, \tag{10.1}$$

where $M^2 = U^2/C^2$, $C^2 = \eta/\lambda\varrho \left(= \frac{G(0)}{\varrho}\right)$ is the wave speed. The characteristics are straight lines which start as lines perpendicular to the flow at $M = 1$ and tilt more toward the free stream as $M > 1$ is increased.

(ii) Our second application is to shear flow. We find, using (9.3), that

$$u = \varkappa y, \quad \tau = \eta\varkappa; \quad \sigma = 2\eta\lambda\varkappa^2, \quad \gamma = v = 0, \tag{10.2}$$

where the shear rate $\varkappa$ is a constant. Inserting the fields (10.2) into (9.10), we find that

$$\begin{aligned}\frac{dy}{dx} &= \frac{\lambda\varkappa}{1+2\lambda^2\varkappa^2-\dfrac{\varkappa^2y^2}{c^2}} \pm \sqrt{\frac{\dfrac{\varkappa^2y^2}{c^2}-\lambda^2\varkappa^2-1}{\left[1+2\lambda^2\varkappa^2-\dfrac{\varkappa^2y^2}{c^2}\right]^2}} \\ &= \frac{1}{\lambda\varkappa \pm \sqrt{\dfrac{\varkappa^2y^2}{c^2}-\lambda^2\varkappa^2-1}}.\end{aligned} \tag{10.3}$$

The vorticity is hyperbolic outside a strip defined by

$$\frac{\varkappa^2y^2}{c^2} > \lambda^2\varkappa^2+1. \tag{10.4}$$

(iii) In extensional flow, we find, using (9.3), that

$$u = sx,\ v = -sy,\quad \tau = 0,\quad \sigma = 2\eta s/(1-2s\lambda),\quad \gamma = -2\eta s/(1+2\lambda s), \tag{10.5}$$

where $0 < s < 1/2\lambda$. The stretch rate s is a positive constant, small enough to keep σ bounded and positive. The unbounded σ at $s = 1/(2\lambda)$ is one of the many undesirable properties the upper-convected Maxwell model. Inserting the field (10.5) into (9.10), we find that

$$\frac{dy}{dx} = \frac{\varrho s^2xy}{\dfrac{\mu}{1-2\lambda s}-\varrho s^2x^2} \pm \frac{\mu}{\sqrt{1-4\lambda^2s^2}}\left\{\frac{x^2}{a^2}+\frac{y^2}{b^2}-1\right\}^{\frac{1}{2}}, \tag{10.6}$$

where

$$a^2 = \frac{\mu}{\varrho s^2(1-2\lambda s)},\qquad b^2 = \frac{\mu}{\varrho s^2(1+2\lambda s)}.$$

We get real characteristics, obeying (10.6), outside the ellipse

$$\frac{x^2}{a^2}+\frac{y^2}{b^2} > 1.$$

(iv) Our fourth application is to sink flow. Here we must work with the polar equations (9.4) and (9.13). For sink flow

$$u = -Q/r,\quad Q > 0 \text{ is the sink strength}, \tag{10.7}$$

$$v = \tau = 0.$$

From (9.4), we find that

$$\frac{d\sigma}{dr}+\frac{2\sigma}{r}-\frac{r\sigma}{\lambda Q}+\frac{2\mu}{r} = 0.$$

We find that

$$\sigma = 2\eta Q/r^2. \tag{10.8}$$

From $(9.4)_3$, we find that

$$\frac{d\gamma}{dr} - \frac{2\gamma}{r} - \frac{r\gamma}{\lambda Q} - \frac{2\mu}{r} = 0.$$

We next introduce ϕ through the change of variables $\gamma = -\mu + \mu\phi/\lambda Q$, where

$$\frac{d\phi}{dr} - \phi\left[\frac{2}{r} + \frac{r}{\lambda Q}\right] + r = 0.$$

We require that ϕ be bounded as $r^2 \to 0$ and find

$$\phi = -r^2 e^{r^2/2\lambda Q} \int \frac{e^{-r^2/2\lambda Q}}{r}\, dr, \tag{10.9}$$

where the last factor is an indefinite integral. For small r, we find that

$$\phi \sim -r^2 \log r > 0.$$

Obviously ϕ is positive at every stationary point. The only stationary point is at $r = \infty$ and ϕ increases monotonically from zero to λQ.

Turning next to (9.13), using (10.7–9), we find that real characteristics exist and are given by (9.13) when

$$(\mu + \gamma)\left[\frac{\varrho Q^2}{r^2} - (\mu + \sigma)\right] = \frac{\mu\phi}{\lambda Q}\left[\frac{\varrho Q^2 - 2\eta Q}{r^2} - \mu\right] > 0. \tag{10.10}$$

The condition for hyperbolicity is satisfied if

$$\varrho Q > 2\eta;$$

that is, when the flow rate is large. In this case the flow is hyperbolic in the circle

$$r < \sqrt{\frac{\varrho Q^2 - 2\eta Q}{\mu}}. \tag{10.11}$$

The differential equations for the net of characteristics covering the hyperbolic circle at the origin is

$$r\frac{d\theta}{dr} = \pm\sqrt{\frac{\mu + \gamma}{\varrho u^2 - (\mu + \sigma)}} = \pm\sqrt{\frac{\mu\phi/\lambda Q\varrho}{\dfrac{Q^2 - 2\eta Q}{r^2} - \mu}}.$$

(v) Our fifth application is to Couette flow outside a rotating circular cylinder of radius a. We suppose that the fluid sticks to the rod, and the rod rotates with an angular frequency Ω. We assume that $v = v(r)$, $u = \sigma = 0$. Then using $(9.4)_5$ and $(9.4)_2$ we find that

$$v = \Omega a^2/r,$$

$$\tau = -2a^2\eta\Omega/r^2,$$

and from $(9.4)_3$

$$\gamma = 2\tau^2/\mu .$$

From (9.13), we find that

$$r\frac{d\theta}{dr} = \pm\sqrt{\frac{v^2}{c^2}\left(1-\frac{4\lambda^2c^2}{r^2}\right)-1}\,.$$

To have real characteristics it is necessary and sufficient

$$\Sigma(r^2) \overset{\text{def}}{=} \frac{a^4\Omega^2}{c^2r^2}\left(1-\frac{4\lambda^2c^2}{r^2}\right)-1>0.$$

Moreover, $\Sigma(\infty) = -1$, and

$$\frac{d\Sigma}{dr^2} = \frac{a^4\Omega^2}{c^2r^4}\left[\frac{8\lambda^2c^2}{r^2}-1\right].$$

Σ first increases, then decreases, with a single maximum at $r^2 = 8\lambda^2c^2$ given by

$$\Sigma_m = \Sigma(8\lambda^2c^2) = \frac{a^4\Omega^2}{16\lambda^2c^4}-1,$$

which is positive when

$$a^4 > \frac{16\lambda^2c^4}{\Omega^2}. \tag{10.12}$$

If (10.12) holds then there are two values

$$r_c^2 = \frac{8\lambda^2c^2}{1\pm\sqrt{1-\dfrac{16\lambda^2c^4}{\Omega^2a^4}}}$$

at which $\Sigma(r_c^2) = 0$. If

$$a^2 > 4\lambda^2c^2 + \frac{c^2}{\Omega^2},$$

then $\Sigma(a^2) > 0$ and the vorticity is hyperbolic with real characteristics given by (9.13) in the annulus

$$a^2 \leqq r^2 \leqq \frac{8\lambda^2c^2}{1-\sqrt{1-\dfrac{16\lambda^2c^4}{\Omega^2a^4}}}.$$

If (10.12) holds and

$$a^2 < 4\lambda^2c^2 + \frac{c^2}{\Omega^2},$$

then $\Sigma(a^2) < 0$ and the vorticity is hyperbolic with real characteristics given by (9.13) in the annulus

$$\frac{8\lambda^2c^2}{1+\sqrt{1-\dfrac{16\lambda^2c^4}{\Omega^2a^4}}} \leqq r^2 \leqq \frac{8\lambda^2c^2}{1-\sqrt{1-\dfrac{16\lambda^2c^4}{\Omega^2a^4}}}.$$

(vi) M. W. JOHNSON (private communication) has studied the flow of an upper convected Maxwell fluid between eccentric rotating cylinders. He assumes a small gap and uses the lubrication approximation. His analysis shows that change of type occurs.

11. Classification of the Quasilinear System of Equations Governing Steady Flows of a Class of Oldroyd Models

The Oldroyd models under discussion are given by (5.4) and (5.5). When $a = 1$, this model reduces to the upper convected Maxwell model which we studied in § 9 and §10.

We have

$$\boldsymbol{D} = \begin{pmatrix} u_x & \frac{1}{2}(u_y + v_x) \\ \frac{1}{2}(u_y + v_x) & v_y \end{pmatrix},$$

$$\boldsymbol{\Omega} = \begin{pmatrix} 0 & \frac{1}{2}(u_y - v_x) \\ \frac{1}{2}(v_x - u_y) & 0 \end{pmatrix}.$$

This gives rise to the following quasilinear system

$$\begin{aligned} &u\sigma_x + v\sigma_y + \tau(v_x - u_y) - a[2\sigma u_x + \tau(u_y + v_x)] - 2\mu u_x = -\sigma/\lambda, \\ &u\tau_x + v\tau_y + \tfrac{1}{2}(\sigma - \gamma)(u_y - v_x) - \frac{a}{2}(\sigma + \gamma)(u_y + v_x) - \mu(u_y + v_x) = -\tau/\lambda, \\ &u\gamma_x + v\gamma_y + \tau(u_y - v_x) - a[2\gamma v_y + \tau(u_y + v_x)] - 2\mu v_y = -\frac{\gamma}{\lambda}, \\ &\varrho(uu_x + vu_y) + p_x - \sigma_x - \tau_y = 0, \\ &\varrho(uv_x + vv_y) + p_y - \tau_x - \tau_y = 0, \\ &u_x + v_y = 0. \end{aligned} \tag{11.1}$$

The analysis of characteristics follows the one used in § 9 exactly. We find characteristic directions, $\alpha = -dy/dx$ from the equation

(11.2)
$$(1 + \alpha^2)(\alpha u + v)^2 \left\{\varrho(\alpha u + v)^2 + \left(\frac{\gamma - \sigma}{2}\right)(\alpha^2 - 1) - 2\tau\alpha - (\alpha^2 + 1) \times \left(\mu + a\left(\frac{\gamma + \sigma}{2}\right)\right)\right\} = 0,$$

corresponding to (9.8). We can prove that the vorticity changes type from elliptic to hyperbolic when the sign of

(11.3)
$$\left[\varrho u^2 + \frac{\gamma}{2}(1 - a) - \frac{\sigma}{2}(1 + a) - \mu\right]\left[(1 + a)\frac{\gamma}{2} + (a - 1)\frac{\sigma}{2} + \mu - \varrho v^2\right] + (\varrho uv - \tau)^2$$

changes from negative to positive.

For shear flows, we find using (11.1) that

$$u = \varkappa y, \quad v = 0, \quad \tau = \frac{\eta\varkappa}{1 + \varkappa^2\lambda^2(1 - a^2)} \stackrel{\text{def}}{=} \frac{\eta\varkappa}{D},$$

$$\sigma = \lambda\varkappa(a + 1)\,\tau,$$

$$\gamma = \lambda\varkappa(a - 1)\,\tau,$$

where $\varkappa > 0$ is the shear rate.

Condition (11.3) now reads

(11.4)

$$\frac{\eta^2\varkappa^2}{D^2} + \left[\varrho\varkappa^2 y^2 - \frac{\varkappa^2\eta\lambda}{2D}(a - 1)^2 - \frac{\varkappa^2\eta\lambda}{2D}(1 + a)^2 - \mu\right]\left[\mu + \frac{a^2 - 1}{D}\varkappa^2\eta\lambda\right] > 0,$$

i.e.

$$\varrho\varkappa^2 y^2 B > \frac{-\eta^2\varkappa^2}{D^2} + B\left[\frac{\varkappa^2\eta\lambda}{2D}(a - 1)^2 + \frac{\varkappa^2\eta\lambda}{2D}(1 + a)^2 + \mu\right],$$

where $B \stackrel{\text{def}}{=} \frac{(a^2 - 1)}{D}\varkappa^2\eta\lambda + \mu = \frac{\mu}{D}$.

If a is in the range $-1 \leqq a \leqq 1$, B is always positive, and (11.4) describes the exterior of a strip in the x,y-plane (it is easy to see that the quantity on the right of the last inequality is always positive).

In extensional flow, we find using (11.1) that

$$u = sx, \quad v = -sy, \quad \tau = 0, \quad \sigma = \frac{2\eta s}{1 - 2a\lambda s}, \quad \gamma = \frac{-2\eta s}{1 + 2a\lambda s},$$

where $0 < s < 1/(2a\lambda)$ for $1 > a > 0$ or $0 < s < 1/(-2a\lambda)$ for $-1 \leqq a < 0$ (no restriction if $a = 0$).

Condition (11.3) is now evaluated as

(11.5)

$$\varrho s^2 x^2\left[\mu + \frac{2\eta s(2a^2\lambda s - 1)}{1 - 4a^2\lambda^2 s^2}\right] + \varrho s^2 y^2\left[\mu + \frac{2\eta s(1 + 2a^2\lambda s)}{1 - 4a^2\lambda^2 s^2}\right]$$
$$> \left[\mu + \frac{2\eta s(1 + 2a^2\lambda s)}{1 - 4a^2\lambda^2 s^2}\right]\left[\mu + \frac{2\eta s(2a^2\lambda s - 1)}{1 - 4a^2\lambda^2 s^2}\right].$$

Let us distinguish two cases

(i) $\mu + \frac{2\eta s(2a^2\lambda s - 1)}{1 - 4a^2\lambda^2 s^2} > 0$ (this is the case for $a = \pm 1$).

Then the region (11.5) is the exterior of the ellipse

$$\frac{x^2}{A^2} + \frac{y^2}{B^2} = 1,$$

where

$$A^2 \stackrel{\text{def}}{=} \frac{\mu + \dfrac{2\eta s(2a^2\lambda s + 1)}{1 - 4a^2\lambda^2 s^2}}{\varrho s^2},$$

$$B^2 \stackrel{\text{def}}{=} \frac{\mu + \dfrac{2\eta s(2a^2\lambda s - 1)}{1 - 4a^2\lambda^2 s^2}}{\varrho s^2}.$$

(ii) $\mu + \dfrac{2\eta s(2a^2\lambda s - 1)}{1 - 4a^2\lambda^2 s^2} < 0.$ Then (11.5) yields

$$\frac{x^2}{A^2} - \frac{y^2}{B^2} < 1,$$

where

$$A^2 = \frac{\mu + \dfrac{2\eta s(1 + 2a^2\lambda s)}{1 - 4a^2\lambda^2 s^2}}{\varrho s^2},$$

$$B^2 = -\frac{\mu + \dfrac{2\eta s(2a^2\lambda s - 1)}{1 - 4a^2\lambda^2 s^2}}{\varrho s^2}.$$

Equation of the vorticity, steady case

One easily finds the equation satisfied by the vorticity $\zeta = -\Delta\psi$, where ψ is the stream function given by $\psi_y = u$, $v = -\psi_x$:

(11.6) $$[-\varrho u^2 + \mu + \tfrac{1}{2}\sigma(1 + a) + \tfrac{1}{2}\gamma(a - 1)]\,\zeta_{xx} + 2(-\varrho uv + \tau)\,\zeta_{xy} + [-\varrho v^2 + \mu + \tfrac{1}{2}\sigma(a - 1) + \tfrac{1}{2}\gamma(a + 1)]\,\zeta_{yy} + \text{terms of lower order} \stackrel{\text{def}}{=} A\zeta_{xx} + 2B\zeta_{xy} + C\zeta_{yy}.$$

This is a mixed elliptic-hyperbolic second-order equation. It is hyperbolic for $\Delta = B^2 - AC > 0$, *i.e.*, when (11.3) holds (see the previous discussion).

Equation for vorticity in the unsteady case

In unsteady planar flow the equation for vorticity is

(11.7) $$\varrho\left(\frac{\partial\zeta}{\partial t} + (\boldsymbol{u}\cdot\nabla)\,\zeta\right) + \lambda\left[\varrho\frac{\partial^2\zeta}{\partial t^2} + 2\varrho(\boldsymbol{u}\cdot\nabla)\frac{\partial\zeta}{\partial t} + \left(\varrho u^2 - \mu - \frac{\sigma}{2}(1 + a) + \frac{\gamma}{2}(1 - a)\right)\frac{\partial^2\zeta}{\partial x^2} + 2(\varrho uv - \tau)\frac{\partial^2\zeta}{\partial x\,\partial y} + \left(\varrho v^2 - \mu + \frac{\sigma}{2}(1 - a) - \frac{\gamma}{2}(a + 1)\right)\frac{\partial^2\zeta}{\partial y^2}\right] = \lambda[l.o.].$$

The term in brackets on the right-hand side is of lower order and does not involve any time derivative. The first term on the left-hand side is irrelevant for the analysis of the type of equation (11.7), but it shows how (11.7) reduces to the unsteady Newtonian case when $\lambda = 0$ (recall that $\mu = \eta/\lambda$).

Equation (11.7) is of evolution type (hyperbolic) when

(11.8)
$$\lambda^2\tau^2 - [\eta - \lambda(\tfrac{1}{2}\gamma(1-a) - \tfrac{1}{2}\sigma(1+a))] \cdot [\eta - \lambda(\tfrac{1}{2}\sigma(1-a) - \tfrac{1}{2}\gamma(1+a))] < 0,$$

(11.9)
$$\lambda[\tfrac{1}{2}\gamma(1-a) - \tfrac{1}{2}\sigma(1+a)] - \eta < 0.$$

12. Quasilinear Systems for Simple Fluids with Fading Memory of the Coleman-Noll Type

The determinate stress $\boldsymbol{\tau}$ in a simple fluid is given by an isotropic functional of the history of the relative Cauchy strain $G(s) = F_t^T(t-s)\,F_t(t-s) - \mathbf{1}$, *i.e.*

(12.1)
$$Q\boldsymbol{\tau}Q^T = F\left[\overset{\infty}{\underset{s=0}{Q}}\ G(s)\,Q^T\right]$$

for all constant orthogonal tensors Q. By taking the material derivative of (12.1), we obtain

(12.2)
$$Q\frac{d\boldsymbol{\tau}}{dt}Q^T = F_1\left[QGQ^T \,\middle|\, Q\frac{dG}{dt}Q^T\right],$$

where F_1 denotes the first functional derivative of F. We have already noted in § 5 that different choices for the domain of the linear functional $F_1[QGQ^T \mid \cdot]$ lead to different representations of F_1. If, following COLEMAN & NOLL, we assume that the functional is defined on a weighted L^2-space, we obtain an integral

(12.3)
$$F_1\left[G\,\middle|\,\frac{dG}{dt}\right] = \int_0^\infty \mathbf{K}(s, G)\frac{dG(s)}{dt}\,ds.$$

Here $\mathbf{K}(s, G)$ is a fourth-order tensor depending on s and on the values $\{G(\sigma), 0 \leqq \sigma < \infty\}$. For the following, we assume that $\mathbf{K}$ and its first derivative (as function of s) are integrable. In particular, it follows that $\mathbf{K}$ is uniformly bounded in s.

The isotropy condition can be written as

(12.4)
$$K_{ijkl}(s, G) = Q_{ai}Q_{bj}Q_{ck}Q_{dl}K_{abcd}(s, QGQ^T).$$

This consequence of isotropy does not appear to lead easily to very explicit representations for $\mathbf{K}$ in situations where G is not confined to special motions. Of course $\mathbf{K}$ is symmetric in the first two indices, and only the symmetric part in the last two indices enters into (12.3).

We next note that the material derivative of G is given by

$$\frac{d}{dt}G(s) = \frac{d}{dt}(F^{-T}(t)\,F^T(t-s)\,F(t-s)\,F^{-1}(t)) \tag{12.5}$$

$$= -F^{-T}(t)\,\dot{F}^T(t)\,G(s) - G(s)\,\dot{F}(t)\,F^{-1}(t) - \frac{d}{ds}G(s)$$

$$= -L^T G(s) - G(s)\,L - \frac{dG(s)}{ds},$$

where $L(\boldsymbol{x}, t) = \nabla \boldsymbol{u}(x, t)$ is the present value of the velocity gradient. Hence, we find

$$\int_0^\infty K_{ijkl}(s, G)\frac{dG_{kl}(s)}{dt}\,ds = -\int_0^\infty (K_{ijkl} + K_{ijlk})\,G_{pl}(s)\,ds \cdot L_{pk}(t) - \int_0^\infty K_{ijkl}(s, G)\frac{dG(s)}{ds}\,ds. \tag{12.6}$$

It follows that

$$\frac{d\tau_{ij}}{dt} = M_{ijkp}(\boldsymbol{x}, t)\,L_{pk}(\boldsymbol{x}, t) + N_{ij}(\boldsymbol{x}, t), \tag{12.7}$$

where

$$M_{ijkp}(\boldsymbol{x}, t) = -\int_0^\infty (K_{ijkl} + K_{ijlk})\,G_{pl}(s)\,ds, \tag{12.8}$$

and

(12.9)

$$N_{ij}(\boldsymbol{x}, t) = \int_0^\infty K_{ijkl}\left(s, \mathop{G}_{s'=0}^{\infty}(s')\right)\frac{dG_{kl}(s)}{ds}\,ds = -\int_0^\infty G_{kl}(s)\frac{d}{ds}K_{ijkl}\left(s, \mathop{G}_{s'=0}^{\infty}(s')\right)ds.$$

The coefficients M_{ijkp} of L_{pk} and the terms $N_{ij}(\boldsymbol{x})$ are of lower order and $N_{ij}(\boldsymbol{x}, t)$ does not enter into analysis of problems of change of type.

We may write (12.9) as

$$\frac{d\tau_{ij}}{dt} = S_{ijkp}\,D_{kp} + A_{ijkp}\Omega_{kp} + N_{ij}, \tag{12.10}$$

where

$$M_{ijkp} = S_{ijkp} + A_{ijkp},$$

and **S** is symmetric and **A** is skew symmetric in k and p. D and Ω denote the symmetric and skew symmetric part of L.

In its general form, (12.10) and the equations of motion are a quasilinear system of first order in the derivatives of p, $\boldsymbol{u}$, $\boldsymbol{\tau}$. We could write out conditions for evolutionary character or for change of type in steady flow as before. However, we cannot in general isolate an equation for the vorticity, as we did for the Oldroyd models.

We can identify a class of models which is more special than the completely general equation (12.10), but much more general than the three-constant Oldroyd model. Let us assume that

$$S_{ijkp} = \tfrac{1}{2}(\delta_{ik}P_{jp} + \delta_{jk}P_{ip} + \delta_{ip}P_{jk} + \delta_{jp}P_{ik}), \tag{12.11}$$

where P is any second order tensor, expressible by integrals of the type (12.8). We need no assumptions at all on the anti-symmetric part A_{ijkp}. Using (12.11) we may reduce (12.10) to

$$\begin{gathered} \frac{d\tau_{ij}}{dt} = P_{ik}\,D_{kj} + P_{jk}\,D_{ki} + A_{ijkp}\,\Omega_{kp} + N_{ij}, \\ \frac{d\boldsymbol{\tau}}{dt} = \boldsymbol{P}\boldsymbol{D} + \boldsymbol{D}\boldsymbol{P}^T + \tfrac{1}{2}(\mathbf{A}\boldsymbol{\Omega} + (\mathbf{A}\boldsymbol{\Omega})^T) + \boldsymbol{N}, \end{gathered} \tag{12.12}$$

where A_{ijkp} is symmetric in ij and skew symmetric in kp.

The Oldroyd model with three constants arises from (12.12) with special choices for the tensors $\mathbf{A}$, $\boldsymbol{P}$ and $\boldsymbol{N}$:

$$\begin{gathered} -2A_{ijkl} = \delta_{ik}\tau_{jl} + \delta_{jk}\tau_{il} - \delta_{il}\tau_{kj} - \delta_{jl}\tau_{ki}, \\ P_{ik} = a\tau_{ik} + \frac{\eta}{\lambda}\,\delta_{ik}, \\ N_{ij} = -\frac{1}{\lambda}\,\tau_{ij}. \end{gathered} \tag{12.13}$$

We now demonstrate that the quasilinear system associated with (12.12) is expressible in terms of the vorticity $\zeta = \operatorname{curl} \boldsymbol{u}$. The equations of motion are

$$\varrho\frac{d\boldsymbol{u}}{dt} = -\nabla p + \operatorname{div}\boldsymbol{\tau}, \qquad \operatorname{div}\boldsymbol{u} = 0. \tag{12.14}$$

We apply the operations curl and $\dfrac{d}{dt}$ to (12.14), and find that to leading order

$$\varrho\frac{d^2}{dt^2}\zeta = \operatorname{curl}\operatorname{div}\frac{d\boldsymbol{\tau}}{dt} + \ldots. \tag{12.15}$$

The question is now whether the right side of (12.15) is expressible in terms of ζ, to leading order. Clearly $\boldsymbol{\Omega}$ is expressible in terms of ζ, and so we need no restrictions on $\mathbf{A}$. Then, working the part with tensor $\boldsymbol{P}$, we calculate

$$\begin{aligned} \varepsilon_{abi}\,\partial_b\partial_j(P_{ik}\,D_{kj} + P_{jk}\,D_{ki}) &= \varepsilon_{abi}[P_{ik}\,\partial_b\partial_j D_{kj} + P_{jk}\,\partial_b\partial_j D_{ki}] + \ldots \\ &= \tfrac{1}{2}\varepsilon_{abi}P_{ik}\,\nabla^2\frac{\partial u_k}{\partial x_b} + \tfrac{1}{2}\varepsilon_{abi}P_{jk}\,\frac{\partial^3 u_i}{\partial x_b\,\partial x_j\,\partial x_k} + \ldots. \end{aligned}$$

Noting now that

$$\nabla^2 \boldsymbol{u} = -\text{curl}\ \boldsymbol{\zeta},$$

we get

$$= -\tfrac{1}{2}\varepsilon_{abi}P_{ik}\,\partial_b\,(\text{curl}\ \zeta)_k + \tfrac{1}{2}\,P_{jk}\,\partial_j\partial_k\zeta_a + \dots$$

It follows now that to leading order (12.15) is a second order quasilinear system of equations for the components of the vorticity.

Acknowledgement. The authors thank M. LUSKIN for useful discussions on the subject of this paper. JOSEPH's work was sponsored by The United States Army under Contract No. DAAG29-82-K0051 and by the Fluid Mechanics Branch of the National Science Foundation; RENARDY's, by the United States Army under Contract No. DAAG29-80-C-0041, using material based upon work supported by the National Science Foundation under Grant Nos. MCS-7927062, Mod. 2, MCS-8210950 and MCS-8215064.

Note added in proof: BERNARD COLEMAN, in an unpublished work, has shown that our equation (12.10) can be consistent with his thermodynamics only if $S_{ijkp} = S_{kpij}$. This implies that the second order tensor $\boldsymbol{P}$ in (12.11) is symmetric.

References

1. S. AGMON, A. DOUGLIS & L. NIRENBERG, Estimates near the boundary for solutions of elliptic partial differential equations satisfying general boundary conditions, Comm. Pure Appl. Math. **12** (1959), 623–727 and **17** (1964), 35–92.
2. M. AHRENS, D. D. JOSEPH, M. RENARDY & Y. RENARDY, Remarks on the stability of viscometric flow, Rheol Acta, forthcoming.
3. U. AKBAY, E. BECKER, S. KROZER & S. SPONAGEL, Instability of slow viscometric flow, Mech. Res. Comm. **7** (1980), 199–204.
4. B. BERNSTEIN, E. A. KEARSLEY & L. J. ZAPAS, A study of stress relaxation with finite strain, Trans. Soc. Rheol. **7** (1963), 391–410.
5. G. BÖHME, Strömungsmechanik nichtnewtonscher Fluide, Teubner Studienbücher Mechanik, B. G. Teubner, Stuttgart 1981.
6. B. D. COLEMAN, M. E. GURTIN & I. HERRERA R., Waves in materials with memory I & II Arch. Rational Mech. Anal. **19** (1965), 1–19; 239–265.
7. B. D. COLEMAN & M. E. GURTIN, On the stability against shear waves of steady flows of non-linear viscoelastic fluids, J. Fluid Mech. **33** (1968), 165–181.
8. B. D. COLEMAN, Necking and drawing in polymeric fibers under tension, Arch. Rational Mech. Anal. **83** (1983), 115–137.
9. B. D. COLEMAN & W. NOLL, An approximation theorem for functionals with applications in continuum mechanics, Arch. Rational Mech. Anal. **6** (1960), 355–370.
10. M. J. CROCHET, A. R. DAVIES & K. WALTERS, Numerical Simulation of Non-Newtonian Flow, Elsevier, 1984.
11. C. F. CURTISS & R. B. BIRD, A kinetic theory for polymer melts, J. Chem. Phys. **74** (1981), 2016–2033.
12. J. L. ERICKSEN, Equilibrium of bars, J. Elasticity **5** (1975), 191–201.
13. A. G. FABULA, An experimental study of grid turbulence in dilute high-polymer solutions, Ph. D. Thesis, Pennsylvania State Univ., University Park 1966.
14. I. M. GELFAND, Some problems in the theory of quasilinear equations, Amer. Math. Soc. Translations **29** (1963), 295–380.

15. H. GIESEKUS, A unified approach to a variety of constitutive models for polymer fluids based on the concept of configuration dependent molecular mobility, Rheol. Acta **21** (1982), 366–375.
16. H. HATTORI, Breakdown of smooth solutions in dissipative nonlinear hyperbolic equations, Quart. Appl. Math. **40** (1982/83), 113–127.
17. J. K. HUNTER & M. SLEMROD, Unstable viscoelastic fluid flow exhibiting hysteretic phase changes, Phys. Fluids **26** (1983), 2345–2351.
18. D. F. JAMES, Laminar flow of dilute polymer solutions around circular cylinders, Ph. D. Thesis, California Institute of Technology, Pasadena 1967.
19. A. JAMESON, Iterative solution of transonic flows over airfoils and wings, including flows at Mach 1, Comm. Pure Appl. Math. **27** (1974), 283–300.
20. A. JAMESON, Transonic flow calculations, in: H. J. WIRZ & J. J. SMOLDEREN (ed.): Numerical Methods in Fluid Dynamics, McGraw Hill/Hemisphere, Washington 1978.
21. M. W. JOHNSON & D. SEGALMAN, A model for viscoelastic fluid behavior which allows non-affine deformation, J. Non-Newtonian Fluid Mech. **2** (1977), 255–270.
22. A. KAYE, Co A Note 134, The college of Aeronautics, Cranfield, Bletchley, England 1962.
23. J. Y. KAZAKIA, The evolution of small rotational perturbations to a uniform rotation of a viscoelastic fluid contained between parallel plates, J. Non-Newtonian Fluid Mech. **15** (1984), 45–60.
24. J. Y. KAZAKIA & R. S. RIVLIN, Run-up and spin-up in a viscoelastic fluid, Rheol. Acta **20** (1981), 111–127; R. S. RIVLIN, II–IV, Rheol. Acta **21** (1982), 107–111 and 213–222, **22** (1983), 275–283.
25. A. I. LEONOV, Nonequilibrium thermodynamics and rheology of viscoelastic polymer media, Rheol. Acta **15** (1976), 85–98.
26. J. L. LIONS & E. MAGENES, Non-homogeneous boundary value problems and applications I, Springer, Berlin-Heidelberg-New York 1972.
27. M. LUSKIN, On the classification of some model equations for viscoelasticity, Rheol. Acta (to appear).
28. R. MALEK-MADANI & J. A. NOHEL, Formation of singularities for a conservation law with memory, (forthcoming) SIAM J. Math. Anal.
29. P. MARKOWICH & M. RENARDY, Lax-Wendroff methods for hyperbolic history value problems, SIAM J. Num. Anal. **21** (1984), 24–51.
30. M. S. MOCK, Systems of conservation laws of mixed type, J. Diff. Eq. **37** (1980), 70–88.
31. E. MURMAN, Analysis of embedded shock waves calculated by relaxation methods, Proc. AIAA Conf. on Computational Fluid Dynamics, Palm Springs 1973, 27–40.
32. E. MURMAN & J. D. COLE, Calculation of plane steady transonic flows, AIAA J. **9** (1971), 114–121.
33. A. NARAIN & D. D. JOSEPH, Linearized dynamics for step jumps of velocity and displacements of shearing flows of a simple fluid, Rheol. Acta **21** (1982), 228–250.
34. J. G. OLDROYD, Non-Newtonian effects in steady motion of some idealized elastico-viscous liquids, Proc. Roy. Soc. London A **245** (1958), 278–297.
35. C. J. S. PETRIE & M. M. DENN, Instabilities in polymer processing, AIChE J. **22** (1976), 200–236.
36. S. I. REGIRER & I. M. RUTKEVICH, Certain singularities of the hydrodynamic equations of non-Newtonian media, J. Appl. Math. Mech. **32** (1968), 962–966.
37. M. RENARDY, Some remarks on the propagation and non-propagation of discontinuities in linearly viscoelastic liquids, Rheol. Acta **21** (1982), 251–264.

38. M. Renardy, Singularly perturbed hyperbolic evolution problems with infinite delay and an application to polymer rheology, SIAM J. Math. Anal. **15** (1984), 333–349.
39. M. Renardy, A local existence and uniqueness theorem for a K-BKZ fluid, (forthcoming) Arch. Rational Mech. Anal.
40. I. M. Rutkevich, The propagation of small perturbations in a viscoelastic fluid, J. Appl. Math. Mech. **34** (1970), 35–50.
41. I. M. Rutkevich, On the thermodynamic interpretation of the evolutionary conditions of the equations of the mechnics of finitely deformable viscoelastic media of Maxwell type. J. Appl. Math. Mech. **36** (1972), 283–295.
42. J. C. Saut & D. D. Joseph, Fading Memory, Arch. Rational Mech. Anal. **81** (1983), 53–95.
43. M. Slemrod, Unstable viscoelastic fluid flows, in: R. L. Sternberg, A. J. Kalinowski & J. S. Papadakis (ed.), Nonlinear Partial Differential Equations in Engineering and Applied Sciences, Marcel Dekker, New York-Basel 1980, 33–43.
44. M. Slemrod, Instability of steady shearing flows in a nonlinear viscoelastic fluid, Arch. Rational Mech. Anal. **68** (1978), 211–225.
45. K. Straub, Zur Untersuchung der Anlaufströmung von viskoelastischen Flüssigkeiten, Rheol. Acta **16** (1977), 385–393.
46. R. I. Tanner, Note on the Rayleigh problem for a viscoelastic fluid, ZAMP **13** (1962), 573–580.
47. J. P. Tordella, Unstable flow of molten polymers, in: F. Eirich, Rheology: Theory and Applications, Vol. 5, Academic Press, New York-London 1969.
48. C. Truesdell & W. Noll, The Nonlinear Field Theories of Mechanics, Handbuch der Physik III/3, ed. S. Flügge, Springer, Berlin-Heidelberg-New York 1965.
49. J. S. Ultman & M. M. Denn, Anomalous heat transfer and a wave phenomenon in dilute polymer solutions, Trans. Soc. Rheology **14** (1970), 307–317.
50. C. Truesdell, A First Course in Rational Continuum Mechanics, Academic Press, New York-San Francisco-London 1977.
51. G. Gripenberg, Nonexistence of smooth solutions for shearing flows in a nonlinear viscoelastic fluid, SIAM J. Math. Anal. **13** (1982), 954–961.
52. C. Truesdell, The meaning of viscometry in fluid mechanics. Annual Reviews of Fluid Mech. **6** (1974), 111–146.

Department of Aerospace Engineering and Mechanics
University of Minnesota, Minneapolis

Department of Mathematics and Mathematics Research Center
University of Wisconsin, Madison

Department of Mathematics
Université de Paris-Sud, Orsay

(Received March 23, 1984; revised June 1 and September 28, 1984)

Basic Theory of Small-Amplitude Waves in a Constrained Elastic Body

P. CHADWICK, A. M. WHITWORTH & P. BOREJKO

In honour of Walter Noll on his 60th birthday

1. Introduction

The effects of one or more internal constraints on the properties of waves in elastic media have been considered by many authors with particular reference to incompressibility and inextensibility, the constraints most widely used in modelling the behaviour of solid materials. The idea of broadening the theoretical framework of such studies by assigning a fully general form to the constraints is seemingly due to SCOTT [1975] who investigated the propagation and growth of acceleration waves of arbitrary shape in an elastic body experiencing one or two constraints. Subsequently BOREJKO & CHADWICK [1980] have discussed the energetics of acceleration waves in an arbitrarily constrained elastic body and, in the same context, an analysis of simple waves has been presented by WHITWORTH [1982].

We extend this field of enquiry in the present paper by developing the linearized dynamics of a non-heat-conducting elastic body subject to a general system of internal constraints and an arbitrary homogeneous pre-strain. The underlying non-linear theory is summarized in Section 2 as a preliminary to the derivation, in the first two parts of Section 3, of the linearized equations governing small-amplitude disturbances of the equilibrium configuration into which the body is carried by the primary deformation.

The remainder of Section 3 is devoted to an examination of energy changes accompanying small-amplitude motions. The linearized theory produces, straightforwardly, an approximation to the energy flux, but the total energy linked to this vector through the energy conservation equation is found not to be the sum of the kinetic and internal energies, as in linear elastodynamics. In place of the internal energy there appears a pseudo-internal energy containing contributions from the constraints and hence peculiar to the linearized theory of constrained and pre-strained elastic media. In the presence of a primary deformation the physical interpretations of the energy flux and total energy are obscured by the occurrence of terms of both leading and next-leading orders in the quantities supposed small in the linearization. We meet this difficulty by dividing the energy flux and the total energy into "interaction" and "incremental" components satisfying individual conservation equations and admitting distinct interpretations. The interaction in question is between the small-amplitude motion and the primary deformation, and the associated parts of the energy flux and total energy are of leading order. The incremental components relate solely to the small-amplitude motion and do not vanish, therefore, in the absence of pre-strain. It is these contributions which are directly relevant to the physics of energy propagation.

In Section 4 we study solutions of the governing equations of the linear theory representing homogeneous plane waves (or body waves) of arbitrary profile. The set of internal constraints is described as being "fully active" in relation to a disturbance of this kind if the vectors resulting from the action of the reaction tensors on the wave normal are linearly independent. The maximum number of fully-active constraints is thus three and the largest number compatible with homogeneous plane-wave propagation is found to be two. The solutions for the cases of one and two fully-active constraints have strong affinities with the corresponding results for acceleration discontinuities and simple waves obtained by Scott [1975] and Whitworth [1982]. They are therefore set down in Section 4b) with a minimum of detail. In a direction in which the constraints are not fully active there exist, in addition to the plane-wave solutions, one or more "ghost" disturbances, so called because they perturb the scalar constraint multipliers without displacing the material. Degenerate solutions of this sort have been shown by Whitworth & Chadwick [1984, 1, 2] to be essential to the construction of surface waves in particular types of constrained elastic bodies.

In Section 4d) the ray velocity of a homogeneous plane wave is shown to coincide with the group velocity when the wave profile is sinusoidal and to be aligned with the ray, or bicharacteristic, of the wave, directed normally to the slowness surface associated with the pre-strained state of the transmitting body. Finally, in Section 4e), we demonstrate that energy relations which apply to homogeneous plane waves in an elastic body free from constraints and pre-strain can be extended to the more general situation envisaged herein. Specifically, we prove that (i) the kinetic energy and the incremental component of the pseudo-internal energy are equipartitioned, (ii) the velocity of propagation of energy (kinetic plus incremental pseudo-internal) equals the ray velocity, and is thus codirectional with the ray, and (iii) the components along the wave normal of the ray and energy propagation velocities are equal to the wave speed.

We employ in this paper both direct and indicial notations, vector and tensor components being referred to an arbitrary orthonormal basis. All subscripts take the values 1, 2, 3 and the summation convention applies to repeated suffixes.

2. Basic Equations

We consider a body $\mathscr{B}$ composed of a homogeneous non-heat-conducting elastic material which is subject to one or more internal constraints. It is assumed that $\mathscr{B}$ possesses a natural undistorted state B_u and an equilibrium configuration B_e which is chosen as the reference state of $\mathscr{B}$. The deformation $B_u \to B_e$ is taken to be homogeneous, and throughout the paper external body forces and heat supply are supposed absent.

a) Kinematics

In a motion of $\mathcal{B}$ the configuration at the current time t is denoted by B_t and the positions of a representative particle in B_u, B_e and B_t by $\boldsymbol{X}$, $\bar{\boldsymbol{x}}$ and $\boldsymbol{x}$ respectively. The gradients of the deformations $B_u \to B_t$, $B_e \to B_t$ and $B_u \to B_e$ are, in turn,

$$\boldsymbol{F} = \partial_{\boldsymbol{X}} \boldsymbol{x}, \qquad \tilde{\boldsymbol{F}} = \partial_{\bar{\boldsymbol{x}}} \boldsymbol{x}, \qquad \bar{\boldsymbol{F}} = \partial_{\boldsymbol{X}} \bar{\boldsymbol{x}}, \tag{2.1}$$

$\bar{\boldsymbol{F}}$ being a constant tensor, and there follow the relations

$$\boldsymbol{F} = \tilde{\boldsymbol{F}} \bar{\boldsymbol{F}}, \qquad J = \tilde{J} \bar{J}, \tag{2.2}$$

where $J = \det \boldsymbol{F}$ *etc.* In this scheme of notation $\bar{\boldsymbol{x}}$ and $\bar{\boldsymbol{F}}$ are the values of $\boldsymbol{x}$ and $\boldsymbol{F}$ in B_e and the overbar is used systematically henceforth to indicate evaluation in B_e.

The internal constraints are given by

$$\lambda^A(\boldsymbol{F}) = 0, \qquad A = 1, 2, \ldots, N, \tag{2.3}$$

and in order that these restrictions on the motion shall not fully specify the strain we stipulate that $N \leqq 5$ (see TRUESDELL & NOLL [1965, p. 73]).

b) Stress

We utilize two measures of stress in B_t, the Cauchy stress $\boldsymbol{\sigma}$ and the nominal stress $\boldsymbol{s}$ relative to B_e. The connection between the two measures is given by

$$\boldsymbol{\sigma} = \tilde{J}^{-1} \tilde{\boldsymbol{F}} \boldsymbol{s} \tag{2.4}$$

(*cf.* CHADWICK [1976, p. 99]).

The constraints (2.3) give rise to internal forces contributing to $\boldsymbol{\sigma}$ a linear combination of reaction tensors $\boldsymbol{N}^A$ which individually do no work. Thus

$$\operatorname{tr}(\boldsymbol{N}^A \dot{\boldsymbol{F}} \boldsymbol{F}^{-1}) = 0, \qquad A = 1, 2, \ldots, N, \tag{2.5}$$

for all admissible motions compatible with the constraints, a dot signifying material time differentiation (with $\bar{\boldsymbol{x}}$ held fixed). The conditions (2.5) determine the reaction tensors as

$$\boldsymbol{N}^A = J^{-1} \boldsymbol{F} (\partial_{\boldsymbol{F}} \lambda^A)^T, \qquad A = 1, 2, \ldots, N, \tag{2.6}$$

and the symmetry of the $\boldsymbol{N}^A$'s is a consequence of the principle of objectivity (cf. TRUESDELL & NOLL [1965, p. 71]).

c) Field Equations

The referential forms of the equation of motion and the energy equation are

$$\bar{\varrho}\ddot{x} = \operatorname{Div} s, \qquad \bar{\varrho}\dot{U} = \operatorname{tr}(s\dot{\tilde{F}}), \tag{2.7}$$

(*cf.* CHADWICK [1976, pp. 101, 112]) where $\bar{\varrho}$ is the uniform mass density of the material in B_e, U is the internal energy per unit mass and the divergence operator is with respect to position $\bar{x}$ in B_e.

On taking the scalar product of each side of $(2.7)_1$ with the velocity $\dot{x}$ and adding the resulting relation to $(2.7)_2$ we obtain the energy conservation equation

$$\bar{\varrho}\dot{\mathscr{E}} = -\operatorname{Div} J, \tag{2.8}$$

with

$$\mathscr{E} = K + U \quad \text{and} \quad K = \tfrac{1}{2}\dot{x} \cdot \dot{x}. \tag{2.9}$$

Here $\mathscr{E}$ is the total and K the kinetic energy per unit mass and

$$J = -s\dot{x} \tag{2.10}$$

is the referential energy flux vector.

d) Constitutive Equations

Since the material under consideration is elastic the internal energy U is a function of F, the deformation gradient from the undistorted state B_u. Moreover,

$$W(F) = \bar{\varrho}\bar{J}U(F), \tag{2.11}$$

the internal energy per unit volume of B_u, generates the extra stress, that is the part of the Cauchy stress wholly determined by the deformation. The remainder of σ, the constraint stress, is a linear combination of the reaction tensors N^A given by (2.6), whence

$$\sigma = J^{-1}F(\partial_F W)^T + \sum_{A=1}^{N} \alpha^A N^A. \tag{2.12}$$

The scalar multipliers α^A in (2.12) are, at this stage, arbitrary: ultimately they must be so chosen that the equation of motion and appropriate boundary conditions are fulfilled. Combining equations (2.4) and (2.12) and making use of (2.2) and (2.6) we obtain the alternative constitutive equations

$$\begin{aligned} s &= \bar{J}^{-1}\bar{F}(\partial_F W)^T + \tilde{J}\tilde{F}^{-1} \sum_{A=1}^{N} \alpha^A N^A \\ &= \bar{J}^{-1}\bar{F}\left(\partial_F W + \sum_{A=1}^{N} \alpha^A \partial_F \lambda^A\right)^T \end{aligned} \tag{2.13}$$

for the nominal stress. It is apparent from (2.12) and $(2.13)_1$ that in B_e, that is when $\tilde{F} = I$ (the identity tensor) and $\alpha^A = \bar{\alpha}^A$, the Cauchy and nominal stresses are equal:

$$\bar{\sigma} = \bar{s}. \tag{2.14}$$

It may readily be verified, with the help of (2.2) and (2.5), that the expressions for U and s provided by (2.11) and $(2.13)_1$ satisfy identically the energy equation $(2.7)_2$.

Because of the presence of constraints, W, which we refer to from now on as the strain energy, is not uniquely specified as a function of $\boldsymbol{F}$: we could, for example, add any linear combination of the constraint functions λ^A. To account fully for the implications of this non-uniqueness we allow W to depend on $\boldsymbol{F}$ both explicitly and through the λ^A's so that

$$W(\boldsymbol{F}) = \hat{W}(\boldsymbol{F}, \lambda^1(\boldsymbol{F}), \lambda^2(\boldsymbol{F}), \ldots, \lambda^N(\boldsymbol{F})). \tag{2.15}$$

Then, with the aid of (2.6),

$$(\partial_{\boldsymbol{F}} W)^T = (\partial_{\boldsymbol{F}} \hat{W})^T + J \sum_{A=1}^{N} \frac{\partial \hat{W}}{\partial \lambda^A} \boldsymbol{F}^{-1} \boldsymbol{N}^A,$$

and α^A is seen to be replaced by

$$\hat{\alpha}^A = \alpha^A + \frac{\partial \hat{W}}{\partial \lambda^A} \tag{2.16}$$

in equations (2.12) and (2.13). In view of the arbitrariness of the scalars α^A, the form of the constitutive equations is therefore unchanged. Further comment on the non-uniqueness of the strain energy is made in Section 3b).

e) *Moduli in B_e*

The linear elasticity tensor $\boldsymbol{B}$ in B_e based on the stress-deformation pair $(\boldsymbol{s}, \boldsymbol{F})$ is defined by the component relation

$$B_{ijkl} = \bar{J}^{-1} \bar{F}_{ip} \bar{F}_{kq} \frac{\partial^2 W}{\partial F_{jp}\, \partial F_{lq}} (\bar{\boldsymbol{F}}) \tag{2.17}$$

(*cf.* CHADWICK & CURRIE [1972, § 2b)]). Analogous moduli $\check{\boldsymbol{B}}^A$, derived from the constraint functions λ^A, have components

$$\check{B}^A_{ijkl} = \bar{J}^{-1} \bar{F}_{ip} \bar{F}_{kq} \frac{\partial^2 \lambda^A}{\partial F_{jp}\, \partial F_{lq}} (\bar{\boldsymbol{F}}), \qquad A = 1, 2, \ldots, N, \tag{2.18}$$

and we set

$$B^*_{ijkl} = B_{ijkl} + \sum_{A=1}^{N} \bar{\alpha}^A \check{B}^A_{ijkl}. \tag{2.19}$$

The symmetry property

$$B^*_{klij} = B^*_{ijkl} \tag{2.20}$$

is a direct consequence of the definitions (2.17) to (2.19), and the quasi-symmetry condition

$$B^*_{jikl} - B^*_{ijkl} = \bar{\sigma}_{jk}\, \delta_{il} - \bar{\sigma}_{ik}\, \delta_{jl} \tag{2.21}$$

may be established as follows. The symmetry of the Cauchy stress implies, via (2.4), $(2.13)_2$ and $(2.2)_1$, that

$$\bar{J}^{-1}F_{ip}\frac{\partial W^*}{\partial F_{jp}} = \bar{J}^{-1}F_{jp}\frac{\partial W^*}{\partial F_{ip}}, \tag{2.22}$$

where $W^* = W + \sum_{A=1}^{N} \alpha^A \lambda^A$. Applying to each side of (2.22) the operator $F_{kq}\,\partial/\partial F_{lq}$, then evaluating in B_e and calling on equations $(2.13)_2$, (2.14) and (2.17) to (2.19) we arrive at (2.21).

In concluding this section we remark that the intermediate configuration B_e has been introduced *ab initio* only to facilitate the later developments. A nonlinear theory of a constrained elastic body which is free from this complication may be secured by putting $\bar{F} = I$, and hence $\tilde{F} = F$, in the above results.

3. Motions of Small Amplitude

We adopt

$$\delta = \|H\| + \sum_{A=1}^{N} |\beta^A/\bar{\alpha}^A| \tag{3.1}$$

as a measure of the departure of the body $\mathscr{B}$ from the equilibrium state B_e, where

$$H = \tilde{F} - I, \quad \|\cdot\| = \{\mathrm{tr}\,((\cdot)^T(\cdot))\}^{\frac{1}{2}}, \quad \beta^A = \alpha^A - \bar{\alpha}^A. \tag{3.2}$$

In a motion of $\mathscr{B}$ for which δ is universally small B_t is always close to B_e and the excursions of the multipliers α^A from their equilibrium values $\bar{\alpha}^A$ (supposed non-zero) are correspondingly limited. Accordingly, we describe a motion of $\mathscr{B}$ as being *of small amplitude about* B_e if $\delta < \delta' \ll 1$ at each particle at all relevant times. Such motions are our exclusive concern from now on.

a) Linearized Constraint and Constitutive Equations

The linearization of the theory formulated in Section 2 on the basis of the small-amplitude property requires the strain energy W and the constraint functions λ^A to be sufficiently smooth in a neighbourhood of $F = \bar{F}$. We assume that W and λ^A are C^2 functions in

$$N_\varepsilon = \{F : \|F - \bar{F}\| < \varepsilon,\ \varepsilon > \delta'\,\|\bar{F}\|\}.$$

In view of the identity

$$F - \bar{F} = H\bar{F}, \tag{3.3}$$

supplied by $(2.2)_1$ and $(3.2)_1$, and the inequality $\|T_1T_2\| \leqq \|T_1\|\,\|T_2\|$, satisfied by arbitrary tensors T_1 and T_2 (*cf.* LANCASTER [1969, p. 202]), it follows from (3.1) that the values of F attained in any small-amplitude motion of $\mathscr{B}$ belong to N_ε.

We can now apply Taylor's theorem about $\boldsymbol{F} = \bar{\boldsymbol{F}}$ to $\lambda^A(\boldsymbol{F})$ and to the representation of the components of $\boldsymbol{s}$ given by $(2.13)_2$. With appeal to equations (3.3), (2.6), (2.14) and (2.17) to (2.19) we find that

$$\lambda^A(\boldsymbol{F}) = \lambda^A(\bar{\boldsymbol{F}}) + \bar{J}\,\mathrm{tr}\,(\bar{\boldsymbol{N}}^A\boldsymbol{H}) + o(\|\boldsymbol{H}\|), \tag{3.4}$$

$$s_{ij} = \bar{\sigma}_{ij} + B^*_{ijrs}H_{sr} + \sum_{A=1}^{N} \bar{N}^A_{ij}\beta^A + o(\delta), \tag{3.5}$$

(see, for example, Flett [1980, p. 198]). Since the internal constraints (2.3) are operative in every configuration of $\mathscr{B}$, $\lambda^A(\boldsymbol{F})$ and $\lambda^A(\bar{\boldsymbol{F}})$ are both zero and (3.4) reduces to

$$\mathrm{tr}\,(\bar{\boldsymbol{N}}^A\boldsymbol{H}) + o(\|\boldsymbol{H}\|) = 0. \tag{3.6}$$

On substituting from (3.5) and $(3.2)_1$ into equation (2.4), and noting that $\tilde{J}^{-1} = 1 - \mathrm{tr}\,\boldsymbol{H} + o(\|\boldsymbol{H}\|)$, we obtain the expression

$$\sigma_{ij} = \bar{\sigma}_{ij}(1 - H_{rr}) + \bar{\sigma}_{pj}H_{ip} + B^*_{ijrs}H_{sr} + \sum_{A=1}^{N} \bar{N}^A_{ij}\beta^A + o(\delta) \tag{3.7}$$

for the components of Cauchy stress in B_t.

Linearized forms of the constraint and constitutive equations appropriate to small-amplitude motions of $\mathscr{B}$ about B_e result from the neglect of the remainder terms in equations (3.5) to (3.7). At the same time we introduce the displacement vector

$$\boldsymbol{u} = \boldsymbol{x} - \bar{\boldsymbol{x}} \tag{3.8}$$

and the notation $(\cdot)_{,i} = \partial(\cdot)/\partial\bar{x}_i$ for partial derivatives with respect to the referential coordinates. Then $H_{ij} = u_{i,j}$, from $(2.1)_2$ and $(3.2)_1$, and the linearized equations read

$$\bar{N}^A_{pq}u_{q,p} = 0, \qquad A = 1, 2, \ldots, N, \tag{3.9}$$

$$s_{ij} = \bar{\sigma}_{ij} + B^*_{ijrs}u_{s,r} + \sum_{A=1}^{N} \bar{N}^A_{ij}\beta^A, \tag{3.10}$$

$$\sigma_{ij} = \bar{\sigma}_{ij}(1 - u_{r,r}) + \bar{\sigma}_{pj}u_{i,p} + B^*_{ijrs}u_{s,r} + \sum_{A=1}^{N} \bar{N}^A_{ij}\beta^A. \tag{3.11}$$

We observe that the symmetry of the right-hand side of equation (3.11) is guaranteed by the condition (2.21).

The application of Taylor's theorem to the internal-energy function $U(\boldsymbol{F})$ yields

$$U(\boldsymbol{F}) = U(\bar{\boldsymbol{F}}) + \bar{\varrho}^{-1}\bar{\sigma}_{pq}u_{q,p} + \tfrac{1}{2}\bar{\varrho}^{-1}B_{pqrs}u_{q,p}u_{s,r} + o(\|\boldsymbol{H}\|^2), \tag{3.12}$$

use being made of equations (2.11), (3.3), (2.12), (3.9) and (2.17). For the corresponding expansion of $\lambda^A(\boldsymbol{F})$ equations (3.6) and (3.9) tell us that the sum of the quadratic and remainder terms is zero*. It does not follow, however, that these terms vanish separately and, in particular, the second-order term $\tfrac{1}{2}\bar{J}\check{\bar{B}}^A_{pqrs}u_{q,p}u_{s,r}$ is generally non-zero.

---

* The argument here is purely kinematic: the fact that the displacement components subsequently have to satisfy linearized equations of motion makes no difference, though, to the conclusion.

b) Linearized Equations of Motion

With the aid of (3.8) and $(2.1)_2$ the field equations (2.7) can be put into the component forms

$$\bar{\varrho}\ddot{u}_i = s_{pi,p}, \quad \bar{\varrho}\dot{U} = s_{pq}\dot{u}_{q,p}. \tag{3.13}$$

On combining $(3.13)_1$ with the linear approximation (3.10) to the components of nominal stress and remembering that $\bar{\boldsymbol{\sigma}}$, $\bar{\boldsymbol{N}}^A$ and $\boldsymbol{B}^*$ are constant tensors, we arrive at the system

$$\bar{\varrho}\ddot{u}_i = B^*_{pirs}u_{s,pr} + \sum_{A=1}^{N} \bar{N}^A_{pi}\beta^A_{,p} \tag{3.14}$$

of linear partial differential equations. In conjunction with the linearized constraints (3.9), (3.14) constitute the basic equations of the theory of small-amplitude motions of a constrained elastic body. Together they furnish $3 + N$ relations between the $3 + N$ unknown functions u_i and β^A.

No approximation has been made in the left-hand side of equation (3.14). The term omitted from the right-hand side is the divergence of the remainder in (3.5) and this is $o(1)$ times the terms retained, the implied limit being $\delta \to 0$. There is no characteristic length in the theory by means of which the absolute magnitude of the error incurred in the derivation of (3.14) could be estimated and, in the final analysis, the accuracy of a solution of the linearized equations must be assessed *a posteriori*, as in classical elastodynamics and acoustics.

To check that the nature of the basic equations (3.9) and (3.14) is unchanged by the substitution for $W(\boldsymbol{F})$ of the generalized strain energy given in (2.15), we return to the definitions (2.17) to (2.19) and obtain, in place of B^*_{ijkl},

$$\hat{B}^*_{ijkl} + \sum_{A=1}^{N} \left\{ \bar{F}_{iq} \frac{\partial^2 \hat{W}}{\partial F_{jq}\, \partial \lambda^A}(\bar{\boldsymbol{F}})\, \bar{N}^A_{kl} + \bar{F}_{kq} \frac{\partial^2 \hat{W}}{\partial F_{lq}\, \partial \lambda^A}(\bar{\boldsymbol{F}})\, \bar{N}^A_{ij} \right\}$$

$$+ \bar{J} \sum_{A=1}^{N} \sum_{B=1}^{N} \frac{\partial^2 \hat{W}}{\partial \lambda^A\, \partial \lambda^B}(\bar{\boldsymbol{F}})\, \bar{N}^A_{ij} \bar{N}^B_{kl},$$

where

$$\hat{B}^*_{ijkl} = \hat{B}_{ijkl} + \sum_{A=1}^{N} \bar{\hat{\alpha}}^A \check{B}^A_{ijkl}, \quad \hat{B}_{ijkl} = \bar{J}^{-1} \bar{F}_{ip} \bar{F}_{kq} \frac{\partial^2 \hat{W}}{\partial F_{jp}\, \partial F_{lq}}(\bar{\boldsymbol{F}}),$$

and $\hat{\alpha}^A$ is given by (2.16). Bearing in mind equations (3.9) we deduce that B^*_{ijkl} and β^A are replaced by $\hat{B}^*_{ijkl}$ and $\hat{\beta}^A$ respectively in (3.14) where

$$\hat{\beta}^A = \beta^A + \bar{F}_{rq} \frac{\partial^2 \hat{W}}{\partial F_{sq}\, \partial \lambda^A}(\bar{\boldsymbol{F}})\, u_{s,r}.$$

The expected form invariance is therefore confirmed.

c) Energetics

In essence the fundamental linear equations (3.9) and (3.14) have been derived by neglecting the squares of the "small quantities" H_{ij} and $\beta^A/|\bar{\hat{\alpha}}^A|$. It is a well-known feature of linear theories of mechanics, however, that terms which are quadratic in quantities considered small are indispensible to a satisfactory treatment of the balance and transport of energy. In the present work both linear and quadratic terms appear in the energy relations and care is needed in handling the interrelationship of the different levels of approximation. A lucid account of the basic issues involved has been given by LIGHTHILL [1978, § 1.3] in connection with sound waves and the discussion which follows reflects its influence.

First we calculate J_i and U by substituting into equations (2.10) and $(3.13)_2$ from (3.8) and (3.10). After simplifying with the aid of (3.9) we obtain

$$J_i = -\bar{\sigma}_{iq}\dot{u}_q - B^*_{iqrs}\dot{u}_q u_{s,r} - \sum_{A=1}^{N} \bar{N}^A_{iq}\dot{u}_q \beta^A, \tag{3.15}$$

$$\dot{U} = \bar{\varrho}^{-1}(\bar{\sigma}_{pq}\dot{u}_{q,p} + B^*_{pqrs}\dot{u}_{q,p}u_{s,r}), \tag{3.16}$$

and it is easily verified, with the help of (2.9), (3.8), (3.9) and (3.14), that these approximations solve the component form of the energy conservation equation (2.8).

By virtue of the symmetry relation (2.20), equation (3.16) can be integrated to give

$$U = \bar{U} + \bar{\varrho}^{-1}\bar{\sigma}_{pq}u_{q,p} + \tfrac{1}{2}\bar{\varrho}^{-1}B^*_{pqrs}u_{q,p}u_{s,r}, \tag{3.17}$$

and there follows from (2.9) and (3.8) the formula

$$\mathscr{E} = \bar{U} + \bar{\varrho}^{-1}\bar{\sigma}_{pq}u_{q,p} + \tfrac{1}{2}\dot{u}_q\dot{u}_q + \tfrac{1}{2}\bar{\varrho}^{-1}B^*_{pqrs}u_{q,p}u_{s,r} \tag{3.18}$$

for the total energy. Evidently the right-hand side of (3.17) does not agree with the expression for $U(\boldsymbol{F})$ obtained by discarding the remainder term from the Taylor expansion (3.12), and it is not possible, therefore, in the present theory, to secure a correct analysis of energy flow by simply representing the internal energy U as a truncated Taylor expansion. It is necessary to introduce a *pseudo-internal energy* by replacing the linear elasticity tensor $\boldsymbol{B}$ in the Taylor expansion (3.12) by the modulus $\boldsymbol{B}^*$ defined by (2.19) and involving the internal constraints*. The concluding remark of Section 3a) states that, while the constraints are workless exactly and to leading order in the small-amplitude approximation, they have non-zero power when quadratic terms are kept. We now infer that the second-order contribution to the rate-of-working of the constraint stress influences the energetics of small-amplitude motions.

Our second main step consists of partitioning the energy flux $\boldsymbol{J}$ and the total energy $\mathscr{E}$ into components to which we apply the descriptions *interaction* and *incremental*. Thus, with reference to equations (3.15) and (3.18), we set

$$J_i = J_i^{\text{int}} + J_i^{\text{incr}}, \qquad \mathscr{E} = \bar{U} + \mathscr{E}^{\text{int}} + \mathscr{E}^{\text{incr}}, \tag{3.19}$$

---

* Equivalently we may say that the particular form $W(\boldsymbol{F}) + \sum_{A=1}^{N} \bar{\alpha}^A \lambda^A(\boldsymbol{F})$ of the generalized strain energy (2.15) must be used in place of $W(\boldsymbol{F})$.

where

$$J_i^{\text{int}} = -\bar{\sigma}_{iq}\dot{u}_q, \tag{3.20}$$

$$J_i^{\text{incr}} = -B^*_{iqrs}\dot{u}_q u_{s,r} - \sum_{A=1}^{N} \bar{N}^A_{iq}\dot{u}_q\beta^A, \tag{3.21}$$

$$\mathscr{E}^{\text{int}} = \bar{\varrho}^{-1}\bar{\sigma}_{pq}u_{q,p}, \tag{3.22}$$

$$\mathscr{E}^{\text{incr}} = \tfrac{1}{2}\dot{u}_q\dot{u}_q + \tfrac{1}{2}\bar{\varrho}^{-1} B^*_{pqrs}u_{q,p}u_{s,r}. \tag{3.23}$$

The interaction components (called by LIGHTHILL [1978, § 1.3] excluded quantities) contain the equilibrium stress $\bar{\boldsymbol{\sigma}}$ and are of leading order in the small quantities. The incremental components, on the other hand, persist when the primary deformation $B_u \to B_e$ is absent, that is when $\bar{\boldsymbol{\sigma}} = \mathbf{0}$.

It follows from the definitions (3.20) to (3.23), with appeal to (2.20) and the basic equations (3.9) and (3.14), that the interaction and incremental components satisfy the individual energy conservation equations

$$\bar{\varrho}\dot{\mathscr{E}}^{\text{int}} = -J^{\text{int}}_{p,p}, \qquad \bar{\varrho}\dot{\mathscr{E}}^{\text{incr}} = -J^{\text{incr}}_{p,p}, \tag{3.24}$$

(*cf.* LIGHTHILL [1978, pp. 14, 15]). Besides corroborating, via (3.19), that $\boldsymbol{J}$ and $\mathscr{E}$ comply with equation (2.8), the relations (3.24) show that $\boldsymbol{J}^{\text{int}}$ and $\boldsymbol{J}^{\text{incr}}$ are the fluxes of the energy densities $\bar{\varrho}\mathscr{E}^{\text{int}}$ and $\bar{\varrho}\mathscr{E}^{\text{incr}}$ respectively. The interaction and incremental components can therefore be given distinct interpretations. The former represent the interplay of the small-amplitude motion with the primary deformation $B_u \to B_e$, and the latter are associated with the self-action of the small-amplitude motion. It is plainly the incremental components which are of primary relevance to the transport of energy in a small-amplitude motion of a constrained elastic body, and it is worth noting, in this connection, that the mean values of $\boldsymbol{J}^{\text{int}}$ and $\boldsymbol{J}^{\text{incr}}$ in a time-harmonic motion of small amplitude are respectively zero and non-zero.

On account of $\boldsymbol{J}^{\text{incr}}$ being homogeneous of degree 1 in $\boldsymbol{H}$ and the β^A's, it is unaltered by transfer from the referential to the spatial description. The right-hand side of equation (3.21) is therefore the incremental component of the spatial energy flux vector

$$\boldsymbol{j} = -\boldsymbol{\sigma}\dot{\boldsymbol{u}}$$

which, from (2.4), (2.10) and (3.8), is related to the referential flux $\boldsymbol{J}$ by

$$\boldsymbol{j} = \tilde{J}^{-1}\tilde{\boldsymbol{F}}\boldsymbol{J}.$$

Since $\boldsymbol{j}$ is the direct expression of the physical concept of energy flux, this observation endorses the significance ascribed to $\boldsymbol{J}^{\text{incr}}$.

4. Homogeneous Plane Waves

We study in this final section solutions of the basic linear equations (3.9) and (3.14) of the type

$$u_i = \phi(\boldsymbol{n}\cdot\bar{\boldsymbol{x}} - vt)\, p_i, \qquad \beta^A = \phi'(\boldsymbol{n}\cdot\bar{\boldsymbol{x}} - vt)\, q^A, \qquad A = 1, 2, \ldots, N. \tag{4.1}$$

These expressions characterize a homogeneous plane wave propagating in the direction defined by the unit vector $\boldsymbol{n}$ with speed $v\ (>0)$. The vector $\boldsymbol{p}$ is the polarization of the wave and the function ϕ, assumed to have non-zero second derivative ϕ'', represents the unvarying profile of the disturbance.

a) The Propagation Conditions

The linear algebraic equations obtained by entering the waveform (4.1) into equations (3.14) and (3.9) and cancelling common factors can be expressed in direct notation as

$$Q(\boldsymbol{n})\,\boldsymbol{p} + \sum_{A=1}^{N} q^A \boldsymbol{c}^A(\boldsymbol{n}) = \bar{\varrho} v^2 \boldsymbol{p}, \tag{4.2}$$

$$\boldsymbol{c}^A(\boldsymbol{n}) \cdot \boldsymbol{p} = 0, \qquad A = 1, 2, \ldots, N, \tag{4.3}$$

where the tensor $Q(\boldsymbol{n})$ (symmetric because of (2.20)) and the vectors $\boldsymbol{c}^A(\boldsymbol{n})$ are defined by the component relations

$$Q_{ij}(\boldsymbol{n}) = B^*_{pirj} n_p n_r, \qquad c_i^A(\boldsymbol{n}) = \overline{N}{}^A_{pi} n_p, \qquad A = 1, 2, \ldots, N. \tag{4.4}$$

We refer to equations (4.2) and (4.3), which determine $\boldsymbol{p}$ and q^A to within a common scalar multiplier, as the *propagation conditions* of the wave (4.1).

In relation to the homogeneous plane wave (4.1), the constraints (2.3) are said to be *fully active* in the direction $\boldsymbol{n}$ if the vectors $\boldsymbol{c}^A(\boldsymbol{n})$ are linearly independent. It is obviously impossible for more than three constraints to be fully active in this sense, and when $N = 3$ and $\boldsymbol{n}$ is a direction in which the constraints are fully active equations (4.2) and (4.3) have only the null solution $\boldsymbol{p} = \boldsymbol{0}$, $q^A = 0$. A homogeneous plane wave can therefore propagate in a direction in which the constraints are fully active only if $N = 1$ or $N = 2$. We examine these cases in subsection b) before dealing in part c) with the possibility of the constraints not being fully active.

No generality is lost in the sequel by taking the polarization $\boldsymbol{p}$ to be a unit vector. The propagation conditions then yield the relation

$$\bar{\varrho} v^2 = B^*_{pqrs} n_p p_q n_r p_s, \tag{4.5}$$

obtained by forming scalar products with $\boldsymbol{p}$ in (4.2) and appealing to (4.3) and $(4.4)_1$.

b) Fully Active Constraints

When $N = 1$ and $\boldsymbol{c}^1(\boldsymbol{n}) \neq \boldsymbol{0}$ equations (4.2) and (4.3) give

$$Q^\dagger(\boldsymbol{n})\,\boldsymbol{p} = \bar{\varrho} v^2 \boldsymbol{p}, \qquad q^1 = -|\boldsymbol{c}^1(\boldsymbol{n})|^{-1}\, \boldsymbol{c}(\boldsymbol{n}) \cdot \{Q(\boldsymbol{n})\,\boldsymbol{p}\}, \tag{4.6}$$

where

$$Q^\dagger(\boldsymbol{n}) = \{\boldsymbol{I} - \boldsymbol{c}(\boldsymbol{n}) \otimes \boldsymbol{c}(\boldsymbol{n})\}\, Q(\boldsymbol{n}) \tag{4.7}$$

and $\boldsymbol{c}(\boldsymbol{n})$ is the unit vector parallel to $\boldsymbol{c}^1(\boldsymbol{n})$. Equations $(4.6)_1$ and (4.7) have been discussed, at the present level of generality, by SCOTT [1975] and WHITWORTH [1982], and the results show that there exist at most two homogeneous plane waves with wave normal $\boldsymbol{n}$. The speeds of propagation of these waves are given by the characteristic equation of $\boldsymbol{Q}^\dagger(\boldsymbol{n})$ which can be written as

$$(\bar{\varrho}v^2)^2 - [\operatorname{tr} \boldsymbol{Q}(\boldsymbol{n}) - \boldsymbol{c}(\boldsymbol{n}) \cdot \{\boldsymbol{Q}(\boldsymbol{n})\, \boldsymbol{c}(\boldsymbol{n})\}]\, \bar{\varrho}v^2 + \boldsymbol{c}(\boldsymbol{n}) \cdot \{\boldsymbol{Q}^{\mathrm{adj}}(\boldsymbol{n})\, \boldsymbol{c}(\boldsymbol{n})\} = 0, \tag{4.8}$$

$\boldsymbol{Q}^{\mathrm{adj}}(\boldsymbol{n})$ being the adjugate of $\boldsymbol{Q}(\boldsymbol{n})$. The definitions (2.19) and $(4.4)_1$ imply that if $\check{\boldsymbol{B}}^1_{pirj} n_p n_r \neq \boldsymbol{0}$*, $\boldsymbol{Q}(\boldsymbol{n})$ depends on the constraint multiplier α^1, in which event strong ellipticity of the linear elasticity tensor $\boldsymbol{B}$ is not a sufficient condition for the existence of two positive real roots of equation (4.8). This circumstance has stability connotations, first noticed by CHEN & GURTIN [1974] in relation to wave propagation in an inextensible elastic body.

When $N = 2$ and $\boldsymbol{c}^1(\boldsymbol{n})$ and $\boldsymbol{c}^2(\boldsymbol{n})$ are linearly independent we deduce from the propagation conditions (4.2) and (4.3) that

$$\boldsymbol{p} = \boldsymbol{d}(\boldsymbol{n}), \qquad \bar{\varrho}v^2 = \boldsymbol{d}(\boldsymbol{n}) \cdot \{\boldsymbol{Q}(\boldsymbol{n})\, \boldsymbol{d}(\boldsymbol{n})\}, \tag{4.9}$$

$$q^A = -\{\boldsymbol{Q}(\boldsymbol{n})\, \boldsymbol{d}(\boldsymbol{n})\} \cdot \sum_{B=1}^{2} \overset{-1}{C}{}^{AB} \boldsymbol{c}^B(\boldsymbol{n}), \qquad A = 1, 2, \tag{4.10}$$

where $\boldsymbol{d}(\boldsymbol{n})$ is the unit vector codirectional with $\boldsymbol{c}^1(\boldsymbol{n}) \wedge \boldsymbol{c}^2(\boldsymbol{n})$ and

$$C^{AB} = \boldsymbol{c}^A(\boldsymbol{n}) \cdot \boldsymbol{c}^B(\boldsymbol{n}).$$

It is evident from these results that a single homogeneous plane wave can travel in the direction $\boldsymbol{n}$ if and only if the right-hand side of equation $(4.9)_2$ is positive, and that when this wave exists the polarization, the speed of propagation and the constraint multiplier increments are given explicitly by (4.9) and (4.10). Again, the existence or non-existence of a homogeneous plane wave solution has a bearing on the stability of $\mathscr{B}$ in the configuration B_e (*cf.* WHITWORTH [1982, § 5(a)].

c) *Inactive and Partially Active Constraints*

Turning now to the situation in which the internal constraints are not fully active, we suppose first that $N = 1$ and $\boldsymbol{c}^1(\boldsymbol{n}) = \boldsymbol{0}$. Equation (4.3) then holds identically, (4.2) fixes $\boldsymbol{p}$ apart from an ambiguity of sign, and q^1 is indeterminate. In addition to the homogeneous plane waves which can propagate in the direction $\boldsymbol{n}$ in the absence of constraints, equations (3.9) and (3.14) thus admit a *ghost* solution (not necessarily wavelike) which perturbs the constraint multiplier α^1 but produces no displacement.

Proceeding to the case $N > 1$, let $M\,(\leq 3)$ be the number of linearly independent vectors in the set $\mathscr{C} = \{\boldsymbol{c}^A(\boldsymbol{n}) : A = 1, 2, \ldots, N\}$. If $\bar{\boldsymbol{c}}^B(\boldsymbol{n})$, $B = 1, \ldots, M$,

---

* Normally, $\check{\boldsymbol{B}}^1_{pirj} n_p n_r \neq \boldsymbol{0}$, but an important exception arises in the case of incompressibility (see WHITWORTH [1982, § 4(a)]).

are linearly independent members of $\mathscr{C}$ there exist scalars γ^{AB}, $A = 1, \ldots, N$, $B = 1, \ldots, M$, such that

$$c^A(n) = \sum_{B=1}^{M} \gamma^{AB} \bar{c}^B(n), \qquad A = 1, 2, \ldots, N,$$

and the matrix (γ^{AB}) is of rank M. Hence, if we put

$$\bar{q}^B = \sum_{A=1}^{N} \gamma^{AB} q^A, \qquad B = 1, 2, \ldots, M, \tag{4.11}$$

the propagation conditions apply with B, M, $\bar{q}^B$ and $\bar{c}^B(n)$ substituted in turn for A, N, q^A and $c^A(n)$. One of the cases of fully active constraints considered above then pertains and p and the M scalars $\bar{q}^B$ are determined by the modified forms of (4.2) and (4.3). The expressions for the q^A's obtained from (4.11) contain $N - M$ parameters, however, so the homogeneous plane-wave solutions ($\leqq 3 - M$ in number) are accompanied by $N - M$ independent ghost disturbances.

d) *Ray and Group Velocities. Slowness Surface*

We are concerned in the remainder of this section with any homogeneous plane wave which can travel in the direction n, regardless of whether the constraints are fully or only partially active in this direction.

The ray velocity C of the plane wave (4.1) is defined by

$$C_i = \frac{\partial v}{\partial n_i}, \tag{4.12}$$

(see, for example, Varley & Cumberbatch [1965, § 2]) where the components of the wave normal n are to be treated as independent variables. We calculate C by generalizing a method applied by Fedorov [1968, p. 124] to a sinusoidal homogeneous plane wave of small amplitude propagating in an anisotropic elastic body free from pre-strain. On differentiating each side of equation (4.5) with respect to n_i, invoking the symmetry relation (2.20) and the propagation conditions (4.2) and (4.3), with (4.4), and recalling that p is a unit vector, we derive from (4.12) the formula

$$C_i = (\bar{\varrho} v)^{-1} \left(B^*_{iqrs} p_q n_r p_s + \sum_{A=1}^{N} \bar{N}^A_{iq} p_q q^A \right). \tag{4.13}$$

The concept of group velocity relates to an ensemble, or superposition, of homogeneous plane waves and can be developed explicitly only for harmonic disturbances of the form

$$\begin{aligned} u_i &= a \exp\{i(k \cdot \bar{x} - \omega t)\}\, p_i, \\ \beta^A &= i(k \cdot k)^{\frac{1}{2}} \exp\{i(k \cdot \bar{x} - \omega t)\}\, b^A, \qquad A = 1, 2, \ldots, N. \end{aligned} \tag{4.14}$$

The scalars a and b^A in (4.14) are complex and the angular frequency $\omega\ (> 0)$

and the wave vector $\boldsymbol{k}$, both real, are connected to the slowness $\boldsymbol{s}$, the speed of propagation v and the wave normal $\boldsymbol{n}$ by

$$\boldsymbol{s} = v^{-1}\boldsymbol{n} = \omega^{-1}\boldsymbol{k}. \tag{4.15}$$

Although the wave (4.14) is non-dispersive, the dependence of v on $\boldsymbol{n}$ implies the possibility of constructive interference between the constitutents of a monochromatic ensemble of harmonic waves travelling in neighbouring directions to form groups progressing at a rate different from the local wave speed. The group velocity $\boldsymbol{G}$ has components

$$G_i = \frac{\partial \omega}{\partial k_i} \tag{4.16}$$

(see LIGHTHILL [1960, App. A], STROH [1962, § 5]).

On account of the equalities (4.15), equation (4.5) can be rewritten in the alternative forms

$$\bar{\varrho}\omega^2 = B^*_{pqrs}k_p p_q k_r p_s \tag{4.17}$$

and

$$H(\boldsymbol{s}) := B^*_{pqrs}s_p p_q s_r p_s - \bar{\varrho} = 0. \tag{4.18}$$

With the further use of the generalized method of FEDOROV, outlined above, in association with equations (4.17) and (4.16) it is found that the ray and group velocities of a harmonic homogeneous plane wave are equal:

$$\boldsymbol{C} = \boldsymbol{G}.$$

In combination with equations (4.2) to (4.4), with $v\boldsymbol{s}$ replacing $\boldsymbol{n}$, (4.18) represents in the Cartesian frame with coordinates s_i the slowness surface corresponding to the configuration B_e. Equations (4.5), (4.17) and (4.18) provide the relations

$$\frac{\partial H}{\partial s_i} = 2\bar{\varrho}\left.\frac{\partial v}{\partial n_i}\right|_{n_i = v s_i} = 2\bar{\varrho}\left.\frac{\partial \omega}{\partial k_i}\right|_{k_i = \omega s_i},$$

from which it is clear that the ray velocity of a homogeneous plane wave and the ray and group velocities of a harmonic disturbance of this kind are directed along the normal to the slowness surface at the point representing the wave. This direction is known as the ray, or bicharacteristic.

e) *Energy Relations*

In the linearized dynamics of a homogeneous anisotropic elastic body, unconstrained and stress free in its undisturbed state, the researches of SCHOUTEN [1951, Chapter VII, § 7], SYNGE [1956] and FEDOROV [1968, § 21] have established, for harmonic homogeneous plane waves, the following results. First, the kinetic and strain energies are equipartitioned. Second, the velocity with which the total energy of the wave propagates is equal to the group velocity. Third, the energy flux produced by the wave is directed along the normal to the slowness surface. And, fourth, the projections of the energy propagation and group velocities on the wave normal equal the speed of propagation of the wave. We prove in conclusion that analogous relationships apply to homogeneous plane waves of arbitrary form in the more general setting of the present paper in which, besides exhibiting material anisotropy, the transmitting body is subject to a primary deformation and a system of internal constraints.

According to equations (2.9), (3.8) and (3.23), the kinetic energy and the incremental component of the pseudo-internal energy in a motion of small amplitude about B_e, each measured per unit mass, are given by

$$K = \tfrac{1}{2}\dot{u}_q\dot{u}_q, \qquad U^{\text{incr}} = \tfrac{1}{2}\bar{\varrho}^{-1}B^*_{pqrs}u_{q,p}u_{s,r}. \tag{4.19}$$

When the small-amplitude motion is the homogeneous plane wave (4.1), equations (4.19) become

$$K = \tfrac{1}{2}v^2\phi'^2, \qquad U^{\text{incr}} = \tfrac{1}{2}\bar{\varrho}^{-1}B^*_{pqrs}n_pp_qn_rp_s\phi'^2, \tag{4.20}$$

the argument of ϕ' being suppressed. We deduce from (4.20) and (4.5) that

$$K = U^{\text{incr}}, \tag{4.21}$$

proving that the kinetic and incremental pseudo-internal energies are equal in a homogeneous plane wave.

The equipartition result (4.21) becomes invalid if the internal energy (3.12) is used (without remainder) instead of the pseudo-internal energy (3.17). A particular instance of this inequality has been encountered by SCOTT & HAYES [1976, § 6] in a study of small-amplitude vibrations of an incompressible and inextensible elastic body.

A natural way of assigning a velocity $\boldsymbol{U}$ to the incremental propagation of energy associated with the wave (4.1) is through the definition

$$U_i = (\bar{\varrho}\mathscr{E}^{\text{incr}})^{-1}\, j_i^{\text{incr}}. \tag{4.22}$$

Since the incremental components of the spatial and referential energy fluxes are equal, as pointed out at the end of Section 3c), equations (3.21) and (4.1) imply the relation

$$j_i^{\text{incr}} = v\left(B^*_{iqrs}p_qn_rp_s + \sum_{A=1}^{N} \bar{N}^A_{iq}p_qq^A\right)\phi'^2. \tag{4.23}$$

Equations (4.20) and (4.21) give

$$\mathscr{E}^{\text{incr}} = K + U^{\text{incr}} = v^2\phi'^2,$$

whence, from (4.22), (4.23) and (4.13), the ray and energy propagation velocities are equal:

$$\boldsymbol{C} = \boldsymbol{U}. \tag{4.24}$$

It follows from equations (4.24) and (4.13), with reference to (4.5), $(4.4)_2$ and (4.3), that

$$\boldsymbol{C}\cdot\boldsymbol{n} = \boldsymbol{U}\cdot\boldsymbol{n} = v,$$

completing the results anticipated at the beginning of this subsection.

The analysis presented in Sections 4d) and 4e) has close parallels with the treatment of the energetics of acceleration waves in a constrained elastic body given by BOREJKO & CHADWICK [1980], and this may appear unsurprising in view of the broad analogies which exist between homogeneous plane waves of small amplitude and acceleration waves in non-heat-conducting elastic media (Truesdell [1961]). The severe restrictions imposed by Borejko and Chadwick on conditions ahead of the acceleration wave have no counterpart in the present investigation, however, and the exactness of the theory of singular surfaces gives no occasion for the subtleties met in Section 3c) in connection with the introduction of a pseudo-internal energy and the decomposition of fields into interaction and incremental components.

Acknowledgements. The authors are grateful to Dr. N. H. Scott for helpful comments on this work. A. M. Whitworth's contribution was made during the tenure of a Studentship awarded by the Science and Engineering Research Council of the U.K., and P. Borejko's partially under the auspices of a Visiting Fellowship at the University of East Anglia.

References

Borejko, P. & P. Chadwick, Wave Motion **2**, 361–374 (1980).
Chadwick, P., Continuum Mechanics. Concise Theory and Problems. London: Allen & Unwin 1976.
Chadwick, P. & P. K. Currie, Arch. Rational Mech. Anal. **49**, 137–158 (1972).
Chen, P. J. & M. E. Gurtin, Int. J. Solids Struct. **10**, 275–281 (1974).
Fedorov, F. I., Theory of Elastic Waves in Crystals. New York: Plenum Press 1968.
Flett, T. M., Differential Analysis. Cambridge etc.: Cambridge University Press 1980.
Lancaster, P., Theory of Matrices. New York and London: Academic Press 1969.
Lighthill, M. J., Phil. Trans. R. Soc. A **252**, 397–430 (1960).
Lighthill, J., Waves in Fluids. Cambridge: Cambridge University Press 1978.
Schouten, J. A., Tensor Analysis for Physicists. Oxford: Clarendon Press 1951.
Scott, N., Arch. Rational Mech. Anal. **58**, 57–75 (1975).
Scott, N. & M. Hayes, Quart. J. Mech. Appl. Math. **29**, 467–486 (1976).
Stroh, A. N., J. Math. Phys. **41**, 77–103 (1962).
Synge, J. L., Proc. R. Irish Acad. Sect. A **58**, 13–21 (1956).
Truesdell, C., Arch. Rational Mech. Anal. **8**, 263–296 (1961).
Truesdell, C. & W. Noll, The Non-Linear Field Theories of Mechanics. Handbuch der Physik, Vol. III/3. Berlin-Heidelberg-New York: Springer 1965.
Varley, E. & E. Cumberbatch, J. Inst. Maths Applics **1**, 101–112 (1965).
Whitworth, A. M., Quart. J. Mech. Appl. Math. **35**, 461–484 (1982).
Whitworth, A. M. & P. Chadwick, Wave Motion **6**, 289–302 (1984, 1).
Whitworth, A. M. & P., Chadwick, Arabian J. Sci. Engng **9**, 67–76 (1984, 2).

School of Mathematics and Physics
University of East Anglia
Norwich, England

and

Institut für Theoretische Mechanik
Universität Karlsruhe
Karlsruhe, West Germany

(Received June 25, 1984)

A Corpuscular Approach to Continuum Mechanics: Basic Considerations

A. I. MURDOCH

Dedicated to Walter Noll on his sixtieth birthday

Communicated by M. E. GURTIN

1. Introduction

The macroscopic behaviour of a material derives ultimately from its microstructure, which at the molecular level and below is of a fundamentally discrete nature. As much is often known of the microstructure, molecular interactions, and thermal motions associated with specific materials, it is of interest to ascertain whether, and to what extent, such information can usefully be incorporated into continuum models. The first step in this direction is to treat discrete fundamental entities (for example, molecules, ions, or atoms) as particles (point masses) and relate continuum theories to corpuscular considerations. This is, of course, an important feature of the kinetic theory of gases, lattice dynamics, and statistical mechanics. The first two theories take account of particle interactions and thermal motions as these pertain to general motions of moderately rarefied gases and vibrations in ordered solids, respectively, but such considerations play no explicit part in standard formulations of statistical mechanics. Many discussions which bear upon the relations between discrete descriptions of materials and continuum mechanics have been given (motivated in the main by considerations of microstructure), such as the contributions of ERICKSEN (1960, 1961), DAHLER & SCRIVEN (1963), ERINGEN & SUHUBI (1964), GREEN & RIVLIN (1964), KRUMHANSL (1965), ERINGEN, KRÖNER, KRUMHANSL, KUNIN, MINDLIN, and RIVLIN in KRÖNER (1968), RIVLIN (1968, 1976), ALBLAS (1976), and CAPRIZ & PODIO-GUIDUGLI (1976, 1977). This work is intended to complement the aforementioned by emphasising the rôles played *in general* by corpuscular interactions and thermal motions in the formulation of continuum theories.

Recently[1] MURDOCH $(1983)_1$ indicated how such aspects of microscopic behaviour can be utilised in the motivation of continuum concepts and fields, and of the balance relations these fields satisfy. Difficulties were there indicated in connection with the motivation of a generalised (tensor) moment of momentum balance (which seems to be necessary adequately to describe certain situations

[1] In what follows this will be referred to as I.

in which micro*structure* is important[1]) and with the generalisation of the notion of binding energy when interactions are not governed by separation-dependent pair-potentials. Further, the effect of diffusion was not adequately taken into account[2], so precluding comparison with kinetic theory. Here two resolutions of the aforementioned difficulties are presented, due attention is paid to diffusion, and the opportunity is taken to eludicate and simplify certain aspects of the previous work. This contribution is intended to be followed by discussions of mixtures and of materials which consist of large, somewhat inflexible, molecules, and accordingly is directed towards this end. Such motivation accounts for the particular emphasis placed upon couple-stress and the correspondingly general forms of moment of momentum and energy balances. The analysis is based upon the equations of motion of individual particles. Four very weak assumptions concerning the nature of corpuscular interactions together with two assumptions expressing the random nature of thermal motions[3] suffice, upon averaging both in space and time, to establish the balance relations sought. Alternative forms of balances of generalised moment of momentum and of energy are indicated consequent upon a further assumption concerning interactions.

The averaging procedures necessary to effect the passage from a discrete to a continuum model are introduced in Section 2, wherein mass density and macroscopic velocity are defined in terms of purely spatial, local averages, the latter in a manner which mandates mass conservation. Local averages in time (necessary to treat fluctuating quantities) and in space-time are discussed, and two aspects of the random nature of thermal motion characterised. Remarks are made on conditions deemed necessary for the existence of local averages. In Section 3 balance relations for a continuous body are derived on the basis of assumptions concerning the nature of corpuscular interactions. In particular, linear momentum balance is established as a consequence of such interactions having essentially short range and yielding negligible net resultant when summed over all particles within regions having smallest characteristic dimensions of order 10^{-5} m or greater[4]. Considerations from statistical mechanics indicate that the traction field upon any hypothetical surface within the body which derives from interactions across this surface is not well-defined as a macroscopic quantity by spatial averaging alone. Here additional smoothing is effected by means of temporal averaging. Accordingly it proves necessary to define all macroscopic (continuum) fields introduced in this section in terms of averages in space-time. The separate contributions to the Cauchy stress field which derive from interactions and from diffusion are rendered explicit. A generalised (tensor-valued) moment of momentum balance relation is established on the further assumption that interactions between particles in any microscopically-large region should give rise to negligible resultant moment about any point. The skew part of this relation corresponds to the usual balance of moment of momentum, but differs from the conventional (non-polar)

---

[1] Indeed, such a balance may become a field equation for a measure of microstructural deformation. *Cf.*, *e.g.*, CAPRIZ & PODIO-GUIDUGLI (1976, 1977).

[2] The author gratefully acknowledges having this drawn to his attention by Professor C. CERCIGNANI.

[3] These are appropriate for both fluid *and* solid phases.

[4] Such regions will be termed *microscopically large*.

form[1] by the presence therein of terms which represent couple-stress and body couple. In the absence of interactions, body couple, and diffusive generalised couple-stress, the local form of the complete balance relation identifies the Cauchy stress tensor with the negative of the pressure tensor of kinetic theory. An additional assumption, involving the interplay between interactions and thermal motions (later interpretable in terms of a balance of heat conduction exchange rates between adjacent, microscopically-large, regions), enables the energy balance relation to be obtained. The internal energy density is shown to be the sum of a purely thermal term (the so-called heat content, or averaged local kinetic energy density associated with thermal motion) and a form of interaction work density. Alternative balance relations for (tensor) moment of momentum and energy are discussed in Section 4, modulo the assumption of a balance of total interaction-related energy exchange rates between adjacent, microscopically-large regions. The appropriate internal energy, the same as that introduced in Section 3 in the absence of interactive couple-stress, has the interaction work density contribution equal to binding energy density when interactions are delivered by separation-dependent pair-potentials. Some general observations concerning radiation, microstructure, material objectivity, and non-local interactions are made in Section 5.

2. Preliminaries

2.1 Volume averages

The existence of local volume-continuous macroscopic fields is based upon the assumption that sums of appropriate microscopic quantities over macroscopically-small regions are proportional to the volumes thereof, provided that these regions are large enough to contain many particles and have, roughly speaking, dimensions of the same order of magnitude in all directions. Further, any region of macroscopic proportions must be divisible into many such disjoint, macroscopically-small, sub-regions. It may also be necessary to assume that such sub-regions preserve their nature when subjected to deformations of the scale associated with continua.

The foregoing may be formalised by defining[2] an **ε-cell centred at $\boldsymbol{x}$** to be a simply-connected region of Euclidean space, with piece-wise smooth boundary, satisfying

$$\tfrac{1}{2}\varepsilon(1+\alpha) < |\boldsymbol{x}-\boldsymbol{y}| < \tfrac{3}{2}\varepsilon(1+\alpha) \text{ with } |\boldsymbol{x}-\bar{\boldsymbol{x}}| \text{ of order } O(\varepsilon^2), \tag{2.1}$$

where $\boldsymbol{y}$ denotes any point on its boundary, $\bar{\boldsymbol{x}}$ its centroid, and $|\alpha| \ll 1$. In order to exclude pathological boundaries it is also assumed that nowhere on any subsurface should curvatures be too large, nor should there be too many component subsurfaces.

Let g_i denote a quantity associated with a particle (point mass) P_i: for example,

---

[1] *Cf., e.g.*, TRUESDELL & TOUPIN (1960, § 196, 205).

[2] This definition is considered preferable to that presented in I.

its mass. Consider $\sum_i g_i/V_\varepsilon$, where the sum is taken over an ε-cell of volume V_ε, centred at $\boldsymbol{x}$. If this volume average is $d(\boldsymbol{x})(1+\beta)$ for some fixed quantity $d(\boldsymbol{x})$, with[1] $|\beta| \ll 1$ over a range of ε values macroscopically small yet microscopically large[2], quite independently of the shape of the cell, then $d(\boldsymbol{x})$ is termed the *ε-limit* of the volume average at $\boldsymbol{x}$ and we write

$$d(\boldsymbol{x}) =: \lim_\varepsilon \left\{ \sum_i{}' g_i/V_\varepsilon \right\}.$$

(Here, and in what follows, a summation sign with superposed prime indicates that the sum is taken over an ε-cell). If such limits exist everywhere in a region R then d will be a field on R. Hereafter, in working with any given ε-limit, $1+\beta$ will be approximated by 1 for any $|\beta| \ll 1$. It follows that any ε-limit, within this approximative scheme, is unique and expressible as a volume average for some cell. These observations justify the formal algebra of ε-limits to be used. If a macroscopic region R, in which d is defined, is subdivided into ε-cells, then the sum $\sum_i g_i$ taken over all particles P_i in R may be written as a sum of individual ε-cell sums. Provided the division is such that the ε-cell sums have volume averages which coincide with the corresponding ε-limits (modulo neglect of β terms) then

$$\sum_i g_i = \sum_{\text{cells}} d(\boldsymbol{x})\, V_\varepsilon(\boldsymbol{x}),$$

where $V_\varepsilon(\boldsymbol{x})$ denotes the volume of the cell centred at $\boldsymbol{x}$. If d is continuous (as is entirely plausible: given points $\boldsymbol{x}_1$ and $\boldsymbol{x}_2$ sufficiently close to each other, the values $d(\boldsymbol{x}_1)$ and $d(\boldsymbol{x}_2)$ will be given by volume averages over practically identical ε-cells which will hence contain almost the same particles) the sum over ε-cells clearly constitutes a Riemann sum which approximates the (Riemann) integral $\int_R d$. Hereafter such sums will always be identified with the corresponding integrals. This is consistent with the approximative convention if d does not vary significantly at the length scale associated with ε.

2.2 *Time averages*

In order properly to discuss the relations between macroscopic fields and microscopic quantities it proves necessary to introduce the notion of a temporal average. If φ denotes a time-dependent quantity then

$$\varphi_\Delta(t) := \frac{1}{\Delta} \int_{t-\Delta}^{t} \varphi(\tau)\, d\tau$$

is termed its *Δ-time average at instant t*.

---

[1] Of course, β will depend upon the quantity in question, $\boldsymbol{x}$, and the cell concerned. What is intended here is that for the range of ε values stated $|\beta|$ is uniformly small.

[2] That is, ε is of order 10^{-5} m ($= 10^5$ Å). In what follows it is to be understood that the term 'ε-cell' refers to cells of this scale.

Quantities which vary appreciably on a microscopic time scale are said to *fluctuate*. It may be possible to smooth out such behaviour, and so obtain a parameter varying appreciably only on a macroscopic time scale, by time averaging over macroscopically-small yet microscopically-large time intervals (for definiteness, Δ of order 10^{-6} s, say). If this average on φ is insensitive to changes in Δ at this level (modulo the degree of accuracy associated with volume averages) then it will be denoted by $\overline{\varphi}$. Symbolically,

$$\overline{\varphi}(t) =: \lim_\Delta \{\varphi_\Delta(t)\},$$

where the 'Δ-limit' is the temporal counterpart of the 'ε-limit' previously discussed. Consistent with the assumed insensitivity of $\varphi_\Delta(t)$ to changes in Δ of the order indicated above, it will be assumed that

$$\overline{(\overline{\varphi})} = \overline{\varphi}. \tag{2.2}$$

Any quantity φ which satisfies $\varphi = \overline{\varphi}$ will be said to *vary macroscopically with time*. All macroscopic fields will be deemed to have this property.

Whenever $\overline{\varphi}$ is meaningful, the **fluctuation** φ' in φ is defined by

$$\varphi' := \varphi - \overline{\varphi}.$$

Clearly (2.2) implies

$$\overline{(\varphi')} = 0. \tag{2.3}$$

If ψ denotes another quantity and $\overline{\psi}$ is meaningful, then consistent with the foregoing is the assumption that

$$\overline{\varphi'\overline{\psi}} = 0, \tag{2.4}$$

and, consequently,

$$\overline{\varphi\psi} = \overline{\varphi}\,\overline{\psi} + \overline{\varphi'\psi'}. \tag{2.5}$$

Relations (2.4) and (2.5) have been expressed in a form appropriate when φ and ψ are scalar-valued. Obvious generalisations follow in the case of tensor products of φ and ψ when these quantities are tensorial in nature.

The observation that for any given Δ,

$$(\dot{\varphi})_\Delta\,(t) = \dot{\widehat{\varphi_\Delta(t)}},$$

suggests it might be reasonable to assume

$$\overline{\dot{\varphi}} = \dot{\overline{\varphi}}, \tag{2.6}$$

but this will not be invoked in what follows.

2.3 Kinematics

If P_i has mass m_i and velocity $\boldsymbol{v}_i$, the **density** $\varrho(\boldsymbol{x}, t)$ and velocity $\boldsymbol{v}(\boldsymbol{x}, t)$ at a geometrical point $\boldsymbol{x}$ at instant t are defined (cf. § 2.1) by

$$\varrho(\boldsymbol{x}, t) := \lim_\varepsilon \left\{ \sum_i{}' m_i/V_\varepsilon \right\} \text{ and } \boldsymbol{v}(\boldsymbol{x}, t) := (\varrho(\boldsymbol{x}, t))^{-1} \lim_\varepsilon \left\{ \sum_i{}' m_i \boldsymbol{v}_i/V_\varepsilon \right\}, \tag{2.7}$$

whenever the right-hand sides (involving sums over particles which lie at instant t within ε-cells centred at $\boldsymbol{x}$) make sense. The mass and linear momentum associated with a macroscopic region R in which ϱ and $\boldsymbol{v}$ are defined are accordingly

$$\int_R \varrho \quad \text{and} \quad \int_R \varrho \boldsymbol{v},$$

respectively. Since ε-limits may be expressed as sums over ε-cells of appropriate size (modulo quantities neglected in the approximative scheme) it follows that

$$\boldsymbol{v}(\boldsymbol{x}, t) = \sum_i{}' m_i \boldsymbol{v}_i / \sum_i{}' m_i$$

for appropriate cells.

The global existence of ϱ requires, at the ε-cell level, that matter essentially be distributed homogeneously. Together with our assumption $(2.1)_2$ concerning the location of the centroid of an ε-cell centred at $\boldsymbol{x}$, this motivates the assumption that for such a cell

$$|\boldsymbol{x}_\varepsilon - \boldsymbol{x}| \textit{ is of order } O(\varepsilon^2), \tag{2.8}$$

where $\boldsymbol{x}_\varepsilon$ denotes the mass centre of the particles in the cell.

As was fully discussed in I, the notions of material point, motion thereof, and material time derivatives of functions defined on the trajectory of a body, can at this stage be made precise, and conservation of mass motivated in the form

$$d/dt \left\{ \int_R \varrho \right\} = - \int_{\partial R} \varrho \boldsymbol{v} \cdot \boldsymbol{n}.$$

It will be assumed that both ϱ and $\boldsymbol{v}$ are of class C^1 on the trajectory of the body, from which the usual local relations

$$\partial p/\partial t + \operatorname{div}(\varrho \boldsymbol{v}) = 0 = \dot{\varrho} + \varrho \operatorname{div} \boldsymbol{v} \tag{2.9}$$

and Reynolds' transport theorem follow.

2.4 Averages in space-time

A given corpuscular quantity g_i may not yield a volume average; that is, the relevant ε-limit may not exist. Even when such a limit exists it may not be identifiable with a macroscopic variable because it does not vary macroscopically with time; that is, it fluctuates. However, in such cases a further *temporal* averaging may yield a macroscopic field: that is, a field which is essentially constant over ε-length and Δ-time scales. More precisely, let

$$s(y, \tau; \varepsilon) := \sum_i{}' g_i$$

denote the appropriate sum, taken over an ε-cell centred at y at instant τ. If the cell boundary is thought to vary with time in such a way as to follow the macroscopic motion of the body as defined by $\boldsymbol{v}$, the corresponding time average at point $\boldsymbol{x}$ at instant t is

$$\hat{d}(\boldsymbol{x}, t; \varepsilon, \Delta) := \frac{1}{\Delta} \int_{t-\Delta}^{t} (s(\hat{\boldsymbol{x}}(\tau), \tau; \varepsilon)/V_\varepsilon(\tau)) \, d\tau .$$

Here $\hat{\boldsymbol{x}}(\tau)$ denotes the location of the material point at instant τ which coincides with $\boldsymbol{x}$ at instant t, and $V_\varepsilon(\tau)$ denotes the volume of the cell at instant τ. If $\hat{d}(\boldsymbol{x}, t; \varepsilon, \Delta)$ varies negligibly with ε and Δ when these parameters are macroscopically small but microscopically large, and is insensitive to cell shape, then for this range of parameter values $\hat{d}$ may essentially be regarded as a function d of $\boldsymbol{x}$ and t. We symbolise this by writing

$$d(\boldsymbol{x}, t) =: \lim_\varepsilon \overline{\{s(\hat{\boldsymbol{x}}(\tau), \tau; \varepsilon)/V_\varepsilon(\tau)\}}$$

or, more directly,

$$d(\boldsymbol{x}, t) = \lim_\varepsilon \left\{ \overline{\sum_i{}' g_i/V_\varepsilon} \right\}.$$

As the variation of V_ε with τ is prescribed by the macroscopic motion, V_ε may be regarded as essentially constant over the time-averaging period. This is emphasized by writing

$$d(\boldsymbol{x}, t) = \lim_\varepsilon \left\{ \overline{\sum_i{}' g_i} \,/V_\varepsilon \right\}.$$

Hereafter all thermodynamical variables to be introduced are necessarily defined in terms of space-time averages. It is thus to be expected that there should be lower limits both in space *and* time for the range of applicability of the corresponding continuum theories. The possibility clearly may exist of increased precision being gained spatially at the expense of temporal accuracy (or vice-versa), somewhat reminiscent of the uncertainty principle of quantum mechanics.

2.5 Thermal motion

In what follows, thermal fields will be related to the fundamentally erratic nature of microscopic motions. While the characters of such motions in fluids and solids are distinctly different, it proves possible to adopt a unified treatment in terms of thermal velocity. The **thermal velocity**[1] $\tilde{\boldsymbol{v}}_i(t)$ of a particle P_i, located at point $\boldsymbol{x}_i$ at instant t, is defined by

$$\tilde{\boldsymbol{v}}_i(t) := \boldsymbol{v}_i(t) - \boldsymbol{v}(\boldsymbol{x}_i, t),$$

where $\boldsymbol{v}_i(t)$ denotes the instantaneous velocity of P_i and $\boldsymbol{v}$ the macroscopic velocity field. If P_i is contained in an ε-cell centred at $\boldsymbol{x}$, then $\tilde{\boldsymbol{v}}_i(t)$ is evidently approximated by the **notional thermal velocity**

$$\hat{\boldsymbol{v}}_i(t) := \boldsymbol{v}_i(t) - \boldsymbol{v}(\boldsymbol{x}, t) - \boldsymbol{L}(\boldsymbol{x}, t)\,(\boldsymbol{x}_i - \boldsymbol{x}), \tag{2.10}$$

[1] In the kinetic theory of gases this is termed the *random*, or *peculiar*, velocity (*cf.* CERCIGNANI, 1975).

where L denotes the velocity gradient. Of course, if $\nabla L(x, t)$ is of order $O(1)$ then $\tilde{v}_i$ and $\hat{v}_i$ may be identified upon neglect of terms of order $O(\varepsilon^2)$. In view of (2.8), it is to be noted that x may be replaced by x_ε in (2.10) within this level of approximation (which will be adopted consistently hereafter) provided L is of order $O(1)$.

The thermal velocity is assumed to be of random character. More specifically, it will be assumed that at any instant, within any ε-cell centred at x, and modulo quantities negligible within the approximative convention adopted[1],

$$\text{T. M. 1.} \quad \sum_i{}' m_i \tilde{v}_i = \mathbf{0}$$

and

$$\text{T. M. 2.} \quad \sum_i{}' (x_i - x) \otimes m_i \tilde{v}_i = \mathbf{0}.$$

2.6 On the existence of limits

In order that local volume averages should give rise to physically meaningful macroscopic fields, it is necessary that the sums involved be additive in the following sense. Let F denote such a field and S_1, S_2, denote relevant sums of microscopic quantities over ε-cells of volumes V_1, V_2, centred at x_1, x_2, respectively. If these cells share a portion of common boundary, but are disjoint, then the sum S computed from the cells taken together should approximate $F(x)$, where x denotes the centroid of the composite volume. If F has a bounded second derivative, then the observation that

$$(x_1 - x) V_1 + (x_2 - x) V_2 = \mathbf{0}$$

motivates, upon neglect of $O(\varepsilon^2)$ terms, the requirement

$$S = S_1 + S_2 .$$

In what follows, only quantities which fulfill this requirement (or a strictly analogous criterion in the case of the surface averages later to be discussed) will be employed. Where fields defined in terms of particle interactions are concerned, additivity is shown to be a consequence of very mild restrictions upon these interactions.

The analysis of the next section is rendered less than straightforward precisely because sums of the form

$$\sum_i{}' \sum_k{}' (x_i - x) \otimes f_{ik} \tag{2.11}$$

(taken over all particles P_i, P_k within an ε-cell centred at x, with f_{ik} denoting the force exerted upon P_i by P_k) are not considered additive. That is, such an assumption (or corresponding assumption concerning the nature of interactions) is

---

[1] It may be noted that T.M.1. follows from the approximation $\tilde{v}_i \cong \hat{v}_i$ together with (2.10), (2.8), and the assumption that L is of order $O(1)$.

considered physically implausible. A measure of the subtlety of the situation is afforded by noting that sums of the form

$$\sum_i{}' \sum_l (\boldsymbol{x}_i - \boldsymbol{x}) \otimes \boldsymbol{f}_{il}$$

are additive. Here the sum involves interactions $\boldsymbol{f}_{il}$ between particles P_i in the cell and *all* other particles P_l, whether inside or outside the cell. In this case additivity is a consequence of interaction assumptions I.1. and I.2. (*cf.* § 3.1) together with the boundedness of $\nabla(\text{div}\, \boldsymbol{T}^-)$.

It was shown in I[1] how the difficulties associated with the above example can be avoided in motivating the conventional balance of *angular* momentum. However, the problem becomes manifest when a *generalised* (tensor) moment of momentum balance is sought. Such a balance relation is most helpful in understanding the inter-relationships between various macroscopic fields, in relating this treatment to kinetic theory, and in laying the foundations for a subsequent study of materials with microstructure in which the relation may become a field equation for microstructural deformation. In its discussion of this generalised balance relation, my previous analysis tentatively introduced a variable $\boldsymbol{J}$ which presupposed the additivity of sums of the form (2.11). This analysis is accordingly now to be regarded as unsatisfactory.

3. Thermodynamic balance relations

3.1 Linear momentum balance

A body is regarded to be a system of interacting particles. The motion of a typical particle P_i is governed (in an inertial frame) by

$$\sum_l \boldsymbol{f}_{il} + \boldsymbol{b}_i = d/dt(m_i \boldsymbol{v}_i), \tag{3.1}$$

where $\boldsymbol{f}_{il}$ denotes its interaction with P_l and $\boldsymbol{b}_i$ represents external force. It is to be emphasized that the nature of interactions may be very general. In particular, $\boldsymbol{f}_{il}$ may depend not only upon the distance between P_i and P_l, but also upon their velocities and the behaviour of *other* particles. In mind here is the awareness that the point masses are models of entities with *structure*, and the 'state' of such an entity (which will influence its interactions with other entities) must be expected to be affected by its near-neighbours. For this reason $\boldsymbol{f}_{ii}$ will not be assumed to vanish in what follows. It suffices for our purposes in this subsection that interactions satisfy two mild restrictions:

I.1. *The net self-force associated with particles in a microscopically-large region is negligible,*

and

I.2. $\sum_l \boldsymbol{f}_{il} = \sum_n \boldsymbol{f}_{in}$, *where* $P_i P_n < \delta \ll \varepsilon$.

[1] *Cf.* Equations (3.18), (3.19) therein.

Remarks

1. The net self-force associated with any region is $\sum_i \sum_k f_{ik}$, where the sum is taken over all P_i, P_k within this region. Since such a sum may be expressed in the form

$$\sum_i f_{ii} + \tfrac{1}{2} \sum_{i \neq k} \sum (f_{ik} + f_{ki}),$$

I.1. is clearly a very much weaker restriction upon interactions than the assumptions that for all P_i, P_k as above

$$f_{ii} = \mathbf{0} \quad \text{and} \quad f_{ik} + f_{ki} = \mathbf{0}.$$

2. Assumption I.1. implies that at the level of net interactions between ε-cells or larger regions, action and reaction are equal and opposite, and that the self-force associated with any macroscopic region is negligible.
3. Assumption I.2, which requires that interactions be *formally* of short range, is much weaker than the *actual* short-range stipulation

$$f_{il} = \mathbf{0} \quad \text{if} \quad P_i P_l \geqq \delta,$$

since *individual* interactions may be admitted which have much greater effective range. The motivation here stems from charged particle interactions, which may have no finite range yet result in I.2., within our approximative scheme, by virtue of judicious book-keeping in balancing the forces on P_i due to particles of opposite charges at distances in excess of δ. Assumption I.2. is consistent with molecular physics, wherein interactions having ranges of order 10^{-7} m (= 1000 Å) are termed very long range (*cf.*, *e.g.*, HIRSCHFELDER, 1967).

Let R denote any region of macroscopic dimensions within that region in which ϱ and $\boldsymbol{v}$ are defined. Here and in what follows it will also be assumed that ∂R is piece-wise smooth, with outward unit normal field $\boldsymbol{n}$ varying negligibly at the ε-length scale on smooth subsurfaces. Consider the sum of equations (3.1) taken over all P_i in R at instant t. As a consequence of I.1. and I.2., the interaction contribution $\sum_i \sum_l f_{il}$ reduces to $\sum_i \sum_j f_{ij}$, where P_i lies inside R, P_j lies outside R, and $P_i P_j < \delta$. Consequently the only non-vanishing interaction contribution derives from particles lying within inner and outer 'δ-envelopes' of ∂R: that is, surfaces parallel to ∂R but distant δ therefrom. This region may be covered by ε-cells whose centres lie upon ∂R. Such cells give rise to what we call **ε-subsurfaces** and **ε-surface cells** (based upon ∂R): an ε-subsurface (ε-surface cell), based upon ∂R and centred at $\boldsymbol{x}$, is the intersection of an ε-cell centred at $\boldsymbol{x} \in \partial R$ with ∂R (with the region between ∂R and the 'inner δ-envelope' of ∂R). For a system of disjoint ε-surface cells whose union is the 'inner' δ-collared region based upon ∂R, the sum

$$\sum_i \sum_j f_{ij} = \sum_{\substack{\text{surface}\\ \text{cells}}} \Big(\sum_i{}' \sum_j f_{ij} \Big),$$

where the superposed prime denotes summation over particles within any individual (surface) cell. It is tempting to write

$$\boldsymbol{t}(\boldsymbol{x}, t) := \lim_\varepsilon \left\{ \sum_i{}' \sum_j \boldsymbol{f}_{ij}/A_\varepsilon \right\},$$

where A_ε denotes the area of the ε-subsurface (centred at $\boldsymbol{x}$) associated with any surface cell, and the limit is taken to exist in a manner precisely analogous to that discussed for volume averages. However, considerations of statistical mechanics suggest that this is *not* the case (*cf.* ALBLAS, 1976, p. 281). *It is for this reason that recourse to time averaging is essential.* Thus, choosing a system of ε-surface cells as above, and allowing these to deform with the motion, it is assumed that the limit

$$\boldsymbol{t}(\boldsymbol{x}, t) := \lim_\varepsilon \left\{ \overline{\sum_i{}' \sum_j \boldsymbol{f}_{ij}/A_\varepsilon} \right\} = \lim_\varepsilon \left\{ \overline{\sum_i{}' \sum_j \boldsymbol{f}_{ij}}/A_\varepsilon \right\} \tag{3.2}$$

exists[1]. We term $\boldsymbol{t}(\boldsymbol{x}, t)$ the **interaction traction** associated with ∂R at $\boldsymbol{x} \in \partial R$ at instant t. The notation and ideas here are an obvious extension of the discussion of § 2.4. The assumption that A_ε be essentially constant over a Δ-interval has been made in reaching the final expression in (3.2). It is to be noted, however, that the double sums in (3.2) cannot be expected to involve the same particles at different instants in the time-averaging period.

The foregoing has indicated that a mere summation of equations (3.1) for particles within R at a given instant is insufficient to establish the existence of a macroscopic traction field. In order to accommodate the necessary averaging in time, let R_τ $(t - \Delta \leqq \tau \leqq t$ with Δ macroscopically small yet microscopically large) denote that region whose boundary deforms with the macroscopic motion and coincides with R at instant t (so that $R_t = R$). For any such τ the sum of equations (3.1) may be computed for all P_i in R_τ and then the Δ-time average of the resulting τ-dependent relation calculated. The contribution from interactions has been discussed and is clearly to be identified with

$$\int_{\partial R} \boldsymbol{t}\,.$$

An alternative expression for the interaction contribution may be obtained by considering a decomposition of R into ε-cells. This induces a corresponding decomposition of R_τ, in terms of which

$$\sum_i \sum_l \boldsymbol{f}_{il} = \sum_{\text{cells}} \left(\sum_i{}' \sum_l \boldsymbol{f}_{il} \right).$$

The first sum is taken over interactions between all particles P_i within R_τ at instant τ and all particles P_l, and the second sum is taken over all cells within R_τ, with particle P_i situated within a given cell and P_l any other particle. The time

[1] Here $\boldsymbol{x}$ denotes the centre of the (deforming) ε-subsurface, at instant t, associated with the deforming surface cell over which the sum is taken.

average of the foregoing yields for the interaction contribution

$$\left(\int_{\partial R} t =\right) \int_R f,$$

where (3.3)

$$f(x, t) := \lim_\varepsilon \left\{ \overline{\sum_i{}' \sum_l f_{il}/V_\varepsilon} \right\}.$$

The cells involved in $(3.3)_2$ have centres at x at instant t.

The contribution from external force is simply identified with

$$\int_R b,$$

where (3.4)

$$b(x, t) := \lim_\varepsilon \left\{ \overline{\sum_i{}' b_i/V_\varepsilon} \right\}.$$

The cells involved in $(3.4)_2$ have centres at x at instant t.

The contribution from inertial terms in (3.1) to the time-averaged sum was argued in I to be

$$d/dt \left\{ \int_{R_t} \varrho v \right\}.$$

However, by oversight this did not take care of diffusion, as will now be shown. The correct expression is formally

$$\frac{1}{\Delta} \int_{t-\Delta}^{t} \left\{ \sum_i{}^{\tau} d/d\tau(m_i v_i) \right\} d\tau, \tag{3.5}$$

where the superposed τ attached to the summation sign is intended to emphasize that the integrand involves a changing particle population: at instant τ only contributions from particles P_i which lie in R_τ are to be summed. In particular, if P_i enters the deforming region at instant t', remains therein until it leaves at instant t'', where $t - \Delta \leqq t' < t'' \leqq t$, then the contribution to (3.5) for this period of residence in the region is

$$\frac{1}{\Delta} m_i \{v_i(t'') - v_i(t')\}.$$

Of course, P_i may subsequently re-enter and leave the deforming region many times during the Δ-time interval, and so yield further contributions to the sum. By consideration of all possibilities it follows that (3.5) may be expressed as

$$\frac{1}{\Delta} \left\{ \sum_{\substack{P_i \\ \text{in } R_t}} m_i v_i - \sum_{\substack{P_i \\ \text{in } R_{t-\Delta}}} m_i v_i \right\} + \frac{1}{\Delta} \sum_j m_j v_j (\tilde{v}_j \cdot n)/|\tilde{v}_j \cdot n|. \tag{3.6}$$

Here the suffix j is intended to label particles P_j instantaneously in the process of crossing the boundary of the region during the time interval considered. A single particle which crosses the boundary more than once will contribute to the second term on each crossing. The outward unit normal field to the boundary of the region has been denoted by $\boldsymbol{n}$, and it is to be noted that $(\tilde{\boldsymbol{v}}_j \cdot \boldsymbol{n})/|\tilde{\boldsymbol{v}}_j \cdot \boldsymbol{n}| = -1\,(+1)$ if the particle enters (leaves) the region. (Recall that $\tilde{\boldsymbol{v}}_j$ denotes the thermal velocity $\boldsymbol{v}_j - \boldsymbol{v}(\boldsymbol{x}_j, \tau)$, where $\boldsymbol{x}_j$ denotes the instantaneous location of P_j). The first term in (3.6) is clearly represented by

$$\frac{1}{\Delta}\left\{\int_{R_t} \varrho \boldsymbol{v} - \int_{R_{t-\Delta}} \varrho \boldsymbol{v}\right\}$$

which is identified with

$$d/dt\left\{\int_{R_t} \varrho \boldsymbol{v}\right\}. \tag{3.7}$$

The second term in (3.6) derives from surface-defined terms. Consider

$$\frac{1}{\Delta}\sum_j{}' \, m_j \boldsymbol{v}_j (\tilde{\boldsymbol{v}}_j \cdot \boldsymbol{n})/|\tilde{\boldsymbol{v}}_j \cdot \boldsymbol{n}| \, A_\varepsilon,$$

where the sum is taken over all particles instantaneously crossing a deforming ε-subsurface based upon the boundary of the deforming region, during the interval $t - \Delta \leqq \tau \leqq t$, of area A_ε and centred at $\boldsymbol{x}$ at instant t. Neglecting the variation of $\boldsymbol{n}$ both with position and time, and of A_ε with time, we identify this expression with

$$\int_{\partial R} \mathbf{d}$$

where (3.8)

$$\mathbf{d}(\boldsymbol{x}, t) := \lim\nolimits_{\varepsilon,\Delta}\left\{\frac{1}{A_\varepsilon \Delta}\left(\sum{}' \, m_j \boldsymbol{v}_j \otimes \tilde{\boldsymbol{v}}_j/|\tilde{\boldsymbol{v}}_j \cdot \boldsymbol{n}|\right)\right\} \cdot \boldsymbol{n}.$$

We term $-\mathbf{d}$ the **diffusive traction** associated with ∂R. Here the nature of the sum has been described in the foregoing. Its identification as a macroscopic field $\mathbf{d}$ requires that it be insensitive to variations both in ε and Δ at the macroscopically-small yet microscopically-large length and time scales, and it is in this sense that the formal 'limit' in $(3.8)_2$ is to be interpreted.

It follows from $(3.3)_1$, $(3.4)_1$, (3.7) and $(3.8)_1$ that the macroscopic form of linear momentum balance associated with the region R at instant t is

$$\int_{\partial R} \boldsymbol{t} + \int_R \boldsymbol{b} = \int_R \{\boldsymbol{f} + \boldsymbol{b}\} = d/dt\left\{\int_{R_t} \varrho \boldsymbol{v}\right\} + \int_{\partial R} \mathbf{d}.$$

The assumption of continuous dependence of $\mathbf{d}$ both upon position and local orientation of R leads, in standard fashion, to the existence of the **diffusive stress tensor** $\mathscr{D}$, in terms of which

$$\mathbf{d} = \mathscr{D}\boldsymbol{n}.$$

Comparison with $(3.8)_2$ leads us to

$$\textit{identify } \mathscr{D}(\boldsymbol{x}, \mathrm{t}) \textit{ with } \lim\nolimits_{\varepsilon,\Delta}\left\{\frac{1}{A_\varepsilon \Delta}\left(\sum_j{}' \, m_j \boldsymbol{v}_j \otimes \tilde{\boldsymbol{v}}_j/|\tilde{\boldsymbol{v}}_j \cdot \boldsymbol{n}|\right)\right\}.$$

The balance relation may now be written as

$$\int_{\partial R} (\boldsymbol{t} - \mathscr{D}\boldsymbol{n}) + \int_R \boldsymbol{b} = d/dt \left\{ \int_{R_t} \varrho \boldsymbol{v} \right\} = \int_R \varrho \boldsymbol{a},$$

where $\boldsymbol{a}$ denotes the acceleration field. If $\boldsymbol{t}$ is assumed to depend continuously both upon position and orientation then it may be shown that

$$\boldsymbol{t} = \boldsymbol{T}^{-}\boldsymbol{n}, \tag{3.9}$$

where $\boldsymbol{T}^{-}$ denotes the **interaction stress tensor.** Thus the balance relation may be expressed in the standard form

$$\int_{\partial R} \boldsymbol{T}\boldsymbol{n} + \int_R \boldsymbol{b} = d/dt \left\{ \int_{R_t} \varrho \boldsymbol{v} \right\},$$

where (3.10)

$$\boldsymbol{T} := \boldsymbol{T}^{-} - \mathscr{D}.$$

Of course, $\boldsymbol{T}$ denotes the **Cauchy stress tensor,** and in the presence of sufficient smoothness $(3.10)_1$ takes the local form

$$\operatorname{div} \boldsymbol{T} + \boldsymbol{b} = \varrho \boldsymbol{a}. \tag{3.11}$$

Equation $(3.10)_2$ makes explicit the separate contributions to the stress which derive from interactions and diffusion.

3.2 Generalised moment of momentum

The generalised moment of momentum balance with respect to an arbitrary fixed point $\boldsymbol{x}_0$ for a macroscopic region R is obtained by tensorially premultiplying (3.1) by $(\boldsymbol{x}_i - \boldsymbol{x}_0)$, summing over all P_i in R_τ (related to R as in § 3.1), and then taking the time average of the result over the interval $t - \Delta \leqq \tau \leqq t$. The discussion of this presented in I omitted the effect of diffusion and indicated certain difficulties. In what follows both diffusion and these difficulties are considered in detail. The latter indicate the subtlety involved when an attempt is made to obtain a general tensor moment balance rather than the usual moment balance (*cf.* TRUESDELL & TOUPIN, 1960, § 205).

A further weak restriction upon interactions is assumed in the form:

I.3. *The net self-couple associated with particles in a microscopically-large region is negligible.*

That is, $\sum_i{}' \sum_k{}' (\boldsymbol{x}_i - \boldsymbol{x}_0) \wedge \boldsymbol{f}_{ik}$, with sums taken over all P_i, P_k in some ε-cell, or union of such cells, is negligible. Of course, from I.1. it follows that such a sum is independent of $\boldsymbol{x}_0$. In the event that $\boldsymbol{f}_{ii} = \boldsymbol{0}$, $\boldsymbol{f}_{ik} + \boldsymbol{f}_{ik} = \boldsymbol{0}$, and $\boldsymbol{f}_{ik}$ is parallel to $(\boldsymbol{x}_i - \boldsymbol{x}_k)$ then I.3. is trivially satisfied.

At instant τ the contribution from interactions to the sum described in the first paragraph of this subsection is

$$\sum_i (\boldsymbol{x}_i - \boldsymbol{x}_0) \otimes \sum_l \boldsymbol{f}_{il} = \sum_i (\boldsymbol{x}_i - \boldsymbol{x}_0) \sim \sum_l \boldsymbol{f}_{il} + \sum_i (\boldsymbol{x}_i - \boldsymbol{x}_0) \wedge \sum_l \boldsymbol{f}_{il}, \tag{3.12}$$

where the sum is over all P_i in R_τ, all particles P_l, and $\boldsymbol{a} \sim \boldsymbol{b}$ $(\boldsymbol{a} \wedge \boldsymbol{b})$ denotes the symmetric (skew) part of the tensor product $\boldsymbol{a} \otimes \boldsymbol{b}$ of any pair of vectors $\boldsymbol{a}, \boldsymbol{b}$. Now P_l may lie inside or outside R_τ: labelling points in R_τ with suffix k, those outside R_τ with suffix j, and upon invoking I.3. for the whole region R_τ we find that

$$\begin{aligned}\sum_i (\boldsymbol{x}_i - \boldsymbol{x}_0) \wedge \sum_l \boldsymbol{f}_{il} &= \sum_i (\boldsymbol{x}_i - \boldsymbol{x}_0) \wedge \left\{\sum_k \boldsymbol{f}_{ik} + \sum_j \boldsymbol{f}_{ij}\right\} \\ &= \sum_i (\boldsymbol{x}_i - \boldsymbol{x}_0) \wedge \sum_j \boldsymbol{f}_{ij}.\end{aligned}$$

As a consequence of I.2., the last sum involves only those particles P_i inside R_τ which lie within a distance δ of ∂R_τ. Accordingly this sum way be expressed as a sum over particles within ε-surface cells based upon ∂R_τ: namely,

$$\sum_{\text{surface cells}} \left\{\sum_i{}' ((\boldsymbol{x}_i - \boldsymbol{x}) + (\boldsymbol{x} - \boldsymbol{x}_0)) \wedge \sum_j \boldsymbol{f}_{ij}\right\}, \tag{3.13}$$

where $\boldsymbol{x}$ denotes the centre of a given surface cell at instant t. Upon taking the time average of this sum, we see that its continuum counterpart is

$$\int_{\partial R} \{\hat{\mathscr{C}} + (\boldsymbol{x} - \boldsymbol{x}_0) \wedge \boldsymbol{t}\},$$

where[1] (3.14)

$$\hat{\mathscr{C}}(\boldsymbol{x}, t) := \lim_\varepsilon \left\{\overline{\sum_i{}' (\boldsymbol{x}_i - \boldsymbol{x}) \wedge \sum_j \boldsymbol{f}_{ij}}/A_\varepsilon\right\}.$$

As now customary, the area A_ε of an ε-subsurface has been assumed to change negligibly during the time-averaging interval.

The remaining term in (3.12) may be broken down into a sum of contributions from ε-cells into which R_τ may be decomposed: namely

$$\begin{aligned}&\sum_{\text{cells}} \left\{\sum_i{}' (\boldsymbol{x}_i - \boldsymbol{x}_0) \sim \boldsymbol{f}_{il}\right\} \\ &\quad = \sum_{\text{cells}} \left\{\sum_i{}' (\boldsymbol{x}_i - \boldsymbol{x}) \sim \sum_l \boldsymbol{f}_{il} + (\boldsymbol{x} - \boldsymbol{x}_0) \sim \sum_i{}' \sum_l \boldsymbol{f}_{il}\right\}.\end{aligned} \tag{3.15}$$

[1] It may be observed that $\boldsymbol{x}$ as it appears in the sums contained in (3.13) and (3.14) may be replaced by $\hat{\boldsymbol{x}}(\tau)$, the centre of the cell in question at instant τ, without significant change: $\hat{\boldsymbol{x}}(\tau)$ varies macroscopically with time (its evolution is governed by the macroscopic field $\boldsymbol{v}$), whence $\varphi(\tau) := \hat{\boldsymbol{x}}(\tau) - \hat{\boldsymbol{x}}(t)$ has negligible time average and thus (2.5) with $\psi := \sum_i{}' \sum_j \boldsymbol{f}_{ij}$ justifies the assertion.

In respect of the last term we notice that as a consequence of $(3.3)_1$ and (3.9) we may

$$\textit{identify } f = \lim_\varepsilon \left\{ \overline{\sum_i{}' \sum_l f_{il}/V_\varepsilon} \right\} \textit{ with } \operatorname{div} T^-, \tag{3.16}$$

and note that in such cases I.1. implies that the sum in $(3.3)_2$ need only be taken over particles P_l *outside* the cell in question.

Defining[1]

$$J(x, t) := \lim_\varepsilon \left\{ \overline{\sum_i{}' (x_i - x) \sim \sum_l f_{il}/V_\varepsilon} \right\}, \tag{3.17}$$

from (3.12)–(3.17) we see that the contribution to generalised moment of momentum balance from interactions is

$$\int_{\partial R} \{\hat{\mathscr{C}} + (x - x_0) \wedge t\} + \int_R \{J + (x - x_0) \sim \operatorname{div} T^-\}. \tag{3.18}$$

The external force contribution to the continuum balance relation is simply obtained as

$$\int_R \{G + (x - x_0) \otimes b\},$$

where (3.19)

$$G(x, t) := \lim_\varepsilon \left\{ \overline{\sum_i{}' (x_i - x) \otimes b_i/V_\varepsilon} \right\}$$

denotes **generalised body couple.**

The remaining sum is

$$\sum_i (x_i - x_0) \otimes d/dt\,(m_i v_i) = \sum_i d/dt\,\{(x_i - x_0) \otimes m_i v_i\} - \sum_i v_i \otimes m_i v_i. \tag{3.20}$$

Decomposing the last sum into a sum over ε-cells within R_τ, writing

$$v_i = \hat{v}_i + v(\hat{x}(\tau), \tau) + L(\hat{x}(\tau), \tau)\,(x_i - \hat{x}(\tau))$$

in each such cell (centred at $\hat{x}(\tau)$), using the approximative convention of neglecting $O(\varepsilon^2)$ terms together with assumptions T. M. 1, 2, and (2.8), identifying $\hat{v}_i$ with $\tilde{v}_i$ (*cf.* § 2.5), and taking the Δ-time average, we find for the corresponding continuum expression

$$\int_R \{K + \varrho v \otimes v\},$$

where (3.21)

$$K(x, t) := \lim_\varepsilon \left\{ \overline{\sum_i{}' \tilde{v}_i \otimes m_i \tilde{v}_i/V_\varepsilon} \right\}.$$

---

[1] As here defined J differs from the quantity similarly labelled in I in two respects: only the symmetric part of the tensor product is here involved, and the sum is taken over *all* P_l, both inside and outside R_τ. *Cf.* § 2.6, in particular the discussion of sums of the form (2.11). In the definition of J, x may be replaced by $\hat{x}(\tau)$ (as in the alternative definition of $\hat{\mathscr{C}}$) without significant change.

The symmetric tensor field K is termed the **thermal tensor** field. The middle sum in (3.20) can be treated in the same way as the terms involving the change of momentum in § 3.1. Indeed, the time average of the sum is

$$\frac{1}{\Delta}\int_{t-\Delta}^{t}\sum_i{}^{\tau}\, d/d\tau\,((\boldsymbol{x}_i-\boldsymbol{x}_0)\otimes m_i\boldsymbol{v}_i)\,d\tau,$$

where the sum is taken over all particles instantaneously within R_τ. Reasoning as in § 3.1., we may express this term as follows:

$$\frac{1}{\Delta}\left\{\sum_{P_i\text{ in }R_t}(\boldsymbol{x}_i-\boldsymbol{x}_0)\otimes m_i\boldsymbol{v}_i-\sum_{P_i\text{ in }R_{t-\Delta}}(\boldsymbol{x}_i-\boldsymbol{x}_0)\otimes m_i\boldsymbol{v}_i\right\}$$
$$+\frac{1}{\Delta}\sum_j(\boldsymbol{x}_j-\boldsymbol{x}_0)\otimes m_j\boldsymbol{v}_j(\tilde{\boldsymbol{v}}_j\cdot\boldsymbol{n})/|\tilde{\boldsymbol{v}}_j\cdot\boldsymbol{n}|. \tag{3.22}$$

The second expression is to be regarded as a sum over particles instantaneously on the point of crossing the boundary at any time in the interval $t-\Delta\leqq\tau\leqq t$: each such crossing contributes a term to this sum. The sums over R_t and $R_{t-\Delta}$ may be broken up into a sum of ε-cell sums. Each such ε-cell sum may be written as

$$\sum_i{}'(\boldsymbol{x}_i-\boldsymbol{x}_0)\otimes m_i\boldsymbol{v}_i=\sum_i{}'(\boldsymbol{x}_i-\boldsymbol{x})\otimes m_i(\hat{\boldsymbol{v}}_i+\boldsymbol{v}+\boldsymbol{L}(\boldsymbol{x}_i-\boldsymbol{x}))$$
$$+(\boldsymbol{x}-\boldsymbol{x}_0)\otimes\sum_i{}'m_i\boldsymbol{v}_i. \tag{3.23}$$

As a consequence of T. M. 1., T. M. 2., and (2.8), (3.23) reduces to $(\boldsymbol{x}-\boldsymbol{x}_0)\otimes\left(\sum_i{}'m_i\right)\boldsymbol{v}$ upon neglect of $O(\varepsilon^2)$ quantities, which also (*cf.* § 2.5) justifies the replacement of $\hat{\boldsymbol{v}}_i$ by $\tilde{\boldsymbol{v}}_i$. Clearly the first term in (3.22) can thus be expressed as

$$\frac{1}{\Delta}\left\{\int_{R_t}(\boldsymbol{x}-\boldsymbol{x}_0)\otimes\varrho\boldsymbol{v}-\int_{R_{t-\Delta}}(\boldsymbol{x}-\boldsymbol{x}_0)\otimes\varrho\boldsymbol{v}\right\}$$

which we identify with

$$d/dt\left\{\int_{R_t}(\boldsymbol{x}-\boldsymbol{x}_0)\otimes\varrho\boldsymbol{v}\right\}. \tag{3.24}$$

The remaining (surface) sum in (3.22) may be written as a sum over deforming ε-subsurface sums, as in § 3.1., which have the form

$$\frac{1}{\Delta}\sum_j{}'((\boldsymbol{x}_j-\boldsymbol{x})+(\boldsymbol{x}-\boldsymbol{x}_0))\otimes m_j\boldsymbol{v}_j(\tilde{\boldsymbol{v}}_j\cdot\boldsymbol{n})/|\tilde{\boldsymbol{v}}_j\cdot\boldsymbol{n}|.$$

In view of $(3.8)_2$, this sum is identified with

$$\int_{\partial R}\{\mathbf{M}+(\boldsymbol{x}-\boldsymbol{x}_0)\otimes\mathscr{D}\boldsymbol{n}\},$$

where (3.25)

$$\mathbf{M}(\boldsymbol{x},t):=\left(\lim_{\varepsilon,\Delta}\left\{\frac{1}{A_\varepsilon\Delta}\sum_j{}'(\boldsymbol{x}_j-\boldsymbol{x})\otimes m_j\boldsymbol{v}_j\otimes\tilde{\boldsymbol{v}}_j/|\tilde{\boldsymbol{v}}_j\cdot\boldsymbol{n}|\right\}\right)\boldsymbol{n}.$$

Taken together, relations (3.18), $(3.19)_1$, $(3.21)_1$, (3.24), $(3.25)_1$ and (3.9) imply that moment of momentum balance takes the form

$$\int_{\partial R} \{\hat{\mathscr{C}} + (\boldsymbol{x} - \boldsymbol{x}_0) \wedge T^{-}\boldsymbol{n}\} + \int_R \{J + (\boldsymbol{x} - \boldsymbol{x}_0) \sim \operatorname{div} T^{-} + G + K + (\boldsymbol{x} - \boldsymbol{x}_0) \otimes \boldsymbol{b}\}$$

$$= \int_R (\boldsymbol{x} - \boldsymbol{x}_0) \otimes \varrho \boldsymbol{a} + \int_{\partial R} \{\mathbf{M} + (\boldsymbol{x} - \boldsymbol{x}_0) \otimes \mathscr{D}\boldsymbol{n}\}. \tag{3.26}$$

Use of (3.11) reduces this to the form

$$\int_{\partial R} \{\hat{\mathscr{C}} - \mathbf{M}\} + \int_R \{J + G + K - \operatorname{sk} T^{-} - \mathscr{D}^T\} = \mathbf{0}. \tag{3.27}$$

This relation may be expressed in an alternative manner upon re-writing (3.12) in the form

$$\sum_i (\boldsymbol{x}_i - \boldsymbol{x}_0) \otimes \sum_l f_{il} = \sum_{\text{cells}} \left(\sum_i{}' (\boldsymbol{x}_i - \boldsymbol{x}) \otimes \sum_l f_{il} + (\boldsymbol{x} - \boldsymbol{x}_0) \otimes \sum_i{}' \sum_l f_{il} \right),$$

where $\boldsymbol{x}$ denotes the centre of any given cell at instant t. The continuum form of the left-hand side is, upon time averaging,

$$\int_R \{\mathscr{G} + (\boldsymbol{x} - \boldsymbol{x}_0) \otimes \boldsymbol{f}\}$$

where

$$\mathscr{G}(\boldsymbol{x}, t) := \lim_\varepsilon \left\{ \overline{\sum_i{}' (\boldsymbol{x}_i - \boldsymbol{x}) \otimes \sum_l f_{il} / V_\varepsilon} \right\}.$$

The analogue of (3.26) is

$$\int_R \{\mathscr{G} + G + K + (\boldsymbol{x} - \boldsymbol{x}_0) \otimes (\boldsymbol{f} + \boldsymbol{b})\}$$

$$= \int_R (\boldsymbol{x} - \boldsymbol{x}_0) \otimes \varrho \boldsymbol{a} + \int_{\partial R} \{\mathbf{M} + (\boldsymbol{x} - \boldsymbol{x}_0) \otimes \mathscr{D}\boldsymbol{n}\}.$$

Use of the appropriate local form of linear momentum balance reduces this to

$$\int_R \{\mathscr{G} + G + K - \mathscr{D}^T\} = \int_{\partial R} \mathbf{M}.$$

In standard fashion the assumption that $\mathbf{M}$ depends continuously upon position and local orientation implies from the last relation that there exists a (rank three) tensor $\mathscr{M}$ such that

$$\mathbf{M} = \mathscr{M}\boldsymbol{n}.$$

Comparison with $(3.25)_2$ leads us to

$$\textit{identify } \mathscr{M}(\boldsymbol{x}, t) \textit{ with } \lim_{\varepsilon,\Delta} \left\{ \overline{\frac{1}{A_\varepsilon \Delta} \sum_j{}' (\boldsymbol{x}_j - \boldsymbol{x}) \otimes m_j \boldsymbol{v}_j \otimes \tilde{\boldsymbol{v}}_j / |\tilde{\boldsymbol{v}}_j \cdot \boldsymbol{n}|} \right\}.$$

Replacing $\mathbf{M}$ by $\mathscr{M}\boldsymbol{n}$ in (3.27) and assuming that $\hat{\mathscr{C}}$ depends continuously upon position and local orientation implies that

$$\hat{\mathscr{C}} = \hat{C}\boldsymbol{n}, \tag{3.28}$$

where $\hat{C}$ denotes the (rank three) **interaction couple-stress tensor.** Relation (3.26) may now be expressed in the form

$$\int_{\partial R} \{C^- n + (x - x_0) \otimes Tn\} + \int_R \{J + K - \operatorname{sym} T^- + G + (x - x_0) \otimes b\}$$
$$= \int_R (x - x_0) \otimes \varrho a,$$

where (3.29)

$$C^- := \hat{C} - \mathcal{M}.$$

After use of (3.11) we find for the local form[1] of $(3.29)_1$

$$\operatorname{div} C^- - \operatorname{sk} T^- - \mathcal{D}^T + J + G + K = \mathbf{0}. \tag{3.30}$$

The skew nature of $\hat{\mathscr{C}}$ (cf. $(3.14)_2$) implies $\operatorname{div} \hat{C}$ is skew-valued. Hence the skew and symmetric parts of (3.30) give rise to the relations

$$\operatorname{div} \hat{C} - \operatorname{sk}(\operatorname{div} \mathcal{M} + T - G) = \mathbf{0}$$

and (3.31)

$$J + K - \operatorname{sym}(\operatorname{div} \mathcal{M} + \mathcal{D} - G) = \mathbf{0}.$$

Remarks

1. From $(3.31)_1$, the stress tensor is symmetric if and only if $\operatorname{div} \hat{C} = \operatorname{sk}(\operatorname{div} \mathcal{M} - G)$. In the event that $\hat{C}$, $\mathcal{M}$, and G are all negligible, then $(3.31)_2$ yields

$$T = \operatorname{sym} T = \operatorname{sym} T^- - \operatorname{sym} \mathcal{D} = \operatorname{sym} T^- - J - K.$$

While both T^- and J are thermally dependent (as a consequence of being defined via time averages of sums of terms which fluctuate because of thermal motion) this relation yields a *purely* thermal contribution to stress in the form of $-K$.

If interactions are negligible to the extent that $\hat{C}$, T^-, and J may be neglected (physically plausible only in the case of non-dense gases), then $(3.10)_2$ and (3.30) yield

$$T = -\mathcal{D} = (\operatorname{div} \mathcal{M})^T - G^T - K.$$

This result may be compared with the expression for the stress computed on the basis of kinetic theory (*cf.* TRUESDELL, 1969, or CERCIGNANI, 1975), namely

$$T = -\varrho \overline{c \otimes c}.$$

The superposed bar here denotes expectation with respect to the molecular density function, and c the thermal velocity. Clearly this stress is to be identified with $-K$ (cf. $(3.21)_2$). Hence the present analysis, upon further neglect of[2] $\mathcal{M}$ and G, can be seen to yield the stress of kinetic theory. Such derivation is, however, somewhat indirect: neglect of interaction stress T^- makes the Cauchy stress diffusive in nature and thence, by the balance of generalised moment of momentum and neglect of C^-, J and G, leads to the result.

---

[1] In Cartesian co-ordinates $(\operatorname{div} A)_{ij} = A_{ijk,k}$ for a tensor field of third rank A.
[2] *Cf.* Remark 4 of this subsection and the final paragraph of § 4.3 in I.

2. It is helpful in what follows to note that (3.12) may be written, upon use of I.1. and I.3., as

$$\sum_i \sum_l (\boldsymbol{x}_i - \boldsymbol{x}_0) \otimes \boldsymbol{f}_{il} = \sum_{\text{cells}} \left\{ \sum_i{}' \sum_l (\boldsymbol{x}_i - \boldsymbol{x}_0) \otimes \boldsymbol{f}_{il} \right\}$$

$$= \sum_{\text{cells}} \left\{ \sum_i{}' \sum_k{}' (\boldsymbol{x}_i - \boldsymbol{x}) \sim \boldsymbol{f}_{ik} + \sum_i{}' \sum_p ((\boldsymbol{x}_i - \boldsymbol{x}) + (\boldsymbol{x} - \boldsymbol{x}_0)) \otimes \boldsymbol{f}_{ip} \right\}, \quad (3.32)$$

where P_i, P_k both lie within any given cell and P_p lies outside this cell. Invoking (3.18) (the continuum version of (3.12)) with (3.28), making the identification (3.16), and recalling (3.17) leads us to

$$\textit{identify} \;\; \lim_\varepsilon \left\{ \overline{\sum_i{}' \sum_p (\boldsymbol{x}_i - \boldsymbol{x}) \wedge \boldsymbol{f}_{ip} / V_\varepsilon} \right\} \;\; \textit{with} \;\; (\operatorname{div} \hat{\boldsymbol{C}} - \operatorname{sk} \boldsymbol{T}^-). \quad (3.33)$$

Of course, the sum in (3.33) denotes the net moment, about the centre of a given cell, of forces exerted thereon by particles outside this cell.

3. The symmetric tensor field $\boldsymbol{K}$ defined in $(3.21)_2$ depends only upon corpuscular masses and thermal velocities, and has thus been called the thermal tensor field. In the balance of energy, to follow, its trace, tr $\boldsymbol{K}$, appears quite naturally. This clearly represents twice the (time-averaged) kinetic energy density, ϱh say, associated with thermal motion. Accordingly

$$h := (\operatorname{tr} \boldsymbol{K})/2\varrho \quad (3.34)$$

is termed the (specific) **heat content.**

4. In the identifications of $\mathscr{D}$ and $\mathscr{M}$ with corpuscular quantities the factor $\boldsymbol{v}_j$ may be replaced by $\tilde{\boldsymbol{v}}_j$ provided that

$$\lim_{\varepsilon,\Delta} \left\{ \frac{1}{A_\varepsilon \Delta} \sum_j{}' m_j \tilde{\boldsymbol{v}}_j \cdot \boldsymbol{n} / |\tilde{\boldsymbol{v}}_j \cdot \boldsymbol{n}| \right\} = 0,$$

$$\lim_{\varepsilon,\Delta} \left\{ \frac{1}{A_\varepsilon \Delta} \sum_j{}' m_j (\boldsymbol{x}_j - \boldsymbol{x}) (\tilde{\boldsymbol{v}}_j \cdot \boldsymbol{n}) / |\tilde{\boldsymbol{v}}_j \cdot \boldsymbol{n}| \right\} = \boldsymbol{0},$$

and

$$\lim_{\varepsilon,\Delta} \left\{ \frac{1}{A_\varepsilon \Delta} \sum_j{}' (\boldsymbol{x}_j - \boldsymbol{x}) \otimes m_j (\boldsymbol{x}_j - \boldsymbol{x}) (\tilde{\boldsymbol{v}}_j \cdot \boldsymbol{n}) / |\tilde{\boldsymbol{v}}_j \cdot \boldsymbol{n}| \right\} = \boldsymbol{0}.$$

The first equation merely expresses the fact that there is no *net* mass transfer across the deforming subsurface. The remaining equations may be regarded as consequences of the spatially-random nature of diffusion across the subsurface. This latter consideration indicates that $\mathscr{M}$ itself is likely to be negligibly small. In what follows $\mathscr{M}$ will be retained for generality.

3.3 Energy balance

Energy balance is obtained by multiplying (3.1) scalarly by $\boldsymbol{v}_i$, adding over all P_i within R_τ, and then taking the time average over the interval $t - \Delta \leqq \tau \leqq t$.

At each instant τ the interaction contribution is

$$S(\tau) := \sum_i \sum_l f_{il} \cdot v_i = \sum_{\text{cells}} \left\{ \sum_i{}' \sum_l f_{il} \cdot v_i \right\}$$

$$= \sum_{\text{cells}} \left\{ \sum_i{}' \left(\sum_k{}' f_{ik} + \sum_p f_{ip} \right) \cdot (\tilde{v}_i + v + L(x_i - x)) \right\},$$

where P_k is a typical particle in the cell centred at $x \ (= \hat{x}(\tau))$ containing P_i, P_p is any particle outside this cell, and v, L denote $v(x, \tau)$ and $L(x, \tau)$ respectively. Use has also been made of (2.10) and the identification of $\hat{v}_i$ with $\tilde{v}_i$. Invoking I.1. and I.3. simplifies the above to[1]

$$S(\tau) = \sum_{\text{cells}} \left\{ \sum_i{}' \left(\sum_k{}' f_{ik} + \sum_p f_{ip} \right) \cdot \tilde{v}_i + \left(\sum_i{}' \sum_p f_{ip} \right) \cdot v \right.$$

$$\left. + \left(\sum_i{}' \sum_l (x_i - x) \sim f_{il} \right) \cdot L^T + \left(\sum_i{}' \sum_p (x_i - x) \wedge f_{ip} \right) \cdot L^T \right\}. \quad (3.35)$$

We now make the further assumption

I.4. $\overline{\sum_i{}' \sum_p f_{ip} \cdot \tilde{v}_i}$ *is additive over microscopically-large regions.*

By this we mean that a time-averaged sum $\overline{\sum_i \sum_j f_{ij} \cdot \tilde{v}_i}$ taken over all particles P_i within any group of adjoining but mutually disjoint ε-cells, with P_j outside this group, may be written as the sum of such expressions as appear in I.4., one for each cell. Of course, the cells are assumed to deform with the macroscopic motion. As will later be clear, this assumption is equivalent to the assumption that any pair of disjoint ε-cells which share a portion of common boundary should have heat conduction interchange rates appropriate to this portion which are equal and opposite. It follows from I.4. that since $\overline{\sum_i{}' \sum_l f_{il} \cdot \tilde{v}_i}$ is additive $\left(\overline{\sum_i{}' \sum_l f_{il} \cdot \tilde{v}_i} \right.$ is clearly additive$\left.\right)$ then so too is $\overline{\sum_i{}' \sum_k{}' f_{ik} \cdot \tilde{v}_i}$. This renders plausible the existence of

$$P(x, t) := \lim_\varepsilon \left\{ \overline{\sum_i{}' \sum_k{}' f_{ik} \cdot \tilde{v}_i / V_\varepsilon} \right\}. \quad (3.36)$$

Further, if P_j denotes any particle outside R_τ then by I.4. and I.2.

$$\sum_{\text{cells}} \left\{ \overline{\sum_i{}' \sum_p f_{ip} \cdot \tilde{v}_i} \right\} = \overline{\sum_i \sum_j f_{ij} \cdot \tilde{v}_i}$$

$$= \sum_{\substack{\text{surface} \\ \text{cells}}} \left\{ \overline{\sum_i{}' \sum_j f_{ij} \cdot \tilde{v}_i} \right\}. \quad (3.37)$$

---

[1] If A, B denote second-rank tensors then their inner product $A \cdot B$ in Cartesian coordinates is $A_{ij} B_{ij}$.

If we define

$$q(\boldsymbol{x}, t) := \lim_\varepsilon \left\{ \overline{\sum_i{}' \sum_j \boldsymbol{f}_{ij} \cdot \tilde{\boldsymbol{v}}_i / A_\varepsilon} \right\}, \tag{3.38}$$

then the time average of $S(\tau)$ as expressed by (3.35) yields, upon invoking (3.36), (3.37), (3.38), (3.17), and making the identifications (3.16) and (3.33), the continuum contribution to energy balance from interactions: namely,

$$\int_{\partial R} q + \int_R \{P + (\operatorname{div} \boldsymbol{T}^-) \cdot \boldsymbol{v} + (\boldsymbol{J} + \operatorname{div} \hat{\boldsymbol{C}} - \operatorname{sk} \boldsymbol{T}^-) \cdot \boldsymbol{L}^T\}. \tag{3.39}$$

Implicit use of (2.5) has also been invoked, in treating the terms involving $\boldsymbol{v}$ and $\boldsymbol{L}$, together with the assumption that these fields vary macroscopically with time.

The contribution of body force to energy balance is straightforwardly found (*cf.* I) to be

$$\int_R \{\mathrm{r} + \boldsymbol{b} \cdot \boldsymbol{v} + \boldsymbol{G} \cdot \boldsymbol{L}^T\},$$

where (3.40)

$$\mathrm{r}(\boldsymbol{x}, t) := \lim_\varepsilon \left\{ \overline{\sum_i{}' \boldsymbol{b}_i \cdot \tilde{\boldsymbol{v}}_i / V_\varepsilon} \right\}.$$

The remaining contribution, from inertial terms, is

$$\frac{1}{\Delta} \int_{t-\Delta}^{t} \sum_i{}^{\tau} d/d\tau \{\tfrac{1}{2} m_i \boldsymbol{v}_i^2\} \, d\tau .$$

Proceeding much as in § 3.1. this yields (*cf.* (3.6) for the interpretation of the suffix j)

$$\frac{1}{\Delta} \left\{ \sum_{\substack{P_i \text{ in} \\ R_t}} \tfrac{1}{2} m_i \boldsymbol{v}_i^2 - \sum_{\substack{P_i \text{ in} \\ R_{t-\Delta}}} \tfrac{1}{2} m_i \boldsymbol{v}_i^2 \right\} + \frac{1}{\Delta} \sum_j \tfrac{1}{2} m_j \boldsymbol{v}_j^2 (\tilde{\boldsymbol{v}}_j \cdot \boldsymbol{n}) / |\tilde{\boldsymbol{v}}_j \cdot \boldsymbol{n}| . \tag{3.41}$$

Using the thermal velocity approximation (2.10), the random nature of thermal velocities as delineated by T. M. 1. and T. M. 2., and omitting $O(\varepsilon^2)$ terms, identifies the continuum counterpart of the first term as (*cf.* (3.34))

$$d/dt \left\{ \int_{R_t} \varrho (h + \boldsymbol{v}^2/2) \right\}. \tag{3.42}$$

Decomposing the last term of expression (3.41) into a sum over deforming ε-subsurfaces of the deforming boundary, writing $\boldsymbol{v}_j$ in terms of $\tilde{\boldsymbol{v}}_j$ using (2.10) and the essential equivalence of $\hat{\boldsymbol{v}}_j$ and $\tilde{\boldsymbol{v}}_j$, and making the assumptions listed in Remark 4 of § 3.2, we obtain the continuum formulation

$$\int_{\partial R} \{\mathrm{k} + \mathscr{D}\boldsymbol{n} \cdot \boldsymbol{v} + \mathscr{M}\boldsymbol{n} \cdot \boldsymbol{L}^T\},$$

where (3.43)

$$\mathrm{k}(\boldsymbol{x}, t) := \left(\lim_{\varepsilon, \Delta} \left\{ \frac{1}{A_\varepsilon \Delta} \sum_j{}' \tfrac{1}{2} m_j (\tilde{\boldsymbol{v}}_j \cdot \tilde{\boldsymbol{v}}_j) \, \tilde{\boldsymbol{v}}_j / |\tilde{\boldsymbol{v}}_j \cdot \boldsymbol{n}| \right\} \right) \cdot \boldsymbol{n} .$$

It follows from (3.39), $(3.40)_1$, (3.42) and $(3.43)_1$ that energy balance takes the form

$$\int_{\partial R} \{q - \mathrm{k} - \mathscr{D}\boldsymbol{n} \cdot \boldsymbol{v} - \mathscr{M}\boldsymbol{n} \cdot \boldsymbol{L}^T\} + \int_{R} \{P + \mathrm{r} + (\operatorname{div} \boldsymbol{T}^- + \boldsymbol{b}) \cdot \boldsymbol{v} + (\boldsymbol{J} + \operatorname{div} \hat{\boldsymbol{C}} - \operatorname{sk} \boldsymbol{T}^- + \boldsymbol{G}) \cdot \boldsymbol{L}^T\} = d/dt \left\{ \int_{R_t} \varrho(h + \tfrac{1}{2}\boldsymbol{v}^2) \right\}. \tag{3.44}$$

An alternative formulation of energy balance, which does not require interaction assumptions I.1.-I.4., may be obtained more simply: namely,

$$\int_{R} \{Q + \mathrm{r} + (\boldsymbol{f} + \boldsymbol{b}) \cdot \boldsymbol{v} + (\mathscr{G} + \boldsymbol{G}) \cdot \boldsymbol{L}^T\} = d/dt \left\{ \int_{R_t} \varrho(h + \boldsymbol{v}^2/2) \right\} + \int_{\partial R} \{\mathrm{k} + \mathscr{D}\boldsymbol{n} \ \boldsymbol{v} + \mathscr{M}\boldsymbol{n} \cdot \boldsymbol{L}^T\},$$

where

$$Q(\boldsymbol{x}, t) := \lim_\varepsilon \left\{ \overline{\sum_i{}' \sum_l \boldsymbol{f}_{il} \cdot \tilde{\boldsymbol{v}}_i / V_\varepsilon} \right\}.$$

It follows that if k is assumed to depend continuously upon position and local orientation then there is a vector field $\boldsymbol{k}$ in terms of which

$$\mathrm{k} = \boldsymbol{k} \cdot \boldsymbol{n}.$$

Comparison with $(3.43)_2$ leads us to

$$\textit{identify } \boldsymbol{k} \textit{ with } \lim_{\varepsilon,\Delta} \left\{ \frac{1}{A_\varepsilon \Delta} \sum_j{}' \tfrac{1}{2} m_j (\tilde{\boldsymbol{v}}_j \cdot \tilde{\boldsymbol{v}}_j) \tilde{\boldsymbol{v}}_j / |\tilde{\boldsymbol{v}}_j \cdot \boldsymbol{n}| \right\}.$$

Replacing k by $\boldsymbol{k} \cdot \boldsymbol{n}$ in (3.44) and assuming q depends continuously upon position and local orientation leads to the existence of a vector field $\boldsymbol{q}^-$ in terms of which

$$q = -\boldsymbol{q}^- \cdot \boldsymbol{n}.$$

The **heat-flux vector** $\boldsymbol{q}$, defined by

$$\boldsymbol{q} := \boldsymbol{q}^- + \boldsymbol{k}, \tag{3.45}$$

allows (3.44), upon invoking $(3.10)_2$, $(3.29)_2$ and appropriate identities, to be expressed in the form[1]

$$\int_{\partial R} \{-\boldsymbol{q} \cdot \boldsymbol{n} + \boldsymbol{T}\boldsymbol{n} \cdot \boldsymbol{v} + \boldsymbol{C}^-\boldsymbol{n} \cdot \boldsymbol{L}^T\} + \int_{R} \{P + \mathrm{r} + \boldsymbol{b} \cdot \boldsymbol{v} + (\boldsymbol{J} - \operatorname{sym} \boldsymbol{T}^- + \boldsymbol{G}) \cdot \boldsymbol{L}^T - \hat{\boldsymbol{C}} \cdot \nabla(\boldsymbol{L}^T)\} = d/dt \left\{ \int_{R_t} \varrho(h + \tfrac{1}{2}\boldsymbol{v}^2) \right\}. \tag{3.46}$$

---

[1] The term $-\hat{\boldsymbol{C}} \cdot \nabla(\boldsymbol{L}^T)$ may be reduced to $+\hat{\boldsymbol{C}} \cdot \nabla \boldsymbol{W}$, where $\boldsymbol{W} := \operatorname{sk} \boldsymbol{L}$, on using symmetry properties both of $\hat{\boldsymbol{C}}$ and $\nabla(\boldsymbol{L}^T)$. Here the inner product $\boldsymbol{A} \cdot \boldsymbol{B} = A_{ijk} B_{ijk}$ for rank three tensors $\boldsymbol{A}$, $\boldsymbol{B}$ has been employed.

Using (3.11), $(3.10)_2$, and $(3.29)_2$, for the local form of (3.46) we obtain

$$\mathrm{r} - \operatorname{div} \boldsymbol{q} + (-\operatorname{sk} \boldsymbol{T}^{-} - \mathscr{D}^T + \operatorname{div} \boldsymbol{C}^{-} + \boldsymbol{J} + \boldsymbol{G}) \cdot \boldsymbol{L}^T - \mathscr{M} \cdot \nabla(\boldsymbol{L}^T) + P = \varrho \dot{h}. \tag{3.47}$$

Relation (3.30) reduces (3.47) to

$$\mathrm{r} - \operatorname{div} \boldsymbol{q} - \boldsymbol{K} \cdot \boldsymbol{D} - \mathscr{M} \cdot \nabla(\boldsymbol{L}^T) + P = \varrho \dot{h} \tag{3.48}$$

after it is noted that (given the symmetric nature of $\boldsymbol{K}$)

$$\boldsymbol{K} \cdot \boldsymbol{L}^T = \boldsymbol{K} \cdot \boldsymbol{D},$$

where $\boldsymbol{D}$ denotes sym $\boldsymbol{L}$.

Remarks

1. The energy balance (3.46) may be displayed in more conventional format as follows. Let

$$\varrho\alpha := -P - (\boldsymbol{J} - \operatorname{sym} \boldsymbol{T}^{-}) \cdot \boldsymbol{D} + \hat{\boldsymbol{C}} \cdot \nabla(\boldsymbol{L}^T), \tag{3.49}$$

and define

$$\beta(\boldsymbol{x}, t) = \beta(\boldsymbol{\chi}_{\varkappa}(\hat{\boldsymbol{x}}, t), t) := \int_{t_0}^{t} \alpha(\boldsymbol{\chi}_{\varkappa}(\hat{\boldsymbol{x}}, \tau), \tau)\, d\tau, \tag{3.50}$$

where $\varkappa$ denotes an arbitrary reference configuration of the body, t_0 an arbitrary reference time ($t_0 < t$), and $\hat{\boldsymbol{x}}$ the geometrical point occupied in configuration $\varkappa$ by that material point located at $\boldsymbol{x}$ at instant t. Clearly (3.49) and (3.50) enable us to put (3.46) into the form

$$\int_{\partial R} \{-\boldsymbol{q} \cdot \boldsymbol{n} + \boldsymbol{T}\boldsymbol{n} \cdot \boldsymbol{v} + \boldsymbol{C}^{-}\boldsymbol{n} \cdot \boldsymbol{L}^T\} + \int_{R} \{\mathrm{r} + \boldsymbol{b} \cdot \boldsymbol{v} + \boldsymbol{G} \cdot \boldsymbol{L}^T\}$$

$$= d/dt \left\{ \int_{R_t} \varrho(\varepsilon + \tfrac{1}{2}v^2) \right\}, \tag{3.51}$$

where the (specific) **internal energy** ε is defined by

$$\varepsilon := \beta + h. \tag{3.52}$$

2. When interactions are negligible (so that, in particular, $\boldsymbol{q}^{-}$ and P may be neglected) then (3.48) yields

$$\mathrm{r} - \operatorname{div} \boldsymbol{k} - \boldsymbol{K} \cdot \boldsymbol{D} - \mathscr{M} \cdot \nabla(\boldsymbol{L}^T) = \varrho \dot{h}. \tag{3.53}$$

In the event that $\mathscr{M}$ and $\boldsymbol{G}$ are also negligible, (3.53) reduces (*cf.* Remark 1 of § 3.2) to

$$\mathrm{r} - \operatorname{div} \boldsymbol{k} - \mathscr{D} \cdot \boldsymbol{D} = \tfrac{3}{2} \varrho \dot{\widehat{(p/\varrho)}},$$

where

$$p := \tfrac{1}{3} \operatorname{tr} \mathscr{D}$$

denotes the negative of the mean stress.

3. If β is negligible, as to be expected when interactions are negligible (that is, in the case of moderately rarefied gases), internal energy coincides with heat content. However, this cannot be expected of liquid or solid phases, since interactions play a key rôle therein.

4. Alternative form of balance relations

4.1 An alternative form of energy balance may be obtained by considering another way of expressing the quantity $S(\tau)$ introduced at the beginning of § 3.3. Clearly

$$S(\tau) := \sum_i \sum_l \boldsymbol{f}_{il} \cdot \boldsymbol{v}_i = \sum_i \sum_j \boldsymbol{f}_{ij} \cdot \boldsymbol{v}_i + \sum_i \sum_k \boldsymbol{f}_{ik} \cdot \boldsymbol{v}_i, \tag{4.1}$$

where P_i, P_k lie in R_τ and P_j lies outside R_τ. In view of I.2. the i, j sum may be expressed as a sum over ε-surface cells based upon ∂R_τ. The thermal velocity approximation (2.10) within each surface cell implies that the corresponding time average should be represented by

$$\int_{\partial R} \{q + \boldsymbol{t} \cdot \boldsymbol{v} + \mathscr{C} \cdot \boldsymbol{L}^T\},$$

where (4.2)

$$\mathscr{C}(\boldsymbol{x}, t) := \lim_\varepsilon \left\{ \overline{\sum_i{}' \sum_j (\boldsymbol{x}_i - \boldsymbol{x}) \otimes \boldsymbol{f}_{ij} / A_\varepsilon} \right\}.$$

As demonstrated in I, the remaining sum,

$$\sum_i \sum_k \boldsymbol{f}_{ik} \cdot \boldsymbol{v}_i$$

is precisely the negative of the time derivative of the (binding) energy necessary to assemble the particles in their instantaneous locations from a state of infinite dispersion, when interactions are governed by separation-dependent pair-potentials. If R_τ be divided into ε-cells and the binding energy associated with interactions between distinct cells ignored (which implies that

$$\sum_i{}' \sum_k{}' \boldsymbol{f}_{ik} \cdot \boldsymbol{v}_i \quad \text{is additive},$$

the sum here being taken over all P_i, P_k within a given cell) then the time average of this sum is expressible as

$$-\int_R \varrho \hat{\alpha}$$

where (4.3)

$$\varrho(\boldsymbol{x}, t)\, \hat{\alpha}(\boldsymbol{x}, t) := -\lim_\varepsilon \left\{ \overline{\sum_i{}' \sum_k{}' \boldsymbol{f}_{ik} \cdot \boldsymbol{v}_i / V_\varepsilon} \right\}.$$

Of course, this conclusion enables us to write the sum in the form

$$-d/dt \left\{ \int_{R_t} \varrho \hat{\beta} \right\}, \tag{4.4}$$

where $\hat{\beta}$ is defined in terms of $\hat{\alpha}$ precisely as in (3.50) and is to be interpreted as the negative of the (specific) binding energy within this interaction context. However, given the much more general nature of the interactions here considered, it is clearly desirable to generalise this concept. This is simply accomplished provided the interaction assumption

I.5. $\overline{\sum_i{}' \sum_k{}' f_{ik} \cdot v_i}$ *is additive over microscopically-large regions*

is valid. In such case the quantities $\hat{\alpha}$ and $\hat{\beta}$ may be defined as in the foregoing. Consequently, the time average of (4.1) yields the continuum representation, on invoking $(4.2)_1$ and (4.4),

$$\int_{\partial R} \{q + \boldsymbol{t} \cdot \boldsymbol{v} + \mathscr{C} \cdot \boldsymbol{L}^T\} - d/dt \left\{ \int_{R_t} \varrho\hat{\beta} \right\}. \tag{4.5}$$

We term $\hat{\beta}$ the **generalised** (specific) **binding energy.** It follows, upon replacing (3.39) by the equivalent (modulo I.5) expression (4.5), that energy balance may be expressed in the form

$$\int_{\partial R} \{-\boldsymbol{q} \cdot \boldsymbol{n} + \boldsymbol{Tn} \cdot \boldsymbol{v} + (\mathscr{C} - \mathscr{M}\boldsymbol{n}) \cdot \boldsymbol{L}^T\} + \int_R \{\mathrm{r} + \boldsymbol{b} \cdot \boldsymbol{v} + \boldsymbol{G} \cdot \boldsymbol{L}^T\}$$

$$= d/dt \left\{ \int_{R_t} \varrho(\hat{\beta} + h + \tfrac{1}{2}\boldsymbol{v}^2) \right\}. \tag{4.6}$$

If $\mathscr{C}$ depends continuously upon position and surface orientation then (4.6) implies it is expressible in terms of a tensor of rank three: that is,

$$\mathscr{C} = \bar{C}\boldsymbol{n}. \tag{4.7}$$

Defining the **alternative generalised couple-stress tensor** C by

$$C := \bar{C} - \mathscr{M}, \tag{4.8}$$

we see that (4.6) has the term $(\mathscr{C} - \mathscr{M}\boldsymbol{n}) \cdot \boldsymbol{L}^T$ expressible as $C\boldsymbol{n} \cdot \boldsymbol{L}^T$. Upon invoking (3.11) we can write the local form of (4.6) as

$$\mathrm{r} - \operatorname{div} \boldsymbol{q} + (\boldsymbol{T}^T + \operatorname{div} C + \boldsymbol{G}) \cdot \boldsymbol{L}^T + C \cdot \nabla(\boldsymbol{L}^T) = \varrho\dot{\hat{\varepsilon}}, \tag{4.9}$$

where

$$\hat{\varepsilon} := \hat{\beta} + h$$

denotes the **alternative** (specific) **internal energy.**

Remarks

1. It is to be noted that (*cf.* $(3.14)_2$) $\mathscr{C}$ has skew part $\hat{\mathscr{C}}$. It follows (*cf.* (4.7) and (3.28)) that

$$\operatorname{sk}(\operatorname{div} \bar{C}) = \operatorname{div} \hat{C}. \tag{4.10}$$

2. Since $\sum_i{}' \sum_l f_{il} \cdot v_i$ is clearly additive when the sum is taken over particles P_i in any region whatsoever, with P_l denoting any particle, whether in this region or not, it follows that I.5. is equivalent to the requirement (*cf.* I.4) that

$$\overline{\sum_i{}' \sum_p f_{ip} \cdot v_i} \quad \text{is additive over microscopically-large regions.}$$

(Here P_i is within the region and P_p outside). Equivalently, the time-averaged power expended (by interactions) by one ε-cell upon another is equal but opposite to the power expended by the latter cell upon the former.

3. The alternative internal energy $\hat{\beta}$ is perhaps to be preferred to the internal energy β of § 3.3, given the interpretation of $\hat{\beta}$ as generalised binding energy. (Of course, in the event that $(\mathscr{C} - \hat{\mathscr{C}})$ is negligible, then no distinction between β and $\hat{\beta}$ need be made.) However, the tensor moment of momentum balance appropriate to energy balance (4.6) needs some subtlety in its formulation, as we now show.

4.2 An alternative form of (3.18), and hence alternative to $(3.29)_1$, as a generalised moment of momentum balance, may be obtained as follows. First we observe that (3.12) may be written as

$$\sum_i \sum_l (x_i - x_0) \otimes f_{il} = \sum_i \sum_j (x_i - x_0) \otimes f_{ij} + \sum_i \sum_k (x_i - x_0) \otimes f_{ik}, \tag{4.11}$$

where the sums are over particles P_i, P_k, P_j as in (4.1). Assumption I.2. allows the i, j sum to be expressed as a sum over surface cells, namely

$$\sum_{\substack{\text{surface}\\ \text{cells}}} \left\{ \sum_i{}' \sum_j ((x_i - x) + (x - x_0)) \otimes f_{ij} \right\}. \tag{4.12}$$

The time average of (4.12), upon recalling $(4.2)_2$ and (3.2), yields as continuum counterpart

$$\int_{\partial R} \{\mathscr{C} + (x - x_0) \otimes t\}. \tag{4.13}$$

Making use of (4.13) in the continuum version of (4.11), which was previously expressed by (3.18), we obtain the identification of

$$\int_{\partial R} \{\hat{\mathscr{C}} + (x - x_0) \wedge t\} + \int_R \{J + (x - x_0) \sim \operatorname{div} T^-\}$$

with (4.14)

$$\int_{\partial R} \{\mathscr{C} + (x - x_0) \otimes t\} + \overline{\sum_i \sum_k (x_i - x_0) \otimes f_{ik}}.$$

It follows, upon using (3.28), (4.7), and (4.10), that the time-averaged sum in (4.14) is to be identified with

$$\int_R \{J - \operatorname{sym}(T^- + \operatorname{div} \bar{C})\}.$$

This clearly motivates the definition of the (symmetric) tensor field

$$\boldsymbol{Z} := \boldsymbol{J} - \operatorname{sym}(\boldsymbol{T}^{-} + \operatorname{div}\overline{\boldsymbol{C}}), \tag{4.15}$$

in terms of which the time average of (4.11) may be identified (upon also recalling (4.12) and (4.13)) with

$$\int_{\partial R} \{\overline{\boldsymbol{C}}\boldsymbol{n} + (\boldsymbol{x} - \boldsymbol{x}_0) \otimes \boldsymbol{T}^{-}\boldsymbol{n}\} + \int_{R} \boldsymbol{Z}. \tag{4.16}$$

Employing this formulation (rather than (3.18)) leads to the generalised moment of momentum balance about $\boldsymbol{x}_0$ taking the form

$$\int_{\partial R} \{\boldsymbol{C}\boldsymbol{n} + (\boldsymbol{x} - \boldsymbol{x}_0) \otimes \boldsymbol{T}\boldsymbol{n}\} + \int_{R} \{\boldsymbol{Z} + \boldsymbol{G} + \boldsymbol{K} + (\boldsymbol{x} - \boldsymbol{x}_0) \otimes \boldsymbol{b}\}$$
$$= \int_{R} (\boldsymbol{x} - \boldsymbol{x}_0) \otimes \varrho\boldsymbol{a}, \tag{4.17}$$

upon using $(3.10)_2$ and (4.8). Invoking (3.11) yields for the local form of (4.17)

$$\operatorname{div}\boldsymbol{C} + \boldsymbol{T}^T + \boldsymbol{Z} + \boldsymbol{G} + \boldsymbol{K} = \boldsymbol{0}. \tag{4.18}$$

Remarks

1. In the absence of assumption I.5. the balances of tensor moment of momentum and energy take the forms exhibited in Section 3. If I.5. is valid then such balance relations are *equivalent* to those given in this section. In the latter context, comparison of (3.46) and (4.6) reveals that

$$\varrho\dot{\beta} = \operatorname{div}((\overline{\boldsymbol{C}} - \hat{\boldsymbol{C}})^{\sim} : \boldsymbol{D}) - \hat{\boldsymbol{C}} \cdot \nabla\boldsymbol{W} - P + (\boldsymbol{T}^{-} - \boldsymbol{J}) \cdot \boldsymbol{D}$$
$$= \operatorname{div}((\overline{\boldsymbol{C}} - \hat{\boldsymbol{C}})^{\sim} : \boldsymbol{D}) + \varrho\dot{\beta}, \tag{4.19}$$

on[1] using (3.49) and (3.50) (*cf.* also Remark 3 of § 4.1, together with (4.7) and (3.28), in respect of $(4.19)_2$).

2. While (modulo assumption I.5.) relation (4.18) is equivalent to (3.30) as a statement of local tensor moment of momentum balance (this is easily verified upon invoking (4.15) and (4.10)), the former relation proves to be more useful when comparison is to be made with theories of generalised continua (*cf.*, *e.g.*, CAPRIZ & PODIO-GUIDUGLI 1976, 1977).

5. Some general observations

5.1 While interactions between atoms, ions, and molecules are well-known to be predominantly electromagnetic in nature, remote influence on such entities will in general derive from the effects both of gravity and electromagnetism. Thus in (3.1), $\boldsymbol{f}_{il}$ is to be regarded as of electromagnetic character and $\boldsymbol{b}_i$ as a composite

[1] If $\boldsymbol{A}$ and $\boldsymbol{B}$ denote, respectively, rank three and rank two tensors then in Cartesian co-ordinates $\boldsymbol{A}^{\sim} : \boldsymbol{B} = (\boldsymbol{A}^{\sim})_{ijk} B_{kj} = A_{kji} B_{kj}$.

of gravitational and electromagnetic forces. Noting that thermal velocity is a fluctuating quantity (in both fluid and solid phases) and recalling (2.5), we can regard the definition $(3.40)_2$ as suggesting the fluctuating part of $\boldsymbol{b}_i$ is paramount in determining r. Since gravitational and certain electromagnetic effects are well-described in terms of macroscopic (and hence nonfluctuating) fields, r is accordingly to be associated with the effect of electromagnetic forces of a fluctuational nature. This justifies (within our simplistic model) *on physical grounds* our identification of r with the effect of radiation, over and above any such formal interpretation as would follow from comparison of the energy balances here exhibited with the standard form (*cf.* TRUESDELL & TOUPIN, 1960, § 241) of energy balance.

5.2 In our analysis no explicit attention has been paid to micro*structure*. However, the fundamental assumptions concerning the nature of corpuscular thermal motions and interactions may reasonably be expected to remain valid when particles are interpreted as appropriate sub-units of large molecules. In particular this is expected to be true of liquid crystalline phases, wherein couple-stress is manifestly significant. This establishes the physical relevance of the emphasis placed here upon the generalised moment of momentum and energy balances. Of course, in liquid crystalline phases (*cf. e.g.*, DE GENNES, 1974) much is known of the geometry of the macromolecules and local common alignment. This information may be utilised to structure the book-keeping of sections 3 and 4 in a fashion which enables terms in the balance relations to be related to such geometry and alignment. This matter will be taken up in a sequel to this work.

5.3 Provided that interactions are objective, and *observer changes*[1] *involve only functions* c *and* Q *which vary macroscopically with time*, it immediately follows from the definitions that the continuum fields $\boldsymbol{T}^-$, $\hat{\mathscr{C}}$, $\boldsymbol{J}$, $\hat{\boldsymbol{C}}$, $\mathscr{C}$, and $\bar{\boldsymbol{C}}$ are objective. The observation that thermal velocities are also objective further implies the objectivity of the fields $\boldsymbol{K}$, h, $\boldsymbol{P}$, q, $\boldsymbol{q}^-$, α, β, and (modulo the assumptions of Remark 4 of § 3.2) $\mathscr{D}$, $\mathscr{M}$, and $\mathscr{k}$, as a consequence of their definitions. It follows that the fields $\boldsymbol{T}$, $\boldsymbol{C}^-$, $\boldsymbol{C}$, and $\boldsymbol{q}$ are also objective, as also is r from (3.48).

As recently pointed out[2], material frame-indifference essentially involves two assumptions: namely, that all observers should agree upon the nature of any given material, and that the physical quantities which characterise a given, material should be described by objective fields. Here it has been indicated that objectivity of these latter fields is conditional upon observers not fluctuating one with respect to another.

5.4 Interaction assumption I.2. required only that the effective range of interactions be markedly less than 10^{-5} m; for example, 10^{-7} m (= 1000 Å). Since nearest-neighbour separations are typically of the order of a few Ångstroms only, in solid and liquid phases, the balance relations here derived are valid for what molecular physicists describe as very long-range interactions. That is, these relations are valid for *microscopically* (highly) *non-local* interactions. Such are to

---

[1] *Cf.* TRUESDELL & NOLL (1965), § 17.

[2] *Cf.* MURDOCH $(1983)_2$.

be distinguished from the *macroscopically non-local* interactions discussed by EDELEN (1976), who, since gravitation is well-understood, seems to have addressed interactions of electromagnetic nature not normally considered.

Acknowledgement. The author would like to thank Professor A. MORRO for many helpful comments and suggestions during the preparation of this work.

References

1960 ERICKSEN, J. L., Theory of anisotropic fluids. Trans. Soc. Rheol. **4**, 29–39.

TRUESDELL, C. & R. A. TOUPIN, The Classical Field Theories. In Handbuch der Physik, Vol. III/1 (ed. S. FLÜGGE). Berlin-Heidelberg-New York: Springer.

1961 ERICKSEN, J. L., Conservation laws for liquid crystals. Trans. Soc. Rheol. **5**, 23–34.

1963 DAHLER, J. S. & L. E. SCRIVEN, Theory of structured continua I. General consideration of angular momentum and polarization. Proc. R. Soc. Lond. A **275**, 504–527.

1964 ERINGEN, A. C. & E. S. SUHUBI, Nonlinear theory of simple microelastic solids I, Int. J. Eng. Sc. **2**, 189–203.

GREEN, A. E. & R. S. RIVLIN, Simple force and stress multipoles. Arch. Rational Mech. Anal. **16**, 325–353.

1965 KRUMHANSL, J. A., Generalized continuum field representations for lattice vibrations. In Lattice Dynamics, Copenhagen 1963 (ed. R. F. WALLIS). Oxford: Pergamon.

TRUESDELL, C. & W. NOLL, The Non-linear Field Theories of Mechanics. In Handbuch der Physik, Vol. III/3 (ed. S. FLÜGGE). Berlin-Heidelberg-New York: Springer.

1967 HIRSCHFELDER, J. O., Study Week on Molecular Forces. Pontificiae Academiae Scientiarum Scripta Varia 31. Amsterdam: North-Holland, New York: Wiley.

1968 KRÖNER, E. (ed.), Mechanics of Generalised Continua, Proceedings of IUTAM Symposium, Freudenstadt and Stuttgart, 1967. Berlin-Heidelberg-New York: Springer.

RIVLIN, R. S., The formulation of theories in generalized continuum mechanics and their physical significance. Ist. Naz. Alta Mat., Symposia Mathematica **1**, 357–373.

1969 TRUESDELL, C., Rational Thermodynamics. New York: McGraw Hill.

1974 DE GENNES, P. G., The Physics of Liquid Crystals. London: Oxford University Press.

1975 CERCIGNANI, C., Theory and application of the Boltzmann equation. Edinburgh and London: Scottish Academic.

1976 ALBLAS, J. B., A note on the physical foundation of the theory of multipole stresses. Arch. Mech. Stos. **28**, 279–298.

CAPRIZ, G. & P. PODIO GUIDUGLI, Discrete and continuous bodies with affine structure. Ann. Mat. Pura Appl. (IV), **111**, 195–217.

EDELEN, D. G. B., Nonlocal Field Theories. In Continuum Physics, Vol. IV (ed. A. C. ERINGEN). New York: Academic.

RIVLIN, R. S., The passage from a particle system to a continuum model. Arch. Mech. Stos. **28**, 549–561.

1977 CAPRIZ, G. & P. PODIO GUIDUGLI, Formal structure and classification of theories of oriented materials. Ann. Mat. Pura Appl. (IV) **115**, 17–39.

1983 MURDOCH, A. I., The motivation of continuum concepts and relations from discrete considerations. Q. J. Mech. Appl. Math. **36**, 163–187.

MURDOCH, A. I., On material frame-indifference, intrinsic spin, and certain constitutive relations motivated by the kinetic theory of gases. Arch. Rational Mech. Anal. **83**, 185–194.

Department of Mathematics
University of Strathclyde
Glasgow

(Received July 16, 1984)

Hydrodynamic Flow with Steady and Preserved Vorticity

WAN-LEE YIN

Dedicated to Professor Walter Noll on his Sixtieth Birthday

1. Introduction

In the absence of non-conservative body force and viscous action, the dynamical equations of an incompressible fluid deliver isochoric circulation-preserving motions. According to CAUCHY'S vorticity formula ([1], p. 173), the vector $JF^{-1}\boldsymbol{\omega}$ is preserved at all material particles in a circulation-preserving motion, where $\boldsymbol{\omega}$ is the present vorticity, $\boldsymbol{F}$ is the deformation gradient relative to a fixed initial configuration, and $J \equiv \det \boldsymbol{F} = 1$ in an isochoric motion. It is obvious that the property of preserving $JF^{-1}\boldsymbol{\omega}$ is generally different from the preservation of vorticity itself. Indeed, under the assumption of the circulation-preserving property, the condition of preserving $J\boldsymbol{\omega}$ is equivalent to the purely spatial criterion* that the velocity vector $\boldsymbol{v}$ be stationary in the direction of the vorticity, $\boldsymbol{\omega} \cdot \nabla \boldsymbol{v} = \boldsymbol{0}$ [1, p. 175]. Since the latter criterion is trivially satisfied in a plane motion, all plane hydrodynamic flows are included in the class of vorticity-preserving hydrodynamic flows**. Yet this class may contain few inherently three-dimensional motions since the condition $\boldsymbol{\omega} \cdot \nabla \boldsymbol{v} = \boldsymbol{0}$ requires the material vortex-lines to proceed in rigid translational motions.

The class of all three-dimensional rotational vector fields $\boldsymbol{v}$ with constant values along individual vector-lines of curl $\boldsymbol{v}$ has been determined in a recent paper [2]. It was found in particular that the vector-lines of curl $\boldsymbol{v}$ are plane curves, and when they are not straight lines, the field $\boldsymbol{v}$ assumes constant values on individual surfaces of a one-parameter family. These purely geometrical results have direct implications on circulation-preserving motions that also preserve $J\boldsymbol{\omega}$. Both the velocity and acceleration fields of such motions belong to the special class of vector fields with constant values along vortex-lines. The differential equa-

---

* One is reminded that the circulation-preserving property itself has a purely spatial formulation, the D'ALEMBERT-EULER condition curl $\dot{\boldsymbol{v}} = \boldsymbol{0}$.

** We use the term "hydrodynamic flow" for circulation-preserving isochoric motions.

tions governing their intrinsic acceleration components in appropriate material coordinate systems were formulated and, excepting one particular case, solved.

In the present paper, attention is confined to the special subclass of isochoric flows with steady vorticity. Restricting the scope of the previous work [2] and applying its geometrical results, I obtain all three-dimensional hydrodynamic flows possessing steady and preserved vorticity. Excepting plane flows with or without superposed rigid translations in the perpendicular direction, all other solutions belong to four simple families and they are steady flows or differ from steady flows by superposed rigid translational motions. Nevertheless, the weaker assumption of steady vorticity introduces considerable analytical subtleties in the study which would not arise in the case of steady flow. A combination of material and spatial descriptions is required to elucidate the implications of various geometrical and kinematical conditions, some referring to instantaneous configurations in space while others to properties preserved by material lines or surfaces. Certain analytical details presented in the work may prove useful in systematic and exact studies of other classes of three-dimensional unsteady motions*. As for the results of the present analysis, they are summarized in the final section of this paper.

2. Governing Equations

If the vorticity vector of a rotational hydrodynamic flow is not only steady in the region of flow but also preserved at each material particle, then the velocity $\boldsymbol{v}$ and the vorticity

$$\boldsymbol{\omega} \equiv \operatorname{curl} \boldsymbol{v} \neq \boldsymbol{0}$$

satisfy the following system of equations

$$\operatorname{div} \boldsymbol{v} = 0, \tag{1}$$

$$\partial \boldsymbol{\omega} / \partial t = \boldsymbol{0}, \tag{2}$$

$$\dot{\boldsymbol{\omega}} = \boldsymbol{v} \cdot \nabla \boldsymbol{\omega} = \boldsymbol{0}, \tag{3}$$

$$\operatorname{curl} \dot{\boldsymbol{v}} = \operatorname{curl} (\boldsymbol{\omega} \times \boldsymbol{v}) = \boldsymbol{0}, \tag{4}$$

where a dot indicates the material derivative with respect to time. Since $\operatorname{div} \boldsymbol{\omega} = 0$, Eqs. (1), (3) and the vector identity

$$\operatorname{curl} (\boldsymbol{\omega} \times \boldsymbol{v}) = \boldsymbol{v} \cdot \nabla \boldsymbol{\omega} - \boldsymbol{\omega} \cdot \nabla \boldsymbol{v} + (\operatorname{div} \boldsymbol{v})\, \boldsymbol{\omega} - (\operatorname{div} \boldsymbol{\omega})\, \boldsymbol{v},$$

imply that the circulation-preserving condition (Eq. (4)) is equivalent to

$$\boldsymbol{\omega} \cdot \nabla \boldsymbol{v} = \boldsymbol{0}. \tag{5}$$

Hence each vortex-line is in rigid translational motion. Equation (5) may also be derived directly from the condition $(J\boldsymbol{\omega})^{\cdot} = \boldsymbol{0}$ by using CAUCHY's vorticity

* Recent works by YIN [3] and STALLYBRASS [4] demonstrated that certain classes of *plane* unsteady rotational flows may be fruitfully investigated by the powerful method of complex variables.

formula, without assuming incompressibility and the steadiness of vorticity [1, p. 175].

According to the results of a previous work [2], Eq. (5) implies that the vortex-lines are plane curves. If they are not straight lines, then the velocity assumes constant values on individual surfaces of a one-parameter family, and these surfaces are vortex-surfaces. The two cases, one with rigidly translating straight vortex-lines and the other with rigidly translating vortex-surfaces, while not mutually exclusive, comprise all rotational solutions of Eq. (5). These two cases are referred to as Case 1 and Case 2, respectively, in the following analysis. In Sections 3 to 5, we show that all solutions of Eqs. (1)–(4) which do not belong to Case 2 are generally unsteady *plane* flows with or without superposed rigid translational motions along the fixed perpendicular direction. The solutions belonging to Case 2 are determined in Sections 6 to 9.

3. Preliminary Analysis of Case 1

The solutions of Case 1 as defined in the preceding section are characterized by rectilinear vortex-lines. In a circulation-preserving motion, the vortex-lines are material lines. Since the velocity vector is constant along each vortex-line (Eq. (5)), the material lines are in rigid translational motion. The steadiness of vorticity implies the steadiness of the spatial configuration of vortex-lines. In the special case when the vortex-lines coincide with the streamlines (Beltrami motions), the material lines translate along themselves. The analysis of this relatively simple case is defered to Section 5. In this and the subsequent section, we investigate those solutions of Case 1 whose streamlines are distinct from the rectilinear vortex-lines. For such flows, Eq. (4) ensures the local existence of a family of Lamb surfaces [5, pp. 404–405], $\xi =$ constant, such that

$$\boldsymbol{\omega} \times \boldsymbol{v} = \nabla \xi, \qquad (\nabla \xi) \neq \mathbf{0}. \tag{6}$$

It is clear from the last equation that the Lamb surfaces are composed of vortex-lines and streamlines. Hence they are vortex-surfaces and stream-surfaces at the same time. Although the vortex-lines are steady, the Lamb surfaces are generally not steady because the vortex-lines which belong to the same Lamb surface at one instant may belong to different Lamb surfaces at later instants. Since div $\boldsymbol{v} = 0$, $\boldsymbol{v}$ may be expressed in a sufficiently small region as the cross-product of the gradients of two scalar fields, and the surfaces on which either one of the scalar functions assumes constant values are stream-surfaces. Such scalar fields are not uniquely defined by $\boldsymbol{v}$, and one scalar field may be an arbitrary function that assumes constant values along individual streamlines, *e.g.*, the function ξ defining the Lamb surfaces. Then the other scalar function is determined by $\boldsymbol{v}$ and ξ to within an additive function of ξ. This yields

$$\boldsymbol{v} = \nabla \xi \times \nabla \zeta. \tag{7}$$

Similarly, Eqs. (6) and div $\boldsymbol{\omega} = 0$ deliver the following representation for the vorticity:

$$\boldsymbol{\omega} = \nabla \xi \times \nabla \eta, \tag{8}$$

where η is a scalar function determined by $\boldsymbol{\omega}$ and ξ to within an additive function of ξ.

Equations (3) and (7) imply that

$$\nabla\xi\times\nabla\zeta\cdot\nabla\boldsymbol{\omega} = 0.$$

This ensures the local validity of the functional relation

$$\boldsymbol{\omega} = \boldsymbol{\omega}(\xi, \zeta). \tag{9}$$

Similarly, Eqs. (5) and (8) deliver

$$\boldsymbol{v} = \boldsymbol{v}(\xi, \eta). \tag{10}$$

Furthermore, Eqs. (6)–(8) yield

$$\nabla\xi = \boldsymbol{\omega}\times\boldsymbol{v} = (\nabla\xi\times\nabla\eta)\times(\nabla\xi\times\nabla\zeta) = (\nabla\xi\times\nabla\eta\cdot\nabla\zeta)\,\nabla\xi,$$

or,

$$\nabla\xi\times\nabla\eta\cdot\nabla\zeta = 1.$$

The last equation implies that the three scalar functons ξ, η and ζ are functionally independent. Hence they may be used as a curvilinear coordinate system in a sufficiently small region. In general, this coordinate system is neither material nor steady.

From Eqs. (8) to (10) we obtain

$$\nabla\xi\times\nabla\eta = \operatorname{curl}\boldsymbol{v}(\xi, \eta) = \nabla\xi\times\boldsymbol{v}_\xi + \nabla\eta\times\boldsymbol{v}_\eta,$$

where the subscripts indicate differentiation with respect to the curvilinear coordinates. The scalar products of the last equation with $\nabla\xi$ and $\nabla\eta$ yield

$$-\boldsymbol{\omega}\cdot\boldsymbol{v}_\xi = \boldsymbol{\omega}\cdot\boldsymbol{v}_\eta = 0. \tag{11}$$

If the derivatives $\boldsymbol{v}_\xi$ and $\boldsymbol{v}_\eta$ are linearly dependent, then there is a scalar function $\alpha(\xi, \eta)$ such that

$$\boldsymbol{v}_\xi = \alpha(\xi, \eta)\,\boldsymbol{v}_\eta.$$

Let $\chi(\xi, \eta)$ be a particular solution of the first-order partial differential equation

$$\partial\chi/\partial\xi - \alpha(\xi, \eta)\,\partial\chi/\partial\eta = 0.$$

Then the last two equations yield

$$(\partial\chi/\partial\xi)\,\boldsymbol{v}_\eta - (\partial\chi/\partial\eta)\,\boldsymbol{v}_\xi = \boldsymbol{0}.$$

Hence the vector-valued function $\boldsymbol{v}(\xi, \eta)$ is functionally dependent on the scalar function $\chi(\xi, \eta)$: $\boldsymbol{v} = \boldsymbol{v}(\chi)$. Since such solutions belong to Case 2, their analysis is deferred to later sections.

If the derivatives $\boldsymbol{v}_\xi$ and $\boldsymbol{v}_\eta$ are linearly independent, then Eq. (11) implies that $\boldsymbol{\omega}$ is parallel to $\boldsymbol{v}_\xi\times\boldsymbol{v}_\eta$:

$$\frac{\boldsymbol{\omega}}{|\boldsymbol{\omega}|} = \pm\frac{\boldsymbol{v}_\xi\times\boldsymbol{v}_\eta}{|\boldsymbol{v}_\xi\times\boldsymbol{v}_\eta|}.$$

Since the left-hand side is independent of η and the right-hand side is independent of ζ, the two sides may represent the same unit vector field $\boldsymbol{r}$ only if $\boldsymbol{r}$ is independent of both η and ζ:

$$\boldsymbol{\omega}/|\boldsymbol{\omega}| = \boldsymbol{r}(\xi).$$

This result implies that the direction of vorticity is constant on each Lamb surface $\xi =$ constant. Consequently, the Lamb surfaces are general cylindrical surfaces with the rectilinear vortex-lines as their generators. Since the velocity $\boldsymbol{v}(\xi, \eta)$ does not depend on ζ, the streamlines on the same cylindrical Lamb surface at a particular instant are parallel curves.

4. Further Analysis of Case 1

We define the unit vector fields $\boldsymbol{n}$ and $\boldsymbol{s}$ by

$$\boldsymbol{n}(\xi, \eta) = \nabla\xi/|\nabla\xi|, \qquad \boldsymbol{s}(\xi, \eta) = \boldsymbol{r}(\xi)\times\boldsymbol{n}(\xi, \eta), \tag{12}$$

and let u and v denote the components of velocity along the directions $\boldsymbol{r}$ and $\boldsymbol{s}$:

$$\boldsymbol{v} = u(\xi, \eta)\, \boldsymbol{r}(\xi) + v(\xi, \eta)\, \boldsymbol{s}(\xi, \eta). \tag{13}$$

Furthermore, we let $W = |\boldsymbol{\omega}|$; then

$$\boldsymbol{\omega} = W(\xi, \zeta)\, \boldsymbol{r}(\xi) \tag{14}$$

and Eq. (6) becomes

$$\nabla\xi = W\boldsymbol{r}\times(u\boldsymbol{r} + v\boldsymbol{s}) = -Wv\boldsymbol{n}. \tag{15}$$

Hence Eqs. (8) and (7) yield, respectively,

$$W\boldsymbol{r} = -Wv\boldsymbol{n}\times\{(\boldsymbol{n}\cdot\nabla\eta)\,\boldsymbol{n} + (\boldsymbol{s}\cdot\nabla\eta)\,\boldsymbol{s}\} = -Wv(\boldsymbol{s}\cdot\nabla\eta)\,\boldsymbol{r},$$

$$u\boldsymbol{r} + v\boldsymbol{s} = -Wv\boldsymbol{n}\times\nabla\zeta = -Wv\{(\boldsymbol{s}\cdot\nabla\zeta)\,\boldsymbol{r} - (\boldsymbol{r}\cdot\nabla\zeta)\,\boldsymbol{s}\}.$$

It follows that

$$\nabla\eta = -(1/v)\,\boldsymbol{s} + (\boldsymbol{n}\cdot\nabla\eta)\,\boldsymbol{n}, \tag{16}$$

$$\nabla\zeta = \boldsymbol{r}/W - u\boldsymbol{s}/(Wv) + (\boldsymbol{n}\cdot\nabla\zeta)\,\boldsymbol{n}.$$

Equation (15) implies that

$$-\nabla\times\nabla\xi = \boldsymbol{0} = W_\zeta v\,\nabla\zeta\times\boldsymbol{n} + Wv_\eta\,\nabla\eta\times\boldsymbol{n} + Wv(\nabla\xi\times\boldsymbol{n}_\xi + \nabla\eta\times\boldsymbol{n}_\eta).$$

The s-component of the last equation and Eqs. (15) and (16) yield

$$vW_\zeta/W + W^2v^2\boldsymbol{r}\cdot\boldsymbol{n}_\xi - vW(\boldsymbol{n}\cdot\nabla\eta)\,\boldsymbol{r}\cdot\boldsymbol{n}_\eta = 0.$$

Differentiating $\boldsymbol{r}(\xi)\cdot\boldsymbol{n}(\xi, \eta) = 0$ with respect to η delivers $\boldsymbol{r}\cdot\boldsymbol{n}_\eta = 0$. Consequently,

$$-W_\zeta/W^3 = v\boldsymbol{r}\cdot\boldsymbol{n}_\xi,$$

where the left-hand side is independent of η and the right-hand side is independent of ζ. Hence both sides must reduce to a function of ξ only. It follows that

$$\boldsymbol{r}' \cdot \boldsymbol{n} = -f(\xi)/v(\xi, \eta), \qquad 1/W^2 = 2f(\xi)\,\zeta + g(\xi). \tag{17}$$

Now, Eqs. (8) and (13) yield

$$\nabla\xi \times \nabla\eta = \operatorname{curl} \boldsymbol{v} = u_\xi \nabla\xi \times \boldsymbol{r} + u_\eta \nabla\eta \times \boldsymbol{r} + u \nabla\xi \times \boldsymbol{r}'$$
$$+ v_\xi \nabla\xi \times \boldsymbol{s} + v_\eta \nabla\eta \times \boldsymbol{s} + v \nabla\xi \times \boldsymbol{s}_\xi + v \nabla\eta \times \boldsymbol{s}_\eta .$$

Taking scalar products with $\nabla\xi$ and $\nabla\eta$ delivers

$$W u_\eta = 0, \qquad -W(u_\xi + v\boldsymbol{r} \cdot \boldsymbol{s}_\xi) = 0 .$$

Hence $u = u(\xi)$ and

$$\boldsymbol{r}' \cdot \boldsymbol{s} = u'(\xi)/v(\xi, \eta).$$

The last equation and Eq. (17b) yield

$$v\boldsymbol{r}' = -f(\xi)\,\boldsymbol{n} + u'(\xi)\,\boldsymbol{s}. \tag{18}$$

Consequently,

$$\{v(\xi, \eta)\}^2\,|\boldsymbol{r}'(\xi)|^2 = \{f(\xi)\}^2 + \{u'(\xi)\}^2 .$$

This result requires that either v be independent of η or $\boldsymbol{r}' = 0$ and $f = u' = 0$.

In the former case Eqs. (18) and (13) imply that the vector fields $\boldsymbol{n}$, $\boldsymbol{\lambda}$ and $\boldsymbol{v}$ are all independent of η. Hence the velocity assumes constant values on individual Lamb surfaces. The flows belong to Case 2 and will be obtained in later sections.

In the latter case the steady direction of vorticity is also constant in space. Let (x, y, z) be a *time-independent* rectangular coordinate system with z-axis parallel to the vortex-lines. Then Eq. (5) implies that $\boldsymbol{v}$ is independent of z. Let

$$\boldsymbol{v} = \boldsymbol{v}(x, y, t) = u\boldsymbol{i} + v\boldsymbol{j} + w\boldsymbol{k}.$$

Then

$$\boldsymbol{\omega} = \operatorname{curl} \boldsymbol{v} = w_y \boldsymbol{i} - w_x \boldsymbol{j} + (v_x - u_y)\,\boldsymbol{k}$$

is parallel to $\boldsymbol{k}$. Hence w is spatially constant. The motion differs from a plane flow by a superposed rigid translational motion along the z-direction:

$$\boldsymbol{v}(x, y, t) = u(x, y, t)\,\boldsymbol{i} + v(x, y, t)\,\boldsymbol{j} + w(t)\,\boldsymbol{k}. \tag{19}$$

The analysis of this and the preceding section yields the conclusion, stated in Section 2, that all solutions of Eq. (1)–(4) which do not belong to Case 2 are generally unsteady plane flows with or without superposed rigid translational motions along the fixed perpendicular direction.

The class of *plane* hydrodynamic flows with steady vorticity has been previously determined [6]. Besides steady plane flows, the class includes only three families of unsteady flows:

(i) Flows of constant vorticity

$$\boldsymbol{v} = -\alpha y \boldsymbol{i} + \nabla\theta(x, y, t), \qquad \nabla^2\theta = 0,$$

(ii) Steady simple shearing motion with superposed rigid translational motion along the direction of streamlines

$$\boldsymbol{v} = \{u(x) + a(t)\}\, \boldsymbol{i},$$

(iii) Steady Couette flow with a superposed unsteady irrotational vortex

$$\boldsymbol{v} = \{u(r) + b(t)/r\}\, \boldsymbol{i}_\theta.$$

5. Beltrami Flows

A motion in which the streamlines coincide with the vortex-lines is called a Beltrami motion. It is easily proved that a Beltrami motion with steady vorticity is necessarily a steady flow ([1], p. 98).

For flows of this type, Eqs. (3) and (5) imply that both $\boldsymbol{\omega}$ and $\boldsymbol{v}$ have constant values along their common vector lines. Let

$$\boldsymbol{v} = V\boldsymbol{e}, \qquad \boldsymbol{\omega} = W\boldsymbol{e}, \qquad |\boldsymbol{e}| = 1.$$

Then $\boldsymbol{e} \cdot \nabla \boldsymbol{\omega} = \boldsymbol{e} \cdot \nabla \boldsymbol{v} = \mathbf{0}$, or,

$$\boldsymbol{e} \cdot \nabla W = \boldsymbol{e}, \nabla V = \boldsymbol{e}, \qquad \boldsymbol{e} \cdot \nabla \boldsymbol{e} = 0. \tag{20}$$

Condition (20c) admits an equivalent expression:

$$\boldsymbol{e} \times \operatorname{curl} \boldsymbol{e} = \mathbf{0}.$$

This and Eq. (20b) imply that

$$\mathbf{0} = \boldsymbol{e} \times \boldsymbol{\omega} = \boldsymbol{e} \times \operatorname{curl}(V\boldsymbol{e}) = \boldsymbol{e} \times (V \operatorname{curl} \boldsymbol{e} + \nabla V \times \boldsymbol{e}) = \nabla V. \tag{21}$$

Hence the steady speed V is also constant in space.

Equations (1) and (21) imply that the rectilinear unit vector field $\boldsymbol{e}$ is solenoidal:

$$\operatorname{div} \boldsymbol{e} = \operatorname{div}(\boldsymbol{e}/V) = \boldsymbol{v} \cdot \nabla(1/V) + (\operatorname{div} \boldsymbol{v})/V = 0.$$

In an early paper [7], I obtained all solenoidal rectilinear vector fields of unit magnitude. It is convenient for the present purpose to present the results in the following manner:

$$\boldsymbol{e} = \boldsymbol{e}(\psi), \qquad \boldsymbol{e} \cdot \nabla \psi = 0, \qquad \boldsymbol{e}'(\psi) \cdot \nabla \psi = 0, \tag{22}$$

where the surfaces $\psi =$ constant are parallel or non-parallel planes. Equations (21) and (22a, b) imply that the velocity , $\boldsymbol{v} = V\boldsymbol{e}$, is tangential to individual ψ-planes and assumes constant values on these planes. Thus the flows of the present case also belong to Case 2. In such steady flows, the various material planes $\psi =$ constant move tangentially and rigidly along different directions $\boldsymbol{e}(\psi)$ with a common speed V. If the planes are not parallel, then they possess an enveloping surface S and Eq. (22c) requires that the direction of the velocity of each plane be parallel to the rule line of S where the plane touches the enveloping surface. *

* I am aware that Mr. Samuel M. Venable, Jr. has independently obtained the Beltrami flows reported in this section. The special case $\boldsymbol{\omega} \times \boldsymbol{v} = \mathbf{0}$ is examined here for the sake of completeness of the analysis.

6. Preliminary Analysis of Case 2

The preceding analysis has shown that, excepting plane flows with or without superposed rigid motions in the perpendicular direction, the flows of Case 1 also belong to Case 2. In the latter case, the velocity field assumes constant values on individual surfaces of a one-parameter family:

$$\boldsymbol{v} = \boldsymbol{v}(\zeta), \quad \nabla\zeta \neq \boldsymbol{0}.$$

Since

$$\boldsymbol{\omega} = \operatorname{curl} \boldsymbol{v} = \nabla\zeta \times \boldsymbol{v}'(\zeta), \tag{23}$$

$v'(\zeta)$ does not vanish in a rotational motion. Hence the condition

$$\operatorname{div} \boldsymbol{v} = \nabla\zeta \cdot \boldsymbol{v}'(\zeta) = 0 \tag{24}$$

implies that the normal vector on a given ζ-surface is everywhere orthogonal to the constant vector $\boldsymbol{v}'(\zeta)$. Therefore, ζ-surfaces are planes or general cylinders with generators parallel to the vectors $\boldsymbol{v}'(\zeta)$. Equation (23) implies that the vortex-lines are parallel lines on planar ζ-surfaces or parallel latitudes on cylindrical ζ-surfaces. In the latter case, the generators of the ζ-cylinders are along the binormal direction of the steady vortex lines. Hence the cylindrical ζ-surfaces are steady surfaces.

The vortex-lines are material lines because the motion is circulation-preserving. These material lines may move tangentially to the ζ-surfaces or they may move across the ζ-surfaces. The latter case is considered first; the former case is examined in Sections 8 and 9.

7. Translation of Vortex-Lines Across ζ-Surfaces

If ζ-surfaces are planes and the rectilinear material vortex-lines, which momentarily occupy ζ-planes, move *across* these planes, then all steady vortex-lines in space are generated by translational motions of the parallel material vortex-lines which at some instant occupy a particular ζ-plane. Hence the steady vortex-line are parallel lines in space. In a time-independent coordinate system with z-axis parallel to the common direction of vortex-lines, the velocity has the expression of Eq. (19). It is a special case of plane flow with possibly a superposed translational motion along the perpendicular direction.

If ζ-surfaces are general cylinders and the material vortex-lines, which at every instant are parallel latitudes of the cylindrical surfaces, proceed in translational motions *across* the surfaces, then all steady vortex-lines in space have the same geometrical shape and the same orientation. Hence the cylindrical ζ-surfaces, which contain the vortex-lines as parallel latitudes, are *steady*, parallel surfaces. Let (x, y, z) be a time-independent rectangular coordinate system with z-axis parallel to the generators of the cylinders. Then

$$\boldsymbol{v} = \boldsymbol{v}(\zeta, t) = u(\zeta, t)\,\boldsymbol{i} + v(\zeta, t)\,\boldsymbol{j} + w(\zeta, t)\,\boldsymbol{k}, \quad \zeta = \zeta(x, y). \tag{25}$$

Equation (24) implies that the vector $\boldsymbol{v}_\zeta$, constant on each ζ-cylinder, is orthogonal to all unit normal vectors of the cylinder. Hence $\boldsymbol{v}_\zeta$ is parallel to the generators of the cylindrical surface:

$$\boldsymbol{v}_\zeta \times \boldsymbol{k} = \boldsymbol{0}.$$

Differentiating Eq. (25) with respect to ζ and substituting the result into the last equation, one finds that the components u and v depend only on time:

$$\boldsymbol{v} = u(t)\,\boldsymbol{i} + v(t)\,\boldsymbol{j} + w(x, y, t)\,\boldsymbol{k}.$$

It follows that

$$\boldsymbol{\omega} = \text{curl}\,\boldsymbol{v} = w_y\boldsymbol{i} - w_x\boldsymbol{j}.$$

Since $\boldsymbol{\omega}$ is steady, its components w_y and $-w_x$ are independent of time. Equation (3) yields

$$u(t)w_{xy} + v(t)\,w_{yy} = 0, \qquad u(t)\,w_{xx} + v(t)\,w_{xy} = 0. \tag{26}$$

Since the vortex-lines are parallel latitudes on curved cylindrical surfaces, $\boldsymbol{\omega} = w_y\boldsymbol{i} - w_x\boldsymbol{j}$ is not a constant vector in the x–y plane. Equation (26) implies that there is at least one steady, non-vanishing vector in the same plane, ∇w_x or ∇w_y, orthogonal to the direction of $u(t)\,\boldsymbol{i} + v(t)\,\boldsymbol{j}$. Hence the latter direction is steady. Let this direction be chosen as the x-direction of a new time-independent coordinate system without changing the z-axis of the original system. Relative to the new coordinates the velocity has the following expression with a new component along the x-direction:

$$\boldsymbol{v} = u(t)\,\boldsymbol{i} + w(x, y, t)\,\boldsymbol{k}.$$

The two equations of (26) are replaced by

$$u(t)\,w_{xy} = u(t)\,w_{xx} = 0.$$

Since the material vortex-lines move across the spatial surfaces of constant velocity, the normal component of velocity $u(t)$ does not vanish. The last equation yields

$$w_{xy} = w_{xx} = 0,$$

or,

$$w(x, y, t) = \alpha x + f(y) + g(t).$$

Hence,

$$\boldsymbol{v} = \{\alpha x + f(y) + g(t)\}\,\boldsymbol{k} + u(t)\,\boldsymbol{i}, \tag{27}$$

where α is a time-independent constant and f and g are arbitrary functions of their respective arguments.

8. Generalized Poiseuille Flow

The analysis in the preceding section yields all solutions of Case 2 whose material vortex-lines move across the spatial surfaces ζ = constant. In the following we investigate solutions with material vortex-lines translating tangentially in ζ-surfaces. In such flows ζ-surfaces are Lamb surfaces.

Since $\boldsymbol{v}$ is constant on each ζ-surface and tangential to the surfaces, the streamlines are parallel straight lines on the surface. Let

$$\boldsymbol{v}(\zeta) = V\boldsymbol{e}, \quad |\boldsymbol{e}(\zeta)| = 1.$$

Then $\boldsymbol{e} \cdot \nabla\zeta = 0$ and

$$\boldsymbol{e} \cdot \nabla V = V'(\zeta)\, \boldsymbol{e} \cdot \nabla\zeta = 0, \quad \operatorname{div} \boldsymbol{e} = (\operatorname{div} \boldsymbol{v})/V = 0. \tag{28}$$

Hence $\boldsymbol{e}$ is a solenoidal rectilinear vector field of unit magnitude. A general representation of such vector fields was given in Eq. (22) in terms of a scalar function ψ, where $\psi =$ constant are planar stream surfaces. The representation shows that the vector field $\boldsymbol{e} = \boldsymbol{e}(\zeta)$, which assumes constant values on cylindrical Lamb surfaces $\zeta =$ constant, also assumes constant values on planar stream-surfaces $\psi =$ constant:

$$\boldsymbol{e} = \boldsymbol{e}(\zeta) = \boldsymbol{e}(\psi).$$

Hence,

$$\operatorname{curl} \boldsymbol{e} = \nabla\zeta \times (d\boldsymbol{e}/d\zeta) = \nabla\psi \times (d\boldsymbol{e}/d\psi),$$

$$\operatorname{div} \boldsymbol{e} = \nabla\zeta \cdot (d\boldsymbol{e}/d\zeta) = \nabla\psi \cdot (d\boldsymbol{e}/d\psi),$$

or,

$$\left(\nabla\zeta - \frac{d\zeta}{d\psi}\nabla\psi\right) \times \frac{d\boldsymbol{e}}{d\psi} = \boldsymbol{0}, \quad \left(\nabla\zeta - \frac{d\zeta}{d\psi}\nabla\psi\right) \cdot \frac{d\boldsymbol{e}}{d\zeta} = 0. \tag{29}$$

Since the two vectors $\nabla\zeta - (d\zeta/d\psi)\,\nabla\psi$ and $d\boldsymbol{e}/d\zeta$ cannot be parallel and orthogonal at the same time, one or the other must vanish.

If $d\boldsymbol{e}/d\zeta = 0$, then $\boldsymbol{e}$ is a constant field and the streamlines are parallel straight lines. Let (x, y, z) be a generally time-dependent rectangular coordinate system such that the streamlines are parallel to the z-axis at every instant. Then $\boldsymbol{e} = \boldsymbol{k} \equiv \nabla z$ and Eq. (28a) implies that

$$\boldsymbol{v} = V(x, y, t)\, \boldsymbol{k}. \tag{30}$$

Hence

$$\omega = V_y \boldsymbol{i} - V_x \boldsymbol{j} \tag{31}$$

The last equation implies that the *steady* vortex-lines are contained in the coordinate planes $z =$ constant. Unless all vortex-lines are parallel straight lines (in which case the flow has the expression of Eq. (19)), there are two linearly independent steady directions orthogonal to the z-axis, and the direction of the z-axis is necessarily steady. Hence a *time-independent* coordinate system may be defined relative to which the velocity and vorticity have the expressions of (30) and (31), respectively. Steadiness of the vorticity requires that $V_{xt} = V_{yt} = 0$, i.e., V_t be a possibly time-dependent constant. Consequently,

$$\boldsymbol{v} = \{w(x, y) + f(t)\}\, \boldsymbol{k}. \tag{32}$$

The motion is a steady generalized Poiseuille flow with a superposed unsteady rigid translational motion along the axial direction.

9. Tangential Translation of Material Planes

If the streamlines are not parallel straight lines, then Eq. (29) implies that

$$\nabla\zeta - (d\zeta/d\psi)\,\nabla\psi = \mathbf{0}.$$

Hence ζ and ψ are functionally dependent scalar fields, and the surfaces of constant velocity, ζ = constant, coincide with ψ-planes. The velocity is a constant tangential vector on each ψ-plane:

$$\boldsymbol{v} = \boldsymbol{v}(\psi), \qquad \boldsymbol{v}\cdot\nabla\psi = 0. \tag{33}$$

Consider first the case of parallel ψ-planes. Let (x, y, z) be a generally time-dependent coordinate system with the z-axis normal to the parallel planes. The velocity and vorticity are given respectively by

$$\boldsymbol{v} = u(z, t)\,\boldsymbol{i} + v(z, t)\,\boldsymbol{j}, \tag{34}$$

$$\boldsymbol{\omega} = -v_z\boldsymbol{i} + u_z\boldsymbol{j}. \tag{35}$$

Unless all vortex-lines are parallel straight lines, there are two linearly independent steady directions of $\boldsymbol{\omega}$. Hence the orthogonal direction $\boldsymbol{k} = \boldsymbol{i}\times\boldsymbol{j}$ is steady. A time-independent coordinate system may be obtained relative to which the expressions (34) and (35) maintain their forms. Then the steadiness of vorticity implies that $v_{zt} = u_{zt} = 0$, or

$$\boldsymbol{v} = u(z)\,\boldsymbol{i} + v(z)\,\boldsymbol{j} + f(t)\,\boldsymbol{i} + g(t)\,\boldsymbol{j}.$$

If ψ-planes are not parallel, then their enveloping surface is a curved developable surface S. The vector fleld $\boldsymbol{e}(\psi)$ in Eq. (22) is obtained by assigning the directions of the rulings of S to the tangent planes of S. Each tangent plane ψ = constant translates rigidly along the direction of the corresponding ruling. Since ψ-planes coincide with ζ-surfaces, they are Lamb surfaces composed of parallel rectilinear vortex-lines. Unless all vortex-lines are parallel (in which case the flow is described by Eq. (19)), different ψ-planes in space are composed of steady, parallel vortex-lines along different directions. Hence these planes are the only planar vortex-surfaces in space. They must be steady planes because they are uniquely determined by the spatial configuration of steady vortex-lines. Hence their enveloping surface S and their corresponding tangent lines on S are steady. This implies that the field $\boldsymbol{e}(\psi)$ is steady. Let

$$\boldsymbol{v} = V(\psi, t)\,\boldsymbol{e}(\psi). \tag{37}$$

Then

$$\boldsymbol{\omega} = V_\psi\,\nabla\psi\times\boldsymbol{e} + V\,\mathrm{curl}\,\boldsymbol{e},$$

where the two vectors on the right-hand side have steady and orthogonal directions. Hence V_ψ is steady, and unless curl $\boldsymbol{e} = 0$, V is also steady. However, if curl $\boldsymbol{e} = \mathbf{0}$, then

$$\nabla\psi\times\boldsymbol{e}'(\psi) = \mathbf{0}, \qquad \mathrm{div}\,\boldsymbol{e} = \nabla\psi\cdot\boldsymbol{e}'(\psi) = 0,$$

so that $e'(\psi)$ is at once parallel and orthogonal to $\nabla\psi$. This would require e to be a constant vector* and the velocity to be of the form of Eq. (32). Thus if neighboring ψ-planes translate along different directions, then the speed V is steady and consequently the flow is steady.

10. Conclusion

The preceding analysis shows that the complete class of hydrodynamic flows with steady and preserved vorticity consists in the following seven families:

(1) Steady plane flows with possible superposition of time-dependent rigid translational motions in the perpendicular direction:

$$\boldsymbol{v}(x, y, t) = u(x, y)\,\boldsymbol{i} + v(x, y)\,\boldsymbol{j} + w(t)\,\boldsymbol{k},$$

$$u = -\psi_y(x, y), \quad v = \psi_x(x, y), \quad \partial(\psi, \nabla^2\psi)/\partial(x, y) = 0,$$

(2) Plane flows of constant vorticity with possible superposition of time-dependent rigid translational motions in the perpendicular direction:

$$\boldsymbol{v}(x, y, t) = -\alpha y\boldsymbol{i} + \nabla\theta(x, y, t) + w(t)\,\boldsymbol{k}, \nabla^2\theta = 0,$$

(3) Steady Couette flow with possible superposition of an unsteady concentric irrotational vortex and a rigid translational motion in the perpendicular direction:

$$\boldsymbol{v}(r, \theta, r) = \{u(r) + b(t)/r\}\,\boldsymbol{i}_\theta + w(t)\,\boldsymbol{k},$$

(4) Steady shearing flow of parallel and congruent cylindrical surfaces with superposed rigid translational motion parallel to and across the cylindrical surfaces,

$$\boldsymbol{v}(x, y, t) = \{\alpha x + f(y) + g(t)\}\,\boldsymbol{k} + u(t)\,\boldsymbol{i},$$

(5) Steady generalized Poiseuille flow with possible superposition of unsteady rigid translational motion along the axial direction

$$\boldsymbol{v}(x, y, t) = \{w(x, y) + f(t)\}\,\boldsymbol{k},$$

(6) Steady tangential translations of parallel planes with or without superposed unsteady tangential rigid motion,

$$\boldsymbol{v}(z, t) = \{u(z) + f(t)\}\,\boldsymbol{i} + \{v(z) + g(t)\}\,\boldsymbol{j},$$

(7) Steady tangential translations of the tangent planes of a curved developable surface along the directions of the respective rulings where the tangent planes touch the developable surface.

It is worthwhile to note that the velocity fields of the first six families depend only on one or two time-independent rectangular coordinates. The steady flows of the last family do not admit superposition of unsteady rigid motions.

---

* The system of equations $\operatorname{curl} \boldsymbol{e} = \boldsymbol{0}$, $\operatorname{div} \boldsymbol{e} = 0$ and $|\boldsymbol{e}| = 1$ yields $\boldsymbol{e} = \nabla\theta$ where $\nabla^2\theta = 0$ and $|\nabla\theta|^2 = 1$. Our conclusion that $\boldsymbol{e}$ is necessarily a constant field corresponds to the analytical theorem that the only solutions of the overdetermined system of partial differential equations $\nabla^2\theta = 0$ and $|\nabla\theta|^2 = 1$ are linear functions of x, y and z.

References

1. C. TRUESDELL, *The Kinematics of Vorticity.* Indiana University Press, Bloomington (1954).
2. W.-L. YIN, "Velocity fields that are constant along every vortex-line", *Arch. Rational Mech. Anal.*, in press.
3. W.-L. YIN, "Circulation-preserving plane flows of incompressible viscous fluids", *Arch. Rational Mech. Anal.* **83**, 169–184 (1983).
4. M. P. STALLYBRASS, "A class of exact solutions of the Navier-Stokes equations—plane unsteady flow", *Intl. J. Engr. Sci.* **21**, 179–186 (1983).
5. C. TRUESDELL & R. TOUPIN, The Classical Field Theories, *Handbuch der Physik*, III/1 Berlin-Göttingen-Heidelberg: Springer (1960).
6. W.-L. YIN, "Plane hydrodynamic flows with steady vorticity: a reduction theorem", *J. Appl. Math. Phys.* (ZAMP) **35**, 430–434 (1984).
7. W.-L. YIN, "Construction of solenoidal rectilinear vector fields of unit magnitude," *Arch. Rational Mech. Anal.* **43**, 321–324 (1971).

School of Engineering Science and Mechanics
Georgia Institute of Technology
Atlanta

(Receiced September 18, 1984)

Some Surface Defects in Unstressed Thermoelastic Solids

J. L. Ericksen

Dedicated to Walter Noll, on the Occasion of His Sixtieth Birthday

I. Introduction

In the literature on crystals, "twinning" is a word used to describe a variety of phenomena involving different but symmetry-related configurations which coexist in crystals, meeting to form surfaces of discontinuity. As is discussed by Pitteri [1], there have been various attempts to formulate a more precise general definition of the word, as it applies to crystals. His discussion makes clear that some types of twinning are outside the scope of thermoelasticity theory. His definition excludes some phenomena which some experts on crystals call twins, like the "rotational twins" described by Barrett & Massalski [2, p. 406], things which seem to me more reminiscent of the multiple births we commonly describe by other words, like triplets or sextuplets. Whatever one calls them, they are of physical interest, as are other somewhat similar phenomena. My purpose is to present elements of thermoelasticity theory for things of this general kind.

II. Thermoelastic Bodies

To abstract features which seem significant, we consider a homogeneous thermoelastic body, referred to a homogeneous reference configuration. For present purposes, it is characterized by one smooth constitutive equation, of the form

$$\phi = \phi(C, \theta), \tag{2.1}$$

with ϕ identified as the Helmholtz free energy per unit mass, θ as absolute temperature and

$$C = F^T F = C^T. \tag{2.2}$$

Here, F is the usual deformation gradient, with

$$\det F > 0, \tag{2.3}$$

implying that C is positive definite. To discuss symmetry-related configurations we want there to be some non-trivial material symmetry, so we require that

$$\phi(H^T C H, \theta) = \phi(C, \theta), \quad H \in G, \tag{2.4}$$

G being some non-trivial subgroup of the unimodular group. For crystals, molecular theory suggests for G a group which can be represented by the unimodular matrices of integers, as is discussed by ERICKSEN [3], and the experience is that this is an apt choice for analyses of twinning, *etc.* The H then representing G need not be these matrices of integers, but are similar to them. Select subgroups can suffice for particular problems. For general discussion, we need not fix any particular choice of G.

For any particular choice of $\theta = \theta_1$, $C = C_1$ is a *natural state* provided the inequality

$$\phi(C, \theta_1) \geqq \phi(C_1, \theta_1) \tag{2.5}$$

holds in the strict sense, at least when $C - C_1$ is small enough. In more physical terms, these are stable or metastable unstressed equilibrium configurations, equilibrium of the simplest kind. With (2.4), for any $H \in G$, $H^T C_1 H$ is also a natural state. Many studies of thermoelasticity theory presume that

$$H^T C_1 H = C_1 \quad \forall\, H \in G. \tag{2.6}$$

The group suggested above for crystals leaves invariant no symmetric, positive tensor so, with this choice of G, (2.6) cannot be satisfied. Even if we ignore this, it is necessary to deny (2.5), to accommodate common examples of twinning. So we want ϕ to be such that, for some $H \in G$, and some natural state C_1,

$$C_2 = H^T C_1 H \neq C_1, \quad \det H = 1, \tag{2.7}$$

giving us at least two different, symmetry-related natural states. Alternatively, if F_1 and F_2 are any deformation gradients corresponding to C_1 and C_2, as indicated by (2.2) and (2.3), there exists a rotation R,

$$R^{-1} = R^T, \quad \det R = 1, \tag{2.8}$$

such that

$$F_2 = R F_1 H. \tag{2.9}$$

One more condition is inherent in notions of phenomena like twinning, that it should be possible for such natural states to coexist, coherently, in the same body. That is, it should be feasible to regard F_1 and F_2 as values of F in adjacent regions, with F the gradient of a continuous deformation. This introduces the requirement that the classical kinematic conditions of compatibility be satisfied where the regions meet, *viz.*

$$F_2 = (1 + a \otimes n)\, F_1. \tag{2.10}$$

Here, n is the unit normal to the surface of discontinuity, in the current configuration, and a is the so-called amplitude vector. In this most elementary formulation, F_1, F_2, R *etc.* are constants; we are interested in the possibilities for patching together homogeneous deformations, with occasional planes of discontinuity, *etc.* What we desire, I think, is a neat way of classifying and picturing all of these, for crystals at least. Certainly, there is interest in like phenomena encountered in stressed crystals, and other types of complications, but I will ignore such ramifications. As it applies to thermoelasticity theory, PITTERI'S [1] definition of twinning requires that

$$R^2 = H^2 = 1, \tag{2.11$_1$}$$

(2.9) then implying that

$$F_1 = RF_2H, \tag{2.11$_2$}$$

reflecting the idea that F_1 and F_2 are related like twins. Actually, one of the conditions (2.11) implies the other, so one can assume less. As he indicates, the theory of such twins in unstressed crystals is rather well understood. Similar remarks apply to the lattice-invariant shears, as defined by ERICKSEN [4], cases where

$$R = 1, \quad H \neq 1. \tag{2.13}$$

Other possibilities include at least some of the "rotational twins" described by BARRETT & MASSALSKI [2, p. 406]. Their description suggests that, for these, we want, for some integer N,

$$H^N = 1 \tag{2.14}$$

and perhaps more; their statements about possible amplitudes, *etc.* seem not to follow from this alone. Later, we will discuss implications of (2.14) in detail. Since I am hardly an expert crystallographer, I could be overlooking other possibilities known to such experts. In any event, it seems sensible to try to understand the full set of mathematical possibilities.

III. A Solution

In the setting indicated, it is rather natural to think of selecting F_1 and H, then trying to calculate values of other quantities involved, when they exist. From (2.9) and (2.10), one can eliminate F_2, to obtain

$$RF_1H = (1 + a \otimes n)\, F_1, \tag{3.1}$$

wherein the rotation R as well as the vectors a and n are regarded as unknowns. We can lump together the known quantities, writing

$$J \stackrel{\text{def}}{=} F_1 H^{-1} F_1^{-1}, \tag{3.2}$$

replacing (3.1) by

$$R = (1 + a \otimes n)\, J. \tag{3.3}$$

From $(2.7)_2$ and (3.2),

$$\det J = \det H^{-1} = 1. \tag{3.4}$$

Whatever other properties J might have depends on the choice of F_1 and H. For crystals, H, hence J is similar to a matrix of integers, so $\operatorname{tr} J$ is an integer, for example and, later, we make a little use of this. With (3.4), we do have, by taking determinants of both sides of (3.3),

$$1 = \det (1 + a \otimes n), \tag{3.5}$$

implying that

$$(1 + a \otimes n)^{-1} = 1 - a \otimes n, \tag{3.6}$$

and, bearing in mind that we already assumed n to be a unit vector,

$$a \cdot n = 0, \quad n \cdot n = 1, \tag{3.7}$$

With (3.3), we can eliminate the rotation R, using

$$\begin{aligned} R^{-1} &= J^{-1}(1 - a \otimes n) \\ &= R^T \\ &= J^T(1 + n \otimes a), \end{aligned}$$

or, equivalently,

$$\begin{aligned} K &\stackrel{\text{def}}{=} JJ^T \\ &= (1 - a \otimes n)(1 - n \otimes a). \end{aligned} \tag{3.8}$$

From its definition and (3.4), we have that

$$K = K^T > 0, \quad \det K = 1. \tag{3.9}$$

Again, any other properties it might have depend on special choices of F_1 and H. Equations of the type (3.8) arise in almost all discussions of simple shearing deformations in nonlinear elasticity theory, from which it is perhaps clear that (3.8) cannot hold for all K satisfying (3.9). To analyze this, let e by a unit vector perpendicular to a and n,

$$e \cdot a = e \cdot n = 0, \quad e \cdot e = 1. \tag{3.10}$$

Then (3.8) implies that

$$Ke = e \tag{3.11}$$

so it is clearly necessary that

$$\det (K - 1) = 0. \tag{3.12}$$

So, we must pick F_1 and H to satisfy this condition, or the equivalent. If we do, we can calculate possible vectors a and n, satisfying (3.8), use (3.3) to calculate R, in brief to solve the basic problem posed.

One way to do the calculation is to introduce the spectral representation of K. Granted that (3.12) and (3.9) hold, it will be of the form

$$K = e \otimes e + \lambda f \otimes f + \lambda^{-1} g \otimes g, \tag{3.13}$$

with e, f and g orthonormal vectors, and $\lambda > 0$. If $\lambda = 1$, then $K = 1$, and it then follows from (3.1) that $a = 0$, a trivial case of no interest. Otherwise, it is clear that the e occurring in (3.11) must be that given in (3.13), so we have not seriously abused notation. Also, from (3.7) and (3.4), we infer that

$$Kn = n - a, \tag{3.14}$$

or

$$a = (1 - K)\, n \tag{3.15}$$

and, using (3.7),

$$n \cdot Kn = n \cdot n = 1. \tag{3.16}$$

So, write

$$n = (\cos \psi) f + (\sin \psi) g, \tag{3.17}$$

and use (3.16) to determine ψ. This gives

$$\lambda \cos^2 \psi + \lambda^{-1} \sin^2 \psi = 1$$

or

$$\cos \psi = \pm 1/\sqrt{\lambda + 1}, \quad \sin \psi = \pm \sqrt{\lambda/(\lambda + 1)}, \tag{3.18}$$

where any choice of the signs can be used. For each choice, (3.15) determines a corresponding value of a. By simple calculation one can verify that the values calculated satisfy (3.8). Nominally, this gives four solutions, but only two are essentially different. That is, we obviously get one solution from another by replacing a by $-a$ and n by $-n$, and this accounts for two of the four. This analysis seems to be new, though special cases have been treated by various writers. One can deduce various conditions which are equivalent to (3.11), but I won't belabor this.

From this view, the only difficulty lies in deciding under what physical circumstances it is likely that (3.12) will be satisfied, in better understanding physical implications of the analysis. Clearly, it only takes an infinitesimal shift in K to violate (3.12). Normally, we think of F_1 as varying smoothly with θ, over some temperature interval, the phenomenon of thermal expansion. Superficially, this makes it seem likely that (3.12) will fail to hold over an interval of temperatures. That there are likely exceptions is known from the theory of twinning. Briefly suppose that $H^2 = 1$. Then from (3.2), it is also true that $J^2 = 1$. With (3.4) and a bit of matrix theory, one can then show that J must be of the form

$$J = -1 + 2c \otimes d, \quad c \cdot d = 1, \tag{3.19}$$

so

$$K = (-1 + 2c \otimes d)(-1 + 2d \otimes c). \tag{3.20}$$

Clearly, (3.11) then holds, with e perpendicular to c and d; it matters little what is F_1, as long as H does not map C_1 to itself. Somewhat similar reasoning applies to the lattice invariant shears, typified by (2.13). Cases to be discussed do restrict C_1, but only a little more, so the superficial impression mentioned above is at least not entirely correct.

A rather curious result is implicit in analyses given by ERICKSEN [5], although one has to examine the special cases to check it. After I noticed it, and mentioned it to RICHARD JAMES, he found and showed me a more direct proof. *Suppose that we have a solution for any case where H is a rotation. Then there exist* 180° *rotations* R_1 *and* R_2, viz.

$$R_1^2 = R_2^2 = 1, \tag{3.21}$$

such that

$$RF_1H = R_1F_1R_2. \tag{3.22}$$

This comes close to saying that, with H orthogonal, all solutions are of the twinning type, basically, although R_2 might not belong to G. In various cases encountered in practice, one can choose a reference configuration so H is orthogonal, although it might not be a choice which intuition suggests is most natural. In this respect, the lattice invariant shears are exceptional, involving H not similar to any rotation. So, (3.22) gives a characterization of a subset of the possibilities, and we might hope for some characterization of other subsets, covering the lot.

IV. Cases with $H^N = 1$

If, as is the case for crystals, G admits non-trivial finite subgroups, we have the possibility of using H in one of these. There will then be a smallest integer such that

$$H^N = 1. \tag{4.1}$$

As indicated before, the case $N = 2$, associated with twins, as defined by PITTERI [1], is pretty well understood, so we are primarily interested in cases with $N > 2$. Let me review some classical arguments, for the reader who might be rusty. First, H leaves invariant at least one symmetric positive definite tensor. One way of determining it is to take any tensor S such that

$$S = S^T > 0, \tag{4.2}$$

and calculate the average

$$\langle S \rangle = \frac{1}{N} \sum_{M=1}^{N} (H^M)^T S H^M. \tag{4.3}$$

It is then easy to verify that

$$\langle S \rangle = \langle S \rangle^T > 0, \quad H^T \langle S \rangle H = \langle S \rangle. \tag{4.4}$$

Also, H must be similar to some rotation,

$$H = L^{-1} \tilde{R} L, \quad \tilde{R}^{-1} = \tilde{R}^T, \quad \det \tilde{R} = 1. \tag{4.5}$$

One way of determining L is to solve

$$L^T L = \langle S \rangle$$

for it. Then, (4.4) gives

$$H^T L^T L H = L^T L \to (L^{-T} H^T L^T)(L H L^{-1}) = 1, \tag{4.6}$$

with this and $(2.7)_2$, (4.5) is immediate. For crystals, the possibilities are

$$N = 1, 2, 3, 4, 6. \tag{4.7}$$

As mentioned before, H should now also be similar to a matrix of integers, so

$$\operatorname{tr} H = \operatorname{tr} \tilde{R} = 1 + 2 \cos \chi = \text{integer}, \tag{4.8}$$

χ being the angle of rotation. Determine the angles possible, and you get (4.7). The result, and arguments used to get it are familiar to crystallographers, since they are used in deducing the crystallographic point groups.

Further, according to the theory of material symmetry, one can subject H to a similarity transformation, by making a change of reference configuration, as is discussed by NOLL [6]. So, there is no real loss of generality in choosing a reference such that

$$H = \tilde{R}, \tag{4.9}$$

I'll assume thus. For crystals, this implies that H then belongs to one of the crystallographic point groups, rather familiar things. However, it will generally not be the point group commonly associated with the natural state considered.

With (4.9), we can employ (3.22) in the form

$$RF_1\tilde{R} = R_1F_1R_2 \tag{4.10}$$

with R_1 and R_2 satisfying (3.21). So, we first consider

$$R_1F_1R_2 = (1 + a \otimes n)\, F_1, \tag{4.11}$$

a standard twinning problem. As is known, and clear from the previous discussion, it will have two solutions. Since R_2 is a 180° rotation, we can represent it in the form

$$R_2 = -1 + 2h \otimes h, \quad h \cdot h = 1, \tag{4.12}$$

h being the axis. Rotating an unstressed body leaves it unstressed. We can introduce a normalizing condition based on this, and do this by taking

$$F_1 = F_1^T = \sqrt{C_1}, \tag{4.13}$$

the usual positive definite symmetric square root. Then, the calculations give, for one solution,

$$\begin{aligned} n &= F_1^{-1}h/\|F_1^{-1}h\|, \\ a &= 2(n - \|F_1^{-1}h\|\, F_1h), \\ R_1 &= -1 + 2n \otimes n, \end{aligned} \tag{4.14}$$

clearly consistent with (3.21). Similarly, the second solution is

$$\begin{aligned} a &= \alpha F_1h, \\ n &= 2\alpha^{-1}(F_1^{-1}h - F_1h/\|F_1h\|^2), \\ R_1 &= -1 + 2a \otimes a/\|a\|^2, \end{aligned} \tag{4.15}$$

the scalar α being fixed by the condition that $n \cdot n = 1$, which gives

$$\alpha^2 = 4(\|F_1^{-1}h\|^2 - \|F_1h\|^{-2}). \tag{4.16}$$

Of course, it is here presumed that R_2 does not leave invariant C_1,

$$R_2C_1R_2 \neq C_1. \tag{4.17}$$

With (4.12) and (4.13), this is equivalent to

$$R_2F_1R_2 \neq F_1, \tag{4.18}$$

and it guarantees that the two solutions are well defined, with $a \neq 0$. This gives the solutions explicitly, for $N = 2$; take $R_2 = \tilde{R}$.

Otherwise, for (4.10) to hold, it is necessary that

$$\tilde{R}^T C_1 \tilde{R} = R_2 C_1 R_2, \tag{4.19}$$

or

$$\hat{R}^T C_1 \hat{R} = C_1, \tag{4.20}$$

where

$$R = \tilde{R} R_2, \quad \tilde{R} = \hat{R} R_2, \tag{4.21}$$

and clearly we don't want to have $\hat{R} = 1$. Mathematically, it is not so hard to satisfy these conditions. For example, any rotation can be written as the product of two 180° rotations, so take $\tilde{R}$ to be what you will; we will it to be a rotation, with angle $2\pi/N$. So represent it, taking one rotation as R_2, the other as $\hat{R}$; these need not belong to G. Then pick C_1 to have the axis of $\hat{R}$ as an eigenvector, and satisfy (4.17). Then (4.20) will hold. Some other forms of $\tilde{R}$ can be accommodated, by letting C_1 have two equal eigenvalues; if all three coincide, (4.17) can't hold, and we must respect it. Physically, we would like C_1 to be a natural state, for some realistic form of the constitutive function, which should be invariant under a group containing $\tilde{R}$. Possibly, some limitation might be deduced from this, but I am not sure of this.

If we satisfy (4.19), we will have some values for R_2 and F_1, information needed to use (4.14) or (4.15), either giving us values for R_1, *etc.* Also, with (4.13), (4.19) is equivalent to

$$\tilde{R}^T F_1 \tilde{R} = R_2 F_1 R_2, \tag{4.22}$$

from which it follows that (4.10) holds, with

$$R = R_1 R_2 \tilde{R}^T. \tag{4.23}$$

To recapitulate briefly, we then have, the quantities determined as indicated,

$$\begin{aligned} R F_1 \tilde{R} &= R_1 R_2 \tilde{R}^T F_1 \tilde{R} \\ &= R_1 R_2^2 F_1 R_2 \\ &= R_1 F_1 R_2 \\ &= (1 + a \otimes n) F_1. \end{aligned} \tag{4.24}$$

Rather obviously, the analysis applies to any case where H is similar to a rotation, not just to cases where H belongs to a finite group. For the crystals, it is known that H is similar to a rotation if and only if it belongs to a finite group and, at least so far, most of the interest in such phenomena has centered around the crystals.

Clearly, the set of possibilities here considered has a rather clear group-theoretic status. With the apparatus given, it is easy enough to construct illustrative examples. I have found some fitting the description of rotational twins given BARRETT & MASSALSKI [2, p. 406], for example, with the property that $R^N = 1$. However, I have not found time for a deeper study of the set.

There remain other logical possibilities, cases where H is not similar to a rotation and $R \neq 1$. One can concoct examples, by composing twinning solutions with suitably related lattice invariant shears, and one might replace the twinning solutions by the other types considered above. I haven't noticed examples of an essentially different kind, and observations known to me suggest nothing else. Lacking good theorems or more illuminating examples, I can say little more. While the basic formulation is quite elementary, the questions it raises are not so easy to settle.

Acknowledgement. This material is based on work supported by the National Science Foundation under Grant No. MEA-8304750.

References

1. PITTERI, M., On the kinematics of mechanical twinning in crystals, *Arch. Rational Mech. Anal.* **88**, 25–58 (1985).
2. BARRETT, C. & MASSALSKI, T. B., *Structure of Metals*, 3rd ed. McGraw Hill, Inc., 1966.
3. ERICKSEN, J. L., "Special Topics in Nonlinear Elastostatics", in Advances in Applied Mechanics (ed. C.-S. YIH), vol. **17**, New York: Academic Press, 1977.
4. ERICKSEN, J. L., *The Cauchy and Born hypotheses for crystals*, MRC Technical Summary Report # 2591, University of Wisconsin, 1983.
5. ERICKSEN, J. L., Stress-free joints, *J. Elasticity* **13**, 3–15 (1983).
6. NOLL, W., A mathematical theory of the mechanical behavior of continuous media, *Arch. Rational Mech. Anal.* **2**, 197–226 (1958).

Department of Aerospace
Engineering and Mechanics
and School of Mathematics
University of Minnesota
Minneapolis

(Received August 7, 1984)

Principal Waves in Orthotropic Elastic Membranes

H. COHEN & C.-C. WANG

Dedicated to W. Noll on the Occasion of his Sixtieth Birthday

1. Introduction

Acceleration waves in isotropic elastic membranes were studied by POP & WANG in [1]. Their results include propagation conditions and amplitude equations for tangential and transverse acceleration waves in general propagating into regions of a membrane at rest in the physical space. Special attention was given to certain waves called principal waves, whose directions of propagation are proper vectors of the characteristic Riemannian metric of the isotropic elastic membrane. Amplitude vectors of principal waves are either parallel or orthogonal to the direction of propagation. Hence principal waves are longitudinal waves, shear waves, or transverse waves.

In this paper we extend results of [1] to waves in orthotropic elastic membranes, especially principal waves.

As noted by POP & WANG [1], their results for waves in isotropic elastic membranes are comparable to those obtained by BOWEN & WANG [2] for waves in isotropic elastic (3-dimensional) material bodies. Later BOWEN & WANG [3] extended the results in [2] to waves in orthotropic elastic material bodies. Our extension of the results of [1] is just like the extension of the results of [2] to those of [3], except that here the waves are in membranes.

As in [1] we denote a membrane by $\mathscr{M}$, which is assumed to be a connected, oriented, 2-dimensional differentiable manifold. We denote the physical space by $\mathscr{E}$ and the physical translation space by $\mathscr{V}$. For simplicity, we pick a fixed origin $\mathbf{0}$ in $\mathscr{E}$ so that any location in $\mathscr{E}$ may be characterized by its position vector relative to $\mathbf{0}$. Then a configuration of $\mathscr{M}$ is given by an embedding $\boldsymbol{r}\colon \mathscr{M} \to \mathscr{V}$, where $\boldsymbol{r}(p)$ denotes the position vector of any membrane point $p \in \mathscr{M}$. As usual, the tangent space of $\mathscr{M}$ at p is denoted by $\mathscr{M}_p$. Then a local configuration of p is given by a (linear) embedding $\boldsymbol{v}_p\colon \mathscr{M}_p \to \mathscr{V}$. In particular, a configuration $\boldsymbol{r}$ gives rise to a field of local configurations, $\boldsymbol{r}_{*p}\colon \mathscr{M}_p \to \mathscr{V}, p \in \mathscr{M}$.

We assume that the membrane points $p \in \mathscr{M}$ are elastic. This means that the stress tensor $\underset{\sim}{\boldsymbol{T}}$ at the position $\boldsymbol{r}(p)$ of p in any configuration $\boldsymbol{r}$ is determined by the local configuration $\boldsymbol{r}_{*p}$. Thus the constitutive equation of p takes the form

$$T = \hat{T}(\boldsymbol{v}_p, p), \tag{1.1}$$

where $\hat{T}(.,p)$ is called the response function of p. In [1], the membrane points $p \in \mathcal{M}$ are required to be isotropic, *i.e.* the symmetry group of p contains all orthogonal tensors with respect to a certain characteristic inner product on $\mathcal{M}_p$. In this paper, we relax the symmetry requirement from isotropy to orthotropy.

As defined in [4] the symmetry group of an orthotropic membrane point p contains reflections of $\mathcal{M}_p$ in two independent directions, say ξ_α, $\alpha = 1, 2$, called the principal axes of p. We define the reflections $\boldsymbol{\pi}_\alpha: \mathcal{M}_p \to \mathcal{M}_p$, $\alpha = 1, 2$, by

$$\boldsymbol{\pi}_1 \xi_1 = -\xi_1, \quad \boldsymbol{\pi}_1 \xi_2 = \xi_2,$$

and

$$\boldsymbol{\pi}_2 \xi_1 = \xi_1, \quad \boldsymbol{\pi}_2 \xi_2 = -\xi_2. \tag{1.2}$$

Then p is an orthotropic elastic membrane point if the response function of p obeys the conditions

$$\hat{T}(\boldsymbol{v}_p \boldsymbol{\pi}_\alpha, p) = \hat{T}(\boldsymbol{v}_p, p), \quad \alpha = 1, 2, \tag{1.3}$$

for all local configurations $\boldsymbol{v}_p$ of p. Clearly, isotropic points are orthotropic, but orthotropic points need not be isotropic.

Let p be an orthotropic membrane point. Then as in [4] we call a local configuration $\boldsymbol{\varkappa}_p$ undistorted if the principal axes ξ_α, $\alpha = 1, 2$, are orthogonal in $\boldsymbol{\varkappa}_p(\mathcal{M}_p)$. In this case, the response functions relative to $\boldsymbol{\varkappa}_p$ are invariant with respect to (orthogonal) reflections in $\boldsymbol{\varkappa}_p(\xi_\alpha)$, $\alpha = 1, 2$. Specifically, let $\boldsymbol{v}_p$ be represented by the deformation gradient $\boldsymbol{F}: \boldsymbol{\varkappa}_p(\mathcal{M}_p) \to \mathcal{V}$ relative to $\boldsymbol{\varkappa}_p$, *viz*,

$$\boldsymbol{v}_p = \boldsymbol{F}\boldsymbol{\varkappa}_p. \tag{1.4}$$

Then the response function $\hat{T}(.,p)$ may be represented by

$$\hat{T}(\boldsymbol{v}_p, p) = \hat{T}(\boldsymbol{F}\boldsymbol{\varkappa}_p, p) = \hat{T}_\varkappa(\boldsymbol{F}, p), \tag{1.5}$$

where $\hat{T}_\varkappa(.,p)$ is called the response function of p relative to $\boldsymbol{\varkappa}_p$. For an undistorted $\boldsymbol{\varkappa}_p$ the reflections $\boldsymbol{P}_\alpha$ in the directions $\boldsymbol{e}_\alpha = \boldsymbol{\varkappa}_p(\xi_\alpha)$ as defined by

$$\boldsymbol{P}_1 \boldsymbol{e}_1 = -\boldsymbol{e}_1, \quad \boldsymbol{P}_1 \boldsymbol{e}_2 = \boldsymbol{e}_2,$$
$$\boldsymbol{P}_2 \boldsymbol{e}_1 = \boldsymbol{e}_1, \quad \boldsymbol{P}_2 \boldsymbol{e}_2 = -\boldsymbol{e}_2, \tag{1.6}$$

are orthogonal tensors on the plane $\boldsymbol{\varkappa}_p(\mathcal{M}_p)$. Substituting (1.5) into (1.3), we see that $\hat{T}_\varkappa(.,p)$ obeys the conditions

$$\hat{T}_\varkappa(\boldsymbol{F}\boldsymbol{P}_\alpha, p) = \hat{T}_\varkappa(\boldsymbol{F}, p), \quad \alpha = 1, 2, \tag{1.7}$$

for all deformation gradients $\boldsymbol{F}$ relative to $\boldsymbol{\varkappa}_p$.

We assume that $\mathcal{M}$ has a configuration $\boldsymbol{r}$ in which the local configurations $\boldsymbol{r}_{*p}$, $p \in \mathcal{M}$, are all undistorted. Then there is an orthogonal surface coordinate system (θ^α) whose natural basis vectors $\boldsymbol{e}_\alpha$, $\alpha = 1, 2$, are principal axes in $\boldsymbol{r}_{*p}(\mathcal{M}_p)$ for all $p \in \mathcal{M}$. A typical principal wave in $\boldsymbol{r}(\mathcal{M})$ is one whose wave fronts are coordinate curves in (θ^α), say $\theta^2 = \text{const.}$, while their normal trajectories are the other coordinate curves, say $\theta^1 = \text{const.}$

Geometry and kinematics of moving membranes are treated in detail in [5], and many useful formulas are summarized in [1, Sec. 2, 3]. We shall make use of those formulas directly in this paper. Compatibility conditions for tensor fields on moving membranes are derived in [1, Sec. 4] and in reference [6, Part B]. We shall make use of those conditions directly in this paper also. We keep the notations for this paper the same as those in [1] as much as possible.

Before closing this section, we remark that the class of orthotropic materials is a fairly large one, including not only isotropic or transversely isotropic materials, but also many types of solid crystals as well as fluid crystals. A recent work [7] on principal waves in elastic subfluids is based entirely on results in [3] for waves in 3-dimensional orthotropic elastic materials. Similar application of results of this paper can be made for waves in various special types of orthotropic elastic membranes, such as subfluid membranes.

2. The Field of Response Functions and Their Gradients

In continuum mechanics it is generally assumed that the response function $\hat{T}(.\,, p)$ of any membrane point $p \in \mathcal{M}$ obeys the following condition of material frame-indifference:

$$\hat{T}(Q\boldsymbol{v}_p, p) = Q\hat{T}(\boldsymbol{v}_p, p)\, Q^T \tag{2.1}$$

for any rotation $Q\colon \boldsymbol{v}_p(\mathcal{M}_p) \to \mathcal{V}$ of $\boldsymbol{v}_p(\mathcal{M}_p)$ in $\mathcal{V}$. Then the response function $\hat{T}_\varkappa(.\,, p)$ relative to $\varkappa_p$ obeys the similar condition:

$$\hat{T}_\varkappa(QF, p) = Q\hat{T}_\varkappa(F, p)\, Q^T. \tag{2.2}$$

Taking the polar decomposition of the deformation gradient F relative to $\varkappa_p$, *viz*,

$$F = RU, \tag{2.3}$$

where $U\colon \varkappa_p(\mathcal{M}_p) \to \varkappa_p(\mathcal{M}_p)$ is positive-definite and symmetric and where $R\colon \varkappa_p(\mathcal{M}_p) \to \boldsymbol{v}_p(\mathcal{M}_p)$ is a rotation, we obtain from (2.2) the representation

$$\hat{T}_\varkappa(F, p) = RT_\varkappa(U, p)\, R^T. \tag{2.4}$$

We can rewrite the representation (2.4) in the form

$$T = FH(C, p)\, F^T, \tag{2.5}$$

where $C = F^TF = U^2$ is the right Cauchy-Green tensor of the deformation gradient F, and where

$$H(C, p) = C^{-\frac{1}{2}}\hat{T}_\varkappa(C^{\frac{1}{2}}, p)\, C^{-\frac{1}{2}}. \tag{2.6}$$

Notice that $C\colon \varkappa_p(\mathcal{M}_p) \to \varkappa_p(\mathcal{M}_p)$ is positive-definite and symmetric, and its square root $C^{\frac{1}{2}}$ is just the right stretch tensor U. From (2.6) it is evident that $H(.\,, p)$ depends on the local reference configuration $\varkappa_p$, but for simplicity we drop

$\varkappa_p$ in the notation $H(., p)$. The meaning of the relative response function $H(\cdot, p)$ may be explained as follows:

Let $\{f_\alpha, \alpha = 1, 2\}$ be any basis for $\varkappa_p(\mathcal{M}_p)$. Then we can characterize a deformation $F: \varkappa_p(\mathcal{M}_p) \to v_p(\mathcal{M}_p)$ in general by the basis

$$e_\alpha = F f_\alpha, \quad \alpha = 1, 2. \tag{2.7}$$

In particular, the covariant components of the tensor C relative to $\{f_\alpha\}$ are given by

$$C_{\alpha\beta} = f_\alpha \cdot C f_\beta = f_\alpha \cdot F^T F f_\beta = F f_\alpha \cdot F f_\beta = e_\alpha \cdot e_\beta. \tag{2.8}$$

Thus $C: \varkappa_p(\mathcal{M}_p) \to \varkappa_p(\mathcal{M}_p)$ has the component form

$$C = (e_\alpha \cdot e_\beta) f^\alpha \otimes f^\beta, \tag{2.9}$$

where $\{f^\alpha\}$ denotes the reciprocal basis of $\{f_\alpha\}$. Like C, the tensor function $H(C, p)$ acts on $\varkappa_p(\mathcal{M}_p)$. Hence $H(C, p)$ has a component form relative to the basis $\{f_\alpha\}$,

$$H(C, p) = H^{\alpha\beta}(C, p) f_\alpha \otimes f_\beta, \tag{2.10}$$

where $H^{\alpha\beta}$ are the contravariant components of H. Substituting (2.10) into (2.5), we obtain

$$T = F(H^{\alpha\beta} f_\alpha \otimes f_\beta) F^T = H^{\alpha\beta} F f_\alpha \otimes F f_\beta = H^{\alpha\beta} e_\alpha \otimes e_\beta, \tag{2.11}$$

which means that $H^{\alpha\beta}$ are also the contravariant components of the stress tensor T relative to the basis $\{e_\alpha\}$ in the deformed configuration $v_p(\mathcal{M}_p)$.

For application to our problem we pick the undistorted configuration r as the reference configuration. More precisely, the situation is as follows: Ahead of the wave front the membrane is at rest in an undistorted configuration r, which is taken to be the reference configuration. Behind the wave front the membrane is moving in space, and the instantaneous configurations are generally not undistorted. A typical instantaneous configuration of the membrane is sketched in the adjoining figure 1. The reference configuration r in our analysis is the configuration r of the membrane before the arrival of the wave. We are primarily interested in the motion of the wave front and in the jump of the acceleration at the wave front, where the membrane points p are undistorted. Since a convected

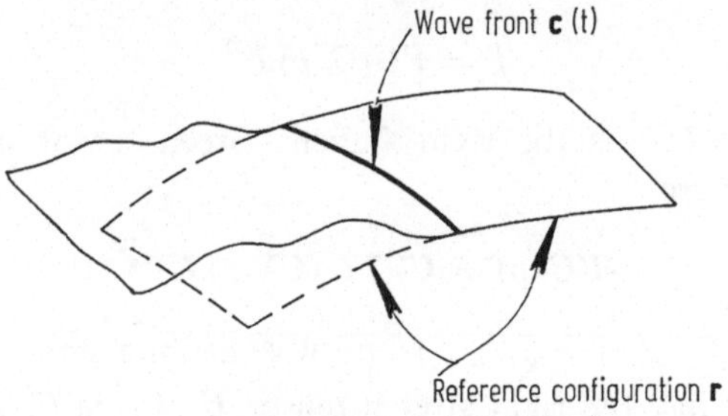

Fig. 1. Wave propagating into membrane at rest

coordinate system (θ^α) is used, the covariant components $C_{\alpha\beta}$ of C are simply the $e_{\alpha\beta}$ components of the surface metric as defined by

$$e_{\alpha\beta} = \boldsymbol{e}_\alpha \cdot \boldsymbol{e}_\beta. \tag{2.12}$$

From (2.11) the stress tensor field has the component form

$$\boldsymbol{T} = H^{\alpha\beta}(e_{\gamma\delta}, \theta^\mu)\, \boldsymbol{e}_\alpha \otimes \boldsymbol{e}_\beta, \tag{2.13}$$

where $e_{\gamma\delta}$ are the values of the argument $C_{\gamma\delta}$ while θ^μ characterize the point $p \in \mathcal{M}$.

It should be noted that the representation (2.13) for the stress tensor field is valid for all elastic membranes, regardless of any material symmetry. For materially uniform isotropic elastic membranes considered in [1] the stress tensor is given by a representation of the form (2.13) such that the component functions $H^{\alpha\beta}$ take the special forms (*cf.* [1, (5.20)])

$$H^{\alpha\beta} = \phi_0(I, II)\, e^{\alpha\beta} + \phi_1(I, II)\, G^{\alpha\beta}, \tag{2.14}$$

where I, II, $e^{\alpha\beta}$, and $G^{\alpha\beta}$ are functions of $e_{\gamma\delta}$ and $g_{\xi\eta}$ which depend on θ^μ. ($g_{\xi\eta}$ are components of the characteristic Riemannian metric which depends on θ^μ). For orthotropic membranes the component functions $H^{\alpha\beta}$ need not take the special form (2.14), but they must obey certain symmetry conditions, as we now explain.

Since the reference configuration $\boldsymbol{r}$ is undistorted, the reflections $\boldsymbol{P}_\alpha$ defined by (1.6) are orthogonal tensors. Hence if the polar decomposition of $\boldsymbol{F}$ is given by (2.3), then that of $\boldsymbol{F}\boldsymbol{P}_\alpha$ is given by

$$\boldsymbol{F}\boldsymbol{P}_\alpha = (\boldsymbol{R}\boldsymbol{P}_\alpha)\,(\boldsymbol{P}_\alpha^T\boldsymbol{U}\boldsymbol{P}_\alpha). \tag{2.15}$$

In (2.15) we can drop the transpose sign T on $\boldsymbol{P}_\alpha$, since $\boldsymbol{P}_\alpha$, being an orthogonal reflection, is symmetric. Substituting (2.15) into (1.7) and using (2.2), we obtain

$$\boldsymbol{P}_\alpha\hat{\boldsymbol{T}}_\varkappa(\boldsymbol{U}, p)\, \boldsymbol{P}_\alpha = \hat{\boldsymbol{T}}_\varkappa(\boldsymbol{P}_\alpha\boldsymbol{U}\boldsymbol{P}_\alpha, p). \tag{2.16}$$

Substituting (2.16) into (2.6), we then get

$$\boldsymbol{P}_\alpha\boldsymbol{H}(\boldsymbol{C}, p)\, \boldsymbol{P}_\alpha = \boldsymbol{H}(\boldsymbol{P}_\alpha\boldsymbol{C}\boldsymbol{P}_\alpha, p). \tag{2.17}$$

This condition is the same as the condition [3, (2.16)] in the 3-dimensional theory of orthotropic material bodies. As a result, $\boldsymbol{H}(\cdot, p)$ and its gradients obey essentially the same restrictions as those in [3].

Specifically, in the reference configuration $\boldsymbol{r}_{*p}(\mathcal{M}_p)$, $\boldsymbol{C}$ is the identity tensor and hence its covariant, contravariant, and mixed components are $e_{\mu\nu}$, $e^{\mu\nu}$, and δ^μ_ν, respectively. The condition (2.17) implies that the contravariant, covariant, and mixed component matrices of $\boldsymbol{T}$ are all diagonal matrices, *e.g.*,

$$[T^{\alpha\beta}] = \operatorname{diag}(T^{11}, T^{22}), \tag{2.18}$$

where T^{11} and T^{22} are given by the values of H^{11} and H^{22} at the arguments θ^μ and $C_{\gamma\delta} = e_{\gamma\delta}$ such that

$$e_{12} = \boldsymbol{e}_1 \cdot \boldsymbol{e}_2 = 0. \tag{2.19}$$

As usual, the physical components of T relative to the orthonormal basis $\{e\langle\alpha\rangle\}$ in the direction of the orthogonal basis $\{e_\alpha\}$ are related to the contravariant components by

$$T\langle 11\rangle = T^{11}e_{11},\, T\langle 22\rangle = T^{22}e_{22}. \tag{2.20}$$

Next, the first gradients of the relative response functions $H^{\alpha\beta}(C_{\gamma\delta}, \theta^\mu)$, evaluated at the undistorted reference configuration, have the following non-zero components:

$$H^{\alpha\alpha\beta\beta} = \frac{\partial H^{\alpha\alpha}}{\partial C_{\beta\beta}}, \tag{2.21}$$

where α and β are free indices, and

$$H^{\alpha\beta\alpha\beta} = \frac{\partial H^{\alpha\beta}}{\partial C^{\alpha\beta}}, \tag{2.22}$$

where α and β are distinct free indices. As usual, the first gradients obey the following symmetry conditions:

$$H^{\alpha\beta\alpha\beta} = H^{\beta\alpha\alpha\beta} = H^{\alpha\beta\beta\alpha} = H^{\beta\alpha\beta\alpha}. \tag{2.23}$$

The arguments of the functions in (2.21)–(2.23) are evaluated at $C_{\gamma\delta} = e_{\gamma\delta}$ and θ^μ, where $e_{\gamma\delta}$ obey the condition (2.19).

We have to be careful in our interpretation of the meaning of the first gradients. First, the partial derivatives in (2.22) are taken from an extension of the functions $H^{\alpha\beta}(C_{\gamma\delta}, \theta^\mu)$ to all values of $C_{\gamma\delta}$. As shown in reference [8], there are extensions such that

$$\frac{\partial H^{\alpha\beta}}{\partial C_{\gamma\delta}} = \frac{\partial H^{\alpha\beta}}{\partial C_{\delta\gamma}}, \tag{2.24}$$

at all symmetric arguments $C_{\gamma\delta}$. Furthermore, for such extensions the partial derivatives in (2.22) and (2.24) at symmetric $C_{\gamma\delta}$ are independent of the choice of the extension. Hence the functions $H^{\alpha\beta\alpha\beta}$ are well defined by (2.22).

Next, we must remember that the basis $\{e_\alpha\}$ is only orthogonal and is generally not orthonormal in the undistorted reference configuration. Hence the first gradients defined by (2.21) and (2.22) are not elasticities in the usual sense. However, the usual elasticities can be calculated in terms of the first gradients, as we now explain.

In the reference configuration $r_{*p}(\mathcal{M}_p)$ we introduce the orthonormal physical basis $\{e\langle\alpha\rangle\}$ associated with the orthogonal basis $\{e_\alpha\}$, *viz*,

$$e\langle\alpha\rangle = e_\alpha/\sqrt{e_{\alpha\alpha}}, \tag{2.25}$$

where α is a free index. Now consider any deformation in the tangent plane $r_{*p}(\mathcal{M}_p)$,

$$F\colon r_{*p}(\mathcal{M}_p) \to r_{*p}(\mathcal{M}_p). \tag{2.26}$$

Such a deformation may be characterized by the physical component from.

$$F = F\langle\alpha\beta\rangle\, e\langle\alpha\rangle \otimes e\langle\beta\rangle, \tag{2.27}$$

where α and β are asummed from 1 to 2 as usual. The deformation $\boldsymbol{F}$ transforms the basis vector $\boldsymbol{e}_\alpha$ to the vector

$$\boldsymbol{F}\boldsymbol{e}_\alpha = \sqrt{e_{\alpha\alpha}}\, F\langle\beta\alpha\rangle\, \boldsymbol{e}\langle\beta\rangle, \tag{2.28}$$

Hence the covariant components of the Cauchy-Green tensor $\boldsymbol{C}$ in the deformed configuration are

$$C_{\alpha\beta} = \boldsymbol{F}\boldsymbol{e}_\alpha \cdot \boldsymbol{F}\boldsymbol{e}_\beta = \sqrt{e_{\alpha\alpha} e_{\beta\beta}}\, F\langle\eta\alpha\rangle\, F\langle\eta\beta\rangle, \tag{2.29}$$

where α and β are free indices while η is summed from 1 to 2. From (2.29), (2.28), and (2.13), the physical components of the stress tensor in the deformed configuration are given by

$$T\langle\tau\eta\rangle = H^{\alpha\beta}\left(\sqrt{e_{\gamma\gamma}e_{\delta\delta}}\, F\langle\lambda\gamma\rangle\, F\langle\lambda\,\delta\rangle, \theta^\mu\right) \sqrt{e_{\alpha\alpha}e_{\beta\beta}}\, F\langle\tau\alpha\rangle\, F\langle\eta\beta\rangle, \tag{2.30}$$

where $\tau, \eta, \gamma, \delta$, and μ are free indices while α, β, and λ are summed. Notice that (2.30) reduces to (2.20) when $\boldsymbol{F} = \boldsymbol{1}$ or, equivalently, when $F\langle\alpha\beta\rangle = \delta\langle\alpha\beta\rangle$. The elasticities in the usual sense are the first gradients of $T\langle\tau\eta\rangle$ with respect to $F\langle\phi\psi\rangle$, evaluated at $\boldsymbol{F} = \boldsymbol{1}$ or at $F\langle\phi\psi\rangle = \delta\langle\phi\psi\rangle$. Specifically, the elasticities are given by

$$\frac{\partial T\langle\tau\eta\rangle}{\partial F\langle\phi\psi\rangle} = 2\sqrt{e_{\tau\tau}e_{\eta\eta}e_{\phi\phi}e_{\psi\psi}}\, H^{\tau\eta\phi\psi} + \sqrt{e_{\psi\psi}e_{\eta\eta}}\, H^{\psi\eta}\, \delta\langle\tau\phi\rangle + \sqrt{e_{\psi\psi}e_{\tau\tau}}\, H^{\psi\tau}\, \delta\langle\eta\phi\rangle, \tag{2.31}$$

where we have used the condition (2.23).

From (2.31) we see that the following elasticities are generally not zero:

$$\frac{\partial T\langle\alpha\alpha\rangle}{\partial F\langle\alpha\alpha\rangle} = 2e_{\alpha\alpha}(e_{\alpha\alpha}H^{\alpha\alpha\alpha\alpha} + H^{\alpha\alpha}), \tag{2.32}$$

where α is a free index,

$$\frac{\partial T\langle\alpha\alpha\rangle}{\partial F\langle\beta\beta\rangle} = 2e_{\alpha\alpha}e_{\beta\beta}H^{\alpha\alpha\beta\beta}, \tag{2.33}$$

where α and β are distinct indices, and

$$\frac{\partial T\langle\alpha\beta\rangle}{\partial F\langle\alpha\beta\rangle} = \frac{\partial T\langle\beta\alpha\rangle}{\partial F\langle\alpha\beta\rangle} = e_{\beta\beta}(2e_{\alpha\alpha}H^{\alpha\beta\alpha\beta} + H^{\beta\beta}), \tag{2.34}$$

where α and β are again distinct free indices. From (2.34) we obtain a universal relation for orthotropic membranes in undistorted states:

$$\frac{\partial T\langle\alpha\beta\rangle}{\partial F\langle\alpha\beta\rangle} - \frac{\partial T\langle\beta\alpha\rangle}{\partial F\langle\beta\alpha\rangle} = e_{\beta\beta}H^{\beta\beta} - e_{\alpha\alpha}H^{\alpha\alpha} = T\langle\beta\beta\rangle - T\langle\alpha\alpha\rangle, \tag{2.35}$$

where we have used the condition (2.20).

It should be noted that the formulas (2.31), (2.32), (2.33), (2.34), and (2.35) have the same forms as [3, (2.15), (2.19), (2.20), (2.21), and (2.22)], respectively, in the 3-dimensional theory. The presence of the metric $e_{\alpha\alpha}$ in the formulas here is due to the fact that the basis $\{\boldsymbol{e}_\alpha\}$ is only orthogonal and is generally not orthonormal. We are forced to use a convected coordinate system (θ^α) here to describe the motion of a membrane, since we cannot define any Eulerian coordinate system for a membrane whose surface is not fixed in the physical space. The orthogonal system (θ^α) is the best we can use in such a circumstance.

By following exactly the same argument as in [3], we obtain for an orthotropic hyperelastic membrane the additional universal conditions,

$$\frac{\partial T\langle\alpha\alpha\rangle}{\partial F\langle\beta\beta\rangle}-\frac{\partial T\langle\beta\beta\rangle}{\partial F\langle\alpha\alpha\rangle}=T\langle\beta\beta\rangle-T\langle\alpha\alpha\rangle, \tag{2.36}$$

cf. [3, (2.45)]. Combining (2.36) with (2.35), we then obtain

$$\frac{\partial T\langle\alpha\alpha\rangle}{\partial F\langle\beta\beta\rangle}-\frac{\partial T\langle\beta\beta\rangle}{\partial F\langle\alpha\alpha\rangle}+\frac{\partial T\langle\beta\alpha\rangle}{\partial F\langle\beta\alpha\rangle}-\frac{\partial T\langle\alpha\beta\rangle}{\partial F\langle\alpha\beta\rangle}=0, \tag{2.37}$$

cf. [3, (2.47)].

Second gradients of the relative response function $H^{\alpha\beta}(C_{\gamma\delta},\theta^\mu)$ also obey essentially the same restrictions as those in [3]. Specifically, the second gradients of $H^{\alpha\beta}(C_{\gamma\delta},\theta^\mu)$, evaluated at the undistorted state with $C_{\gamma\delta}=e_{\gamma\delta}$ satisfying (2.19), have the following non-zero components

$$H^{\alpha\alpha\beta\beta\gamma\gamma}=\frac{\partial^2 H^{\alpha\alpha}}{\partial C_{\beta\beta}\,\partial C_{\gamma\gamma}}, \tag{2.38}$$

where α, β, γ are free indices, and

$$H^{\alpha\alpha\alpha\beta\alpha\beta}=\frac{\partial^2 H^{\alpha\alpha}}{\partial C_{\alpha\beta}\,\partial C_{\alpha\beta}},\qquad H^{\alpha\beta\alpha\beta\alpha\alpha}=\frac{\partial^2 H^{\alpha\beta}}{\partial C_{\alpha\beta}\,\partial C_{\alpha\alpha}}, \tag{2.39}$$

where α and β are distinct free indices. As usual the second gradients obey the following symmetry conditions:

$$H^{\alpha\beta\gamma\delta\mu\nu}=H^{\beta\alpha\gamma\delta\mu\nu}=H^{\alpha\beta\delta\gamma\mu\nu}=H^{\alpha\beta\gamma\delta\nu\mu}=H^{\alpha\beta\mu\nu\gamma\delta}. \tag{2.40}$$

By the same remark as before, the second gradients $H^{\alpha\beta\gamma\delta\mu\nu}$ of the relative response functions $H^{\alpha\beta}(C_{\gamma\delta},\theta^\mu)$ are not second order elasticities in the usual sense, but we can calculate the usual second order elasticities in terms of them. Specifically, we take second partial derivatives of (2.30) and evaluate the result at $\boldsymbol{F}=\mathbf{1}$ or at $F\langle\alpha\beta\rangle=\delta\langle\alpha\beta\rangle$, obtaining

$$\begin{aligned}\frac{\partial^2 T\langle\tau\eta\rangle}{\partial F\langle\phi\psi\rangle\,\partial F\langle\omega\sigma\rangle}&=4\sqrt{e_{\tau\tau}e_{\eta\eta}e_{\phi\phi}e_{\psi\psi}e_{\omega\omega}e_{\sigma\sigma}}\,H^{\tau\eta\phi\psi\omega\sigma}\\&+2\sqrt{e_{\tau\tau}e_{\eta\eta}e_{\psi\psi}e_{\sigma\sigma}}\,\delta\langle\phi\omega\rangle\,H^{\tau\eta\psi\sigma}+2\sqrt{e_{\sigma\sigma}e_{\eta\eta}e_{\phi\phi}e_{\psi\psi}}\,\delta\langle\tau\omega\rangle\,H^{\sigma\eta\phi\psi}\\&+2\sqrt{e_{\tau\tau}e_{\sigma\sigma}e_{\phi\phi}e_{\psi\psi}}\,\delta\langle\eta\omega\rangle\,H^{\tau\sigma\phi\psi}+2\sqrt{e_{\eta\eta}e_{\psi\psi}e_{\omega\omega}e_{\sigma\sigma}}\,\delta\langle\tau\phi\rangle\,H^{\eta\psi\omega\sigma}\\&+2\sqrt{e_{\tau\tau}e_{\psi\psi}e_{\omega\omega}e_{\sigma\sigma}}\,\delta\langle\eta\phi\rangle\,H^{\tau\psi\omega\sigma}\\&+\sqrt{e_{\psi\psi}e_{\sigma\sigma}}\,H^{\psi\sigma}(\delta\langle\tau\phi\rangle\,\delta\langle\eta\omega\rangle+\delta\langle\eta\phi\rangle\,\delta\langle\tau\omega\rangle).\end{aligned} \tag{2.41}$$

This formula has the form as [3, (2.32)] except that the metric $e_{\tau\tau}$ is present here because $\{e_\alpha\}$ is only an orthogonal basis. By virtue of (2.41) the non-zero second order elasticities are

$$\frac{\partial^2 T\langle\alpha\alpha\rangle}{\partial F\langle\alpha\alpha\rangle\,\partial F\langle\alpha\alpha\rangle} = 2e_{\alpha\alpha}(2e_{\alpha\alpha}e_{\alpha\alpha}H^{\alpha\alpha\alpha\alpha\alpha\alpha} + 5e_{\alpha\alpha}H^{\alpha\alpha\alpha\alpha} + H^{\alpha\alpha}), \tag{2.42}$$

where α is a free index, and

$$\frac{\partial^2 T\langle\alpha\alpha\rangle}{\partial F\langle\alpha\alpha\rangle\,\partial F\langle\beta\beta\rangle} = 4e_{\alpha\alpha}e_{\beta\beta}(e_{\alpha\alpha}H^{\alpha\alpha\alpha\alpha\beta\beta} + H^{\alpha\alpha\beta\beta}), \tag{2.43}$$

$$\frac{\partial^2 T\langle\alpha\alpha\rangle}{\partial F\langle\alpha\beta\rangle\,\partial F\langle\alpha\beta\rangle} = 2e_{\beta\beta}(2e_{\alpha\alpha}e_{\alpha\alpha}H^{\alpha\alpha\alpha\beta\alpha\beta} + e_{\alpha\alpha}H^{\alpha\alpha\beta\beta} + 4e_{\alpha\alpha}H^{\alpha\beta\alpha\beta} + H^{\beta\beta}), \tag{2.44}$$

$$\frac{\partial^2 T\langle\alpha\alpha\rangle}{\partial F\langle\alpha\beta\rangle\,\partial F\langle\beta\alpha\rangle} = 4e_{\alpha\alpha}e_{\beta\beta}(e_{\alpha\alpha}H^{\alpha\alpha\alpha\beta\alpha\beta} + H^{\alpha\beta\alpha\beta}), \tag{2.45}$$

$$\frac{\partial^2 T\langle\alpha\alpha\rangle}{\partial F\langle\beta\alpha\rangle\,\partial F\langle\beta\alpha\rangle} = 2e_{\alpha\alpha}e_{\alpha\alpha}(2e_{\beta\beta}H^{\alpha\alpha\alpha\beta\alpha\beta} + H^{\alpha\alpha\alpha\alpha}), \tag{2.46}$$

$$\frac{\partial^2 T\langle\alpha\beta\rangle}{\partial F\langle\alpha\beta\rangle\,\partial F\langle\alpha\alpha\rangle} = 2e_{\alpha\alpha}e_{\beta\beta}(2e_{\alpha\alpha}H^{\alpha\beta\alpha\beta\alpha\alpha} + H^{\beta\beta\alpha\alpha} + 2H^{\alpha\beta\alpha\beta}), \tag{2.47}$$

$$\frac{\partial^2 T\langle\alpha\beta\rangle}{\partial F\langle\beta\alpha\rangle\,\partial F\langle\alpha\alpha\rangle} = e_{\alpha\alpha}(4e_{\alpha\alpha}e_{\beta\beta}H^{\alpha\beta\alpha\beta\alpha\alpha} + 2e_{\alpha\alpha}H^{\alpha\alpha\alpha\alpha} + 2e_{\beta\beta}H^{\alpha\beta\alpha\beta} + H^{\alpha\alpha}), \tag{2.48}$$

where α and β are distinct free indices.

3. Propagation Conditions

As in [1] and [6] we define the jump of a field Φ across a singular curve $\boldsymbol{c}$ in a configuration $\boldsymbol{r}(\mathcal{M})$ by

$$[\Phi] = \Phi^- - \Phi^+, \tag{3.1}$$

where Φ^- and Φ^+ denote the limits of Φ at $\boldsymbol{c}$ taken, respectively, from the negative and the positive regions neighboring $\boldsymbol{c}$ in $\boldsymbol{r}(\mathcal{M})$. An acceleration wave is a singular curve $\boldsymbol{c}(t)$ for the second partial derivatives of the position vector $\boldsymbol{r}(t, \theta^\alpha)$. Thus the position vector $\boldsymbol{r}$, the velocity vector $\boldsymbol{v}$, the surface basis vectors $\boldsymbol{e}_\alpha$, and the surface unit normal

$$\boldsymbol{e}_3 = \boldsymbol{e}_1 \times \boldsymbol{e}_2 / \|\boldsymbol{e}_1 \times \boldsymbol{e}_2\|, \tag{3.2}$$

are continuous across the wave, but the acceleration vector $\boldsymbol{a}$, the velocity gradient

$$\boldsymbol{L} = \boldsymbol{v}_{,\alpha} \otimes \boldsymbol{e}^\alpha, \tag{3.3}$$

and the surface gradients $\boldsymbol{e}_{i,\alpha}$, as well as their derivatives, suffer jump discontinuities across $\boldsymbol{c}(t)$. In (3.3) $\{\boldsymbol{e}^\alpha\}$ denotes the reciprocal basis of $\{\boldsymbol{e}_\alpha\}$ and a comma denotes a partial derivative, *viz*, $,\alpha = \partial/\partial\theta^\alpha$. We introduce a fixed rectangular

Cartesian coordinate system (x^i) in $\mathscr{E}$ with origin $\mathbf{0}$ and Cartesian basis $\{\boldsymbol{h}_i\}$. Then we denote the partial derivative with respect to x^i by $,i$, *viz*, $,i = \partial/\partial x^i$, also. Of course, partial derivatives of vectors or tensors are taken with respect to the Euclidean structure on $\mathscr{E}$.

As in [1] we define the amplitude vector $\boldsymbol{\sigma}$ of an acceleration wave by

$$\boldsymbol{\sigma} = [\boldsymbol{a}] = [a^\alpha]\, \boldsymbol{e}_\alpha + [a^3]\, \boldsymbol{e}_3 . \tag{3.4}$$

Since we assume that the region of the membrane ahead of the wave is at rest in $\mathscr{E}$, the amplitude vector is simply the limit $\boldsymbol{a}^-$ of the acceleration field taken from the region of the membrane behind the wave. The amplitude vector $\boldsymbol{\sigma}$ obeys the propagation condition which may be derived from the equations of motion as follows.

From dynamical principles governing the motions of a membrane the stress tensor field must obey the field equations:

$$T^{\alpha\beta}_{\ \ |\alpha} + \varrho f^\beta = \varrho a^\beta, \tag{3.5}$$

and

$$\Omega_{\alpha\beta} T^{\alpha\beta} + \varrho f^3 = \varrho a^3, \tag{3.6}$$

where ϱ denotes the mass density (mass per unit surface area), f^i denote components of the body force relative to the basis $\{\boldsymbol{e}_i\}$, and $\Omega_{\alpha\beta}$ denote the second fundamental form of the membrane surface in the coordinate system (θ^μ). The vertical stroke in the leading term of (3.5) denotes the surface covariant derivative with respect to the surface metric tensor $e_{\gamma\delta}$. Specifically,

$$T^{\alpha\beta}_{\ \ |\alpha} = T^{\alpha\beta}_{\ \ ,\alpha} + T^{\gamma\beta} \left\{ {\alpha \atop \gamma\alpha} \right\} + T^{\alpha\gamma} \left\{ {\beta \atop \gamma\alpha} \right\}, \tag{3.7}$$

where $\left\{ {\beta \atop \gamma\alpha} \right\}$ denote the surface Christoffel symbols in (θ^μ). Of course, the balance of moment of momentum implies that the stress tensor $T^{\alpha\beta}$ is symmetric, and the conservation of mass implies that ϱ satisfies the continuity equation:

$$\dot{\varrho} + \varrho L^\alpha_\alpha = 0, \tag{3.8}$$

where the superimposed dot in the leading term denotes the material time derivative, *i.e.* partial derivative with respect to time at fixed convected coordinates θ^μ.

Using the constitutive equation (2.13), we can calculate the leading term on the right hand side of (3.7) by

$$T^{\alpha\beta}_{\ \ ,\alpha} = H^{\alpha\beta\gamma\delta} e_{\gamma\delta,\alpha} + H^{\alpha\beta}_{\ \ ,\alpha} . \tag{3.9}$$

The partial derivatives $e_{\gamma\delta,\alpha}$ may be calculated as usual by the condition

$$0 = e_{\gamma\delta|\alpha} = e_{\gamma\delta,\alpha} - e_{\lambda\delta} \left\{ {\lambda \atop \gamma\alpha} \right\} - e_{\gamma\lambda} \left\{ {\lambda \atop \delta\alpha} \right\}. \tag{3.10}$$

In [1] the jump of the Christoffel symbols and the second fundamental form are derived, *cf.* [1, (6.6)]

$$\left[\left\{ {\lambda \atop \alpha\beta} \right\} \right] = \frac{1}{U^2} \sigma^\lambda n_\alpha n_\beta, \qquad [\Omega_{\alpha\beta}] = \frac{1}{U^2} \sigma^3 n_\alpha n_\beta, \tag{3.11}$$

where U denotes the normal speed and n_α denotes the positive unit normal vector of the wave. Using (3.10) and (3.11), we can calculate the jump of $e_{\gamma\delta,\alpha}$ by

$$[e_{\gamma\delta,\alpha}] = \frac{1}{U^2}(\sigma_\delta n_\gamma + \sigma_\gamma n_\delta)\, n_\alpha. \tag{3.12}$$

Then from (3.9) the jump of $T^{\alpha\beta}_{\ ,\alpha}$ is given by

$$[T^{\alpha\beta}_{\ ,\alpha}] = \frac{2}{U^2} H^{\alpha\beta\gamma\delta}\sigma_\gamma n_\delta n_\alpha, \tag{3.13}$$

where we have used the symmetry condition (2.23). Notice that there is no jump for the term $H^{\alpha\beta}_{\ ,\alpha}$ since the relative response functions $H^{\alpha\beta}(\cdot, \theta^\mu)$ are assumed to depend smoothly on the convected surface coordinates θ^μ, and since the arguments $C_{\gamma\delta} = e_{\gamma\delta}$ of $H^{\alpha\beta}_{\ ,\alpha}$ are continuous. Taking the jump of (3.7) and using (3.11) and (3.13), we obtain

$$[T^{\alpha\beta}_{\ |\alpha}] = \frac{1}{U^2}(2H^{\alpha\beta\gamma\delta} + H^{\delta\beta}e^{\alpha\gamma} + H^{\alpha\delta}e^{\beta\gamma})\, \sigma_\gamma n_\delta n_\alpha. \tag{3.14}$$

Now taking the jump of the field equations (3.5) and (3.6) and using the preliminary results (3.14) and (3.11), we obtain the propagation conditions:

$$(2H^{\alpha\beta\gamma\delta} + H^{\delta\beta}e^{\alpha\gamma} + H^{\alpha\delta}e^{\beta\gamma})\, n_\delta n_\alpha \sigma_\gamma = \varrho U^2 \sigma^\beta, \tag{3.15}$$

and

$$H^{\alpha\beta} n_\alpha n_\beta \sigma^3 = \varrho U^2 \sigma^3. \tag{3.16}$$

In view of these conditions we define the surface acoustic tensor $A(\boldsymbol{n})$ by

$$A(\boldsymbol{n}) = (2H^{\alpha\beta\gamma\delta} + H^{\delta\beta}e^{\alpha\gamma} + H^{\alpha\delta}e^{\beta\gamma})\, n_\delta n_\alpha \boldsymbol{e}_\beta \otimes \boldsymbol{e}_\gamma. \tag{3.17}$$

and the spatial acoustic tensor $Q(\boldsymbol{n})$ by

$$Q(\boldsymbol{n}) = A(\boldsymbol{n}) + H^{\alpha\beta} n_\alpha n_\beta \boldsymbol{e}_3 \otimes \boldsymbol{e}_3. \tag{3.18}$$

Then the propagation conditions may be rewritten as

$$Q(\boldsymbol{n})\, \boldsymbol{\sigma} = \varrho U^2 \boldsymbol{\sigma}, \tag{3.19}$$

which means that the amplitude vector $\boldsymbol{\sigma}$ is a proper vector of the acoustic tensor $Q(\boldsymbol{n})$, and that ϱU^2 is the corresponding proper number.

As remarked in [1] the representation (3.18) for $Q(\boldsymbol{n})$ implies that the surface unit normal $\boldsymbol{e}_3$ is a possible amplitude vector provided that

$$H^{\alpha\beta} n_\alpha n_\beta > 0. \tag{3.20}$$

Such an acceleration wave is called a transverse wave. The other possible amplitude vectors of $Q(\boldsymbol{n})$ are contained in the tangent plane $\boldsymbol{r}_{*p}(\mathcal{M}_p)$ of the membrane surface and are proper vectors of the surface acoustic tensor $A(\boldsymbol{n})$. Such acceleration waves are called tangential waves. As usual, a tangential wave is called a longitudinal wave if $\boldsymbol{\sigma}$ is parallel to the unit normal $\boldsymbol{n}$ of the wave $c(t)$ and is called a shear wave if $\boldsymbol{\sigma}$ is parallel to the unit tangent vector $\boldsymbol{m}$ of $c(t)$.

As mentioned before in this paper we are primarily interested in principal waves in orthotropic membranes. Hence $\boldsymbol{n}$ is parallel to a principal axis, say $\boldsymbol{e}_2$. For such a wave the wave front $c(t)$ is a coordinate curve $\theta^2 = \text{const.}$ while the normal trajectory is a coordinate curve $\theta^1 = \text{const.}$ The unit normal $\boldsymbol{n}$ of such a principal wave has the component form

$$\boldsymbol{n} = n^2 \boldsymbol{e}_2 = n_2 \boldsymbol{e}^2, \tag{3.21}$$

where

$$n^2 n^2 = e^{22}, \quad n^2 n_2 = 1, \quad n_2 n_2 = e_{22}. \tag{3.22}$$

In (3.21) and (3.22) we have used the fact that $\{\boldsymbol{e}_\alpha\}$ is an orthogonal basis. Substituting (3.21) into (3.17) and using the restrictions on $H^{\alpha\beta}$ and $H^{\alpha\beta\gamma\delta}$ developed in the preceding section, we see that the surface acoustic tensor is a symmetric tensor with $\{\boldsymbol{e}\langle\alpha\rangle\}$ as its principal basis, *viz*, $\boldsymbol{A}(n^2\boldsymbol{e}_2)$ has the spectral form

$$\boldsymbol{A}(n^2\boldsymbol{e}_2) = A\langle 11\rangle\, \boldsymbol{e}\langle 1\rangle \otimes \boldsymbol{e}\langle 1\rangle + A\langle 22\rangle\, \boldsymbol{e}\langle 2\rangle \otimes \boldsymbol{e}\langle 2\rangle \tag{3.23}$$

where

$$A\langle 11\rangle = e_{22}(2e_{11}H^{1212} + H^{22}), \tag{3.24}$$

$$A\langle 22\rangle = 2e_{22}(e_{22}H^{2222} + H^{22}). \tag{3.25}$$

Consequently, a principal wave may be a shear wave with

$$\varrho U_{21}^2 = e_{22}(2e_{11}H^{1212} + H^{22}), \tag{3.26}$$

a longitudinal wave with

$$\varrho U_{22}^2 = 2e_{22}(e_{22}H^{2222} + H^{22}), \tag{3.27}$$

or a transverse wave with

$$\varrho U_{23}^2 = e_{22}H^{22}. \tag{3.28}$$

Comparing (3.26), (3.27) and (3.28) with (2.34), (2.32) and (2.20), we see that the speed of a principal shear wave is characterized by a shear elasticity at the undistorted state:

$$\varrho U_{21}^2 = \frac{\partial T\langle 12\rangle}{\partial F\langle 12\rangle}, \tag{3.29}$$

the speed of a principal longitudinal wave is characterized by a longitudinal elasticity at the undistorted state:

$$\varrho U_{22}^2 = \frac{\partial T\langle 22\rangle}{\partial F\langle 22\rangle}, \tag{3.30}$$

while the speed of a principal transverse wave is characterized by a principal stress at the undistorted state:

$$\varrho U_{23}^2 = T\langle 22\rangle. \tag{3.31}$$

Clearly, the formulas (3.29) and (3.30) are consistent with the formula [3, (3.14)] in the 3-dimensional theory, while the formulas (3.29), (3.30) and (3.31) generalize the formulas [1, (6.30), (6.29), and (6.31)] from principal waves in isotropic membranes to those in orthotropic membranes. As in the 3-dimensional theory, the speeds of principal shear waves obey the universal relation:

$$\varrho(U_{12}^2 - U_{21}^2) = T\langle 11\rangle - T\langle 22\rangle. \tag{3.32}$$

In view of (3.31) the speeds of principal shear waves and principal transverse waves are related by the universal relation:

$$U_{12}^2 - U_{21}^2 = U_{13}^2 - U_{23}^2. \tag{3.33}$$

These relations are valid for principal waves at all undistorted states in all orthotropic elastic membranes.

Since the principal wave speeds do not involve elasticities of the form (2.33), they do not give any information about validity of the condition (2.36) for hyperelasticity. This situation is the same as in the 3-dimensional theory. Elasticities of the form (2.33) will enter into the amplitude equations, however, as we shall see in the next section.

4. Amplitude Equations

To obtain the amplitude equations for tangential or transverse waves we start from the formulas:

$$2\frac{D\sigma^\alpha}{\delta t} = \frac{\sigma^\alpha}{U}\frac{DU}{\delta t} + [\dot{a}^\alpha] - U^2\hat{v}^\alpha \tag{4.1}$$

and

$$2\frac{D\sigma^3}{\delta t} = \frac{\sigma^3}{U}\frac{DU}{\delta t} + [\dot{a}^3] - U^2\hat{v}^3, \tag{4.2}$$

which are derived in detail in [1, Sec. 7], *cf.* [1, (7.13) and (7.14)]. In (4.1) and (4.2) $\hat{v}^\alpha$ and $\hat{v}^3$ are defined by

$$\hat{v}^\alpha = [v^\alpha_{|\beta\gamma}]\, n^\beta n^\gamma, \qquad \hat{v}^3 = [v^3_{|\beta\gamma}]\, n^\beta n^\gamma, \tag{4.3}$$

cf. [1, (4.21)], and $D/\delta t$ denotes the normal time derivative which is defined in general by [1, (3.35)], *viz*,

$$\frac{D\sigma^\alpha}{\delta t} = \frac{\delta(\sigma^\alpha)}{\delta t} + \sigma^\lambda \xi^\alpha_\lambda, \qquad \frac{D\sigma^3}{\delta t} = \frac{\delta(\sigma^3)}{\delta t}, \tag{4.4}$$

where $\delta/\delta t$ denotes the displacement derivative (*i.e.* time derivative along the normal trajectory of the wave), and where

$$\xi^\alpha_\lambda = \boldsymbol{e}^\alpha \cdot \frac{\delta \boldsymbol{e}_\lambda}{\delta t} = v^\alpha_{,\lambda} + \left\{{\alpha \atop \lambda\beta}\right\} U n^\beta, \tag{4.5}$$

cf. [1, (3.34)]. As noted in [1], (4.4) and (4.5) imply that the normal time derivative is just the surface projection of the displacement derivative for any surface tensor such as $\sigma^\alpha \boldsymbol{e}_\alpha$, and for the special case of a surface scalar such as σ^3 the normal time derivative coincides with the displacement derivative. Like ξ^α_λ, ξ^3_λ denotes the normal projection of the displacement derivative $\delta e_\lambda/\delta t$, *viz.*

$$\xi^3_\lambda = e^3 \cdot \frac{\delta e_\lambda}{\delta t} = v^3_{,\lambda} + \Omega_{\lambda\alpha} U n^\alpha, \tag{4.6}$$

where e^3 is the same as e_3, *cf.* (3.2). It should be noted that both ξ^α_λ and ξ^3_λ are continuous across the wave, *viz,*

$$[\xi^\alpha_\lambda] = 0, \qquad [\xi^3_\lambda] = 0, \tag{4.7}$$

cf. [1, (4.22) and (4.23)].

In view of (4.1) and (4.2) derivation of the amplitude equations concerns mainly calculation of the jumps $[\dot{a}^\alpha]$ and $[\dot{a}^3]$. These jumps may be obtained from the material time derivatives (*i.e.* partial time derivatives when the convected coordinates θ^μ are held constant) of the field equations (3.5) and (3.6). Specifically,

$$\varrho[\dot{a}^\beta] = \overline{[T^{\alpha\beta}_{|\alpha}]}^{\cdot} + \overline{[\varrho f^\beta]}^{\cdot} - \sigma^\beta[\dot{\varrho}], \tag{4.8}$$

and

$$\varrho[\dot{a}^3] = \overline{[\Omega_{\alpha\beta} T^{\alpha\beta}]}^{\cdot} + \overline{[\varrho f^3]}^{\cdot} - \sigma^3[\dot{\varrho}], \tag{4.9}$$

where we have used the product rule for $\overline{[\varrho a^\alpha]}^{\cdot}$ and $\overline{[\varrho a^3]}^{\cdot}$, and the condition that the region ahead of the wave is at rest.

As noted in [1] the jump $[\dot{\varrho}]$ may be obtained by taking the jump of the continuity equation (3.8)

$$[\dot{\varrho}] = \frac{\varrho}{U} \sigma^\xi n_\xi. \tag{4.10}$$

Then the jump $\overline{[\varrho f^i]}^{\cdot}$ may be calculated by the product rule

$$\overline{[\varrho f^i]}^{\cdot} = f^i[\dot{\varrho}] + \varrho[\dot{f}^i], \tag{4.11}$$

where f^i may be determined by its limit from the positive side of the wave front $c(t)$, *viz*

$$f^\beta = -\frac{1}{\varrho} T^{\alpha\beta +}_{|\alpha}, \qquad f^3 = -\frac{1}{\varrho} H^{\alpha\beta} \Delta_{\alpha\beta}, \tag{4.12}$$

where $\Delta_{\alpha\beta}$ denotes the positive limit of the second fundamental form $\Omega_{\alpha\beta}$, *i.e.*,

$$\Delta_{\alpha\beta} = \Omega^+_{\alpha\beta}. \tag{4.13}$$

From (3.7), (3.9), and (3.10) the positive limit $T^{\alpha\beta +}_{|\alpha}$ is given by

$$T^{\alpha\beta +}_{|\alpha} = e_{\gamma\lambda}(2H^{\alpha\beta\gamma\delta} + H^{\delta\beta} e^{\alpha\gamma} + H^{\alpha\delta} e^{\beta\gamma})\, \Gamma^\lambda_{\delta\alpha} + H^{\alpha\beta}_{,\alpha}, \tag{4.14}$$

where $\Gamma^\lambda_{\beta\alpha}$ denote the positive limit of the surface Christoffel symbols $\left\{ {\lambda \atop \beta\alpha} \right\}$, *i.e.*,

$$\Gamma^\lambda_{\beta\alpha} = \left\{ {\lambda \atop \beta\alpha} \right\}^+. \tag{4.15}$$

Notice that the combination of terms in the parentheses on the right hand side of (4.14) is the same as that on the right hand side of (3.17).

Next the jumps $[\dot{f}^i]$ may be calculated in the same way as in [1]. As usual we assume that

$$[f] = [\dot{f}] = \mathbf{0}. \tag{4.16}$$

Then using the kinematical formula [1, (2.27)] for the material time derivative of $\boldsymbol{f}$, we obtain

$$[\dot{f}^\beta] + f^\lambda [v^\beta_{|\lambda}] - f^3 [v^3_{,\lambda}]\, e^{\lambda\beta} = 0 \tag{4.17}$$

and

$$[\dot{f}^3] + f^\lambda [v^3_{,\lambda}] = 0, \tag{4.18}$$

cf. [1, (7.21) and (7.22)]. Hence from (4.12) we obtain

$$\varrho[\dot{f}^\beta] = -\frac{1}{U} T^{\alpha\lambda+}_{|\alpha} \sigma^\beta n_\lambda, \quad \varrho[\dot{f}^3] = -\frac{1}{U} T^{\alpha\lambda+}_{|\alpha} \sigma^3 n_\lambda, \tag{4.19}$$

where we have used the condition that the region ahead of the wave is at rest so that

$$[v^\lambda_{|\alpha}] = [v^\lambda_{,\alpha}] = -\frac{1}{U} \sigma^\lambda n_\alpha, \quad [v^3_{,\alpha}] = -\frac{1}{U} \sigma^3 n_\alpha, \tag{4.20}$$

and in $(4.19)_1$ we have used the condition that the amplitude vector $\boldsymbol{\sigma}$ is either a tangential vector

$$\boldsymbol{\sigma} = \sigma^\alpha \boldsymbol{e}_\alpha = \sigma_\alpha \boldsymbol{e}^\alpha, \quad \sigma_3 = \sigma^3 = 0, \tag{4.21}$$

or a transverse vector

$$\boldsymbol{\sigma} = \sigma^3 \boldsymbol{e}_3 = \sigma_3 \boldsymbol{e}^3, \quad \sigma_\alpha = \sigma^\alpha = 0. \tag{4.22}$$

The formulas $(4.19)_1$ or $(4.19)_2$, like the formulas $(4.12)_1$ or $(4.12)_2$, will be used for amplitude vector $\boldsymbol{\sigma}$ having the form (4.21) or (4.22), respectively.

Having calculated the last two terms in (4.8) and (4.9), we now turn our attention to the leading term on the right hand side of (4.8). The calculations leading to that term are very long, since we have to take the jump of the material time derivative of (3.7) with the leading terms on the right hand side given by (3.9). Using the chain rule, we have

$$\overline{T^{\alpha\gamma\{\beta \atop \gamma\alpha\}}}^{\cdot} = H^{\alpha\gamma\mu\nu} \dot{e}_{\mu\nu} \{{\beta \atop \gamma\alpha}\} + H^{\alpha\gamma} \overline{\{{\beta \atop \gamma\alpha}\}}^{\cdot}, \tag{4.23}$$

which has the jump

$$\begin{aligned} [\overline{T^{\alpha\gamma\{\beta \atop \gamma\alpha\}}}^{\cdot}] &= H^{\alpha\gamma\mu\nu} (\Gamma^\beta_{\gamma\alpha} + [\{{\beta \atop \gamma\alpha}\}])\, [\dot{e}_{\mu\nu}] + H^{\alpha\gamma} [\overline{\{{\beta \atop \gamma\alpha}\}}^{\cdot}] \\ &= -\frac{2}{U} H^{\alpha\gamma\mu\nu} \left(\Gamma^\beta_{\gamma\alpha} + \frac{1}{U^2} \sigma^\beta n_\gamma n_\alpha \right) \sigma_\mu n_\nu + H^{\alpha\gamma} [v^\beta_{|\gamma\alpha}], \end{aligned} \tag{4.24}$$

where we have used the kinematical formulas [1, (2.15) and (2.21)] for $\dot{e}_{\mu\nu}$ and $\overline{\{{\beta \atop \gamma\alpha}\}}^{\cdot}$, the product rule for jumps, the symmetry condition (2.23), the condition (4.21) for a tangential wave, the compatibility conditions (3.11) and (4.20), and the condition that the region ahead of the wave is at rest. Similarly, we have

$$\begin{aligned} [\overline{T^{\gamma\beta\{\alpha \atop \gamma\alpha\}}}^{\cdot}] &= H^{\gamma\beta\mu\nu} (\Gamma^\alpha_{\gamma\alpha} + [\{{\alpha \atop \gamma\alpha}\}])\, [\dot{e}_{\mu\nu}] + H^{\gamma\beta} [\overline{\{{\alpha \atop \gamma\alpha}\}}^{\cdot}] \\ &= -\frac{2}{U} H^{\gamma\beta\mu\nu} \left(\Gamma^\alpha_{\gamma\alpha} + \frac{1}{U^2} \sigma^\alpha n_\gamma n_\alpha \right) \sigma_\mu n_\nu + H^{\gamma\beta} [v^\alpha_{|\gamma\alpha}]. \end{aligned} \tag{4.25}$$

Next we calculate the jump of the leading term on the right hand side of (3.7). That term is given by the formula (3.9). Using the chain rule, we have

$$\dot{\overline{H^{\alpha\beta}_{,\alpha}}} = H^{\alpha\beta\mu\nu}_{,\alpha}\dot{e}_{\mu\nu}, \tag{4.26}$$

which has the jump

$$[\dot{\overline{H^{\alpha\beta}_{,\alpha}}}] = -\frac{2}{U} H^{\alpha\beta\mu\nu}_{,\alpha}\sigma_\mu n_\nu, \tag{4.27}$$

where we have used conditions such as (2.23), (4.21), and (4.20) mentioned before. Substituting (3.10) into (3.9) and using the symmetry condition (2.23), we can express the leading term on the right hand side of (3.9) as

$$H^{\alpha\beta\gamma\delta}e_{\gamma\delta,\alpha} = 2H^{\alpha\beta\gamma\delta}e_{\lambda\gamma}\left\{{\lambda \atop \delta\alpha}\right\}. \tag{4.28}$$

Using the chain rule, we then have

$$\dot{\overline{H^{\alpha\beta\gamma\delta}e_{\gamma\delta,\alpha}}} = 2H^{\alpha\beta\gamma\delta\mu\nu}\dot{e}_{\mu\nu}e_{\lambda\gamma}\left\{{\lambda \atop \delta\alpha}\right\} + 2H^{\alpha\beta\gamma\delta}\dot{e}_{\lambda\gamma}\left\{{\lambda \atop \delta\alpha}\right\} + 2H^{\alpha\beta\gamma\delta}e_{\lambda\gamma}\dot{\overline{\left\{{\lambda \atop \delta\alpha}\right\}}}, \tag{4.29}$$

which has the jump

$$\begin{aligned}
[\dot{\overline{H^{\alpha\beta\gamma\delta}e_{\gamma\delta,\alpha}}}] &= 2H^{\alpha\beta\gamma\delta\mu\nu}e_{\lambda\gamma}\left(\Gamma^\lambda_{\delta\alpha} + \left[\left\{{\lambda \atop \delta\alpha}\right\}\right]\right)[\dot{e}_{\mu\nu}] + 2H^{\alpha\beta\gamma\delta}\left(\Gamma^\lambda_{\delta\alpha} + \left[\left\{{\lambda \atop \delta\alpha}\right\}\right]\right)[\dot{e}_{\lambda\gamma}] \\
&\quad + 2H^{\alpha\beta\gamma\delta}e_{\lambda\gamma}\left[\dot{\overline{\left\{{\lambda \atop \delta\alpha}\right\}}}\right] \\
&= -\frac{4}{U}H^{\alpha\beta\gamma\delta\mu\nu}e_{\lambda\gamma}\left(\Gamma^\lambda_{\delta\alpha} + \frac{1}{U^2}\sigma^\lambda n_\delta n_\alpha\right)\sigma_\mu n_\nu \\
&\quad - \frac{2}{U}H^{\alpha\beta\gamma\delta}\left(\Gamma^\lambda_{\delta\alpha} + \frac{1}{U^2}\sigma^\lambda n_\delta n_\alpha\right)(\sigma_\lambda n_\gamma + \sigma_\gamma n_\lambda) + 2H^{\alpha\beta\gamma\delta}e_{\lambda\gamma}[v^\lambda_{|\delta\alpha}],
\end{aligned} \tag{4.30}$$

where we have used the same set of conditions as mentioned before in the derivation of (4.24), (4.25), and (4.27).

Now we have calculated all the terms contained in the expression for the leading term on the right hand side of (4.8). Hence (4.8) reduces to

$$\begin{aligned}
\varrho[\dot{a}^\beta] &= (2H^{\alpha\beta\gamma\delta} + H^{\delta\beta}e^{\alpha\gamma} + H^{\alpha\delta}e^{\beta\gamma})\,e_{\lambda\gamma}[v^\lambda_{|\delta\alpha}] + \frac{\varrho}{U}\sigma^\beta\sigma^\zeta n_\zeta \\
&\quad + \frac{1}{U}(2H^{\alpha\beta\gamma\delta} + H^{\delta\beta}e^{\alpha\gamma} + H^{\alpha\delta}e^{\beta\gamma})\,e_{\lambda\gamma}\Gamma^\lambda_{\delta\alpha}\sigma^\zeta n_\zeta + \frac{1}{U}H^{\alpha\beta}_{,\alpha}\sigma^\zeta n_\zeta \\
&\quad - \frac{1}{U}(2H^{\alpha\lambda\gamma\delta} + H^{\delta\lambda}e^{\alpha\gamma} + H^{\alpha\delta}e^{\lambda\gamma})\,e_{\gamma\xi}\Gamma^\xi_{\delta\alpha}\sigma^\beta n_\lambda - \frac{1}{U}H^{\alpha\lambda}_{,\alpha}\sigma^\beta n_\lambda \\
&\quad - \frac{4}{U}H^{\alpha\beta\gamma\delta\mu\xi}e_{\lambda\gamma}\left(\Gamma^\lambda_{\delta\alpha} + \frac{1}{U^2}\sigma^\lambda n_\delta n_\alpha\right)\sigma_\mu n_\xi - \frac{2}{U}H^{\alpha\beta\gamma\delta}_{,\alpha}\sigma_\gamma n_\delta \\
&\quad - \frac{2}{U}H^{\alpha\beta\gamma\delta}\left(\Gamma^\lambda_{\delta\alpha} + \frac{1}{U^2}\sigma^\lambda n_\delta n_\alpha\right)(\sigma_\lambda n_\gamma + \sigma_\gamma n_\lambda) \\
&\quad - \frac{2}{U}H^{\alpha\gamma\mu\xi}\left(\Gamma^\beta_{\gamma\alpha} + \frac{1}{U^2}\sigma^\beta n_\gamma n_\alpha\right)\sigma_\mu n_\xi - \frac{2}{U}H^{\gamma\beta\mu\xi}\left(\Gamma^\alpha_{\gamma\alpha} + \frac{1}{U^2}\sigma^\alpha n_\gamma n_\alpha\right)\sigma_\mu n_\xi.
\end{aligned} \tag{4.31}$$

As in [1] we now apply the second geometric condition of compatibility [1, (4.19)] to the velocity field $\boldsymbol{v}$, obtaining

$$[v^{\lambda}_{|\delta\alpha}] = \hat{v}^{\lambda} n_{\delta} n_{\alpha} - \frac{D}{\partial s}\left(\frac{\sigma^{\lambda}}{U}\right)(n_{\delta} m_{\alpha} + m_{\delta} n_{\alpha}) + \frac{\eta}{U}\sigma^{\lambda} m_{\delta} m_{\alpha}, \tag{4.32}$$

where $D/\partial s$ denotes the surface covariant derivative with respect to the arc length s along the wave front $\boldsymbol{c}(t)$, and where η denotes the geodesic curvature of $\boldsymbol{c}(t)$. Substituting (4.32) into (4.31) and then into (4.1), we finally obtain the amplitude equation for a tangential wave:

$$\begin{aligned}
2\varrho\frac{D\sigma^{\beta}}{\delta t} &= \{(2H^{\alpha\beta\gamma\delta} + H^{\delta\beta}e^{\alpha\gamma} + H^{\alpha\delta}e^{\beta\gamma})\, n_{\delta} n_{\alpha} e_{\lambda\gamma} - \varrho U^2\, \delta^{\beta}_{\lambda}\}\, \hat{v}^{\lambda} + \frac{\varrho\sigma^{\beta}}{U}\frac{DU}{\delta t} \\
&\quad - (2H^{\alpha\beta\gamma\delta} + H^{\delta\beta}e^{\alpha\gamma} + H^{\alpha\delta}e^{\beta\gamma})\, e_{\lambda\gamma}\left\{\frac{D}{\partial s}\left(\frac{\sigma^{\lambda}}{U}\right)(m_{\alpha} n_{\delta} + m_{\delta} n_{\alpha}) - \frac{\eta}{U}\sigma^{\lambda} m_{\delta} m_{\alpha}\right\} \\
&\quad - \frac{1}{U}(2H^{\alpha\beta\gamma\delta} + H^{\delta\beta}e^{\alpha\gamma} + H^{\alpha\delta}e^{\beta\gamma})\, e_{\lambda\gamma}\Gamma^{\lambda}_{\delta\alpha}\,\sigma^{\zeta} n_{\zeta} - \frac{1}{U} H^{\alpha\beta}_{,\alpha}\sigma^{\zeta} n_{\zeta} \\
&\quad - \frac{1}{U}(2H^{\alpha\lambda\gamma\delta} + H^{\delta\lambda}e^{\alpha\gamma} + H^{\alpha\delta}e^{\lambda\gamma})\, e_{\gamma\xi}\Gamma^{\xi}_{\delta\alpha}\sigma^{\beta} n_{\lambda} - \frac{1}{U} H^{\alpha\lambda}_{,\alpha}\sigma^{\beta} n_{\lambda} \\
&\quad - \frac{4}{U} H^{\alpha\beta\gamma\delta\mu\xi} e_{\lambda\gamma}\left(\Gamma^{\lambda}_{\delta\alpha} + \frac{1}{U^2}\sigma^{\lambda} n_{\delta} n_{\alpha}\right)\sigma_{\mu} n_{\xi} - \frac{2}{U} H^{\alpha\beta\gamma\delta}_{,\alpha}\sigma_{\gamma} n_{\delta} \\
&\quad - \frac{2}{U} H^{\alpha\beta\gamma\delta}\left(\Gamma^{\lambda}_{\delta\alpha} + \frac{1}{U^2}\sigma^{\lambda} n_{\delta} n_{\alpha}\right)(\sigma_{\lambda} n_{\gamma} + \sigma_{\gamma} n_{\lambda}) \\
&\quad - \frac{2}{U} H^{\alpha\gamma\mu\xi}\left(\Gamma^{\beta}_{\gamma\alpha} + \frac{1}{U^2}\sigma^{\beta} n_{\gamma} n_{\alpha}\right)\sigma_{\mu} n_{\xi} \\
&\quad - \frac{2}{U} H^{\gamma\beta\mu\xi}\left(\Gamma^{\alpha}_{\gamma\alpha} + \frac{1}{U^2}\sigma^{\alpha} n_{\gamma} n_{\alpha}\right)\sigma_{\mu} n_{\xi} - \frac{\varrho}{U}\sigma^{\beta}\sigma^{\zeta} n_{\zeta}.
\end{aligned} \tag{4.33}$$

As it stands the amplitude equation is very complicated. By virtue of (3.17), the leading term on the right hand side may be rewritten as

$$\{A^{\beta}_{\lambda}(\boldsymbol{n}) - \varrho U^2\, \delta^{\beta}_{\lambda}\}\, \hat{v}^{\lambda}, \tag{4.34}$$

which may be eliminated by multiplying the equation (4.33) with a left proper vector of the surface acoustic tensor $\boldsymbol{A}(\boldsymbol{n})$. Then the third term on the right hand side may be combined with the normal time derivative on the left hand side to form a time derivative along certain special oblique trajectories called the rays, or the bicharacteristics of the wave. In this paper we are primarily interested in principal waves, for which the acoustic tensors are symmetric and the rays are normal trajectories. Hence we leave the amplitude equation for a tangential wave in the general form (4.33) without further reductions as remarked.

We consider now the special case in which the wave is a principal shear wave having its wave front given by a coordinate curve $\theta^2 = \text{const.}$ and its normal trajectory given by a coordinate curve $\theta^1 = \text{const.}$ For such a principal wave the unit normal vector $\boldsymbol{n}$ is given by (3.21), *viz*,

$$\boldsymbol{n} = n^2 \boldsymbol{e}_2 = n_2 \boldsymbol{e}^2 = \boldsymbol{e}\langle 2\rangle, \qquad n_1 = n^1 = 0, \tag{4.35}$$

where

$$n^2 n^2 = e^{22}, \qquad n^2 n_2 = 1, \qquad n_2 n_2 = e_{22}, \tag{4.36}$$

the unit tangent vector $\boldsymbol{m}$ is given by

$$\boldsymbol{m} = m^1 \boldsymbol{e}_1 = m_1 \boldsymbol{e}^1 = \boldsymbol{e}\langle 1\rangle, \qquad m_2 = m^2 = 0, \tag{4.37}$$

where

$$m^1 m^1 = e^{11}, \qquad m^1 m_1 = 1, \qquad m_1 m_1 = e_{11}, \tag{4.38}$$

and the amplitude vector $\boldsymbol{\sigma}$ is given by

$$\boldsymbol{\sigma} = \sigma^1 \boldsymbol{e}_1 = \sigma_1 \boldsymbol{e}^1 = \sigma \boldsymbol{e}\langle 1\rangle = \sigma \boldsymbol{m}, \tag{4.39}$$

where σ denotes the amplitude of the wave, *i.e.*,

$$\sigma = \sigma^1 \sqrt{e_{11}} = \sigma_1 \sqrt{e^{11}}. \tag{4.40}$$

Since the region ahead of the wave is assumed to be undistorted and (θ^α) is an orthogonal coordinate system with basis vectors $\boldsymbol{e}_\alpha$ pointing in the direction of the principal axes, the Christoffel symbols $\left\{ {\alpha \atop \gamma\beta} \right\}$ are given by

$$\begin{aligned}
&\left\{ {1 \atop 11} \right\} = \tfrac{1}{2} e^{11} e_{11,1}, && \left\{ {2 \atop 22} \right\} = \tfrac{1}{2} e^{22} e_{22,2}, \\
&\left\{ {1 \atop 21} \right\} = \left\{ {1 \atop 12} \right\} = \tfrac{1}{2} e^{11} e_{11,2}, && \left\{ {2 \atop 11} \right\} = -\tfrac{1}{2} e^{22} e_{11,2}, \\
&\left\{ {2 \atop 21} \right\} = \left\{ {2 \atop 12} \right\} = \tfrac{1}{2} e^{22} e_{22,1}, && \left\{ {1 \atop 22} \right\} = -\tfrac{1}{2} e^{11} e_{22,1}.
\end{aligned} \tag{4.41}$$

It follows from $(4.41)_{2,3}$ that

$$e_{11}\left\{ {1 \atop 12} \right\} + e_{22}\left\{ {2 \atop 11} \right\} = 0, \qquad e_{22}\left\{ {2 \atop 12} \right\} + e_{11}\left\{ {1 \atop 22} \right\} = 0. \tag{4.42}$$

The geodesic curvature η of the wave front $\mathbf{c}(t)$ is given (*cf.* [1, (3.11), and (3.14)]) in general by

$$\eta = \boldsymbol{n} \cdot \frac{d\boldsymbol{m}}{ds} = -\boldsymbol{m} \cdot \frac{d\boldsymbol{n}}{ds}, \tag{4.43}$$

where s denotes the arc length along $\mathbf{c}(t)$. In view of (4.35) and (4.37), we can express η as

$$\eta = n_\alpha \Gamma^\alpha_{\beta\gamma} m^\beta m^\gamma = e^{11} \sqrt{e_{22}}\, \Gamma^2_{11} = -\sqrt{e^{22}}\, \Gamma^1_{12}, \tag{4.44}$$

where the metric and the Christoffel symbols are evaluated on $\mathbf{c}(t)$, *cf.* (4.15). In terms of the metric η is given by

$$\eta = -\tfrac{1}{2} e^{11} \sqrt{e^{22}}\, \overset{+}{e}_{11,2}. \tag{4.45}$$

Let the propagation of the wave front be given by

$$\theta^2 = \theta^2(t). \tag{4.46}$$

Then the normal velocity of the wave is

$$u = \frac{d\theta^2}{dt} e_2 = \frac{d\theta^2}{dt} \sqrt{e_{22}}\, n. \tag{4.47}$$

Hence the normal speed U is given by

$$U = \sqrt{e_{22}}\, \frac{d\theta^2}{dt}. \tag{4.48}$$

As a result, the displacement derivative and the normal time derivative can both be expressed in terms of the partial derivative with respect to θ^2. For the amplitude equation of the principal shear wave, we need an expression for the normal time derivative of ϱU^2. From (3.26)

$$\varrho U^2 = e_{22}(2e_{11}H^{1212} + H^{22}). \tag{4.49}$$

Using the chain rule based on (4.48), we then have

$$\begin{aligned}\frac{\delta(\varrho U^2)}{\delta t} = \frac{D(\varrho U^2)}{\delta t} &= \frac{U}{\sqrt{e_{22}}} \frac{\partial}{\partial\theta^2}\{e_{22}(2e_{11}H^{1212} + H^{22})\} \\ &= \frac{U}{\sqrt{e_{22}}}\{\overset{+}{e}_{22,2}(2e_{11}H^{1212} + H^{22}) + e_{22}(2H^{1212}\overset{+}{e}_{11,2} \\ &\quad + 2H^{121211}\overset{+}{e}_{11,2}e_{11} + 2H^{121222}\overset{+}{e}_{22,2}e_{11} + 2H^{1212}_{,2}e_{11} \\ &\quad + H^{2211}\overset{+}{e}_{11,2} + H^{2222}\overset{+}{e}_{22,2} + H^{22}_{,2})\},\end{aligned} \tag{4.50}$$

where the displacement derivative coincides with the normal time derivative since ϱU^2 is a scalar, and where the partial derivatives of the metric may be expressed in terms of the Christoffel symbols $\Gamma^{\alpha}_{\beta\gamma}$ through limits of the relations (4.41).

Now we have derived all the preliminary formulas needed for the reduction of the amplitude equation (4.33) for the special case of a principal shear wave. Using the kinematical formula [1, (3.38)] for the normal time derivative of the unit tangent vector m, we can express the left hand side of (4.33) as

$$2\varrho \frac{D(\sigma m^\beta)}{\delta t} = 2\varrho m^\beta \frac{D\sigma}{\delta t} + 2\varrho\sigma \frac{DU}{\partial s} n^\beta. \tag{4.51}$$

Hence multiplying (4.33) by m_β and summing on β, we obtain an expression for the normal time derivative of the amplitude σ:

$$2\varrho \frac{D\sigma}{\delta t} = m_\beta \{A^\beta_\gamma(n) - \varrho U^2\, \delta^\beta_\gamma\}\, \hat{v}^\lambda + \frac{\varrho\sigma}{U}\frac{DU}{\delta t} + \Sigma_1, \tag{4.52}$$

where the leading term on the right hand vanishes since the acoustic tensor for a principal wave is given by (3.23). Among terms included in the sum Σ_1 there is one involving the tangential derivative of the amplitude vector σ. Using (4.39) and (4.43), we can calculate that term by

$$\frac{D}{\partial s}\left(\frac{\sigma^\lambda}{U}\right) = \frac{D}{\partial s}\left(\frac{\sigma}{U} m^\lambda\right) = \frac{\sigma}{U}\eta n^\lambda + m^\lambda \frac{D}{\partial s}\left(\frac{\sigma}{U}\right), \tag{4.53}$$

cf. [1, (8.24)]. Now using the preliminary formulas (4.35)–(4.51) and (4.53), we can reduce the sum Σ_1 in (4.52) to

$$\Sigma_1 = -\frac{\sigma}{U^2}\frac{D(\varrho U^2)}{\delta t} + \eta\frac{\sigma}{U}(\varrho U^2)$$

$$= -\frac{\varrho\sigma}{U}\frac{DU}{\delta t} - \frac{\sigma}{U}\frac{D(\varrho U)}{\delta t} + \eta\sigma\varrho U. \tag{4.54}$$

Substituting (4.54) into (4.52), we finally obtain the amplitude equation for a principal shear wave:

$$2\varrho\frac{D\sigma}{\delta t} = -\frac{\sigma}{U}\frac{D(\varrho U)}{\delta t} + \eta\varrho U\sigma. \tag{4.55}$$

This equation has exactly the same form as [1, (8.29)] for a principal shear wave in an isotropic elastic membrane and is comparable to [3, (4.38)] which governs the amplitude for a principal shear wave in an orthotropic elastic material. The only difference between (4.55) and [3, (4.38)] is that the geodesic curvature η of the wave front $\mathbf{c}(t)$ replaces the mean curvature K of the wave surface in the 3-dimensional body.

Next we consider the special case when the wave is a principal longitudinal wave. For such a principal wave $\boldsymbol{n}$ and $\boldsymbol{m}$ are still given by (4.35) and (4.37), respectively, but $\boldsymbol{\sigma}$ is given by

$$\boldsymbol{\sigma} = \sigma^2\mathbf{e}_2 = \sigma_2\mathbf{e}^2 = \sigma\mathbf{e}\langle 2\rangle = \sigma\boldsymbol{n}, \tag{4.56}$$

where

$$\sigma = \sigma^2\sqrt{e_{22}} = \sigma_2\sqrt{e^{22}}. \tag{4.57}$$

Using the chain rule based on (4.48), we now have

$$\frac{\delta(\varrho U^2)}{\delta t} = \frac{D(\varrho U^2)}{\delta t} = \frac{U}{\sqrt{e_{22}}}\frac{\partial}{\partial\theta^2}\{2e_{22}(e_{22}H^{2222} + H^{22})\}$$

$$= \frac{U}{\sqrt{e_{22}}}\{2\overset{+}{e}_{22,2}(e_{22}H^{2222} + H^{22}) + 2e_{22}(\overset{+}{e}_{22,2}H^{2222}$$

$$+ e_{22}H^{222211}\overset{+}{e}_{11,2} + e_{22}H^{222222}\overset{+}{e}_{22,2} + e_{22}H^{2222}_{,2}$$

$$+ H^{2211}\overset{+}{e}_{11,2} + H^{2222}\overset{+}{e}_{22,2} + H^{22}_{,2}\}. \tag{4.58}$$

Having derived the additional preliminary formulas (4.56)–(4.58), we are now in a position to derive the reduction of the amplitude equation (4.33) for the special case of a principal longitudinal wave. First, using the kinematical formula [1, (3.39)] for the normal time derivative of the unit normal vector $\boldsymbol{n}$, we can express the left hand side of (4.33) as

$$2\varrho\frac{D(\sigma n^\beta)}{\delta t} = 2\varrho n^\beta\frac{D\sigma}{\delta t} - 2\varrho\sigma\frac{DU}{\partial s}m^\beta. \tag{4.59}$$

Hence multiplying (4.33) by n_β and summing on β, we obtain an expression for the normal time derivative of the amplitude σ:

$$2\varrho \frac{D\sigma}{\delta t} = n_\beta\{A^\beta_\gamma(\boldsymbol{n}) - \varrho U^2 \delta^\beta_\lambda\} \hat{v}^\lambda + \frac{\varrho\sigma}{U}\frac{DU}{\delta t} + \Sigma_2, \tag{4.60}$$

where the leading term on the right hand side again vanishes since the acoustic tensor for a principal wave is given by (3.23). Among terms included in the sum Σ_2 there is one involving the tangential derivative of the amplitude vector $\boldsymbol{\sigma}$. Using (4.56) and (4.43), we can express that term as follows:

$$\frac{D}{\partial s}\left(\frac{\sigma^\lambda}{U}\right) = \frac{D}{\partial s}\left(\frac{\sigma}{U} n^\lambda\right) = -\frac{\sigma}{U}\eta m^\lambda + n^\lambda \frac{D}{\partial s}\left(\frac{\sigma}{U}\right), \tag{4.61}$$

cf. [1, (8.31)]. Now using the preliminary formulas (4.35)–(4.38), (4.41)–(4.48), (4.56)–(4.59), (4.61), (2.32)–(2.34), and (2.42), we can reduce the sum Σ_2 in (4.60) to

$$\Sigma_2 = -\frac{(\sigma)^2}{U^3}\frac{\partial^2 T\langle 22\rangle}{\partial F\langle 22\rangle\,\partial F\langle 22\rangle} - \frac{\sigma}{U^2}\frac{D(\varrho U^2)}{\delta t}$$
$$+ \eta\frac{\sigma}{U}\left(\frac{\partial T\langle 22\rangle}{\partial F\langle 22\rangle} + \frac{\partial T\langle 22\rangle}{\partial F\langle 11\rangle} - \frac{\partial T\langle 11\rangle}{\partial F\langle 22\rangle} + \frac{\partial T\langle 12\rangle}{\partial F\langle 12\rangle} - \frac{\partial T\langle 21\rangle}{\partial F\langle 21\rangle}\right). \tag{4.62}$$

Substituting (4.62) into (4.60), we finally obtain the amplitude equation for a principal longitudinal wave:

$$2\varrho\frac{D\sigma}{\delta t} = -\frac{(\sigma)^2}{U^3}\frac{\partial^2 T\langle 22\rangle}{\partial F\langle 22\rangle\,\partial F\langle 22\rangle} - \frac{\sigma}{U}\frac{D(\varrho U)}{\delta t}$$
$$+ \eta\frac{\sigma}{U}\left(\frac{\partial T\langle 22\rangle}{\partial F\langle 22\rangle} + \frac{\partial T\langle 22\rangle}{\partial F\langle 11\rangle} - \frac{\partial T\langle 11\rangle}{\partial F\langle 22\rangle} + \frac{\partial T\langle 12\rangle}{\partial F\langle 12\rangle} - \frac{\partial T\langle 21\rangle}{\partial F\langle 21\rangle}\right). \tag{4.63}$$

This equation generalizes the amplitude equation [1, (8.17)] from principal longitudinal waves in isotropic elastic membranes to principal longitudinal waves in orthotropic elastic membranes. Also, (4.63) is comparable to the amplitude equation [3, (4.32)] which governs the amplitude for a principal wave in an orthotropic elastic material. The only difference between (4.63) and [3, (4.32)] is that the geodesic curvature η of the wave front $\boldsymbol{c}(t)$ replaces the normal curvatures $\Omega\langle 22\rangle$ and $\Omega\langle 33\rangle$ of the wave surface in the 3-dimensional body.

If the orthotropic elastic membrane is hyperelastic, then by (2.37) the sum of the last four terms inside the parentheses on the right hand side of (4.63) vanishes. In that case the amplitude equation for a principal longitudinal wave reduces to

$$2\varrho\frac{D\sigma}{\delta t} = -\frac{(\sigma)^2}{U^3}\frac{\partial^2 T\langle 22\rangle}{\partial F\langle 22\rangle\,\partial F\langle 22\rangle} - \frac{\sigma}{U}\frac{D(\varrho U)}{\delta t} + \eta\varrho U\sigma. \tag{4.64}$$

This equation differs from the equation [3, (4.44)] by only a single term: the geodesic curvature η of the wave front $\boldsymbol{c}(t)$ replaces the mean curvature K of the wave surface in the 3-dimensional body.

Having derived the amplitude equations for principal tangential waves, we now turn our attention to the leading term on the right hand side of (4.9) which governs the amplitude of a transverse wave in general. Using the product rule, we have

$$\overline{[\Omega_{\alpha\beta}T^{\alpha\beta}]}^{\boldsymbol{\cdot}} = [\Omega_{\alpha\beta}]\,[\dot{T}^{\alpha\beta}] + \Delta_{\alpha\beta}[\dot{T}^{\alpha\beta}] + H^{\alpha\beta}[\dot{\Omega}_{\alpha\beta}], \tag{4.65}$$

where $\Delta_{\alpha\beta}$ is defined by (4.13). As before, we can calculate the material time derivative $\dot{T}^{\alpha\beta}$ by the chain rule:

$$\dot{T}^{\alpha\beta} = H^{\alpha\beta\gamma\delta}\dot{e}_{\gamma\delta}. \tag{4.66}$$

Then the jump of $\dot{T}^{\alpha\beta}$ can be expressed in terms of the jump of $\dot{e}_{\gamma\delta}$ which is given by the kinematical formula [1, (2.15)]. However, for a transverse wave the amplitude vector σ is given by (4.22) which implies that

$$[\dot{e}_{\gamma\delta}] = 0. \tag{4.67}$$

As a result, we have also

$$[\dot{T}^{\alpha\beta}] = 0, \tag{4.68}$$

as in the case of a transverse wave in an isotropic elastic membrane; *cf.* [1, (7.37)].

Similarly, the jump of $\dot{\Omega}_{\alpha\beta}$ may be obtained by taking the jump of the kinematical formula [1, (2.22)], and the result is

$$[\dot{\Omega}_{\alpha\beta}] = [v^3_{|\alpha\beta}]; \tag{4.69}$$

cf. [1, (7.38)], where we have used the condition (4.22). Substituting (4.68) and (4.69) into (4.65) and then using the previous results (4.10), (4.11), (4.14), and (4.19), we obtain

$$\varrho[\dot{a}^3] = H^{\alpha\beta}[v^3_{|\alpha\beta}] - \frac{1}{U}(2H^{\alpha\lambda\gamma\delta} + H^{\delta\lambda}e^{\alpha\gamma} + H^{\alpha\delta}e^{\lambda\gamma})\,e_{\gamma\xi}\Gamma^{\xi}_{\delta\alpha}\sigma^3 n_\lambda - \frac{1}{U}H^{\alpha\lambda}_{,\alpha}\sigma^3 n_\lambda. \tag{4.70}$$

As in [1] we apply the second geometric condition of compatibility [1, (4.20)] to the velocity field $\boldsymbol{v}$, obtaining

$$[v^3_{|\alpha\beta}] = \hat{v}^3 n_\alpha n_\beta - \frac{D}{\partial s}\left(\frac{\sigma^3}{U}\right)(n_\alpha m_\beta + n_\beta m_\alpha) + \frac{\eta}{U}\sigma^3 m_\alpha m_\beta, \tag{4.71}$$

where v^3 is defined by (4.3). Substituting (4.71) into (4.70) and then into (4.2), we obtain the amplitude equation for a transverse wave:

$$\begin{aligned} 2\varrho\frac{D\sigma^3}{\delta t} = {} & \{H^{\alpha\beta}n_\alpha n_\beta - \varrho U^2\}\,\hat{v}^3 + \frac{\varrho\sigma^3}{U}\frac{DU}{\delta t} - 2H^{\alpha\beta}\frac{D}{\partial s}\left(\frac{\sigma^3}{U}\right)n_\alpha m_\beta \\ & + H^{\alpha\beta}\frac{\eta}{U}\sigma^3 m_\alpha m_\beta - \frac{1}{U}(2H^{\alpha\lambda\gamma\delta} + H^{\delta\lambda}e^{\alpha\gamma} + H^{\alpha\delta}e^{\lambda\gamma})\,e_{\gamma\xi}\Gamma^{\xi}_{\delta\alpha}\sigma^3 n_\lambda \\ & - \frac{1}{U}H^{\alpha\lambda}_{,\alpha}\sigma^3 n_\lambda. \end{aligned} \tag{4.72}$$

Notice that by virtue of the propagation condition (3.19) with $Q(\boldsymbol{n})$ given by (3.18) and with $\boldsymbol{\sigma}$ given by (4.22), the leading term on the right hand side of (4.72) vanishes. However, we have kept that term in (4.72) in order to show the similarity of (4.72) and (4.33).

Next we specialize the amplitude equation (4.72) to that governing the amplitude of a principal transverse wave. For this case $\boldsymbol{n}$ and $\boldsymbol{m}$ are given by (4.35)–(4.38). Since $\boldsymbol{e}_3$ is a unit vector (*cf.* (3.2)), the amplitude σ is given directly by

$$\sigma = \sigma^3 = \sigma_3. \tag{4.73}$$

Using the chain rule based on (4.48), we have

$$\begin{aligned}\frac{\partial(\varrho U^2)}{\delta t} = \frac{D(\varrho U^2)}{\delta t} &= \frac{U}{\sqrt{e_{22}}}\frac{\partial}{\partial\theta^2}\{e_{22}H^{22}\} \\ &= \frac{U}{\sqrt{e_{22}}}\{\overset{+}{e}_{22,2}H^{22} + e_{22}H^{2211}\overset{+}{e}_{11,2} + e_{22}H^{2222}\overset{+}{e}_{22,2} + e_{22}H^{22}_{,2}\}.\end{aligned} \tag{4.74}$$

Having derived the needed preliminary formula (4.74), we write the amplitude equation (4.72), specialized by (4.35)–(4.38) and (4.73), in the form

$$2\varrho\frac{D\sigma}{\delta t} = \{Q^3_3(n) - \varrho U^2\}\,\hat{v}^3 + \frac{\varrho\sigma}{U}\frac{DU}{\delta t} + \Sigma_3, \tag{4.75}$$

where the leading term on the right hand side actually vanishes. Using the preliminary formulas (4.35)–(4.38), (4.41)–(4.48), (4.22), (4.73), and (4.74), we can reduce the sum Σ_3 in (4.75) to

$$\Sigma_3 = -\frac{\sigma}{U^2}\frac{D(\varrho U^2)}{\delta t} + \frac{\sigma}{U}(\varrho U^2), \tag{4.76}$$

which has exactly the same form as Σ_1, *cf.* (4.54). As a result, the amplitude equation for a principal transverse wave is

$$2\varrho\frac{D\sigma}{\delta t} = -\frac{\sigma}{U}\frac{D(\varrho U)}{\delta t} + \eta\varrho U\sigma, \tag{4.77}$$

which has exactly the same form as the amplitude equation for a principal shear wave; *cf.* (4.55).

The fact that the amplitude equations for a principal shear wave and for a principal transverse wave have exactly the same form has been remarked in [1, Sec. 8] for principal waves in isotropic elastic membranes. Here we generalize that remark to principal waves in orthotropic elastic membranes.

Since the amplitude equations (4.55), (4.64), and (4.77) for principal shear waves, principal longitudinal waves, and principal transverse waves, respectively, in orthotropic elastic membranes have the same forms as corresponding amplitude equations for principal waves in isotropic elastic membranes, their solutions for special cases, such as those enjoying cylindrical symmetry, conical symmetry, or spherical symmetry, may be obtained in the same way as in [1]. Hence there is no need for us to repeat the details here.

Acknowledgement. The authors acknowledge the support in part of the National Sciences and Engineering Research Council of Canada.

References

1. POP, J. J., & C.-C. WANG. Acceleration Waves in Isotropic Elastic Membranes. *Arch. Rational Mech. Anal.* **77**, 47–93 (1981).
2. BOWEN, R. M., & C.-C. WANG. Acceleration Waves in Inhomogeneous Isotropic Elastic Bodies. *Arch. Rational Mech. Anal.* **38**, 13–45 (1970).
3. BOWEN, R. M., & C.-C. WANG. Acceleration Waves in Orthotropic Elastic Materials. *Arch. Rational Mech. Anal.* **47**, 149–170 (1972).
4. COHEN, H., & C.-C. WANG. On the Response and Symmetry of Elastic and Hyperelastic Membrane Points. *Arch. Rational Mech. Anal.* **85**, 355–391 (1984).
5. NAGHDI, P. M. The Theory of Shells and Plates. Handbuch der Physik, Vol. VIa/2. Berlin-Heidelberg-New York: Springer 1972.
6. COHEN, H., & C.-C. WANG. On Compatibility Conditions for Singular Surfaces. *Arch. Rational Mech. Anal.* **80**, 205–261 (1982).
7. COHEN, H., & C.-C. WANG. Principal Waves in Elastic Subfluids. *Arch. Rational Mech. Anal.* **83**, 139–168 (1983).
8. COHEN, H., & C.-C. WANG. A Note on Hyperelasticity. *Arch. Rational Mech. Anal.* **85**, 213–236 (1984).

Department of Civil Engineering,
University of Manitoba,
Winnipeg, Manitoba

and

Department of Mathematical Sciences,
Rice University,
Houston, Texas

(Received November 12, 1984)

Continua with Latent Microstructure

GIANFRANCO CAPRIZ

Dedicated to Walter Noll on his sixtieth birthday

1. Introduction

In [1] a general dynamic model of a continuum with microstructure was proposed which is inspired by (and contains as special cases) known models for continua with voids, liquid crystals, multipolar media, *etc.* The elementary background which supports the proposal was described at length in a course of lectures [2] at the Banach Institute, Warsaw, and reported later in a paper [3]. Further arguments in favour of the model can be found in [4], where, in particular, the case of a fluid with microstructure is studied in some detail.

Here I suggest that a related concept of latent microstructure can be useful in offering an interpretation of constitutive prescriptions involving displacement gradients of higher order, an interpretation which allows one to circumvent a known incompatibility of some of those prescriptions with the Clausius-Duhem inequality. The suggestion is in line with an early remark of TOUPIN (see [5]) on the possible identification of certain materials of second order with Cosserat continua whose microrotations are constrained.

I give complete developments only for a few special cases, where the essential ideas are easily illustrated, avoiding too complex formal developments. Also, in those cases, developments are strictly linked with recent work by DUNN & SERRIN; and, actually, direct inspiration for this paper came from their report [6]. The introduction they propose of an additional term (the interstitial working) in the equation which expresses the balance of energy finds below another justification. Their end result (a subcase of Korteweg fluids is compatible with the Clausius-Duhem inequality) finds also support; but, besides, here a path is open to discuss more complex cases excluded by their premisses.

In Section 2 and 3 I recall briefly the approach proposed in [3] and [4] in a form most suitable for the present developments. In Section 4 I illustrate the concept of latent microstructure and deduce some immediate consequences. In Section 5 I introduce the simplest case of internal constraints and derive reduced balance equations, to mirror closely the developments of DUNN & SERRIN. Finally, in Section 6 and 7 I consider two special cases of elastic fluids.

At a recent meeting (Oberwolfach, Jan. 1984) Professor MAUGIN asserted that the subject matter of Sections 2 and 3 is exhaustively treated in some of his papers, in particular in [7].

2. Continua with Lagrangian microstructure

A body $\mathfrak{B}$ (a set whose members $\mathfrak{X}$ are material elements) is said to be a *continuum with microstructure* if the following properties apply.
(P1) There is a differentiable manifold $\mathcal{Q}$ of finite dimension q whose members $\underset{\sim}{\nu}$ are the microstates (or sets of order parameters). Over $\mathcal{Q}$ a group operation of rigid rotation is defined; *i.e.*, given any proper orthogonal tensor Q and any $\underset{\sim}{\nu} \in \mathcal{Q}$ a unique $\underset{\sim}{\nu}^{(Q)}$ is determined with the properties.

$$\underset{\sim}{\nu}^{(Q^{(1)}Q^{(2)})} = (\underset{\sim}{\nu}^{(Q^{(2)})})^{(Q^{(1)})}, \quad \underset{\sim}{\nu}^{(1)} = \underset{\sim}{\nu},$$

where $Q^{(1)}, Q^{(2)} \in \mathrm{Orth}^+$ and 1 is the identity tensor.
(P2) A class $\mathscr{C}$ of mappings (called here *complete placements*) $\mathfrak{B} \to \mathscr{E} \times \mathcal{Q}$ ($\mathscr{E}$, three-dimensional Euclidean space) exists

$$x = x(\mathfrak{X}), \quad \underset{\sim}{\nu} = \underset{\sim}{\nu}(\mathfrak{X}), \quad x \in \mathscr{E}, \quad \underset{\sim}{\nu} \in \mathcal{Q},$$

such that:

(i) the *apparent placement* $x = x(\mathfrak{X})$ is a one-to-one mapping of $\mathfrak{B}$ into $\mathscr{E}$, as in the case of placements of ordinary continua; the range $\mathscr{B} = x(\mathfrak{B})$ is an open subset of $\mathscr{E}$;

(ii) any couple of apparent placements $x'(\mathfrak{X})$, $x''(\mathfrak{X})$ is such that the induced bijection of $\mathscr{B}' = x'(\mathfrak{B})$ onto $\mathscr{B}'' = x''(\mathfrak{B})$ is smooth, again as in the case of ordinary continua;

(iii) each complete placement $(x, \underset{\sim}{\nu}) \in \mathscr{C}$ is such that the mapping $\underset{\sim}{\nu} \circ x^{-1}$ of $\mathscr{B} = x(\mathfrak{B})$ into $\mathcal{Q}$ is smooth;

(iv) if $(x, \underset{\sim}{\nu}) \in \mathscr{C}$, then also all the placements $(x^{(Q)}, \underset{\sim}{\nu}^{(Q)})$ belong to $\mathscr{C}$, if

$$x^{(Q)}(\mathfrak{X}) = x(\mathfrak{X}) + c + Q(x(\mathfrak{X}) - x(\mathfrak{X}')), \tag{2.1}$$

with c any vector, $\mathfrak{X}'$ a fixed element of $\mathfrak{B}$, $Q \in \mathrm{Orth}^+$, and $\underset{\sim}{\nu}^{(Q)}$ is specified in accordance with the rule under P1.

Remark 1. Any manifold $\mathcal{Q}'$ which can be put into a sufficiently smooth one-to-one correspondence with $\mathcal{Q}$ can be used to describe the microstates. Correspondingly the way different observers relate their measures of the order parameters (*i.e.*, the specification of the operation of rigid rotation) must be appropriately adapted. The topological properties of $\mathcal{Q}$ are of the essence when, for instance, defects in the microstructure are classified; these matters are not pursued here.

Remark 2. Examples exist of different continua, for all of which the microstructure may be described by the same manifold $\mathcal{Q}$, but to each of which a different group of rigid rotations over $\mathcal{Q}$ is associated.

Examples:

(i) Continua with voids. For $\mathcal{Q}$ the interval [0, 1) of real numbers may be taken; $\underset{\sim}{\nu}$ is the scalar which specifies the void fraction. Rigid rotations leave the microstate unaltered: $\underset{\sim}{\nu}^{(Q)} = \underset{\sim}{\nu}$.

(ii) Standard model for nematic liquid crystals. For $\mathcal{Q}$ the set of diads $c \otimes c$ (c, a unit vector) is taken; $\mathcal{Q}$ is isomorphic to the projective plane, or to a spherical surface where the antipodal points are identified. As a consequence of a rotation specified by a tensor Q the diad $c \otimes c$ goes into $Q(c \otimes c)\, Q^T$.

For the developments below, the introduction of further notation is appropriate. Let r be the vector associated with the orthogonal tensor Q, so that $Q = \exp \mathbf{e}\, r$ ($\mathbf{e}$, the Ricci commutator); then I write $\underset{\sim}{\nu}^{(r)}$ as an alternative to $\underset{\sim}{\nu}^{(Q)}$, and call $\aleph(\underset{\sim}{\nu})$ the Fréchet derivative of $\underset{\sim}{\nu}^{(r)}$ at $\underset{\sim}{\nu}$:

$$\aleph(\underset{\sim}{\nu})\, r := \left. \frac{d\underset{\sim}{\nu}^{(r)}}{dr} \right|_{r=0} [r], \tag{2.2}$$

so that

$$\underset{\sim}{\nu}^{(r)} = \underset{\sim}{\nu} + \aleph(\underset{\sim}{\nu})\, r + o(|r|). \tag{2.3}$$

$\aleph(\underset{\sim}{\nu})$ is a linear map of $\mathscr{V}$ (the translation space of $\mathscr{E}$) into the tangent space $\mathscr{T}\mathcal{Q}_\nu$ at $\underset{\sim}{\nu}$.

Most developments involve only values of $\underset{\sim}{\nu}$ in a 'small' subset of $\mathcal{Q}$, which can be imagined covered by one chart only. Thus I use freely a set of local coordinates ν_α ($\alpha = 1, 2, \dots q$) and also, at each $\underset{\sim}{\nu}$, a corresponding set $\{r_\alpha\}$ of vectors such that

$$(\aleph(\underset{\sim}{\nu})\, r)_\alpha = r_\alpha \cdot r. \tag{2.4}$$

A motion of $\mathfrak{B}$ is a mapping of a real interval $[\tau_0, \tau_1)$ into $\mathscr{C}$:

$$x = x(\mathfrak{X}, \tau), \qquad \underset{\sim}{\nu} = \underset{\sim}{\nu}(\mathfrak{X}, \tau).$$

Velocity v and generalized velocity $\underset{\sim}{v}$ in that motion are

$$v := \dot{x}(\mathfrak{X}, \tau), \qquad \underset{\sim}{v} := \dot{\underset{\sim}{\nu}}(\mathfrak{X}, \tau);$$

notice that v may happen to be any element of $\mathscr{V}$, whereas $\underset{\sim}{v}(\mathfrak{X}, \tau)$ must belong to $\mathscr{T}\mathcal{Q}_\nu(\mathfrak{X}, \tau)$.

Of course, velocity and generalized velocity, at a given instant τ, can be also considered as fields on $\mathscr{B}_\tau := x(\mathfrak{B}, \tau)$, as tacitly implied by most formulae below. The connected notion of virtual velocities, as fields on $\mathscr{B}_\tau$, will also be of use later.

A velocity distribution is rigid (with translational speed $\dot{c}$ and angular speed w) if it has the form

$$\begin{aligned} v^{(R)}(\mathfrak{X}, \tau) &= \dot{c}(\tau) + w(\tau) \times (x(\mathfrak{X}, \tau) - x(\mathfrak{X}', \tau)), \\ \underset{\sim}{v}^{(R)}(\mathfrak{X}, \tau) &= \aleph(\underset{\sim}{\nu}(\mathfrak{X}, \tau))\, w(\tau). \end{aligned} \tag{2.5}$$

When a local chart is introduced and the notation (2.4) is used, the components of $\underset{\sim}{v}^{(R)}$ can be written as follows

$$v_\alpha^{(R)} = r_\alpha \cdot w. \tag{2.6}$$

As any ordinary continuum, $\mathfrak{B}$ is assumed to have a mass, which can be expressed as a total of a mass density ϱ and which is preserved; during any motion the usual equation of balance of mass

$$\dot{\varrho} + \varrho \operatorname{div} v = 0 \tag{2.7}$$

is presumed valid.

On the contrary one can expect that micromotions contribute to the the kinetic energy of the body; it is a matter of convenience to define the order parameters in such a way that the kinetic energy of $\mathfrak{B}$ reduces to the classical expression when $\underset{\sim}{v}$ vanishes. Here I assume explicitly that the kinetic energy is measured, in general, by

$$\int_{\mathscr{B}} \varrho(\tfrac{1}{2}v^2 + \varkappa(\underset{\sim}{\nu}, \underset{\sim}{v})), \tag{2.8}$$

where $\varkappa$ is a mapping of the tangent bundle $\mathscr{T}\mathscr{Q}$ of $\mathscr{Q}$ into $\mathbb{R}^+$, such that

$$\varkappa(\underset{\sim}{\nu}, \underset{\sim}{v}) \geqq 0, \quad \varkappa(\underset{\sim}{\nu}, \mathit{0}) = 0.$$

Usually $\varkappa$ is assumed to be a positive (semi) definite quadratic form on each $\mathscr{T}\mathscr{Q}_{\underset{\sim}{\nu}}$; in terms of local coordinates

$$\varkappa = \tfrac{1}{2} \sum_{1}^{q}{}_{\alpha\beta}\, \mu^{\alpha\beta}(\underset{\sim}{\nu})\, v_\alpha v_\beta. \tag{2.9}$$

Mechanical actions on $\mathfrak{B}$ comprise the usual external body force of density ϱb (per unit volume of $\mathfrak{B}$) and the usual surface traction t. Besides, one must account for actions on the microstructure; they can be appropriately described around any element $\mathfrak{X}$ by members of the cotangent space $\mathscr{T}^*\mathscr{Q}_{\underset{\sim}{\nu}}(\mathfrak{X})$: an external body force $\varrho\underset{\sim}{\beta}$, a resultant of internal actions $-\underset{\sim}{\zeta}$ (both per unit volume of $\mathscr{B}$) and a surface traction $\underset{\sim}{\tau}$ (with components, respectively, $\varrho\beta^\alpha$, $-\zeta^\alpha$, τ^α).

Mechanical balance is expressed by the usual equation for momentum

$$\left(\int_{\mathscr{E}} \varrho v\right)^{\cdot} = \int_{\partial\mathscr{E}} t + \int_{\mathscr{E}} \varrho b, \tag{2.10}$$

(valid for any subset $\mathscr{E}$ of $\mathscr{B}$ with smooth boundary $\partial\mathscr{E}$) and by an equation modelled on the Lagrange equation

$$\left(\int_{\mathscr{E}} \varrho \frac{\partial\varkappa}{\partial \underset{\sim}{v}}\right)^{\cdot} - \int_{\mathscr{E}} \varrho \frac{\partial\varkappa}{\partial \underset{\sim}{\nu}} = \int_{\partial\mathscr{E}} \underset{\sim}{\tau} + \int_{\mathscr{E}} (\varrho\underset{\sim}{\beta} - \underset{\sim}{\zeta}). \tag{2.11}$$

Remark 3. Eqn. (2.11) is correct only if $\varkappa$ is a quadratic form in $\underset{\sim}{v}$; otherwise the function $\varkappa$ appearing in it must not be identified with the kinetic energy density; rather the latter is the Legendre transform $\varkappa'$ of the former, or in terms of components

$$\varkappa' := \sum_{1}^{q}{}_{\alpha} \frac{\partial\varkappa}{\partial v_\alpha} v_\alpha - \varkappa.$$

Below, as in (2.11), we presume that $\varkappa'$ and $\varkappa$ coincide.

Sufficient regularity of t and $\underset{\sim}{\tau}$ implies the existence of the tensor of stress T and of a linear operator from $\mathscr{V}$ into $\mathscr{T}^*\mathscr{Q}_{\underset{\sim}{\nu}}$, the microstress ב, such that

$$t = Tn, \qquad \underset{\sim}{\tau} = ב n, \tag{2.12}$$

where n is the unit vector of the exterior normal to $\partial\mathscr{C}$. Local balance equations ensue: Cauchy's balance equation

$$\varrho\ddot{x} = \varrho b + \operatorname{div} T, \tag{2.13}$$

and a new balance equation for micromomentum

$$\varrho\left(\left(\frac{\partial \varkappa}{\partial \underset{\sim}{v}}\right)^{\cdot} - \frac{\partial \varkappa}{\partial \underset{\sim}{v}}\right) = \varrho\underset{\sim}{\beta} - \zeta + \operatorname{div} ב; \tag{2.14}$$

on a local chart the latter can be written

$$\varrho\left(\left(\frac{\partial \varkappa}{\partial v_\alpha}\right)^{\cdot} - \frac{\partial \varkappa}{\partial v_\alpha}\right) = \varrho\beta^\alpha - \zeta^\alpha + \operatorname{div} t^\alpha, \tag{2.15}$$

where t^α is defined so that (see (2.12))

$$\tau^\alpha = t^\alpha \cdot n. \tag{2.16}$$

Under appropriate circumstances (2.13), (2.14) can be deduced also from a variational principle; I refer to [4] for details.

3. Balance of moment of momentum, energy and entropy

A kinetic energy theorem follows from (2.13), (2.14); if $\mathscr{C}$ is any portion of $\mathscr{B}$ where the motion is smooth, then

$$\begin{aligned}\left(\int_{\mathscr{C}} \varrho\left(\tfrac{1}{2}v^2 + \varkappa\right)\right)^{\cdot} &= \int_{\mathscr{C}} \varrho(b \cdot v + \underset{\sim}{\beta} \cdot \underset{\sim}{v}) + \int_{\partial\mathscr{C}} (t \cdot v + \underset{\sim}{\tau} \cdot \underset{\sim}{v}) + \\ &\quad - \int_{\mathscr{C}} (T \cdot \operatorname{grad} v + ב \cdot \operatorname{grad} \underset{\sim}{v} + \underset{\sim}{\zeta} \cdot \underset{\sim}{v}).\end{aligned} \tag{3.1}$$

The first two terms in the right-hand side express the power of external actions; thus one is led to interpret the quantity

$$-(T \cdot \operatorname{grad} v + ב \cdot \operatorname{grad} v + \underset{\sim}{\zeta} \cdot \underset{\sim}{v}) \tag{3.2}$$

as the density per unit volume of the power of internal actions.

In simple continua the symmetry of T is a necessary and sufficient condition for the vanishing of that density for all virtual rigid velocity distributions. In the present context that condition can be expressed using formulae (2.5) and reads (with an obvious interpretation of the transposition symbol)

$$\mathbf{e}\, T = \aleph^T \underset{\sim}{\zeta} + (\operatorname{grad} \aleph^T)\, ב. \tag{3.3}$$

Alternatively, and with the use of a local chart, eqn. (3.3) can be written as follows

$$\operatorname{skw} T = \tfrac{1}{2}\,\mathbf{e}\left(\sum_{1}^{q}{}_{\alpha}\,(\zeta^{\alpha} r_{\alpha} + (\operatorname{grad} r_{\alpha})\,t^{\alpha})\right). \tag{3.4}$$

This equation is interpreted as the expression of the balance of moment of momentum for $\mathfrak{B}$; I refer to [7] and to earlier papers cited there for the derivation of conditions of the type (3.3) in many specific theories of non-simple continua.

With the use of (3.3), the power density (3.2) can be put into a form whose invariance under superposed rigid body motions is immediately apparent

$$-(T \cdot D + \underset{\sim}{\zeta} \cdot (\underset{\sim}{v} - \aleph s) + ב \cdot (\operatorname{grad} \underset{\sim}{v} - (\operatorname{grad} \aleph)^{t} s)), \tag{3.5}$$

or

$$-\left(T \cdot D + \sum_{1}^{q}{}_{\alpha}\,(\zeta^{\alpha}(v_{\alpha} - r_{\alpha} \cdot s) + t^{\alpha} \cdot (\operatorname{grad} v_{\alpha} - (\operatorname{grad} r_{\alpha})^{T} s)\right);$$

here D is the strain rate tensor and s is the spin in the macromotion:

$$D := \operatorname{sym} \operatorname{grad} v, \qquad s := -\tfrac{1}{2}\,\mathbf{e}\,(\operatorname{skw} \operatorname{grad} v), \tag{3.6}$$

and the exponent t affixed to grad $\aleph$ indicates transposition of the last two of the three indices.

The balance of energy is expressed by an appropriate modification of the classical relation, which involves the extra kinetic energy and the power of actions on the microstructure

$$\left(\int_{\mathscr{E}} \varrho(\varepsilon + \tfrac{1}{2}v^{2} + \varkappa)\right)^{\cdot} = \int_{\partial\mathscr{E}} (t \cdot v + \underset{\sim}{\tau} \cdot \underset{\sim}{v} + q \cdot n) + \int_{\mathscr{E}} \varrho(b \cdot v + \underset{\sim}{\beta} \cdot v + \chi); \tag{3.7}$$

here ε is internal energy per unit mass, q is heat flux into $\mathscr{E}$, χ is radiant heating per unit mass.

When the functions involved are sufficiently smooth one can deduce from (3.7) the local relation

$$\varrho\dot{\varepsilon} = T \cdot \operatorname{grad} v + \underset{\sim}{\zeta} \cdot \underset{\sim}{v} + ב \cdot \operatorname{grad} \underset{\sim}{v} + \operatorname{div} q + \varrho\chi. \tag{3.8}$$

Use will also be made later of some consequences of the Clausius-Duhem inequality, which will be accepted here in the standard form

$$\left(\int_{\mathscr{E}} \varrho\eta\right)^{\cdot} \geqq \int_{\partial\mathscr{E}} \frac{q \cdot n}{\theta} + \int_{\mathscr{E}} \varrho\frac{\chi}{\theta}, \tag{3.9}$$

where traditional notation is used for entropy η and temperature θ. The local consequence of (3.9) will be called upon below

$$\varrho\dot{\eta} \geqq \operatorname{div}\frac{q}{\theta} + \varrho\frac{\chi}{\theta}. \tag{3.10}$$

If one eliminates χ in (3.10) using (3.8) and expresses ε in terms of the Helmholtz free energy ψ

$$\psi := \varepsilon - \theta\eta,$$

then one obtains the reduced inequality

$$\varrho\,(\dot{\psi}+\eta\dot{\theta}) - T\cdot \operatorname{grad} v - \underset{\sim}{\zeta}\cdot \underset{\sim}{v} - \beth\cdot \operatorname{grad} \underset{\sim}{v} - \frac{q \operatorname{grad}\theta}{\theta} \leqq 0. \tag{3.11}$$

Referential versions of (2.14), (3.3) and (3.11) can be easily given by the introduction of Piola-type stresses involving the use of the deformation gradient F and its determinant $\iota := \det F$, *i.e.*

$$P := \iota T(F^{-1}),^{T} \quad \underset{\sim}{\vartheta} := \iota\underset{\sim}{\zeta}, \quad \daleth := \iota\,\beth(F^{-1})^{T}, \tag{3.12}$$

and the operators Div and Grad on the reference placement, so that, in particular,

$$\iota \operatorname{div}\beth = \operatorname{Div}\daleth.$$

The equations are $(\varrho^* = \varrho\iota)$:

$$\varrho_* \left(\left(\frac{\partial\varkappa}{\partial \underset{\sim}{v}}\right)^{\cdot} - \frac{\partial\varkappa}{\partial \underset{\sim}{v}}\right) = \varrho_*\underset{\sim}{\beta} - \underset{\sim}{\vartheta} + \operatorname{Div}\beth, \tag{3.13}$$

$$\mathbf{e}\,(PF^{T}) = \aleph^{T}\underset{\sim}{\vartheta} + (\operatorname{Grad}\aleph)'\,\gimel, \tag{3.14}$$

$$\varrho_*(\dot{\psi}+\eta\dot{\theta}) - P\cdot\dot{F} - \underset{\sim}{\vartheta}\cdot\underset{\sim}{v} - \daleth\cdot \operatorname{Grad}\underset{\sim}{v} - \frac{(F^{-1}q)\cdot \operatorname{Grad}\theta}{\theta} \leqq 0. \tag{3.15}$$

Remark. In [4] the existence also of a thermal microstructure is conjectured. Here I keep to less shifting ground.

4. Latent microstructure

I say that the microstructure is latent when, though its effects are felt in the balance equations, all relevant quantities can be expressed in terms of geometric and kinematic quantities pertaining to apparent placements.

More precisely I suppose that hypotheses (i)–(iv) below are satisfied which are introduced in order of increasing severity so that consequences of corresponding specificity can be deduced. The first two hypotheses are:

(i) there is no inertia connected with the microstructure, *i.e.*, $\varkappa$ vanishes identically:

(ii) there are no external body actions on the microstructure, *i.e.*, $\underset{\sim}{\beta}$ vanishes identically;

These hypotheses alone have several consequences which I explore first; under (i) and (ii) the balance equation (2.14) reduces to

$$\underset{\sim}{\zeta} = \operatorname{div}\beth, \tag{4.1}$$

or, on a local chart,

$$\zeta^{\alpha} = \operatorname{div} t^{\alpha},$$

and, in referential form,

$$\underset{\sim}{\vartheta} = \operatorname{Div}\daleth. \tag{4.2}$$

When constitutive prescriptions are assigned for $\underset{\sim}{\zeta}$ and $\beth$, eqn. (4.1) may take the analytical form of an evolution equation for the order parameters $\underset{\sim}{\nu}$.

As a consequence of (4.1) the balance equation of moment of momentum (3.3) becomes more compact

$$\mathbf{e}\, T = \operatorname{div}\,(\aleph^T \beth) \tag{4.3}$$

or

$$\mathbf{e}\, T = \operatorname{div}\left(\sum_1^q {}_\alpha\, r_\alpha \otimes t^\alpha\right),$$

and, in referential form,

$$\mathbf{e}\,(PF^T) = \operatorname{Div}\,(\aleph^T \text{ר}). \tag{4.4}$$

Formula (4.3) generalizes the Cosserats' moment equation obtained, within their context, under assumptions akin to (i) and (ii).

There is another interesting consequence of (4.1): the power density of internal actions due to the microstructure can be written as the divergence of a vector u:

$$\begin{aligned} \underset{\sim}{\zeta} \cdot \underset{\sim}{v} + \beth \cdot \operatorname{grad} \underset{\sim}{v} &= \operatorname{div} u, \\ u := \beth^T \underset{\sim}{v} &\left(= \sum_1^q {}_\alpha\, t^\alpha v_\alpha \right), \end{aligned} \tag{4.5}$$

so that (3.8) can be given the apparently standard form

$$\varrho \dot{\varepsilon} = T \cdot \operatorname{grad} v + \operatorname{div} q' + \varrho \chi, \tag{4.6}$$

where a modified flux vector q' appears

$$q' = q + u. \tag{4.7}$$

DUNN & SERRIN suggest, for their modified Neuman equation, the form (4.6), and call u the interstitial work flux. The introduction of different flux vectors, q' and q respectively, in the Neumann equation and in the Clausius-Duhem inequality has been suggested repeatedly by I. MÜLLER; it is, perhaps, worth remarking that, when only hypotheses (i) and (ii) are accepted, u and hence q' need not be objective.

If, instead, one assumes that:

(iii) the interstitial work flux is objective, then u must not change when $\underset{\sim}{v}$ is altered by the addition of a term of the form $\aleph w$, with w any vector; as a consequence (see $(4.5)_2$)

$$\beth^T \aleph = 0; \tag{4.8}$$

then condition (4.3) requires further that

$$T \in \mathrm{Sym}. \tag{4.9}$$

The last hypothesis is less precise; it is rather suggestive of many specific subcases, the simpler of which I study in the following Sections:

(iv) a set of frictionless holonomic or anholonomic constraints expresses either the order parameters $\underset{\sim}{\nu}$ in terms of displacement gradient F and, perhaps, its gradients or the generalized velocities $\underset{\sim}{v}$ in terms of grad v and, perhaps, higher gradients of v.

5. The elastic materials of Dunn & Serrin

I observe in this Section that the elastic materials studied by DUNN & SERRIN in [6] can be considered as special continua with latent microstructure when, on the one hand, hypothesis (iv) is rendered more specific as follows:
(iv)′ the internal parameters are constrained by a condition:

$$\underset{\sim}{\nu} = \hat{\underset{\sim}{\nu}}(F), \tag{5.1}$$

and, on the other hand, it is also presumed that
(v) free energy, entropy, heat flux, the stress and microstress are functions of F, Grad F, θ and grad θ.

The implications of (iv)′ and (v) are less cumbersome to obtain if Cartesian components (specified below, with latin letters for indices) are used (repeated indices are to be summed). From (5.1) it follows that

$$v_\alpha = \frac{\partial \hat{v}_\alpha}{\partial F_{aA}} v_{a,b} F_{bA} \tag{5.2}$$

and, in a rigid motion with angular velocity w,

$$v_\alpha^{(R)} = \mathbf{e}_{abc} \frac{\partial \hat{v}_\alpha}{\partial F_{aA}} F_{cA} w_b, \tag{5.3}$$

so that

$$\aleph_{\alpha a} = \mathbf{e}_{abc} \frac{\partial \hat{v}_\alpha}{\partial F_{cA}} F_{bA}. \tag{5.4}$$

Condition (4.8) now requires that the third-order tensor

$$\mathbf{s}_{ijk} := \sum_{1}^{q} {}_\alpha \overset{\alpha}{\beth}_i \frac{\partial \hat{v}_\alpha}{\partial F_{jA}} F_{kA} \tag{5.6}$$

be symmetric in the last two indices

$$\mathbf{s}_{ijk} = \mathbf{s}_{ikj}. \tag{5.7}$$

The condition that the constraint (5.1) is frictionless is interpreted in the usual sense, rendered precise in a general setting by GURTIN & PODIO-GUIDUGLI (see [7]):

(a) stress and microstresses are each the sum of two terms, active and reactive respectively,

$$T = \overset{a}{T} + \overset{r}{T}, \qquad \underset{\sim}{\zeta} = \overset{a}{\underset{\sim}{\zeta}} + \overset{r}{\underset{\sim}{\zeta}}, \qquad \beth = \overset{a}{\beth} + \overset{r}{\beth}; \tag{5.8}$$

(b) the power of the reactive parts of stress and microstress vanishes for all velocity fields allowed by the constraints; this condition is more conveniently expressed in terms of Piola-type stresses:

$$\overset{r}{P} \cdot \dot{F} + \overset{r}{\underset{\sim}{\vartheta}} \cdot \underset{\sim}{v} + \overset{r}{\daleth} \cdot \operatorname{Grad} \underset{\sim}{v} = 0, \tag{5.9}$$

for all F and all v given by (5.2).

It is easy to check that (5.9) implies

$$\overset{r}{P}_{iA} + \sum_1^q {}_\alpha\left(\overset{r}{\vartheta}{}^\alpha \frac{\partial \hat{v}_\alpha}{\partial F_{iA}} + \overset{r}{\daleth}{}^\alpha_B \left(\frac{\partial \hat{v}_\alpha}{\partial F_{iA}}\right)_{,B}\right) = 0, \tag{5.10}$$

$$\sum_1^q {}_\alpha \overset{r}{\daleth}{}^\alpha_A \frac{\partial \hat{v}_\alpha}{\partial F_{iB}} + \sum_1^q {}_\alpha \overset{r}{\daleth}{}^\alpha_B \frac{\partial \hat{v}_\alpha}{\partial F_{iA}} = 0. \tag{5.11}$$

On the other hand property (a) above and (5.11) imply

$$\sum_1^q {}_\alpha \overset{a}{\daleth}{}^\alpha_A \frac{\partial \hat{v}_\alpha}{\partial F_{iB}} - \sum_1^q {}_\alpha \overset{a}{\daleth}{}^\alpha_B \frac{\partial \hat{v}_\alpha}{\partial F_{iA}} = 0. \tag{5.12}$$

The interpretation *à la* COLEMAN-NOLL of the dissipation inequality and an ensuing well-known train of reasoning, together with hypothesis (v) and (5.12), exclude a dependence of ψ and η on grad θ, impose the usual relation between these two functions and, besides, lead to the relations

$$\overset{a}{P}_{iA} + \sum_1^q {}_\alpha\left(\overset{a}{\vartheta}{}^\alpha \frac{\partial \hat{v}_\alpha}{\partial F_{iA}} + \left(\overset{a}{\daleth}{}^\alpha_B \frac{\partial \hat{v}_\alpha}{\partial F_{iA}}\right)_{,B} - \overset{a}{\daleth}{}^\alpha_{B,B} \frac{\partial \hat{v}_\alpha}{\partial F_{iA}}\right) = \varrho_* \frac{\partial \psi}{\partial F_{iA}}, \tag{5.13}$$

$$\sum_1^q {}_\alpha \left(\overset{a}{\daleth}{}^\alpha_B \frac{\partial \hat{v}_\alpha}{\partial F_{iA}}\right) = \varrho_* \frac{\partial \psi}{\partial F_{iA,B}}; \tag{5.14}$$

observe that the tensor on the right-hand side of (5.14) is symmetric in the indices A, B.

Summing up term by term (5.10) and (5.13) and using (5.14) one arrives at the relation

$$P_{iA} = \varrho_* \frac{\partial \psi}{\partial F_{iA}} - \left(\varrho_* \frac{\partial \psi}{\partial F_{iA,B}}\right)_{,B} - \left(\sum_1^q {}_\alpha \overset{r}{\daleth}{}^\alpha_B \frac{\partial \hat{v}_\alpha}{\partial F_{iA}}\right)_{,B}. \tag{5.15}$$

To render explicit in terms of the derivatives of ψ the last term in (5.15) it suffices to exploit (5.7) in the manner shown by DUNN & SERRIN in Appendix A of their paper and so obtain

$$P_{iA} = \varrho_* \frac{\partial \psi}{\partial F_{iA}} - \left(\varrho_* \frac{\partial \psi}{\partial F_{iA,B}}\right)_{,B} - \left(\varrho_* F_{iB}\left(\frac{\partial \psi}{\partial F_{jB,C}} F^{-1}_{Aj} - \frac{\partial \psi}{\partial F_{jB,A}} F^{-1}_{Cj}\right)\right)_{,C}, \tag{5.16}$$

which is DUNN & SERRIN's formula (3.4).

As observed by DUNN & SERRIN, the symmetry of PF^T is ensured automatically because ψ is an objective scalar.

6. A subcase of the Korteweg fluid

The reduction of (5.16) to the special case when ψ depends (on the temperature θ,) on $\iota = \det F$ and its first gradient only has already been performed by DUNN & SERRIN and so needs not be reconsidered here. However, that case can be viewed also from a different angle, and therefore a few remarks are in order.

Suppose that the microstructure be such that $\underset{\sim}{v}$ (as happens when $\underset{\sim}{\nu}$ is determined through ι alone) be not affected by superposed rigid rotations ($\aleph \equiv 0$); then the interstitial work flux is automatically objective and assertion (iii) becomes a theorem.

If, in addition, there is only one internal parameter (then the notation can reduce to the usual letters without superscript and wavy underlining: ν, ζ, t) and that parameter is constrained to coincide with $\imath$, $\nu \equiv \iota$, then one does not even need referential arguments. One can take into account Euler's formula for $\dot{\iota}$

$$\dot{\iota} = \iota \operatorname{div} v$$

and its consequence

$$\operatorname{grad} \dot{\iota} = (\operatorname{grad} \iota)^{\cdot} + (\operatorname{grad} v)^T \operatorname{grad} \iota,$$

to put expression (3.2) for the power density into the form

$$A \cdot \operatorname{dev} (T + (\operatorname{grad} \iota) \otimes t) + \alpha (\operatorname{tr} T + 3\iota\zeta + (\operatorname{grad} \iota) \cdot t) + t \cdot (\operatorname{grad} \iota)^{\cdot}, \tag{6.1}$$

$$A := \operatorname{dev} \operatorname{grad} v, \qquad \alpha = \tfrac{1}{3} \operatorname{div} v, \tag{6.2}$$

to obtain directly, on the one hand,

$$\overset{r}{t} = 0, \quad \operatorname{dev} \overset{r}{T} = 0, \quad \operatorname{tr} \overset{r}{T} + 3\iota\overset{r}{\zeta} = 0 \tag{6.3}$$

and hence

$$\operatorname{tr} \overset{a}{T} - 3\iota\overset{a}{\zeta} = 0,$$

and, on the other hand,

$$\overset{a}{t} = \varrho \frac{\partial \psi}{\partial (\operatorname{grad} \iota)}, \tag{6.4}$$

$$\operatorname{div} (\overset{a}{T} + (\operatorname{grad} \iota) \otimes \overset{a}{t}) = 0,$$

$$\operatorname{tr} \overset{a}{T} + 3\iota\overset{a}{\zeta} + (\operatorname{grad} \iota) \cdot \overset{a}{t} = 3\iota\varrho \frac{\partial \psi}{\partial \iota},$$

together with (see (4.1))

$$\overset{a}{\zeta} + \overset{r}{\zeta} = \operatorname{div} \overset{a}{t}. \tag{6.5}$$

The final expression for T,

$$T = \left(\varrho\iota \frac{\partial \psi}{\partial \iota} - \iota \operatorname{div} \left(\varrho \frac{\partial \psi}{\partial (\operatorname{grad} \iota)}\right)\right) 1 - (\operatorname{grad} \iota) \otimes \varrho \frac{\partial \psi}{\partial (\operatorname{grad} \iota)}, \tag{6.6}$$

coincides with the expression (1.25) given by DUNN & SERRIN. Notice again that the symmetry of T is assured, because the dependence of ψ on grad ι may occur only through the modulus $|\operatorname{grad} \iota|$, as ψ is an objective scalar.

7. A more general elastic fluid of the Korteweg class

Suppose that $\mathcal{Q}$ coincides with $\mathcal{V}\times\mathbb{R}$ ($q = 4$), that ν_1, ν_2, ν_3 are Cartesian components of a vector d and ν_4 is a scalar ν; then we can put

$$\underset{\sim}{\zeta} \equiv \begin{pmatrix} z \\ \zeta \end{pmatrix} \quad \text{and} \quad \daleth \equiv \begin{pmatrix} V \\ t \end{pmatrix},$$

with $\zeta \in \mathbb{R}$, $z, t \in \mathcal{V}$ and $V \in \mathrm{Lin}$ and write the balance equation of moment of momentum in the form

$$z = \operatorname{div} V, \qquad \zeta = \operatorname{div} t. \tag{7.1}$$

Suppose further that

(iv)″ appropriate constraints bind d and ν to ι and grad ι; more precisely

$$d = \operatorname{grad} \varrho, \qquad \nu = \varrho. \tag{7.2}$$

Remark. Of course, ϱ is in one-to-one correspondence with ι and one could, perhaps more appropriately, accept an alternative direct definition of d and ν in terms of ι and grad ι. The choice (7.2) brings the following developments more strictly in line with those of DUNN & SERRIN.

When the notation (6.2) is used and account is taken of obvious consequences of the constraints (7.2)

$$\dot{d} = -3 \operatorname{grad}(\varrho\alpha) - (A + \alpha 1) \operatorname{grad} \varrho, \qquad \dot{\nu} = -3\varrho\alpha,$$

the power density of internal actions

$$T \cdot \operatorname{grad} v + z \cdot \dot{d} + V \cdot \operatorname{grad} \dot{d} + t \cdot \operatorname{grad} \dot{\nu} + \zeta\dot{\nu},$$

can be given the expression

$$\begin{aligned} &\alpha(\operatorname{tr} T - (4z + 3t) \cdot \operatorname{grad} \varrho - 4V \cdot \operatorname{grad}^2 \varrho - 3\varrho\zeta) \\ &- 3 (\operatorname{grad} \alpha) \cdot (\varrho(z + t) + 3 (\operatorname{sym} V) \operatorname{grad} \varrho) \\ &- 3\varrho (\operatorname{grad}^2 \alpha) \cdot V + (\operatorname{grad}^2 v) \cdot ((\operatorname{grad} \varrho) \otimes V) \\ &+ A \cdot \operatorname{dev} (T - (\operatorname{grad}^2 \varrho) V^T - (\operatorname{grad} \varrho) \otimes z). \end{aligned}$$

It follows that the reactive components of stress and microstress satisfy the conditions

$$\begin{aligned} \overset{r}{T} &= \varrho\overset{r}{\zeta}\, 1 + (\operatorname{grad} \varrho) \otimes \overset{r}{z} - (\operatorname{grad}^2 \varrho) \overset{r}{V}, \\ \overset{r}{z} + \overset{r}{t} &= 0, \qquad \overset{r}{V} \in \mathrm{Skw}. \end{aligned} \tag{7.3}$$

Assume now that

(v)′ free energy, entropy, heat flux, stress and microstress are functions of ϱ, grad ϱ, grad^2 ϱ, θ and grad θ.

Because

$$\begin{aligned} (\operatorname{grad}^2 \varrho)^{\cdot} = &-5\alpha \operatorname{grad}^2 \varrho - 6 \operatorname{sym} ((\operatorname{grad} \varrho) \otimes (\operatorname{grad} \alpha)) \\ &- 3\varrho \operatorname{grad}^2 \alpha - 2 \operatorname{sym} ((\operatorname{grad}^2 \varrho) A) - B, \end{aligned}$$

with

$$B_{ij} = \varrho_{,h} v_{h,ij}, \qquad B \in \mathrm{Sym},$$

the total time derivative of ψ is given by

$$\dot{\psi} = \frac{\partial \psi}{\partial \theta} \dot{\theta} + \frac{\partial \psi}{\partial \theta_{,i}} (\theta_{,i})^{\cdot} - \left(3\varrho \frac{\partial \psi}{\partial \varrho} + 4 \frac{\partial \psi}{\partial \varrho_{,i}} \varrho_{,i} + 5 \frac{\partial \psi}{\partial \varrho_{,ij}} \varrho_{,ij}\right) \alpha$$

$$- \left(3\varrho \frac{\partial \psi}{\partial \varrho_{,i}} + 6\varrho_{,j} \frac{\partial \psi}{\partial \varrho_{,ij}}\right) \alpha_{,i} - 3\varrho \frac{\partial \psi}{\partial \varrho_{,ij}} \alpha_{,ij}$$

$$- \left(\frac{\partial \psi}{\partial \varrho_{,j}} \varrho_{,i} + \frac{\partial \psi}{\partial \varrho_{,kj}} \varrho_{,ik} + \frac{\partial \psi}{\partial \varrho_{,ik}} \varrho_{,jk}\right) A_{ij} - \frac{\partial \psi}{\partial \varrho_{,ij}} B_{ij}.$$

Then, again, the argument first introduced by COLEMAN & NOLL leads to the usual conclusion that ψ does not depend on θ and grad θ, and to the following relations for the active components of stress and microstress:

$$\overset{a}{V} = \varrho \frac{\partial \psi}{\partial (\mathrm{grad}^2 \varrho)}, \qquad \overset{a}{z} + \overset{a}{t} = \varrho \frac{\partial \psi}{\partial (\mathrm{grad}\, \varrho)},$$

$$\overset{a}{T} - (\mathrm{grad}\, \varrho) \otimes \overset{a}{z} - \varrho \overset{a}{\zeta}\, 1 = \tag{7.4}$$

$$-\varrho^2 \frac{\partial \psi}{\partial \varrho} 1 - \varrho\, (\mathrm{grad}\, \varrho) \otimes \frac{\partial \psi}{\partial (\mathrm{grad}\, \varrho)} - \varrho\, (\mathrm{grad}^2\, \varrho) \left(\frac{\partial \psi}{\partial (\mathrm{grad}\ ^2 \varrho)}\right).$$

Formulae (7.3), (7.4) together lead to

$$V = \varrho \frac{\partial \psi}{\partial (\mathrm{grad}^2 \varrho)} + \overset{r}{V}, \qquad \overset{r}{V} \in \mathrm{Skw},$$

$$z + t = \varrho \frac{\partial \psi}{\partial (\mathrm{grad}\, \varrho)}, \tag{7.5}$$

$$T = \varrho \left(\zeta - \varrho \frac{\partial \psi}{\partial \varrho}\right) 1 - (\mathrm{grad}\, \varrho) \otimes t - (\mathrm{grad}^2\, \varrho)\, V.$$

The balance equations (7.1) allow us to eliminate ζ and t from $(7.5)_3$

$$T = -\varrho \left(\varrho \frac{\partial \psi}{\partial \varrho} + \mathrm{div}\left(\mathrm{div}\left(\varrho \frac{\partial \psi}{\partial (\mathrm{grad}^2 \varrho)}\right) - \varrho \frac{\partial \psi}{\partial (\mathrm{grad}\, \varrho)}\right)\right) 1$$

$$+ (\mathrm{grad}\, \varrho) \otimes \mathrm{div}\, V - (\mathrm{grad}^2\, \varrho)\, V - \varrho\, (\mathrm{grad}\, \varrho) \otimes \frac{\partial \psi}{\partial (\mathrm{grad}\, \varrho)}. \tag{7.6}$$

Finally to express the reactive part of V in terms of the derivatives of ψ, we exploit the assumption of objectivity of the interstitial working. In a rigid motion with rotational speed w

$$\dot{\varrho} = 0, \qquad (\mathrm{grad}\, \varrho)^{\cdot} = (\mathrm{grad}\, \varrho) \times w,$$

so that the assumption mentioned is equivalent to the condition

$$((\mathrm{grad}\, \varrho) \otimes V)\, v \in \mathrm{Sym} \qquad \forall\, v \in \mathscr{V}. \tag{7.7}$$

Following now the argument, already cited, of DUNN & SERRIN, we obtain from (7.5) and (7.7)

$$\varrho_{,k} \overset{r}{V}_{ij} = \varrho \left(\varrho_{,i} \frac{\partial \psi}{\partial \varrho_{,jk}} - \varrho_{,j} \frac{\partial \psi}{\partial \varrho_{,ki}} \right),$$

with the conclusion that

$$\begin{aligned} T_{ij} = &-\varrho \left(\varrho \frac{\partial \psi}{\partial \varrho} + \left(\left(\varrho \frac{\partial \psi}{\partial \varrho_{,hk}} \right)_{,h} - \varrho \frac{\partial \psi}{\partial \varrho_{,k}} \right)_{,k} \delta_{ij} \right) \cdot \\ &+ \varrho_{,i} \left(\left(\varrho \frac{\partial \psi}{\partial \varrho_{,jk}} \right)_{,k} - \varrho \frac{\partial \psi}{\partial \varrho_{,j}} \right) - \varrho_{,ik} \varrho \frac{\partial \psi}{\partial \varrho_{,kj}} \qquad (7.8) \\ &+ \varrho \left(\varrho_{,j} \frac{\partial \psi}{\partial \varrho_{,ki}} - \varrho_{,k} \frac{\partial \psi}{\partial \varrho_{,ij}} \right)_{,k}, \end{aligned}$$

a formula which generalizes (6.6) and may be the starting point for ampler analyses regarding the class of elastic fluids of the Korteweg type. The formula does not fall within the subclass considered by DUNN & SERRIN; one of their hypotheses on the interstitial working excludes a dependence of ψ on $\text{grad}^2 \varrho$.

Acknowledgements. This work is part of a research program under the auspices of the Italian Ministry of Education. I thank also the Department of Mathematics of Heriot-Watt University in Edinburgh for support given me for a short visit, during which I had the leisure to write the final draft.

References

1. G. CAPRIZ & P. PODIO-GUIDUGLI, Materials with finite-dimensional structure, in A. P. S. SELVADURAI, ed., *Mechanics of Structured Media.* Elsevier (1981), 255–268.
2. G. CAPRIZ, Introductory remarks to the dynamics of continua with microstructure, im *Mathematical Methods and Models in Mechanics.* Banach Center Publ. **15** (1985), 85–109.
3. G. CAPRIZ & P. PODIO-GUIDUGLI, Structured continua from a Lagrangian point of view. *Ann. Mat. Pura Appl.* **135** (1983), 1–25.
4. G. CAPRIZ, Continui con microstruttura. *Boll. Un. Mat. Ital.* **3A** (1984), 181–195.
5. R. A. TOUPIN, Theories of elasticity with couple-stress. *Arch. Rational Mech. An.* **17** (1964), 85–112.
6. J. E. DUNN & J. SERRIN, On the thermomechanics of interstitial working. IMA Preprint 24 (1983).
7. G. A. MAUGIN, The method of virtual power in continuum mechanics. *Acta Mechanica* **35** (1980), 1–70.
8. M. GURTIN & P. PODIO-GUIDUGLI, The thermodynamics of constrained materials. *Arch. Rational Mech. An.* **51** (1973), 192–208.

Università di Pisa
Italy

(Received October 19, 1984)

A Skin Effect Approximation for Eddy Current Problems

R. C. MacCamy & Ernst Stephan

Dedicated to Walter Noll

1. Introduction

This paper concerns an *interface* problem in the plane. It arises in the study of *eddy currents* in electromagnetic theory as outlined in section two. Let Γ be a simple, closed curve in E^2 with interior Ω_- and exterior Ω_+. For any $\chi \in C(\bar{\Omega}_\pm)$ let $\chi^\pm$ denote the limits from $\Omega_\pm$. Let $\boldsymbol{n}$ be the exterior normal to Γ and write χ_n for $\operatorname{grad} \chi \cdot \boldsymbol{n}$.

Problem $P(\alpha)$. Given $\alpha > 0$ and a, constants, and $f, g \in C(\Gamma)$ find u such that;

$$\begin{aligned} &\Delta u = i\alpha^2 u \text{ in } \Omega_-; \quad \Delta u = 0 \text{ in } \Omega_+ \\ &u^- = u^+ + f, \quad u_n^- = u_n^+ + g \text{ on } \Gamma \\ &u(\boldsymbol{x}) - a \log |\boldsymbol{x}| = O(1) \text{ as } |\boldsymbol{x}| \to \infty. \end{aligned} \tag{1.1}$$

We will also be concerned with the boundary-value problem:

Problem $P(\infty)$. Given a, constant, and $f \in C(\Gamma)$ find u such that;

$$\begin{aligned} &\Delta u = 0 \text{ in } \Omega_+; \quad u^+ + f = 0 \text{ on } \Gamma \\ &u(\boldsymbol{x}) - a \log |\boldsymbol{x}| = O(1) \text{ as } |\boldsymbol{x}| \to \infty. \end{aligned} \tag{1.2}$$

Our concern is with the behavior of the solution of $P(\alpha)$ as α tends to infinity and the work continues that in [2]. It is shown in [2] that $P(\alpha)$ and $P(\infty)$ each have unique solutions. Further, a formal asymptotic expansion for the solution of $P(\alpha)$ was given in [2]. This had the form,

$$u \sim \begin{cases} \sum_{k=0}^{\infty} u_k^+ (\sqrt{i}\alpha)^{-k} \text{ in } \Omega^+ \\ \chi \sum_{k=1}^{\infty} u_k^- (\sqrt{i}\alpha)^{-k}, \quad \chi = e^{-\sqrt{i}\alpha\tau} \text{ in } \Omega^-. \end{cases} \tag{1.3}$$

Here τ is the distance measured along the inner normal into Ω_- from Γ. The coefficient u_0^+ is the solution of $P(\infty)$. The subsequent $u_k^\pm$ are computed recursively; the u_k^+ by solving problems of the same form as $P[\infty]$ and the u_k^- are determined by quadrature. A version of this procedure for three dimensional problems is described in [5].

The construction of (1.3) is reviewed in section three. The main new result, given in section four, is a proof of the validity of (1.3). The procedure (1.3), essentially reduces $P(\alpha)$ to $P(\infty)$, a considerable saving for numerical work as discussed in [2]. It is also of interest in that the term χ represents the well known *skin-effect* for eddy current problems.

Numerical experiments presented in [2] indicate that the skin effect approximation yields quite accurate results over a large range of values of α.

2. Physical Background

In the general eddy current problem one has an initial electromagnetic field in all space (thought of as air, a dielectric). One inserts into this field a metallic obstacle. This distorts the initial field and induces *eddy currents* in the obstacle.

We make the following assumptions:

(i) Conduction current in air and displacement current in metal can be neglected.

(ii) The metal is non-ferromagnetic, isotropic and homogeneous.

(iii) The metal and air have the same magnetic permeability.

(iv) The initial field is *transverse* magnetic; the electric field is in a fixed direction and the magnetic field orthogonal to it and all fields are uniform in the direction of the electric field.

(v) The obstacle is a uniform cylinder in the electric field direction.

(vi) All fields start from rest.

It follows from analysis in [2] that a mathematical model of the above situation has the following form. For any region $S \subset E^2$ write $\boldsymbol{S}$ for $(0,\infty)\times S$ and for χ on $\boldsymbol{S}$ let $\dot{\chi}$ be its t-derivative. Let Ω_- be the cylinder's cross section and $\Gamma = \partial\Omega_-$.

Problem $\mathscr{P}(\gamma,\beta)$. Given $\gamma > 0$, $\beta > 0$, constants, and $F, G \in C(\boldsymbol{\Gamma})$ find U such that

$$\begin{gathered} \Delta U = \gamma\dot{U} \text{ in } \boldsymbol{\Omega}_-; \quad \Delta U = \beta\ddot{U} \text{ in } \boldsymbol{\Omega}_+ \\ U^- = U^+ + F, \quad U_n^- = U_n^+ + G \text{ on } \boldsymbol{\Gamma} \\ U(0,\boldsymbol{x}) \equiv 0 \text{ in } \Omega_-; \quad U(0,x) \equiv \dot{U}(0,\boldsymbol{x}) \equiv 0 \quad \text{in } \Omega_+. \end{gathered} \tag{2.1}$$

A scaling has been performed in (2.1) so that all quantities are dimensionless. The (dimensionless) electric field $\boldsymbol{E}$ and magnetic displacements $\boldsymbol{B}$ are given by,

$$\boldsymbol{E} = -U\boldsymbol{e}_3; \quad \boldsymbol{B} = (\operatorname{grad} U)^\perp, \tag{2.2}$$

where $\boldsymbol{e}_3$ is in the direction of the cylinder and $(m\boldsymbol{e}_1 + n\boldsymbol{e}_2)^\perp = n\boldsymbol{e}_1 - m\boldsymbol{e}_2$. $(2.1)_1$ represents Faraday's and Ampère's laws while $(2.1)_2$ corresponds to continuity of tangential electric and magnetic fields across the interface. In (2.1) U represents the total field in Ω_- and the scattered field in Ω_+. The quantities F and G are U^0 and U_n^0 on Γ where U^0 generates the initial field.

There are two standard approximations in the study of eddy currents.

I. Quasi-static approximations. The dimensionless parameter β is usually very small numerically and a standard engineering procedure is to take it to be zero. There is a technical difficulty in this as described in [2]. In meaningful physical problems when $\beta \to 0$ $\dot{U}^0$ tends to a limit function which grows logarithmically as $|x| \to \infty$. One must accordingly allow the scattered field U in Ω_+ to grow the same way in order to keep the total field finite. One obtains then:

Problem $\mathscr{P}(\gamma, 0)$. Given $\gamma > 0$, constant, $F, G \in C(\Gamma)$ and $A \in C[0, \infty)$ find U such that

$$\begin{gathered} \Delta U = \gamma \dot{U} \text{ in } \Omega_-; \quad \Delta U = 0 \text{ in } \Omega_+ \\ U^- = U^+ + F; \quad U_n^- = U_n^+ + G \text{ on } \Gamma \\ U(0, x) = 0 \text{ in } \Omega_-; \quad U(t, x) - A(t) \log |x| = O(1) \text{ as } |x| \to \infty. \end{gathered} \tag{2.3}$$

(Notice that one omits the initial conditions in Ω_+.)

II. Perfect conductor approximations. It is often true that the dimensionless parameter γ is large numerically. (It depends linearly on the conductivity.) Another standard practice is to let $\gamma \to \infty$ in (2.3). Here one sets $U \equiv 0$ in Ω_- and neglects the second of $(2.3)_2$. This yields:

Problem $\mathscr{P}(\infty, 0)$. Given $F \in C(\Gamma)$ and $A \in C[0, \infty)$ find U such that:

$$\begin{gathered} \Delta U = 0 \text{ in } \Omega^+; \quad 0 = U^+ + F \text{ on } \Gamma \\ U(t, x) - A(t) \log |x| = O(1) \text{ as } |x| \to \infty. \end{gathered} \tag{2.4}$$

(Notice that here t appears only as a parameter.)

The final step which leads to the problems of section one is to apply Laplace transforms. Let u, f, g, a represent the transforms of the functions U, F, G, A for the transform variable $s = i\eta$. Then one verifies that for $\mathscr{P}(\infty, 0)$ u satisfies $P(\infty)$ and for $\mathscr{P}(\gamma, 0)$ u satisfies $P(\sqrt{\gamma\eta})$. Conversely, if $u(\eta, x)$ represents the solution of $P(\infty)$ or $P(\sqrt{\gamma\eta})$ one can, formally, recover the solutions of $\mathscr{P}(\infty, 0)$ or $\mathscr{P}(\gamma, 0)$ from the inversion integrals,

$$U(t, x) = (2\pi)^{-1} \int_{-\infty}^{+\infty} e^{i\eta t} u(\eta; x)\, d\eta. \tag{2.5}$$

The present status of the various problems, under appropriate smoothness conditions, can be summarized as follows.

Theorem 2.1. (i) *There is at most one solution of* $\mathscr{P}(\gamma, \beta)$, $\mathscr{P}(\gamma, 0)$, $P(\alpha)$ *and* $P(\infty)$.
(ii) *There are solutions* u_α, u_∞ *of* $P(\alpha)$ *and* $P(\infty)$.

The proofs for $P(\alpha)$ and $P(\infty)$ are given in [2]. The uniqueness proofs for $\mathscr{P}(\gamma, \beta)$ and $\mathscr{P}(\gamma, 0)$ are easy consequences of Green's theorem. The result in section four shows that $u_\alpha \to u_\infty$ (in certain norms) as $\alpha \to \infty$. We make the following conjectures:

(i) There are solutions $U_{\gamma\beta}$ and $U_{\gamma 0}$ of $\mathscr{P}(\gamma, \beta)$ and $\mathscr{P}(\gamma, 0)$.
(ii) $U_{\alpha\beta} \to U_{\gamma 0}$ as $\beta \to 0$, $\quad U_{\gamma 0} \to u_\infty$ as $\gamma \to \infty$.

Remark 2.1. The problem $P(\sqrt{\gamma\eta})$ also arises if the incident field, and hence F and G, are time periodic with frequency η and one looks for limit solutions of $\mathscr{P}(\gamma, 0)$ of the same form.

Remark 2.2. It is clear that the expansion (1.3) greatly facilitates the approximate calculation of the integral (2.5) for $\mathscr{P}(\gamma, 0)$ for it can be used to approximate that portion of the integral where η is large.

Remark 2.3. The reduction of $\mathscr{P}(\gamma, \beta)$ to $\mathscr{P}(\gamma, 0)$, justified above on physical scaling grounds, also has an interesting interpretation in terms of *history* problems. In order to be tractable numerically the problems $\mathscr{P}(\gamma, \beta)$ must somehow be reduced to a problem on a finite domain. To this end let us consider the inner boundary problem,

$$\Delta U = \gamma \dot{U} \text{ in } \boldsymbol{\Omega}_-; \quad U_n^- = \Psi_- \text{ on } \boldsymbol{\Gamma}; \quad U(0, \boldsymbol{x}) \equiv 0 \text{ in } \Omega_-. \tag{2.6}$$

If Ψ_- were known this would be a standard problem, but Ψ_- is not known. Consider, however, the corresponding exterior problem:

$$\Delta U = \beta^2 \ddot{U} \text{ in } \boldsymbol{\Omega}_+; \quad U_n^+ = \Psi_+ \text{ on } \boldsymbol{\Gamma}; \quad U(0, \boldsymbol{x}) \equiv \dot{U}(0, \boldsymbol{x}) \equiv 0 \text{ in } \Omega_+. \tag{2.7}$$

The solution of (2.7), at $(t, \boldsymbol{x})$, is a linear functional of the history $\Psi_+^t(\Psi_+^t(\tau, \cdot) = \Psi_+(t - \tau, \cdot),\ 0 \leqq \tau \leqq t)$ of the values of Ψ_+ on Γ. Thus

$$U(t, \boldsymbol{x}) = \mathscr{I}_\beta(\Psi_+^t(\cdot, \cdot))(\boldsymbol{x}). \tag{2.8}$$

If we let $\boldsymbol{x}$ tend to Γ we obtain then,

$$U^+(t, \boldsymbol{x}) = \mathscr{I}_\beta^+(\Psi_+^t(\cdot, \cdot))(\boldsymbol{x}) \text{ on } \boldsymbol{\Gamma}. \tag{2.9}$$

Now (2.6), (2.7) and the second interface condition yield $\Psi_- = \Psi_+ + G$. Hence if we substitute into (2.9) and the first interface condition we have,

$$U^-(t, \boldsymbol{x}) = \mathscr{I}_\beta^+(\Psi_-^t(\cdot, \cdot))(\boldsymbol{x}) + F(t, \boldsymbol{x}) - \mathscr{I}_\beta^+(G^t(\cdot, \cdot))(\boldsymbol{x}) \text{ on } \boldsymbol{\Gamma}. \tag{2.10}$$

The above calculation shows that we can attempt to solve $\mathscr{P}(\gamma, \beta)$ by seeking the pair (U, Ψ_-) satisfying (2.6) and (2.10). Note that condition (2.10) is non-local with respect to both space and time. Suppose, however, we apply the same ideas to $\mathscr{P}(\gamma, 0)$. We have,

$$\Delta U = 0 \text{ in } \boldsymbol{\Omega}_+; \quad U_n^+ = \Psi_+ \text{ on } \boldsymbol{\Gamma}; \quad U(t, \boldsymbol{x}) - A(t) \log |\boldsymbol{x}| = O(1). \tag{2.11}$$

Now the solution at $(t, \boldsymbol{x})$ depends only on the present time values of Ψ_+ and A. Then (2.9) is replaced by,

$$U^+(t, \boldsymbol{x}) = \mathcal{I}_0^+(\Psi_-(t, \cdot))(\boldsymbol{x}) + A(t)\, k(\boldsymbol{x}) \tag{2.12}$$

where $k(\boldsymbol{x})$ is known. The analog of (2.10) is non-local only with respect to space.

We observe that (2.9) is a realization in $\mathcal{P}(\infty, \beta)$ of the general idea of an *environmental* operator introduced by NOLL in [6]. Moreover, the simplified condition (2.12) is closely related to the idea of COLEMAN & NOLL ([1]) of approximating functionals on history space. $\beta \ll 1$ corresponds to having "slowly varying fields" in one time scale.

3. The Formal Expansion

We begin this section by giving a solution procedure for $P(\infty)$. This procedure was introduced in [3] and analyzed further in [4]. It is particularly well adapted to our purposes.

We introduce the simple layer $\mathcal{S}[\sigma]$,

$$\mathcal{S}[\sigma](\boldsymbol{x}) = (2\pi)^{-1} \int_\Gamma \sigma(y) \log|\boldsymbol{x} - y|\, ds_y. \tag{3.1}$$

For a smooth Γ and smooth σ it is well known that $\mathcal{S}[\sigma] \in C(E^2)$, with value $S(\sigma]$ on γ and $\Delta\mathcal{S}[\sigma] \equiv 0$ in $\Omega_- \cup \Omega_+$. Its normal derivative, however, jumps across Γ according to the relation,

$$(\mathcal{S}[\sigma])_n^\pm(\boldsymbol{x}) = \pm \tfrac{1}{2}\sigma(\boldsymbol{x}) + N[\sigma](\boldsymbol{x}) \tag{3.2}$$

where N is an integral operator with smooth kernel. The result from [3] is the following, again under appropriate smoothness assumptions:

Theorem 3.1. *For any f and a there exists a unique σ on Γ and constant C such that,*

$$S[\sigma] + C + f = 0 \text{ on } \Gamma; \quad (2\pi)^{-1} \int_\Gamma \sigma(y)\, ds_y = a \tag{3.3}$$

and

$$u = \mathcal{S}[\sigma] + C \tag{3.4}$$

is a solution of $P(\infty)$.

Note that for this solution,

$$u_n^+ = \tfrac{1}{2}\sigma + N[\sigma]. \tag{3.5}$$

Recall that in the application of $P(\alpha)$ to $\mathcal{P}(\gamma, 0)$, f, g, and a are the Laplace transforms of F, G, A. If F is very smooth with respect to t, f will have an expansion for large η of the form,

$$f(\eta, \boldsymbol{x}) \sim \sum_{n=1}^{\infty} \left(\frac{\partial}{\partial t}\right)^n f(t, \boldsymbol{x})\big|_{t=0} (i\eta)^{-n} \tag{3.6}$$

with similar expressions for g and a. This means that we should study $P(\alpha)$ under the assumption that f, g and a have asymptotic expansions in powers of α^{-1}. In the interest of simplicity we will assume f, g, and a are independent of α. The extension to the more general case corresponding to (3.6) will be clear.

We avoid complications on smoothness by assuming that the curve Γ is C^∞ and that the functions f and g are C^∞ in $\boldsymbol{x}$.

The first step in our procedure is to introduce a local coordinate system near Γ. Let $\boldsymbol{x} = \boldsymbol{X}(\xi)$, $0 \leqq \xi \leqq L$, be the parametric form for Γ ($\boldsymbol{X}$ a C^∞ function) and let $\boldsymbol{n}(\xi)$ be the unit *inner* normal. Then we can introduce the local coordinate system,

$$\boldsymbol{x} = \boldsymbol{X}(\xi) + \tau \boldsymbol{n}(\xi) \tag{3.7}$$

near Γ. This coordinate system is orthogonal with the form

$$Q^2(d\xi)^2 + (d\tau)^2, \quad Q(\xi, \tau) = (1 - \tau k(\xi)) \tag{3.8}$$

where $k(\xi)$ is the curvature. $Q > 0$ for τ small. If $\boldsymbol{t}(\xi)$ is the unit tangent then the gradient and Laplacian have the form,

$$\operatorname{grad} = \boldsymbol{t} Q^{-1} \frac{\partial}{\partial \xi} + \boldsymbol{n} \frac{\partial}{\partial \tau}; \quad \Delta = Q^{-1}\left(\frac{\partial}{\partial \xi}\left(Q^{-1}\frac{\partial}{\partial \xi}\right) + \frac{\partial}{\partial \tau}\left(Q\frac{\partial}{\partial \tau}\right)\right). \tag{3.9}$$

In the expansion (1.3) we consider χ and the u_k^- as functions of ξ and η. We set $\delta = \sqrt{i}\,\alpha$ and then we have,

$$\operatorname{grad} \chi = -\delta \chi \boldsymbol{n} \quad \text{and} \quad \Delta \chi = (\delta^2 - \delta Q^{-1} Q_\tau)\,\chi. \tag{3.10}$$

Let us write $(1.3)_2$ as $u \sim \chi V$. Then we have $\Delta u = \Delta \chi V + 2 \operatorname{grad} \chi \cdot \operatorname{grad} V + \chi \Delta V$. We substitute $(1.3)_2$ and equate the result to $\delta^2 \chi V$. Then we cancel χ and equate powers of δ. It is easy to verify that this process yields the equations:

$$\left.\begin{aligned} &2\frac{\partial}{\partial \tau} u_1^- + Q^{-1} Q_\tau u_1^- = 0 \\ &2\frac{\partial}{\partial \tau} u_m^- + Q^{-1} Q_\tau u_m^- = \Delta u_{m-1}^-, \quad m \geqq 2. \end{aligned}\right\} \text{ in } \Omega_- \tag{3.11}$$

From (1.2) and (1.3) we obtain, on equating powers of δ,

$$\begin{gathered} \Delta u_m^+ = 0 \quad \text{in } \Omega_+ \\ u_0(\boldsymbol{x}) - a \log |\boldsymbol{x}| = O(1), \quad u_m(\boldsymbol{x}) = O(1) \text{ as } |\boldsymbol{x}| \to \infty, \quad m > 0. \end{gathered} \tag{3.12}$$

We turn next to the interface conditions. If we substitute (1.3) into $u^- = u^+ + f$ and equate coefficients of δ^{-m} we find,

$$\begin{gathered} u_0^+(\boldsymbol{X}(\xi)) + f(\boldsymbol{X}(\xi)) = 0 \\ u_m^-(\xi, 0) = u_m^+(\boldsymbol{X}(\xi)), \quad m \geqq 1. \end{gathered} \tag{3.13}$$

We note that the derivative in the external normal direction is minus the τ derivative on Γ. Accordingly, when we substitute (1.3) into $u_n^- = u_n^+ + g$ we find,

$$\begin{gathered} u_1^-(\xi, 0) = \frac{\partial}{\partial n} u_0^+(\boldsymbol{X}(\xi)) + g(\boldsymbol{X}(\xi)) \\ u_m^-(\xi, 0) = \frac{\partial}{\partial n} u_{m-1}^+(\boldsymbol{X}(\xi)) - \frac{\partial}{\partial \tau} u_{m-1}^-(\xi, 0), \quad m \geqq 2. \end{gathered} \tag{3.14}$$

We can now describe the process of determining $u_n^\pm$.

Step (0). We determine u_0^+ from (3.12) and (3.13) for $m = 0$, using Theorem 3.1. Observe that u_0^+ is the solution of $P(\infty)$.

Step (0)'. Compute $u_1^-(\xi, 0)$ from (3.14) and the value of u_0^+ from Step (0). Then solve (3.11), for $m = 1$, with this initial value. For each ξ this is a linear ordinary differential equation in τ. This equation can be written $(\sqrt{Q(\xi, \tau)}\, u_1^-(\xi, \tau))_\tau = 0$ and $Q(\xi, 0) \equiv 1$; hence we have,

$$u_1^-(\xi, \tau) = (Q(\xi, \tau))^{-1/2}\, u_1^-(\xi, 0). \tag{3.15}$$

Step (1). By (3.12) and (3.13) for $m = 1$,

$$\begin{gathered} \Delta u_1^+ = 0 \text{ in } \Omega_+, \quad u_1^+(x) = O(1) \text{ as } |x| \to \infty \\ u_1^+(X(\xi)) = u_1^-(X(\xi)). \end{gathered} \tag{3.16}$$

This is of the same form as $P(\infty)$ but with $a = 0$ and f replaced by u_1^-. (Observe that (3.14) yields the boundary value of u_1^- in terms of u_0^+ so that to determine u_1 we do not actually have to solve for u_1^-.) We calculate u_1 by Theorem 3.1.

Step (1)'. Compute the quantity $\frac{\partial}{\partial n} u_1^+(X(\xi)) - \frac{\partial}{\partial \tau} u_1^-(\xi, 0)$ from the above steps, hence determining $u_2^-(\xi, 0)$ by (3.14), $m = 2$. Compute Δu_1^- from (3.9) and the preceding values. Then we can write (3.11) for $m = 2$ in the form

$$\frac{\partial}{\partial \tau} (\sqrt{Q(\xi, \tau)}\, u_2^-(\xi, \tau)) = \tfrac{1}{2} \sqrt{Q(\xi, \tau)}\, \Delta u_1(\xi, \tau).$$

Hence,

$$u_2^-(\xi, \tau) = \tfrac{1}{2} (Q(\tau, \xi))^{-1/2} \int_0^\tau \sqrt{Q(\xi, \tau)}\, \Delta u_1(\xi, \tau)\, d\tau) + Q(\xi, \tau)^{-1/2}\, u_2^-(\xi, 0). \tag{3.17}$$

The process can be continued. For each m we have Step (m) in which we obtain u_m^+ by solving a problem of the form $P(\infty)$ with $a = 0$ and $u_m^+(X(\xi))$ determined by the $u_k^\pm$ for $k \leqq m - 1$. This is done with Theorem 3.1. Then we have Step $(m)'$ in which we calculate u_{m+1}^- from a formula like (3.17).

Remark 3.1. After Step (m) we will have to calculate $(u_m^+)_n^+$. Observe that this easy to do with the procedure in Theorem 3.1 by using (3.5). This is what makes the solution procedure numerically tractable as discussed in [2].

The above process yields the formal expansion. It can be carried out as long as we restrict our attention to values of τ small enough that the coordinate system is well defined and Q remains positive. Let us assume this is true for $\tau \leqq \bar{\tau}$. For any $\tau_0 < \bar{\tau}$ set

$$\Omega_-(\tau_0) = \{x : \tau(x) \leqq \tau_0\}.$$

Then our procedure is valid for $x \in \Omega_-(\tau_0)$.

An analysis of the method will reveal that as we increase m we need an increasing amount of smoothness on the data. Recall that we assumed that both the data and Γ were C^∞. Regularity results in [4] for the procedure in Theorem 3.1 combined with the formulas like (3.17) can be used to show that the $u_m^\pm$ for $m \geqq 1$ are all in $C^\infty(\bar{\Omega}_\pm)$, respectively.

We set

$$U_m = \begin{cases} \sum_{k=0}^{m} u_k^+ \delta^{-k} & \text{in } \Omega_+ \\ & \qquad m \geqq 0. \\ \chi \sum_{k=1}^{m+1} u_k^- \delta^{-k} & \text{in } \Omega_-(\tau_0) \end{cases} \tag{3.18}$$

We call these the mth-order asymptotic approximations. In the next section we will compare them with the actual solution. For this purpose we will need the following result which follows readily from the definition of the $u_k^\pm$.

Theorem 3.2. *For each* $m \geqq 0$,

$$\begin{aligned} \Delta U_m - i\alpha^2 U_m &= \chi\, \Delta u_{m+1}^-(\delta)^{-m-1} \equiv \hat{F}_m \quad \text{in} \quad \Omega_-(\tau_0) \\ \Delta U_m &= 0 \quad \text{in } \Omega_+ \\ U_m^- - U_m^+ &= f - (u_{m+1}^-)^- (\delta)^{-m-1} \equiv f + f_m \\ (U_m)_n^- - (U_m)_n^+ &= g - \frac{\partial}{\partial \tau}(u_{m+1}^-)|_{\tau=0} (\delta)^{-m-1} \equiv g + g_m . \end{aligned} \tag{3.19}$$

$$U_0(\boldsymbol{x}) - a \log|\boldsymbol{x}| = O(1), \qquad U_m(\boldsymbol{x}) = O(1), \qquad m > 0, \qquad \text{as} \quad |\boldsymbol{x}| \to \infty.$$

4. Convergence

Let us set v_m, $m = 0, 1, 2, \ldots,$ equal to the difference between the solution u of $P(\alpha)$ and the mth approximation U_m of (3.19). The goal of this section is to show that, in some sense, $v_m = 0(\alpha^{-m-1})$. We describe the results more precisely. The first is an estimate in Sobolev norms. For $S \subset \Omega_-$ we set

$$\begin{aligned} |||v_m|||^2(S) = \|v\|_1^2(S) + \int_{\Omega_+} \|\nabla v_m(\boldsymbol{x})\|^2\, d\boldsymbol{x} + \|v_m^-\|_{1/2}^2(\Gamma) \\ + \|(v_m)_n^-\|_{-1/2}^2(\Gamma) + \|v_m^+\|_{1/2}^2(\Gamma) + \|(v_m)_n^+\|_{-1/2}^2(\Gamma). \end{aligned} \tag{4.1}$$

Here $\|v\|_1(S)$ is the Sobolev norm, $\|v\|_1^2(S) = \int_S (|\operatorname{grad} v(\boldsymbol{x})|^2 + v(\boldsymbol{x})^2)\, d\boldsymbol{x}$.

Theorem 4.1. *Given any integer* $m \geqq 0$ *there is a constant* $L(m) > 0$ *such that,*

$$|||v_m|||\,(\Omega_-(\tau/2)) \leqq L(m)\, \alpha^{-m-1} \qquad \text{for any } \alpha > 1. \tag{4.2}$$

We can also obtain pointwise estimates in Ω_+. We use the usual notation

$$D^\gamma = \left(\frac{\partial}{\partial x_1}\right)^{\gamma_1} \left(\frac{\partial}{\partial x_2}\right)^{\gamma_2} \quad \text{for } \gamma = (\gamma_1, \gamma_2),\ |\gamma| = \gamma_1 + \gamma_2,\ \gamma_i \text{ non-negative integers.}$$

Theorem 4.2. *For any integers* $m \geqq 0$, $k \geqq 0$ *and any positive* ϱ *there is a constant* $L(m, k, \varrho) > 0$ *such that if* $\Omega_+(\varrho) = \{\boldsymbol{x} : \boldsymbol{x} \in \Omega_+, d(\boldsymbol{x}, \Gamma) > \varrho\}$ *then*

$$\|D^\gamma v_m\|_{L_\infty}(\Omega_+(\varrho)) \leqq L(m, k, \varrho)\, \alpha^{-m-1} \quad \text{for any } \alpha > 1 \text{ and } \gamma \text{ with } |\gamma| \leqq k. \tag{4.3}$$

Since U_m is only defined in the strip $\Omega_-(\bar{\tau})$ we cannot expect to estimate v_m in all of Ω_-. Our final result shows that the solution u is very small in $\Omega_-\backslash\Omega_-(\bar{\tau}/2)$ anyway, when α is large.

Theorem 4.3. *For any integer* $k \geqq 0$ *there is a constant* $L(k)$ *such that*

$$\|D^\gamma u\|_{L_\infty}(\Omega_-\backslash\Omega_-(\bar{\tau}/2)) \leqq L(k)\, e^{-\bar{\tau}\alpha/2} \text{ for any } \alpha > 1 \text{ and } \gamma \text{ with } |\gamma| \leqq k. \quad (4.4)$$

The key element in the proofs of all these results is a set of *a priori* estimates for the *problem* $P(\alpha; F, f, g)$ of finding v such that

$$\begin{aligned} \Delta v &= i\alpha^2 v + F \quad \text{in } \Omega_-; \quad \Delta v = 0 \quad \text{in } \Omega_+ \\ v^- &= v^+ + f; \quad v_n^- = v_n^+ + g \quad \text{on } \Gamma \\ v &= O(1) \quad \text{as } |\boldsymbol{x}| \to \infty. \end{aligned} \quad (4.5)$$

Let $\|\cdot\|_1'$ denote the norm in the space H_1', the dual of H_1 with respect to $L_2(\Omega)$, and set,

$$M(F, f, g, a) = \|F\|_1'(\Omega_-) + \|f\|_{1/2}(\Gamma) + \|g\|_{-1/2}(\Gamma) + |a|. \quad (4.6)$$

Lemma 4.1. *There is a constant* $k > 0$ *such that if* v *is a solution of* $P(\alpha : F, f, g)$ *then* $|||v|||\ (\Omega_-) \leq kM(F, f, g, a)$ *for any* $a > 0$.

Proof: First we apply Green's theorem to Ω and obtain

$$\int_{\Omega_-} \|\text{grad } v(\boldsymbol{x})\|^2\, d\boldsymbol{x} + i\alpha^2 \int_{\Omega} |v(\boldsymbol{x})|^2\, dx = \int_{\Gamma} v_n^- \bar{v}^- ds - \int_{\Omega} F(\boldsymbol{x})\bar{v}(\boldsymbol{x})\, d\boldsymbol{x}. \quad (4.7)$$

For $\Delta v = 0$ in Ω_+ and $v = O(1)$ as $|\boldsymbol{x}| \to \infty$ one has grad $v = O(|\boldsymbol{x}|^{-2})$. Thus one can do an exterior Green's theorem calculation, with a limiting argument, and obtain

$$\int_{\Omega_+} \|\text{grad } v(\boldsymbol{x})\|^2\, d\boldsymbol{x} = -\int_{\Gamma} v_n^+ \bar{v}^+ ds. \quad (4.8)$$

From the interface conditions we have $v_n^+\bar{v}^+ = v_n^-\bar{v}^- - g\bar{v}^- - v_n^-\bar{f} + g\bar{f}$. Thus adding (4.7) and (4.8) yields,

$$Q(v) = A(F, v) + B(g, v) + C(f, v) + D(g, f), \quad (4.9)$$

where,

$$\begin{aligned} Q(v) &= \int_{\Omega_-} \|\text{grad } v(\boldsymbol{x})\|^2\, d\boldsymbol{x} + i\alpha^2 \int_{\Omega_-} |v(\boldsymbol{x})|^2\, d\boldsymbol{x} + \int_{\Omega_+} \|\text{grad } v(\boldsymbol{x})\|^2\, dx \\ A(F, v) &= -\int_{\Omega_-} F(\boldsymbol{x})\, \bar{v}(\boldsymbol{x})\, d\boldsymbol{x}, \quad B(g, v) = \int_{\Gamma} g\bar{v}^-\, ds \\ C(f, v) &= \int_{\Gamma} v_n^- \bar{f}\, ds, \quad D(g, f) = -\int_{\Gamma} g\bar{f}\, ds. \end{aligned} \quad (4.10)$$

We now estimate. From (4.10) we see that for $\alpha \geqq 1$ we have

$$Q(v) \geqq \|v\|_1^2(\Omega_-) + \int_{\Omega_+} \|\text{grad } v(\boldsymbol{x})\|^2\, d\boldsymbol{x}. \quad (4.11)$$

We have, for all α,

$$|A(F,v)| \leqq \|F\|_{-1}(\Omega_-)\,\|v\|_1(\Omega); \qquad |D(g,f)| \leqq \|g\|_{-1/2}(\Gamma)\,\|f\|_{1/2}(\Gamma) \tag{4.12}$$

and by the trace theorem there is a k_1, independent of α, such that

$$B(g,v) \leqq \|g\|_{-1/2}(\Gamma)\,\|v^-\|_{1/2} \leqq k_1\|g\|_{-1/2}(\Gamma)\,\|v\|_1(\Omega_-). \tag{4.13}$$

We need to estimate C and for this we have to consider the solution formula of Th. 3.1. It is shown in [4] that for that formula there is a constant k_2 such that

$$\|\sigma\|_{-1/2} + |C| \leqq k_2(\|f\|_{1/2} + |a|).$$

Since the kernel in N of (3.5) is smooth, and $a=0$ for v_n, it follows that for some k_3,

$$\|v_n^-\|_{-1/2}(\Gamma) \leqq \|v_n^+\|_{-1/2}(\Gamma) + \|g\|_{-1/2}(\Gamma) \leqq k_3(\|f\|_{1/2}(\Gamma) + \|g\|_{-1/2}(\Gamma)).$$

Thus we can use the trace theorem again to conclude that

$$|C(f,v)| \leqq \|v_n^-\|_{-1/2}(\Gamma)\,\|f\|_{1/2}(\Gamma) \leqq k_3\|f\|_{1/2}(\Gamma)(\|f\|_{1/2}(\Gamma) + \|g\|_{-1/2}(\Gamma)). \tag{4.14}$$

The conclusion of the lemma follows from (4.11)–(4.14).

An argument similar to that in Lemma 4.1 can be used to show that the solution u of $P(\alpha)$ satisfies an estimate of the form,

$$|||u||| \leqq kM(0,f,g,a) \qquad \text{for any } \alpha \geqq 1. \tag{4.15}$$

We would like to apply the lemma to v_m but before doing so we must extend U_m to all of Ω_-. Let ζ be a C^∞ function which has its support in $\bar\Omega_-(\tau')$ for some τ', $\bar\tau/2 < \tau' < \bar\tau$, and which is identically one in $\bar\Omega_-(\bar\tau/2)$. Put $Z_m = \zeta U_m$. Then $Z_m \equiv U_m$ in $\bar\Omega_-(\bar\tau/2)$, $Z_m \equiv 0$ in $\Omega_-\backslash\Omega_-(\bar\tau)$ and,

$$\Delta Z_m - i\alpha^2 Z_m \begin{cases} = \hat F_m \text{ in } \Omega_-(\bar\tau/2) \\ = 0 \text{ in } \Omega_-\backslash\Omega_-(\bar\tau) \\ = \zeta\hat F_m + 2\,\operatorname{grad}\zeta\cdot\operatorname{grad}U_m + \Delta\zeta U_m \text{ in } \Omega'_-. \end{cases} \tag{4.16}$$

Here $\hat F_m$ is as in (3.19) and Ω'_- is the strip $\Omega'_- = \{x : \bar\tau/2 < \tau(x) < \bar\tau\}$.

Lemma 4.2. *For any integer $m \geqq 0$ there is a constant $K_m > 0$ such that if $\Delta v_m - i\alpha^2 v_m = F_m$ then*

$$\|F_m\|_1'(\Omega_-) \leqq K\alpha^{-m-1} \qquad \text{for any } \alpha > 0. \tag{4.17}$$

Proof. From (3.19) we see that $\|\hat F_m\|_{L_\infty}(\Omega_-(\bar\tau)) \leqq K_1\alpha^{-m-1}$, K_1 independent of α. F_m is zero in $\Omega_-\backslash\Omega_-(\bar\tau)$ so we need only consider the strip Ω'_-. We observe that differentiations with respect to x of U_m produce powers of α through χ. We note, however, that χ is decaying exponentially so that for x in Ω'_- we will have $\|D^\gamma\chi\|_{L_\infty}(\Omega'_-)$ bounded independently of α. Hence the same is true of U_m and, by (4.16), of F_m and (4.16) yields the result of the Lemma.

Proof of Theorem 4.1. From (3.19) we see that if $v_m' = u - Z_m$ then v_m' is a solution of $P(\alpha; F_m, f_m, g_m)$ for $m \geqq 0$. We see immediately from (3.19) that there are constants $K_1(m)$ such that $\|f_m\|_{1/2}(\Gamma)$, $\|g_m\|_{-1/2}(\Gamma) \leqq K_1(m)\,\alpha^{-m-1}$. This fact together with Lemma 4.2 shows that we may apply Lemma 4.1 to v_m' and obtain $|||v_m'|||(\Omega_-) \leqq L(m)\,\alpha^{-m-1}$ for some $L(m)$. The estimate (4.2) follows.

Proof of Theorem 4.2. We have $\Delta v_m = 0$ in Ω_+ with $v_m^- = f_m$ on Γ and $v_m = O(1)$ as $|\boldsymbol{x}| \to \infty$. Thus by Th. 3.1 we can represent v_m in Ω_+ as

$$v_m = \mathcal{S}[\sigma_m] + C_m,$$

where, as before, $\|\sigma_m\|_{-1/2}(\Gamma) \leqq k_2(\|f_m\|_{1/2}(\Gamma))$. Also we have $\int_\Gamma \sigma_m\, ds = 0$. We observe that all derivatives of the logarithm in (3.1) are smooth and uniformly bounded on any set $d(\boldsymbol{x}, \Gamma) \geqq \varrho$. It follows that for any ϱ and γ, with $|\gamma| \neq 0$, there is a constant $K(\gamma, \varrho)$ such that,

$$\| D^\gamma v_m \|_{L_\infty}(\Omega_+(\varrho)) \leqq K(\gamma, \varrho) \|f_m\|_{1/2}(\Gamma). \tag{4.18}$$

Since $\int_\Gamma \sigma_m\, ds = 0$ we can write

$$v_m(\boldsymbol{x}) = \int_\Gamma \sigma_m(y)\,\{\log|\boldsymbol{x} - \boldsymbol{y}| - \log|\boldsymbol{x}|\}\, dy + C_m$$

and conclude as before that

$$\| v_m \|_{L_\infty}(\Omega_+(\varrho)) \leqq K(\varrho)\, (\|f_m\|_{1/2}(\Gamma). \tag{4.19}$$

Now we have only to observe that by (3.19) $f_m = 0(\alpha^{-m-1})$ to obtain the conclusion.

Proof of Theorem 4.3. We use the integral representation of the solution of $P(\alpha)$. Let

$$g(\boldsymbol{x}, \boldsymbol{y}) = -\frac{i}{4} H_0^{(1)}(\sqrt{i}\,\alpha\,|\boldsymbol{x} - \boldsymbol{y}|) \tag{4.20}$$

be the fundamental solution for $\Delta v - i\alpha^2 v = 0$. $H_0^{(1)}(z)$ is the Hankel function of order zero. Then one has the standard representation of u

$$u(\boldsymbol{x}) = \int_\Gamma \left(u^-(\boldsymbol{y}) \frac{\partial}{\partial n} g(\boldsymbol{x}, \boldsymbol{y}) - u_n^-(\boldsymbol{y})\, g(\boldsymbol{x}, \boldsymbol{y}) \right) d\boldsymbol{y} \quad \text{in } \Omega_-. \tag{4.21}$$

Now the function $H_0^{(1)}(z)$ has the property that it and all its derivatives are of order e^{iz}. By (4.15) and (4.1) we see that $\|u^-\|_{1/2}(\Gamma)$ and $\|u_n^-\|_{-1/2}(\Gamma)$ are bounded independently of α so that the conclusion of Th. 4.3 follows immediately from (4.21), together with derivatives of that formula.

Acknowledgement. MacCamy was supported by the National Science Foundation under Grant MCS-8219675. Some of the work was done while he was visiting Technische Hochschule, Darmstadt, under the sponsorship of the German Academic Exchange Service.

References

1. Coleman, B. D., & Noll, W., An approximation theorem for functionals, with applications in continuum mechanics, *Arch. Rational Mech. and Anal* **6** (1960), 355–370.
2. Hariharan, S., & MacCamy, R. C., Integral equation procedures for eddy current problems, *J. Comp. Phys.* **45** (1982), 80–89.
3. Hsiao, G. C., & MacCamy, R. C., Solution of boundary value problems by integral equations of the first kind, SIAM *Review* **15** (1973), 687–705.
4. Hsiao, G. C., & Wendland, W. L., A finite element method for some integral equations of first kind, *J. Math. Anal. and Appl.* **55** (1977), 449–481.
5. MacCamy, R. C., & Stephan, E., Solution procedures for three-dimensional eddy current problems, *J. Math. Anal. and Appl.* **10** (1984), No. 2 348–379.
6. Noll, W., A general framework for problems in the statics of finite elasticity, in *Contemporary Developments in Continuum Mechanics and Partial Differential Equations, North-Holland Mathematics Studies*, 1978.

Carnegie-Mellon University
Pittsburgh
&
Georgia Institute of Technology
Atlanta

(Received December 26, 1984)

Algebraic Theories for Continuous Semilattices

OSWALD WYLER

Dedicated to Walter Noll on his Sixtieth Birthday

Continuous lattices were introduced by D. SCOTT in [5]; they have been studied intensively and found numerous applications We refer to the Compendium [1] for a comprehensive survey of continuous lattices and related topics. In some applications, *e.g.* in K. KEIMEL'S study of compact convex sets in locally convex topological vector spaces [2], continuous sup semilattices have been useful.

Like continuous lattices, continuous semilattices have a very rich structure. They are *inter alia* the algebras for five algebraic theories, on four base categories. It is the main purpose of this paper to construct these algebraic theories.

Traditionally, finitary algebraic theories on sets have been constructed from operations and formal laws. Category theory has shown that finitary and infinitary algebraic theories, not only on sets but on arbitrary base categories, can be described by monads, and that "algebraic" functors between categories of algebras correspond to morphisms of monads. Thus this paper is concerned with the algebras for the nine monads which appear in the following diagram of monads and morphisms of monads.

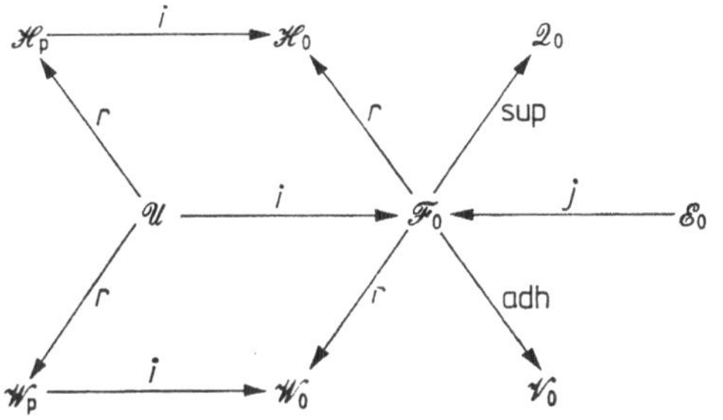

The monads will be constructed in the paper; i stands for inclusion, r for restriction, and adh for adherence. For five of the monads, algebras are continuous sup semilattices; we show that the algebraic functors induced by the four morphisms with domain $\mathscr{F}_0$ are isomorphisms of categories. This is proved by four applications of a general categorical theorem which we state and prove in Section 0.

Only basic categorical concepts are used in this paper; we refer to [3] for these concepts and their properties. Categorical concepts used in this paper, but not defined in [3], are defined in Section 0.

The order for continuous lattices will be as in [2] and [8a], dual to the order of the Compendium [1] and of [8b]. This may cause minor inconveniences, but it makes possible coherent orders for various ordered structures. Powersets are ordered by set inclusion, and all natural maps of ordered structures, except for complementation, preserve order.

0. Categorical background

0.1. If $\mathscr{T} = (T, \eta, \mu)$ and $\mathscr{S} = (S, h, m)$ are monads on categories $\mathscr{A}$ and $\mathscr{B}$, then a *morphism* $(R, \pi): \mathscr{S} \to \mathscr{T}$ consists of a functor $R: \mathscr{A} \to \mathscr{B}$ (note the contravariance) and a natural transformation $\pi: SR \to RT$ such that $\pi \cdot hR = R\eta$, and $\pi \cdot mR = R\mu \cdot \pi T \cdot S\pi$. Monads and their morphisms form a category, with composition $(R, \pi) \cdot (\bar{R}, \bar{\pi}) = (R\bar{R}, R\bar{\pi} \cdot \pi\bar{R})$.

For the categories $\mathscr{A}^{\mathscr{T}}$ and $\mathscr{B}^{\mathscr{S}}$ of algebras, an *algebraic functor* over $R: \mathscr{A} \to \mathscr{B}$ is a functor $\bar{R}: \mathscr{A}^{\mathscr{T}} \to \mathscr{B}^{\mathscr{S}}$ such that $U^{\mathscr{S}}\bar{R} = RU^{\mathscr{T}}$ for the forgetful functors $U^{\mathscr{T}}: \mathscr{A}^{\mathscr{T}} \to \mathscr{A}$ and $U^{\mathscr{S}}$. The condition $U^{\mathscr{S}}\bar{R} = RU^{\mathscr{T}}$ determines $\bar{R}$ for homomorphisms.

A morphism $(R, \pi): \mathscr{S} \to \mathscr{T}$ of monads induces an algebraic functor $(R, \pi)^*: \mathscr{A}^{\mathscr{T}} \to \mathscr{B}^{\mathscr{S}}$ over R, with $(R, \pi)^*(A, \alpha) = (RA, R\alpha \cdot \pi_A)$ for a $\mathscr{T}$-algebra (A, α). Conversely, every algebraic functor $\bar{R}: \mathscr{A}^{\mathscr{T}} \to \mathscr{B}^{\mathscr{S}}$ over $R: \mathscr{A} \to \mathscr{B}$ is obtained in this way, with $\pi_A = \bar{\mu}_A \cdot SR\eta_A$ for an object A of $\mathscr{A}$ with $\bar{R}(TA, \mu_A) = (RTA, \bar{\mu}_A)$. We omit the straightforward proof.

We have obtained a syntax/semantics duality between morphisms of monads and algebraic functors. If (R, π) is a morphism of monads, then the induced algebraic functor $(R, \pi)^*$ may be denoted simply by π^* if the context permits.

We note, without proof, that the natural transformation $\pi: SR \to RT$ lifts to $\pi: F^{\mathscr{S}}R \to (R, \pi)^* F^{\mathscr{T}}$ for the free algebra functors, if $(R, \pi): \mathscr{S} \to \mathscr{T}$ is a morphism of monads.

0.2. We shall deal repeatedly with a morphism of monads $(R, \pi): \mathscr{S} \to \mathscr{T}$ which satisfies the following conditions.

(i) R is faithful, and all morphisms π_A are epimorphic.

(ii) There is a functor $\Delta: \mathscr{B}^{\mathscr{S}} \to \mathscr{A}$ such that $R\Delta = U^{\mathscr{S}}$ and $\Delta(R, \pi)^* = U^{\mathscr{T}}$.

(iii) Every morphism $S\pi_A$ is epimorphic, and the structure of an $\mathscr{S}$-algebra (B, β) always factors $\beta = u\pi_A$, with $A = \Delta(B, \beta)$ and $u: RTA \to B$ in $\mathscr{B}$.

Theorem. *If a morphism* (R, π) *of monads satisfies* (i) *and* (ii) *above, then the algebraic functor* $(R, \pi)^*$ *is full and faithful, and injective on objects. If* (R, π) *also satisfies* (iii), *then* $(R, \pi)^*$ *is an isomorphism of categories.*

Proof. We write π^* for $(R, \pi)^*$. This functor is faithful if R is. If (ii) is also valid, then $A = \Delta\pi^*(A, \alpha)$ for a $\mathscr{T}$-algebra (A, α), and then $R\alpha \cdot \pi_A$ determines α uniquely. Thus π^* is injective on objects.

If $g: \pi^*(A, \alpha) \to \pi^*(C, \gamma)$ and $f = \Delta g: A \to C$, then $g = Rf$, and $Rf \cdot R\alpha \cdot \pi_A = R\gamma \cdot \pi_C \cdot SRf = R\gamma \cdot RTf \cdot \pi_A$, using naturality of π in the second step. Now $f: (A, \alpha) \to (C, \gamma)$, and $g = \pi^* f$, if (i) and (ii) are valid.

For the last part, we must only show that π^* is surjective on objects if (iii) is also valid. Thus consider an $\mathscr{S}$-algebra (B, β) and put $A = \Delta(B, \beta)$. If $\beta = u\pi_A$, we must show that $u = R\alpha$ for a $\mathscr{T}$-algebra structure α of A.

For this, we consider the following diagram:

$$\begin{array}{ccccc}
SSRA & \xrightarrow{S\pi_A} & SRTA & \xrightarrow{Su} & SRA \\
\Big\downarrow & & \Big\downarrow{\scriptstyle \pi_{TA}} & & \Big\downarrow{\scriptstyle \pi_A} \\
{\scriptstyle m_{RA}} & & RTTA & \xrightarrow{RT\alpha} & RTA \\
\Big\downarrow & & \Big\downarrow{\scriptstyle R\mu_A} & & \Big\downarrow{\scriptstyle u} \\
SRA & \xrightarrow{\pi_A} & RTA & \xrightarrow{u} & B
\end{array} .$$

The outer square and the lefthand rectangle commute by hypothesis. Since $S\pi_A$ is epi, $u: \pi^* F^{\mathscr{T}} A \to (B, \beta)$ follows, *i.e.* the righthand rectangle commutes. Now $u = R\alpha$ for $\alpha = \Delta u$, and the upper righthand square commutes by naturality of π. Since π_{TA} is epi, the lower righthand square commutes, and hence $\alpha\mu_A = \alpha \cdot T\alpha$ by (i). Finally, $R\alpha \cdot R\eta_A = u\pi_A e_{RA} = \beta e_B = \mathrm{id}_{RA}$, and $\alpha\eta_A = \mathrm{id}_A$ follows by (i) □

0.3. A contravariant functor G from $\mathscr{B}$ to $\mathscr{A}$ will be shown as a (covariant) functor $G: \mathscr{B}^{\mathrm{op}} \to \mathscr{A}$, and we recall that contravariant functors $F: \mathscr{A}^{\mathrm{op}} \to \mathscr{B}$ and $G: \mathscr{B}^{\mathrm{op}} \to \mathscr{A}$ are called *adjoint on the right* if F^{op} is left adjoint to G, in symbols $F^{\mathrm{op}} \dashv G$. An adjunction on the right has two *units* $\eta: \mathrm{Id}_{\mathscr{A}} \to GF^{\mathrm{op}}$ and $\varepsilon: \mathrm{Id}_{\mathscr{B}} \to FG^{\mathrm{op}}$, with η the unit and $\varepsilon^{\mathrm{op}}$ the counit for $F^{\mathrm{op}} \dashv G$.

We now consider a not necessarily commutative diagram

$$\begin{array}{ccc}
\mathscr{B}^{\mathrm{op}} & \xrightarrow{S^{\mathrm{op}}} & \mathscr{B}_1^{\mathrm{op}} \\
{\scriptstyle F^{\mathrm{op}}}\Big\downarrow\Big\uparrow{\scriptstyle G} & & {\scriptstyle F_1^{\mathrm{op}}}\Big\downarrow\Big\uparrow{\scriptstyle G_1} \\
\mathscr{A} & \xrightarrow{R} & \mathscr{A}_1
\end{array}$$

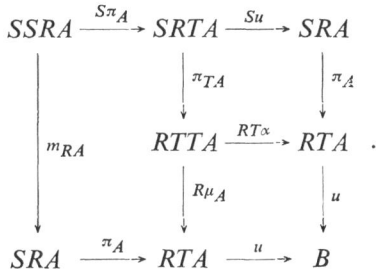

of categories and functors, with adjunctions on the right for the vertical arrows. This induces a bijective correspondence between natural transformations $\varkappa: RG \to G_1 S^{\mathrm{op}}$ and $\lambda: SF \to F_1 R^{\mathrm{op}}$ as follows. $\varkappa$ and λ correspond to each other, and are called *adjoint*, if $\varkappa_B \cdot Rg$ and $\lambda_A \cdot Sf$ are adjoint for $F_1^{\mathrm{op}} \dashv G_1$ whenever $g: A \to GB$ and $f: B \to FA$ are adjoint for $F^{\mathrm{op}} \dashv G$.

Let $\mathcal{T}$ on $\mathcal{A}$ and $\mathcal{S}$ on $\mathcal{A}_1$ be the monads induced by the adjunctions on the right, and let $K: \mathcal{B}^{\mathrm{op}} \to \mathcal{A}^{\mathcal{T}}$ and $K_1: \mathcal{B}_1^{\mathrm{op}} \to \mathcal{A}_1^{\mathcal{S}}$ be the comparison functors, with $U^{\mathcal{T}}K = G$ and $KF^{\mathrm{op}} = F^{\mathcal{T}}$, and with $U^{\mathcal{S}}K_1 = G_1$ and $K_1F_1^{\mathrm{op}} = F^{\mathcal{S}}$. There are two situations in which adjoint natural transformations produce a morphism of monads.

(a) If all $G_1\lambda_A$ factor $G_1\lambda_A = \varkappa_{FA} \cdot \pi_A$, with every $\varkappa_{FA}$ mono, then the π_A define a morphism $(R, \pi): \mathcal{S} \to \mathcal{T}$, and $\varkappa$ lifts to a natural transformation $\varkappa: (R, \pi)^* K \to K_1 S^{\mathrm{op}}$.

(b) If $\mathcal{A}_1 = \mathcal{A}$ and $R = \mathrm{Id}_{\mathcal{A}}$, and if all $\varkappa_{FA}$ factor $\varkappa_{FA} = G_1\lambda_A \cdot \pi_A$, with every $G_1\lambda_A$ mono, then the π_A define a morphism $(\mathrm{Id}, \pi): \mathcal{T} \to \mathcal{S}$, and $\varkappa$ lifts to a natural transformation $\varkappa: K \to (\mathrm{Id}, \pi)^* K_1 S^{\mathrm{op}}$.

We omit the straightforward proofs.

0.4. We recall that a *concrete category* is a category $\mathcal{A}$ equipped with a faithful functor $|\ |: \mathcal{A} \to \mathrm{ENS}$ to sets. In this situation, a morphism g of $\mathcal{A}$ usually is identified with the underlying mapping $|g|$, and one often uses the same notation for an object A of $\mathcal{A}$ and the underlying set $|A|$.

We also recall that the contravariant powerset functor P on sets, with $(Pu)(T) = u^{\leftarrow}(T)$ for a mapping $u: A \to B$ and $T \subset B$, is self-adjoint on the right.

A contravariant functor $F: \mathcal{A}^{\mathrm{op}} \to \mathcal{B}$ between concrete categories will be called a *preimage functor* if $|FA| \subset P|A|$ for every object A of $\mathcal{A}$, and always $(Fu)(T) = u^{\leftarrow}(T)$ for $u: A \to A'$ in $\mathcal{A}$ and $T \in |FA'|$. In this situation, subset insertions define a natural transformation $|\ \ |F \to P|\ \ |^{\mathrm{op}}$ which we call the *natural inclusion* of F.

We say that preimage functors $F: \mathcal{A}^{\mathrm{op}} \to \mathcal{B}$ and $G: \mathcal{B}^{\mathrm{op}} \to \mathcal{A}$ are *exponentially adjoint* if F and G are adjoint on the right, and the natural inclusions $|\ \ |F \to P|\ \ |^{\mathrm{op}}$ of F and $|\ \ |G \to P|\ \ |^{\mathrm{op}}$ of G are adjoint. This means that $x \in f(y) \Leftrightarrow y \in g(x)$, for $x \in |A|$ and $y \in |B|$, defines a bijection between morphisms $f: B \to FA$ of $\mathcal{B}$ and $g: A \to GB$ of $\mathcal{A}$, for all objects A of $\mathcal{A}$ and B of $\mathcal{B}$.

Example: the contravariant powerset functor P on sets is a preimage functor, exponentially adjoint to itself.

If preimage functors $F: \mathcal{A}^{\mathrm{op}} \to \mathcal{B}$ and $G: \mathcal{B}^{\mathrm{op}} \to G$ are exponentially adjoint, then the units $\eta: \mathrm{Id}_{\mathcal{A}} \to GF^{\mathrm{op}}$ and $\varepsilon: \mathrm{Id}_{\mathcal{B}} \to FG^{\mathrm{op}}$ are given by $S \in \eta_A(x) \Leftrightarrow x \in S$, for $x \in |A|$ and $S \in |FA|$, and $T \in \varepsilon_B(y) \Leftrightarrow y \in T$, for $y \in |B|$ and $T \in |GB|$.

1. The proper filter monad on sets

1.1. We denote by MSL_0 the category of meet semilattices with 0 (and 1). Morphisms of MSL_0 preserve finite meets and 0. The contravariant powerset functor on sets obviously lifts to a preimage functor $P_0: \mathrm{ENS}^{\mathrm{op}} \to \mathrm{MSL}_0$, with P_0A the powerset of A for a set A, ordered by set inclusion and regarded as meet semilattice with 0.

If $f: A \to PL$ and $g: L \to PA$ are exponentially adjoint, *i.e.* always $a \in f(x) \Leftrightarrow x \in g(a)$ if $x \in A$ and $a \in L$, for a set A and an object L of MSL_0, then g is a morphism $g \cdot L \to P_0 A$ of MSL_0 iff every $f(x)$ is a proper filter in L. It follows immediately that we have a preimage functor $G_0 : \mathrm{MSL}_0^{\mathrm{op}} \to \mathrm{ENS}$, exponentially adjoint to P_0 (see 0.4), with $G_0 L$ the set of all proper filters in L for an object L of MSL_0.

We denote by $\mathscr{F}_0$ the monad on sets obtained from this adjunction; this is the *proper filter monad.* The *proper filter functor* $F_0 = G_0 P_0^{\mathrm{op}}$ assigns to every set A the set of all proper filters on A.

1.2. We denote by LAT the category of lattices, with 0 and 1, and by P_p: $\mathrm{ENS}^{\mathrm{op}} \to \mathrm{LAT}$ the preimage functor which assigns to every set its powerset, ordered by set inclusion and considered as a lattice. This functor has an exponential adjoint (see 0.4) $G_p : \mathrm{LAT}^{\mathrm{op}} \to \mathrm{ENS}$, with $G_p L$ the set of all prime filters in L for a lattice L.

The resulting monad on sets is the *ultrafilter monad* which we denote by $\mathscr{U}$, with functor part $U = G_p P_p^{\mathrm{op}}$ the *ultrafilter functor* on sets. As is well known, $\mathscr{U}$-algebras are compact Hausdorff spaces; the $\mathscr{U}$-algebra structure of a compact Hausdorff space X assigns to every ultrafilter on X its limit for X.

If $S: \mathrm{LAT} \to \mathrm{MSL}_0$ is the inclusion functor, then clearly $SP_p = P_0$. Adjoint to the resulting identity natural transformation is $\varkappa: G_p \to G_0 S^{\mathrm{op}}$ given by inclusions. By 0.3.(b), this produces a morphism $i = (\mathrm{Id}, \varkappa P_p^{\mathrm{op}}): \mathscr{U} \to \mathscr{F}_0$ of monads. Thus every $\mathscr{F}_0$-algebra (L, α) has an underlying compact Hausdorff space $i^*(L, \alpha)$, with the restriction of α to ultrafilters as convergence of ultrafilters. Morphisms of $\mathscr{F}_0$-algebras are continuous for the underlying compact topologies thus obtained.

1.3. We define a *sup semilattice* as an ordered set L, such that every non-empty subset of L has a supremum in L, and we denote by SSL the category of sup semilattices, with morphisms which preserve suprema of non-empty subsets.

We denote by $\underline{E}_0$ the free sup semilattice functor on sets, left adjoint to the underlying set functor $|\ |: \mathrm{SSL} \to \mathrm{ENS}$. It is well known that $\underline{E}_0 A$, for a set A, is the set of all non-empty subsets of A, with set unions as suprema, and that $\underline{E}_0$ for mappings is given by direct images.

We denote by $\mathscr{E}_0$ the *powerset monad* on sets which results from the adjunction $\underline{E}_0 \dashv |\ |$, with functor part $E_0 = |\ | \underline{E}_0$. Algebras for $\mathscr{E}_0$ are sup semilattices; the $\mathscr{E}_0$-algebra structure of a sup semilattice is given by suprema of non-empty subsets.

Every sup semilattice L has a one-point extepsion to a complete lattice $\tilde{L}$, obtained by adding a zero o_L to L, and a map $f: L \to L'$ of SSL extends to $\tilde{f}: \tilde{L} \to \tilde{L}'$ preserving suprema, with $\tilde{f}(o_L) = o_{L'}$. We obtain a functor D_0: $\mathrm{SSL}^{\mathrm{op}} \to \mathrm{MSL}_0$ by putting $D_0 L = \tilde{L}$, considered as object of MSL_0, and putting $x \leqq (D_0 f)(x') \Leftrightarrow \tilde{f}(x) \leqq x'$, for a map $f: L \to L'$ of SSL and $(x, x') \in \tilde{L} \times \tilde{L}'$.

The functor $D_0 E_0^{\mathrm{op}}$ clearly is naturally isomorphic to the functor P_0. Adjoint to this isomorphism is a natural transformation $\varkappa: |\ | \to G_0 D_0^{\mathrm{op}}$; one sees easily that $\varkappa_L(x) = \uparrow x$, for a sup semilattice L and $x \in L$. We note that proper filters in $D_0 L$ are filters in L.

By 0.3.(b), the natural transformation $\varkappa : | \ | \to G_0 D_0^{op}$ induces a morphism $j = (\mathrm{Id}, \varkappa \underline{E}_0) : \varepsilon_0 \to \mathscr{F}_0$ of monads, and hence an algebraic functor j^*, from $\mathscr{F}_0$-algebras to SSL, which preserves underlying sets and mappings. Thus an $\mathscr{F}_0$-algebra (L, α) has an underlying sup semilattice $j^*(L, \alpha)$, with $\sup S = \alpha(\uparrow S)$ for a non-empty subset S of L. Homomorphisms of $\mathscr{F}_0$-algebras preserve these suprema.

1.4. By the general theory, the functor $G_0 : \mathrm{MSL}_0^{op} \to \mathrm{ENS}$ lifts to a contravariant comparison functor K_0 to $\mathscr{F}_0$-algebras, with $K_0 L = (G_0 L, G_0 \varepsilon_L)$ for an object L of MSL_0, and $K_0 f = G_0 f : K_0 L \to K_0 M$ for $f : L \to M$ in MSL_0.

By 0.4, $a \in (G_0 \varepsilon_L)(\Phi) \Leftrightarrow \varepsilon_L(a) \in \Phi$, for $a \in L$ and a proper filter Φ on $G_0 L$, with $\varphi \in \varepsilon_L(a) \Leftrightarrow a \in \varphi$ for a proper filter φ in L. Using this for $\Phi = \uparrow S$, we see that $\sup S$ in $j^* K_0 L$ is the set intersection of all $\varphi \in S$. Thus the order of filters in $K_0 L$ is the natural order of filters, dual to set inclusion.

For an ultrafilter Ψ on $G_0 L$ with limit φ in $i^* K_0 L$, we get $a \in \varphi \Leftrightarrow \varphi \in \varepsilon_L(a) \Leftrightarrow \varepsilon_L(a) \in \Psi$, hence also $\varphi \notin \varepsilon_L(a) \Leftrightarrow G_0 L \setminus \varepsilon_L(a) \in \Psi$. Thus $i^* K_0 L$ is a Stone space, with the coarsest topology for which all sets $\varepsilon_L(a)$ are clopen.

1.5. Proposition. *If (L, α) is an $\mathscr{F}_0$-algebra, with underlying compact space $X = i^*(L, \alpha)$, then $\alpha(\Phi) = \sup \mathrm{adh}_X \Phi$ for a proper filter Φ on L, with supremum in $j^*(L, \alpha)$. Morphisms of $\mathscr{F}_0$-algebras are all morphisms of the underlying sup semilattices which are continuous for the underlying compact topologies.*

Proof. All morphism of $\mathscr{F}_0$-algebras, including an algebra structure α, are continuous and preserve suprema as stated. In the free algebra $K_0 P_0 L$, a filter Φ on L is the supremum of all finer ultrafilters Ψ; thus $\alpha(\Phi)$ is the supremum of the limits $\alpha(\Psi)$. These limits form the adherence $\mathrm{adh}_X \Phi$ of Φ; thus $\alpha(\Phi) = \sup \mathrm{adh}_X \Phi$.

Now if $f : X \to Y$ for $X = i^*(L, \alpha)$ and $Y = i^*(M, \beta)$, and $f : j^*(L, \alpha) \to j^*(M, \beta)$ in SSL, then

$$f(\sup \mathrm{adh}_X \Phi) = \sup f(\mathrm{adh}_X \Phi) = \sup \mathrm{adh}_Y (F_0 f)(\Phi)$$

for a proper filter Φ on L, since $f : X \to Y$ preserves adherences of filters for compact Hausdorff spaces X and Y. Thus $f : (L, \alpha) \to (M, \beta)$x is a homomorphism of $\mathscr{F}_0$-algebras □

1.6. Proposition. *$\mathscr{F}_0$-algebras can be embedded into compact join semilattices (without 0) as a full subcategory. If (L, α) is an $\mathscr{F}_0$-algebra, then $\alpha(\Phi) = \inf \{\sup S : S \in \Phi\}$ for a proper filter Φ on L.*

Proof. We have $\vee \cdot (\alpha \times \alpha) = \alpha \cdot \vee$ for joins in $j^*(L, \alpha)$ and in $j^* K_0 P_0 L$. Since joins in $j^* K_0 P_0 L$ are set intersections, we have $\vee^{\leftarrow}(S^\#) = S^\# \times S^\#$ for $S \subset L$ and $S^\# = \varepsilon_{P_0 L}(S)$, with $\Phi \in S^\# \Leftrightarrow S \in \Phi$. This shows that $\vee$ is continuous for the compact topology of $K_0 P_0 L$. As α and $\alpha \times \alpha$ are continuous for the compact topologies and surjective, hence topological quotient maps, it follows that $\vee$ for (L, α) is continuous for $i^*(L, \alpha)$. Thus $\mathcal{F}_0$-algebras are compact join semilattices.

Compact join semilattices are compact ordered spaces [4]; thus filter bases have infima, and dual filter bases suprema, which are topological limits. It follows that maps of compact join semilattices preserve infima of filter bases, and suprema of dual filter bases. Since these maps preserve finite non-empty joins, they preserve all non-empty joins. With 1.5, it follows that $\mathcal{F}_0$-algebras and their induced compact join semilattices have the same maps.

Now if (L, α) is an $\mathcal{F}_0$-algebra and Φ a proper filter on L, then Φ is the infimum in K_0P_0L of all filters $\uparrow S$ with $S \in \Phi$. These filters form a filter basis in K_0P_0L; thus $\alpha(\Phi)$ is the infimum of all $\alpha(\uparrow S) = \sup S$ with $S \in \Phi$ □

2. Continuous sup semilattices

2.1. A sup semilattice L (1.3) is called *complete* if every filter base in L has an infimum in L.

For elements a, b of a complete sup semilattice L, we say that b is *way above* a, and we write $b \gg a$, if b is in every filter φ in L with $\inf \varphi \leqq a$. Since $\uparrow a$ is such a filter, $a \leqq b$ follows. The elements way above a in L form a filter in L which we denote by $\Uparrow a$, with $\inf \Uparrow a \geqq a$. We say that L is a *continuous sup semilattice* (continuous-complete sup semilattice in [2]) if $\inf \Uparrow a = a$ for all $a \in L$.

Morphism of complete and continuous sup semilattices are mappings which preserve non-empty suprema, and infima of filter bases or equivalently of filters.

2.2. We use again the notations of 1.3. For a sup semilattice L, putting $q_L(x) = \downarrow x$ for $x \in L$ provides a map $q_L : L \to P|L|$. With $q_L(o_L) = \emptyset$, this clearly becomes a morphism $q_L : D_0L \to P_0|L|$ in MSL_0, and it is easily verified that this morphism q_L is natural in L.

Proper filters in a meet semilattice M with 0, ordered dually to set inclusion, form a sup semilattice, with set intersections as suprema. Maps G_0f, for morphisms f of MSL_0, clearly preserve these suprema; thus we can lift G_0 to a functor $G_0 : \mathrm{MSL}_0 \to \mathrm{SSL}$, with $|\ |G_0 = G_0$.

Proposition. *The functors* D_0 *and* G_0 *are adjoint on the right, with* $q : D_0 \to P_0|\ |^{\mathrm{op}}$ *and* $\mathrm{id}(G_0)$ *adjoint.*

Proof. For objects L of SSL and M of MSL_0, we know that $a \in f(x) \Leftrightarrow x \in g_1(a)$, for $x \in L$ and $a \in M$, provides a bijection between morphisms $g_1 : M \to P_0L$ of MSL_0 and mappings $f : |L| \to G_0M$. In this situation, $a \in f(\sup x_i) \Leftrightarrow \sup x_i \in g_1(a)$, and $a \in \sup f(x_i) \Leftrightarrow x_i \in g_1(a)$ for all i. Thus f preserves non-empty suprema iff each $g_1(a)$ is a principal dual filter $\downarrow g(a)$, *i.e.* iff g_1 factors $g_1 = q_L \cdot g$. Since q_L is a natural embedding, this gives the desired bijection between maps $f : L \to G_0M$ in SSL and $g : M \to D_0L$ in MSL_0, and this bijection is natural in L as well as in M. The proof also shows that q and id (G_0) are adjoint □

2.3. We denote by $\mathcal{Q}_0$ the monad on SSL obtained from the adjunction $D_0^{\mathrm{op}} \dashv G_0$ of 2.2, with functor part $Q_0 = G_0 D_0^{\mathrm{op}}$. Since $f: L \to G_0 M$ and $g: M \to D_0 L$ are adjoint iff always $a \in f(x) \Leftrightarrow x \leqq g(a)$, both units of this adjunction are principal filter maps. We note that $Q_0 L$ is the sup semilattice of all filters in L; these are the proper filters in $D_0 L$.

Theorem. $\mathrm{SSL}^{\mathcal{Q}_0}$ *is the category of continuous sup semilattices, with algebra structures* $\inf_L : Q_0 L \to L$.

Proof. If (L, γ) is a $\mathcal{Q}_0$-algebra, then $\gamma(\varphi) \leqq \gamma(\uparrow a) = a$ for a filter φ in L and $a \in \varphi$. On the other hand, if $x \leqq a$ for all $a \in \varphi$, then $\uparrow x \leqq \varphi$, and $x \leqq \gamma(\varphi)$ follows. Thus L is complete, and $\gamma(\varphi) = \inf_L \varphi$ for all φ in $Q_0 L$.

Now a $\mathcal{Q}_0$-algebra is a complete sup semilattice L such that $\inf: Q_0 L \to L$ preserves suprema and satisfies the formal laws for monadic algebras. These laws require that always $\inf \uparrow x = x$, which is valid, and $\inf \inf \Phi = \inf (Q_0 \inf)(\Phi)$ for a filter Φ in $Q_0 L$. This is also valid since $\inf \Phi$ is the set union of all $\varphi \in \Phi$, and the filter $Q_0 \inf(\Phi)$ has the elements $\inf \varphi$, for $\varphi \in \Phi$, as a basis.

The map $\inf_L : Q_0 L \to L$ preserves suprema iff there is a mapping $\sigma: L \to Q_0 L$ such that $\inf_L \varphi \leqq x \Leftrightarrow \varphi \leqq \sigma(x)$ for all $\varphi \in Q_0 L$ and $x \in L$. Then $\sigma(x)$ must be the supremum of all φ with $\inf_L \varphi \leqq x$, and $\inf_L \sigma(x) \leqq x$. But then $\sigma(x) = \Uparrow x$; thus $\inf_L$ preserves suprema, and L is a $\mathcal{Q}_0$-algebra, with structure $\inf_L$, iff L is a continuous sup semilattice.

Morphisms of $\mathcal{Q}_0$-algebras must preserve suprema of non-empty subsets, and $\mathcal{Q}_0$-algebra structures $\inf_L$; thus they are the maps of continuous sup semilattices □

2.4. By 2.2 and 0.3.(a), we have a morphism of monads $\sup = (|\ |, G_0 q^{\mathrm{op}}): \mathcal{F}_0 \to \mathcal{Q}_0$. For a proper filter Φ on a sup semilattice L, the filter $(G_0 g_L)(\Phi)$ in L consists of all $x \in L$ with $\downarrow x \in \Phi$, and hence of all $\sup S$ with $S \in \Phi$.

Theorem. *The algebraic functor* $\sup^*$ *from continuous sup semilattices to* $\mathcal{F}_0$*- algebras is an isomorphism of categories, preserving underlying sets and mappings, and with* $j^* \sup^*$ *the forgetful functor from continuous sup semilattices to* SSL.

Proof. $\sup^*$ clearly preserves underlying sets and mappings. For a continuous sup semilattice L and $\alpha = \inf_L \cdot G_0 g_L$, we have $(G_0 g_L)(\uparrow S) = \uparrow \sup S$ for non-empty $S \subset L$, hence $\alpha(\uparrow S) = \sup S$. Thus $j^* \sup^* L$ is the underlying sup semilattice of L.

Now $\sup^*$ has the factorization property of 0.2.(ii), with $\Delta = j^*$, and 0.2.(i) is also satisfied since every filter φ in a sup semilattice L satisfies $\varphi = (G_0 g_L)(\Phi)$ for the filter Φ on L with the sets $\downarrow x$, $x \in \varphi$, as filter basis. 0.2.(iii) is satisfied by 1.6; thus $\sup^*$ is an isomorphism by 0.2 □

3. The proper Vietoris monad

3.1. For a compact Hausdorff space X, we denote by V_0X the set of all non-empty closed subsets of X, ordered by set inclusion and provided with the coarsest topology such that the sets $\downarrow S = \{A \in V_0X : A \subset S\}$ are closed for $S \subset X$ closed, and open for $S \subset X$ open. This is the *Vietoris space* [6] of X.

V_0X is a Hausdorff space, for if A, B are closed in X with $A \not\subset B$, then there are disjoint open sets R and S in X with $B \subset S$, and R and A not disjoint. Then $V_0X \setminus \uparrow(X \setminus R)$ and $\uparrow S$ are disjoint neighborhoods of A and B in V_0X.

3.2. We denote by CH the category of compact Hausdorff spaces, and by $|\ |: \mathrm{CH} \to \mathrm{ENS}$ its underlying set functor. For a compact Hausdorff space X, adherences of proper filters on X provide a mapping $\mathrm{adh}_X : F_0\,|X| \to |V_0X|$.

Proposition. $\mathrm{adh}_X : K_0P_0\,|X| \to V_0X$ *is continuous and preserves suprema.*

Corollary. V_0X *is a compact Hausdorff space.*

Proof. For $S \subset X$ closed and a filter Φ on X, we have $\mathrm{adh}_X \Phi \subset S$ iff all neighborhoods of S in X are in Φ. Thus $\mathrm{adh}_X^{\leftarrow}(\downarrow S) = \wedge R^{\#}$, with $R^{\#} = \varepsilon_L(R)$ for $L = P_0\,|X|$ (see 1.4), for all neighborhoods R of S in X. This set is closed in $i^*\,K_0P_0\,|X|$.

For $S \subset X$ open, we have $\mathrm{adh}_X \Phi \subset S$ iff S contains a closed neighborhood of $\mathrm{adh}_X \Phi$, with $R \in \Phi$. Thus $\mathrm{adh}_X^{\leftarrow}(\downarrow S) = \vee R^{\#}$ for closures R of open sets contained in S. This set is open in $i^*K_0P_0\,|X|$; thus adh_X is continuous.

For filters Φ_i on X and $S \subset X$ closed, we have $\mathrm{adh}_X \sup \Phi_i \subset S \Leftrightarrow$ all neighborhoods of S are in $\sup \Phi_i \Leftrightarrow$ all neighborhoods of S are in every Φ_i $\Leftrightarrow \mathrm{adh}_X \Phi_i \subset S$ for every $i \Leftrightarrow \sup \mathrm{adh}_X \Phi_i \subset S$, with the last supremum in V_0X. This shows that adh_X preserves suprema.

Since $\mathrm{adh}_X \uparrow S = S$ for S closed in X, adh_X is surjective. As adh_X is continuous and $K_0P_0\,|X|$ compact, V_0X is compact □

3.3. For $f: X \to Y$ in CH, let V_0f be the restriction of $f^{\rightarrow}$ to V_0X and V_0Y. Then $(V_0f)^{\leftarrow}(\downarrow T) = \uparrow f^{\leftarrow}(T)$ for $T \subset Y$, and $V_0f: V_0X \to V_0Y$ is continuous. Thus we have a functor $V_0 : \mathrm{CH} \to \mathrm{CH}$.

For a compact Hausdorff space X, we denote by $s_X : X \to V_0X$ the singleton map, with $s_X(x) = \{x\}$ for $x \in X$. Since $s_X{\leftarrow}(\downarrow S) = S$ for $S \subset X$, the map s_X is continuous.

L. Vietoris proved in [7] (see also *e.g.* [8b], 5.4) that the set union $\vee K$ is closed in X for K closed in V_0X, if X is a compact Hausdorff space. Thus set unions define a map $u_X : V_0V_0X \to V_0X$. Clearly $u_X{\leftarrow}(\downarrow S) = \downarrow\downarrow S$ for $S \subset X$; thus u_X is continuous.

It is easily seen that adh_X, s_X, u_X are natural in X.

Proposition. *The functor* V_0, *and the natural transformations obtained above, define a monad* $\mathscr{V}_0 = (V_0, s, u)$ *on* CH, *and a morphism* $(|\ |, \mathrm{adh}) : \mathscr{F}_0 \to \mathscr{V}_0$ *of monads.*

We call $\mathscr{V}_0$ the *proper Vietoris monad.*

Proof. The monadic identities for s and u are easily verified; we omit details. We must show that $\text{adh} \cdot \eta|\ | = |\ | s$, and

$$\text{adh} \cdot \mu|\ | = |\ | u \cdot \text{adh}\, V_0 \cdot F_0\, \text{adh},$$

if $\mathscr{F}_0 = (F_0, \eta, \mu)$. The first of these is obvious; the point filter $\eta_{|X|}(x)$ has $\{x\}$ as its adherence. For the second one, put $Z = i^* K_0 P_0 |X|$. Since $\mu_{|X|}$ is the $\mathscr{F}_0$-algebra structure of $K_0 P_0 |X|$, we have a diagram

$$\begin{array}{ccccc}
F_0|Z| & \xrightarrow{\text{adh}_Z} & |V_0 Z| & \xrightarrow{\text{sup}} & |Z| \\
\downarrow{\scriptstyle F_0\, \text{adh}_X} & & \downarrow{\scriptstyle V_0\, \text{adh}_X} & & \downarrow{\scriptstyle \text{adh}_X,} \\
F_0|V_0 X| & \xrightarrow{\text{adh}_{V_0 X}} & |V_0 V_0 X| & \xrightarrow{u_X} & |V_0 X|
\end{array}$$

with the factorization of $\mu_{|X|}$ by 1.5 on top. The lefthand square commutes by naturality of adh_X, and the righthand square since adh_X preserves suprema. Thus the diagram commutes □

3.4. Theorem. *The algebraic functor* $(|\ |, \text{adh})^*$, *from* $\mathscr{V}_0$*-algebras to* $\mathscr{F}_0$*-algebras, is an isomorphism of categories, preserving underlying sets and mappings, with* $i^*(|\ |, \text{adh})^*$ *the forgetful functor from* $\mathscr{V}_0$*-algebras to compact Hausdorff spaces.*

Proof. Put $(|\ |, \text{adh})^* = \text{adh}^*$; this preserves underlying sets and mappings. If (X, ξ) is a $\mathscr{V}_0$-algebra and Ψ an ultrafilter on X, with limit x, then $\xi(\text{adh}_X \Psi) = \xi(\{x\}) = x$. Thus x is the limit of Ψ in $i^* \text{adh}^*(X, \xi)$, *i.e.* $i^* \text{adh}^* (X, \xi) = X$. Since $i^* \text{adh}^*$ and $|\ | i^*$ preserve underlying sets and mappings, we have 0.2.(ii) for adh^*, with $\Delta = i^*$. Since $\text{adh}_X \uparrow S = S$ for S closed in X, 0.2.(i) also is satisfied. Finally, 0.2.(iii) is satisfied by 1.5, so that 0.2 applies □

4. The closed proper filter monad

4.1. We denote by TOP the category of topological spaces and continuous maps, and by $R: \text{TOP} \to \text{ENS}$ its underlying set functor. For purposes of this paper, objects of TOP could be restricted to be T_0 spaces or sober spaces, or the super-sober spaces of the Compendium [1].

For a topological space X, an object L of MSL_0, and adjoint maps $f: RX \to G_0 L$ and $g: L \to P_0 RX$, we have $g(a) = f^{\leftarrow}(a^\sharp)$, for $a \in L$ and $a^\sharp = \varepsilon_L(a) = \{\varphi \in G_0 A : a \in \varphi\}$. If $\Gamma_0 X$ is the set of all closed sets of X, ordered by set inclusion and considered as object of MSL_0, and $\Sigma_0 L$ the set $G_0 L$ with the coarsest topology such that all sets $a^\#$ are closed, it follows that $f: X \to \Sigma_0 L$ is continuous iff g maps L into $\Gamma_0 X$. Thus we have exponentially adjoint preimage functors (0.4) $\Gamma_0: \mathrm{TOP}^{\mathrm{op}} \to \mathrm{MSL}_0$ and $\Sigma_0: \mathrm{MSL}_0^{\mathrm{op}} \to \mathrm{TOP}$.

By this construction, $R\Sigma_0 = G_0$, and set inclusions define a natural transformation $\lambda: \Gamma_0 \to P_0 R^{\mathrm{op}}$, adjoint to id (G_0). The monad on TOP obtained from the adjunction $\Gamma_0^{\mathrm{op}} \dashv \Sigma_0$ is the *proper closed filter monad*. We denote it by $\mathscr{W}_0$, with functor part $W_0 = \Sigma_0 \Gamma_0^{\mathrm{op}}$.

By 0.2.(a), we have a morphism $r = (R, G_0\lambda^{\mathrm{op}}): \mathscr{F}_0 \to \mathscr{W}_0$ of monads, with $(G_0\lambda_X)(\Phi)$ the restriction of Φ to closed sets for a proper filter Φ on a topological space X.

4.2. If L is a lattice, then prime filters in L form a subspace $\Sigma_p L$ of $\Sigma_0 L$; this defines a preimage functor $\Sigma_p: \mathrm{LAT}^{\mathrm{op}} \to \mathrm{TOP}$, exponentially adjoint to the functor $\Gamma_p: \mathrm{TOP}^{\mathrm{op}} \to \mathrm{LAT}$ with $\Gamma_p X$ the lattice of closed sets of a space X. If $S: \mathrm{LAT} \to \mathrm{MSL}_0$ is the inclusion functor, then $\Gamma_0 = S\Gamma_p$, and subspace inclusions define a natural transformation $\varkappa: \Sigma_p \to \Sigma_0 S^{\mathrm{op}}$, with $\varkappa$ and id(Γ_0) clearly adjoint.

We denote by $\mathscr{W}_p$ the monad on TOP resulting from the adjunction $\Gamma_p^{\mathrm{op}} \dashv \Sigma_p$, with functor part $W_p = \Sigma_p \Gamma_p^{\mathrm{op}}$. This is the *prime closed filter monad* on TOP, and $W_p X$ is the *prime Wallman compactification* of X for a topological space X. We have an inclusion morphism $i = (\mathrm{Id}, \varkappa\Gamma_p^{\mathrm{op}}): \mathscr{W}_p \to \mathscr{W}_0$, and a restriction morphism $r: \mathscr{U} \to \mathscr{W}_p$.

$\mathscr{W}_p$-algebras were studied in [9]; they are compact ordered spaces. If $(Z, \leqq)$ is a compact ordered space, then $Z = r^*(X, \alpha)$ for a unique $\mathscr{W}_p$-algebra (X, α), where r^* is the algebraic functor induced by the restriction morphism $r: \mathscr{U} \to \mathscr{W}_p$. In this situation, X is Z with the upper topology, *i.e.* open sets of X are increasing open sets of Z, the topology of Z is the patch topology of X [1; 9], and Z has the induced order of X, with $x \leqq y$ in the order of Z iff $x \in \mathrm{cl}_X\{y\}$.

4.3. By 4.1 and 4.2, a $\mathscr{W}_0$-algebra (X, α) has an induced compact ordered space $i^*(X, \alpha)$, and an induced continuous sup semilattice, or $\mathscr{F}_0$-algebra, $r^*(X, \alpha)$. Since the diagram at the beginning of this paper commutes, both have the same compact topology, the patch topology of X. They also have the same order:

Proposition. *If (X, α) is a $\mathscr{W}_0$-algebra, then the order of the induced continuous sup semilattice $r^*(X, \alpha)$ is the induced order of X.*

Proof. We note first that the induced order of a space $\Sigma_0 L$ is the natural order of filters in L, dual to set inclusions, since $\psi \leqq \varphi$ in the induced order of $\Sigma_0 L$ iff ψ is in every basic closed set $\cup a_i^\#$ with φ in $\cup a_i^\#$.

Now let (X, α) be a $\mathscr{W}_0$-algebra, with induced $\mathscr{F}_0$-algebra structure $\alpha \cdot \mathrm{cl}_X$, where $\mathrm{cl}_X \Phi$ is the restriction of a filter Φ to closed sets. If $x \leqq y$ in the induced order of X, then $x \vee y = \alpha(\uparrow\mathrm{cl}_X\{x, y\}) = \alpha(\uparrow\mathrm{cl}_X\{y\}) = y$ in $r^*(X, \alpha)$. Conversely, $\uparrow\mathrm{cl}_X\{x\} \leqq \uparrow\mathrm{cl}_X\{x, y\}$ in the induced order of $W_0 X$. The continuous map α preserves the induced order; thus $x \leqq x \vee y$ in the induced order of X, and $x \leqq y$ in this order if $x \vee y = y$ □

4.4. For an $\mathscr{F}_0$-algebra (L, α), let $U(L, \alpha)$ be L provided with the upper topology for the induced compact ordered space of (L, α). This clearly defines a functor U, from $\mathscr{F}_0$-algebras to TOP, which preserves underlying sets and mappings, with RU the forgetful functor from $\mathscr{F}_0$-algebras to sets. Note that our upper topology is the lower topology of the Compendium [1].

Theorem. *The algebraic functor r^* from $\mathscr{W}_0$-algebras to $\mathscr{F}_0$-algebras is an isomorphism of categories, preserving underlying sets and mappings, and with Ur^* the forgetful functor from $\mathscr{W}_0$-algebras to* TOP.

Proof. r^* clearly preserves underlying sets and mappings. If (X, α) is a $\mathscr{W}_0$-algebra, then X has the upper topology of the induced compact ordered space which by 4.3 is also the induced compact ordered space of $r^*(X, \alpha)$. Thus $Ur^*(X, \alpha) = X$. Since Ur^* preserves underlying mappings, it follows that Ur^* is the forgetful functor for $\mathscr{W}_0$-algebras. Thus 0.2.(ii) is satisfied with $\Delta = U$.

0.2.(i) holds; every proper filter of closed sets on a topological space X is the restriction of a proper filter on RX to closed sets.

It remains to obtain the factorization of 0.2.(iii), *i.e.* if Φ is a proper filter on L for an $\mathscr{F}_0$-algebra (L, α), then $\alpha(\Phi) = \inf\{\sup S : S \in \Phi\}$ depends only upon the decreasing closed sets in Φ, *i.e.* the sets closed for $U(L, \alpha)$. This is the case since $\sup S = \sup \mathrm{cl}_Z S = \sup \downarrow\mathrm{cl}_Z S$ for $S \subset Z$ if $Z = i^*(L, \alpha)$, considered as compact ordered space, with $\downarrow\mathrm{cl}_Z S$ closed for Z and decreasing, and in Φ if $S \in \Phi$ □

5. The proper open filter monad

5.1. We denote by $\mathcal{O}_0 X$ and $\mathcal{O}_p X$ the set of open sets of a topological space X, ordered by set inclusion and regardes as an object of MSL_0 and a lattice respectively. In the other direction, we denote by $\Pi_0 L$ the set $G_0 L$ of proper filters in a meet semilattice L with 0, provided with the topology for which the sets $a^\# = \varepsilon_L(a) = \{\varphi \in G_0 L : a \in \varphi\}$, for $a \in L$, form a basis of open sets. If L is a lattice, then $\Pi_p L$ will be the subspace of $\Pi_0 L$ consisting of all prime filters in L.

As in 4.1 and 4.2, this defines exponentially adjoint preimage functors $\mathcal{O}_0 : \mathrm{TOP}^{\mathrm{op}} \to \mathrm{MSL}_0$ and $\Pi_0 : \mathrm{MSL}_0^{\mathrm{op}} \to \mathrm{TOP}$, and exponentially adjoint preimage functors $\mathcal{O}_p : \mathrm{TOP}^{\mathrm{op}} \to \mathrm{LAT}$ and $\Pi_p : \mathrm{LAT}^{\mathrm{op}} \to \mathrm{TOP}$. We denote by $\mathscr{H}_0$ the *proper open filter monad* on TOP, and by $\mathscr{H}_p$ the *prime open filter monad* on TOP, resulting from these adjunctions, with functor parts $H_0 = \Pi_0 \mathcal{O}_0^{\mathrm{op}}$ and $H_p = \Pi_p \mathcal{O}_p^{\mathrm{op}}$.

5.2. If $R : \mathrm{TOP} \to \mathrm{ENS}$ is the underlying set functor, then $R\Pi_0 = G_0$ and $R\Pi_p = G_p$, and we have natural inclusions $\lambda : \mathcal{O}_0 \to P_0 R^{\mathrm{op}}$ and $\lambda : \mathcal{O}_p \to P_p R^{\mathrm{op}}$, adjoint to id (G_0) and id (G_p). For $\pi = G_0 \lambda^{\mathrm{op}}$ and $\pi = G_p \lambda^{\mathrm{op}}$, restricting proper filters or ultrafilters to their open sets, and $r = (R, \pi)$, we have morphisms of monads $r : \mathscr{F}_0 \to \mathscr{H}_0$ and $r : \mathscr{U} \to \mathscr{H}_p$.

We have $\mathcal{O}_0 = S\mathcal{O}_p$ for the inclusion functor S of 4.2, and subspace inclusions define $\varkappa : \Pi_p \to \Pi_0 S^{\mathrm{op}}$, adjoint to id $(\mathcal{O}_0)$. Thus we have an inclusion morphism $i = \mathrm{Id}, \varkappa\mathcal{O}_p^{\mathrm{op}}) : \mathcal{H}_p \to \mathcal{H}_0$.

Let $\Delta : \mathrm{LAT} \to \mathrm{LAT}$ denote the dual lattice functor which reverses order in every lattice and preserves underlying sets and mappings. Complements of closed sets define a natural isomorphism $\varrho : \Delta\Gamma_p \to \mathcal{O}_p$. For a lattice L, the complement $L \setminus \varphi$ of a prime filter φ in L is a prime filter in ΔL; it is easily seen that complements of prime filters define a natural homeomorphism $\sigma : \Sigma_p \to \Pi_p \Delta^{\mathrm{op}}$, with σ and ϱ adjoint. It follows by 0.3 that $\sigma\Gamma_p^{\mathrm{op}} = \Pi_p \varrho^{\mathrm{op}} \cdot \pi$ for an isomorphism of monads $(\mathrm{Id}, \pi) : \mathcal{W}_p \to \mathcal{H}_p$. We note that $X \setminus A \in \pi_X(\Psi) \Leftrightarrow A \notin \Psi$, for a prime filter Ψ of closed sets of a topological space X and a closed set A of X, and that the maps ϱ_X are the only order reversing natural maps encountered in this paper.

5.3. If h is the unit of $\mathcal{H}_0$ or of $\mathcal{H}_p$, then $h_X(x)$ is the filter of open neighborhoods of x in X, for a topological space X and $x \in X$. We define the *dual induced order* of X, dual to the induced order of X, by putting $x \leqq y$, for x, y in X, iff $h_X(x) \leqq h_X(y)$.

For φ in a space $\Pi_0 L$ or $\Pi_p L$, the sets $a^\sharp$ with $a \in \varphi$ form a base of neighborhoods of φ; it follows that the dual induced order for $\Pi_0 L$ or $\Pi_p L$ is the natural order for filters in L, dual to set inclusion.

By 5.2, we have algebraic functors r^*, from $\mathcal{H}_p$-algebras to compact Hausdorff spaces, and from $\mathcal{H}_0$-algebras to $\mathcal{F}_0$-algebras. These functors preserve underlying sets and mappings.

Theorem. *If* $(Z, \leqq)$ *is a compact ordered space* [4], *then* $Z = r^*(X, \alpha)$ *for a unique* $\mathcal{H}_p$*-algebra* (X, α). *In this situation,* X *has the lower topology of* Z *and the order of* Z *is the dual induced order of* X. *Every* $\mathcal{H}_p$*-algebra is obtained in this way, and morphisms of* $\mathcal{H}_p$*-algebras are morphisms of compact ordered spaces.*

Proof. If $Z = r^*(X, \alpha)$ for an $\mathcal{H}_p$-algebra (X, α), then it is seen as in [9], 1.7, that the topology of Z is finer than the topology of X. If φ is an ultrafilter on X and $\bar{\varphi}$ the prime filter of open sets in φ, then $\alpha(\bar{\varphi})$ is the limit of φ for Z; thus φ converges to all $x \geqq \ (\bar{\varphi})$ for the dual induced order of X. Conversely, if φ converges to x for X, then $h_X(x) \geqq \bar{\varphi}$ in $H_p X$; thus $x \geqq \alpha(\bar{\varphi})$ for the dual induced order of X since the continuous map α preserves this order.

If $x \nleqq y$ in X, so that $y \notin \mathrm{cl}_X\{x\}$, then $h_X(y)$ is not in the closed set $\alpha^{\leftarrow}(\mathrm{cl}_X\{x\})$ in $H_p X$; thus there is a basic open set $V^\sharp$ in $H_p X$, with V an open neighborhood of y in X, disjoint from $\alpha^{\leftarrow}(\mathrm{cl}_X\{x\})$. It follows that $X \setminus V$ is in every ultrafilter with limit x for Z; thus $X \setminus V$ is a neighborhood of x in Z. Now $(X \setminus V) \times V$ is a neighborhood of (x, y) in $Z \times Z$, disjoint from the graph of $\leqq$ since V is decreasing for the dual induced order; thus $(Z, \leqq)$ is a compact ordered space. Ultrafilters on X have the same limits for X as for the lower topology of $(Z, \leqq)$; thus X has this lower topology.

The remainder of the proof follows the proof of the corresponding results of [9], with only minor changes □

5.4. Remark. Since the monads $\mathscr{H}_p$ and $\mathscr{W}_p$ are isomorphic, the same topological spaces have algebra structures for the two monads. These spaces are the super-sober spaces of the Compendium [1]. The compact ordered spaces obtained in this way from a super-sober space X are dual. They have the same topology, the patch topology of X [1; 9], but dual orders.

5.5. The morphisms of monads of 5.2 provide algebraic functions i^* and r^*, from $\mathscr{H}_0$-algebras to $\mathscr{H}_p$-algebras and to $\mathscr{F}_0$-algebras. If (X, α) is an $\mathscr{H}_0$-algebra, then the $\mathscr{F}_0$-algebra $r^*(X, \alpha)$ and the compact ordered space obtained from $i^*(X, \alpha)$ provide us with the same compact Hausdorff topology, by commutativity of the diagram at the beginning of this paper. We now show that they also provide us with the same order.

Proposition. *If (X, α) is an $\mathscr{H}_0$-algebra, then the order of the $\mathscr{F}_0$-algebra $r^*(X, \alpha)$ is the dual induced order of X.*

Proof. We have $r = (R, \pi)$ with π_X restricting a proper filter on X to its open sets. If (X, α) is an $\mathscr{H}_0$-algebra, then $x \vee y = \alpha(\pi_X(\uparrow\{x, y\}))$ in $r^*(X, \alpha)$, for x, y in X. If $x \leqq y$ in the dual induced order of X, then $\pi_X(\uparrow(\{x, y\}) = h_X(y)$; thus $x \vee y = y$. Conversely, $h_X(x) \leqq \pi_X(\uparrow\{x, y\})$ in the dual induced order of H_0X, and α preserves this order. Thus $x \leqq y$ in the dual induced order of X if $x \vee y = y$ in $r^*(X, \alpha)$ □

5.6. The lower topology of a continuous sup semilattice L is the Scott topology, with $U \subset L$ open iff U is decreasing and meets every filter φ in L with $\inf_L \varphi$ in U. Scott topologies provide a functor S, from $\mathscr{F}_0$-algebras to TOP, which preserves underlying sets and mappings. It follows that RS is the forgetful functor from $\mathscr{F}_0$-algebras to sets.

Theorem. *The algebraic functor r^* from $\mathscr{H}_0$-algebras to $\mathscr{F}_0$-algebras is an isomorphism of categories, preserving underlying sets and mappings, and with Sr^* the forgetful functor from $\mathscr{H}_0$-algebras to* TOP *for the Scott topology functor S.*

Proof. 0.2.(i) and 0.2.(ii), with $\Delta = S$, are verified as in the proof of 4.4. To obtain 0.2.(iii) for an $\mathscr{F}_0$-algebra (L, α), we must show that $\alpha(\Phi) = \sup \operatorname{adh}_Z \Phi = \sup \downarrow\operatorname{adh}_Z \Phi$, for $Z = i^*(L, \alpha)$ and a proper filter Φ on L, depends only on the decreasing open sets in Φ. Restricting Φ to these sets can only increase $\downarrow\operatorname{adh}_Z \Phi$ and $\alpha(\Phi)$. On the other hand, if $\uparrow x$ and $\operatorname{adh}_Z \Phi$ are disjoint, then $\operatorname{adh}_Z \Phi$ has a decreasing open neighborhood V with $X \setminus V$ a neighborhood of $\uparrow x$. Then $V \in \Phi$, and $\uparrow x$ and $\operatorname{cl}_Z V$ are disjoint. Thus restricting Φ to decreasing open sets does not increase $\downarrow\operatorname{adh}_Z \Phi$ □

References

1. G. GIERZ, K. H. HOFMANN, K. KEIMEL, J. D. LAWSON, M. MISLOVE, D. S. SCOTT, A Compendium of Continuous Lattices. Springer-Verlag, 1980.

2. K. Keimel, Continuous Lattices, General Convexity Spaces, and a Fixed Point Theorem. Notes, December 1983.
3. S. MacLane, Categories for the Working Mathematician. Springer-Verlag, 1971.
4. L. Nachbin, Topology and Order. Princeton, 1965.
5. D. Scott, Continuous lattices. Toposes, Algebraic Geometry and Logic, Lecture Notes in Math. **274** (1972), 93–136.
6. L. Vietoris, Bereiche zweiter Ordnung. Monatsh. für Math. und Physik **32** (1922), 258–280.
7. L. Vietoris, Kontinua zweiter Ordnung. Monatsh. für Math. und Physik **33** (1923), 49–62.
8. O. Wyler, Algebraic Theories of Continuous Lattices.
   a. Technical Report, Carnegie-Mellon University, December, 1976.
   b. Continuous Lattices, Lecture Notes in Math. **871** (1981), 390–413.
9. O. Wyler, Compact Ordered Spaces and Prime Wallman Compactifications. Categorical Topology, Sigma Series in Pure Mathematics 5, pp. 618–635. Heldermann Verlag Berlin (1984).

Department of Mathematics
Carnegie-Mellon University
Pittsburgh, Pennsylvania

(Received October 29, 1984)

The Existence of the Flux Vector and the Divergence Theorem for General Cauchy Fluxes

MIROSLAV ŠILHAVÝ

To Professor W. Noll on his 60th birthday

Contents

Abstract

A new proof of the existence of the flux vector is given for general Cauchy fluxes. The proof is based on an approximation theorem in the theory of functions rather than on the classical tetrahedron argument. This enables us to replace the usual assumptions of Lipschitz continuity with respect to area and volume by less restrictive assumptions so as to produce the flux vector fields with possible singularities. The classical expression for the area density of the flux is proved and the flux is shown to satisfy an appropriate version of the divergence theorem.

1. Introduction

There are many physical quantities that, at a given instant, can be associated with each surface in a body. Examples of such quantities are the contact force and the heat conducted in that body. As these quantities usually can be interpreted as fluxes through the surfaces, a mathematical object describing them is called the Cauchy flux in recent papers [1], [2], [3].

The physical interpretation requires that the flux behave additively on compatible material surfaces. Further, it is natural to assume that if two surfaces differ by a set of zero area, then the values of the flux on these two surfaces coincide. Finally, the fluxes encountered in physics usually satisfy a classical balance law which states that the flux corresponding to the boundary of any part of the body is equal to some quantity that is additive and volume-continuous with respect to parts of the body. A central result in the theory of Cauchy fluxes states that the above properties, when rendered precise and augmented with suitable technical assumptions, imply the existence of the flux vector field whose scalar product with the normal to the surface yields the surface density of the flux. The first result of this type was established by CAUCHY [4] in 1823 under an additional non-technical assumption that the density depends only on the normal. NOLL (1957) [5] showed that this assumption is essentially a consequence of the other assumptions on the flux. This stimulated new research, and the works of NOLL [6, 7], GURTIN & WILLIAMS [8], GURTIN, MIZEL & WILLIAMS [9], GURTIN & MARTINS [1], and ZIEMER [3] contain further important developments.

The most general proofs available in the literature can be carried out only under the assumption of global Lipschitz continuity with respect to the area of the surface and with respect to the volume of the part of the body. The subsequent sections contain a precise statement of these conditions. They are rather restrictive because they imply the global boundedness of the field of the flux vector and the global boundedness of its divergence. On the other hand, it seems reasonable, both from physical and mathematical points of view, to allow singularities of the flux vector at certain points. This is perhaps best motivated when the flux is visualized as the contact force: neither does the presently available theory of partial differential equations afford tools sufficient to guarantee the boundedness of the stress tensor, nor does that boundedness seem natural from the point of view of applications.

In this paper I give a new argument to prove the existence of the flux vector under less restrictive additional assumptions which, I believe, cover more situations encountered in mechanics, including those in which unbounded and singular stresses occur. The proof is based on the following observation: if the flux vector $\boldsymbol{q}$ exists and is sufficiently smooth, then by the divergence theorem,

$$\begin{aligned}\int_{\partial P} v\boldsymbol{q}\cdot\boldsymbol{n}\,dA &= \int_P \operatorname{div}(v\boldsymbol{q})\,dV\\ &= \int_P (v\cdot\operatorname{div}\boldsymbol{q}+\nabla v\cdot\boldsymbol{q})\,dV\end{aligned}\tag{1.1}$$

for every smooth (in fact, Lipschitz continuous) function v and every part P of the body. Here ∂P is the boundary of the part P and $\boldsymbol{n}$ the exterior normal to ∂P, dA is the element of the area of ∂P and dV is the element of volume of P. Now, the first expression in (1.1) can be given immediate meaning provided the Cauchy flux behaves as a measure on ∂P:

$$\int_{\partial P} v\boldsymbol{q}\cdot\boldsymbol{n}\,dA = \int_{\partial P} v\,dQ \tag{1.2}$$

and we note that the right-hand side of (1.2) does not contain the flux vector $\boldsymbol{q}$. On the other hand, the last expression in (1.1) can be used to *define* both the flux vector and its divergence. In other words, we seek to prove the existence of a vector field q and a scalar field b such that

$$\int_{\partial P} v \, dQ = \int_{P} (v \cdot b + \nabla v \cdot q) \, dV, \tag{1.3}$$

for every part P of the body and every Lipschitz continuous function v. Obvious candidates for the components of q are the surface densities of the flux with respect to planar material surfaces perpendicular to the coordinate axes. That with such a definition of q (and with an appropriate definition of b) the formula (1.3) holds can be proved by a surprisingly simple computation if v is affine or piecewise affine. The validity of (1.3) is then extended to general v by approximating v by a sequence of piecewise affine functions.

From (1.3) we deduce that the divergence div q of q in the sense of distributions satisfies $\operatorname{div} q = b$ and this in turn implies that

$$\int_{\partial P} v \, dQ = \int_{P} \operatorname{div} (vq) \, dV \tag{1.4}$$

which is the **divergence theorem** for the Cauchy flux Q. The validity of this theorem is important in considerations about energy in mechanics (q = stress tensor, v = velocity) and in manipulations with the Clausius-Duhem inequality (q = heat flux vector, $v = 1/\theta$ = the reciprocal of the absolute temperature).

Next we use the divergence theorem (1.4) to establish the usual expression for the surface densities of the flux in terms of the flux vector. A set N_0 of exceptional points of the body emerges in the proof. N_0 is small in the sense that its volume is zero, and the formula for the density of the flux can be proved only for surfaces whose intersection with N_0 has area at most zero. Whether or not N_0 is empty depends on whether or not the flux vector can be changed on a set of zero Lebesgue measure to produce a function whose Lebesgue set [10] is the whole region occupied by the body. The latter condition is certainly satisfied if the components of the flux vector can be represented by continuous functions, in which case the exceptional set is empty and the surface density is given by the usual expression for every surface. This is the position of the classical result within the present approach. Also the other results known can be recovered by using the present methods, and the present approach often permits slight generalizations of them.

The present study hence shows that each Cauchy flux satisfying the additional technical assumption gives rise to a vector field with divergence in the sense of distributions of class L^1. I do not know if the converse is also true. It can be proved that any bounded measurable vector field with bounded divergence gives rise to a Cauchy flux satisfying the conditions of Lipschitz continuity. The general vector fields with divergence of class L^1 induce certain power functionals close to the Cauchy fluxes. However, it is not clear whether the power functionals can be identified with the Cauchy fluxes.

2. Bodies, parts, and material surfaces

Throughout, we identify the N-dimensional Euclidean space with the space $\mathbb{R}^N$ of N-tuples of real numbers. We further denote by $\mathbb{S}^{N-1}$ the unit sphere in $\mathbb{R}^N$;

$\mathbb{S}^{N-1} = \{n \in \mathbb{R}^N : |n| = 1\}$, where $|\cdot|$ denotes the Euclidean norm. V denotes the Lebesgue measure in $\mathbb{R}^N$ and A the $(N-1)$-dimensional Hausdorff measure in $\mathbb{R}^N$. The values of these measures on a Borel set S are denoted by $V(S)$ and $A(S)$, respectively.

We identify the body with a bounded open region $B \subset \mathbb{R}^N$. No regularity of the boundary is assumed. To stress the generality of the present approach, I interpret the parts of the body B as sets of finite perimeter contained in B. However, the results of the present paper can be established for smaller collections of parts of B as well. An extensive treatment of sets of finite perimeter is contained in FEDERER's book [11], and I also refer to ZIEMER [3, § 2] for a brief introduction to this topic. The treatment of sets of finite perimeter in continuum mechanics can be found in BANFI & FABRIZIO [12, 13] and ZIEMER [3].

A part P of B is any Borel subset of B of finite perimeter whose boundary ∂P in the sense of measure theory is contained in B. We denote by $\mathcal{P}$ the set of all parts of B. Each set P of finite perimeter has a well defined exterior normal $\boldsymbol{n}^P$ which is defined A-almost everywhere (A-a.e.) on ∂P. The class $\mathcal{P}$ forms a ring of subsets of B, and this ring generates the σ-ring of Borel subsets of B. Note also that actually it is natural to identify the sets $P_1, P_2 \in \mathcal{P}$ if their symmetric difference has Lebesgue measure zero and to consider them as forming the same part of the body B. That means that rather than with the ring $\mathcal{P}$ one should work with the quotient of $\mathcal{P}$ modulo the ideal of the sets of Lebesgue measure zero, as BIRKHOFF & VON NEUMANN proposed in a different context [14]. The same applies to the class of all surfaces and the ideal of sets of Hausdorff measure (area) zero. I do not follow this possibility here, although I note that in this way certain requirements of absolute continuity would be automatically satisfied.

A material surface S is any pair $S = (\hat{S}, \boldsymbol{n})$, where $\hat{S}$ is a Borel subset of B and $\boldsymbol{n} : \hat{S} \to \mathbb{R}^N$ a Borel measurable function such that there is a part $P \in \mathcal{P}$ with $\hat{S} \subset \partial P$ and

$$\boldsymbol{n}(x) = \boldsymbol{n}^P(x)$$

for all $x \in \hat{S}$ for which the exterior normal $\boldsymbol{n}^P(x)$ is defined and

$$\boldsymbol{n}(x) = 0$$

otherwise. The function $\boldsymbol{n}$ is the normal to the surface S; it orients S. The opposite of the material surface $S = (\hat{S}, \boldsymbol{n})$ is the pair $-S$ given by $-S = (\hat{S}, -\boldsymbol{n})$. It may happen that $-S$ is not a material surface. However, if the material surface S is contained in some compact subset of B, then $-S$ is a material surface. We shall frequently identify the material surface S with the corresponding underlying set $\hat{S}$. For instance, if P is a part of the body, then ∂P will denote both the set ∂P and the oriented surface $(\partial P, \boldsymbol{n}^P)$. If S_1, S_2 are two material surfaces, then the inclusion $S_1 \subset S_2$ will be understood to mean not only that the underlying sets satisfy the corresponding relation, but also that S_1 has the same orientation as S_2 on S_1. The set of all material surfaces is denoted $\mathcal{S}$.

A countable family $S_1, S_2, \ldots \in \mathcal{S}$ of material surfaces is said to be compatible if there is a part $P \in \mathcal{P}$ such that $S_i \subset \partial P$ $(i = 1, 2, \ldots)$. A compatible family $S_1, S_2, \ldots$ is said to be disjoint if the underlying sets of all members of the family are pairwise disjoint. In this case, we denote by $\bigcup_{i=1}^{\infty} S_i$ the (oriented) union of the family, *i.e.*, the material surface whose underlying set is the union of the underlying sets of the members of the family and whose orientation is the same as that of each member of the family.

The symbol $L^1(B, V)$ denotes the usual Lebesgue space of all measurable functions f defined on B which satisfy

$$\int_B |f| \, dV < \infty .$$

Similarly, if S is a material surface, then $L^1(S, A)$ denotes the space of all Borel functions f defined on S which satisfy

$$\int_S |f| \, dA < \infty .$$

3. Cauchy fluxes

Before defining formally the concept of the Cauchy flux, we briefly discuss the definitions employed in the earlier papers [1], [2], [3]. As in these works, also here the Cauchy flux will be a function $Q : \mathscr{S} \to \mathbb{R}$ which assigns to each material surface S a number $Q(S)$. In [1], [2], [3], the function Q is subject to the following requirements: (a) additivity:

$$Q(S_1 \cup S_2) = Q(S_1) + Q(S_2) \tag{3.1}$$

whenever S_1 and S_2 are two disjoint compatible material surfaces, and, (b) Lipschitz continuity with respect to area, *i.e.*,

$$|Q(S)| \leqq C\, A(S) \tag{3.2}$$

for every $S \in \mathscr{S}$, where C is a constant independent of S. The additivity (3.1) is well justified. In contrast, I wish to discuss the role of (3.2). Examination of the proofs reveals that (3.2) serves two purposes in the theory. First, this condition is used, in conjunction with additivity, to prove [8] that Q can be extended to a countably additive measure on each material surface, defined on all Borel subsets of that surface, and absolutely continuous with respect to the area measure on the surface. Hence Q has a density on each material surface. Were this the only reason to impose (3.2), then of course, to generalize it would be a routine excercise in measure theory. (This exercise is stated as Proposition 1, below.) However, there are deeper reasons to impose (3.2), namely, certain parts of the argument proving the dependence of the density on the normal [3] cannot be carried out without (3.2); also the tetrahedron argument, even in its refined form in [1], rests on this assumption.

Now we wish to avoid (3.2) but the countable additivity and the absolute continuity drawn from (3.2) seem to be reasonable. (This in particular applies to the condition of absolute continuity, because the boundaries of disjoint parts add only to within sets of area zero.) Hence we introduce a definition of the Cauchy flux such that countable additivity and absolute continuity are satisfied but (3.2) need not hold. However, the existence of the flux vector cannot be established in this generality and certain additional assumption, much weaker than (3.2), will have to be added in Section 5.

Several equivalent versions of the present definition of the Cauchy flux can be given, and it is hard to select one of them as the basic one. The following proposition, a routine consequence of measure theory (FUGLEDE [15]) lists the equivalent versions of the definition.

Proposition 1. *For a real-valued function Q defined on $\mathcal{S}$ the following four conditions are equivalent:*

(1) *For every $S \in \mathcal{S}$ there is a Borel function $q^S \in L^1(S, A)$ such that*

$$Q(S') = \int_{S'} q^S \, dA$$

for every $S' \in \mathcal{S}$, $S' \subset S$.

(2) *The function Q is countably additive on compatible material surfaces*, i.e.,

$$Q\left(\bigcup_{i=1}^{\infty} S_i\right) = \sum_{i=1}^{\infty} Q(S_i)$$

for every disjoint compatible family $S_1, S_2, \ldots \in \mathcal{S}$, and moreover,

$$Q(S) = 0$$

whenever $A(S) = 0$.

(3) *The function Q is additive on compatible material surfaces*, i.e.,

$$Q(S_1 \cup S_2) = Q(S_1) + Q(S_2)$$

for every pair S_1, S_2 of compatible disjoint material surfaces, and moreover, for each $S \in \mathcal{S}$ there is a non-negative Borel function $h^S \in L^1(S, A)$ such that

$$|Q(S')| \leqq \int_{S'} h^S \, dA$$

for every $S' \in \mathcal{S}$, $S' \subset S$.

(4) *The function Q is additive on compatible material surfaces and, moreover, for every $\varepsilon > 0$ there is a $\delta > 0$ such that*

$$\sum_{i=1}^{l} |Q(S_i)| < \varepsilon$$

for every finite number $S_1, \ldots, S_l$ of compatible disjoint material surfaces satisfying

$$\sum_{i=1}^{l} A(S_i) < \delta .$$

Any function Q satisfying the above four equivalent conditions is called a **Cauchy flux**, and the function q^S as in (1) is called the **surface density** of Q. Contact with the definitions of the Cauchy flux given in the earlier papers [1], [2], [3] is established through Condition (3), as obviously every function satisfying (3.1), (3.2) satisfies also (3) with $h^S = C =$ const. However, the present definition is more general, as the converse need not hold.

4. Weakly balanced Cauchy fluxes

The basic motivation for the concept of a weakly balanced Cauchy flux is the fact that the fluxes encountered in physics satisfy the integral form of the classical balance law. One possible definition of a weakly balanced Cauchy flux is precisely the condition expressing this, and it is listed below as Condition (1) in Theorem 1. However, using the countable additivity of Cauchy fluxes, we can adopt seemingly weaker but actually equivalent conditions. These are listed as the remaining two conditions in Theorem 1.

Theorem 1. *For a Cauchy flux the following three conditions are equivalent:*
(1) *There is a function* $b \in L^1(B, V)$ *such that*

$$Q(\partial P) = \int_P b\, dV \tag{4.1}$$

for every part P.
(2) *There is a function* $k \in L^1(B, V)$ *such that* $k \geqq 0$ *and*

$$|Q(\partial P)| \leqq \int_P k\, dV$$

for every part P.
(3) *For every* $\varepsilon > 0$ *there is a* $\delta > 0$ *such that*

$$\sum_{i=1}^{l} |Q(\partial P_i)| < \varepsilon$$

for every finite number $P_1, \ldots, P_l$ *of disjoint parts for which*

$$\sum_{i=1}^{l} V(P_i) < \delta.$$

We say that the Cauchy flux Q is **weakly balanced** if it satisfies the three equivalent conditions of the preceding theorem. The function b as in Condition (1) is called the **volume density of the flux** Q. Of the three equivalent conditions, Condition (1) is most easily handled. It will be precisely this condition that, together with Proposition 2, below, will be used in the following section to establish the existence of the flux vector. Note, however, that the important feature of Conditions (2), (3) is that they do not postulate the additivity of $Q(\partial P)$ on disjoint parts.

The present definition of weak balancing is less restrictive than the definitions in [1], [2], [3], wherein the following stronger version of Condition (2) is adopted as the definition: there is a constant C such that

$$|Q(\partial P)| \leqq CV(P) \tag{4.2}$$

for every part of the body.

Proposition 2. *If* Q *is a weakly balanced Cauchy flux, then*

$$Q(-S) = -Q(S) \tag{4.3}$$

for every $S \in \mathcal{S}$ *with* $-S \in \mathcal{S}$. *Hence the surface density satisfies*

$$q^{-S}(x) = -q^{S}(x) \tag{4.4}$$

for A-a.e. $x \in S$.

Proof of Theorem 1 & Proposition 2. The implication (1) $\Rightarrow$ (2) is trivial, and (2) $\Rightarrow$ (3) is routine in measure theory. Concerning the implication (3) $\Rightarrow$ (1), we note that (3) is a general necessary and sufficient condition under which an additive set function defined on a ring generating the measurable sets can be represented as in (1); see [15]. In the present situation we apply this condition to the set function M, given by

$$M(P) = Q(\partial P), \quad P \in \mathcal{P}.$$

However, we must verify that M is additive on $\mathcal{P}$. In order to do that, we first prove that if (3) holds, then the conclusion of Proposition 2 holds. This proof is easy but long at the present level of generality, for non-smooth material surfaces may occur. I shall only sketch the basic idea, or, rather, I shall indicate how the proof can be reduced to the verification of (4.3), (4.4) for smooth material surfaces only. Namely, if S is smooth and its relative $(N-2)$-dimensional boundary is sufficiently regular, then we can use almost the same argument as in [8], [1] to prove (4.3). I note that the proofs in [8], [1] are given under assumptions stronger than the present concerning Q, but in this case the generalization is easy. Now the validity of (4.3) for every smooth material surface with sufficiently regular relative boundary suffices to prove, by using the derivatives with respect to regular families shrinking to a point $x \in S$, the validity of (4.4) for A-a.e. point of a smooth surface S. By integrating over a Borel subset S' of S, we prove the validity of (4.3) for a general S' contained in a smooth material surface. Finally, the proof of (4.3) for a general non-smooth material surface S is completed by using the nontrivial fact that A-almost all of any $S \in \mathcal{S}$ can be covered by a countable family of smooth material surfaces S [11]. This implies that, to within a possible set of area zero, S can be written as a union of a compatible, disjoint, countable family of material surfaces that are contained in appropriate smooth material surfaces. Then, by using the countable additivity of Q, absolute continuity, and the validity of (4.3) for surfaces contained in smooth surfaces, we establish (4.3) generally.

Note. Another proof of Proposition 2 is possible (under the assumption that Condition (3) is valid) which does not use the non-trivial result about covering a general S by smooth material surfaces. It is based on the observation that if P is any part of the body, then for A-a.e. $x \in \partial P$ the following quantities tend to 0 as $r \to 0+$:

$$r^{-N+1} Q(\partial B(x, r)), \quad r^{-N+1} Q(\partial(P \cap B(x, r))), \quad r^{-N+1} Q(\partial(P^c \cap B(x, r))),$$

where $B(x, r)$ is the open ball of radius r centered at x, and P^c is the complement of P in B. This is proved from Condition (3) by using directly the definition of Hausdorff measure, the equality $V(\partial P) = 0$, and an argument similar to the one given in FEDERER [11], pp. 179–181. The details are omitted. Next one splits the boundary of $P \cap B(x, r)$ into two parts, one being the portion of the boundary contained in $B(x, r)$ and the other being the portion contained in $\partial B(x, r)$. The same is done for $P^c \cap B(x, r)$. The portions of the boundaries contained in $B(x, r)$ are opposite each to other in the sense defined in section 2. Then, using the additivity of Q, and the fact that the limits of the quantities indicated above are 0, we may evoke the basic result on the differentiation of set functions to prove (4.4) for a general boundary P. The result then follows.

Now the conclusion of Proposition 2 is used to find that the set function M is indeed additive: if P_1 and P_2 are two disjoint parts, then the contributions to the sum $Q(\partial P_1) + Q(\partial P_2)$ from the overlapping parts of the boundaries cancel in view of (4.3) and their opposite orientations.The proof is complete.

5. The flux vector and the divergence theorem

To prove the existence of a flux vector that satisfies the divergence theorem, we have to impose a further condition on the Cauchy flux. This condition is embodied by the definition of the Cauchy flux of class L^1, below. We first introduce the following terminology. We say that a material surface $S = (\hat{S}, \boldsymbol{n})$ is planar if there is a hyperplane H_0 in $\mathbb{R}^N$ for which $\hat{S} \subset H_0$. Given a unit vector $\boldsymbol{n}^* \in \mathbb{S}^{N-1}$, we say that a planar material surface $S = (\hat{S}, \boldsymbol{n})$ is perpendicular to $\boldsymbol{n}^*$ if $\boldsymbol{n}(x) = \boldsymbol{n}^*$ for every $x \in \hat{S}$. A Cauchy flux Q is said to be **of class L^1** if there are N linearly independent vectors $\boldsymbol{n}_1, \ldots, \boldsymbol{n}_N \in \mathbb{S}^{N-1}$ and a non-negative Borel function $h \in L^1(B, V)$ such that

$$|Q(S)| \leqq \int_S h \, dA \tag{5.1}$$

for every planar material surface S perpendicular to one of the vectors $\boldsymbol{n}_1, \ldots, \boldsymbol{n}_N$. Since h is non-negative and Borel measurable, the surface integral in (5.1) is meaningful, but it can be infinite for certain surfaces S. Such a situation is not excluded in the above definition. However, using the condition $h \in L^1(B, V)$ and Fubini's theorem, one can prove that $h \in L^1(H_0 \cap B, A)$ for "almost every" hyperplane H_0 perpendicular to one of the vectors $\boldsymbol{n}_1, \ldots, \boldsymbol{n}_N$. For material surfaces contained in such hyperplanes the finiteness of the integral in (5.1) is then guaranteed. The condition (5.1) thus introduces some uniformity on the variations of the Cauchy flux on hyperplanes perpendicular to the vectors $\boldsymbol{n}_1, \ldots, \boldsymbol{n}_N$.

If the Cauchy flux satisfies the assumption of area Lipschitz continuity (3.2) employed in [1], [2], [3], then Q is of class L^1 and satisfies (5.1) for every material surface, not only for the planar material surfaces perpendicular to the vectors $\boldsymbol{n}_i$. It is precisely the passage from (3.2) to (5.1) that enables us to obtain a larger class of flux vector fields including the unbounded ones. Actually the results to be given below can be established under even less restrictive assumption of local summability, defined in an appropriate way.

The main conclusion in the present section deals with vector fields of class L^1 over B. A vector field $\boldsymbol{q}$ of class L^1 on B is an N-tuple $\boldsymbol{q} = (q_1, \ldots, q_N)$ of measurable functions each of which belongs to $L^1(B, V)$; we write $\boldsymbol{q} \in L^1(B, V)$

in this case. Let $C_0^\infty(B)$ be the set of all infinitely differentiable functions φ: $\mathbb{R}^N \to \mathbb{R}$ with compact support in B. The divergence in the sense of distributions of the vector field $\boldsymbol{q} \in L^1(B, V)$ is a linear functional $\operatorname{div} \boldsymbol{q}$ on $C_0^\infty(B)$ whose value $(\operatorname{div} \boldsymbol{q}, \varphi)$ on a general $\varphi \in C_0^\infty(B)$ is given by

$$(\operatorname{div} \boldsymbol{q}, \varphi) = -\int_B \boldsymbol{q} \cdot \nabla\varphi \, dV,$$

where $\nabla\varphi$ is the gradient of φ. We say that the vector field $\boldsymbol{q}$ has divergence of class L^1 if there is a function $b \in L^1(B, V)$ such that

$$(\operatorname{div} \boldsymbol{q}, \varphi) = \int_B b\varphi \, dV$$

for every $\varphi \in C_0^\infty(B)$. The function b is defined uniquely to within to values on a set of Lebesgue measure zero, and we identify it with the divergence,

$$\operatorname{div} \boldsymbol{q} = b.$$

The fact that $\boldsymbol{q}$ has divergence of class L^1 can be expressed by writing $\operatorname{div} \boldsymbol{q} \in L^1(B, V)$. We easily verify the following: if $\boldsymbol{q}$ has divergence of class L^1 and if v is a Lipschitz continuous function on B with compact support, then also $v\boldsymbol{q}$ has divergence of class L^1 and

$$\operatorname{div}(v\boldsymbol{q}) = v \operatorname{div} \boldsymbol{q} + \nabla v \cdot \boldsymbol{q} \tag{5.2}$$

where ∇v is the gradient of the function v, which, according to the theorem of Rademacher, exists V-a.e. on B.

The definition of Cauchy flux implies that Q induces a Borel measure on each material surface S; if v is a bounded Borel function defined on S, then the symbol $\int_S v \, dQ$ will denote the integral of v with respect to the measure induced by Q on S:

$$\int_S v \, dQ = \int_S v \, q^S \, dA,$$

where q^S is the surface density of the flux.

Theorem 2. *Let Q be a weakly balanced Cauchy flux of class L^1. Then there is a Borel-measurable vector field $\boldsymbol{q}$ with $\boldsymbol{q} \in L^1(B, V)$ and $\operatorname{div} \boldsymbol{q} \in L^1(B, V)$ such that*

$$\int_{\partial P} v \, dQ = \int_P \operatorname{div}(v\boldsymbol{q}) \, dV \tag{5.3}$$

for every part P and every Lipschitz continuous function v on B. The field $\boldsymbol{q}$ also satisfies the local form of the balance law:

$$\operatorname{div} \boldsymbol{q} = b \tag{5.4}$$

where b is the volume density of Q.

Any function $\boldsymbol{q}$ satisfying the assertions of the above theorem will be called a **flux vector** for Q. Any two flux vectors differ at most on a set of Lebesgue measure zero.

Under stronger assumptions about the flux ZIEMER [3] proves that div $\boldsymbol{q} \in L^1(B, V)$ (in fact, under his assumptions, div $\boldsymbol{q} \in L^\infty(B, V)$), but his method of proof, following essentially the traditional line, does not lead to the divergence theorem (5.3).

Proof. By using a suitable affine transformation, one can assume that the vectors occurring in the definition of the Cauchy flux of class L^1 are the natural basis vectors $\boldsymbol{e}_1, \ldots, \boldsymbol{e}_N$ in $\mathbb{R}^N$.

We shall first prove that for each i $(1 \leqq i \leqq N)$ there is a Borel function $q_i \in L^1(B, V)$ such that

$$\int_{\partial P} x_i^* \, dQ = \int_P (x_i^* b + q_i) \, dV \tag{5.5}$$

for every $P \in \mathscr{P}$, where $x_i^* : \mathbb{R}^N \to \mathbb{R}$ is a natural coordinate function, given by

$$x_i^*(x) = x_i, \qquad x = (x_1, \ldots, x_N) \in \mathbb{R}^N.$$

Then we shall define a vector field $\boldsymbol{q} \in L^1(B, V)$ by

$$\boldsymbol{q}(x) = (q_1(x), \ldots, q_N(x)), \qquad x \in B, \tag{5.6}$$

and prove that

$$\int_{\partial P} v \, dQ = \int_P (vb + \nabla v \cdot \boldsymbol{q}) \, dV \tag{5.7}$$

for every $P \in \mathscr{P}$ and every Lipschitz continuous function v on B.

To prove (5.5), let i $(1 \leqq i \leqq N)$ be fixed. Consider, for each $t \in \mathbb{R}$, the closed half-space $R(t)$ in $\mathbb{R}^N$ given by $R(t) = \{x \in \mathbb{R}^N : x_i^*(x) \leqq t\}$, the corresponding open half-space $R^0(t) = \{x \in \mathbb{R}^N : x_i^*(x) < t\}$, and the boundary hyperplane $N(t) = \{x \in \mathbb{R}^N : x_i^*(x) = t\}$.

Let P be any part of the body. Denote by $P(t)$ the intersection of P with the closed half-space $R(t)$,

$$P(t) = P \cap R(t).$$

It can be shown that $P(t)$ is a set of finite perimeter; hence $P(t) \in \mathscr{P}$ for every $t \in \mathbb{R}$. The boundary of $P(t)$ in the sense of measure theory is given by

$$\partial P(t) = S(t) \cup T(t),$$

where $S(t)$ is the portion of the boundary ∂P of the original set P that is contained in the open half-space $R^0(t)$,

$$S(t) = \partial P \cap R^0(t),$$

and $T(t)$ is the remaining part of $\partial P(t)$; one has, for almost every $t \in \mathbb{R}$,

$$T(t) = P \cap N(t) \tag{5.8}$$

to within a set of zero area. All these essentially geometric facts are intuitively clear and can be verified formally by using the definition of a set of finite perimeter. Note also that most of the above facts are special cases of a general method of "slicing" described in FEDERER [11].

We put

$$F(t) = Q(S(t)),$$
$$G(t) = Q(T(t)),$$
$$H(t) = \int_{P(t)} b\, dV.$$

The boundedness of P and (5.8) imply that G vanishes outside some compact interval. Applying Condition (1) of Theorem 1 to $P(t)$ yields

$$F(t) + G(t) = H(t) \tag{5.9}$$

for every $t \in \mathbb{R}$. Applying (5.2) to $T(t)$ yields

$$|G(t)| \leqq \int_{T(t)} h\, dA. \tag{5.10}$$

Now F is the distribution function of the function x_i^* with respect to the measure induced by Q on ∂P and H is the distribution function of x_i^* with respect to the measure $L \mapsto \int_{L \cap P} b\, dV$, $L \subset \mathbb{R}^N$, a Borel set. Hence F and H have bounded variation. (5.9) implies that also G has bounded variation. The general properties of distribution functions enable us to express the integrals of x_i^* with respect to the indicated measures through the corresponding distribution functions as the "first moments":

$$\int_{\partial P} x_i^*\, dQ = \int_R t\, dF(t),$$
$$\int_P x_i^* b\, dV = \int_R t\, dH(t).$$

The last two formulas, (5.9), integration by parts, and the fact that G vanishes outside some compact interval justify the following computation:

$$\begin{aligned}\int_{\partial P} x_i^*\, dQ - \int_P x_i^* b\, dV &= \int_{\mathbb{R}} t\, d(F - H)(t)\\ &= -\int_{\mathbb{R}} t\, dG(t)\\ &= \int_{\mathbb{R}} G(t)\, dt\end{aligned}$$

Inequality (5.10), relation (5.8) and Fubini's theorem establish the inequality

$$\begin{aligned}\left|\int_{\partial P} x_i^*\, dQ - \int_P x_i^* b\, dV\right| &\leqq \int_{\mathbb{R}} |G(t)|\, dt\\ &\leqq \int_{\mathbb{R}} \left(\int_{T(t)} h\, dA\right) dt\\ &= \int_{\mathbb{R}} \left(\int_{P \cap N(t)} h\, dA\right) dt\\ &= \int_P h\, dV,\end{aligned}$$

i.e.,

$$\left| \int_{\partial P} x_i^* \, dQ - \int_P x_i^* b \, dV \right| \leqq \int_P h \, dV \tag{5.11}$$

for every part $P \in \mathscr{P}$.

It is now observed that the set function

$$P \mapsto \int_{\partial P} x_i^* \, dQ - \int_P x_i^* b \, dV, \qquad P \in \mathscr{P}, \tag{5.12}$$

is additive. Indeed, it is obvious that the volume integral is additive. The additivity of the surface integral in (5.12) is verified by using (4.4) to cancel the contributions to the sum from the overlapping parts of the boundaries.

To summarize, the set function (5.12) is additive and satisfies (5.11). The result of FUGLEDE [15] then implies the existence of a Borel function $q_i \in L^1(B, V)$ such that (5.5) holds.

We now define $\boldsymbol{q}$ by (5.6) and prove (5.7). Noting that $\nabla x_i^* = \boldsymbol{e}_i$, one sees that (5.5) is precisely (5.7) for $v = x_i^*$. Further, (5.7) is also satisfied by constant functions, for in this case the gradient vanishes and (5.7) reduces to the already established equality (4.1) of Theorem 1. But this proves that (5.7) holds for all affine functions since an affine function is a linear combination of $x_1^*, \ldots, x_N^*$ and of constants.

Next, if v is piecewise affine, we establish the validity of (5.7) for a general part P by dividing it into smaller parts R_k on which v is affine, and using the already established validity of (5.7) for affine functions. Then adding the equalities (5.7) for each R_k and using the additivity of the expressions on both sides of the equality, we find that (5.7) holds. (To prove the additivity of the left-hand side, we must evoke (4.4) to cancel the contributions of the overlapping parts of the boundaries of R_k's.)

Finally, if v is a general Lipschitz continuous function on B, then there is a sequence of piecewise affine functions v_n such that

$$v_n \to v \qquad \text{uniformly on } B$$

$$\nabla v_n \to \nabla v \qquad V\text{-a.e. on } B$$

and

$$|\nabla v_n| \leqq C \qquad V\text{-a.e. on } B,$$

with C a constant independent of n. (See *e.g.* EKELAND & TEMAM [16].) Applying (5.7) to v_n and letting n tend to ∞ then yields (5.7) in the general case.

We now prove (5.4). For every $\varphi \in C_0^\infty(B)$ there is a part P of the body such that the support of φ is contained in the interior of P. Hence the function φ vanishes on the boundary of P, and applying (5.7) to P and the function φ yields

$$0 = \int_{\partial P} \varphi \, dQ = \int_P \varphi b + \nabla \varphi \cdot \boldsymbol{q} \, dV = \int_B \varphi b + \nabla \varphi \cdot \boldsymbol{q} \, dV.$$

But this equality means precisely that the divergence of $\boldsymbol{q}$ in the sense of distributions is of class L^1 and equals b.

Finally, (5.4) and (5.2) enable us to reduce the right-hand side of (5.7) to the right-hand side of (5.3), and the proof is complete.

6. The surface densities

In this section we consider a group of results associated with the expression for the surface density in terms of the flux vector.

Theorem 3. *Let Q be a weakly balanced Cauchy flux of class L^1 with the corresponding flux vector $\boldsymbol{q}$. Then there is a Borel subset $N_0 \subset B$ of Lebesgue measure zero such that*

$$Q(S) = \int_S \boldsymbol{q} \cdot \boldsymbol{n}\, dA \tag{6.1}$$

for every material surface $S = (\hat{S}, \boldsymbol{n})$ satisfying

$$A(\hat{S} \cap N_0) = 0. \tag{6.2}$$

In other words, the surface density corresponding to any such a material surface is given by

$$q^S(x) = \boldsymbol{q}(x) \cdot \boldsymbol{n}(x) \qquad A\text{-a.e. on } S. \tag{6.3}$$

Proof. Let $\varphi : \mathbb{R}^N \to \mathbb{R}$ be a non-negative, infinitely differentiable, spherically symmetric function with compact support satisfying

$$\int_{\mathbb{R}^{N-1}} \varphi(y_1, \dots, y_{N-1}, 0)\, dy_1 \dots dy_{N-1} = 1, \tag{6.4}$$

and let P be any part of the body. In this situation, the following lemma holds.

Lemma 1. *For A-a.e. $x \in \partial P$,*

$$\lim_{r \to 0+} r^{-N} \int_P \nabla\varphi(r^{-1}(x - y))\, dV(y) = \boldsymbol{n}^P(x) \tag{6.5}$$

and

$$\lim_{r \to 0+} r^{-N+1} \int_{\partial P} \varphi(r^{-1}(x - y))\, dA(y) = 1. \tag{6.6}$$

The first assertion of the lemma is verified by a direct computation if P is a half-space. The general case is then verified by using this special case and the definition of the normal in the sense of measure theory. The details are omitted. By applying the Gauss-Green theorem (FEDERER [11]), we can transform the volume integral in (6.5) into a surface integral to restate (6.5) in the form

$$\lim r^{-N+1} \int_{\partial P} \varphi(r^{-1}(x - y))\, \boldsymbol{n}^P(y)\, dA(y) = \boldsymbol{n}^P(x)$$

(from now on I omit the symbol $r \to 0+$ in the limits.) This statement implies (6.6) at every Lebesgue point of $\boldsymbol{n}^P$ relative to A. The lemma is proved.

The proof of Theorem 3 is now easily completed. Let N_0 be the complement (in B) of the set of all Lebesgue points x for $\boldsymbol{q}$ that satisfy

$$\lim r^{-N+1} \int_B \varphi(r^{-1}(x-y))\, b(y)\, dV(y) = 0 \tag{6.7}$$

We easily find that (6.7) is satisfied at every Lebesgue point x for b and since V-a.e. point of B is simultaneously a Lebesgue point for both $\boldsymbol{q}$ and b, we have $V(N_0) = 0$.

Now let S satisfy (6.2). Then A-a.e. point $x \in S$ is a Lebesgue point for $\boldsymbol{q}$ and satisfies (6.7). In virtue of this and in view of (6.5) then, if P is any part with $S \subset \partial P$,

$$\lim r^{-N} \int_P \nabla\varphi(r^{-1}(x-y)) \cdot \boldsymbol{q}(y)\, dV(y) = \boldsymbol{q}(x) \cdot \boldsymbol{n}^P(x) \tag{6.8}$$

and similarly, by (6.6),

$$\lim r^{-N+1} \int_{\partial P} \varphi(r^{-1}(x-y))\, q^{\partial P}(y)\, dA(y) = q^{\partial P}(x) \tag{6.9}$$

for A-a.e. $x \in S$. Application of Theorem 2 (in the form of the equality (5.7)) to the function $v(y) = r^{-N+1}\varphi(r^{-1}(x-y))$ and then use of (6.7), (6.8) and (6.9) will reduce (5.7) with our special choice of v to

$$q^{\partial P}(x) = \boldsymbol{q}(x) \cdot \boldsymbol{n}^P(x) \qquad \text{for } A\text{-a.e. } x \in S,$$

and the results follow.

Unfortunately it is not known whether one can change the flux vector on a set of zero Lebesgue measure so as to make (6.1) hold for every material surface. The set N_0 was defined as the complement of the set of all Lebesgue points for $\boldsymbol{q}$ that satisfy (6.7). The following proposition gives some information about the size of the set of points x that satisfy (6.7).

Lemma 2. *If* $b \in L^1(B, V)$, *then*

$$\lim r^{-N+1} \int_{B(x,r)} |b(y)|\, dV(y) = 0 \tag{6.10}$$

for A-a.e. *point* $x \in B$.

According to this lemma, not only the "volume", but also the "area" of the set of points that do not satisfy (6.10) is zero. Note that if b is the volume density of a weakly balanced Cauchy flux, then the lemma implies that

$$\lim r^{-N+1} Q(\partial B(x,r)) = 0$$

for A-a.e. $x \in B$ and this information is a strengthened version of the information obtained in the "Note" at the end of Section 4 concerning the vanishing of the limits of the quantities indicated there. In that note we derived this information from Condition (3) of Theorem 1 without having b at our disposal.

Proof. Let M be the set of all points of B where (6.10) holds. As has been pointed out in the proof of Theorem 3, the complement of M satisfies

$$V(B - M) = 0. \tag{6.11}$$

We apply FEDERER [11, Thm. 2.10.18, item (2)] to the present situation. Namely, we identify his space X with our B, his measure μ with the indefinite integral

$$\mu(Z) = \int_{Z \cap B} |b| \, dV$$

for any Borel subset $Z \subset B$, choose his function ζ to be given by $\zeta(S) = (\operatorname{diam} S)^{N-1}$, the family C to be the family of all balls, and the set A to be our M. The measure ψ occurring in the indicated theorem then must be our measure A. With this choice of the objects the theorem asserts that

$$\lim r^{-N+1} \int_{M \cap B(x,r)} |b| \, dV = 0$$

for A-a.e. $x \in B - M$. In virtue of (6.11) the limit in the last equality is the same as the limit in (6.10), and hence (6.10) holds for A-a.e. $x \in B - M$. However, since according to our definition of M (6.10) holds at *no* point of $B - M$, we conclude that $A(B - M) = 0$, and the proof is complete.

According to this lemma, we may concentrate our attention entirely on the complement of the set of Lebesgue points of $\boldsymbol{q}$ since adding the points that do not satisfy (6.7) increases negligeably the size of the set N_0 from the point of view of surface integration. The set of all Lebesgue points of $\boldsymbol{q}$ will be the whole of B if the flux vector is continuous, and that in turn is true if the flux Q is continuous in the following sense. A Cauchy flux Q of class L^1 is said to be **continuous** if there are N linearly independent vectors $\boldsymbol{n}_1, \ldots, \boldsymbol{n}_N \in \mathbb{S}^{N-1}$ and continuous functions $\boldsymbol{q}_1^*, \ldots, \boldsymbol{q}_N^*$ on B such that

$$Q(S) = \int_S \boldsymbol{q}_i^* \, dA$$

for every planar material surface element S perpendicular to $\boldsymbol{n}_i$.

Theorem 4. *Let Q be a weakly balanced continuous Cauchy flux. Then its flux vector is continuous, and* (6.1) *holds for every material surface.*

This is essentially a result of NOLL [5] under slightly weaker assumptions. (Note that Lemma 2 can be avoided completely if Q satisfies the conditions of Lipschitz continuity with respect to volume.)

We conclude with another theorem generalizing a known result (NOLL [5], ZIEMER [3]). To state the result, we say that a Cauchy flux is **of class L^∞** if there are N linearly independent vectors $\boldsymbol{n}_1, \ldots, \boldsymbol{n}_N \in \mathbb{S}^{N-1}$ and a constant C such that

$$|Q(S)| \leqq CA(S)$$

for every planar material surface S perpendicular to one of the vectors $\boldsymbol{n}_1, \ldots, \boldsymbol{n}_N$.

Theorem 5. *Let Q be a weakly balanced Cauchy flux of class L^∞. Then there is a bounded function* $q^*: B \times \mathbb{S}^{N-1} \to \mathbb{R}$ *such that*

$$Q(S) = \int_S q^*(x, \boldsymbol{n}(x))\, dA$$

for every material surface S.

The proof is based on essentially the same idea as the proof of Theorem 3 and will be omitted. Note again that the present version of the result generalizes the existing ones since the condition of Lipschitz continuity with respect to volume is not imposed and Lipschitz continuity with respect to area is postulated only for planar material surfaces perpendicular to the vectors $\boldsymbol{n}$. Also it is worth stressing that the previous works use this result as an important intermediate step in proving the more concrete results of the type of Theorems 3, 4, while here we have found a way to avoid use of the function $q^*(x, \boldsymbol{n})$. Theorem 5 is included for completeness and to indicate that no information was lost by following the path indicated in this paper.

Acknowledgements. I thank Professors WALTER NOLL and MARIO PITTERI for discussions on a previous draft of the paper during my stays at CNUCE, the institute of CNR in Pisa, and at the University of Padova. These discussions helped me in particular to make a final choice between the two possible constructions of the flux vector I had. I also thank Professor CLIFFORD TRUESDELL for improving the English of the paper and for a discussion on the historical position of Cauchy's postulate.

References

1. M. E. GURTIN & L. C. MARTINS, Cauchy's theorem in classical physics, *Arch. Rational Mech. Anal.* **60** (1976), 305–324.
2. L. C. MARTINS, On Cauchy's theorem in classical physics: some counterexamples, *Arch. Rational Mech. Anal.* **60** (1976), 325–328.
3. W. P. ZIEMER, Cauchy flux and sets of finite perimeter, *Arch. Rational Mech. Anal.* **84** (1983), 189–201.
4. A. L. CAUCHY, Recherches sur l'équilibre et le mouvement intérieur des corps solides ou fluides, élastiques ou non-élastiques, *Bull. Soc. Philomath.* (1823), 9–13.
5. W. NOLL, The foundations of classical mechanics in the light of recent advances in continuum mechanics, pp. 266–281 of *The Axiomatic Method, with Special Reference to Geometry and Physics* (Symposium at Berkley, 1957). Amsterdam: North-Holland Publishing Co. 1959.
6. W. NOLL, Lectures on the foundations of continuum mechanics. *Arch. Rational Mech. Anal.* **52** (1973), 62–92.
7. W. NOLL, The foundations of mechanics, in *Nonlinear Continuum Theories*, C.I.M.E. Lectures, Roma 1966.
8. M. E. GURTIN & W. O. WILLIAMS, An axiomatic foundation for continuum thermodynamics, *Arch. Rational. Mech. Anal.* **26** (1967), 83–117.
9. M. E. GURTIN, V. J. MIZEL, & W. O. WILLIAMS, A note on Cauchy's stress theorem, *J. Math. Anal. Appl.* **22** (1968), 398–401.
10. W. RUDIN, *Real and Complex Analysis*, McGraw-Hill, New York 1966.
11. H. FEDERER, Geometric measure theory, Springer-Verlag, New York 1969.

12. C. BANFI & M. FABRIZIO, Sul concetto di sottocorpo nella meccanica dei continui, *Rend. Acc. Naz. Lincei*, **66** (1979), 136–142.
13. C. BANFI & M. FABRIZIO, Global theory for thermodynamic behaviour of a continuous medium, *Ann. Univ. Ferrara* **27** (1981), 181–199.
14. G. BIRKHOFF & J. VON NEUMANN, The logic of quantum mechanics, *Annals of Math.* **37** (1936), 823–843.
15. B. FUGLEDE, On a theorem of F. Riesz, *Math. Scand.* **3** (1955) 283–302.
16. I. EKELAND & R. TEMAM, Convex Analysis and Variational Problems, Amsterdam, North-Holland Publishing Co. 1976.

Mathematical Institute
Czechoslovak Academy of Sciences
Prague

(Received December 28, 1984)

Isochoric Circulation-Preserving Motions with Stream-Lines of a Potential Motion

A. W. MARRIS

Dedicated to Professor Walter Noll on his sixtieth birthday

Introduction

We consider the class of steady isochoric circulation-preserving motions. These motions are defined kinematically by the conditions

$$\operatorname{curl}(\boldsymbol{\omega} \times \boldsymbol{v}) = \mathbf{0}, \qquad \text{I.1}$$

and

$$\operatorname{div} \boldsymbol{v} = 0, \qquad \text{I.2}$$

where

$$\boldsymbol{v} = v\boldsymbol{s} \qquad \text{I.3}$$

is the velocity, $\boldsymbol{s}$ being the unit vector tangent to the stream-line, and where

$$\boldsymbol{\omega} = \operatorname{curl} \boldsymbol{v} \qquad \text{I.4}$$

is the vorticity.

In this paper we prove the following.

Theorem 1. *The only steady isochoric rotational circulation-preserving motion whose stream-line pattern is that of the isochoric irrotational motion* $\boldsymbol{v}_1$ *is a complex-lamellar motion whose velocity magnitude bears a constant value on a stream-line, that is, a motion defined by* I.1 *and the conditions*

$$\boldsymbol{s} \cdot \operatorname{grad} v \equiv \frac{\delta v}{\delta s} = 0, \qquad \text{I.5}$$

$$\operatorname{div} \boldsymbol{s} = 0, \qquad \text{I.6}$$

and

$$\Omega = \boldsymbol{s} \cdot \operatorname{curl} \boldsymbol{s} = 0. \qquad \text{I.7}$$

Since the steady isochoric irrotational motion $\boldsymbol{v}_1$ has the same stream-line pattern as the circulation-preserving motion, we have

$$\boldsymbol{v}_1 = v_1 \boldsymbol{s} \qquad \text{I.8}$$

and by I.6

$$\frac{\delta v_1}{\delta s} = 0, \quad \text{I.9}$$

and we have

Corollary 1. *The steady isochoric irrotational motion $\boldsymbol{v}_1$ is a motion whose velocity magnitude bears a constant value on a stream-line.*

In (1937) Hamel presented what in its published form was an incomplete proof of the theorem that the vector lines of the motion $\boldsymbol{v}_1$ of Corollary 1 must be parallel straight lines, concentric circles or circular helices mounted on concentric circular cylinders. This theorem was proved by the writer in (1973).

From Theorem 1 and Corollary 1 we deduce

Theorem 2. *The only steady rotational isochoric circulation-preserving motion whose stream-line pattern is that of a steady isochoric irrotational motion is a motion $\boldsymbol{v} = v\boldsymbol{s}$, where the vector lines of the unit vector $\boldsymbol{s}$ are parallel straight lines, concentric circles or circular helices mounted on concentric circular cylinders.*

In 1952, Prim (1952, p. 483), published the following theorem:

Let $\boldsymbol{v}$ be a steady, isochoric, complex-lamellar circulation-preserving motion such that the velocity magnitude v is constant along a stream-line, then there is a steady, isochoric, lamellar motion for which the velocity magnitude is constant along a stream-line, having the same stream-lines as the complex-lamellar motion.

Prim thus showed that if the steady isochoric circulation-preserving motion is required to satisfy I.5 and I.6 *a priori*, then the stream-lines must be those of the helical motion of Hamel.

The present theorem differs from Prim's in that we postulate that the stream-lines of the circulation-preserving motion must be those of an isochoric irrotational motion and then show that the stream-lines must be of the above kind.

Summary of Background Material

We employ the basis $\boldsymbol{s}$, $\boldsymbol{n}$ and $\boldsymbol{b}$, where $\boldsymbol{s}$ is the unit vector tangent to the stream-line $\boldsymbol{n}$ is the unit principal normal and $\boldsymbol{b} = \boldsymbol{s} \times \boldsymbol{n}$ is the unit bi-normal.* We refer to (1970 Appendix) and quote the representations

$$\operatorname{curl} \boldsymbol{s} = \Omega \boldsymbol{s} + \varkappa \boldsymbol{b}, \quad (1.1)$$

$$\operatorname{curl} \boldsymbol{n} = -\operatorname{div} \boldsymbol{b}\boldsymbol{s} + \Omega_n \boldsymbol{n} + \psi \boldsymbol{b}, \quad (1.2)$$

$$\operatorname{curl} \boldsymbol{b} = (\varkappa + \operatorname{div} \boldsymbol{n})\, \boldsymbol{s} - \theta \boldsymbol{n} + \Omega_b \boldsymbol{b}, \quad (1.3)$$

* If the vector-lines of $\boldsymbol{s}$ are rectilinear, the unit vector $\boldsymbol{n}$ may be taken in any direction perpendicular to $\boldsymbol{s}$, then $\boldsymbol{b}$ is defined by $\boldsymbol{b} = \boldsymbol{s} \times \boldsymbol{n}$.

where Ω, Ω_n, Ω_b are the abnormalities of the vector fields of $\boldsymbol{s}$, $\boldsymbol{n}$ and $\boldsymbol{b}$, related to the torsion τ of the vector-line of $\boldsymbol{s}$ by

$$\Omega_n + \Omega_b = \Omega - 2\tau. \tag{1.4}$$

Also $\varkappa$ is the curvature of the s-lines and θ and ψ are defined by

$$\theta = \boldsymbol{b} \cdot \operatorname{grad} \boldsymbol{s} \cdot \boldsymbol{b}, \tag{1.5}$$

$$\psi = \boldsymbol{n} \cdot \operatorname{grad} \boldsymbol{s} \cdot \boldsymbol{n},$$

and

$$\operatorname{div} \boldsymbol{s} = \theta + \psi. \tag{1.6}$$

The identity

$$\operatorname{curl} \operatorname{grad} F = \mathbf{0}$$

applied to the scalar point function F yields*

$$\frac{\delta^2 F}{\delta b\, \delta n} - \frac{\delta^2 F}{\delta n\, \delta b} = \Omega \frac{\delta F}{\delta s} - \operatorname{div} \boldsymbol{b} \frac{\delta F}{\delta n} + (\varkappa + \operatorname{div} \boldsymbol{n}) \frac{\delta F}{\delta b},$$

$$\frac{\delta^2 F}{\delta s\, \delta b} - \frac{\delta^2 F}{\delta b\, \delta s} = \Omega_n \frac{\delta F}{\delta n} - \theta \frac{\delta F}{\delta b}, \tag{1.7}$$

$$\frac{\delta^2 F}{\delta n\, \delta s} - \frac{\delta^2 F}{\delta s\, \delta n} = \varkappa \frac{\delta F}{\delta s} + \psi \frac{\delta F}{\delta n} + \Omega_b \frac{\delta F}{\delta b}.$$

For the motion I.3 we have

$$\operatorname{div} \boldsymbol{v} = \frac{\delta v}{\delta s} + v \operatorname{div} \boldsymbol{s},$$

$$= \frac{\delta v}{\delta s} + v(\theta + \psi) = 0, \tag{1.8}$$

by (1.6), and

$$\boldsymbol{\omega} = \operatorname{curl} \boldsymbol{v} = \operatorname{grad} v \times \boldsymbol{s} + v \operatorname{curl} \boldsymbol{s} = \Omega v \boldsymbol{s} + \frac{\delta v}{\delta b} \boldsymbol{n} + \left(\varkappa v - \frac{\delta v}{\delta n}\right) \boldsymbol{b}, \tag{1.9}$$

by (1.1). We shall also use the notation

$$\omega_s = \Omega v, \qquad \omega_n = \frac{\delta v}{\delta b}, \qquad \omega_b = \varkappa v - \frac{\delta v}{\delta n}. \tag{1.10}$$

We then have

$$\boldsymbol{\omega} \times \boldsymbol{v} = -v\omega_n \boldsymbol{b} + v\omega_b \boldsymbol{n},$$

and

$$\operatorname{curl} (\boldsymbol{\omega} \times \boldsymbol{v}) = -\operatorname{grad} (v\omega_n) \times \boldsymbol{b} - v\omega_n \operatorname{curl} \boldsymbol{b}$$

$$+ \operatorname{grad} (v\omega_b) \times \boldsymbol{n} + v\omega_b \operatorname{curl} \boldsymbol{n}. \tag{1.11}$$

---

* We employ the symbol $\dfrac{\delta F}{\delta s}$ to denote $\boldsymbol{s} \cdot \operatorname{grad} F$, then $\dfrac{\delta^2 F}{\delta n\, \delta s}$ denotes $\boldsymbol{n} \cdot \operatorname{grad} (\boldsymbol{s} \cdot \operatorname{grad} F)$ and so on.

With a little manipulation, using (1.1), (1.2), (1.3), the commutation formula (1.7) applied to the velocity magnitude v, the formulae $(1.10)_{2,3}$, and finally the identity

$$\operatorname{div} \operatorname{curl} \boldsymbol{s} = \frac{\delta \Omega}{\delta s} + \Omega \operatorname{div} \boldsymbol{s} + \frac{\delta \varkappa}{\delta b} + \varkappa \operatorname{div} \boldsymbol{b} = 0, \quad \text{by (1.1)},$$

we obtain the representation [1969, p. 129],

$$\begin{aligned}\operatorname{curl}(\boldsymbol{\omega}\times\boldsymbol{v}) &= v[\operatorname{div}(\Omega \boldsymbol{v}) - 2\varkappa\omega_{\mathrm{n}}]\, \boldsymbol{s} \\ &+ \left[\frac{\delta}{\delta s}(v\omega_{\mathrm{n}}) + v(\theta\omega_{\mathrm{n}} + \Omega_{\mathrm{n}}\omega_{\mathrm{b}})\right] \boldsymbol{n} \\ &+ \left[\frac{\delta}{\delta s}(v\omega_{\mathrm{b}}) + v(\psi\omega_{\mathrm{b}} - \Omega_{\mathrm{b}}\omega_{\mathrm{n}})\right] \boldsymbol{b}.\end{aligned} \tag{1.12}$$

In arriving at (1.12) we may note that the expressions for $\boldsymbol{n}\cdot\operatorname{curl}(\boldsymbol{\omega}\times\boldsymbol{v})$ and $\boldsymbol{b}\cdot\operatorname{curl}(\boldsymbol{\omega}\times\boldsymbol{v})$ follow directly from (1.2), (1.3) and (1.11), the reduction is required to obtain the above form of the component $\boldsymbol{s}\cdot\operatorname{curl}(\boldsymbol{\omega}\times\boldsymbol{v})$.

Proof.

For the isochoric circulation-preserving motion $\boldsymbol{v} = v\boldsymbol{s}$, we have by I.1, I.2, (1.6), (1.8), $(1.10)_1$ and (1.12)

$$v\frac{\delta \Omega}{\delta s} - 2\varkappa\omega_{\mathrm{n}} = 0, \tag{2.1}$$

$$\frac{\delta \omega_{\mathrm{n}}}{\delta s} - \psi\omega_{\mathrm{n}} + \Omega_{\mathrm{n}}\omega_{\mathrm{b}} = 0, \tag{2.2}$$

$$\frac{\delta \omega_{\mathrm{b}}}{\delta s} - \theta\omega_{\mathrm{b}} - \Omega_{\mathrm{b}}\omega_{\mathrm{n}} = 0. \tag{2.3}$$

Now let $\boldsymbol{v}_1 = v_1\boldsymbol{s}$ be the isochoric irrotational motion having the same streamlines as the circulation-preserving motion.

Since $\operatorname{div} \boldsymbol{v}_1 = 0$, we have by (1.8)

$$\frac{\delta}{\delta s}\log v_1 = -\operatorname{div} \boldsymbol{s} = -(\theta + \psi). \tag{2.4}$$

Since $\operatorname{curl} \boldsymbol{v}_1 = \boldsymbol{0}$, we have by (1.9)

$$\Omega = 0, \tag{2.5}$$

$$\frac{\delta v_1}{\delta b} = 0, \tag{2.6}$$

$$\frac{\delta v_1}{\delta n} = \varkappa v_1. \tag{2.7}$$

Applying $(1.7)_2$ to $\log v_1$, we have by (2.6)

$$\frac{\delta^2}{\delta b\, \delta s}\log v_1 = -\Omega_{\mathrm{n}}\frac{\delta}{\delta n}\log v_1 = -\varkappa\Omega_{\mathrm{n}}, \quad \text{by (2.7)}$$

so by (2.4)

$$\frac{\delta}{\delta b} \operatorname{div} \boldsymbol{s} = \varkappa \Omega_{\mathrm{n}} . \tag{2.8}$$

Applying $(1.7)_3$ to $\log v_1$ we get from (2.4) and (2.7)

$$\frac{\delta^2}{\delta n\, \delta s} \log v_1 = -\frac{\delta}{\delta n} \operatorname{div} \boldsymbol{s},$$

$$\frac{\delta^2}{\delta s\, \delta n} \log v_1 = \frac{\delta \varkappa}{\delta s},$$

so, by $(1.7)_3$, (2.4), (2.6) and (2.7),

$$\frac{\delta}{\delta n} \operatorname{div} \boldsymbol{s} + \frac{\delta \varkappa}{\delta s} = \varkappa \operatorname{div} \boldsymbol{s} - \psi \varkappa = \theta \varkappa, \quad \text{by (1.6)}. \tag{2.9}$$

Using $(1.7)_1$ in a similar manner, we obtain

$$\frac{\delta \varkappa}{\delta b} + \varkappa \operatorname{div} \boldsymbol{b} = 0, \tag{2.10}$$

a result which follows from (1.1) using $\operatorname{div} \operatorname{curl} \boldsymbol{s} = 0$, $\Omega = 0$.

The conditions (2.5), (2.8), (2.9) and (2.10) must hold for the circulation-preserving motion given by (2.1), (2.2), (2.3).

From $(1.10)_2$ and (2.2)

$$\frac{\delta^2 v}{\delta s\, \delta b} = \psi \omega_{\mathrm{n}} - \Omega_{\mathrm{n}} \left(\varkappa v - \frac{\delta v}{\delta n} \right). \tag{2.11}$$

But by $(1.7)_2$ and $(1.10)_2$,

$$\begin{aligned} \frac{\delta^2 v}{\delta b\, \delta s} &= \frac{\delta^2 v}{\delta s\, \delta b} - \Omega_{\mathrm{n}} \frac{\delta v}{\delta n} + \theta \omega_{\mathrm{n}} \\ &= \operatorname{div} \boldsymbol{s} \omega_{\mathrm{n}} - \Omega_{\mathrm{n}} \varkappa v, \quad \text{by (2.11)}. \end{aligned} \tag{2.12}$$

By (1.8)

$$\begin{aligned} \frac{\delta^2 v}{\delta b\, \delta s} &= -\frac{\delta}{\delta b} (v \operatorname{div} \boldsymbol{s}) = -\omega_{\mathrm{n}} \operatorname{div} \boldsymbol{s} - v \frac{\delta}{\delta b} \operatorname{div} \boldsymbol{s} \\ &= -\operatorname{div} \boldsymbol{s} \omega_{\mathrm{n}} - \Omega_{\mathrm{n}} \varkappa v, \quad \text{by (2.8)}. \end{aligned} \tag{2.13}$$

It follows from (2.12) and (2.13) that

$$\operatorname{div} \boldsymbol{s} \omega_{\mathrm{n}} = 0. \tag{2.14}$$

Again from $(1.10)_3$ and (2.3)

$$\frac{\delta}{\delta s} \left(\varkappa v - \frac{\delta v}{\delta n} \right) - \theta \omega_{\mathrm{b}} - \Omega_{\mathrm{b}} \frac{\delta v}{\delta b} = 0, \tag{2.15}$$

and, by (1.18)

$$\frac{\delta \varkappa}{\delta s} v - \varkappa v \operatorname{div} \boldsymbol{s} - \frac{\delta^2 v}{\delta s\, \delta n} - \theta \omega_{\mathrm{b}} - \Omega_{\mathrm{b}} \frac{\delta v}{\delta b} = 0. \tag{2.16}$$

By $(1.7)_3$ and (1.8)

$$\frac{\delta^2 v}{\delta s\,\delta n} = \frac{\delta^2 v}{\delta n\,\delta s} - \varkappa\frac{\delta v}{\delta s} - \psi\frac{\delta v}{\delta n} - \Omega_b\frac{\delta v}{\delta b}$$

$$= -\frac{\delta}{\delta n}(v \operatorname{div} s) + \varkappa v \operatorname{div} s - \psi\frac{\delta v}{\delta n} - \Omega_b\frac{\delta v}{\delta b}$$

$$= \omega_b \operatorname{div} s - v\frac{\delta}{\delta n}\operatorname{div} s - \psi\frac{\delta v}{\delta n} - \Omega_b\frac{\delta v}{\delta b}, \tag{2.17}$$

by (1.6) and (1.10). Eliminating $\frac{\delta^2 v}{\delta s\,\delta n}$ from (2.16) and (2.17), we get

$$v\left(\frac{\delta\varkappa}{\delta s} + \frac{\delta}{\delta n}\operatorname{div} s\right) - \varkappa v \operatorname{div} s - \theta\omega_b - \omega_b \operatorname{div} s + \psi\frac{\delta v}{\delta n} = 0,$$

or, by (2.9)

$$\theta\varkappa v - \operatorname{div} s\,\varkappa v - \theta\omega_b - \omega_b \operatorname{div} s + \psi\frac{\delta v}{\delta n} = 0,$$

which, by (1.6) and (1.10), gives

$$\operatorname{div} s\,\omega_b = 0. \tag{2.18}$$

From (2.5), (2.14) and (2.18) either the circulation-preserving motion is irrotational or $\operatorname{div} s = 0$. Since the circulation-preserving motion is postulated to be rotational the latter condition must hold and so, by (1.8), $\frac{\delta v}{\delta s} = 0$. Thus Theorem 1 is proved.

We note that from (2.1) and (2.5), that either $\varkappa$ or ω_n is zero. The result follows in any case from (2.18).

References

1937 HAMEL, G., Potentialströmungen mit konstanter Geschwindigkeit. *Sitzgsber. preuss. Akad. Wiss., phys.-math. Kl.* 5–20.

1952 PRIM, R. C., Steady rotational flow of ideal gases. *J. Rational Mech. Anal.* **1**, 425–497.

1969 MARRIS, A. W., On steady three-dimensional motions. *Arch. Rational Mech. Anal.* **35**, 122–168.

1970 MARRIS, A. W., & C.-C. WANG, Solenoidal screw fields of constant magnitude, Appendix, *Arch. Rational Mech. Anal.* **39**, 227–244.

1973 MARRIS, A. W., Hamel's theorem, *Arch. Rational Mech. Anal.* **51**, 85–105.

Georgia Institute of Technology
Atlanta

(Received October 9, 1984)

On Shear Bands in Ductile Materials

BERNARD D. COLEMAN & MARION L. HODGDON

Dedicated to Walter Noll on the occasion of his 60th birthday

Synopsis

Spatially non-uniform solutions are found here for the field relations that govern the behavior in shear of a class of rigid-plastic materials that have rate-independent response and, after a certain amount of plastic flow, exhibit strain softening, *i.e.*, a decline of yield stress with further flow. As strain softening can destabilize homogeneous configurations and result in the concentration of strain in narrow bands, the constitutive relations have been chosen to account for a possibly important influence upon the stress of the spatial variation of the accumulated strain. It is shown that for simple non-steady shearing flows, the shear strain, as a function of time and position, can be expressed in terms of easily evaluated elliptic functions. The theory yields strain fields showing shear bands that are remarkably similar to those observed in ballistic tests of metals under conditions in which inertial forces are negligible but the combined effects of adiabatic heating and thermal softening result in apparent strain softening. Although quantitative comparison with observation is not yet possible, it is expected that the theory will be applicable to slow deformations of geological materials that are ductile at elevated pressures and temperatures and are often softer after flowing as the result of an accumulation of internal damage with deformation.

1. Introduction

We here present a theory of the development of shear bands for a class of ductile materials with rate-independent response.

For the motions we discuss, the flow is in the y-direction of a fixed Cartesian system, and the displacement u in that direction is a function of x and t, with t the time which has elapsed since the body was in an undistorted state with $u(x, 0) \equiv 0$. At each x and t, the shear strain γ is $u_x(x, t)$, the rate of shear $\dot{\gamma}$ is $\gamma_t(x, t) = u_{tx}(x, t)$,[1] and the *accumulated shear strain* $\bar{\gamma}$ is, by definition, the

[1] The subscripts indicate partial derivatives.

total variation of the function $\xi \mapsto \gamma(x, \xi)$ on the interval $[0, t]$. The shear stress τ is the (xy)-component, T^{yx}, of the Cauchy stress tensor.[2] Each material of the class we consider can support a shear stress τ without further deformation provided τ is less than a function φ of the accumulated strain; *i.e.*,

$$\dot{\gamma} = 0 \quad \text{if} \quad |\tau| < \varphi(\bar{\gamma}); \tag{1.1}$$

φ is called the *yield function* of the material; its value, $\varphi(\bar{\gamma})$, called the *yield stress*, is assumed to be positive for small $\bar{\gamma}$. The rate of change of the accumulated strain $\bar{\gamma}_t$ is never negative, because $\bar{\gamma}_t = |\dot{\gamma}|$; moreover,[3]

$$\varphi(\bar{\gamma})_t = \varphi'(\bar{\gamma})\,\bar{\gamma}_t = \varphi'(\bar{\gamma})\,|\dot{\gamma}|. \tag{1.2}$$

Thus, during deformation, *i.e.*, when $\dot{\gamma} \neq 0$, the yield stress increases or decreases in accord with whether φ' is positive or negative. We say that a material shows *strain hardening* when φ' is positive and *strain softening* when φ' is negative.

We are here interested in materials that harden in the early stages of deformation, but eventually soften as they flow. Thus, we suppose that there is a critical value of $\bar{\gamma}$, $\bar{\gamma}_m$, at which $\varphi'(\bar{\gamma}_m) = 0$, and $\varphi'(\bar{\gamma}_m)$ is positive for $\bar{\gamma}$ less than $\bar{\gamma}_m$, and negative for $\bar{\gamma}$ greater than $\bar{\gamma}_m$, as depicted in Figure 1.

Metals generally strain harden in isothermal deformation, but show a decrease in yield stress with an increase in temperature; hence a rigid-plastic metal that shows only moderate isothermal strain hardening can have an adiabatic behavior that does not differ greatly from that shown in Figure 1.[4] When rocks and aggre-

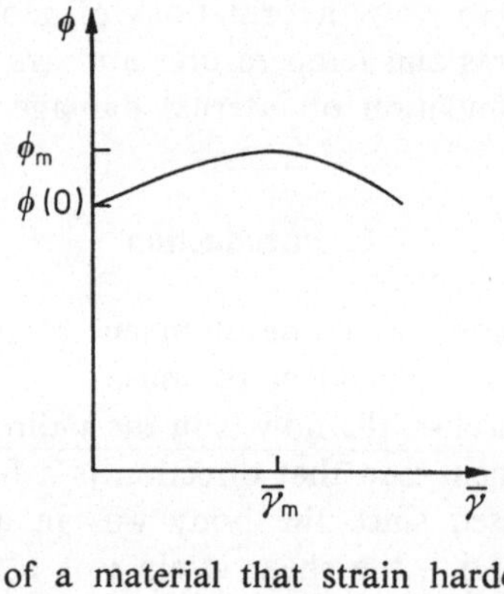

Fig. 1. The yield function of a material that strain hardens when the accumulated shear strain $\bar{\gamma}$ is small, but softens when $\bar{\gamma}$ exceeds a critical value $\bar{\gamma}_m$.

---

[2] See equations (2.1)–(2.6).

[3] $\varphi' = d\varphi/d\bar{\gamma}$.

[4] See, *e.g.*, recent studies of mild steel by COSTIN, CRISMAN, HAWLEY, & DUFFY [1979, 1] and ELEICHE [1981, 1]. Examples of the effects of thermal softening in nearly adiabatic deformations of metals are given in the survey of ROGERS [1979, 3]. Among the many references in that survey is an early paper of ZENER & HOLLOMAN [1944, 1] recognizing that, in deformations of metals at speeds high enough to be approximately adiabatic, thermal softening can lead to a concentration of shear strain in narrow bands.

gates are ductile,[5] it is not unusual for them to show, at large values of the accumulated strain, a decrease of yield stress with increasing strain; such softening can be the result of internal damage induced by shearing.

The conventional theory of rigid-plastic materials (without viscous stresses but with a von Mises yield condition at each value of the accumulated strain) here would give the following constitutive relation for the stress during shearing:

$$\tau = \varphi(\bar{\gamma}) \frac{\dot{\gamma}}{|\dot{\gamma}|} \quad \text{when} \quad \dot{\gamma} \neq 0. \tag{1.3}$$

When the yield function is as in Figure 1, the relation (1.2) yields, for $\bar{\gamma}$ in excess of $\bar{\gamma}_m$, a decrease in shear stress with an increase in shear strain and hence a loss of stability for homogeneous configurations. This is not, by itself, unphysical; in fact, we expect strain softening to result in the occurrence of regions in which $\bar{\gamma}$ varies rapidly with position. However, as no real material is truly simple in the sense that only the history of the first gradient of the displacement influences the stress, we expect spatial derivatives of $\bar{\gamma}$ to influence τ when they are large. These considerations suggest that we should add to (1.3) a term that depends on derivatives of $\bar{\gamma}$. As we expect such a term to have physical effects similar to those of the capillary forces that act to oppose the growth of new surfaces in crack growth and phase transformation, the new term should be compatible with rate-independent response and, without totally suppressing inhomogeneities, should tend to moderate them. Of course, the term must be in accord with the principle of material frame indifference. The simplest term with these properties appears to be one that, for a shearing flow, takes the form $-c\bar{\gamma}_{xx}\dot{\gamma}/|\dot{\gamma}|$, with c a positive constant. Such heuristic reasoning leads us to consider the following modification of (1.3):

$$\tau = [\varphi(\bar{\gamma}) - c\bar{\gamma}_{xx}] \frac{\dot{\gamma}}{|\dot{\gamma}|} \quad \text{when} \quad \dot{\gamma} \neq 0. \tag{1.4}$$

[5] Several materials of geological importance that are brittle under the conditions of everyday experience exhibit ductile behavior at high confining pressure or elevated temperature. A pressure induced transition from brittle to ductile behavior that occurs in carbonates and several sedimentary rocks (see, *e.g.*, HANDIN & HAGER [1957, 1]) was studied in detail by VON KÁRMÁN in his often cited experimental work on Carrara marble [1911, 1]. In compressive triaxial tests at room temperature, that material shows, as expected, brittle failure at low confining pressures p but is fully ductile for p above 500 bars; when $p = 685$ bars, the material can flow to strains of over 7%; in such flow marble is transformed into a chalky material (*i.e.*, a firmly compacted powder) that is very different in appearance from undamaged marble. Sandstone is ductile at room temperature when p is above 1400 bars (see, *e.g.*, MURRELL [1965, 1]). GRIGGS, TURNER, & HEARD [1960, 1] showed that, at a confining pressure of 5 kilobars, basalt and granite are fully ductile at a temperature of 500 °C and can there exhibit strains greater than 10%. These and other examples of ductility of common rocks at elevated pressures and temperatures are discussed in the text of JAEGER & COOK [1979, 2]. A glance at a symposium volume edited by GRIGGS & HANDIN (*e.g.*, [1960, 1]) shows that there is a great variety to the internal structural changes that can accompany the ductile deformation of rocks.

We make no claim to having derived (1.4) from general principles and self-evident assumptions. A theory in which the flow rule (1.4) is used with the yield condition (1.1) can be accepted as a useful, albeit approximate, model of a real material only if its implications can be shown to be in accord with observations of the behavior of the material.

In the next section we shall state our assumptions about material response in a more complete way. We shall there show that our constitutive relations for shearing flows can be viewed as the specialization to such unidimensional motions of a properly invariant statement about the behavior of three-dimensional bodies. In the third and longest section of the article, we shall show that when φ is quadratic, *i.e.*, has the form $\varphi(\bar{\gamma}) = \varphi_m - \alpha(\bar{\gamma} - \bar{\gamma}_m)^2$ with α, like $\bar{\gamma}_m$ and φ_m, a positive constant, our constitutive assumptions give rise to a theory of strain localization in which the field relations can be solved explicitly. We shall there present several graphs that we hope will enable experimenters in solid mechanics and field workers in geology to compare our results with their observations of the form and growth of shear bands.[6]

2. Constitutive Assumptions

Let x, y, z be the Cartesian coordinates at time t of the material point whose coordinates are X, Y, Z when the body is in its reference configuration $\mathscr{R}$. We are here concerned with shearing motions for which

$$\begin{aligned} x &= X, \\ y &= Y + u(X, t) = Y + u(x, t), \qquad (2.1) \\ z &= Z. \end{aligned}$$

The x-derivative of the displacement u is the *shear strain* γ, *i.e.*,

$$\gamma(x, t) = u_x(x, t), \qquad (2.2)$$

[6] Illustrative calculations of the type shown in Figures 4–8 and 10–13 require the assignment of numerical values to the material parameters c, φ_m, α, and $\bar{\gamma}_m$. For the calculations we report, the dimensionless quantity $\bar{\gamma}_m$ was taken to be 0.05. Although that value appears appropriate for several ductile geological materials, it is small for the adiabatic deformation of mild steel (starting at room temperature) with $\dot{\gamma} = 10^3\ \text{sec}^{-1}$. Nevertheless, the graphs of displacement u *versus* position x seen in Figure 10 are remarkably similar in form to patterns formed by scribe lines on tubes of cold rolled mild steel that, in the experiments of COSTIN, CRISMAN, HAWLEY, & DUFFY [1979, 1], showed localization of plastic flow when torsionally loaded (at various initial temperatures, including room temperature) so as to undergo adiabatic shearing at nominal strain rates of *circa* $10^3\ \text{sec}^{-1}$. However, when, as appears to be the case for the steel employed in these experiments, the yield function of a plastic material has a dependence on rate of deformation beyond the dependence accounted for by thermal softening and a change from isothermal to adiabatic deformation, the constitutive relations of the present theory, as they are rate-independent, are at best approximate.

and, as $x = X$, the material time-derivative $\dot{\gamma} = \partial\gamma(X, t)/\partial t$ here equals the spatial time-derivative $\gamma_t = \partial\gamma(x, t)/\partial t$. We assume that the body is in the configuration $\mathscr{R}$ at time $t = 0$, and hence

$$u(x, 0) = 0 \quad \text{and} \quad \gamma(x, 0) = 0, \tag{2.3}$$

for all x. For each $t \geqq 0$ the *accumulated shear strain up to time t* (at x) is denoted by $\bar{\gamma}(x, t)$ and is defined to be the total variation of the shear strain considered a function $\gamma(x, \cdot)$ on $[0, t]$. Thus,

$$\bar{\gamma}(x, t) = \int_0^t |\dot{\gamma}(x, \sigma)| \, d\sigma = \int_0^t \dot{\gamma}(x, \sigma) \, s(\dot{\gamma}(x, \sigma)) \, d\sigma, \tag{2.4}$$

where s denotes the signum function:

$$s(\dot{\gamma}) = \begin{cases} 1 & \text{for} \quad \dot{\gamma} > 0, \\ 0 & \text{for} \quad \dot{\gamma} = 0, \\ -1 & \text{for} \quad \dot{\gamma} < 0. \end{cases} \tag{2.5}$$

We suppose that the matrix of the Cartesian components of the Cauchy stress tensor $\boldsymbol{T}$ has the form

$$[\boldsymbol{T}] = \begin{bmatrix} -p & \tau & 0 \\ \tau & -p & 0 \\ 0 & 0 & -p \end{bmatrix}, \tag{2.6}$$

and we make the following assumption relating τ to $\dot{\gamma}$ and $\bar{\gamma}$. The material composing the body under consideration is characterized by a *yield function* φ, defined and positive on an interval $[0, A)$, $0 < A \leqq \infty$, and a positive constant c, such that if

$$\dot{\gamma} = 0, \quad \text{then} \quad \tau^2 \leqq \varphi(\bar{\gamma})^2; \tag{2.7a}$$

if $$\dot{\gamma} \neq 0, \quad \text{then} \quad \tau^2 \geqq \varphi(\bar{\gamma})^2 \quad \text{and} \quad \tau = [\varphi(\bar{\gamma}) - c\bar{\gamma}_{xx}] \, s(\dot{\gamma}). \tag{2.7b}$$

The assumption (2.7b) yields not only (1.4) but also (1.1), because (1.1) is equivalent to the assertion that $|\tau| \geqq \varphi(\bar{\gamma})$ when $\dot{\gamma} \neq 0$. The two parts of (2.7b), taken together, tell us that $\bar{\gamma}_{xx}$ cannot be positive wherever $\dot{\gamma}$ is not zero and agrees in sign with τ.[7] We note the following consequences of (2.7):

if $$\tau^2 < \varphi(\bar{\gamma})^2, \text{ then } \dot{\gamma} = 0; \tag{2.8a}$$

if $\tau^2 > \varphi(\bar{\gamma})^2$ and $\dot{\gamma}\tau \geqq 0$, then $\dot{\gamma} \neq 0$, $|\tau| = |\varphi(\bar{\gamma}) - c\bar{\gamma}_{xx}|$, and $\bar{\gamma}_{xx} < 0$. (2.8b)

[7] Whereas the implication, $\dot{\gamma} \neq 0 \Rightarrow |\tau| \geqq \varphi(\bar{\gamma})$, is familiar in theories of plasticity, the implication, $\dot{\gamma} \neq 0 \;\&\; \tau\dot{\gamma} \geqq 0 \Rightarrow \bar{\gamma}_{xx} \leqq 0$, linking the ability to deform further in the direction of the applied stress to a local concavity in x of the accumulated strain, is new and may appear strange. In Section 3 we shall see that this constraint upon flow, although unfamiliar, does not preclude the existence of physically interesting solutions of the field relations that result when our constitutive assumptions are combined with balance of forces.

One cannot consider meaningful a constitutive assumption proposed for a unidimensional class of motions unless it is made clear that the relation is the specialization of a properly invariant statement about material response in the three-dimensional world in which actual bodies exist. Fortunately, as we shall show now, our assumption (2.7) is the form taken in shearing motions by at least one class of frame-indifferent three-dimensional constitutive relations.

In terms of the velocity $\boldsymbol{v}$, expressed as a function of spatial position $\boldsymbol{x}$ and time t, the *velocity gradient* $\boldsymbol{L}$, the *stretching tensor* $\boldsymbol{D}$, and the *rate of distortion tensor* $\boldsymbol{D}_{(d)}$ are defined as

$$\begin{aligned} \boldsymbol{L} &= \operatorname{grad} \boldsymbol{v}, \\ \boldsymbol{D} &= \tfrac{1}{2}[\boldsymbol{L} + \boldsymbol{L}^T], \\ \boldsymbol{D}_{(d)} &= \boldsymbol{D} - \tfrac{1}{3}(\operatorname{tr} \boldsymbol{D})\,\boldsymbol{1}, \end{aligned} \tag{2.9}$$

with $\boldsymbol{L}^T$ the transpose of $\boldsymbol{L}$, $\boldsymbol{1}$ the unit tensor, and $\operatorname{tr}\boldsymbol{D}$ the trace of $\boldsymbol{D}$; $\boldsymbol{D}_{(d)}$ is the deviatoric part of $\boldsymbol{D}$. The deviatoric part $\boldsymbol{T}_{(d)}$ of the Cauchy stress $\boldsymbol{T}$ is defined by the relation,

$$\boldsymbol{T} = \boldsymbol{T}_{(d)} - p\boldsymbol{1}, \tag{2.10}$$

in which $p = -\tfrac{1}{3}\operatorname{tr}\boldsymbol{T}$ is the *pressure*. For the norm $\|\boldsymbol{A}\|$ of a tensor $\boldsymbol{A}$, we use the definition

$$\|\boldsymbol{A}\| = [\operatorname{tr}(\boldsymbol{A}\boldsymbol{A}^T)]^{\frac{1}{2}}. \tag{2.11}$$

For a body that is initially in an undistorted reference configuration, we define the *accumulated distortion up to time t* to be

$$\bar{\varepsilon} = \bar{\varepsilon}(\boldsymbol{x}, t) = \sqrt{2} \int_0^t \|\boldsymbol{D}_{(d)}(\boldsymbol{x}, \xi)\|\, d\xi. \tag{2.12}$$

Each (three-dimensional) material of the class we now define is characterized by a positive material parameter c and a function φ that is positive on an interval $[0, A)$. Our constitutive assumption is:

if $$\boldsymbol{D}_{(d)} = \boldsymbol{0}, \quad \text{then} \quad \|\boldsymbol{T}_{(d)}\|^2 \leqq 2\varphi(\bar{\varepsilon})^2; \tag{2.13a}$$

if $\boldsymbol{D}_{(d)} \neq \boldsymbol{0}$, then $\|\boldsymbol{T}_{(d)}\|^2 \geqq 2\varphi(\bar{\varepsilon})^2$ and

$$\boldsymbol{T} = -p\boldsymbol{1} + \sqrt{2}\,[\varphi(\bar{\varepsilon}) - c\,\Delta\bar{\varepsilon}]\frac{\boldsymbol{D}_{(d)}}{\|\boldsymbol{D}_{(d)}\|}. \tag{2.13b}$$

Here Δ denotes the Laplacian, *i.e.*, $\Delta\bar{\varepsilon} = \operatorname{div}\operatorname{grad}\bar{\varepsilon}$. It is clear that the relations (2.13) are compatible with NOLL's principle of material frame-indifference.[8]

[8] [1958, 1]; there it is called the "principle of objectivity of material properties". The principle is explained further in [1965, 2], and an elementary discussion is given in [1966, 1].

Now, for shearing motions it follows from (2.1) that the matrices of the Cartesian components of $\boldsymbol{D}$ and $\boldsymbol{D}_{(d)}$ are

$$[\boldsymbol{D}_{(d)}] = [\boldsymbol{D}] = \tfrac{1}{2}\begin{bmatrix} 0 & \dot{\gamma} & 0 \\ \dot{\gamma} & 0 & 0 \\ 0 & 0 & 0 \end{bmatrix}, \tag{2.14}$$

and hence (2.11) and (2.12) yield

$$\|\boldsymbol{D}_{(d)}\| = |\dot{\gamma}|/\sqrt{2}, \quad \bar{\varepsilon}(\boldsymbol{x}, t) = \bar{\gamma}(x, t), \quad \Delta\bar{\varepsilon} = \bar{\gamma}_{xx} \tag{2.15}$$

with $\bar{\gamma}$ as in (2.4). Thus (2.13b) tells us that if $\dot{\gamma} \neq 0$, the matrix of the Cartesian components of the stress tensor has the form seen in (2.6) with

$$\tau = [\varphi(\bar{\varepsilon}) - c\,\Delta\bar{\varepsilon}]\frac{\sqrt{2}\,\dot{\gamma}/2}{|\dot{\gamma}|/\sqrt{2}} = [\varphi(\bar{\gamma}) - c\bar{\gamma}_{xx}]\,s(\dot{\gamma}), \tag{2.16}$$

and

$$\tau^2 \geqq \varphi(\bar{\varepsilon})^2. \tag{2.17}$$

Moreover, when $[\boldsymbol{T}]$ is as in (2.6) and (2.1) holds with $\dot{\gamma} = 0$, (2.13a) yields

$$\tau^2 \leqq \varphi(\bar{\varepsilon})^2. \tag{2.18}$$

Thus, for shearing motions with $[\boldsymbol{T}]$ as in (2.6), the constitutive assumption (2.13) reduces to (2.7) with φ and c unaltered. Moreover, (2.13) implies that $[\boldsymbol{T}]$ must have the assumed form (2.6) in shearing motions with $\dot{\gamma} \neq 0$.

We have just shown that (2.7) is the form taken in shearing motions by (2.13). Of course, we do not claim that (2.13) is the only possible extension of (2.7) to three-dimensional bodies.

It appears worth mentioning that (2.13) implies the following extension of (2.8):

if $\quad \|\boldsymbol{T}_{(d)}\|^2 < 2\varphi(\bar{\varepsilon})^2$, then $\boldsymbol{D}_{(d)} = \boldsymbol{0}$; (2.19a)

if $\quad \|\boldsymbol{T}_{(d)}\|^2 > 2\varphi(\bar{\varepsilon})^2$, then $\boldsymbol{D}_{(d)} \neq \boldsymbol{0}$ and

$$\boldsymbol{T}_{(d)} = \sqrt{2}\,[\varphi(\bar{\varepsilon}) - c\,\Delta\bar{\varepsilon}]\frac{\boldsymbol{D}_{(d)}}{\|\boldsymbol{D}_{(d)}\|}; \tag{2.19b}$$

if $\quad \|\boldsymbol{T}_{(d)}\|^2 > 2\varphi(\bar{\varepsilon})^2$ and $\boldsymbol{T}_{(d)} \cdot \boldsymbol{D}_{(d)} \geqq 0$,

$$\text{then} \quad \boldsymbol{T}_{(d)} \cdot \boldsymbol{D}_{(d)} > 0, \quad \frac{\boldsymbol{T}_{(d)}}{\|\boldsymbol{T}_{(d)}\|} = \frac{\boldsymbol{D}_{(d)}}{\|\boldsymbol{D}_{(d)}\|}, \quad \text{and} \quad \Delta\bar{\varepsilon} < 0. \tag{2.19c}$$

When there are no body forces and when inertial effects are negligible, balance of forces is equivalent to the assertion that, for all $\boldsymbol{x}$ and t,

$$\operatorname{div} \boldsymbol{T} = \boldsymbol{0}. \tag{2.20}$$

We shall seek shearing motions (2.1) that satisfy this equation, obey the constitutive assumption (2.7), and are such that the pressure p appearing in (2.6) is constant in space. For such motions the normal thrusts, per unit of area, are equal on all planes of the form $x =$ constant and $z =$ constant, including any that may bound the body. Spatial uniformity for p may be realized in several ways, two of which appear important: (i) When the material is incompressible [which does not contradict (2.13)], only isochoric motions are possible, *i.e.*, $\operatorname{div} \boldsymbol{v} = 0$, or, equivalently, $\boldsymbol{D}_{(d)} = \boldsymbol{D}$, a condition met by the shearing motions (2.1), and as p then is not controlled by a constitutive relation, p can be taken to be constant in $\boldsymbol{x}$. (ii) When the material is compressible, p in (2.13b) and (2.6) may be given by a function of the mass density ϱ, and as the motions (2.1) are isochoric, if ϱ is constant in space at time $t = 0$, it will remain so for $t > 0$ and p will then be constant in both $\boldsymbol{x}$ and t. For such a compressible material, one would expect c and the yield function φ to depend on ϱ and hence on the pressure p.[9] Our analysis of (2.20) will be compatible with an influence of p on c and φ.

The constitutive relations (2.7) tell us that in a shearing motion τ is a function of x and t wherever $\dot{\gamma}$ is not zero. When $\dot{\gamma}$ is zero, these relations place no restrictions on τ other than the bound $|\tau| \leqq \varphi(\bar{\gamma})$. We shall assume that τ is a function of x and t even in regions where $\dot{\gamma}$ vanishes. Clearly, when such is the case and p is constant in space, (2.20) reduces to $\tau_x = 0$, *i.e.*, the shearing stress τ is given by a function τ° of time alone, and we have the equation

$$\tau(x, t) = \tau^\circ(t). \tag{2.21}$$

We seek to understand the properties of motions that obey this equation when φ is not monotone but is instead, as shown in Figure 1, a smooth function that attains a maximum φ_m at a point $\bar{\gamma}_m$ and has $\varphi(0) > 0$, $\varphi'(0) > 0$, $\varphi''(\bar{\gamma}_m) < 0$ and, of course, $\varphi'(\bar{\gamma}_m) = 0$. The simplest such function has the form

$$\varphi(\bar{\gamma}) = \varphi_m - \alpha(\bar{\gamma} - \bar{\gamma}_m)^2, \tag{2.22}$$

with

$$\alpha > 0, \quad \bar{\gamma}_m > 0, \quad \varphi_m > \alpha\bar{\gamma}_m^2. \tag{2.23}$$

We shall assume that φ is as in (2.22). There is then a value of $\bar{\gamma}$, which we call $\bar{\gamma}_F$, for which

$$\varphi(\bar{\gamma}_F) = 0; \tag{2.24}$$

indeed,

$$\bar{\gamma}_F = \bar{\gamma}_m + \sqrt{\varphi_m/\alpha}\,. \tag{2.25}$$

It follows from (2.7a) that as $\bar{\gamma}$ approaches $\bar{\gamma}_F$, the maximum magnitude $\varphi(\bar{\gamma})$ of the shear stress τ that the material can sustain without flowing approaches zero. In other words, a body in which $\bar{\gamma}$ attains the value $\bar{\gamma}_F$ at a plane $x =$ constant would appear to have zero strength for shear in that plane, and an engineer would say that the body "has failed" or has "lost cohesion" across that plane. We shall take the domain of φ to be the interval $[0, \bar{\gamma}_F)$. We shall look for solutions $u = u(x, t)$

---

[9] Such a dependence is in accord with many studies of the effects of confining pressure on the ductility of rocks and aggregates. See the following references from footnote 5: [1911, 1], [1957, 1], [1965, 1], [1979, 2].

of (2.20) that obey (2.3), and we shall follow such solutions up to a time t_F which is the "time of failure" in the sense that, for some value x_F of x,[10]

$$\lim_{t \to t_F} \bar{\gamma}(x_F, t) = \bar{\gamma}_F. \tag{2.26}$$

It follows from (2.7) and (2.22) that (2.20) is equivalent to the following restriction on the displacement field u: for each pair (x, t), with $0 \leqq t < t_F$,

$$\text{if} \quad \dot{\gamma} = 0, \quad \text{then} \quad |\tau^\circ(t)| \leqq \varphi_m - \alpha(\bar{\gamma} - \bar{\gamma}_m)^2; \tag{2.27a}$$

$$\text{if} \quad \dot{\gamma} \neq 0, \quad \text{then} \quad |\tau^\circ(t)| \geqq \varphi_m - \alpha(\bar{\gamma} - \bar{\gamma}_m)^2 \quad \text{and}$$

$$\tau^\circ(t) = [\varphi_m - \alpha(\bar{\gamma} - \bar{\gamma}_m)^2 - c\bar{\gamma}_{xx}]\, s(\dot{\gamma}). \tag{2.27b}$$

3. Solution of the Field Relations

We here study the equation (2.21) in the form (2.27) for a body of such depth that the range of x can be taken to be $(-\infty, \infty)$. We assume that when $t = 0$ the body is in its reference configuration $\mathscr{R}$ which implies that (2.3) holds. In view of (2.4), we have

$$\bar{\gamma}(x, 0) = 0 \tag{3.1}$$

for all x. We suppose that for $t \geqq 0$, the function $t \mapsto |\tau^\circ(t)|$ is continuous with a piecewise continuous derivative and

$$\tau^\circ(0) = 0. \tag{3.2}$$

We seek displacement fields $u = u(x, t)$ on $(-\infty, \infty) \times [0, t_F)$ for which $\dot{\gamma}$ agrees in sign with τ° for each x and t, *i.e.*,

$$\tau^\circ(t)\, \dot{\gamma}(x, t) \geqq 0, \tag{3.3}$$

and that are smooth in the sense that, for each x, $t \mapsto \gamma(x, t)$ is continuous and $t \mapsto \dot{\gamma}(x, t)$ is piecewise continuous [which, by (2.4), implies that $t \mapsto \bar{\gamma}(x, t)$ is continuous with a piecewise continuous derivative], and, at each t, $x \mapsto \bar{\gamma}(x, t)$ is bounded and continuous while $x \mapsto \bar{\gamma}_x(x, t)$ and $x \mapsto \bar{\gamma}_{xx}(x, t)$ are bounded and piecewise continuous [which, of course, implies the existence of left-hand and right-hand limits, $\bar{\gamma}_{xx}(x^-, t)$ and $\bar{\gamma}_{xx}(x^+, t)$, for $x \mapsto \bar{\gamma}_{xx}(x, t)$ at all points x]. It follows from (2.4) that the function $t \mapsto \bar{\gamma}(x, t)$ never decreases.

As $\bar{\gamma}_t(x, t) = |\dot{\gamma}(x, t)|$, (2.7a) here asserts that at (x, t),

$$\varphi(\bar{\gamma}(x, t)) \geqq |\tau^\circ(t)| \quad \text{if} \quad \bar{\gamma}_t(x, t) = 0. \tag{3.4}$$

[10] The time t_F of failure in a general motion is the earliest time at which the accumulated distortion $\bar{\varepsilon}$ reaches, at some point in the body, the value $\bar{\varepsilon}_F$ for which $\varphi(\bar{\varepsilon}_F) = 0$; when φ is as in (2.22), $\bar{\varepsilon}_F = \bar{\gamma}_m + \sqrt{\varphi/\alpha}$. It follows from (2.19b) that wherever $\bar{\varepsilon} = \bar{\varepsilon}_F$ no deviatoric stress can be sustained without the material giving way to it. That is, when $\bar{\varepsilon} = \bar{\varepsilon}_F$ and $\boldsymbol{T}_{(d)}$ is not zero, no matter how small the positive number $\|\boldsymbol{T}_{(d)}\|$, the rate of distortion $\boldsymbol{D}_{(d)}$ is not zero.

It follows from (3.3) that wherever $\dot{\gamma} \neq 0$, $\tau^\circ(t)/s(\dot{\gamma}) = |\tau^\circ(t)|$, and hence (2.7b) here becomes

$$\varphi(\bar{\gamma}(x,t)) \leqq |\tau^\circ(t)| \quad \text{and} \quad \varphi(\bar{\gamma}(x,t)) - c\bar{\gamma}_{xx}(x,t) = |\tau^\circ(t)|, \quad \text{if} \quad \bar{\gamma}_t(x,t) > 0, \tag{3.5}$$

and as c is positive,

$$\bar{\gamma}_{xx}(x,t) \leqq 0, \quad \text{if} \quad \bar{\gamma}_t(x,t) > 0. \tag{3.6}$$

As we assume (3.3), the implications (3.4) and (3.5), taken together, are equivalent to the field equation (2.21) and the constitutive assumption (2.7).

It is clear from (2.8a) that γ, and hence $\bar{\gamma}$, will remain equal to zero until the time t_0 at which $|\tau^\circ|$ first reaches the value $\varphi(0)$ which, by (2.22), is $\varphi_m - \alpha\bar{\gamma}_m^2$.

Let t_m be the time at which $|\tau^\circ|$ first reaches the value φ_m. Suppose $|\tau^\circ(t)|$ increases montonically with t for $t_0 \leqq t \leqq t_m$. The relations (3.4) and (3.5) are satisfied in this time interval if the motion is one of homogeneous shearing with, for all x, $\bar{\gamma}(x,t) = \bar{\gamma}^\circ(t)$, where $\bar{\gamma}^\circ(t)$ is the smaller of the two positive roots of the equation

$$\varphi(\bar{\gamma}^\circ(t)) = |\tau^\circ(t)| \tag{3.7}$$

which, by (2.22), yields

$$\bar{\gamma}(x,t) = \bar{\gamma}_m - \left(\frac{\varphi_m - |\tau^\circ(t)|}{\alpha}\right)^{\frac{1}{2}} = \bar{\gamma}^\circ(t) \quad \text{for} \quad t_0 \leqq t \leqq t_m, \tag{3.8}$$

and, in particular, $\bar{\gamma}(x, t_m) = \bar{\gamma}_m$.

One may satisfy (3.4) and (3.5) for $t > t_m$ with a homogeneous motion by letting $|\tau^\circ(t)|$ have a maximum at $t = t_m$ and putting $\bar{\gamma}^\circ(t)$ equal to the larger root of equation (3.7) for $t > t_m$. But, as homogeneous solutions are not expected to be stable when τ° decreases with increasing $\bar{\gamma}^\circ$, we shall look for solutions of (3.4) and (3.5) that obey (3.8) for $t_0 \leqq t \leqq t_m$ and thereafter are not homogeneous.[11]

Before turning to the problem of constructing non-homogeneous extensions of solutions obeying (3.8), we should point out that, for motions initially homogeneous, the assumption of monotonicity for $|\tau^\circ|$ on $[t_0, t_m]$ causes no serious loss of generality. Indeed, let (t_1, t_2) be the first subinterval of $[t_0, t_m]$ on which $|\tau^\circ(t)|$ is strictly decreasing, and suppose (3.8) holds on $[t_0, t_1]$. As $\varphi(\bar{\gamma})$ is monotone increasing in $\bar{\gamma}$ for $\bar{\gamma}$ in $(0, \bar{\gamma}_m)$ and $\bar{\gamma}$ never decreases in time, for each x we can assert that if, at a time $\hat{t}$ in (t_1, t_2), $\bar{\gamma}(x, \hat{t})$ is less than $\bar{\gamma}_m$, then $|\tau^\circ(\hat{t})| < |\tau^\circ(t_1)| = \varphi(\bar{\gamma}^\circ(t_1)) \leqq \varphi(\bar{\gamma}(x, \hat{t}))$, and (2.8a) yields $\dot{\gamma}(x, \hat{t}) = 0$; thus, at each x, $\dot{\gamma}(x,t) = 0$ for all t in (t_1, t_2). A simple extension of this argument to later time-intervals on which $|\tau^\circ|$ may be decreasing shows that no shearing occurs anywhere in the body while $|\tau^\circ|$ is decreasing, unless $|\tau^\circ|$ has first attained the value φ_m.

Whatever shearing occurs after time t_m must result in an increase in $\bar{\gamma}$ beyond $\bar{\gamma}_m$ in some region of the body. If, as we shall suppose, there are, after time t_m,

---

[11] By examining the solutions of the equation $|\tau^\circ(t)| = \varphi(\bar{\gamma}) - c\bar{\gamma}_{xx}$ and employing an argument given by COLEMAN [1983, 1] in a different context, one can prove that the boundedness and continuity of $\bar{\gamma}$ *imply* that $\bar{\gamma}$ be independent of x for $t_0 \leqq t \leqq t_m$.

regions in the body where $\dot{\gamma} \neq 0$ and regions where $\dot{\gamma} = 0$, then $|\tau^\circ(t)|$ cannot increase beyond the value φ_m it attains at t_m, because as $\varphi_m \geqq \varphi(\bar{\gamma})$ for all $\bar{\gamma}$, (3.4) implies that, for the places where $\dot{\gamma}(x, t) = 0$, we have $|\tau^\circ(t)| \leqq \varphi(\bar{\gamma}(x, t)) \leqq \varphi_m$.

For $t > t_m$, let x^*, which will generally depend on t, *i.e.*, $x^* = x^*(t)$, be a boundary point between regions where $\dot{\gamma} = 0$ and $\dot{\gamma} \neq 0$. Let $\bar{\gamma}^*$ be the value of $\bar{\gamma}$ at (x^*, t), *i.e.*, $\bar{\gamma}^* = \bar{\gamma}^*(t) = \bar{\gamma}(x^*(t), t)$ and let $\bar{\gamma}^*_{xx} = \bar{\gamma}^*_{xx}(t)$ be the limit of $\bar{\gamma}_{xx}(x, t)$ as x approaches $x^*(t)$ from the side of $x^*(t)$ on which $\dot{\gamma}(x, t) \neq 0$. It follows from (3.4), (3.5), the continuity of the functions φ and $x \mapsto \bar{\gamma}(x, t)$, and the piecewise continuity of the function $x \mapsto \bar{\gamma}_{xx}(x, t)$ that

$$\varphi(\bar{\gamma}(x^*, t)) = |\tau^\circ(t)| \tag{3.9}$$

and

$$\bar{\gamma}^*_{xx} = 0. \tag{3.10}$$

Now, (3.5) tells us that at each time $t > t_m$, in the regions where $\dot{\gamma} \neq 0$, the function $x \mapsto \bar{\gamma}(x, t)$ obeys the ordinary differential equation $|\tau^\circ(t)| = \varphi(\bar{\gamma}) - c\bar{\gamma}_{xx}$, which, by (2.22), can be written

$$\delta = (\bar{\gamma} - \bar{\gamma}_m)^2 + \frac{c}{\alpha}\bar{\gamma}_{xx}, \tag{3.11}$$

where

$$\delta = \frac{\varphi_m - |\tau^\circ(t)|}{\alpha} = \delta(t) \tag{3.12}$$

is non-negative. With the definitions[12]

$$\eta = \frac{\bar{\gamma} - \bar{\gamma}_m}{\sqrt{\delta}}, \qquad \xi = \sqrt{\frac{2\alpha}{c}}\,\delta^{\frac{1}{4}}x, \tag{3.13}$$

(3.11) becomes

$$2\eta_{\xi\xi} = 1 - \eta^2, \tag{3.14}$$

which is equivalent to the following first-order system of equations for the pair (η, ψ) with $\psi = \eta_\xi$:

$$\begin{aligned} \eta_\xi &= \psi, \\ \psi_\xi &= \tfrac{1}{2}(1 - \eta^2). \end{aligned} \tag{3.15}$$

This system is of Hamiltonian type with

$$H = H(\eta, \psi) = \tfrac{1}{6}\eta^3 - \tfrac{1}{2}\eta + \tfrac{1}{2}\psi^2 \tag{3.16}$$

---

[12] The transformation (3.13) and our method of solving the reduced equation (3.14) were suggested to us by T. W. WRIGHT, who, in the course of research with COLEMAN on related approaches to shear band formation, expressed the general solution (3.23) of (3.14) in terms of elliptic integrals and derived the present equations (3.24) [valid for $-1/3 < H < 1/3$] and (3.27) [the homoclinic orbit corresponding to $H = 1/3$, which is familiar in soliton theory].

constant on each of its solutions. The phase-plane portrait for the system (3.15) is shown in Figure 2. Each curve there was obtained by setting H equal to a constant and solving (3.16) for ψ; thus,

$$\psi = \eta_\xi = \pm\sqrt{-\tfrac{1}{3}P(\eta; H)}\,, \tag{3.17}$$

with $P(\cdot; H)$ the cubic polynomial,

$$P(\eta; H) = \eta^3 - 3\eta - 6H. \tag{3.18}$$

We are interested in solutions $\bar{\gamma}$ of (3.11) that attain their maximum values at a preassigned value of x which, for convenience, we take to be $x = 0$. Such functions $\bar{\gamma}$ correspond to solutions (η, ψ) of (3.14) for which

$$\eta(0) = \eta_0, \qquad \psi(0) = 0, \tag{3.19}$$

where η_0, the maximum value of $\eta(\xi)$, depends on H; when we wish to emphasize this dependence we write

$$\eta_0 = \eta_0[H]. \tag{3.20}$$

The relations $(3.19)_2$ and (3.17) imply that $\eta_0[H]$ is the largest real root of the cubic equation

$$P(\eta; H) = 0. \tag{3.21}$$

As $\bar{\gamma}$ never decreases, at each time after t_m the maximum value of $\bar{\gamma}(x, t)$ must exceed $\bar{\gamma}_m$, which means that $\eta_0[H]$ must be positive. For $H < -1/3$ there are

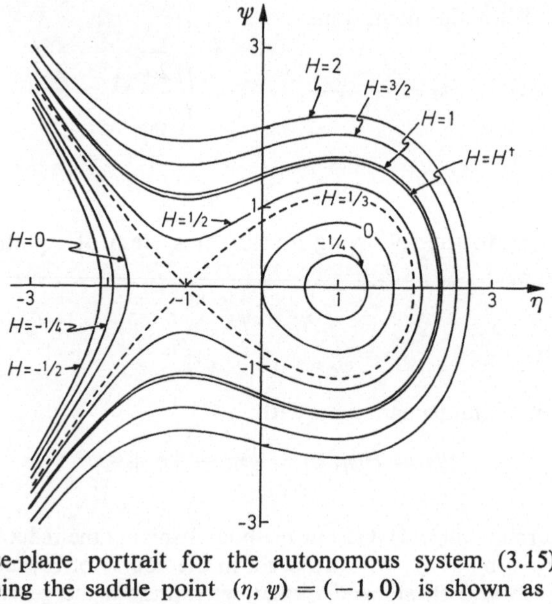

Fig. 2. The phase-plane portrait for the autonomous system (3.15). The invariant manifold containing the saddle point $(\eta, \psi) = (-1, 0)$ is shown as a dashed curve; it corresponds to $H = 1/3$; the homoclinic orbit is the portion of that manifold for which $-1 < \eta \leqq 2$. The orbit with $H = H^\dagger$ will play a special role in the theory. To six figures, the value of $H^\dagger$ is 0.930496.

no branches in the plane with $\eta_0[H] > 0$; but for each $H > -1/3$ there is precisely one curve in the phase plane on which η attains positive values. Thus for each choice of $H > -1/3$ the system (3.15) has a unique solution with

$$\psi(0) = 0 \quad \text{and} \quad \eta(0) = \eta_0[H] > 0; \tag{3.22}$$

this solution,

$$\xi = \pm\sqrt{3} \int_{\eta}^{\eta_0[H]} \frac{d\nu}{\sqrt{-(\nu^3 - 3\nu - 6H)}}, \tag{3.23}$$

can be expressed in terms of Jacobian elliptic functions.[13]

When $-1/3 < H < 1/3$ the cubic (3.21) has, in addition to η_0, two other real roots, η_1 and η_2, with $\eta_0 > \eta_1 > \eta_2$, and (3.23) can be written

$$\xi = \pm\sqrt{3} \int_{\eta}^{\eta_0} \frac{d\nu}{\sqrt{-(\nu - \eta_0)(\nu - \eta_1)(\nu - \eta_2)}}, \tag{3.24}$$

which yields

$$\eta = \eta_1 + (\eta_0 - \eta_1)\,\mathrm{cn}^2\left(\frac{\lambda}{\sqrt{3}}\xi \,\middle|\, m\right) \tag{3.25}$$

with

$$\lambda = \tfrac{1}{2}(\eta_0 - \eta_2)^{\frac{1}{2}}, \qquad m = \frac{\eta_0 - \eta_1}{\eta_0 - \eta_2}. \tag{3.26}$$

When $H = 1/3$ the roots of (3.21) are $\eta_0 = 2$ and $\eta_1 = \eta_2 = -1$, and the solution of (3.15) obeying (3.22) is the homoclinic orbit

$$\eta = 3\,\mathrm{sech}^2 \frac{\xi}{2} - 1. \tag{3.27}$$

In the present case (3.26) gives $\lambda = \frac{1}{2}\sqrt{3}$ and $m = 1$, and as $\mathrm{cn}(\zeta \mid 1) = \mathrm{sech}\,\zeta$, (3.25) yields (3.27) in the limit as H approaches $1/3$ from below.

When $H > 1/3$, the cubic (3.21) has precisely one real root η_0, and (3.23) yields

$$\eta = \eta_0 - \lambda^2 \frac{1 - \mathrm{cn}\left(\frac{\lambda}{\sqrt{3}}\xi \,\middle|\, m\right)}{1 + \mathrm{cn}\left(\frac{\lambda}{\sqrt{3}}\xi \,\middle|\, m\right)} \tag{3.28}$$

with

$$\lambda = 3^{\frac{1}{4}}(\eta_0^2 - 1)^{\frac{1}{4}}, \qquad m = \tfrac{1}{2} + \tfrac{3}{4}\frac{\eta_0}{\lambda^2}. \tag{3.29}$$

---

[13] We use below what appears to be the standard notation for elliptic functions and integrals; see, *e.g.*, Milne-Thomson [1950, 1], [1964, 1].

The solution (3.28) is unbounded and periodic with period[14] $4\sqrt{3}K(m)/\lambda$ and can be written

$$\eta = \eta_0 - 3(\eta_0^2 - 1)\left[2^{\frac{2}{3}}\mathscr{P}\left(\frac{\xi}{2^{\frac{2}{3}}\sqrt{3}}\right) + \eta_0\right]^{-1}, \tag{3.30}$$

where $\mathscr{P}$ is the Weierstrass elliptic function defined by

$$\int_{\mathscr{P}(\zeta)}^{\infty} \frac{dv}{\sqrt{4v^3 - 3 \cdot 2^{\frac{2}{3}}v + 6H}} = \zeta. \tag{3.31}$$

To complement the phase-plane portrait of Figure 2, we show in Figure 3 the graph of the homoclinic orbit (3.27) along with graphs of a bounded periodic solution (3.25) and an unbounded periodic solution (3.28).

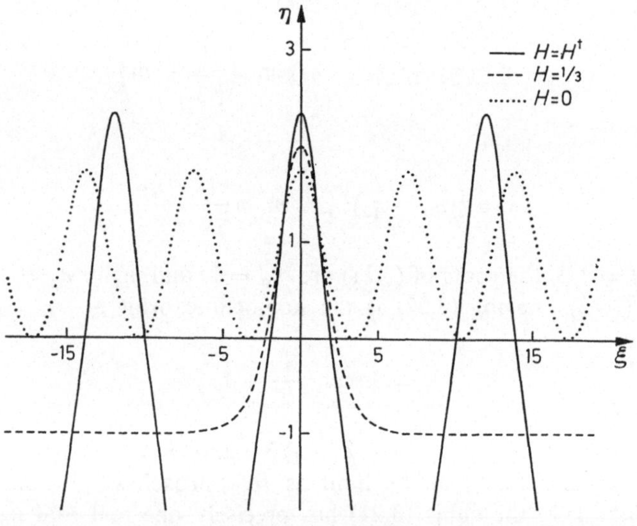

Fig. 3. Representative solutions $\xi \mapsto \eta(\xi)$ of (3.15) with maxima at $\xi = 0$. Each solution corresponds to a distinct value of the Hamiltonian H: ···· a bounded periodic solution of the form (3.25); – – – – the homoclinic orbit (3.27); ——— an unbounded periodic solution of the form (3.28).

We have been seeking solutions $\bar{\gamma}$ of (3.11) that attain their maxima as $x = 0$ and for which there are intervals of the form $-h < x < h$ on which $\bar{\gamma}$ exceeds $\bar{\gamma}_m$. It follows from (3.13) and the solutions given in equations (3.25)–(3.29) for

[14] $K(m)$ is the complete elliptic integral of the first kind with parameter m, *i.e.* with modulus $k = \sqrt{m}$. See, *e.g.*, [1964, 1].

the system (3.15) that the sought solutions of (3.11) are:

$$\bar{\gamma}(x,t) = \hat{\gamma}(x,\delta,H)$$

$$= \begin{cases} \bar{\gamma}_m + \sqrt{\delta}\left[\eta_1 + (\eta_0-\eta_1)\,\mathrm{cn}^2\left(\frac{\lambda}{\sqrt{3}}\delta^{\frac{1}{4}}\sqrt{\frac{2\alpha}{c}}x\,\middle|\,m\right)\right] & \text{for} \quad -\frac{1}{3} < H < \frac{1}{3}, \\ \bar{\gamma}_m + \sqrt{\delta}\left[3\,\mathrm{sech}^2\left(\frac{\delta^{\frac{1}{4}}}{2}\sqrt{\frac{2\alpha}{c}}x\right) - 1\right] & \text{for} \quad H = \frac{1}{3}, \\ \bar{\gamma}_m + \sqrt{\delta}\left[\eta_0 - \lambda^2\frac{1-\mathrm{cn}\left(\frac{\lambda}{\sqrt{3}}\delta^{\frac{1}{4}}\sqrt{\frac{2\alpha}{c}}x\,\middle|\,m\right)}{1+\mathrm{cn}\left(\frac{\lambda}{\sqrt{3}}\delta^{\frac{1}{4}}\sqrt{\frac{2\alpha}{c}}x\,\middle|\,m\right)}\right] & \text{for} \quad H > \frac{1}{3}. \end{cases} \tag{3.32}$$

Here $\delta = (\varphi_m - |\tau^\circ|)/\alpha = \delta(t)$; $\eta_0 > \eta_1 > \eta_2$ are the real roots of $\eta^3 - 3\eta - 6H = 0$ for $-1/3 < H < 1/3$; η_0 is the unique real root of the same cubic equation for $H > 1/3$;

$$\lambda = \begin{cases} \frac{1}{2}(\eta_0 - \eta_2)^{\frac{1}{2}} & \text{for} \quad -1/3 < H < 1/3, \\ 3^{\frac{1}{4}}(\eta_0^2 - 1)^{\frac{1}{2}} & \text{for} \quad H > 1/3; \end{cases} \tag{3.33}$$

and

$$m = \begin{cases} (\eta_0 - \eta_1)/(\eta_0 - \eta_2) & \text{for} \quad -1/3 < H < 1/3, \\ \frac{1}{2} + 3^{\frac{1}{2}}\eta_0(\eta_0^2 - 1)^{-\frac{1}{2}}/4 & \text{for} \quad H > 1/3. \end{cases} \tag{3.34}$$

To see if our theory is compatible with the formation of shear bands, we now suppose that $d|\tau^\circ(t)|/dt$ is (strictly) negative for $t \geqq t_m$, and we attempt to use the results just stated to construct a function $\bar{\gamma}$ on $(-\infty, \infty) \times [t_m, t_F]$ that, in addition to obeying (3.4) and (3.5) and having the smoothness assumed early in this section, meets the following conditions: (i) $\bar{\gamma}(x, t_m) = \bar{\gamma}_m$ for all x; (ii) for each t in (t_m, t_F), $\bar{\gamma}(x,t)$ attains its maximum at $x = 0$; (iii) for each t in (t_m, t_F), the set of values of x where $\bar{\gamma}_t(x,t) \neq 0$ is an interval of the form $-x^*(t) < x < x^*(t)$ with $0 < x^*(t) < \infty$. For x in the interval $(-x^*(t), x^*(t))$ the function $\bar{\gamma}$ must be of the form $\bar{\gamma}(x,t) = \hat{\gamma}(x, \delta(t), H)$, with $\hat{\gamma}$ as in (3.32) and $\delta(t) = (\varphi_m - |\tau^\circ(t)|)/\alpha$. We here have $\dot{\delta}(t) > 0$ and $\lim_{t \to t_m^+} \delta(t) = 0$. We seek an appropriate value of H, independent of t.[15]

The monotonicity of $\bar{\gamma}$ in t implies that, for $t > t_m$,

$$\bar{\gamma}(x,t) = \sup_{t_m \leqq \tilde{t} \leqq t} \bar{\gamma}(x, \tilde{t}) \quad \text{for all} \quad x, \tag{3.35}$$

---

[15] At this stage in our construction it is not obvious that there is an *H independent of t* that gives $\bar{\gamma}$ the desired properties on $(-\infty, \infty) \times [t_m, t_F)$. Fortunately, we are able to find such an H.

and hence, when x^* is a non-increasing function of t and H is independent of t,

$$\bar{\gamma}(x, t) = \sup_{0 \leqq \tilde{\delta} \leqq \delta(t)} \hat{\gamma}(x, \tilde{\delta}, H) \quad \text{for} \quad -x^*(t) \leqq x \leqq x^*(t). \tag{3.36}$$

It is not difficult to show that for each fixed $H > -1/3$ there is a positive function $\delta \mapsto x^\#(\delta)$ with $dx^\#(\delta)/d\delta < 0$ such that

$$\hat{\gamma}_\delta(x, \delta, H) \begin{cases} > 0 & \text{for} \quad -x^\#(\delta) < x < x^\#(\delta), \\ = 0 & \text{for} \quad x = \pm x^\#(\delta), \\ < 0 & \text{for} \quad x^\#(\delta) < |x| < x^\#(\delta) + v, \end{cases} \tag{3.37}$$

where $\hat{\gamma}_\delta = \partial\hat{\gamma}/\partial\delta$ and $v = v(\delta) > 0$. As $dx^\#/d\delta > 0$, we have $\hat{\gamma}_{\tilde{\delta}}(x, \tilde{\delta}, H) > 0$ for $0 < \tilde{\delta} < \delta$ and $-x^\#(\delta) \leqq x \leqq x^\#(\delta)$; in particular,

$$\sup_{0 \leqq \tilde{\delta} < \delta} \hat{\gamma}(x, \tilde{\delta}, H) = \hat{\gamma}(x, \delta, H) \quad \text{for} \quad -x^\#(\delta) \leqq x \leqq x^\#(\delta). \tag{3.38}$$

As we are assuming, for $t > t_m$, that $d|\tau^\circ(t)|/dt > 0$, we here have $\dot{\delta}(t) > 0$. When $\bar{\gamma}(x, t)$ is as in (3.32), $\bar{\gamma}_t(x, t) = \hat{\gamma}_\delta(x, \delta(t), H)\,\dot{\delta}(t)$, and (3.37) tells us that $\bar{\gamma}_t(x, t)$ is positive if $|x| < x^\#(\delta(t))$ and negative if $x^\#(\delta(t)) < |x| < x^\#(\delta(t)) + v$. Of course, the fact that (3.32) and $|x| < x^\#$ together yield $\bar{\gamma}_t(x, t) > 0$ is compatible with our derivation of (3.32) for an interval about $x = 0$ in which $\bar{\gamma}_t \neq 0$. However, as $\bar{\gamma}_t$ is never negative, the observation that (3.32) and $x^\# < |x| < x^\# + v$ yield $\bar{\gamma}_t(x, t) < 0$ tells us that (3.32) cannot hold where $x^\# < |x| < x^\# + v$. Therefore, $x^\#$ is the largest value of x for which (3.32) can hold. We shall attempt to choose H so that

$$x^\#(\delta(t)) = x^*(t), \tag{3.39}$$

where $(-x^*(t), x^*(t))$ is the interval of x-values for which $\bar{\gamma}_t(x, t) \neq 0$. As (3.32) holds on this interval, (3.10), (3.37), and (3.39) imply that

$$\hat{\gamma}_{xx}(x^*(t), \delta(t), H) = \hat{\gamma}_\delta(x^*(t), \delta(t), H) = 0 \tag{3.40}$$

for each $t > t_m$.

Up to this point we have said no more about the number H than that it exceeds $-1/3$ and that we intend to give it a value independent of t. The question arises as to whether this is possible. The answer is yes; in fact, there is a unique value $H^\dagger$ of H for which the positive-valued function $\delta \mapsto x^\#(\delta)$ obeying (3.37) with $H = H^\dagger$ also obeys $\hat{\gamma}_{xx}(x^\#(\delta), \delta, H^\dagger) = 0$. To six figures,

$$H^\dagger = 0.930496, \tag{3.41}$$

whatever the values of the material parameters α, c, $\bar{\gamma}_m$, and φ_m. To prove the claimed existence and uniqueness of $H^\dagger$, for each H and δ we let x^* be the smallest positive solution of the equation $\hat{\gamma}_{xx}(x^*, \delta, H) = 0$ and then seek a number $H^\dagger > -1/3$ for which $\hat{\gamma}_\delta(x^*, \delta, H) = 0$ when $H = H^\dagger$, independently of the choice of δ. By (3.32) and the formula $\frac{\partial}{\partial \zeta} \operatorname{cn}(\zeta \mid m) = -\operatorname{sn}(\zeta \mid m) \operatorname{dn}(\zeta \mid m)$, we have

$$2\sqrt{\delta}\,\hat{\gamma}_\delta(x, \delta, H) =$$

$$\begin{cases} \eta - (\eta_0 - \eta_1)\dfrac{\lambda\xi}{\sqrt{3}} \operatorname{cn}\left(\dfrac{\lambda\xi}{\sqrt{3}}\,\middle|\, m\right) \operatorname{sn}\left(\dfrac{\lambda\xi}{\sqrt{3}}\,\middle|\, m\right) \operatorname{dn}\left(\dfrac{\lambda\xi}{\sqrt{3}}\,\middle|\, m\right) & \text{for } -1/3 < H < 1/3, \\[2ex] \eta - \dfrac{3\xi}{2} \operatorname{sech}^2\left(\dfrac{\xi}{2}\right) \tanh\left(\dfrac{\xi}{2}\right) & \text{for } H = 1/3, \\[2ex] \eta - \dfrac{\lambda^3\xi}{\sqrt{3}} \dfrac{\operatorname{sn}\left(\dfrac{\lambda\xi}{\sqrt{3}}\,\middle|\, m\right) \operatorname{dn}\left(\dfrac{\lambda\xi}{\sqrt{3}}\,\middle|\, m\right)}{\left[1 + \operatorname{cn}\left(\dfrac{\lambda\xi}{\sqrt{3}}\,\middle|\, m\right)\right]^2} & \text{for } H > 1/3, \end{cases} \tag{3.42}$$

where η and ξ are as in (3.13), the numbers η_i are as explained earlier in this section, λ is as in (3.33), and m is as in (3.34). Now, let $\xi^* = \xi(x^*) = \sqrt{\frac{2\alpha}{c}}\,\delta^{\frac{1}{4}} x^*$, and let $\eta^* = \eta(\xi^*)$, *i.e.*, $\eta^* = \delta^{-\frac{1}{2}}[\hat{\gamma}(x^*, \delta, H) - \gamma_m]$. As $\hat{\gamma}_{xx} = 0$ at x^*, we have $\eta_{\xi\xi} = 0$ at ξ^*, and, as η^* is positive, (3.14) yields $\eta^* = 1$, *i.e.*,

$$\hat{\gamma}(x^*, \delta, H) = \bar{\gamma}_m + \sqrt{\delta}\,. \tag{3.43}$$

By evaluating (3.42) and (3.23) at the point $x = x^*$ where $\xi = \xi^*$ and $\eta = \eta^* = 1$, we find that

$$2\sqrt{\delta}\,\hat{\gamma}_\delta(x^*, \delta, H) =$$

$$\begin{cases} 1 - (\eta_0 - \eta_1)\dfrac{\lambda\xi^*}{\sqrt{3}} \operatorname{cn}\left(\dfrac{\lambda\xi^*}{\sqrt{3}}\,\middle|\, m\right) \operatorname{dn}\left(\dfrac{\lambda\xi^*}{\sqrt{3}}\,\middle|\, m\right) & \text{for } -1/3 < H < 1/3, \\[2ex] 1 - \dfrac{2}{\sqrt{3}} \cosh^{-1}\sqrt{\dfrac{3}{2}} & \text{for } H = 1/3, \\[2ex] 1 - \dfrac{\lambda^3\xi^*}{\sqrt{3}} \dfrac{\operatorname{sn}\left(\dfrac{\lambda\xi^*}{\sqrt{3}}\,\middle|\, m\right) \operatorname{dn}\left(\dfrac{\lambda\xi^*}{\sqrt{3}}\,\middle|\, m\right)}{\left[1 + \operatorname{cn}\left(\dfrac{\lambda\xi^*}{\sqrt{3}}\,\middle|\, m\right)\right]^2} & \text{for } H > 1/3, \end{cases} \tag{3.44}$$

and

$$\xi^* = \sqrt{3} \int_1^{\eta_0[H]} \frac{dv}{\sqrt{6H + 3v - v^3}} = \frac{\sqrt{3}}{\lambda} F(\Phi \mid m), \tag{3.45}$$

where λ is as in (3.33), m is as in (3.34), and

$$\cos\Phi = \begin{cases} (1-\eta_1)^{\frac{1}{2}}/(\eta_0-\eta_1)^{\frac{1}{2}} & \text{for} \quad -1/3 < H \leqq 1/3 \\ (\lambda^2-\eta_0+1)/(\lambda^2+\eta_0-1) & \text{for} \quad H > 1/3. \end{cases} \tag{3.46}$$

In (3.45) F, the elliptic integral of the first kind, is related to the elliptic function cn by

$$F(\Phi \mid m) = \mathrm{cn}^{-1}(\cos\Phi \mid m). \tag{3.47}$$

When $H = 1/3$, (3.45) reduces to $\xi^* = 2\cosh^{-1}\sqrt{3/2}$. It is clear from (3.44)–(3.46) that the function $H \mapsto 2\sqrt{\delta}\,\hat{\gamma}_\delta(x^*, \delta, H)$ is independent of δ and the material parameters α, c, γ_m, and φ_m. It is not difficult to show that for all $H > -1/3$ this function is strictly monotone decreasing, and hence there is at most one $H^\dagger$ with $\hat{\gamma}_\delta(x^*, \delta, H^\dagger) = 0$. The graph seen in Figure 4 was calculated using (3.44)–(3.46) and standard methods of evaluating elliptic functions and integrals.[16] It shows that $\hat{\gamma}_\delta(x^*, \delta, H^\dagger) = 0$ does have a root, and it occurs with $H^\dagger > -1/3$. A careful numerical calculation gives us the value 2.324236 for $\eta_0[H^\dagger]$, *i.e.*, for the value of η_0 that makes the last line on the right in (3.44) vanish when ξ^* is as in (3.45); this value of $\eta_0[H^\dagger]$ yields, by (3.16), (3.19), and (3.45)–(3.47), 0.930496 for $H^\dagger$ and 1.25797 for ξ^*. As $x^* = \delta^{-\frac{1}{4}}\xi^*\sqrt{c/(2\alpha)}$, for each $t > t_m$ we have

$$x^* = 1.25797\sqrt{\frac{c}{2\alpha}}\,\delta^{-\frac{1}{4}}. \tag{3.48}$$

When x^* is chosen in this way, (3.42) yields not only $\hat{\gamma}_\delta(x^*, \delta, H^\dagger) = 0$, but also $\hat{\gamma}_\delta(x, \delta, H^\dagger) < 0$ when $|x|$ is less than x^*, and $\hat{\gamma}_\delta(x, \delta, H^\dagger) > 0$ when $|x|$ is greater than but near to $|x^*|$. Thus the dependence of x^* on δ, *i.e.*, the function $x^\#$ defined as $\delta \mapsto x^\#(\delta) = x^*$ with x^* in (3.48), is the function we have been seeking: that for which (3.37), with $H = H^\dagger$, holds *and* $\hat{\gamma}_{xx}(x^\#(\delta), \delta, H^\dagger) = 0$. This proves, among other things, the existence and uniqueness of $H^\dagger$.

Let us now turn to the function $t \mapsto x^*(t)$ generated by (3.48):

$$x^*(t) = 1.25797\sqrt{\frac{c}{2\alpha}}\,\delta(t)^{-\frac{1}{4}} = 0.889518\,c^{\frac{1}{2}}\alpha^{-\frac{1}{4}}\,[\varphi_m - |\tau^\circ(t)|]^{-\frac{1}{4}}. \tag{3.49}$$

As $\hat{\gamma}_{xx}(0, \delta, H^\dagger) < 0$, and $x^*(t)$ is the *smallest* positive number for which $\hat{\gamma}_{xx}(x^*, \delta(t), H^\dagger) = 0$, we have

$$\hat{\gamma}_{xx}(x, \delta(t), H^\dagger) < 0 \quad \text{for} \quad -x^*(t) < x < x^*(t). \tag{3.50}$$

Hence, if we put

$$\bar{\gamma}(x,t) = \hat{\gamma}(x, \delta(t), H^\dagger) = \bar{\gamma}_m + \sqrt{\delta(t)}\left[\eta_0 - \lambda^2\,\frac{1 - \mathrm{cn}\left(\dfrac{\lambda}{\sqrt{3}}\,\delta(t)^{\frac{1}{4}}\sqrt{\dfrac{2\alpha}{c}}\,x \,\Big|\, m\right)}{1 + \mathrm{cn}\left(\dfrac{\lambda}{\sqrt{3}}\,\delta(t)^{\frac{1}{4}}\sqrt{\dfrac{2\alpha}{c}}\,x \,\Big|\, m\right)}\right]$$

$$\text{for} \quad -x^*(t) \leqq x \leqq x^*(t), \tag{3.51a}$$

[16] *E.g.*, [1964, 1].

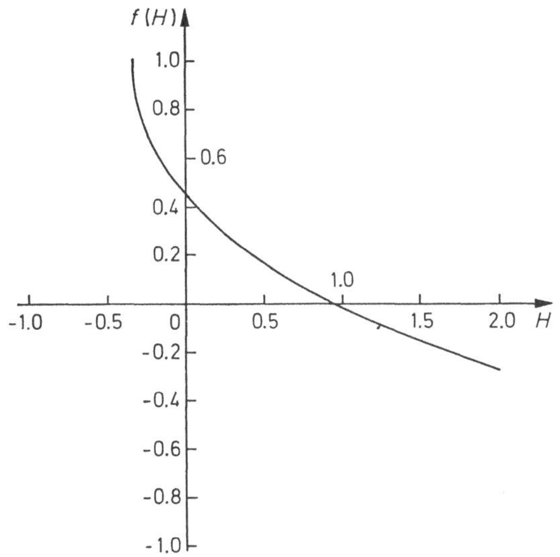

Fig. 4. Graph of $f(H) = 2\sqrt{\delta}\,\hat{\gamma}_\delta(x^*, \delta, H)$ *versus* H. The ordinate is given by (3.44), with λ, m, and ξ^* determined by H through the equations (3.33), (3.34), and (3.45). The graph crosses the abscissa at $H = H^\dagger$.

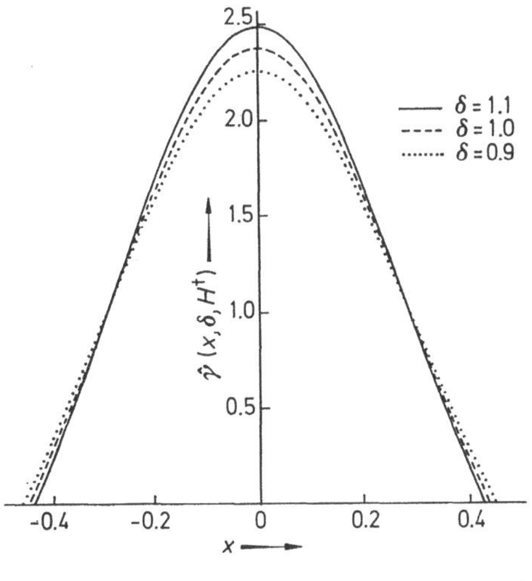

Fig. 5. Illustration of the phenomenon behind (3.37). Here H is set at a fixed value, and δ is varied slightly. The way the curves intersect makes it clear that there is a positive number $x^\# = x^\#(\delta)$ such that $\hat{\gamma}(x, \delta, H)$ increases with δ for $|x|$ less than $x^\#(\delta)$ and decreases with δ for $|x|$ greater than but near to $x^\#(\delta)$. For the case shown, $H = H^\dagger$. The graphs are approximately straight in the region in which they cross, because $\hat{\gamma}_{xx}(x^\#(\delta), \delta, H^\dagger) = 0$.

with

$$\delta(t) = \frac{\varphi_m - |\tau^\circ(t)|}{\alpha}, \qquad \eta_0 = \eta_0[H^\dagger] = 2.324236,$$
$$\lambda = \lambda[H^\dagger] = 1.90632, \qquad m = m[H^\dagger] = 0.979681, \tag{3.51 b}$$

then, as $\hat{\gamma}$ is a solution of the differential equation (3.11), we have for $-x^*(t) < x < x^*(t)$,

$$|\tau^\circ(t)| = \varphi(\bar{\gamma}(x, t)) - c\bar{\gamma}_{xx}(x, t), \tag{3.52 a}$$

and

$$\varphi(\bar{\gamma}(x, t)) < |\tau^\circ(t)|, \tag{3.52 b}$$

which implies that (3.4) and (3.5) are satisfied with $\bar{\gamma}_t(x, t) > 0$ for $-x^*(t) < x < x^*(t)$.

Equation (3.51 a) gives us our solution $\bar{\gamma}(x, t)$ to (3.4) and (3.5) for (x, t) in $[-x^*(t), x^*(t)] \times (t_m, t_F)$. As $x^*(t)$ here obeys (3.39), the relation (3.38) tells us that (3.51 a) does obey (3.36) (with $H = H^\dagger$).

The formula (3.49) for $x^*(t)$ implies that when $|\tau^\circ(t)|$ is very close to φ_m, *i.e.*, when $\delta(t)$ is very small, the band about $x = 0$ in which $\dot{\gamma} \neq 0$ is a wide band; in fact, in the limit $t \mapsto t_m^+$ this band extends from $x = -\infty$ to $x = +\infty$. As $|\tau^\circ(t)|$ drops below φ_m, the band of active shearing narrows rapidly, but as $|\tau^\circ(t)|$ continues to drop the width of the band, $2x^*$, eventually diminishes slowly, for x^* varies with τ° as $[\varphi_m - \tau^\circ]^{-\frac{1}{4}}$. [An example of a graph of $x^*(t)$ *versus* $|\tau^\circ(t)|/\varphi_m$ is seen in Figure 6; there, $\bar{\gamma}_m = 0.05$, $\alpha = 1$, $\varphi_m = 10$, and $c = 0.1$.] While the shear band is evolving (*i.e.*, for $t > t_m$), the accumulated shear at the center of the band, $\bar{\gamma}(0, t)$, is increasing with time as $[\varphi_m - |\tau^\circ(t)|]^{\frac{1}{2}}$. Indeed, by $(3.13)_1$,

$$\bar{\gamma}(0, t) - \bar{\gamma}_m = \eta_0[H^\dagger]\, \delta(t)^{\frac{1}{2}} = 2.32424\alpha^{-\frac{1}{2}}[\varphi_m - |\tau^\circ(t)|]^{\frac{1}{2}}. \tag{3.53}$$

As $\bar{\gamma}(0, t)$ is the maximum value attained in the body by $\bar{\gamma}$ at time t, it follows from (2.25) and (3.53) that at the time t_F of failure, *i.e.*, at the time at which the maximum value of $\bar{\gamma}(x, t)$ is equal to $\bar{\gamma}_F$, the absolute value of the shearing stress $\tau^\circ(t)$ is

$$|\tau^\circ(t_F)| = (1 - \eta_0^{-2})\, \varphi_m = 0.8149\varphi_m. \tag{3.54}$$

Equation (3.49) gives us $x^*(t)$, and (3.51) gives us $\bar{\gamma}(x, t)$ for $|x| \leqq x^*(t)$. We now seek $\bar{\gamma}(x, t)$ for $|x| > x^*(t)$.

For each t in (t_m, t_F) and each x with $|x| > x^*(t)$, there is a time $\tilde{t} = \tilde{t}(|x|)$ in (t_m, t) for which $|x| = x^*(\tilde{t})$. At that time $\tilde{t}$,

$$\bar{\gamma}_t(\zeta, \tilde{t}) = \begin{cases} > 0 & \text{for } |\zeta| < x, \\ = 0 & \text{for } |\zeta| > x; \end{cases} \tag{3.55}$$

in view of the results obtained above, $\bar{\gamma}(\zeta, t) = \hat{\gamma}(\zeta, \delta(\tilde{t}), H^\dagger)$, for each ζ with $|\zeta| \leqq x$ and, in particular, for $\zeta = x$. Moreover, for each $\tilde{\tilde{t}}$ in $(\tilde{t}, t)$, we have $|x| > x^*(\tilde{\tilde{t}})$ and hence $\bar{\gamma}_t(x, \tilde{\tilde{t}}) = 0$, which yields

$$\bar{\gamma}(x, t) = \bar{\gamma}(x, \tilde{t}(x)) \quad \text{for } |x| > x^*(t). \tag{3.56}$$

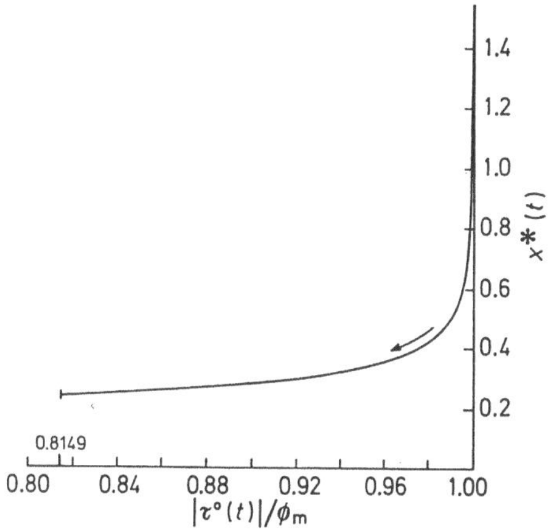

Fig. 6. The half-width, $x^*(t)$, of the band of active shearing [according to (3.49)] *versus* $|\tau^\circ(t)|/\varphi_m$, as t increases from t_m to t_F, *i.e.*, as $|\tau^\circ(t)|/\varphi_m$ decreases from 1 to $1-\eta_0^{-2}=0.8149$. The arrow shows the sense in which the curve is traversed as t increases.

Therefore,

$$\bar{\gamma}(x,t)=\hat{\gamma}(x,\tilde{\delta}(|x|),H^\dagger)\quad\text{for}\quad |x|>x^*(t),\tag{3.57}$$

where $\hat{\gamma}$ is as in (3.51), and $\tilde{\delta}(|x|)=\delta(\tilde{t}(|x|))$ is the value of δ at the instant $\tilde{t}$ when $|x|=x^*(\tilde{t})$. By (3.49) with t replaced by $\tilde{t}$, $x^*(\tilde{t})=\xi^*c^{\frac{1}{2}}(2\alpha)^{-\frac{1}{2}}\,\delta(\tilde{t})^{-\frac{1}{4}}$, and hence

$$\tilde{\delta}(|x|)=\frac{c^2\xi^{*4}}{4\alpha^2|x|^4}=\frac{0.62606c^2}{\alpha^2|x|^4}.\tag{3.58}$$

As $\tilde{\delta}(|x|)$ does not change in time once $x^*(t)$ has decreased below $|x|$, (3.57) yields $\bar{\gamma}_t(x,t)=0$. To show that (3.57) is compatible with (3.4), we must show that (3.57) yields $\tau^\circ(t)\leqq\varphi(\bar{\gamma}(x,t))$. But such is the case, because, by the definition of $\tilde{t}=\tilde{t}(|x|)$: $\bar{\gamma}_{xx}(x,\tilde{t})=0$, (3.56) holds, $|\tau^\circ(t)|<|\tau^\circ(\tilde{t})|$, and (3.52a) holds at x with t replaced by $\tilde{t}$, and, therefore,

$$|\tau^\circ(t)|<|\tau^\circ(\tilde{t})|=\varphi(\bar{\gamma}(x,\tilde{t}))-c\bar{\gamma}_{xx}(x,\tilde{t})=\varphi(\bar{\gamma}(x,t))\qquad\text{for}\qquad |x|>x^*(t).$$

Now, at the instant $\tilde{t}$ at which $|x|=x^*$ and $\delta(\tilde{t})=\tilde{\delta}(|x|)$, we have $\bar{\gamma}_{xx}(x,\tilde{t})=0$, and, by (3.11), $\delta(\tilde{t})^{\frac{1}{2}}=\bar{\gamma}(x,\tilde{t})-\gamma_m$. Thus, (3.57) is equivalent to the following remarkably simple expression for $\bar{\gamma}$ in the "rigid region" where $\bar{\gamma}_t=0$:

$$\bar{\gamma}(x,t)=\bar{\gamma}_m+\tilde{\delta}(|x|)^{\frac{1}{2}}=\bar{\gamma}_m+\frac{c\xi^{*2}}{2\alpha|x|^2}=\bar{\gamma}_m+\frac{0.79124c}{\alpha|x|^2},$$

$$\text{for}\quad x\geqq x^*(t)\quad\text{and for}\quad -x\leqq -x^*(t).\tag{3.59}$$

In summary: The equations (3.51) and (3.59), with x^* given by (3.49) and $\tilde{\delta}$ by (3.58), determine $\bar{\gamma}(x,t)$ for $t>t_m$ in such a way that $\bar{\gamma}_t(x,t)>0$ for

$|x| < x^*(t)$, $\bar{\gamma}_t(x, t) = 0$ for $|x| \geqq x^*(t)$, and (3.4) and (3.5) hold for all x in $(-\infty, \infty)$ and all t in (t_m, t_F). If we put

$$\hat{\delta}(x, t) = \begin{cases} \delta(t), & \text{for} \quad |x| \leqq x^*(t), \\ \tilde{\delta}(|x|), & \text{for} \quad |x| > x^*(t), \end{cases} \tag{3.60}$$

then we can write, for each t in $(t_m, t_F]$,

$$\bar{\gamma}(x, t) = \hat{\gamma}(x, \hat{\delta}(x, t), H^\dagger), \quad \text{for} \quad -\infty < x < \infty; \tag{3.61}$$

of course, the right-hand side of this equation is given by (3.51a) with $\delta(t)$ replaced by $\hat{\delta}(x, t)$; when $|x|$ exceeds $x^*(t)$, the equation reduces to (3.59).

In Figure 7 we present graphs $\bar{\gamma}(x, t)$ *versus* x, based on (3.51) and (3.59), *i.e.*, on (3.61), for times $t > t_m$ at which $|\tau^\circ(t)|/\varphi_m = 0.99$, 0.90, and $0.8149 \simeq |\tau^\circ(t_F)|/\varphi_m$. For these calculations, as for those of Figures 5, 6, 8, and 10–13, the values of the material parameters are taken to be $\bar{\gamma}_m = 0.05$, $\alpha = 1$, $\varphi_m = 10$, and $c = 0.1$.[17] A corresponding graph of $\hat{\delta}(x, t)$ *versus* x, for the time t at which $\tau^\circ(t)/\varphi_m = 0.90$, is shown in Figure 8.

If we assume that the velocity is zero in the plane $x \equiv 0$, then by (2.2) and $(2.3)_1$,

$$u(x, t) = \int_0^x \gamma(\zeta, t)\, d\zeta. \tag{3.62}$$

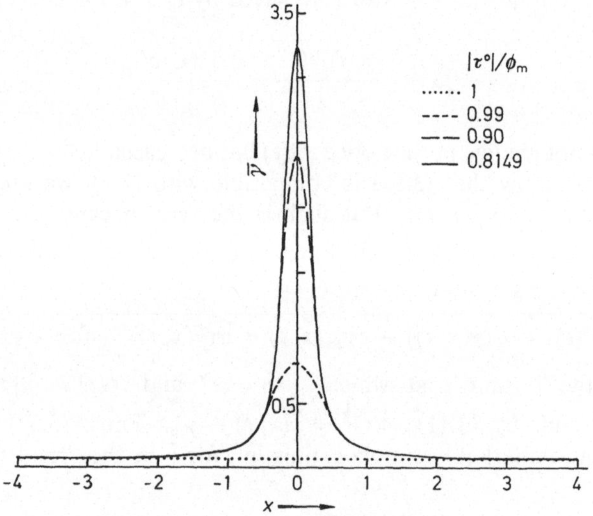

Fig. 7. The accumulated shear strain $\bar{\gamma}$ as a function of x at various times $t \geqq t_m$: $\cdots\cdot$ $t = t_m$, *i.e.*, $|\tau^\circ(t)|/\varphi_m = 1$; ---- $|\tau^\circ(t)|/\varphi_m = 0.99$; — — — $|\tau^\circ(t)|/\varphi_m = 0.90$; —— $t = t_F$, *i.e.*, $|\tau^\circ(t)|/\varphi_m = 0.8149$.

[17] Figure 10 contains graphs of stress *versus* nominal strain for a case in which $c = 0.1$ and also cases in which $c = 0.01$ and 1.0.

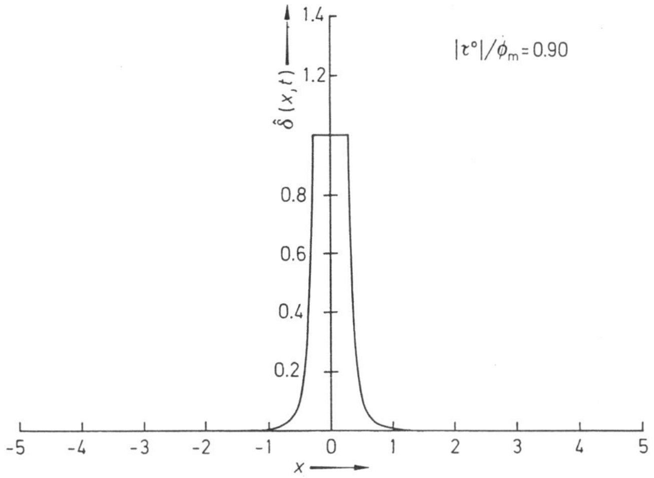

Fig. 8. $\hat{\delta}(x, t)$ as a function of x for the case in Figure 7 in which $|\tau^\circ(t)|/\varphi_m = 0.90$. The calculation is based on (3.60) with $\delta(t)$ as in (3.12) and $\tilde{\delta}(|x|)$ as in (3.58).

We can determine the shear strain at x at time t, $\gamma(x, t)$, from knowledge of the past history of the accumulated shear strain at x, *i.e.*, from knowledge of $\bar{\gamma}(x, \sigma)$ for each $\sigma < t$, provided we know also the sign of $\dot{\gamma}(x, \sigma)$ for each $\sigma < t$. Indeed, by (2.4),

$$\dot{\gamma}(x, t) = \bar{\gamma}_t(x, t)\, s(\dot{\gamma}(x, t)), \tag{3.63}$$

and clearly $(2.3)_2$ yields

$$\gamma(x, t) = \int_0^t \dot{\gamma}(x, \sigma)\, d\sigma. \tag{3.64}$$

It follows from (3.3) that (3.63) can be written

$$\dot{\gamma}(x, t) = \bar{\gamma}_t(x, t)\, s(\tau^\circ(t)), \tag{3.65}$$

and, of course,

$$\bar{\gamma}(x, t) = \int_0^t \bar{\gamma}_t(x, \sigma)\, d\sigma. \tag{3.66}$$

Solutions for which the Shear Stress is of Fixed Sign

The simplest case is that in which τ° does not change sign. If τ° is never negative, the equations (3.64), (3.65), and (3.66) yield $\gamma(x, t) = \bar{\gamma}(x, t)$. In that case, the graph of $\tau^\circ(t)$ *versus* t coincides with a graph of $|\tau^\circ(t)|$ *versus* t of the type seen in Figure 9A, and the graphs of Figure 8 can be considered plots of $\gamma(x, t)$ *versus* x. Strain fields $x \mapsto \gamma(x, t)$ obtained in that way yield, by evaluation of the integral (3.62), the displacement fields $x \mapsto u(x, t)$ shown in Figure 10, where x is the ordinate and $u(x, t)$ the abscissa. The graphs of Figure 10 may be thought of as pictures of a scribe line in the (x, y)-plane that lies along the x-axis at the time $t = 0$ that the body is in an undistorted reference configuration. The dotted line corresponds to the time t_m at which $\tau^\circ(t_m) = \varphi_m$; at that time the scribe line is straight, as it is at earlier times; at later times it is curved as seen in the two graphs drawn with dashes, which show the scribe line at times t for which $\tau^\circ(t)/\varphi_m = 0.99$ and 0.90; the solidly drawn curve shows the shape of the line at the time t_F of failure.

It follows from (3.51) and (3.59) that in the present case in which $\gamma = \bar{\gamma}$, (3.62) yields, for each t in (t_m, t_F),

$$u(x,t) = \begin{cases} \bar{\gamma}_m x + \sqrt{\delta(t)}\left(\eta_0 x - \lambda^2 x + 2\sqrt{3}\,\lambda\delta(t)^{-\frac{1}{4}}\sqrt{\dfrac{c}{2\alpha}}\left[E\left(\dfrac{\lambda}{\sqrt{3}}\delta(t)^{\frac{1}{4}}\sqrt{\dfrac{2\alpha}{c}}x\,\middle|\,m\right)\right.\right. \\ \qquad \left.\left. - \dfrac{\operatorname{sn}\left(\dfrac{\lambda}{\sqrt{3}}\delta(t)^{\frac{1}{4}}\sqrt{\dfrac{2\alpha}{c}}x\,\middle|\,m\right)\operatorname{dn}\left(\dfrac{\lambda}{\sqrt{3}}\delta(t)^{\frac{1}{4}}\sqrt{\dfrac{2\alpha}{c}}x\,\middle|\,m\right)}{1+\operatorname{cn}\left(\dfrac{\lambda}{\sqrt{3}}\delta(t)^{\frac{1}{4}}\sqrt{\dfrac{2\alpha}{c}}x\,\middle|\,m\right)}\right]\right), \\ \qquad\qquad \text{for} \quad 0 \le x \le x^*, \\ \delta(t)^{\frac{1}{4}}\sqrt{\dfrac{c}{2\alpha}}\left([\bar{\gamma}_m\,\delta(t)^{-\frac{1}{2}} + \eta_0 - \lambda^2]\,\xi^* + 2\sqrt{3}\,\lambda E\left(\dfrac{\lambda}{\sqrt{3}}\xi^*\,\middle|\,m\right)\right. \\ \qquad \left. - \dfrac{6}{\lambda^2\xi^*}\left[1+\operatorname{cn}\left(\dfrac{\lambda}{\sqrt{3}}\xi^*\,\middle|\,m\right)\right]\right), \quad \text{for} \quad x = x^*, \\ u(x^*,t) + \bar{\gamma}_m[x - x^*] + \dfrac{c\xi^{*2}}{2\alpha}\left[\dfrac{1}{x^*} - \dfrac{1}{x}\right], \quad \text{for} \quad x \ge x^*, \end{cases} \tag{3.67a}$$

with $\delta(t)$, λ, η_0 and m as in (3.51b), with $\xi^* = 1.25797$, and with $E(\cdot|m)$ given by an elliptic integral of the second kind, *i.e.*,

$$E(w|m) = \int_0^w dn^2(v|m)\,dv = \int_0^\psi \sqrt{1 - m\sin^2\theta}\,d\theta, \tag{3.67b}$$

where

$$\psi = \sin^{-1}(\operatorname{sn}(w|m)). \tag{3.67c}$$

Because γ is symmetric about $x = 0$, we here have $u(-x,t) = -u(x,t)$.

If the depth d of the body is finite, but large enough that the solutions obtained here under the assumption that the range of x is $(-\infty, \infty)$ hold to a good approximation on $(-d, d)$, then at each instant t the average, or "nominal", shear strain in the sample is

$$\langle\gamma\rangle(t) = \frac{u(\frac{1}{2}d,t) - u(-\frac{1}{2}d,t)}{d} = \frac{2}{d}u(\tfrac{1}{2}d,t). \tag{3.68}$$

Plots of $\tau^\circ(t)$ *versus* $\langle\gamma\rangle(t)$ are shown in Figure 11; the values of $\bar{\gamma}_m$, x, and φ_m employed are those used for the calculations shown in Figures 6–8 and 10, and d is taken to be 20; results are shown for c equal to 0.01, 0.1, and 1.0. For a homogeneous deformation, $\langle\gamma\rangle = \gamma$ and hence the graph of τ° *versus* $\langle\gamma\rangle$ coincides with the graph of φ *versus* $\bar{\gamma}$ until $\bar{\gamma}$ reaches $\bar{\gamma}_m$. Eventually, however, the decrease in τ° with an increase in $\langle\gamma\rangle$ is greater than the decrease in φ with an increase in $\bar{\gamma}$. It is interesting that this effect is enhanced by taking c small.

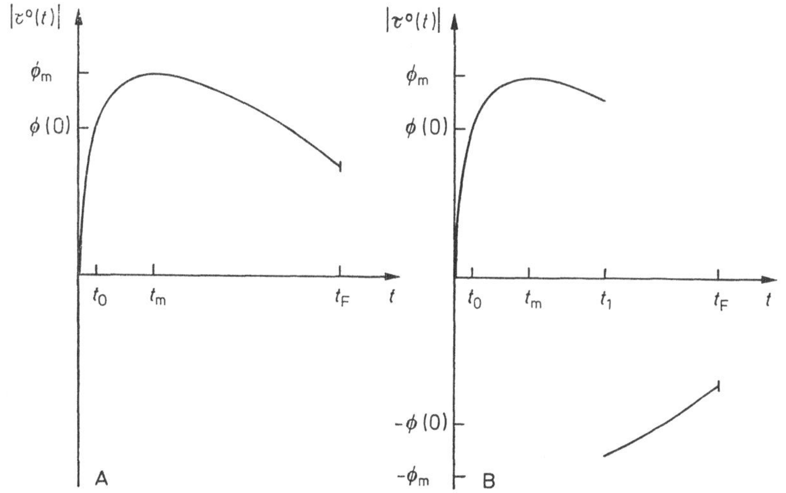

Fig. 9. Schematic representation of the time-dependence of the shear stress for motions analyzed here. (A) Typical case of $|\tau^\circ|$ *versus* t; for the displacement field seen in Figure 10 such a graph would give also τ° as a function of t. (B) τ° *versus* t for motions that obey (3.70) and are illustrated in Figures 12 and 13.

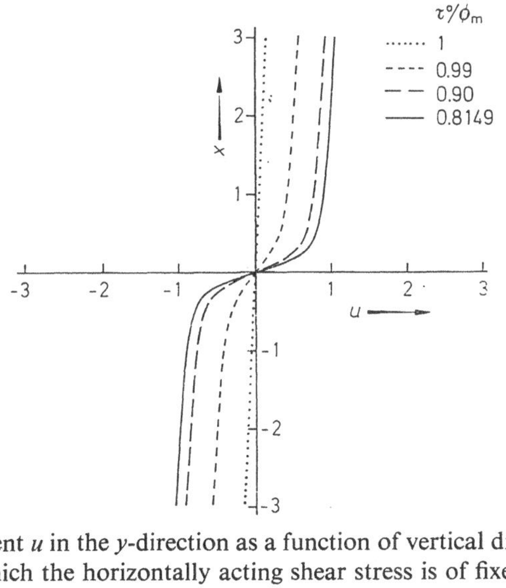

Fig. 10. Displacement u in the y-direction as a function of vertical distance x for motions in which the horizontally acting shear stress is of fixed sign.

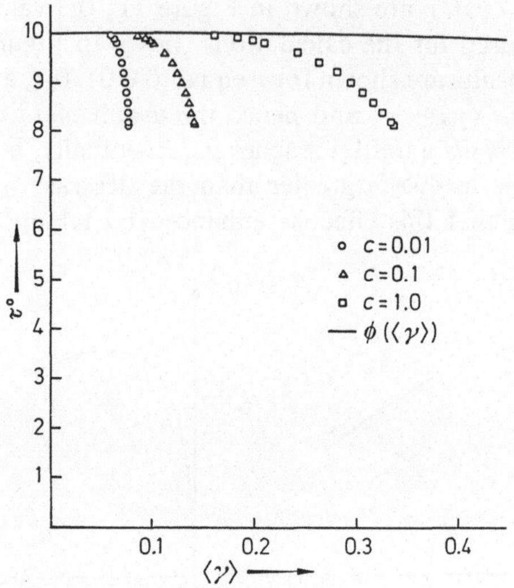

Fig. 11. Shear stress τ° as a function of nominal shear strain $\langle\gamma\rangle$ for a sample of thickness $d = 20$ undergoing a motion in which τ° is always positive. The values of α, φ_m, and $\bar{\gamma}_m$ are as in Figures 5–8, 10, 12, and 13. In the case $c = 0.1$, a scribe line initially along the x-axis would take the shapes seen in Figure 10.

Solutions with Reversal of the Shear Stress

The assumption that $|\tau^\circ|$ is a continuous function of time with a graph of the general form seen in Figure 9A does not preclude the occurrence of times t_i at which τ° reverses sign and abruptly changes from $\tau^\circ(t_i^-)$ to $\tau^\circ(t_i^+) = -\tau^\circ(t_i^-)$. By (3.65), at such times the strain rate $\dot{\gamma}$ reverses sign and changes from $\dot{\gamma}(x, t_i^-)$ to $\dot{\gamma}(x, t_i^+) = -\dot{\gamma}(x, t_i^-)$. Even if the shear stress and the rate of shearing reverse sign one or more times in this way, the function $(x, t) \mapsto \bar{\gamma}(x, t)$ of (2.4) and (3.61) remains a solution of the basic relations (3.4) and (3.5). It follows from (3.64) and (3.65) that, at each time t in the interval (t_i, t_{i+1}) between the ith and the $(i+1)^{\text{th}}$ reversal of sign of τ° and $\dot{\gamma}$, the shear strain γ is related as follows to the accumulated shear strain $\bar{\gamma}$:

$$\gamma(x, t) = \gamma(x, t_i) + [\bar{\gamma}(x, t) - \bar{\gamma}(x, t_i)]\, s(\tau^\circ(t)). \tag{3.69}$$

When, as in Figure 9B, τ° reverses sign once, say at time t_1, with $\tau^\circ(t) > 0$ for $0 < t < t_1$ and $\tau^\circ(t) < 0$ for $t_1 < t < t_F$, we have, at each x,

$$\gamma(x, t) = \begin{cases} \bar{\gamma}(x, t) & \text{for} \quad 0 \leqq t \leqq t_1, \\ 2\bar{\gamma}(x, t_1) - \bar{\gamma}(x, t) & \text{for} \quad t_1 < t < t_F. \end{cases} \tag{3.70}$$

Figures 12 and 13 show calculations based on (3.70) and (3.61) when t_1 in (3.70) is such that $\tau^\circ(t_1^-)/\varphi_m = 0.975$, and the material parameters are as in Figures 6, 7, 8, and 10. As in Figure 10, the graphs of $u(x, t)$ *versus* x of Figure 13 may be thought of as views of a scribe line in the (x, y)-plane that lies along the x-axis when $t = 0$. The line is shown at time t_1, at times t in (t_1, t_F) at which $\tau^\circ(t)/\varphi_m = -0.95$ and -0.90, and at the time t_F.

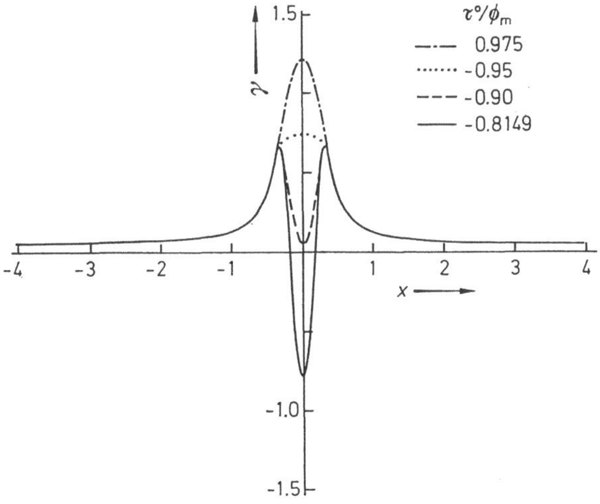

Fig. 12. Shear strain γ as a function of x at various times $t \geqq t_m$ for a motion in which $\tau^\circ(t)$ is initially positive but changes sign when $\tau^\circ(t)/\varphi_m = 0.975$; —·— moment of reversal of shear stress; ····· $\tau^\circ(t)/\varphi_m = -0.95$; – – – – $\tau^\circ(t)/\varphi_m = -0.90$; ——— $t = t_F$, *i.e.*, $\tau^\circ(t)/\varphi_m = -0.8149$.

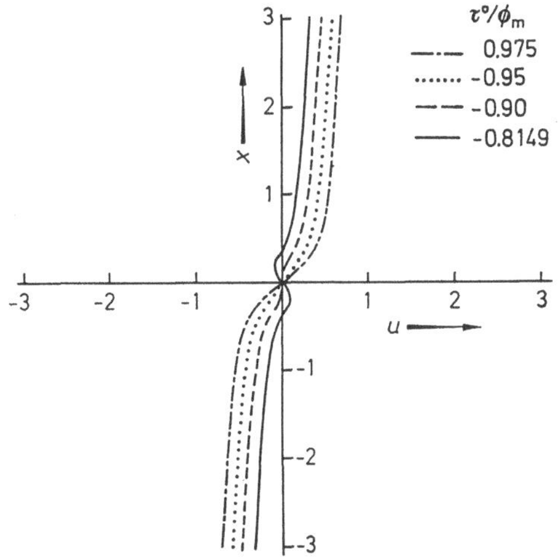

Fig. 13. Displacement u in the y-direction as a function of vertical distance x for a motion with the strain field shown in Figure 12.

Remark. All of the analytical results presented here remain valid if $\bar{\gamma}_m$ is taken to be zero and (2.22), (2.23), and (2.25) are replaced by

$$\varphi(\bar{\gamma}) = \varphi_m - \alpha\bar{\gamma}^2, \tag{2.22'}$$

$$\alpha > 0, \quad \varphi_m > 0, \tag{2.23'}$$

and

$$\bar{\gamma}_F = \sqrt{\varphi_m/\alpha}. \tag{2.25'}$$

When γ_m is set equal to zero in the solutions to (3.4) and (3.5) that we have presented, t_0 and t_m become equal, and the preliminary homogeneous deformation (3.8) disappears from the motion; the body starts to deform when $|\tau^\circ|$ attains the value $|\tau^\circ(t_m)| = \varphi(0) = \varphi_m$; at that moment $|\tau^\circ|$ begins its decline, and a shear band develops with $\bar{\gamma}(x, t)$ given by (3.61) [with $\bar{\gamma}_m = 0$ in (3.51) and $H^\dagger$ again as in (3.41)]; this shear band is such that for each t in (t_m, t_F), $\lim_{x\to\infty} \gamma(x, t) = \lim_{x\to-\infty} \gamma(x, t) = 0$, *i.e.*, points far from the center of the band suffer no shear at all.

Acknowledgments. We are grateful to THOMAS W. WRIGHT for valuable discussions about shear bands in ductile materials. Dr. WRIGHT's criticism of an earlier attempt to construct a theory of the phenomenon led us to propose the constitutive relations studied here, and his correspondence with COLEMAN about related problems materially helped our analysis of the differential equation (3.11).

We are grateful also to STEPHEN L. PASSMAN and TIMOTHY G. TRUCANO for their patient explanations to us of experimental observations of shear bands in metals and for their help in our preliminary work on the subject.

This research was supported in part by the National Science Foundation and Sandia National Laboratories.

References

1911 [1] KÁRMÁN, Th. VON, Festigkeitsversuche unter allseitigem Druck, *Z. Vereines Deut. Ing.* **55**, 1749–1757.

1944 [1] ZENER, C., & J. H. HOLLOMAN, Effect of strain rate upon plastic flow of steel, *J. Appl. Phys.* **15**, 22–32.

1950 [1] MILNE-THOMSON, L. M., *Jacobian Elliptic Function Tables*, Dover (New York).

1957 [1] HANDIN, J., & R. V. HAGER, Jr., Experimental deformation of sedimentary rocks under confining pressure: tests at room temperature on dry samples, *Bull. Am. Assoc. Petrol. Geol.* **41**, 1–50.

1958 [1] NOLL, W., A mathematical theory of the mechanical behavior of continuous media, *Arch. Rational Mech. Anal.* **2**, 197–226.

1960 [1] GRIGGS, D. T., F. J. TURNER, & H. C. HEARD, Deformation of rocks at 500° to 800°C in *Rock Deformation*, eds.: D. T. GRIGGS & J. HANDIN, Geological Society of America Memoir 79, Waverly Press (Baltimore) pp. 39–105; particularly 62–65.

1964 [1] MILNE-THOMSON, L. M., § 16, Jacobian elliptic functions and theta functions; § 17, Elliptic integrals, in *Handbook of Mathematical Functions*, eds.: M. ABRAMOWITZ & L. A. STEGUN, National Bureau of Standards (Washington), Appl. Math. Series, Vol. 55, pp. 567–626.

1965 [1] MURRELL, S. A. F., The effect of triaxial stress systems on the strength of rocks at atmospheric temperature, *Geophys. J.* **10**, 231–281.

[2] TRUESDELL, C., & W. NOLL, *The Non-Linear Field Theories of Mechanics*, Encyclopedia of Physics, Vol. III/3, Springer (Berlin, Heidelberg, New York).

1966 [1] COLEMAN, B. D., H. MARKOVITZ, & W. NOLL, *Viscometric Flows of Non-Newtonian Fluids*, Springer Tracts in Natural Philosophy, Vol. 5, Springer (Berlin, Heidelberg, New York).

1979 [1] COSTIN, L. S., E. E. CRISMAN, R. H. HAWLEY, & J. DUFFY, *On the Localization of Plastic Flow in Mild Steel Tubes under Dynamic Torsional Loading*, Technical Report (January 1979), Division of Engineering, Brown University (Providence).

[2] JAEGER, J. C., & N. G. W. COOK, *Fundamentals of Rock Mechanics*, 3rd Edition, Chapman & Hall (London).

[3] ROGERS, H. C., Adiabatic plastic deformation, *Ann. Rev. Mater. Sci.* **9**, 283–311.

1981 [1] ELEICHE, A.-S. M., Strain-rate history and temperature effects on the torsional-shear behavior of a mild steel, *Experiment. Mech.* **21**, 285–294.

1983 [1] COLEMAN, B. D., Necking and drawing in polymeric fibers under tension, *Arch. Rational Mech. Anal.* **83**, 115–137.

Department of Mathematics
Carnegie-Mellon University
Pittsburgh, Pennsylvania

(Received November 26, 1984)

Dynamic Admissible States, Negative Absolute Temperature, and the Entropy Maximum Principle

CHI-SING MAN

To Walter Noll, Author of Modern Classics in the Foundations of Mechanics and Thermodynamics

§ 1

GIBBS ([1], p. 55) began his major memoir on thermostatics by the dictum of CLAUSIUS: "The energy of the universe is constant. The entropy of the universe tends to a maximum." Replacing the "universe" of CLAUSIUS by a down-to-earth "isolated material system", GIBBS saw the basis for what we now call his Entropy Maximum Principle, which he developed "as a foundation for the general theory of thermodynamic equilibrium" ([2], p. 354). GIBBS's statement of his entropy principle was vague, but he also applied it (to be precise, here I should have said its companion Energy Minimum Principle, which GIBBS "preferred in general to use") to a variety of systems. A general principle will assume appropriate specific forms in different concrete problems. On the other hand, after seeing how a vaguely worded, general principle has been used in various situations, we should be able to discern that which is in common. Thence it is no accident that VAN DER WAALS ([3], § 2) and COLEMAN & NOLL ([4], § 16) should independently give the same formulation of the Entropy Maximum Principle for the same type of "isolated systems", for what they gave is in essence GIBBS's original formulation and they expressed it in the language of fields. The present essay is meant to be a critique of this formulation.

First let me raise two questions about GIBBS's formulation. For definiteness let us consider an instance of "the simplest kind", namely, that of a linearly viscous fluid "enclosed in a rigid and fixed envelop, which is impermeable ... and perfectly non-conducting to heat", neither "complicated by the action of gravity" nor by "capillarity tensions" ([1], p. 62). Let Ω be the region enclosed by the given envelop. We shall assume that the fluid always occupies the entire region Ω.[1] Let M be the mass of the fluid body and E its total energy when the system is prepared.

---

[1] Since we ignore the effect of surface tensions here, this assumption is more realistic than to allow vacuous cavitations.

Let S be the "thermodynamic surface" of the fluid in question. It is a surface in the u–ϱ–s space and is often assumed to be the graph of a function $s = s(u, \varrho)$; here u is the internal energy density per unit volume, ϱ is the mass density and s is the entropy density per unit volume. GIBBS's original formulation of the Entropy Maximum Principle asserts that among all "static states" $(u, \varrho, s)\colon \Omega \to \mathrm{S}$ which satisfy the constraints $\int_\Omega u \, dV = E$ and $\int_\Omega \varrho \, dV = M$, an equilibrium state renders the total entropy $S \equiv \int_\Omega s \, dV$ a maximum; here the integrals are volume integrals.[2] (*Cf.* VAN DER WAALS [3], § 2; COLEMAN & NOLL [4], § 16; COLEMAN & GREENBERG [5], § 1; DUNN & FOSDICK [6], § 3.)

The formulation above can be motivated as follows (see COLEMAN & GREENBERG [5], § 2; DUNN & FOSDICK [6], §§ 2–3): Let r be the radiant heating per unit mass, $\boldsymbol{b}$ the body force per unit mass, $\boldsymbol{v}$ the velocity and $\boldsymbol{q}$ the heat-flux vector. For thermodynamic processes "compatible with isolation" of the fluid body, *e.g.*, those for which $r = 0$, $\boldsymbol{b} = \boldsymbol{0}$, $\boldsymbol{v}|_{\partial\Omega} = \boldsymbol{0}$ and $\boldsymbol{q} \cdot \boldsymbol{n}|_{\partial\Omega} = 0$ (here $\partial\Omega$ denotes the boundary of the region Ω; $\boldsymbol{n}$ is the outward unit-normal field on $\partial\Omega$; see [5], Eq. (2.22)), it follows from the First and the Second Law of continuum thermodynamics that

$$\frac{d}{dt} \int_\Omega \left(u + \tfrac{1}{2}\varrho \boldsymbol{v} \cdot \boldsymbol{v}\right) dV = 0, \tag{1}$$

$$\frac{d}{dt} \int_\Omega s \, dV \geqq 0; \tag{2}$$

here d/dt denotes the derivative with respect to the time t. Provided that the velocity $\boldsymbol{v} = \boldsymbol{0}$ when the system comes to thermodynamic equilibrium, by Eq. (1) the internal energy density u should satisfy the constraint $\int_\Omega u \, dV = E$ at equilibrium. Since the mass of the fluid body must remain always the same, at equilibrium the mass density must satisfy the constraint $\int_\Omega \varrho \, dV = M$. Eq. (2) suggests that an equilibrium state would be one after the attainment of which the total entropy could no longer increase.

The foregoing arguments are persuasive, especially when we just look at the "isolated systems" considered by GIBBS. But GIBBS's memoir, grand as it is, does not exhaust the subject of isolated systems. The classical model of a celestial body as a liquid mass subject to no forces but self-gravitation is a perfect example of an isolated system; the permanently rotating state of a star is a good candidate for a state of maximum entropy. Starting from some state of motion, such a fluid mass must preserve unchanged its total linear momentum and its total angular momentum; thence its velocity field cannot altogether vanish at a state of thermo-

[2] All integrals in this essay are Lebesgue integrals.

dynamic equilibrium.[3] It should now be clear that GIBBS's formulation of the Entropy Maximum Principle, by setting $\boldsymbol{v} = \mathbf{0}$ from the outset, can hardly be sufficiently general. Moreover, even for "isolated systems" considered by GIBBS, we shall see by counter-example (§§ 2–3 below) that the kinetic energy will play a role in the thermostatics and should not be ignored outright. Indeed I contend that except for systems that are rigid or without inertia GIBBS's formulation is reproachable as a variational criterion, for it has unduly restricted the competitors or the admissible states to "static" ones. The Entropy Maximum Principle should be reformulated so that states with non-vanishing velocities are also admissible.

The counter-example that we shall discuss concerns negative absolute temperature. (*Cf.* Appendix 3 of DUNN & FOSDICK [6], and see Remark 3.9 below). If the material constituent of an "isolated system" can assume negative absolute temperatures, GIBBS's formulation of the Entropy Maximum Principle will not rule out the following possibility: when the "isolated system" in question is in equilibrium, its absolute temperature is negative. Indeed in § 2 we shall study an example for which this possibility is realized. The foregoing possibility vanishes, however, when the Entropy Maximum Principle is reformulated so that not only static states but dynamic ones are also admissible. Theorems proved in §§ 3–5 support the following assertion: *For a broad class of materials the Entropy Maximum Principle, when properly formulated, requires the equilibrium absolute temperature of an isolated system to be positive.* In particular the conclusion of the foregoing assertion is valid for the example studied in § 2, contradicting the result proved under GIBBS's formulation. (See Corollary 3.2 and Remark 3.5.)

The other question that I should like to raise concerns boundary conditions. Let us go back to the "isolated system" of "the simplest kind", discussed earlier. To obtain Eqs. (1) and (2) from the First and Second Laws, we must prescribe conditions pertaining to the isolation of the fluid body. Among these conditions are boundary conditions, for example, $\boldsymbol{v}|_{\partial\Omega} = \mathbf{0}$ and $\boldsymbol{q} \cdot \boldsymbol{n}|_{\partial\Omega} = 0$. Thence, when we apply a variational criterion of equilibrium, the admissible states should include only those which are compatible with the boundary conditions. All boundary conditions, however, have been left out in GIBBS's formulation of the Entropy Maximum Principle for "isolated systems" of "the simplest kind".

A physicist of the nineteenth century could hardly forget the all-important boundary conditions; GIBBS certainly would not.[4] That boundary conditions have been left out in the formulation above may be explained as follows: For the particular problem at hand, we can legitimately forget about the boundary con-

---

[3] "It is evident that a mass of any ordinary liquid ..., if left to itself in any state of motion, must preserve unchanged its moment of momentum ... But the viscosity, or internal friction, ... will, if the mass remain continuous, ultimately destroy all relative motion among its parts; so that it will ultimately rotate as a rigid solid." (THOMSON & TAIT [7], § 777). In § 5 below we shall study in detail the special instance of a compressible, linearly viscous and heat-conducting liquid. For such a liquid we shall show that THOMSON & TAIT's observation is borne out as a consequence of the properly reformulated Entropy Maximum Principle, provided that the "ultimate" states are those for which the total entropy of the liquid mass can no longer increase.

[4] See, for example, GIBBS [1], p. 192, Eq. (381), which "expresses the mechanical condition of equilibrium for a surface where [a] solid meets a fluid".

ditions; or, to use present-day jargon, those conditions can be "relaxed". GIBBS's formulation pertains to the "relaxed problem" (without boundary conditions), not the original problem (with boundary conditions). For the particular problem at issue, a solution of the "relaxed problem" also solves the original problem and *vice versa*. Obviously such a happy state of affairs will not always happen. In § 4 we shall apply the Entropy Maximum Principle to a model system; there a boundary condition plays an essential role in the proof of Theorem 4.1. § 4, its intrinsic interest notwithstanding, is also meant as a reminder for thermodynamicists of traditional training, who are often oblivious of boundary conditions. (See § 6 for an example of such heedlessness.)

In § 5 I shall formulate the Entropy Maximum Principle for the "ultimate" states of an isolated liquid mass in unconfined motion, a system for which GIBBS's original formulation is clearly inadequate. By "isolated" is here meant what follows: the liquid mass $\mathscr{B}$ is subjected to no body-forces except self-gravitation, the contact loading on the boundary $\partial\mathscr{B}$ of $\mathscr{B}$ is exerted by surface tension alone, the radiant heating r is null, and the influx of heating on $\partial\mathscr{B}$ is null. Usually, when the effects of self-gravitation are important (*e.g.*, for rotating stars), those due to surface tension will be negligible; and *vice versa* (*e.g.*, for liquid drops). By dropping the appropriate term our formulation will be appropriate to the more familiar situations in which one or the other of the two effects can be ignored. Much has been written on rotating liquid masses. It will be interesting to see what the Entropy Maximum Principle can add to the foundations of the subject. Discourse beyond formulation, study into consequences of the entropy principle, will lead us away from the purpose of this essay. Nevertheless we shall read off some easy results. In particular we shall show that the "ultimate" states must be states of permanent rotation with positive absolute temperature.

The present essay would not be complete, should we silently pass over other formulations of the Entropy Maximum Principle. My formulation is anticipated by the work of STUECKELBERG DE BREIDENBACH and SCHEURER ([8], [9], [10]). Their formulation can be regarded as intermediate between GIBBS's and mine. § 6 is devoted to their work, both as critique and as appraisal. Finally in § 7 we shall discuss the form the Entropy Maximum Principle assumes in TISZA's "macroscopic thermodynamics of equilibrium" [11]. I, for one, had my first encounter with Gibbsian thermostatics through this transmogrified version, thanks to the textbook by CALLEN [12].[5] The criticism in § 7 applies to the entire "composite systems" approach.[6]

The present essay focuses on two questions raised regarding GIBBS's original formulation, namely, the problem of dynamic admissible states and that of boundary conditions. Of course, we can ask other questions. Indeed some good questions were posed by COLEMAN & GREENBERG [5], and they have led to contributions of permanence (besides [5], see COLEMAN [14], and DUNN & FOSDICK [6]). The Entropy Maximum Principle in concrete use is a mathematical principle. Thence

[5] I have taken the names "Entropy Maximum Principle" and "Energy Minimum Principle" from CALLEN's book ([12], p. 88).

[6] According to TISZA ([11], p. 41), "the role of composite systems in dispelling conceptual confusion ... was particularly emphasized by Planck [13]."

there is also the question of mathematical formalism. The language of fields need not always be the most convenient and the most elegant, as NOLL [15] demonstrated for GIBBS's "isolated systems" of "the simplest kind" ([1], p. 62) that contain reacting mixtures. Problems other than the two discussed at great length above, however, are beyond the scope of this essay.

§ 2

For an "isolated system" of "the simplest kind" (see [1], p. 62; *cf.* also § 1 above), GIBBS's formulation of the Entropy Maximum Principle generally does not rule out the following possibility: when the system is in equilibrium, its absolute temperature is negative. It is easy to construct an example for which this possibility is realized. (*Cf.* Appendix 3 of DUNN & FOSDICK [6].)

Let us consider the instance discussed in § 1, namely, an "isolated system" that consists in a body of linearly viscous fluid enclosed in a rigid and fixed envelop. The reader should consult § 1 for definitions, notation and GIBBS's formulation of the Entropy Maximum Principle for that system. Let $s = s(u, \varrho)$ define the "thermodynamic surface" of the fluid in question. Let this function be of class C^1 and let its domain be an open set $\mathscr{D}$ in R^2; in this essay the symbol R denotes the reals. Suppose further that the function $s = s(u, \varrho)$ is strictly concave and $\partial s/\partial u = 1/\theta < 0$ over a non-empty subset $\mathscr{D}^-$ of the domain $\mathscr{D}$; here θ denotes the absolute temperature. Let M and E be the mass and the total energy of the fluid body, respectively. Let V be the volume of the region Ω. Suppose the point defined by the pair $(E/V, M/V)$ falls within $\mathscr{D}^-$. By GIBBS's formulation of the Entropy Maximum Principle we conclude that at equilibrium $u(\boldsymbol{x}) = E/V$, $\varrho(\boldsymbol{x}) = M/V$ for each place $\boldsymbol{x}$ in Ω. Thence the equilibrium temperature $\theta = \left[\frac{\partial s}{\partial u}(E/V, M/V)\right]^{-1} < 0$ over Ω.

The foregoing conclusion is false under a properly reformulated Entropy Maximum Principle. (See Corollary 3.2 and Remark 3.5 below.)

§ 3

We shall reformulate the Entropy Maximum Principle for an "isolated system" that consists in a body of fluid enclosed in a rigid envelop which is impermeable and does not conduct heat. We assume that the rigid envelop is at rest in an inertial frame of reference, the radiant heating r is identically zero and the body force $\boldsymbol{b}$ is conservative with potential Φ. As we shall see, the reformulated entropy principle requires that equilibrium temperature of such "isolated systems" be positive. Although we shall prove the foregoing assertion for two particular types of fluids, it is easy to see that the same conclusion will be valid for a great variety of materials, fluids and solids alike. (See Remark 3.6 below.)

Let us now consider an "isolated system" of the kind at issue: suppose that inside the rigid envelop is a fluid of the type first studied by VAN DER WAALS [3].

We assume that the fluid always occupies the entire region Ω enclosed by the envelop; moreover Ω is mathematically a connected, bounded open set with Lipschitz boundary $\partial\Omega$. Let $\bar{\Omega} \equiv \Omega \cup \partial\Omega$. At any place $\boldsymbol{x}$ in $\bar{\Omega}$ and at any instant t, the entropy density s and absolute temperature θ are given by the constitutive relations

$$s = s(u, \varrho, \nabla\varrho), \tag{3}$$

$$\frac{\partial s}{\partial u} = \frac{1}{\theta}; \tag{4}$$

here $\nabla\varrho$ denotes the spatial gradient of ϱ. We assume that the entropy function $s = s(u, \varrho, \nabla\varrho)$ of the given fluid is of class C^1 and its domain is an open set $\mathcal{D}$ in R^5; moreover, it satisfies the following adscititious conditions:

$$\varrho > 0, \text{ and } u_1 < u_2 \text{ implies} \frac{\partial s}{\partial u}(u_1, \varrho, \nabla\varrho) > \frac{\partial s}{\partial u}(u_2, \varrho, \nabla\varrho). \tag{5}$$

To render our discussion non-vacuous we suppose $\partial s/\partial u = 1/\theta > 0$ over a non-empty proper subset of the domain $\mathcal{D}$.

For each instant t after our "isolated system" has been prepared, we shall call the triple $(u(\cdot, t), \varrho(\cdot, t), \boldsymbol{v}(\cdot, t)) : \bar{\Omega} \to R^5$ the state of the "isolated system" at time t. Suppose there is an instant t_0 after which the "isolated system" is in thermodynamic equilibrium. We denote the equilibrium states by $(u_\circ(\cdot, t), \varrho_\circ(\cdot, t), \boldsymbol{v}_\circ(\cdot, t))$; here t is an instant in $[t_0, \infty)$.

Before we can formulate the Entropy Maximum Principle, we must delineate the class of admissible states. Roughly speaking, admissible states are those compatible with isolation of the given system. More precisely, an admissible state $(u, \varrho, \boldsymbol{v}) : \bar{\Omega} \to R^5$ must satisfy the following requirements:[7]

(i) Boundary conditions. An admissible state of the system should satisfy boundary conditions that *imply* what follows:

$$\boldsymbol{q} \cdot \boldsymbol{n}|_{\partial\Omega} = 0, \quad \boldsymbol{v} \cdot \boldsymbol{n}|_{\partial\Omega} = 0, \quad \boldsymbol{v} \cdot \boldsymbol{T}\boldsymbol{n}|_{\partial\Omega} = 0, \tag{6}$$

where $\boldsymbol{T}$ is the Cauchy stress. (*Cf.* COLEMAN & GREENBERG [5], Definition 2.3.) For the discussion in this section, we can leave the boundary conditions otherwise implicit.

(ii) Energy constraint. Let E be the total energy of the "isolated system" when it is prepared. An admissible state of the system satisfies the constraint

$$\int_\Omega (u + \tfrac{1}{2}\varrho \boldsymbol{v} \cdot \boldsymbol{v})\, dV + \int_\Omega \varrho \Phi(\boldsymbol{x})\, dV + \int_{\partial\Omega} \Psi(u, \varrho, \nabla\varrho)\, dA = E. \tag{7}$$

In Eq. (7) the term $\int_{\partial\Omega} \Psi(u, \varrho, \nabla\varrho)\, dA$ delivers the contact energy between the

---

[7] Admissible states are virtual competitors; the time t does not appear in their definition.

fluid and the solid envelop;[8] the term $\int_\Omega \varrho\Phi(\boldsymbol{x})\,dV$ arises because $\boldsymbol{b} = -\nabla\Phi$ and $\boldsymbol{v}\cdot\boldsymbol{n}|_{\partial\Omega} = 0$ imply that $\int_\Omega \boldsymbol{v}\cdot\varrho\boldsymbol{b}\,dV = -d\left(\int_\Omega \varrho(\boldsymbol{x},t)\,\Phi(\boldsymbol{x})\,dV\right)/dt$, should the fields in question be sufficiently smooth. $\int_\Omega \boldsymbol{v}\cdot\varrho\boldsymbol{b}\,dV$ appears as a term in the equation of energy balance.

(iii) Mass constraint. Let M be the total mass of the fluid body. An admissible state must satisfy the equality

$$\int_\Omega \varrho\,dV = M. \tag{8}$$

(iv) Mathematical conditions. An admissible state should satisfy mathematical conditions such that Eqs. (7), (8), and the boundary conditions (which are still left implicit) make sense, the composite function $s(u, \varrho, \nabla\varrho)(\cdot)$ is well defined and integrable over Ω, *etc.*

Let $S[u, \varrho, \boldsymbol{v}] \equiv \int_\Omega s(u, \varrho, \nabla\varrho)\,dV$ denote the total entropy corresponding to the triple $(u, \varrho, \boldsymbol{v})$. We formulate the *Entropy Maximum Principle* for the "isolated system" at issue as follows:

Let t_0 be the instant after which the "isolated system" is in thermodynamic equilibrium. For each t in $[t_0, \infty)$,

$$S[u_\circ(\cdot,t), \varrho_\circ(\cdot,t), \boldsymbol{v}_\circ(\cdot,t)] \geqq S[u, \varrho, \boldsymbol{v}], \tag{9}$$

for all admissible states $(u, \varrho, \boldsymbol{v})$. Moreover, the total entropy S is constant over the interval of time $[t_0, \infty)$.

The formulation above is as yet abstract and unamenable to mathematical analysis. Even with the conditions stated earlier, the totality of admissible states is still not well defined. To proceed, we need additional information. For the proof of Theorem 3.1, it suffices to lay down the following

Basic Assumptions:

(I) For each instant t in $[t_0, \infty)$, $(u_\circ(\cdot,t), \varrho_\circ(\cdot,t), \boldsymbol{v}_\circ(\cdot,t))$ is an admissible state. In particular the equilibrium state in question satisfies Eq. (7), Eq. (8) and the appropriate boundary conditions, whatever they may be.

(II) The following assertion is valid for each instant t in $[t_0, \infty)$: Let $C_0^\infty(\Omega)$ be the set of C^∞ functions with compact support in Ω. Given any h and $\tilde{h}$ in $C_0^\infty(\Omega)$ and any $\boldsymbol{k}$ in $[C_0^\infty(\Omega)]^3$, there is a neighborhood $\mathcal{N}$ of (0, 0) in R^2 ($\mathcal{N}$ generally dependent upon h, $\tilde{h}$, $\boldsymbol{k}$ and the equilibrium state in question) such that for all (α, β) in $\mathcal{N}$, the triple $(u_\circ + \alpha h + \beta\tilde{h}, \varrho_\circ, \boldsymbol{v}_\circ + \alpha\boldsymbol{k})$ defines an admissible state of the system if it satisfies the energy constraint Eq. (7) and if the composite map $s(u_\circ + \alpha h + \beta\tilde{h}, \varrho_\circ, \nabla\varrho_\circ)(\cdot)$ is well defined and is integrable over Ω.

---

[8] Cahn [16] postulated a similar contact-energy term. As long as the contact energy is a surface integral over the boundary $\partial\Omega$, the analysis below will be independent of the exact form of this term.

Remark 3.1. Because $(u_\circ(\cdot, t), \varrho_\circ(\cdot, t), \boldsymbol{v}_\circ(\cdot, t))$ is an admissible state of the "isolated system", and because $\tilde{h}$, h and $\boldsymbol{k}$ have compact support over Ω, a triple of the form $(u_\circ + \alpha h + \beta \tilde{h}, \varrho_\circ, \boldsymbol{v}_\circ + \alpha \boldsymbol{k})$ satisfies Eq. (8) and the boundary conditions, whatever they be.

In GIBBS's thermostatics the internal energy density and the mass density need not be continuous when the system in question is in equilibrium; indeed they should have a jump discontinuity across a liquid–vapor interface. In VAN DER WAALS's "theory of capillarity", however, the basic hypothesis is to surmise a "continuous variation of density" even across a liquid–vapor interface. Since we are studying VAN DER WAALS's gradient-type fluids, we shall adopt assumptions of continuity and smoothness accordingly.

Theorem 3.1. *Let the "isolated system" at issue be in thermodynamic equilibrium throughout the interval of time* $[t_0, \infty)$. *Let* $(u_\circ(\cdot, t), \varrho_\circ(\cdot, t), \boldsymbol{v}_\circ(\cdot, t))$ *denote the equilibrium state at the instant t. For each t, assume that* $u_\circ(\cdot, t)$ *and* $\boldsymbol{v}_\circ(\cdot, t)$ *are continuous and* $\varrho_\circ(\cdot, t)$ *is of class* C^1 *over* $\bar{\Omega}$. *For each* $\boldsymbol{x}$ *in* $\bar{\Omega}$, *assume that* $u_\circ(\boldsymbol{x}, \cdot)$, $\boldsymbol{v}_\circ(\boldsymbol{x}, \cdot)$, $\varrho_\circ(\boldsymbol{x}, \cdot)$ *and* $\nabla\varrho_\circ(\boldsymbol{x}, \cdot)$ *are continuous over* $[t_0, \infty)$. *Let* $\theta_\circ(\cdot, t)$ *be the temperature field corresponding to the equilibrium state* $(u_\circ(\cdot, t), \varrho_\circ(\cdot, t), \boldsymbol{v}_\circ(\cdot, t))$. *The Entropy Maximum Principle dictates that throughout* $[t_0, \infty)$ *either* (i) $\theta_\circ = \infty$ *or* (ii) $\boldsymbol{v}_\circ = \boldsymbol{0}$ *and* $\theta_\circ = \text{constant} > 0$.

Proof. Fix an instant t in $[t_0, \infty)$. We shall suppress the dependence on t in all expressions when no confusion should arise. Let $\boldsymbol{k}$ be in $[C_0^\infty(\Omega)]^3$, h and $\tilde{h}$ be in $C_0^\infty(\Omega)$, and $\tilde{h}$ satisfy the condition that $\int_\Omega \tilde{h}\, dV \neq 0$. Let $\mathcal{N}$ be the neighborhood of $(0, 0)$ in R^2 guaranteed by Basic Assumption (II). By hypothesis $(u_\circ, \varrho_\circ, \nabla\varrho_\circ)(\bar{\Omega})$ is a compact subset of $\mathcal{D}$. Thence there are positive real numbers a and b such that $(-a, a)\times(-b, b)$ is contained in $\mathcal{N}$ and $(u_\circ + \alpha h + \beta\tilde{h}, \varrho_\circ, \nabla\varrho_\circ)(\bar{\Omega}) \subset \mathcal{D}$ for all α in $(-a, a)$ and β in $(-b, b)$. By Basic Assumption (II), for each (α, β) in $(-a, a)\times(-b, b)$, the triple $(u_\circ + \alpha h + \beta\tilde{h}, \varrho_\circ, \boldsymbol{v}_\circ + \alpha\boldsymbol{k})$ will define an admissible state of the "isolated system" if it satisfies Eq. (7).

Let $\mathcal{S}(\alpha, \beta) \equiv \int_\Omega s(u_\circ + \alpha h + \beta\tilde{h}, \varrho_\circ, \nabla\varrho_\circ)\, dV$, $\mathcal{E}(\alpha, \beta) \equiv \int_\Omega (u_\circ + \alpha h + \beta\tilde{h} + \frac{1}{2}\varrho_\circ(\boldsymbol{v}_\circ + \alpha\boldsymbol{k})\cdot(\boldsymbol{v}_\circ + \alpha\boldsymbol{k}))\, dV$, and $\mathcal{E}_\circ \equiv E - \int_\Omega \varrho_\circ\, \Phi(\boldsymbol{x})\, dV - \int_{\partial\Omega} \Psi(u_\circ, \varrho_\circ, \nabla\varrho_\circ)\, dA$. Since s is of class C^1, by the rule on "differentiation under the integral sign"[9] the functions $\mathcal{S}$ and $\mathcal{E}$ are of class C^1 over $(-a, a)\times(-b, b)$. At $(\alpha, \beta) = (0, 0)$ $(\partial\mathcal{E}/\partial\alpha, \partial\mathcal{E}/\partial\beta) = \left(\int_\Omega (h + \varrho_\circ\boldsymbol{v}_\circ \cdot \boldsymbol{k})\, dV, \int_\Omega \tilde{h}\, dV\right) \neq (0, 0)$. By the Entropy Maximum Principle, $\mathcal{S}(\alpha, \beta)$ assumes a maximum at $(0, 0)$ when (α, β) is constrained to the set defined by the equation $\mathcal{E}(\alpha, \beta) = \mathcal{E}_\circ$. Thence, by the Lagrange multi-

[9] See LANG [17], Ch. XIV, § 4, Lemma 1 and Lemma 2. The cited version will suffice also for our later use.

plier theorem, there is a real number $\lambda(t)$ such that at $(\alpha, \beta) = (0, 0)$, $(\partial\mathscr{S}/\partial\alpha) + \lambda(t)(\partial\mathscr{E}/\partial\alpha) = 0$, $(\partial\mathscr{S}/\partial\beta) + \lambda(t)(\partial\mathscr{E}/\partial\beta) = 0$, *i.e.*,

$$\int_{\Omega} \left[\left(\frac{1}{\theta_\circ} + \lambda(t)\right) h + \lambda(t)\,\varrho_\circ \boldsymbol{v}_\circ \cdot \boldsymbol{k}\right] dV = 0, \tag{10}$$

$$\int_{\Omega} \left(\frac{1}{\theta_\circ} + \lambda(t)\right) \tilde{h}\, dV = 0. \tag{11}$$

We let h and $\boldsymbol{k}$ run over $C_0^\infty(\Omega)$ and $[C_0^\infty(\Omega)]^3$, respectively, and repeat the above analysis for each pair of h and $\boldsymbol{k}$, holding $\tilde{h}$ fixed. We see that Eq. (10) is valid for any h in $C_0^\infty(\Omega)$ and any $\boldsymbol{k}$ in $[C_0^\infty(\Omega)]^3$, and the multiplier $\lambda(t) = -\int_\Omega (1/\theta_\circ)\,\tilde{h}\,dV/\int_\Omega \tilde{h}\,dV$ is independent of h and $\boldsymbol{k}$. By the "fundamental lemma" of the calculus of variations,[10] we conclude that over Ω

$$(1/\theta_\circ) + \lambda(t) = 0, \tag{12}$$

$$\lambda(t)\,\varrho_\circ \boldsymbol{v}_\circ = \boldsymbol{0}.\text{[11]} \tag{13}$$

By hypothesis, for each $\boldsymbol{x}$ in $\bar{\Omega}$, $(1/\theta_\circ)(\boldsymbol{x}, \cdot)$ is continuous over $[t_0, \infty)$. It follows from Eq. (12) that $\lambda(\cdot)$ is continuous over $[t_0, \infty)$. Let $\mathscr{T} \equiv \{t \in [t_0, \infty): \lambda(t) \neq 0\}$. If $\mathscr{T}$ is empty, Eq. (12) dictates that $\theta_\circ = \infty$ throughout $[t_0, \infty)$. Now suppose $\mathscr{T}$ is not empty; then either (a) $\mathscr{T} = [t_0, \infty)$ or (b) $\mathscr{T} \neq [t_0, \infty)$. (a) If $\mathscr{T} = [t_0, \infty)$, by Eq. (13) $\boldsymbol{v}_\circ = \boldsymbol{0}$ over $[t_0, \infty)$. Thence for each $\boldsymbol{x}$ in $\bar{\Omega}$, $\varrho_\circ(\boldsymbol{x}, \cdot)$ and $\nabla\varrho_\circ(\boldsymbol{x}, \cdot)$ are constant over $[t_0, \infty)$. We claim that $\lambda(\cdot)$ is constant over $[t_0, \infty)$. Indeed, if there are instants t^* and $t^\#$ such that $\lambda(t^*) > \lambda(t^\#)$, it will follow from the constitutive hypothesis Eq. $(5)_2$ that $u_\circ(\boldsymbol{x}, t^*) > u_\circ(\boldsymbol{x}, t^\#)$ for each $\boldsymbol{x}$ in $\bar{\Omega}$. By the same hypothesis and the entropy principle, $0 = S(t^*) - S(t^\#) < \int_\Omega [(\partial s/\partial u)(u_\circ(\boldsymbol{x},t^\#), \varrho_\circ(\boldsymbol{x}), \nabla\varrho_\circ(\boldsymbol{x}))]\,[u_\circ(\boldsymbol{x}, t^*) - u_\circ(\boldsymbol{x}, t^\#)]\,dV = -\lambda(t^\#)\int_\Omega [u_\circ(\boldsymbol{x}, t^*) - u_\circ(\boldsymbol{x}, t^\#)]\,dV$; similarly, $0 < -\lambda(t^*)\int_\Omega [u_\circ(\boldsymbol{x}, t^\#) - u_\circ(\boldsymbol{x}, t^*)]\,dV$. Since $\lambda(t^*)$ and $\lambda(t^\#)$ have the same sign by continuity of $\lambda(\cdot)$, we have a contradiction. Therefore $\lambda(\cdot)$ is constant over $[t_0, \infty)$. (b) If $\mathscr{T} \neq [t_0, \infty)$, there will be a $\tilde{t} \neq t_0$ such that $\lambda(\tilde{t}) = 0$ and $\tilde{t}$ is the end-point of some sub-interval $\mathscr{T}_\alpha$ over which $\lambda(t) \neq 0$. By the argument used in case (a), $\lambda(\cdot)$ is constant over the sub-interval $\mathscr{T}_\alpha$. Thence $\lambda(\cdot)$ has a discontinuity at $t = \tilde{t}$, which is impossible. In summary, we have two

---

[10] If $f: \Omega \to R^1$ is locally integrable and $\int_\Omega f(\boldsymbol{x})\,h(\boldsymbol{x})\,dV = 0$ for all h in $C_0^\infty(\Omega)$, then $f = 0$ almost everywhere. See Yosida [18], Ch. I, § 8, pp. 48–49.

[11] This paragraph is just a slight modification of Weierstrass's proof as regards "Euler's rule" for isoperimetric problems (see Bolza [19], § 39). It is included here for completeness. Henceforth whenever similar reasoning is called for, we shall omit the details and just appeal to "Euler's rule".

possibilities, namely, either (i) $\theta_\circ = \infty$ throughout $[t_0, \infty)$ or (ii) $\boldsymbol{v}_\circ = \mathbf{0}$ and $\theta_\circ = \text{constant} \neq \infty$ throughout $[t_0, \infty)$.

Let us proceed to show that $\theta_\circ > 0$ under case (ii). Fix an instant t in $[t_0, \infty)$. Pick a specific function $\boldsymbol{k}$ in $[C_0^\infty(\Omega)]^3$ such that $\int\limits_\Omega \frac{1}{2}\varrho_\circ \boldsymbol{k} \cdot \boldsymbol{k}\, dV \neq 0$. Choose and fix a function $\tilde{h}$ in $C_0^\infty(\Omega)$ such that $\int\limits_\Omega \tilde{h}\, dV = -\int\limits_\Omega \frac{1}{2}\varrho_\circ \boldsymbol{k} \cdot \boldsymbol{k}\, dV$. Since $\varrho_\circ(\boldsymbol{x}) > 0$, $\int\limits_\Omega \tilde{h}\, dV < 0$.

For $\varepsilon > 0$, a triple of the form $(u_\circ + \varepsilon \tilde{h}, \varrho_\circ, \mathbf{0} + \varepsilon^{\frac{1}{2}} \boldsymbol{k})$ satisfies Eq. (7). Thence by Basic Assumption (II) there is a positive number c such that for each ε in $[0, c)$ the triple $(u_\circ + \varepsilon \tilde{h}, \varrho_\circ, \mathbf{0} + \varepsilon^{\frac{1}{2}} \boldsymbol{k})$ defines an admissible state of the "isolated system". Let $\not{f}(\varepsilon) \equiv \int\limits_\Omega s(u_\circ + \varepsilon \tilde{h}, \varrho_\circ, \nabla \varrho_\circ)\, dV$. The function $\not{f}$ is of class C^1 over $[0, d)$ for some d in $(0, c)$. By the Entropy Maximum Principle, $\not{f}$ assumes a maximum at $\varepsilon = 0$. Hence $\not{f}'(0) \leqq 0$. Since $\not{f}'(0) = \int\limits_\Omega (1/\theta_\circ)\, \tilde{h}\, dV = (1/\theta_\circ) \int\limits_\Omega \tilde{h}\, dV$ and $\int\limits_\Omega \tilde{h}\, dV < 0$, we conclude that $\theta_\circ > 0$. □

Remark 3.2. The constitutive assumption $\varrho > 0$ is essential in the proof above. If it were true that $\varrho < 0$, the conclusion for case (ii) would be $\boldsymbol{v}_\circ = \mathbf{0}$ and $\theta_\circ = \text{constant} < 0$. It was STUECKELBERG DE BREIDENBACH who first asserted that "mass density ... has the sign of absolute temperature". See § 6 for a critique of his work.

Remark 3.3. Let us discuss case (ii) a little further. We arrive at the conclusion without recourse to boundary conditions. We must ensure, however, that our conclusion is compatible with Eq. (6), the boundary conditions of "isolation". Since $\boldsymbol{v}_\circ = \mathbf{0}$, Eqs. $(6)_2$ and $(6)_3$ are obviously satisfied. A sufficient condition for Eq. $(6)_1$ to be observed is that for the fluid in question uniform temperature leads to zero heat-flux.

Remark 3.4. The conclusion of Theorem 3.1 remains valid for an interval of time $[t_0, t_1]$ if the Entropy Maximum Principle is replaced by what follows: (1.) The total entropy $S[u_\circ(\cdot, t), \varrho_\circ(\cdot, t), \boldsymbol{v}_\circ(\cdot, t)]$ is constant over $[t_0, t_1]$. (2.) Given any t in $[t_0, t_1]$, any h, $\tilde{h}$ in $C_0^\infty(\Omega)$ and $\boldsymbol{k}$ in $[C_0^\infty(\Omega)]^3$, $S[u_\circ(\cdot, t), \varrho_\circ(\cdot, t), \boldsymbol{v}_\circ(\cdot, t)] \geqq S[u, \varrho, \boldsymbol{v}]$ for all those admissible $(u, \varrho, \boldsymbol{v})$ that are of the form $(u_\circ + \alpha h + \beta \tilde{h}, \varrho_\circ, \boldsymbol{v}_\circ + \alpha \boldsymbol{k})$ if α and β are sufficiently small. Thence we can say that the conclusion of Theorem 3.1 applies also to "metastable" states.

To obtain the counter-example mentioned in §§ 1–2, let us consider a slightly different situation: suppose the "isolated" body in question is a linearly viscous fluid. Since no confusion should arise, we shall use the same terminology and the same set of symbols as before. Everything we have said about VAN DER WAALS's gradient-type fluid applies almost verbatim to the linearly viscous fluid, with just one exception: we can no longer require $u_\circ(\cdot, t)$, $\varrho_\circ(\cdot, t)$ to be continuous, because the linearly viscous fluid falls within GIBBS's constitutive scheme for substances. Before we get to the final result in Corollary 3.2, let us go over the preliminaries briefly, emphasizing only those aspects different from the preceding instance.

The relevant constitutive relations are now

$$s = s(u, \varrho), \qquad \partial s/\partial u = 1/\theta. \tag{14}$$

Again we assume that the entropy function $s = s(u, \varrho)$ is of class C^1 and its domain $\mathscr{D}$ is open in R^2, that $\varrho > 0$, that $s(\cdot, \varrho)$ is strictly concave for each ϱ, and that $\theta < 0$ over a proper, non-empty subset of $\mathscr{D}$.

The general conditions (i) to (iv) regarding admissible states, the basic assumptions (I) and (II) about equilibrium states, and the formulation of the Entropy Maximum Principle remain formally the same as before except that $\nabla\varrho$ and $\nabla\varrho_\circ$ should be deleted from all expressions. For example, the term that expresses contact energy in Eq. (7) should be $\int_{\partial\Omega} \Psi(u, \varrho)\, dA$ and the total entropy corresponding to the state $(u, \varrho, \boldsymbol{v})$ should now read $S = \int_\Omega s(u, \varrho)\, dV$.

Let $L^\infty(\Omega)$ and $L^2(\Omega)$, be respectively, the spaces of essentially bounded and of square integrable functions over Ω. We shall use these spaces (with their respective Banach-space norms) to specify the additional conditions which we presume the equilibrium states to obey.

Corollary 3.2. *For each t in $[t_0, \infty)$, assume that $u_\circ(\cdot, t)$, $\varrho_\circ(\cdot, t)$ are in $L^\infty(\Omega)$ and $\boldsymbol{v}_\circ(\cdot, t)$ is in $[L^2(\Omega)]^3$. Assume, moreover, that the mappings $t \mapsto u_\circ(\cdot, t)$ and $t \mapsto \varrho_\circ(\cdot, t)$, both from $[t_0, \infty)$ to $L^\infty(\Omega)$, are continuous. Suppose there is an instant τ in $[t_0, \infty)$ such that the following conditions hold*: (a) *For almost every $\boldsymbol{x}$ in Ω, $((u_\circ(\boldsymbol{x}, \tau), \varrho_\circ(\boldsymbol{x}, \tau))$ lies in some compact subset $\mathscr{K}$ of $\mathscr{D}$*; (b) $\theta_\circ(\cdot, \tau)$ *is essentially bounded over some subset of Ω with non-zero measure. The Entropy Maximum Principle dictates that $\theta_\circ$ = constant > 0 almost everywhere in Ω throughout the interval of time $[t_0, \infty)$.*

Proof. A brief outline should suffice here, since the argument is essentially the same as that of Theorem 3.1. For brevity we shall suppress the qualification "almost everywhere in Ω" in this proof. Consider the instant $t = \tau$. Because s is of class C^1, and because $u_\circ(\cdot, \tau)$ and $\varrho_\circ(\cdot, \tau)$ are in $L^\infty(\Omega)$, condition (a) implies that given h and $\tilde{h}$ in $C_0^\infty(\Omega)$ the composite maps $s(u_\circ + \alpha h + \beta\tilde{h}, \varrho_\circ)(\cdot, \tau)$ and $\frac{\partial s}{\partial u}(u_\circ + \alpha h + \beta\tilde{h}, \varrho_\circ)(\cdot, \tau)$ are in $L^\infty(\Omega)$ if α and β are sufficiently small. We can follow the argument in Theorem 3.1 to conclude that $\boldsymbol{v}_\circ(\cdot, \tau) = \boldsymbol{0}$ and $\theta_\circ(\cdot, \tau) = \text{constant} > 0$ ($\theta_\circ(\cdot, \tau) = \infty$ is eliminated by condition (b).) By hypothesis the maps $t \mapsto u_\circ(\cdot, t)$ and $t \mapsto \varrho_\circ(\cdot, t)$ are continuous. Thence conditions (a) and (b) must be valid for all instants t in some interval $\mathscr{T}$ that contains τ. By argument similar to that used in the proof of Theorem 3.1, we can show that $\boldsymbol{v}_\circ = \boldsymbol{0}$ and $\theta_0 = \text{constant} > 0$ over $\mathscr{T}$, and that $\mathscr{T} = [t_0, \infty)$. □

Remark 3.5. Let us compare the result above with that of § 2. Corollary 3.2 certainly remains valid when the constitutive function $s = s(u, \varrho)$ is strictly concave and the potentials $\Phi \equiv 0, \Psi \equiv 0$. Moreover the process defined by $u(\cdot, t) = E/V$, $\varrho(\cdot, t) = M/V$ for each t in $[t_0, \infty)$, where $(E/V, M/V)$ is in $\mathscr{D}^-$ (see § 2 for notation), satisfies all the mathematical hypotheses of Corollary 3.2. In GIBBS's formulation of the Entropy Maximum Principle, $(u, \varrho) = (E/V, M/V)$ is the equilibrium state for the example considered in § 2. In our present formulation, since $\theta(E/V, M/V) < 0$, Corollary 3.2 implies that $(u(\cdot, t), \varrho(\cdot, t)) = (E/V, M/V)$ for t in $[t_0, \infty)$ cannot be an equilibrium process. Thus the two formulations lead to mutually contradictory conclusions.

Remark 3.6. Although Theorem 3.1 and Corollary 3.2 refer to specific types of fluids, it is easy to see that similar results will be valid for a great variety of materials. For example, so long as Eq. (4) is valid, adding higher gradients of ϱ as independent variables in Eq. (3) clearly will not make much difference in the subsequent discussion. Likewise, nothing much needs be changed, should we replace ϱ by the deformation gradient $\boldsymbol{F}$ in Eq. (3) or Eq. (14) and impose the zero-displacement boundary condition. Finally, since we are discussing a criterion of thermodynamic equilibrium, GIBBS's constitutive scheme Eq. (14) when given the right interpretation will cover many materials. In this regard the reader is referred to the discussions of COLEMAN & GREENBERG [5], COLEMAN [14], and DUNN & FOSDICK [6].

Remark 3.7. "Isolated systems" studied in this section are of the type that GIBBS considered in his memoir. With results such as Theorem 3.1 and Corollary 3.2 in mind, we may attempt to salvage GIBBS's original formulation (in which the time t and the velocity $\boldsymbol{v}$ do not appear) by a simple operation, namely, to replace the constraint of equal energy by a constraint of inequality. For example, for the "isolated system" of "the simplest kind" discussed in § 1, the energy constraint would read $\int_\Omega u\,dV \leqq E$ instead of $\int_\Omega u\,dV = E$. Similarly Eq. (7) would become $\int_\Omega (u + \varrho\Phi)\,dV + \int_{\partial\Omega} \Psi\,dA \leqq E$. Moreover, we add a further agreement: should $(u_\circ, \varrho_\circ)$ be such as to render the left-hand side of the constraint equal to E, we would infer that $\boldsymbol{v}_\circ = \boldsymbol{0}$, although the velocity $\boldsymbol{v}$ does not appear explicitly in this formulation. With these modifications of GIBBS's formulation, we can now prove that at equilibrium either (i) $\theta_\circ = \infty$ or (ii) $\theta_\circ = \text{constant} > 0$ and $\boldsymbol{v}_\circ = \boldsymbol{0}$. As for proof, we can follow lines like those of Theorem 3.1. For example, let us consider the system studied in Theorem 3.1. Because the formulation here involves neither the velocity $\boldsymbol{v}$ nor the time t, there will not be an analog of Eq. (13) and the multiplier λ will be a constant real number. By appealing to the Kuhn-Tucker necessary conditions for problems that involve inequality constraints (see, for example, BAZARAA & SHETTY [20], Theorem 4.2.10), we shall obtain the inequality $\lambda \leqq 0$ in addition to the equation $(1/\theta_\circ) + \lambda = 0$. (*Cf.* Eq. (12).) Again from the Kuhn-Tucker conditions we know that for $\theta_\circ > 0$, we have $\int_\Omega (u_\circ + \varrho_\circ\Phi)\,dV + \int_{\partial\Omega} \Psi\,dA = E$, from which we conclude that $\boldsymbol{v}_\circ = 0$. Thence the conclusions under this formulation are the same as those of Theorem 3.1. Indeed the reader can easily convince himself that for the type of "isolated systems" considered in this section, the modifications proposed above will suffice to redeem the honor of GIBBS's formulation by making it a viable alternative to our formulation which involves the velocity and the time explicitly. There is, however, no such easy way to save GIBBS's formulation from disgrace when we consider isolated liquid masses in unconfined motion (see § 5). Not that nothing can be done, but the changes will be enormous and the outcome will be artificial.

Remark 3.8. Physicists have long agreed that some nuclear spin systems can assume negative absolute temperatures. Do our results contradict this? Not at all. The theorems here refer to "isolated systems" in equilibrium; nothing is said about nonequilibrium states of these systems. Moreover our results arise from the kinetic-energy term in the energy constraint equation. For systems without inertia and for rigid bodies subject to boundary conditions of adherence or null displacement, the kinetic energy is annulled and so are the conclusions of our theorems. Physicists usually separate out the spins and treat them as a system in itself, which is without inertia. Then they talk about spin-lattice relaxation, in which the temperature of the spins can be positive or negative but the temperature of the "lattice" is always positive. Our results are entirely compatible with their practice. Indeed some physicists might sneer at the present efforts; to them the conclusions above are everyman's knowledge or are nothing but obvious. What follows is extracted from an undergraduate text on magnetic resonance: "Negative absolute temperature is possible only if the system has the property that the highest energy level of the system is bounded. Clearly excluded by this requirement are systems involving kinetic energy, since in such systems the total energy of a system at negative absolute temperature would be infinite" (SCHUMACHER [21], p. 103). The reasoning is debatable; nonetheless this extract supports my contention that the results above will not do any violence to the intuition of physicists.

Remark 3.9. GIBBS, like everyone else in his time, believed that absolute temperature must be positive. Most later authors who adopted GIBBS's formulation of thermostatics made the same assumption on absolute temperature. An exception was DUNN & FOSDICK [6]. The reader who is interested in seeing how negative absolute temperatures fare under GIBBS's original formulation should consult their work [6], in which they discuss or mention negative absolute temperature at the following places: pp. 5, 6, 39, 50, 51, 53, 54, 56, 57, and Appendix 3. Compare in particular our Corollary 3.2 with their Theorem 13 ([6], pp. 56–57), a gem which refers to states of negative absolute temperature with non-zero kinetic energy.

§ 4

Boundary conditions remain implicit in the proof of Theorem 3.1 and Corollary 3.2. In this section we shall study an example which has a different flavor. We shall prove the analog of Theorem 3.1 for a model system; one boundary condition will play a prominent role in the proof.

The "isolated system" that we shall consider is of the same type as those studied in § 3; only the fluid inside the rigid envelop is different. Since no confusion should arise, we shall use the same terminology and the same symbols as in § 3. Here we shall consider a fluid whose relevant constitutive relations are as follows:

$$s(\Theta, \nabla\Theta, \varrho, \nabla\varrho) = \tilde{s}(\Theta, \varrho, \nabla\varrho) + \tilde{a}(\varrho, \nabla\varrho)\,|\nabla\Theta|^2, \tag{15}$$

$$u(\Theta, \nabla\Theta, \varrho, \nabla\varrho) = \tilde{u}(\Theta, \varrho, \nabla\varrho) + \tilde{b}(\varrho, \nabla\varrho)\,|\nabla\Theta|^2, \tag{16}$$

$$\frac{\partial\tilde{s}}{\partial\Theta} = \Theta\frac{\partial\tilde{u}}{\partial\Theta}, \tag{17}$$

$$\varrho > 0, \quad \partial\tilde{u}/\partial\Theta < 0, \quad \tilde{a} \leqq 0, \quad \text{and} \quad \tilde{b} \geqq 0; \tag{18}$$

here $\Theta \equiv 1/\theta$ is the coldness; the functions s and u are defined over an open set $\mathscr{D}$ in R^8; $\tilde{s}$, $\tilde{u}$, $\tilde{a}$, and $\tilde{b}$ are of class C^1; $\Theta < 0$ over a non-empty proper subset of $\mathscr{D}$.

For each instant t after the "isolated system" is prepared, we shall call the triple $(\Theta(\cdot, t), \varrho(\cdot, t), \boldsymbol{v}(\cdot, t))$ its state at time t. Suppose the "isolated system" is in thermodynamic equilibrium over the interval of time $[t_0, \infty)$. We denote the equilibrium states by $(\Theta_\circ(\cdot, t), \varrho_\circ(\cdot, t), \boldsymbol{v}_\circ(\cdot, t))$. Since we have chosen the coldness Θ to replace the internal energy density u as independent variable, admissible states are now triples of functions $(\Theta, \varrho, \boldsymbol{v})\colon \bar{\Omega} \to R^5$.

The general conditions (i) to (iv) regarding admissible states, the basic assumptions (I) and (II) about equilibrium states, and the formulation of the Entropy Maximum Principle are largely the same as those for Theorem 3.1. The only major difference is that now we explicitly require an admissible state to satisfy the boundary condition

$$\nabla\Theta \cdot \boldsymbol{n}|_{\partial\Omega} = 0, \tag{19}$$

where $\boldsymbol{n}$ is the unit outward normal field on $\partial\Omega$. Otherwise the modifications are obvious. For instance the contact energy in Eq. (7) should now be taken as $\int_{\partial\Omega} \Psi(\Theta, \nabla\Theta, \varrho, \nabla\varrho)\, dA$.

An analog of Theorem 3.1 is valid for the "isolated system" in question. Essential use will be made of boundary condition Eq. (19) in the proof.

Theorem 4.1. *Assume that $\boldsymbol{v}_\circ$ is continuous and $\Theta_\circ$, $\varrho_\circ$ are of class C^1 over $\bar{\Omega}\times [t_0, \infty)$. Let $\theta_\circ(\cdot, t) \equiv 1/\Theta_\circ(\cdot, t)$ be the temperature field corresponding to the equilibrium state $(\Theta_\circ(\cdot, t), \varrho_\circ(\cdot, t), \boldsymbol{v}_\circ(\cdot, t))$. The Entropy Maximum Principle dictates that throughout $[t_0, \infty)$ either* (i) $\theta_\circ = \infty$ *or* (ii) $\boldsymbol{v}_\circ = \boldsymbol{0}$ *and* $\theta_\circ =$ *constant* > 0.

Proof. First we fix an instant τ in $[t_\circ, \infty)$. We shall suppress the dependence on τ in all expressions when no confusion should arise. We introduce the following notations:

$$\frac{\delta S}{\delta\Theta} \equiv \frac{\partial\tilde{s}}{\partial\Theta}(\Theta_\circ, \varrho_\circ, \nabla\varrho_\circ) - \nabla\cdot(2\tilde{a}(\varrho_\circ, \nabla\varrho_\circ)\,\nabla\Theta_\circ), \tag{20}$$

$$\frac{\delta U}{\delta\Theta} \equiv \frac{\partial\tilde{u}}{\partial\Theta}(\Theta_\circ, \varrho_\circ, \nabla\varrho_\circ) - \nabla\cdot(2\tilde{b}(\varrho_\circ, \nabla\varrho_\circ)\,\nabla\Theta_\circ). \tag{21}$$

We claim that $\delta U/\delta\Theta$ does not vanish identically in $\bar{\Omega}$. Indeed, should $\delta U/\delta\Theta = 0$ everywhere in $\bar{\Omega}$, the boundary condition Eq. (19) would imply $\int_\Omega (\partial\tilde{u}/\partial\Theta)\, dV = 0$, which contradicts the hypothesis that $\partial\tilde{u}/\partial\Theta < 0$. Thence we can choose an $\tilde{h}$ in $C_0^\infty(\Omega)$ such that $\int_\Omega (\delta U/\delta\Theta)\, \tilde{h}\, dV \neq 0$. Let h be a function in $C_0^\infty(\Omega)$ and let $\boldsymbol{k}$ be in $[C_0^\infty(\Omega)]^3$.

Let $\mathscr{S}(\alpha, \beta) \equiv \int_\Omega (\tilde{s}(\Theta_\circ + \alpha h + \beta\tilde{h}, \varrho_\circ, \nabla\varrho_\circ) + \tilde{a}(\varrho_\circ, \nabla\varrho_\circ)\, |\nabla\Theta_\circ + \alpha\nabla h + \beta\nabla\tilde{h}|^2)\, dV$, $\mathscr{E}(\alpha, \beta) \equiv \int_\Omega (\tilde{u}(\Theta_\circ + \alpha h + \beta\tilde{h}, \varrho_\circ, \nabla\varrho_\circ) + \tilde{b}(\varrho_\circ, \nabla\varrho_\circ)\, |\nabla\Theta_\circ + \alpha\nabla h + \beta\nabla\tilde{h}|^2 + \frac{1}{2}\varrho_\circ\, |\boldsymbol{v}_\circ + \alpha\boldsymbol{k}|^2)\, dV$, and $\mathscr{E}_\circ \equiv E - \int_\Omega \varrho_\circ\Phi(\boldsymbol{x})\, dV - \int_{\partial\Omega} \Psi(\Theta_\circ, \nabla\Theta_\circ, \varrho_\circ, \nabla\varrho_\circ)\, dV$. The functions $\mathscr{S}$ and $\mathscr{E}$ are defined and are of class C^1 in a neighborhood of (0, 0) in R^2; moreover, at $(\alpha, \beta) = (0, 0)$, $\mathscr{E}(0, 0) = \mathscr{E}_\circ$ and $(\partial\mathscr{E}/\partial\alpha, \partial\mathscr{E}/\partial\beta) = \left(\int_\Omega ((\delta U/\delta\Theta)\, h + \varrho_\circ\boldsymbol{v}_\circ \cdot \boldsymbol{k})\, dV, \int_\Omega (\delta U/\delta\Theta)\, \tilde{h}\, dV\right) \neq (0, 0)$. By the Entropy Maximum Principle, $\mathscr{S}$ assumes a maximum at (0, 0) when (α, β) is constrained to the set defined by the equation $\mathscr{E}(\alpha, \beta) = \mathscr{E}_\circ$. We can appeal to "Euler's rule" (*cf.* the proof of Theorem 3.1 for details) to deduce what follows: There is a real number $\lambda(\tau)$ such that

$$\lambda(\tau)\, \varrho_\circ\boldsymbol{v}_\circ = \boldsymbol{0}, \tag{22}$$

$$\frac{\delta S}{\delta\Theta} + \lambda(\tau)\frac{\delta U}{\delta\Theta} = 0. \tag{23}$$

It follows from Eq. (22) that either (a.) $\lambda(\tau) = 0$ or (b.) $\lambda(\tau) \neq 0$ and $\boldsymbol{v}_\circ(\cdot, \tau) = \boldsymbol{0}$. Let us consider the two possibilities in turn.

(a.) $\lambda(\tau) = 0$. Eq. (23) then implies that $\delta S/\delta\Theta = 0$ or $\Theta_\circ(\partial\tilde{u}/\partial\Theta) - \nabla \cdot (2\tilde{a}\,\nabla\Theta_\circ) = 0$. Since $\tilde{a}(\varrho_\circ, \nabla\varrho_\circ) \leqq 0$ and $\nabla\Theta_\circ \cdot \boldsymbol{n}|_{\partial\Omega} = 0$, we have $0 \geqq \int_\Omega 2\tilde{a}\, |\nabla\Theta_\circ|^2\, dV = \int_\Omega \nabla \cdot (\Theta_\circ(2\tilde{a}\,\nabla\Theta_\circ))\, dV - \int_\Omega \Theta_\circ\nabla \cdot (2\tilde{a}\,\nabla\Theta_\circ)\, dV = -\int_\Omega \Theta_\circ^2(\partial\tilde{u}/\partial\Theta)\, dV \geqq 0$. Thence we conclude that $\Theta_\circ(\cdot, \tau) = 0$ or $\theta_\circ(\cdot, \tau) = \infty$.

(b.) $\lambda(\tau) \neq 0$ and $\boldsymbol{v}_\circ(\cdot, \tau) = \boldsymbol{0}$. First we show that $\lambda(\tau) < 0$. We choose and fix a function $\boldsymbol{k}$ in $[C_0^\infty(\Omega)]^3$ and an $\tilde{h}$ in $C_0^\infty(\Omega)$ such that $\int_\Omega (\delta U/\delta\Theta)\, \tilde{h}\, dV > 0$. Let $\mathscr{F}(\alpha, \beta) \equiv g(\beta) + c\alpha^2$, where $g(\beta) \equiv \int_\Omega (\tilde{u}(\Theta_\circ + \beta\tilde{h}, \varrho_\circ, \nabla\varrho_\circ) + \tilde{b}(\varrho_\circ, \nabla\varrho_\circ)\, |\nabla\Theta_\circ + \beta\nabla\tilde{h}|^2)\, dV - \mathscr{E}_\circ$, and $c \equiv \int_\Omega \frac{1}{2}\varrho_\circ\boldsymbol{k} \cdot \boldsymbol{k}\, dV$. The function $\mathscr{F}$ is defined and is of class C^1 in a neighborhood of (0, 0) in R^2; moreover, at $(\alpha, \beta) = (0, 0)$, $\mathscr{F}(0, 0) = 0$ and $\partial\mathscr{F}/\partial\beta = \int_\Omega (\delta U/\delta\Theta)\, \tilde{h}\, dV > 0$. By the implicit function theorem, the set defined by the equation $\mathscr{F}(\alpha, \beta) = 0$ is given in a neighborhood of (0, 0) by a C^1 function $\beta = \beta(\alpha)$, which is defined in a neighborhood of zero in R^1. Moreover, since $\mathscr{F}(\alpha, \beta) = 0$ implies $g(\beta) = -c\alpha^2 \leqq 0$, it follows from $g(0) = 0$ and $g'(0) = \int_\Omega (\delta U/\delta\Theta)\, \tilde{h}\, dV > 0$ that $\beta(\alpha) \leqq 0$. Thence there is a sufficiently small $d > 0$ such that for β in $(-d, 0]$ and for $\alpha = (-g(\beta)/c)^{\frac{1}{2}}$, the triple $(\Theta_\circ + \beta\tilde{h}, \varrho_\circ, \mathbf{0} + \alpha\mathbf{k})$ defines an admissible state. Let $f(\beta) \equiv \int_\Omega (\tilde{s}(\Theta_\circ + \beta\tilde{h}, \varrho_\circ, \nabla\varrho_\circ) + \tilde{a}(\varrho_\circ, \nabla\varrho_\circ)\,|\nabla\Theta_\circ + \beta\,\nabla\tilde{h}|^2)\,dV$; the function f is of class C^1 over $(-d, 0]$. By the Entropy Maximum Principle, $f'(0) \geqq 0$. Since $f'(0) = \int_\Omega (\delta S/\delta\Theta)\,\tilde{h}\,dV = -\lambda(\tau) \int_\Omega (\delta U/\delta\Theta)\,\tilde{h}\,dV$, we conclude that $\lambda(\tau) < 0$.

Let us proceed to show that $\Theta_\circ(\cdot, \tau) = -\lambda(\tau) > 0$. Indeed, since $\tilde{a}(\varrho_\circ, \nabla\varrho_\circ) \leqq 0$, $\tilde{b}(\varrho_\circ, \nabla\varrho_\circ) \geqq 0$, and $\partial\tilde{u}/\partial\Theta < 0$, we can appeal to Eq. (23) and boundary condition Eq. (19) to obtain what follows: $0 \geqq \int_\Omega 2\tilde{a}\,|\nabla\Theta_\circ|^2\,dV + \int_\Omega 2\lambda(\tau)\,\tilde{b}\,|\nabla\Theta_\circ|^2\,dV = -\int_\Omega (\Theta_\circ + \lambda(\tau))\,(\nabla\cdot(2\tilde{a}\,\nabla\Theta) + \lambda(\tau)\,\nabla\cdot(2\tilde{b}\,\nabla\Theta_\circ))\,dV = -\int_\Omega (\Theta_\circ + \lambda(\tau))^2\,(\partial\tilde{u}/\partial\Theta)\,dV \geqq 0$. Hence, $\int_\Omega (\Theta_\circ + \lambda(\tau))^2\,(\partial\tilde{u}/\partial\Theta)\,dV = 0$, and our claim follows from the hypothesis that $\partial\tilde{u}/\partial\Theta < 0$.

Now we can use arguments similar to those used in Theorem 3.1 to show that in fact either (i) $\theta_\circ = \infty$ over $\bar{\Omega}\times[t_0, \infty)$ or (ii) $\theta_\circ = \text{constant} > 0$ over $\bar{\Omega}\times[t_0, \infty)$. □

§ 5

Discourse on the figures of equilibrium of celestial bodies, which were modelled as liquid masses subjected to no forces but self-gravitation, already had had a long history when GIBBS published the first part of his long memoir on thermostatics. Then the MACLAURIN spheroids and the JACOBI ellipsoids were well known as possible figures of relative equilibrium for rotating masses of incompressible liquids; THOMSON & TAIT, almost a decade earlier, had remarked upon the "ultimate" rotation which dissipation would dictate on an isolated mass of liquid (see Footnote 3 above). Self-gravitating fluid masses are perfect examples of isolated systems; it is strange that they did not seem to figure anywhere in GIBBS's mind when he wrote down his entropy principle. The Entropy Maximum Principle clearly should be valid for the equilibrium states of such unconfined isolated systems; yet it is equally apparent that the specific version meant for "isolated systems" of the type studied by GIBBS (see §§ 3–4 above), namely those which are "isolated" bodies confined in rigid enclosures, will now be unsuitable. The purpose of this section is to formulate the Entropy Maximum Principle for isolated liquid masses in unconfined motion.

Henceforth we shall use a fixed inertial frame of reference to describe motion. We consider an isolated liquid mass $\mathscr{B}$ which is in unconfined motion with respect to the chosen frame. We assume that the body $\mathscr{B}$ remains a continuous mass during the course of motion, the radiant heating r is null, the influx of heating on the boundary $\partial\mathscr{B}$ is null, the contact loading on $\partial\mathscr{B}$ is exerted by surface tension alone and the liquid mass $\mathscr{B}$ is subjected to no body-forces except self-gravitation. For ease of comparison with systems studied in §§ 3–4 and comparison with earlier work by STUECKELBERG and SCHEURER (see § 6) we shall take the liquid in question as compressible, linearly viscous, and heat-conducting (in particular, $q = 0$ if $\nabla\Theta = 0$, where $\Theta = 1/\theta$ is the coldness). The constitutive assumptions most relevant to our discussion below are given in the paragraph that contains Eq. (14) in § 3.

Let $\Omega(t)$ be the region occupied by the body $\mathscr{B}$ at the instant t. Since there is now no envelop to confine the liquid mass to a fixed region of space, in general we have $\Omega(t_1) \neq \Omega(t_2)$ if $t_1 \neq t_2$. Suppose there is an instant t_0 after which the body $\mathscr{B}$ is in thermodynamic equilibrium. We denote the equilibrium states by $(u_\circ(\cdot, t), \varrho_\circ(\cdot, t), \boldsymbol{v}_\circ(\cdot, t))$, where t is an instant in $[t_0, \infty)$. Unlike the "isolated systems" studied in §§ 3–4, equilibrium states in the present context need not have the same domain of definition. For each instant t in $[t_0, \infty)$, we assume that $\Omega(t)$ is a connected bounded open set in space and $\partial\Omega(t)$ is a compact C^2 manifold-without-boundary.

The statement of the Entropy Maximum Principle remains, at least formally, the same as before (see the paragraph that contains Eq. (9) in § 3). Only the class of competitors is defined differently. An admissible state is a triple of functions $(u, \varrho, \boldsymbol{v})\colon \bar{\Omega} \to R^5$; here Ω is *some* connected, bounded, open set in space, the boundary $\partial\Omega$ of which is a compact C^2 manifold-without-boundary. It should be emphasized that Ω need not be equal to $\Omega(t)$ for any instant t; in other words it need not be a region once actually occupied by the liquid mass during the course of motion. We require the admissible states to satisfy the following conditions:

(i) Integral constraints. Let M, E, and $\mathscr{L}$ be the total mass, total energy, and total linear momentum of the liquid mass $\mathscr{B}$ at any instant t, respectively; let $\mathscr{M}$ be the total angular momentum of $\mathscr{B}$ about the origin of the chosen frame of reference. An admissible state of the isolated liquid mass satisfies the mass constraint Eq. (8) and the integral constraints

$$\int_\Omega \varrho\boldsymbol{v}\, dV = \mathscr{L}, \tag{24}$$

$$\int_\Omega (\boldsymbol{x} \times \varrho\boldsymbol{v})\, dV = \mathscr{M}, \tag{25}$$

$$\int_\Omega (u + \tfrac{1}{2}\varrho\boldsymbol{v}\cdot\boldsymbol{v})\, dV + \tfrac{1}{2}\int_\Omega \varrho\phi(\boldsymbol{x})\, dV + \int_{\partial\Omega} \sigma(u, \varrho)\, dA = E; \tag{26}$$

in Eq. (26) the symbol σ denotes the coefficient of surface tension and

$$\phi(\boldsymbol{x}) \equiv -G \int_\Omega \frac{\varrho(\boldsymbol{x}')}{|\boldsymbol{x} - \boldsymbol{x}'|}\, dV', \tag{27}$$

where G is the constant of universal gravitation. Let $\boldsymbol{x}_c(\tau) \equiv \left(\int_{\Omega(\tau)} \varrho_\circ(\boldsymbol{x}, \tau)\, \boldsymbol{x}\, dV\right)\Big/ M$ be the center of mass of $\mathscr{B}$ at the instant τ. Eq. (25) is obviously equivalent to the equation

$$\int_\Omega (\boldsymbol{x} - \boldsymbol{x}_c(\tau)) \times \varrho\boldsymbol{v}\, dV = \mathscr{M} - \boldsymbol{x}_c(\tau) \times \mathscr{L}. \tag{28}$$

(ii) Boundary conditions. An admissible state satisfies Eq. (19). Although we have made the physical assumption that the contact loading on $\partial\mathcal{B}$ is exerted by surface tension alone, we do not require the competitors to satisfy a boundary condition to that effect. As a result of our formulation, we can deduce certain natural boundary conditions which an equilibrium state must satisfy. Our assumption about surface tension will reappear as one of those conditions.

(iii) Mathematical conditions. An admissible state should satisfy mathematical conditions such that all the requirements above make sense, the composite function $s(u, \varrho)(\cdot)$ is well defined and is integrable over Ω, *etc.*

To illustrate what the Entropy Maximum Principle will imply, we shall prove an analog of Theorem 3.1 for the isolated liquid mass in unconfined motion. We adopt the following
Basic Assumptions:
(*i*) For each instant t in $[t_0, \infty)$, $(u_\circ(\cdot, t), \varrho_\circ(\cdot, t), \boldsymbol{v}_\circ(\cdot, t))$ is an admissible state.
(*ii*) The following assertion is valid for each instant t in $[t_0, \infty)$: Given any h and $\tilde{h}$ in $C_0^\infty(\Omega(t))$, any $\boldsymbol{k}$ in $[C_0^\infty(\Omega(t))]^3$, and any vector field $\boldsymbol{w}$ on $\bar{\Omega}(t)$ that can be taken as the snapshot velocity field of some rigid motion (*i.e.*, $\boldsymbol{w} = \boldsymbol{a} + \boldsymbol{b} \times (\boldsymbol{x} - \boldsymbol{x}_c(t))$ for some vectors $\boldsymbol{a}$ and $\boldsymbol{b}$), there is a neighborhood $\mathcal{N}$ of (0, 0) in R^2 such that for all (α, β) in $\mathcal{N}$, the triple $(u_\circ + \alpha h + \beta\tilde{h}, \varrho_\circ, \boldsymbol{v}_\circ + \alpha\boldsymbol{k} + \boldsymbol{w}) : \bar{\Omega}(t) \to R^5$ defines an admissible state of the isolated liquid mass if it satisfies the integral constraints Eqs. (24), (25) and (26), and if the composite map $s(u_\circ + \alpha h + \beta\tilde{h}, \varrho_\circ)(\cdot)$ is well-defined and is integrable over $\Omega(t)$.

For the *liquid* mass $\mathcal{B}$, we need not allow $u_\circ$ and $\varrho_\circ$ to have discontinuities.

Theorem 5.1. *Let the isolated liquid mass $\mathcal{B}$ be in thermodynamic equilibrium throughout the interval of time $[t_0, \infty)$. For each t in $[t_0, \infty)$, we assume that $u_\circ(\cdot, t)$, $\varrho_\circ(\cdot, t)$, and $\boldsymbol{v}_\circ(\cdot, t)$ are continuous over $\bar{\Omega}(t)$. For each material point X of the liquid mass, we assume that $u_\circ(X, \cdot)$, $\varrho_\circ(X, \cdot)$, and $\boldsymbol{v}_\circ(X, \cdot)$ are continuous over $[t_0, \infty)$. Let $\theta_\circ(\cdot, t)$ be the temperature field corresponding to the equilibrium state $(u_\circ(\cdot, t), \varrho_\circ(\cdot, t), \boldsymbol{v}_\circ(\cdot, t))$. The Entropy Maximum Principle dictates that throughout $[t_0, \infty)$ either* (i) $\theta_\circ = \infty$ *or* (ii) *the body $\mathcal{B}$ is in rigid motion and $\theta_\circ = constant > 0$.*

Proof. First we fix an instant τ in $[t_0, \infty)$. Let $\boldsymbol{x}_c(\tau)$ denote the center of mass of body $\mathcal{B}$ at the instant τ; $\boldsymbol{x}_c(\tau)$ is a place in $\Omega(\tau)$. Let $\boldsymbol{e}_1(\tau), \boldsymbol{e}_2(\tau), \boldsymbol{e}_3(\tau)$ be an orthonormal triad which defines the principal axes of inertia of $\mathcal{B}$ at time τ. Henceforth we shall suppress the dependence on τ when no confusion should arise. Let $\tilde{h}$ be a function in $C_0^\infty(\Omega(\tau))$ such that $\int_\Omega \tilde{h}\, dV \neq 0$. Let $\boldsymbol{k}_i = \boldsymbol{e}_i$, $\boldsymbol{k}_{3+i} = \boldsymbol{e}_i \times (\boldsymbol{x} - \boldsymbol{x}_c)$ for $i = 1, 2, 3$. Given a function h in $C_0^\infty(\Omega(\tau))$, $\boldsymbol{k}$ in $[C_0^\infty(\Omega(\tau))]^3$, and six real numbers γ_j $(j = 1, \ldots, 6)$, by Basic Assumption (*ii*) the triple $(u_\circ + \alpha h + \beta\tilde{h}, \varrho_\circ, \boldsymbol{v}_\circ + \alpha\boldsymbol{k} + \sum_j \gamma_j\boldsymbol{k}_j) : \bar{\Omega}(\tau) \to R^5$ will define an admissible state of the liquid mass when α and β are sufficiently small, provided that the triple in question satisfies Eqs. (24), (26) and (28). We can appeal to the Entropy Maximum Principle and "Euler's rule" to conclude what follows (details of argument are similar to those in the proof of Theorem 3.1): There are vectors $\zeta(\tau)$, $\xi(\tau)$ and a scalar $\lambda(\tau)$ such that

$$\frac{\partial s}{\partial u}(u_\circ, \varrho_\circ) + \lambda(\tau) = 0, \tag{29}$$

$$\lambda(\tau)\, \boldsymbol{v}_\circ + \zeta(\tau) + \xi(\tau) \times (\boldsymbol{x} - \boldsymbol{x}_c(\tau)) = \mathbf{0}. \tag{30}$$

Eqs. (29) and (30) are valid for each instant τ in $[t_0, \infty)$. There are two possibilities: (i) $\lambda(t) = 0$ for all t in $[t_0, \infty)$; (ii) there is an instant t in $[t_0, \infty)$ such that $\lambda(t) \neq 0$. Under case (ii) we can show by arguments similar to those used in the proof of Theorem 3.1 that $\lambda(\cdot) = \text{constant} \neq 0$ and the body $\mathscr{B}$ is in rigid motion during the interval of time $[t_0, \infty)$. The velocity field of $\mathscr{B}$ at time t, namely $\boldsymbol{v}_\circ(\cdot, t)\colon \overline{\Omega}(t) \to R^3$, is given by $\boldsymbol{v}_\circ(\boldsymbol{x}, t) = (\mathscr{L}/M) + \boldsymbol{\omega}(t) \times (\boldsymbol{x} - \boldsymbol{x}_c(t))$, where $\boldsymbol{\omega}(t) = -\xi(t)/\lambda(t)$ is the angular velocity at time t.

Let us proceed to show that $\lambda(\cdot) = \text{constant} < 0$ in case (ii). We fix again an instant τ in $[t_0, \infty)$. Let $\boldsymbol{k}_0$ be a non-trivial function in $[C_0^\infty(\Omega(\tau))]^3$. Let $\boldsymbol{k}_j$ $(j = 1, \ldots, 6)$ be defined as above. There are constants α_j $(j = 1, \ldots, 6)$ such that $\int_{\Omega(\tau)} \varrho_\circ \left(\boldsymbol{k}_0 + \sum_j \alpha_j \boldsymbol{k}_j\right) dV = \mathbf{0}$ and $\int_{\Omega(\tau)} (\boldsymbol{x} - \boldsymbol{x}_c(\tau)) \times \varrho_\circ \left(\boldsymbol{k}_0 + \sum_j \alpha_j \boldsymbol{k}_j\right) dV = \mathbf{0}$. Let $\boldsymbol{k} \equiv \boldsymbol{k}_0 + \sum_j \alpha_j \boldsymbol{k}_j$; clearly $\boldsymbol{k}$ does not vanish identically over $\overline{\Omega}(\tau)$. We choose a function $\tilde{h}$ in $C_0^\infty(\Omega(\tau))$ such that $\int_{\Omega(\tau)} \tilde{h}\, dV = -\int_{\Omega(\tau)} \tfrac{1}{2}\varrho_\circ\, \boldsymbol{k} \cdot \boldsymbol{k}\, dV < 0$. Since $\int_{\Omega(\tau)} \boldsymbol{v}_\circ(\boldsymbol{x}, \tau) \cdot \varrho_\circ\, \boldsymbol{k}\, dV = 0$, there is a constant $c > 0$ such that for each ε in $[0, c)$, the triple $(u_\circ + \varepsilon \tilde{h}, \varrho_\circ, \boldsymbol{v}_\circ + \varepsilon^{\frac{1}{2}} \boldsymbol{k})\colon \overline{\Omega}(\tau) \to R^5$ defines an admissible state of the body $\mathscr{B}$. We can appeal to the Entropy Maximum Principle to deduce that $\lambda(\tau) < 0$. (*Cf.* the proof of Theorem 3.1 for details.) □

Remark 5.1. By the preceding theorem a mass of compressible, linearly viscous and heat-conducting liquid "if left to itself in any state of motion ... will ultimately rotate as a rigid solid" in an inertial frame with respect to which its center of mass is at rest. While the quoted remark of THOMSON & TAIT originally refers to "a mass of any ordinary liquid" (see Footnote 3 above), we have demonstrated that at least for the type of liquid under our consideration their observation is borne out as a consequence of the Entropy Maximum Principle.

Of course more can be deduced from the Entropy Maximum Principle than Theorem 5.1. However, to study all the implications of the entropy principle for some particular system is not the purpose of the present essay. Let me finish this section by citing some further results, which will be useful in § 6.

In deriving Theorem 5.1 we have considered, for a fixed instant τ in $[t_0, \infty)$, variations in u and $\boldsymbol{v}$ only; the equilibrium density $\varrho_\circ$ and the region $\Omega(\tau)$ are kept fixed. Should we consider admissible variations in u, $\boldsymbol{v}$, ϱ and $\Omega(\tau)$ altogether (and suitably adjust the basic mathematical assumption (*ii*)), we shall obtain in addition to Theorem 5.1 two more equations when $\theta_\circ > 0$. The first equation is

$$\nabla p_\circ + \varrho_\circ \nabla(\phi_\circ - \tfrac{1}{2}|\boldsymbol{v}_\circ|^2) = \boldsymbol{0}, \tag{31}$$

where $p_\circ \equiv -u_\circ + \theta_\circ s(u_\circ, \varrho_\circ) - \theta_\circ\varrho_\circ \frac{\partial s}{\partial \varrho}(u_\circ, \varrho_\circ)$ is the equilibrium pressure, and $\phi_\circ$ is obtained when ϱ is replaced by $\varrho_\circ$ in Eq. (27). The second one is the natural boundary condition

$$p_\circ(\boldsymbol{x}, \tau)|_{\partial\Omega(\tau)} = 2\sigma H, \tag{32}$$

where $H(\boldsymbol{x}, \tau)$ is the mean curvature at the place $\boldsymbol{x}$ on $\partial\Omega(\tau)$. Eq. (32) emerges because admissible states need not have $\overline{\Omega}(\tau)$ as their domain of definition. When the effects of surface tension are ignored (*i.e.*, $\sigma \equiv 0$), the natural boundary condition Eq. (32) reduces to

$$p_\circ(\boldsymbol{x}, \tau)|_{\partial\Omega(\tau)} = 0, \tag{33}$$

which is usually adopted in the classical theory of rotating stars.

Since the motion of $\mathscr{B}$ during the interval of time $[t_0, \infty)$ should also obey the balance of linear momentum, we have

$$\varrho_\circ \boldsymbol{a}_\circ = -\nabla p_\circ - \varrho_\circ \nabla\phi_\circ, \tag{34}$$

where $\boldsymbol{a}_\circ$ is the acceleration field in question. Comparing Eq. (34) with Eq. (31), we see that

$$\boldsymbol{a}_\circ = \nabla(-\tfrac{1}{2}|\boldsymbol{v}_\circ|^2), \tag{35}$$

which implies

$$\nabla \times \boldsymbol{a}_\circ = 2\frac{d}{dt}\boldsymbol{\omega}(t) = \boldsymbol{0}. \tag{36}$$

Thence the "ultimate" rotation must be steady. By "Euler's theorem" we conclude that the liquid mass $\mathscr{B}$ is ultimately in a steady rigid rotation about a principal axis of inertia through the center of mass, which moves at a constant velocity. rn fact, as a consequence of the Entropy Maximum Principle, the body $\mathscr{B}$ must Iotate around an axis associated with the largest moment of inertia.

Remark 5.2. For the type of liquid under consideration, the Entropy Maximum Principle has thus led us to the starting point of the classical theory of figures of celestial bodies, the "fundamental problem" of which concerns an isolated self-gravitating mass of fluid in steady rigid-body rotation about a fixed axis in an inertial frame (see, for example, JARDETZKY [22], Ch. I). It will be interesting to study what the Entropy Maximum Principle can add to the foundations of this classical subject. Nevertheless, I do not wish here to delve into the theory of figures of celestial bodies.

Remark 5.3. In the discussion above we have taken the liquid in question as linearly viscous. This assumption is clearly not essential. Replacing the linearly viscous fluid by VAN DER WAALS's gradient-type fluid (see Eqs. (3) and (4)) will not invalidate Theorem 5.1. Neither will occurrence of higher gradients of ϱ invalidate that theorem, so long as Eq. (4) remains valid. The results cited after Theorem 5.1 are more special. They rely on the assumption that GIBBS's constitutive scheme for substances be valid (*i.e.*, we have Eq. (14) plus the Cauchy stress $T = -pI$, where $p = -u + \theta s - \theta \varrho(\partial s/\partial \varrho)$ is the thermodynamic pressure). The range of GIBBS's scheme is in fact quite broad (see COLEMAN & GREENBERG [5], COLEMAN [14], and DUNN & FOSDICK [6]). Of course, for consistency the liquid mass should also possess some internal dissipative mechanism which will ultimately bring it to thermodynamic equilibrium; thence THOMSON & TAIT referred to "ordinary" liquid in their assertion (quoted in Footnote 3 above).

§ 6

In a series of papers that extended for almost two decades, the Swiss physicist E. C. G. STUECKELBERG DE BREIDENBACH set himself the task to determine through "phenomenological thermodynamics ... the sign of all local state functions occuring in the equations of motion of a fluid [non-relativistic (Galilei covariant) ..., relativistic (restricted = Lorentz covariant, including time reversal) ... and in general relativity ...] in terms of the sign of absolute temperature ... and of the signature of a positive or negative definite space metric ..." ([9], p. 888; the explanation in square brackets is STUECKELBERG & SCHEURER's). His efforts in non-relativistic thermomechanics, which concern us here, were helped by his "disciple and collaborator" P. B. SCHEURER. The critique below is based on SCHEURER's doctoral thesis (published as a joint paper [9] with the mentor) and the book written by them [10].

STUECKELBERG & SCHEURER's paper [9] was mainly devoted to applying the Entropy Maximum Principle, which they called "the 2nd law part b", to a compressible, linearly viscous fluid that obeys Fourier's law of conduction. Both their formulation of the Entropy Maximum Principle and the means by which they exploited it require careful comment. For ease of comparison, in what follows I shall silently convert their notation to that we have adopted above.

They started with the problem of an isolated fluid mass that "is free to move in space (no 'container')"; self-gravitation and surface tension were not considered. As we shall see, ignoring both self-gravitation and surface tension will lead to disastrous results, but let us set aside the impending calamity for the moment. At this point their problem would be essentially the same as that which we have considered in § 5, should we set $G \equiv 0$ and $\sigma \equiv 0$ in our formulation. However STUECKELBERG & SCHEURER quickly announced, "[i]n all cases considered ..., the equilibrium is stationary (more exactly static)", by which they meant the liquid mass occupies "a time independent region" Ω when it comes to thermodynamic equilibrium and the equilibrium "local state functions $s_0(\boldsymbol{x})$, $\varrho_0(\boldsymbol{x})$ and $\boldsymbol{v}_0(\boldsymbol{x})$" are all independent of time. "It is an open question to us whether non-stationary equilibria $\partial \varrho_0(\boldsymbol{x}, t)/\partial t \neq 0$ ... exist [for fluids]", they remarked in Footnote 17 of their paper [9]. In their variational formulation of the Entropy Maximum Principle, admissible states are triples of functions $(s, \varrho, \boldsymbol{v})$ that are defined on the same *fixed* region $\overline{\Omega}$ and satisfy integral constraints given by our Eqs. (8), (24), (25) and Eq. (26) with $G \equiv 0$ and $\sigma \equiv 0$. They did not explicitly require the admissible states to satisfy any boundary condition.

For a fluid mass "that is free to move in space", their assumption of "stationary equilibrium" and their formulation of the Entropy Maximum Principle are totally unacceptable. For them, the inertial frame of reference must be chosen so that the total linear momentum $\mathscr{L} = \mathbf{0}$, an obvious trifle which STUECKELBERG & SCHEURER did not even bother to mention. Granted that $\mathscr{L}$ is null in the chosen frame, the assumption that the fluid mass will ultimately occupy a time-independent region of space is still too narrow. This assumption is invalid even for a Jacobi ellipsoid that rotates steadily about its shortest axis. Restricting admissible states to those defined on the same fixed region $\bar{\Omega}$ is even worse. It will preclude the deduction of natural boundary conditions and render the variational formulation incomplete. In their theory the equilibrium pressure p_0 should satisfy the boundary condition

$$p_\circ(\mathbf{x})|_{\partial\Omega} = 0; \tag{37}$$

Eq. (37), however, did not appear in their work. As a consequence STUECKELBERG & SCHEURER overlooked a deadly defect of their theory. They deduced the following relation for the chemical potential $\mu_\circ$ (see [9], Eq. (0.46)): $\mu_\circ(\mathbf{x}) = \mu_{\circ\circ} + \frac{1}{2}|\mathbf{v}_\circ(\mathbf{x})|^2$, where $\mu_{\circ\circ}$ is a constant. The foregoing relation is equivalent to our Eq. (31) when $G \equiv 0$, *i.e.*,

$$\nabla p_\circ + \varrho_\circ \nabla(-\tfrac{1}{2}|\mathbf{v}_\circ|^2) = \mathbf{0}. \tag{38}$$

The boundary-value problem set by Eq. (38) and Eq. (37) has no solution unless the region Ω is an infinite cylinder whose generator coincides with the axis of steady rotation, the pressure $p_\circ$ is everywhere negative inside the infinite cylinder and the total mass of the liquid is infinite. Thence all that STUECKELBERG & SCHEURER deduced, including their conclusion that "mass density ... has the sign of absolute temperature", was built upon the false premiss of the non-existent.

The Entropy Maximum Principle can be formulated in such away that the effects of self-gravitation and surface tension are included and STUECKELBERG & SCHEURER's assumption of "stationary equilibrium" is entirely unnecessary, as our discussion in § 5 should have amply demonstrated.

Let us now turn to their mathematics. STUECKELBERG & SCHEURER's major tool was the second-order condition of a mathematical assertion that involved Lagrange multipliers. Their condition pertains to the maximum of a functional of the "density type", whose unknown state variables are constrained by a finite number of equations expressed by functionals of the same type. By a functional of the "density type" they meant an integral whose integrand is a function of the unknown state variables and the spatial variable $\mathbf{x}$ only, all the variables being defined over a fixed region of space. For example, when the unknown state variables are u, ϱ, and $\mathbf{v}$, which are defined on the fixed region $\bar{\Omega}$, the most general functional of the "density type" has the form $\int_\Omega f(\mathbf{x}, u, \varrho, \mathbf{v})\,dV$, where the integrand f is some given function of $\mathbf{x}$, u, ϱ, and $\mathbf{v}$. With Θ and $\mathbf{v}$ as unknown state variables, the functional

$\int_{\Omega} (\tilde{s}(\Theta, \varrho_{\circ}, \nabla\varrho_{\circ}) + \tilde{a}(\varrho_{\circ}, \nabla\varrho_{\circ})\ |\nabla\Theta|^2)\, dV$ in Theorem 4.1, where $\varrho_{\circ}$ is a given function, is *not* of the "density type"; likewise, when the effects of self-gravitation and surface tension are included, the total energy is strictly speaking *not* a functional that falls under their "density type". The condition STUECKELBERG & SCHEURER used is the analog of what BOLZA called "the second necessary condition for the isoperimetric problem" ([19], § 40, p. 217). Undoubtedly it will be valid under suitable mathematical hypotheses. Their problems are to find those hypotheses and to determine whether the proposed hypotheses are physically acceptable. In this regard their record is deplorable. STUECKELBERG DE BREIDENBACH first discussed the condition at issue in the appendix to a paper of 1962 [8]. Five years later STUECKELBERG & SCHEURER [9] wrote, "the [earlier] proof ... was too short and left some ambiguity. Thus we devote the present paper to a more rigorous proof which should satisfy [the] physicist. Another paper, based on well-known mathematical methods is in preparation." The paper [23] which STUECKELBERG & SCHEURER promised to write with "the mathematician Dr. J. Poncet" and to contain "a rigorous proof of our theorem" came out four more years later. I find the earlier "proofs" (see [8], [9]) totally incomprehensible, and I cannot understand why their final "rigorous proof" was stuffed with mathematical assumptions so strong as to make the theorem, even if correct, almost physically useless.

I have not used STUECKELBERG'S mathematical assertion in proving the theorems in §§ 3–5. So far as showing the relation between the sign of $\varrho_{\circ}$ and $\theta_{\circ}$ is concerned, my method clearly has two advantages over STUECKELBERG'S: (1.) It is not restricted to systems which involve only functionals of the "density type". See, for example, the model system studied in § 4, to which STUECKELBERG'S mathematical assertion is not applicable. (2.) In all my proofs the constitutive relation for the entropy density, *e.g.*, $s = s(u, \varrho)$, $s = s(u, \varrho, \nabla\varrho)$, *etc.*, is assumed to be of class C^1. STUECKELBERG'S, being a second-order condition, will require stronger assumptions of smoothness. Readers who are familiar with problems in the foundations of thermodynamics will understand the enormous difference between assuming C^1-smoothness and C^2-smoothness of the entropy density.

In their book ([10], § 4.1) STUECKELBERG DE BREIDENBACH & SCHEURER considered also the "isolated system" of "the simplest kind" which we have discussed in § 1 (*cf.* also § 3 above). There they made a grave mistake. After they obtained the result $\boldsymbol{v}_{\circ}(\boldsymbol{x}) = \boldsymbol{0}$ (their Eq. (4.1.3.v)), they remarked, "consequently we can replace the [total] energy with the [total] internal energy." With these words they set $\boldsymbol{v} = \boldsymbol{0}$ in the energy constraint equation and reverted to GIBBS'S original formulation of the Entropy Maximum Principle. They spent much of the rest of that section discussing stability conditions both for $\theta_{\circ} > 0$ and for $\theta_{\circ} < 0$. Thus our Corollary 3.2, a version of which they should have hit upon, completely escaped their notice.

In summary, STUECKELBERG DE BREIDENBACH deserves credit for having had the physical insight to include the velocity in formulating the Entropy Maximum Principle. But, to develop this basic idea requires mathematics and careful logical analysis, for which physical intuition is no substitute. It is lapses of logic and weakness in mathematics that give rise to errors and difficulties in the work of STUECKELBERG & SCHEURER.

§ 7

"If an isolated system is not in equilibrium, we can associate no entropy with it." (TISZA [11], p. 41) For physicists who uphold this ultra-cautious creed,[12] GIBBS's criteria of equilibrium, at any rate his original statements of them, will appear as "lack[ing] precision to the point of being paradoxical" ([11], p. 41). GIBBS spoke of "all possible variations of the state of [an isolated] system which do not alter its energy" and talked about "the energy and the entropy of the system *in its varied state*" ([1], p. 56). What can these varied states be, if they differ from the original equilibrium state and if entropy can be assigned to them without violating the creed?

According to TISZA ([11], p. 41), "many authors have grappled with this dilemma until a satisfactory solution was found in terms of the *composite system*." For definiteness let us restrict our discussion to the "isolated system" of "the simplest kind" considered in § 1. The ultra-cautious physicist will proceed as follows: The region Ω is divided into spatially disjoint subregions by adiabatic and impermeable partitions. When each subsystem is assigned an equilibrium state, the original system, now a composite, is said to be in a state of "constrained equilibrium" because "internal constraints" (here, the adiabatic and impermeable partitions) are required to maintain such a state. Given a state of "constrained equilibrium", the total entropy, internal energy and mole number of the "composite system" are taken as the corresponding sum of values assumed by the subsystems, respectively. In the present context the Entropy Maximum Principle is formulated as follows: The values of internal energy and mole number assumed by the subsystems in the absence of internal constraints are those that maximize the entropy of the "composite system" over the totality of constrained equilibrium states which have the given total internal energy and mole number. (*Cf.* CALLEN [12], p. 24, Postulate II.)

TISZA, who strove to give "a new postulational basis" to "Gibbsian thermodynamics" by axiomatizing the use of composite systems [25], attributed the origin of the concept itself to PLANCK (see Footnote 6 in § 1, above). Presumably from PLANCK's influence, "[a]t present the concept [of composite systems] is in general use" ([11], p. 41); indeed, the textbook literature has provided sufficient evidence.

Let us examine the formulation above more closely. It is in fact not much different from what we call GIBBS's original formulation in § 1. If in GIBBS's field formulation we require the unknown state variables u and ϱ to be step functions, integrals will become finite sums and we shall obtain the ultra-cautious formulation. Conversely the phase rule will ensure that consideration of step functions alone would suffice, should pathological cases be ignored. Indeed, when GIBBS considered "isolated systems" of "the simplest kind", in which he included also an enclosed "mass of matter" naving variable composition, he studied "conditions relating to the equilibrium between the initially existing homogeneous parts of the given mass" ([1], pp. 62–70). There he adopted a formulation mathematically the same as that of the ultra-cautious, just without the horde of linguistic (or

[12] After P. GERMAIN ([24], § 17.3), who used the word "cautious" to label "those theories which start with the assumption of a local state".

conceptual, as the ultra-cautious physicist will retort) gadgets such as "composite system", "internal constraints" and the like. But there are numerous situations under which the ultra-cautious formulation is untenable, *e.g.*, when gravity is present. GIBBS used the field formulation for all those. He certainly did not subscribe to the ultra-cautious creed. (See [26], p. 39, where GIBBS associated entropy with a body which "is not in a state of thermodynamic equilibrium" and whose energy includes "*the vis viva of sensible motions*".)

The main drawback of the composite-system approach comes hand in hand with what its promotors would regard as its strength, namely, a complete divorce from the dynamics of processes. Notions such as "constrained equilibrium", "adiabatic and impermeable partitions", *etc.*, may help the ultra-cautious physicist psychologically to allow himself in effect to associate entropy with a class of non-equilibrium states of the "isolated system". But the same liguistic or conceptual gadgets will simutaneously make it impossible for him even to think of questions such as dynamical significance of the Entropy Maximum Principle or materials such as those of the gradient type, which do not fall within GIBBS's constitutive scheme for substances.

Since the velocity field is antithetical to the ultra-cautious formulation, all the theorems in this essay, including Corollary 3.2, are *a priori* barred from those who subscribe to the composite-system approach. Unfortunately that approach and its variants are so widely adopted that erroneous assertions abound when negative absolute temperature is discussed in textbooks.[13]

§ 8

When GIBBS formulated his Entropy Maximum Principle, he had in mind a particular class of isolated systems; moreover, everyone then believed that absolute temperature must be positive. GIBBS's formulation is perfectly all right for the intended class of isolated systems, if the masses of matter in question can assume only positive absolute temperatures. It may become inadequate when those premisses fail to apply, *e.g.*, when we consider isolated liquid masses in unconfined motion or when negative absolute temperature is a possibility. In this essay I have proposed a new version of the Entropy Maximum Principle in which the time and the velocity appear explicitly. For the two circumstances just mentioned, the reformulated entropy principle remains valid while GIBBS's original formulation fails. During our discussion I have also pointed out, with concrete examples in illustration, that boundary conditions can be important to thermostatic analysis, a truism which traditional thermodynamicists are prone to forget.

Acknowledgment. This essay has its origin in Appendix A to Part 2 of my M. Phil. thesis [28]. I am grateful to Dr. C. L. CHAN, my adviser then at the University of Hong Kong, both for his encouragement and for his giving me the opportunity to study for the

[13] For example, "the conditions for equilibrium stability of states of a system with negative absolute temperatures are the same as at positive absolute temperatures" (BAZAROV [27], p. 215), and the preceding assertion is supposed to be valid also for a fluid.

degree. The manuscript of this essay has undergone numerous changes since then. The most significant revision took place after I presented §§ 1–3 at the Institute for Mathematics and its Applications, University of Minnesota, in September 1984. An objection against my comments on the inequality-constraint formulation (see Remark 3.7), which Professor J. L. ERICKSEN raised during my presentation, prompted me to rewrite much of what I had written and to add what is now § 5 of this essay. After my talk Professor J. B. SERRIN kindly suggested to me problems for further work. § 4 has grown from the work he suggested. I am also thankful to Dr. J. E. DUNN and Professor R. L. FOSDICK for their comments on draught of this essay and for helpful discussion on the motivations for GIBBS's formulation of the Entropy Maximum Principle.

Much of the research reported here was done at the Institute for Mathematics and its applications, University of Minnesota, Minneapolis.

References

1. J. W. GIBBS, "On the equilibrium of heterogeneous substances", pp. 55–353 of *The Scientific Papers of J. Willard Gibbs*, Volume 1, Dover Publications, New York, 1961.
2. J. W. GIBBS, "Abstract of the 'Equilibrium of heterogeneous substances'", pp. 354–371 of *The Scientific Papers of J. Willard Gibbs*, Volume 1, Dover Publications, New York, 1961.
3. J. D. VAN DER WAALS, "Théorie thermodynamique de la capillarité, dans l'hypothèse d'une variation continue de densité", *Archives Néerlandaises des Sciences Exactes et Naturelles* **28** (1895), 121–209; trans. into English with an introduction by J. S. ROWLINSON, *Journal of Statistical Physics* **20** (1979), 197–244.
4. B. D. COLEMAN & W. NOLL, "On the thermostatics of continuous media", *Arch. Rational Mech. Anal.* **4** (1959), 97–128.
5. B. D. COLEMAN & J. M. GREENBERG, "Thermodynamics and the stability of fluid motion", *Arch. Rational Mech. Anal.* **25** (1967), 321–341.
6. J. E. DUNN & R. L. FOSDICK, "The morphology and stability of material phases", *Arch. Rational Mech. Anal.* **74** (1980), 1–99.
7. W. THOMSON & P. G. TAIT, *Treatise on Natural Philosophy*, Clarendon Press, Oxford, 1867.
8. E. C. G. STUECKELBERG, "Relativistic thermodynamics III: velocity of elastic waves and related problems", *Helv. Phys. Acta* **35** (1962), 568–591.
9. E. C. G. STUECKELBERG DE BREIDENBACH & P. B. SCHEURER, "Phenomenological thermodynamics V: the 2nd law applied to extensive functionals with the use of Lagrange multipliers", *Helv. Phys. Acta* **40** (1967), 887–906.
10. E. C. G. STUECKELBERG DE BREIDENBACH & P. B. SCHEURER, *Thermocinétique Phénoménologique Galiléenne*, Birkhäuser Verlag, Basel *etc.*, 1974.
11. L. TISZA, *Generalized Thermodynamics*, The M.I.T. Press, Cambridge (Massachusetts) 1966.
12. H. B. CALLEN, *Thermodynamics*, John Wiley & Sons, New York *etc.*, 1960.
13. M. PLANCK, "Bemerkungen über Quantitätsparameter, Intensitätsparameter und stabiles Gleichgewicht", *Physica* **2** (1935), 1029–1032; reprinted as pp. 681–684 of *Physikalische Abhandlungen und Vorträge*, Band II, Friedr. Vieweg & Sohn, Braunschweig, 1958.
14. B. D. COLEMAN, "On the stability of equilibrium states of general fluids", *Arch. Rational Mech. Anal.* **36** (1970), 1–32.

15. W. NOLL, "On certain convex sets of measures and on phases of reacting mixtures", *Arch. Rational Mech. Anal.* **38** (1970), 1–12.
16. J. W. CAHN, "Critical point wetting", *J. Chem. Phys.* **66** (1977), 3667–3672.
17. S. LANG, *Real Analysis*, Addison-Wesley Publishing Company, Reading (Massachusetts) *etc.*, 1969.
18. K. YOSIDA, *Functional Analysis*, Second Edition, Springer-Verlag, Berlin *etc.*, 1968.
19. O. BOLZA, *Lectures on the Calculus of Variations*, Third Edition, Chelsea Publishing Company, New York, 1973.
20. M. S. BAZARAA & C. M. SHETTY, *Nonlinear Programming*, John Wiley & Sons, New York *etc.*, 1979.
21. R. T. SCHUMACHER, *Introduction to Magnetic Resonance*, W. A. Benjamin, Inc., New York, 1970.
22. W. S. JARDETZKY, *Theories of Figures of Celestial Bodies*, Interscience Publishers, Inc., New York, 1958.
23. J. PONCET, E. C. G. STUECKELBERG DE BREIDENBACH & P. B. SCHEURER, "Sur une question de calcul de variations sous contraintes", *Helv. Phys. Acta* **44** (1971), 522–529.
24. P. GERMAIN, "The role of thermodynamics in continuum mechanics", pp. 317–333 of *Foundations of Continuum Thermodynamics*, J. J. D. DOMINGOS *et al.* (eds.), Halsted Press, New York, 1973.
25. L. TISZA, "The thermodynamics of phase equilibrium", *Annals of Physics* **13** (1961), 1–92; reprinted as pp. 102–193 of *Generalized Thermodynamics*, see ref. [11] above.
26. J. W. GIBBS, "A method of geometric representation of the thermodynamic properties of substances by means of surfaces", pp. 33–54 of *The Scientific Papers of J. Willard Gibbs*, Volume 1, Dover Publications, 1961.
27. I. P. BAZAROV, *Thermodynamics*, Pergamon Press, Oxford etc., 1964.
28. C.-S. MAN, *Critical Studies in some Thermodynamic Problems*, unpublished M. Phil. thesis, accepted by the University of Hong Kong, 1976.

Department of Mathematics
University of Kentucky
Lexington

(Received April 2, 1985)

One-dimensional Variational Problems whose Minimizers do not Satisfy the Euler-Lagrange Equation

J. M. BALL & V. J. MIZEL

Dedicated to Walter Noll

§ 1. Introduction

In this paper we consider the problem of minimizing

$$I(u) = \int_a^b f(x, u(x), u'(x))\, dx \tag{1.1}$$

in thc set $\mathscr{A}$ of absolutely continuous functions $u: [a, b] \to \mathbb{R}$ satisfying the end conditions

$$u(a) = \alpha, \qquad u(b) = \beta, \tag{1.2}$$

where α and β are given constants. In (1.1), $[a, b]$ is a finite interval, $'$ denotes $\dfrac{d}{dx}$, and the integrand $f = f(x, u, p)$ is assumed to be smooth, nonnegative and to satisfy the *regularity condition*

$$f_{pp} > 0. \tag{1.3}$$

The significance of the regularity condition (1.3) is that, as is well known, it ensures the existence of at least one absolute minimizer for I in $\mathscr{A}$, provided f also satisfies an appropriate growth condition with respect to p. Further, it implies that any Lipschitz solution u of the integrated form

$$f_p = \int_a^x f_u\, dy + \text{const.} \qquad \text{a.e. } x \in [a, b] \tag{IEL}$$

of the Euler-Lagrange equation is in fact smooth in $[a, b]$. Notwithstanding these facts and the status of (IEL) as a classical necessary condition for a minimizer, we present a number of examples in which I attains a minimum at some $u \in \mathscr{A}$ but u is *not* smooth and does *not* satisfy (IEL).

To see where the classical argument leading to (IEL) may break down, recall that the argument relies on calculating the derivative

$$\frac{d}{dt} I(u + t\varphi)\big|_{t=0}$$

$$= \lim_{t \to 0} \int_a^b \frac{f(x, u(x) + t\varphi(x), u'(x) + t\varphi'(x)) - f(x, u(x), u'(x))}{t} \, dx \tag{1.4}$$

for φ a smooth function satisfying $\varphi(a) = \varphi(b) = 0$, and concluding that since $I(u + t\varphi)$ is minimized at $t = 0$ the derivative is zero; *viz.*

$$\int_a^b [f_u \varphi + f_p \varphi'] \, dx = 0. \tag{1.5}$$

If $u \in W^{1,\infty}(a, b)$ this argument is clearly valid, since by the mean value theorem the integrand on the right-hand side of (1.4) is uniformly bounded independently of small t and consequently one may pass to the limit $t \to 0$ using the bounded convergence theorem. However, if it is known only that the minimizer u belongs to $\mathscr{A}$, the only readily available piece of information which may aid passing to the limit in (1.4) is that $I(u) < \infty$. Consequently one is typically forced into making assumptions on the derivatives of f, these assumptions being unnecessary for the existence of a minimizer, so as to pass to the limit. More alarmingly, a difficulty may arise at an earlier stage in the argument to due the possibility that near some $u \in \mathscr{A}$ with $I(u) < \infty$ there may be functions $v \in \mathscr{A}$ with $I(v) = \infty$; in fact, in two of our examples we are able to show that for a large class of $\varphi \in C_0^\infty(a, b)$ the minimizers u are such that $I(u + t\varphi) = \infty$ for *all* $t \neq 0$.

The possibility that a minimizer u of I in $\mathscr{A}$ might be singular was envisaged by TONELLI, who proved a striking and little known partial regularity theorem to the effect that u is a smooth solution of the Euler-Lagrange equation on the complement of a closed subset E of $[a, b]$ of measure zero, and that $|u'(x)| = \infty$ for all $x \in E$. He then gave a number of criteria ensuring that "the set E does not exist" and thus that $u \in C^\infty([a, b])$. Remarks in TONELLI [32] suggest that he did not know of any examples in which E is nonempty, and we believe that our examples are the first of this type. A precise statement and proof of a version of the partial regularity theorem is given in § 2, where we also gather together a number of results concerning the existence of minimizers and first order necessary conditions. In this connection we mention that we are unaware of any integral form of a first order necessary condition that is satisfied by every minimizer u in the absence of additional hypotheses on f.

Our first example, given in § 3, is that of minimizing

$$I(u) = \int_0^1 [(x^2 - u^3)^2 (u')^{14} + \varepsilon (u')^2] \, dx \tag{1.6}$$

subject to

$$u(0) = 0, \quad u(1) = k, \tag{1.7}$$

where $\varepsilon > 0$, $k > 0$. (As we point out at the end of § 5, the power 14 is the lowest for which singular minimizers of (1.6) exist.) Note that if $0 < k \leqq 1$ and $\varepsilon = 0$ then the minimum of I is attained by $u(x) = \min(x^{\frac{2}{3}}, k)$; the results summarized below show that the singularity of u at $x = 0$ is not destroyed provided $\varepsilon > 0$ is sufficiently small. The integrand in (1.6) has a scale-invariance property which allows one to transform the Euler-Lagrange equation to an autonomous ordinary differential equation in the plane, and this makes it possible to give a very detailed and complete description of the absolute minimizers u of (1.6), (1.7) for all ε and k. Some of the main conclusions are the following (see especially Theorem 3.12). There exist numbers $\varepsilon_0 = .002474\ldots$, $\varepsilon^* = .00173\ldots$ such that (a) for $0 < \varepsilon < \varepsilon_0$ there exist two elementary solutions $\bar{k}_1(\varepsilon)\, x^{\frac{2}{3}}$, $\bar{k}_2(\varepsilon)\, x^{\frac{2}{3}}$ of the Euler-Lagrange equation on (0, 1]; (b) if $0 < \varepsilon < \varepsilon^*$ and k is sufficiently large I attains an absolute minimum at a unique function u which satisfies $u(x) \sim \bar{k}_2(\varepsilon)\, x^{\frac{2}{3}}$ as $x \to 0+$, $u \in C^\infty((0, 1])$ and $f_u(\cdot, u(\cdot), u'(\cdot)) \notin L^1(0, 1)$, so that (IEL) does not hold: if $k = \bar{k}_2(\varepsilon)$ then $u(x) = \bar{k}_2(\varepsilon)\, x^{\frac{2}{3}}$; (c) if $0 < \varepsilon < \varepsilon^*$ and k is sufficiently large (for example, $k \geqq 1$) there is no smooth solution of the Euler-Lagrange equation on [0, 1] satisfying the end conditions (1.7), and hence I does not attain a minimum among Lipschitz functions; (d) if $\varepsilon > \varepsilon^*$ then there is exactly one u that minimizes I and it is the unique smooth solution of the Euler-Lagrange equation on [0, 1] satisfying (1.7). The detailed structure of the phase portrait that leads to these conclusions would have been extremely difficult to determine without the aid of computer plots, though these do not form part of the proofs. Since the singular minimizers are smooth for $x > 0$ their "Tonelli set" E consists in the single endpoint $\{0\}$ and they *do* satisfy the Euler-Lagrange equation in the sense of distributions, *i.e.* in its "weak" form.

In § 4 we consider the case when $f = f(u, p)$ does not depend on x. We first construct an $f \in C^\infty(\mathbb{R}^2)$ satisfying (in addition to (1.3))

$$|p| \leqq f(u, p) \leqq \text{const.}\,(1 + p^2), \qquad (u, p) \in \mathbb{R}^2, \tag{1.8}$$

and

$$\frac{f(u, p)}{|p|} \to \infty \quad \text{as} \quad |p| \to \infty \quad \text{for each} \quad u \neq 0, \tag{1.9}$$

such that

$$I(u) = \int_{-1}^{1} f(u, u')\, dx \tag{1.10}$$

attains an absolute minimum subject to the end conditions

$$u(-1) = k_1, \qquad u(1) = k_2 \tag{1.11}$$

(for suitable k_1, k_2), at a unique function u_0 whose Tonelli set E is a single interior point $x_0 \in (-1, 1)$ and which satisfies

$$f_u(u_0, u_0') \notin L^1_{\text{loc}}(-1, 1); \tag{1.12}$$

hence (IEL) does not hold, with integration in the Lebesgue sense, and neither is the weak form of the Euler-Lagrange equation satisfied. Next we construct, for *any* preassigned closed set $E \subset [-1, 1]$ of measure zero, a similar function

$f=f^E$ satisfying (1.8) such that for suitable k_1, k_2, I attains an absolute minimum subject to (1.11) at a unique function u_0 whose Tonelli set is precisely E. Again (1.12) holds. These two examples demonstrate the optimality of Corollary 2.12 and the Tonelli partial regularity theorem (Theorem 2.7), respectively. Awareness of conditions necessary for the validity of chain rule calculations ([34], [30], [27], [28]) influenced our initial construction of those examples, thoug the proofs presented here avoid this issue.

In § 5 we consider the problem of minimizing

$$I(u) = \int_{-1}^{1} [(x^4 - u^6)^2 \, |u'|^s + \varepsilon(u')^2] \, dx \tag{1.13}$$

in the set $\mathscr{A}$ of absolutely continuous functions on $[-1, 1]$ (*i.e.* functions in $W^{1,1} = W^{1,1}(-1, 1)$) satisfying the end conditions

$$u(-1) = k_1, \quad u(1) = k_2, \tag{1.14}$$

where $s>3$ and $\varepsilon>0$. (We allow s to take nonintegral values, even though the integrand is smooth only if s is an even integer.) We show (Theorem 5.1) that if $s \geqq 27$ then, provided $-1 \leqq k_1 < 0 < k_2 \leqq 1$ and ε is sufficiently small, every minimizer u_0 of I in $\mathscr{A}$ is such that $u_0(x) \sim |x|^{\frac{2}{3}} \operatorname{sign} x$ as $x \to 0$, $u_0 \in C^\infty([-1, 0) \cup (0, 1])$ and $u_0 \in W^{1,p}$ for $1 \leqq p < 3$. It follows that $E=\{0\}$ and that u_0 does not satisfy the Euler-Lagrange equation either in its weak or its integrated form. Furthermore, if $3 \leqq q \leqq \infty$,

$$\inf_{v \in W^{1,q} \cap \mathscr{A}} I(v) > \inf_{v \in \mathscr{A}} I(v) = I(u_0). \tag{1.15}$$

This remarkable fact is known as the *Lavrentiev phenomenon* (*cf.* LAVRENTIEV [22], MANIÀ [25], CESARI [11]), and its occurrence in a *regular* problem has not previously been noted; in the cited references only the case $q=\infty$ is considered. If $s>27$ then an equally surprising property holds (Theorem 5.5), namely that for any sequence $\{v_m\} \subset W^{1,q} \cap \mathscr{A}$ such that $v_m(x) \to u_0(x)$ for each x in some set containing arbitrarily small positive and negative numbers one has $I(v_m) \to \infty$ as $m \to \infty$. In particular, no minimizing sequence for I in $W^{1,q} \cap \mathscr{A}$ can converge to u_0. Since conventional finite-element methods for minimizing I yield such sequences, it follows that they cannot in general detect singular minimizers. Similarly, if v_η is a minimizer of, for example, an apparently innocuous penalized functional such as

$$I_\eta(u) = \int_{-1}^{1} [(x^4 - u^6)^2 \, |u'|^s + \varepsilon(u')^2 + \eta \, |u'|^{3+\gamma}] \, dx \tag{1.16}$$

in $\mathscr{A}$, where $\gamma>0$, then v_η cannot converge to u_0 as $\eta \to 0+$. Motivated by numerical experiments of BALL & KNOWLES [6] we show also that if $s>27$, $3 \leqq q \leqq \infty$ and $\varepsilon>0$, k_1, k_2 are arbitrary then (Theorem 5.8) I attains a minimum in $W^{1,q} \cap \mathscr{A}$ and any such minimizer u_1 is a smooth solution of the Euler-Lagrange equation on $[-1, 1]$. (Note that such "pseudominimizers" do not in general exist for (1.6), (1.7).) The pseudominimizers can be regarded as being "admissible" minimizers of I with respect to various penalty methods such as (1.16). Finally, we show (Theorem 5.9) that for $s < 26$ all minimizers of I in $\mathscr{A}$ are smooth, and that, at least for the corresponding problem posed on (0, 1), singular minimizers not satisfying the Lavrentiev phenomenon may exist for $26 \leqq s < 27$.

In all the examples considered we analyze whether or not the minimizers satisfy the weak or integrated forms of the DuBois-Reymond equation

$$\frac{d}{dx}(f - u' f_p) = f_x. \tag{DBR}$$

The examples in this paper were motivated by attempts to prove that minimizers of the total energy

$$I(u) = \int_\Omega W(x, Du(x))\,dx \tag{1.17}$$

of an elastic body subject to appropriate boundary conditions are weak solutions of the corresponding Euler-Lagrange equations

$$\frac{\partial}{\partial x^\alpha}\frac{\partial W}{\partial A^i_\alpha} = 0, \quad i = 1, \dots, n. \tag{1.18}$$

Here we have assumed that the body occupies the bounded open subset $\Omega \subset \mathbb{R}^n$ in a reference configuration and that there are no external forces. The particle at $x \in \Omega$ in the reference configuration is displaced to $u(x) \in \mathbb{R}^n$, and $Du(x)$ denotes the gradient of u at x. One of the complications of the problem, which is still open, is that the stored-energy function $W(x, A)$ of the material is defined only for $\det A > 0$ and is typically assumed to satisfy $W(x, A) \to \infty$ as $\det A \to 0+$. The existence of minimizers in appropriate subsets of the Sobolev space $W^{1,1} = W^{1,1}(\Omega; R^n)$ is established in BALL [2] for a class of realistic functions W, and conditions guaranteeing that these minimizers satisfy other first order necessary conditions are announced in BALL [5]. It is known (BALL [3], BALL & MURAT [8]) that even when W satisfies favorable constitutive hypotheses such as strong ellipticity, I may not attain its minimum within the class of smooth functions, and in fact that if $n \leqq q \leqq \infty$ then

$$\inf_{\substack{v \text{ smooth} \\ v|_{\partial\Omega} = \bar{u}|_{\partial\Omega}}} I(v) = \inf_{\substack{v \in W^{1,q} \\ v|_{\partial\Omega} = \bar{u}|_{\partial\Omega}}} I(v) > \inf_{\substack{v \in W^{1,1} \\ v|_{\partial\Omega} = \bar{u}|_{\partial\Omega}}} I(v) \tag{1.19}$$

can occur for appropriate boundary displacements $\bar{u}$. Of course (1.19) is a higher-dimensional version of the Lavrentiev phenomenon. The deformations responsible here for LAVRENTIEV'S gap are those for which cavitation occurs, that is, holes form in the body. Cavitation cannot occur if W satisfies the growth condition

$$W(x, A) \geqq \text{const.}\,|A|^p \quad \text{for } \det A > 0, \tag{1.20}$$

for some $p > n$, by the Sobolev embedding theorem (nor, in fact, if $p = n$). An intriguing possibilty raised by our one-dimensional examples is that singular minimizers and the Lavrentiev phenomenon may occur for (1.17) even when (1.20) holds, and that the singularities of Du might be connected with the initiation of fracture. More work needs to be done to decide whether this can happen under realistic hypotheses on W. Similar considerations may be relevant for other nonlinear elliptic systems (see, for example, Giaquinta [17] and several articles in Ball [4]).

In view of the potential physical significance of singular minimizers and the Lavrentiev phenomenon in elasticity and perhaps other fields, our general view is that they should be studied rather than exorcised. However, it is of course also interesting to determine conditions under which this behavior cannot occur. We mention in particular the theorem of Angell [1] concerning a sufficient condition for nonoccurrence of the Lavrentiev phenomenon, which generalizes earlier results of Tonelli [32], Cinquini [12] and Manià [25]. Angell's theorem is presented in Cesari [11], who gives a wealth of related results. We also refer the reader to the result of Giaquinta & Giusti [18] (see also Giaquinta [17, p. 267]) giving conditions on f for minimizers of (1.1) to be smooth in the case when f satisfies $\lambda p^2 \leqq f(x, u, p) \leqq \Lambda p^2$ for all x, u, p, where $\lambda > 0$.

Many of the results in this paper were announced in Ball & Mizel [7] and Ball [5].

We conclude the introduction with a remark concerning an abuse of notation in which we indulge. If, for example, we write $u \in W^{1,q}(0, \delta) \cap W^{1,2}(0, 1)$, where $0 < \delta < 1$, we mean that $u \in W^{1,2}(0, 1)$ and that u restricted to $(0, \delta)$ belongs to $W^{1,q}(0, \delta)$.

§ 2. Review of positive results concerning minimizers and first order necessary conditions

We consider integrals of the form

$$I(u) = \int_a^b f(x, u(x), u'(x))\, dx,$$

where $-\infty < a < b < \infty$, and where the competing functions $u: [a, b] \to \mathbb{R}$. We discuss the problem of minimizing I in the set

$$\mathcal{A} = \{u \in W^{1,1}(a, b) : u(a) = \alpha, u(b) = \beta\},$$

where α, β are given real constants. By an appropriate choice of representatives, $W^{1,1}(a, b)$ can be identified with the set of absolutely continuous functions $u: [a, b] \to \mathbb{R}$ and we shall henceforth assume this to have been done. To avoid getting enmeshed in technical hypotheses that are unnecessary for our purposes, we make the standing assumptions that $f = f(x, u, p)$ is C^3 in its arguments and bounded below; the reader interested in optimal regularity hypotheses or the case $u: [a, b] \to \mathbb{R}^n$ can consult the cited references. Our aim in this section is to summarize for later reference the available information concerning the existence of minimizers and first order necessary conditions satisfied by them.

Theorem 2.1. (Tonelli's existence theorem). *Suppose* $f_{pp} \geqq 0$ *and* $f(x, u, p) \geqq \varphi(|p|)$, $x \in [a, b]$, $(u, p) \in \mathbb{R}^2$, *where* φ *is bounded below and satisfies* $\dfrac{\varphi(t)}{t} \to \infty$ *as* $t \to \infty$. *Then* I *attains an absolute minimum on* $\mathcal{A}$.

For the proof see, for example, CESARI [11, pp. 112, 372], HESTENES [20], or EKELAND & TÉMAM [16]. The original proof (for the case $\varphi(t) = t^p$, $p > 1$) can be found in TONELLI [31 II, p. 282], and in TONELLI [33] for the general case. TONELLI [31 II, pp. 287, 296] and [33] also proved that minimizers exist when f has superlinear growth in p except in the neighborhood of finitely many points or absolutely continuous curves; significant extensions of some of these results, together with a more complete bibliography, are described in MCSHANE [24], and CESARI [11, Chapter 12]. These results imply, for example, that the functionals I considered in § 4 attain a minimum, but are not needed in our development there since the minimizer is constructed explicitly.

Definitions 2.2. A function $u \in \mathscr{A}$ is a *weak relative minimizer* of I if $I(u) < \infty$ and there exists $\delta > 0$ such that $I(u) \leqq I(v)$ for all $v \in \mathscr{A}$ with $\operatorname{ess\,sup}_{x \in [a,b]} [|u(x) - v(x)| + |u'(x) - v'(x)|] \leqq \delta$. We say that $u \in \mathscr{A}$ is a *strong relative minimizer* of I if there exists $\delta > 0$ such that $I(u) \leqq I(v)$ for all $v \in \mathscr{A}$ with $\max_{x \in [a,b]} |u(x) - v(x)| \leqq \delta$.

We consider the following forms of classical first order necessary conditions for a minimum. The *Euler-Lagrange equation* is

$$\frac{d}{dx} f_p = f_u. \tag{EL}$$

A function $u \in \mathscr{A}$ satisfies the *weak form of the Euler-Lagrange equation* if $f_u, f_p \in L^1_{\text{loc}}(a, b)$ and (EL) holds in the sense of distributions, *i.e.*

$$\int_a^b [f_p \varphi' + f_u \varphi]\, dx = 0 \quad \text{for all} \quad \varphi \in C_0^\infty(a, b). \tag{WEL}$$

A function $u \in \mathscr{A}$ satisfies the *integrated form of the Euler-Lagrange equation* provided $f_u \in L^1(a, b)$ and

$$f_p(x, u(x), u'(x)) = \int_a^x f_u\, dy + \text{const.} \quad \text{a.e. } x \in [a, b]. \tag{IEL}$$

The *DuBois-Reymond equation* is

$$\frac{d}{dx}(f - u' f_p) = f_x. \tag{DBR}$$

A function $u \in \mathscr{A}$ satisfies the *weak form of the DuBois-Reymond equation* if $f - u' f_p$, $f_x \in L^1_{\text{loc}}(a, b)$ and (DBR) holds in the sense of distributions, *i.e.*

$$\int_a^b [(f - u' f_p) \varphi' + f_x \varphi]\, dx = 0 \quad \text{for all } \varphi \in C_0^\infty(a, b). \tag{WDBR}$$

A function $u \in \mathscr{A}$ satisfies the *integrated form of the DuBois-Reymond equation* provided $f_x \in L^1(a, b)$ and

$$f(x, u(x), u'(x)) - u'(x) f_p(x, u(x), u'(x)) = \int_a^x f_x\, dy + \text{const.} \quad \text{a.e. } x \in [a, b]. \tag{IDBR}$$

Of course, if u satisfies (IEL) (respectively (IDBR)) then u satisfies (WEL) (respectively (WDBR)). We will see later that the converse is false in general; what *is* true is that, by the fundamental lemma of the calculus of variations, (WEL) is equivalent to

$$f_p(x, u(x), u'(x)) = \int_c^x f_u \, dy + \text{const.} \quad \text{a.e. } x \in [a, b],$$

for any $c \in (a, b)$, a similar statement holding for (WDBR).

Theorem 2.3.

(i) *Let* $u \in \mathcal{A}$ *be a weak relative minimizer of* I *and suppose that* $f_u(\cdot, \bar{u}(\cdot), u'(\cdot)) \in L^1(a, b)$ *whenever* $\bar{u} \in L^\infty(a, b)$ *with* $\operatorname*{ess\,sup}_{x \in [a,b]} |u(x) - \bar{u}(x)|$ *sufficiently small. Then* u *satisfies* (IEL).

(ii) *Let* $u \in \mathcal{A}$ *be a strong relative minimizer of* I *and suppose that* $f_x(\bar{x}(\cdot), u(\cdot), u'(\cdot)) \in L^1(a, b)$ *whenever* $\bar{x} \in L^\infty(a, b)$ *with* $\operatorname*{ess\,sup}_{x \in [a,b]} |\bar{x}(x) - x|$ *sufficiently small. Then* u *satisfies* (IDBR).

Proof.

(i) For $\delta > 0$ sufficiently small and $G \subset \mathbb{R}$ closed define

$$\gamma_G(x) = \sup_{t \in [-\delta, \delta] \cap G} |f_u(x, u(x) + t, u'(x))|,$$

$$E(x) = \{t \in [-\delta, \delta] : |f_u(x, u(x) + t, u'(x))| = \gamma_{\mathbb{R}}(x)\}.$$

We consider the set-valued mapping $E : x \mapsto E(x)$. Clearly $E(x)$ is closed for a.e. $x \in [a, b]$. Furthermore, for any closed $G \subset \mathbb{R}$ the set

$$\{x \in [a, b] : E(x) \cap G \text{ nonempty}\} = \{x \in [a, b] : \gamma_G(x) - \gamma_{\mathbb{R}}(x) = 0\}$$

is measurable (since $\gamma_G - \gamma_{\mathbb{R}}$ is a measurable function). By a standard measurable selection theorem (*cf.* Cesari [11, p. 283ff]) there exists a measurable function $(x \mapsto t(x)$ with $t(x) \in E(x)$ a.e. $x \in [a, b]$. Hence $\gamma_{\mathbb{R}}(x) = |f_u(x, u(x) + t(x), u'(x))|$ a.e. $x \in [a, b]$, so that our hypothesis is equivalent to the existence of $\gamma \in L^1(a, b)$ such that

$$|f_u(x, \bar{u}(x), u'(x))| \leqq \gamma(x) \quad \text{a.e. } x \in [a, b]$$

for all $\bar{u} \in L^\infty(a, b)$ with $\operatorname*{ess\,sup}_{x \in [a,b]} |u(x) - \bar{u}(x)|$ sufficiently small. The result now follows from Tonelli [31] (see also Cesari [11, p. 61ff], Hestenes [20, p. 196ff]).

(ii) This follows in a similar way from Tonelli [31] (see also Cesari [11, p. 61ff]). Alternatively, one can deduce (ii) from (i) by a reduction based on the idea that $\varphi \equiv 0$ is a weak relative minimum of

$$J(\varphi) = \int_a^b f(x, u_\varphi(x), u'_\varphi(x)) \, dx$$

subject to $\varphi(a) = \varphi(b) = 0$, where $u_\varphi(x) \stackrel{\text{def}}{=} u(z)$, $z + \varphi(z) = x$. □

Corollary 2.4. *Let* $f = f_1(x, u) + f_2(x, p)$. *If* $u \in \mathscr{A}$ *is a weak relative minimizer of* I *then* u *satisfies* (IEL).

Proof. If $\bar{u} \in L^\infty(a, b)$ then $f_u(x, \bar{u}(x), u'(x)) = (f_1)_u(x, \bar{u}(x))$ is uniformly bounded. □

Corollary 2.5. *Let* $f = f_1(x, u) + f_2(u, p)$. *If* $u \in \mathscr{A}$ *is a strong relative minimizer of* I *then* u *satisfies* (IDBR).

Proof. If $\bar{x} \in L^\infty(a, b)$ then $f_x(\bar{x}(x), u(x), u'(x)) = (f_1)_x(\bar{x}(x), u(x))$ is uniformly bounded. □

The above results are notable for the lack of any convexity assumptions on f. The growth assumptions are also considerably weaker than those of corresponding theorems known for multiple integrals. For example, in Theorem 2.3(i) there is no hypothesis on f_p; that the result is true without such a hypothesis is suggested by the fact that f_p is bounded for any solution of (IEL). We are not aware of any counterexamples to Theorem 2.3 if the integrability hypotheses are weakened to read in part (i) $f_u(\cdot, u(\cdot), u'(\cdot)) \in L^1(a, b)$, and in part (ii) $f_x(\cdot, u(\cdot), u'(\cdot)) \in L^1(a, b)$.

We now describe results in which f is assumed convex with respect to p.

Theorem 2.6. *Let* $u \in W^{1,\infty}(a, b)$ (= *Lipschitz continuous functions on* $[a, b]$) *be a weak relative minimizer of* I, *and suppose that* $f_{pp}(x, u(x), p) > 0$ *for all* $x \in [a, b], p \in \mathbb{R}$. *Then* $u \in C^3([a, b])$ *and satisfies* (EL).

Proof. This is standard and can be found in CESARI [11, p. 57ff]. □

Let $\overline{\mathbb{R}} = \mathbb{R} \cup \{-\infty\} \cup \{+\infty\}$ denote the extended real line with its usual topology. We define $C^1([a, b]; \overline{\mathbb{R}})$ to be the set of continuous functions $u: [a, b] \to \mathbb{R}$ such that for all $x \in [a, b]$

$$u'(x) \stackrel{\text{def}}{=} \lim_{h \to 0} \frac{u(x + h) - u(x)}{h} \tag{2.1}$$

exists as an element of $\overline{\mathbb{R}}$ (with the appropriate one-sided limit being taken if $x = a$ or $x = b$), and such that $u': [a, b] \to \overline{\mathbb{R}}$ is continuous.

Theorem 2.7 (TONELLI'S *partial regularity theorem*). *Let* $f_{pp} > 0$. *If* $u \in \mathscr{A}$ *is a strong relative minimizer of* I *then* $u \in C^1([a, b]; \overline{\mathbb{R}})$.

Before proving Theorem 2.7 we note some consequences. Clearly $u'(x)$ as defined in (2.1) coincides almost everywhere with the derivative of u in the sense of distributions. Therefore under the hypotheses of the theorem the *Tonelli set* E defined by

$$E = \{x \in [a, b] : |u'(x)| = \infty\}$$

is a closed set of measure zero. The complement $[a, b] \setminus E$ is a union of disjoint relatively open intervals D_j. By the optimality principle and Theorem 2.6, u

is a C^3 solution of (EL) on each D_j. By Theorem 2.7, $u'(x)$ tends to $+\infty$ or $-\infty$ as x tends to the end-points of every such interval (unless $a \in D_j$ or $b \in D_j$). These consequences of Theorem 2.7 constitute TONELLI's statement of his theorem (TONELLI [31 II, p. 359]); our formulation includes the extra remark that u' is continuous. The proof we give, like TONELLI's, uses the local solvability of (EL), but we avoid his construction of auxiliary integrands by applying the field theory of the calculus of variations. Recently, CLARKE & VINTER [13, 14] have presented certain extensions of TONELLI's theorem to the cases when f is not smooth and $u:[a,b] \to \mathbb{R}^n$. They have also shown [15] that if f is a polynomial then the Tonelli set E is at most countable with finitely many points of accumulation.

Lemma 2.8. *Let* $A \subset \mathbb{R}^2$ *be bounded, and let* $M > 0$, $\delta > 0$. *There exists* $\varepsilon > 0$ *such that if* $(x_0, u_0) \in A$, $|\alpha| \leqq M$, $|\beta| \leqq M$, *the solution* $u(x; \alpha, \beta)$ *of* (EL) *satisfying the initial conditions*

$$u(x_0; \alpha, \beta) = u_0 + \alpha, \qquad u'(x_0; \alpha, \beta) = \beta, \tag{2.2}$$

exists for $|x - x_0| \leqq \varepsilon$, *is unique, and is such that*

(a) *u and u' are C^1 functions of x, α, β in the set*

$$S \stackrel{\text{def}}{=} \{(x, \alpha, \beta) : |x - x_0| \leqq \varepsilon, |\alpha| \leqq M, |\beta| \leqq M\},$$

(b)
$$|u'(x; \alpha, \beta) - \beta| < \delta, \tag{2.3}$$

$$\frac{\partial u}{\partial \alpha}(x; \alpha, \beta) > 0, \qquad \operatorname{sign} \frac{\partial u}{\partial \beta}(x; \alpha, \beta) = \operatorname{sign}(x - x_0), \tag{2.4}$$

for all $(x, \alpha, \beta) \in S$, *where* sign t *takes the values* $-1, 0, 1$ *for* $t < 0$, $t = 0$, $t > 0$, *respectively.*

Proof. Because $f_{pp} > 0$, solving (EL) is equivalent to solving the equation

$$u'' = F(x, u, u'),$$

where $F(x, u, p) \stackrel{\text{def}}{=} (f_u - f_{px} - p f_{pu})/f_{pp}$. Our hypotheses imply that $F \in C^1(\mathbb{R}^3)$. The existence, uniqueness and smoothness assertions follow from standard results (see, for example, HARTMAN [19, Chapter 5]). Furthermore, the derivatives appearing in (2.3), (2.4) depend continuously on x_0, u_0. That $\varepsilon > 0$ can be chosen sufficiently small for (b) to hold follows by a simple compactness argument, using the relations

$$u'(x_0; \alpha, \beta) = \beta, \qquad \frac{\partial u}{\partial \alpha}(x_0; \alpha, \beta) = 1, \qquad \frac{\partial u}{\partial \beta}(x_0; \alpha, \beta) = 0,$$

$$\left(\frac{\partial u}{\partial \beta}\right)'(x_0; \alpha, \beta) = 1. \quad \square$$

Proposition 2.9. (TONELLI [31 II, p. 344ff]). *Let* $m > 0$, $\varrho > 0$, $M_1 > 0$. *Then there exists* $\varepsilon > 0$ *such that if* $x_0, x_1 \in [a, b]$, $0 < x_1 - x_0 \leqq \varepsilon$, $|u_0| \leqq m$ *and* $\left|\frac{u_1 - u_0}{x_1 - x_0}\right| \leqq M_1$ *there is a unique solution* $\tilde{u} \in C^2([x_0, x_1])$ *of* (EL) *satisfying*

$\tilde{u}(x_0) = u_0$, $\tilde{u}(x_1) = u_1$ *and* $\max_{x \in [x_0, x_1]} |\tilde{u}(x) - u_0| \leqq \varrho$, *and* $\tilde{u}$ *is the unique absolute minimizer of*

$$\tilde{I}(u) = \int_{x_0}^{x_1} f(x, u(x), u'(x))\, dx$$

over the set

$$\tilde{\mathscr{A}} = \{u \in W^{1,1}(x_0, x_1) : u(x_0) = u_0, \quad u(x_1) = u_1, \quad \max_{x \in [x_0, x_1]} |\tilde{u}(x) - u_0| \leqq \varrho\}.$$

Proof. Let $\sigma = m + \varrho$, $A = [a, b] \times [-\sigma, \sigma]$, $M > \max(M_1, 2\varrho)$ and let $0 < \delta < M - M_1$. Let $\varepsilon > 0$ be chosen as in Lemma 2.8, and suppose in addition that $3M\varepsilon < \varrho$. Let $x_0, x_1 \in [a, b]$, $0 < x_1 - x_0 \leqq \varepsilon$, $|u_0| \leqq m$ and $\left|\dfrac{u_1 - u_0}{x_1 - x_0}\right| \leqq M_1$. Note that by integrating (2.3) we have that

$$|u(x; \alpha, \beta) - u_0 - \alpha - \beta(x - x_0)| \leqq \delta(x - x_0), \qquad x \in [x_0, x_1]. \tag{2.5}$$

Therefore

$$u(x_1; 0, M) \geqq u_0 + M_1(x_1 - x_0) + (M - M_1 - \delta)(x_1 - x_0) > u_1,$$

$$u(x_1; 0, -M) \leqq u_0 - M_1(x_1 - x_0) - (M - M_1 - \delta)(x_1 - x_0) < u_1.$$

Since $\dfrac{\partial u}{\partial \beta}(x_1; 0, \beta) > 0$ for $\beta \in [-M, M]$ there is a unique $\beta_0 \in [-M, M]$ such that $u(x_1; 0, \beta_0) = u_1$. Define $\tilde{u}(x) = u(x; 0, \beta_0)$. Setting $x = x_1$ in (2.5) we obtain

$$|\beta_0| \leqq \delta + M_1. \tag{2.6}$$

Therefore, again by (2.5), for $x \in [x_0, x_1]$

$$\begin{aligned} |\tilde{u}(x) - u_0| &\leqq (\delta + |\beta_0|)(x - x_0) \\ &\leqq (2\delta + M_1)\varepsilon < \varrho. \end{aligned}$$

Now suppose that $v \in C^2([x_0, x_1])$ is also a solution of (EL) satisfying $v(x_0) = u_0$, $v(x_1) = u_1$ and $\max_{x \in [x_0, x_1]} |v(x) - u_0| \leqq \varrho$. Then $v'(\bar{x}) = \dfrac{u_1 - u_0}{x_1 - x_0}$ for some $\bar{x} \in (x_0, x_1)$ and $(\bar{x}, v(\bar{x})) \in A$, and so applying (2.3) with $(\bar{x}, v(\bar{x}))$ replacing (x_0, u_0) we deduce that

$$\left|v'(x) - \frac{u_1 - u_0}{x_1 - x_0}\right| \leqq \delta \qquad \text{for } x \in [x_0, x_1].$$

In particular,

$$|v'(x_0)| \leqq M_1 + \delta < M.$$

By the uniqueness of β_0 we therefore have that $v'(x_0) = \beta_0$, and thus $v = \tilde{u}$.

To show that $\tilde{u}$ minimizes $\tilde{I}$ in $\tilde{\mathscr{A}}$, we consider the one-parameter family of solutions $\{u(\cdot; \alpha, \beta_0), |\alpha| \leqq M\}$. By (2.5), (2.6) we have

$$u(x; M, \beta_0) - u_0 \geqq M + (\beta_0 - \delta)(x - x_0) \geqq M - (2\delta + M_1)\varepsilon > \varrho$$

and

$$u(x; -M, \beta_0) - u_0 \leqq -M + (\beta_0 + \delta)(x - x_0) \leqq -M + (2\delta + M_1)\varepsilon < -\varrho,$$

for $x \in [x_0, x_1]$. Since $\frac{\partial u}{\partial \alpha}(x; \alpha, \beta_0) > 0$ it follows that $\tilde{u}$ is embedded in a field of extremals that simply covers the region $[x_0, x_1] \times [u_0 - \varrho, u_0 + \varrho]$. Since $f_{pp} > 0$ it follows from Weierstrass's formula (*e.g.* BOLZA [9, p. 91], CESARI [11, p. 72]) that

$$\tilde{I}(u) > \tilde{I}(\tilde{u})$$

for all $u \in \tilde{\mathcal{A}}$, with equality if and only if $u = \tilde{u}$, which concludes the proof. □

Proof of Theorem 2.7. Let $u \in \mathcal{A}$ be a strong relative minimizer of I; thus there exists $\delta_1 > 0$ such that $I(u) \leqq I(v)$ for all $v \in \mathcal{A}$ with $\max_{x \in [a,b]} |u(x) - v(x)| \leqq \delta_1$. Let $\bar{x} \in [a, b]$, and suppose that

$$M(\bar{x}) \overset{\text{def}}{=} \liminf_{\substack{x \to \bar{x} \\ x \neq \bar{x}, x \in [a,b]}} \left| \frac{u(x) - u(\bar{x})}{x - \bar{x}} \right| < \infty. \tag{2.7}$$

Suppose that $\bar{x} \neq b$ and take $\bar{x}_1 > \bar{x}$ with $\bar{x}_1 - \bar{x}$ sufficiently small that $\max_{x \in [\bar{x}, \bar{x}_1]} |u(x) - u(\bar{x})| \leqq \frac{\delta_1}{2}$. Choose $M_1 > M(\bar{x})$. By (2.7) we can apply Proposition 2.9 with $x_0 = \bar{x}$, $u_0 = u(\bar{x})$, $\varrho = \frac{\delta_1}{2}$, $u_1 = u(x_1)$, where $x_1 \in (\bar{x}, \bar{x}_1)$ satisfies

$$x_1 - \bar{x} < \varepsilon, \qquad \left| \frac{u(x_1) - u(\bar{x})}{x_1 - \bar{x}} \right| \leqq M_1.$$

Let $\tilde{u}$ be the corresponding solution of (EL). Let $\hat{u} \in \mathcal{A}$ be defined by $\hat{u}(x) = \tilde{u}(x)$ if $x \in [\bar{x}, x_1]$, $\hat{u}(x) = u(x)$ otherwise. Then $\max_{x \in [a,b]} |\hat{u}(x) - u(x)| \leqq \frac{\delta_1}{2} + \frac{\delta_2}{2} = \delta_1$ and so $I(\hat{u}) - I(u) = \tilde{I}(\tilde{u}) - \tilde{I}(u) \geqq 0$. Since $\tilde{u}$ is the unique minimizer of $\tilde{I}$ in $\tilde{\mathcal{A}}$ it follows that $\tilde{u} = u$ in $[\bar{x}, x_1]$ and hence that $u \in C^2([\bar{x}, x_1])$. Similarly, if $\bar{x} \neq a$ then $u \in C^2([x_0, \bar{x}])$ for some $x_0 < \bar{x}$. In particular u is Lipschitz in the neighborhood of any $\bar{x} \in [a, b]$ with $M(\bar{x}) < \infty$, and thus by Theorem 2.6 is C^3 in a neighborhood of any such $\bar{x}$. Since u is differentiable almost everywhere in $[a, b]$ it follows that $D \overset{\text{def}}{=} \{x \in [a, b] : M(x) < \infty\}$ is a relatively open subset of $[a, b]$ of full measure, and that $u \in C^3(D)$.

Let $E = [a, b] \setminus D$, and let $x_0 \in E$, so that $M(x_0) = \infty$. Suppose that $x_0 \in (a, b)$. By an appropriate reflection of the variables x and/or u we can suppose without loss of generality that there exist points $y_j \to x_0-$ with

$$\lim_{j \to \infty} \frac{u(x_0) - u(y_j)}{x_0 - y_j} = +\infty.$$

Let $M > 0$, $\delta > 0$ be arbitrary and apply Lemma 2.8 with $u_0 = u(x_0)$. The solutions $\{u(\cdot; \alpha, M) : |\alpha| \leqq M\}$ of (EL) form a field of extremals simply covering some neighborhood of (x_0, u_0) in $\mathbb{R}^2$. Thus, for $|x - x_0|$ sufficiently small there exists a unique $\alpha(x)$ with $|\alpha(x)| \leqq M$ such that $u(x) = u(x; \alpha(x), M)$, and by the implicit function theorem and (2.4) α depends continuously on x. Clearly $\alpha(x_0) = 0$. We claim that $\alpha(x)$ is nondecreasing near x_0. In fact suppose there exist sequences $a_j \to x_0$, $b_j \to x_0$, $c_j \to x_0$ with $a_j < b_j < c_j$ and $\alpha(a_j) = \alpha(c_j) \neq \alpha(b_j)$. Then for large enough j the solution $v_j(x) \overset{\text{def}}{=} u(x; \alpha(a_j), M)$, $a_j \leqq x \leqq c_j$, satisfies $v_j(a_j) = u(a_j)$, $v_j(b_j) \neq u(b_j)$, $v_j(c_j) = u(c_j)$ and $\max\limits_{x \in [a_j, c_j]} |u(x) - v_j(x)| \leqq \delta_1$. Since v_j is embedded in a field of extremals, Weierstrass's formula gives

$$\int_{a_j}^{c_j} f(x, u(x), u'(x))\, dx > \int_{a_j}^{c_j} f(x, v_j(x), v_j'(x))\, dx,$$

contradicting our hypothesis that u is a strong relative minimizer. Thus α is either nondecreasing or nonincreasing near x_0; the latter possibility is excluded by noting that by integrating (2.3) (*cf.* (2.5)) we obtain

$$\frac{\alpha(y_j)}{x_0 - y_j} \leqq \delta + M - \frac{u(x_0) - u(y_j)}{x_0 - y_j},$$

so that $\alpha(y_j) < 0$ for j sufficiently large. This proves our claim. Now let $x_j \to x_0$, $z_j \to x_0$ with $x_j > z_j$. Then for large enough j,

$$\begin{aligned}
\frac{u(x_j) - u(z_j)}{x_j - z_j} &= \frac{u(x_j; \alpha(x_j), M) - u(z_j; \alpha(z_j), M)}{x_j - z_j} \\
&\geqq \frac{u(x_j; \alpha(z_j), M) - u(z_j; \alpha(z_j), M)}{x_j - z_j} \\
&= u'(w_j; \alpha(z_j), M) \\
&\geqq M - \delta,
\end{aligned}$$

where $x_j \geqq w_j \geqq z_j$ and we have used (2.3). Thus, since M, δ are arbitrary,

$$\lim_{\to\infty} \frac{u(x_j) - u(z_j)}{x_j - z_j} = +\infty. \tag{2.8}$$

In particular $u'(x_0)$ exists in the sense of (2.1) and equals $+\infty$. A similar argument applies if $x_0 = a$ or $x_0 = b$. We have thus shown that $u'(x)$ exists in the sense of (2.1) for all $x \in [a, b]$. The continuity of u' at x_0 is obvious if $x_0 \in D$, and follows simply from (2.8) otherwise. □

As an application of Theorem 2.7 we prove the following version of results of TONELLI [31, Vol. II, pp. 361, 366], which should be compared with Theorem 2.3.

Theorem 2.10. *Let* $f_{pp} > 0$ *and suppose that*

$$\lim_{|p| \to \infty} \frac{f(x, u, p)}{|p|} = \infty \quad \text{for each } x \in [a, b],\ u \in \mathbb{R}.$$

Let $u(\cdot)\in\mathcal{A}$ *be a strong relative minimizer of* I *and suppose either that* $f_u(\cdot, u(\cdot), u'(\cdot))\in L^1(a, b)$ *or that* $f_x(\cdot, u(\cdot), u'(\cdot))\in L^1(a, b)$. *Then* $u\in C^3([a, b])$ *and satisfies* (EL) *and* (DBR) *on* $[a, b]$.

Proof. Let D_1 be a maximal relatively open interval in $D=[a, b]\setminus E$. By Theorem 2.7, $u\in C^3(D_1)$ and satisfies (EL) and thus (DBR) on D_1. If $f_u(\cdot, u(\cdot), u'(\cdot))\in L^1(a, b)$ then by (EL)

$$|f_p(x, u(x), u'(x))|\leqq \text{const.},\qquad x\in D_1. \tag{2.9}$$

If $f_x(\cdot, u(\cdot), u'(\cdot))\in L^1(a, b)$ then by (DBR)

$$|u'(x) f_p(x, u(x), u'(x)) - f(x, u(x), u'(x))|\leqq \text{const.},\qquad x\in D_1. \tag{2.10}$$

By the following lemma, either (2.9) or (2.10) implies that u' is bounded in D_1, and thus that $D_1=D=[a, b]$. □

Lemma 2.11. *Let* f *satisfy the hypotheses of Theorem 2.10. Then*

$$|f_p(x, u, p)|\to\infty,\quad pf_p(x, u, p)-f(x, u, p)\to\infty$$

as $|p|\to\infty$, *uniformly for* $x\in[a, b]$ *and for* u *in compact sets of* $\mathbb{R}$.

Proof. By the convexity of $f(x, u, \cdot)$ we have that

$$f(x, u, 0)\geqq f(x, u, p)-pf_p(x, u, p),$$

and hence, for $p\neq 0$,

$$\frac{p}{|p|}f_p(x, u, p)\geqq\frac{f(x, u, p)}{|p|}-\frac{f(x, u, 0)}{|p|}.$$

Therefore, for fixed x, u,

$$\lim_{p\to\infty} f_p(x, u, p)=\infty,\qquad \lim_{p\to-\infty} f_p(x, u, p)=-\infty. \tag{2.11}$$

But $f_p(x, u, p)$ is increasing in p. Thus if $x_j\to x$, $u_j\to u$, $p_j\to\infty$ we have for $p_j\geqq M$,

$$f_p(x_j, u_j, p_j)\geqq f_p(x_j, u_j, M),$$

and so

$$\liminf_{j\to\infty} f_p(x_j, u_j, p_j)\geqq f_p(x, u, M).$$

Letting $M\to\infty$ we deduce that the first limit in (2.11) is uniform for x, u in compact sets; otherwise there would exist a convergent sequence (x_j, u_j) and a sequence $p_j\to\infty$ such that $\liminf f_p(x_j, u_j, p_j)<\infty$. The case $p\to-\infty$ is treated similarly.

To prove the second assertion of the lemma we note that

$$f(x, u, 1)\geqq f(x, u, p)-(p-1) f_p(x, u, p),$$

and hence, provided $p > 1$,

$$pf_p(x, u, p) - f(x, u, p) \geqq \frac{f(x, u, p)}{p} \cdot \frac{p}{p-1} - f(x, u, 1) \cdot \frac{p}{p-1}.$$

Therefore, for fixed x, u,

$$\lim_{p \to \infty} [pf_p(x, u, p) - f(x, u, p)] = \infty. \tag{2.12}$$

That the limit in (2.12) is uniform for x, u in compact sets follows as above using the fact that $pf_p(x, u, p) - f(x, u, p)$ is increasing in p for $p > 0$. The case $p \to -\infty$ is handled similarly. □

Corollary 2.12. *Let* $f = f(u, p)$ *satisfy* $f_{pp} > 0$ *and*

$$\lim_{|p| \to \infty} \frac{f(u, p)}{|p|} = \infty \quad \text{for each } u \in \mathbb{R}. \tag{2.13}$$

If $u(\cdot) \in \mathscr{A}$ *is a strong relative minimizer of* I *then* $u(\cdot) \in C^3([a, b])$ *and satisfies* (EL) *and* (DBR) *on* $[a, b]$.

Finally, we remark that if $1 < q < \infty$ then Theorem 2.7 still holds (with the same proof) if we replace $\mathscr{A}$ by $\mathscr{A} \cap W^{1,q}(a, b)$ both in the statement of the theorem and in the definition of a strong relative minimizer. This is perhaps of interest since in § 5 we show that minimizers in $\mathscr{A}$ and $\mathscr{A} \cap W^{1,q}(a, b)$ may be different.

§ 3. An integral with a scale invariance property

In this section we consider the problem of minimizing

$$I(u) = \int_0^1 [(x^2 - u^3)^2 (u')^{14} + \varepsilon(u')^2] \, dx \tag{3.1}$$

subject to

$$u(0) = 0, \quad u(1) = k, \tag{3.2}$$

where $\varepsilon > 0$ and $k > 0$ are given.

Note that the integrand

$$f(x, u, p) = (x^2 - u^3)^2 p^{14} + \varepsilon p^2 \tag{3.3}$$

in (3.1) satisfies

$$f_{pp} \geqq 2\varepsilon > 0. \tag{3.4}$$

The Euler-Lagrange equation corresponding to (3.1) is

$$\frac{d}{dx} (7(x^2 - u^3)^2 (u')^{13} + \varepsilon u') = -3u^2(x^2 - u^3) (u')^{14}. \tag{3.5}$$

It is easily verified that (3.5) has an exact solution $u = \bar{k}x^{\frac{2}{3}}$ on (0, 1] provided

$$\varepsilon = \left(\frac{2\bar{k}}{3}\right)^{12} (1 - \bar{k}^3)\,(13\bar{k}^3 - 7). \tag{3.6}$$

Define

$$\theta(\tau) = (\tfrac{2}{3})^{12}\,\tau^4(1-\tau)\,(13\tau - 7).$$

Differentiating θ we see that θ attains its maximum in the interval $(\frac{7}{13}, 1)$ at the point $\tau^* = \dfrac{25 + \sqrt{79}}{39} = .868928\ldots$, and that $\theta'(\tau) > 0$ for $\frac{7}{13} < \tau < \tau^*$, $\theta'(\tau) < 0$ for $\tau^* < \tau < 1$. Define

$$\varepsilon_0 = \theta(\tau^*) = .002474\ldots$$

We have thus proved

Proposition 3.1. *If* $0 < \varepsilon < \varepsilon_0$ *the Euler-Lagrange equation* (3.5) *has exactly two solutions in* (0, 1] *of the form* $u = \bar{k}x^{\frac{2}{3}}$, $\bar{k} > 0$; *the corresponding values of* $\bar{k}$ *satisfy* $\frac{7}{13} < \bar{k}_1(\varepsilon)^3 < \tau^* < \bar{k}_2(\varepsilon)^3 < 1$. *If* $\varepsilon = \varepsilon_0$ *there is just one such solution, namely* $u = (\tau^*)^{\frac{1}{3}}\, x^{\frac{2}{3}}$; *if* $\varepsilon > \varepsilon_0$ *there are no such solutions.*

The integrand f in (3.3) satisfies the scale invariance property

$$f(\lambda x, \lambda^{\gamma} u, \lambda^{\gamma-1} p) = \lambda^{\varrho} f(x, u, p) \tag{3.7}$$

for all $\lambda > 0$ and all (x, u, p), where $\gamma = \frac{2}{3}$ and $\varrho = -\frac{2}{3}$. We exploit this by making the change of variables

$$v = u^{1/\gamma}, \qquad z = \frac{v}{x}, \qquad q = v', \qquad x = e^t. \tag{3.8}$$

Setting $\lambda = 1/x$ in (3.7) we obtain

$$f(x, u, p) = x^{\varrho} F(z, q), \tag{3.9}$$

where

$$F(z, q) \stackrel{\text{def}}{=} f(1, z^{\gamma}, \gamma z^{\gamma-1} q). \tag{3.10}$$

It is easily verified that, for any smooth integrand satisfying (3.7), (EL) is transformed into the *autonomous* system

$$\begin{aligned} \frac{dz}{dt} &= q - z, \\ \frac{dF_q}{dt} &= F_z - \varrho F_q. \end{aligned} \tag{3.11}$$

More precisely, if $0 < a < b < \infty$ and u is a smooth solution of (EL) on (a, b) satisfying $u(x) > 0$ for all $x \in (a, b)$, then

$$q(t) = \gamma^{-1}[u(e^t)]^{(1-\gamma)/\gamma}\, u'(e^t), \qquad z(t) = e^{-t}[u(e^t)]^{1/\gamma} \tag{3.12}$$

is a smooth solution of (3.11) for $\log a < t < \log b$. Conversely, if (q, z) is a smooth solution of (3.11) defined for $\alpha < t < \beta$ and satisfying $z(t) < 0$ for all $x \in (\alpha, \beta)$ then

$$u(x) = [x \cdot z(\log Ax)]^{\gamma} \tag{3.13}$$

is a smooth solution of (EL) for $e^{\alpha} < Ax < e^{\beta}$, where $A > 0$ is arbitrary. The arbitrary constant in (3.13) arises from the fact that, since (3.11) is autonomous, if $z(t)$ is a solution so is $z(t + \log A)$; equivalently, if $u(x)$ is a positive solution of (EL) so is $A^{-\gamma} u(Ax)$. Note that (3.11 b) is the Euler-Lagrange equation for the integral

$$\hat{I}(v) = \int_0^1 x^{\varrho} F\left(\frac{v(x)}{x}, v'(x)\right) dx,$$

obtained by making the change of variables (3.8) in (3.1). As has been pointed out to us by P. J. OLVER, the fact that the scale invariance property (3.7) implies the existence of a change of variables making (EL) autonomous is a consequence of the theory of Lie groups (*cf.* INCE [21, Chap. 4]). We remark that the above reduction to an autonomous system is used in BALL [3] as a tool for studying the radial equation of nonlinear elasticity in n space dimensions, the appropriate values of γ, ϱ being $\gamma = 1$, $\varrho = n - 1$.

From now on we assume that f is given by (3.3), although it will be apparent to the reader that much of what we have to say applies to a general class of integrands satisfying (3.7) for suitable γ, ϱ. For later use we note that since

$$F(z, q) = (\tfrac{2}{3})^{14} (1 - z^2)^2 z^{-14/3} q^{14} + (\tfrac{2}{3})^2 \varepsilon z^{-2/3} q^2, \tag{3.14}$$

(3.11) takes the form

$$\begin{aligned} \frac{dz}{dt} &= q - z, \\ \frac{dq}{dt} &= G(z, q), \end{aligned} \tag{3.15}$$

$$\begin{aligned} G(z, q) &\overset{\text{def}}{=} \frac{F_z + \frac{2}{3} F_q - (q - z) F_{qz}}{F_{qq}} \\ &= \frac{q^2}{3z} \left[\frac{(\frac{2}{3})^{12} (1 - z^2) [13q(7 - z^2) - 84z] q^{11} + \varepsilon z^4}{91 (\frac{2}{3})^{12} (1 - z^2)^2 q^{12} + \varepsilon z^4}\right]. \end{aligned} \tag{3.16}$$

We study (3.15) in the first quadrant of the (z, q) plane. Note that solutions of (3.15) in the first quadrant correspond to positive solutions u of (3.5) with $u'(x) \geqq 0$. It is clear that any minimizer of (3.1), (3.2) satisfies $u'(x) \geqq 0$ a.e. $x \in [0, 1]$, since otherwise the value of I could be reduced by making u constant on some interval.

Before proceeding with the details of our phase-plane analysis, the reader may wish to look at Figure 3.1 so as to see where we are heading.

We begin by examining the rest points of (3.15) in $z > 0$, $q > 0$. From (3.15), (3.16) these are easily seen to be given by $q = z = \bar{k}^{-\frac{3}{2}}$, where $\bar{k} > 0$ satisfies

(3.6), and correspond to the solutions $u = \bar{k}x^{\frac{2}{3}}$ discussed in Proposition 3.1. Thus, for $0 < \varepsilon < \varepsilon_0$, there are precisely two rest points, namely $q = z = \bar{k}_1(\varepsilon)^{\frac{3}{2}}$ and $q = z = \bar{k}_2(\varepsilon)^{\frac{3}{2}}$, with $\frac{7}{13} < \bar{k}_1(\varepsilon)^3 < \bar{k}_2(\varepsilon)^3 < 1$. We denote these points by P_1 and P_2 respectively. We study the nature of the rest points by linearization. Thus let P denote a rest point $q = z = \bar{k}^{\frac{3}{2}}$. Setting $z = \bar{k}^{\frac{3}{2}} + a$, $q = \bar{k}^{\frac{3}{2}} + b$ gives (3.15) the form

$$\frac{d}{dt}\begin{pmatrix} a \\ b \end{pmatrix} = A\begin{pmatrix} a \\ b \end{pmatrix} + O(|a|^2 + |b|^2), \tag{3.17}$$

where

$$A \stackrel{\text{def}}{=} \begin{pmatrix} -1 & 1 \\ \sigma(\bar{k}) & \frac{2}{3} \end{pmatrix},$$

and

$$\sigma(\bar{k}) \stackrel{\text{def}}{=} \frac{2}{9}\,\frac{(31\bar{k}^3 - 28)}{(1 - \bar{k}^3)(14 - 13\bar{k}^3)}.$$

The eigenvalues of A are given by

$$\lambda_\pm = \tfrac{1}{6}\left(-1 \pm \sqrt{25 + 36\sigma(\bar{k})}\right).$$

Thus,

(i) if $\sigma(\bar{k}) < -\frac{25}{36}$, λ_+, λ_- are complex,

(ii) if $\sigma(\bar{k}) = -\frac{25}{36}$, $\lambda_+, \lambda_- = -\frac{1}{6}$ and A has a double elementary divisor,

(iii) if $-\frac{25}{36} < \sigma(\bar{k}) < -\frac{2}{3}$, $\lambda_- < \lambda_+ < 0$,

(iv) if $\sigma(\bar{k}) = -\frac{2}{3}$, $\lambda_- = -\frac{1}{3}, \lambda_+ = 0$,

(v) if $-\frac{2}{3} < \sigma(\bar{k})$, $\lambda_- < 0 < \lambda_+$.

As is well known (*cf.* HARTMAN [19, p. 212, ff.]), cases (i)–(iii) correspond to P being a sink, and case (v) to a saddle-point. Case (iv) is a critical case where the stability is determined by the nonlinear terms in (3.17), and we discuss this presently. In case (i), P is a focus. In case (ii) P is an improper node, all solutions of (3.15) near P approaching P with slope $\frac{5}{6}$ as $t \to \infty$. In case (iii) P is an improper node with a single pair of solutions approaching P with slope $\lambda_- + 1 \in (\frac{2}{3}, \frac{5}{6})$ as $t \to \infty$, and all other nearby solutions approaching P with slope $\lambda_+ + 1 \in (\frac{5}{6}, 1)$ as $t \to \infty$. In case (v) the slope of the stable manifold of P at P is $\lambda_- + 1 < \frac{2}{3}$, that of the unstable manifold $\lambda_+ + 1 > 1$. We now note that $\sigma(\bar{k}) > -\frac{2}{3}$

Fig. 1. The phase-plane diagram for (3.15). Shown in particular are the smooth solution orbit, which leaves the origin with slope 3/2, and the stable and unstable manifolds of P_2. The absolute minimizers of I correspond to appropriate portions of the dashed curves (see Theorem 3.12).

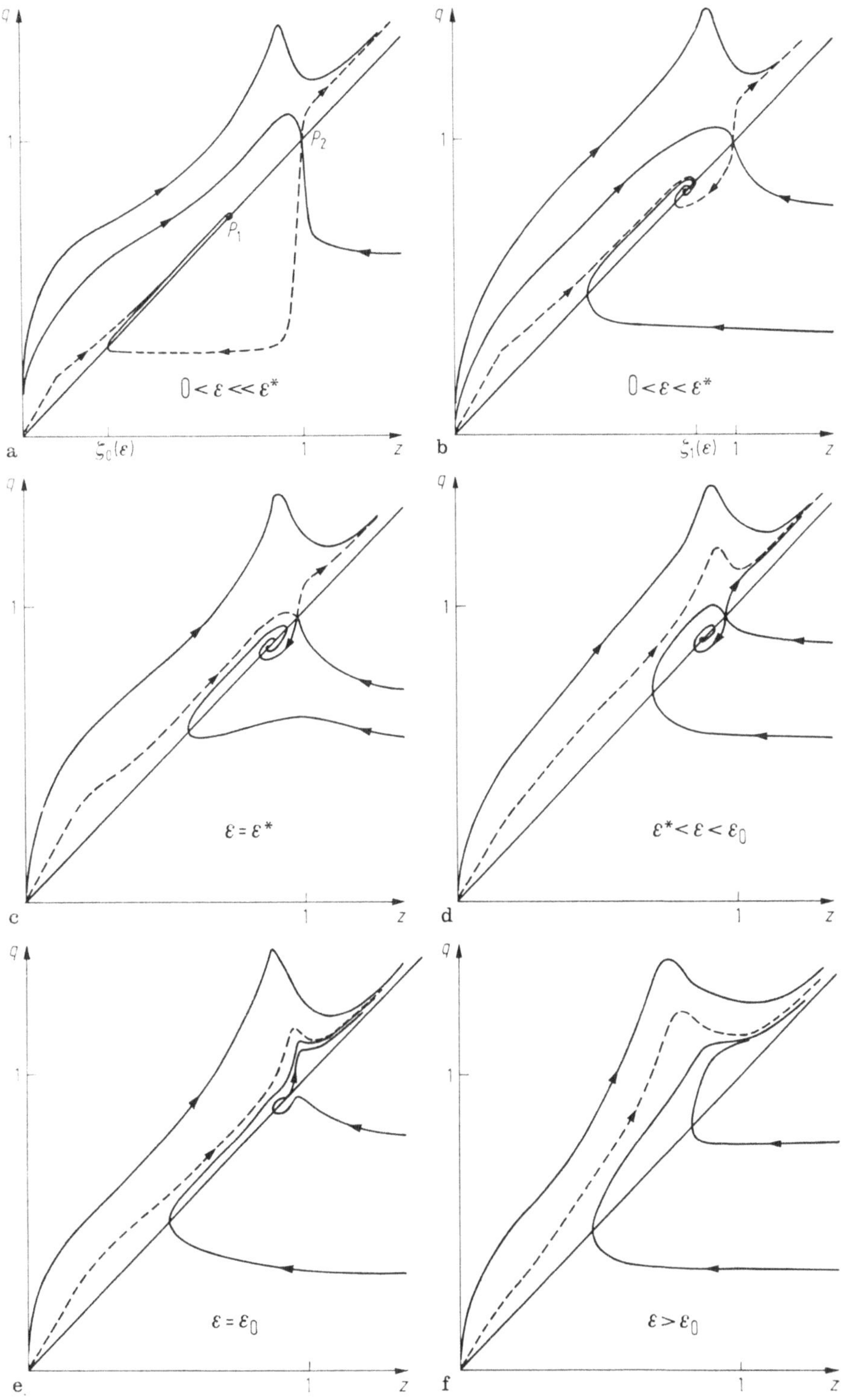

q
1
P_2
P_1
$0<\varepsilon\ll\varepsilon^*$
a
$\zeta_0(\varepsilon)$
1
z
q
1
$0<\varepsilon<\varepsilon^*$
b
$\zeta_1(\varepsilon)$
1
z
q
1
$\varepsilon=\varepsilon^*$
c
1
z
q
1
$\varepsilon^*<\varepsilon<\varepsilon_0$
d
1
z
q
1
$\varepsilon=\varepsilon_0$
e
1
z
q
1
$\varepsilon>\varepsilon_0$
f
1
z

(respectively $\sigma(\bar{k}) < -\frac{2}{3}$) if and only if

$$39\tau^2 - 50\tau + 14 > 0 \quad \text{(respectively } <0\text{)}$$

where $\tau = \bar{k}^3$, and since $\dfrac{25 - \sqrt{79}}{39} < \frac{7}{13}$ this holds if and only if $\tau > \tau^* = \dfrac{25 + \sqrt{79}}{39}$ (respectively $\tau < \tau^*$). The case $\sigma(\bar{k}) = -\frac{2}{3}$ corresponds to $\tau = \tau^*$. Similarly, $\sigma(\bar{k}) > -\frac{25}{36}$ (respectively $\sigma(\bar{k}) < -\frac{25}{36}$) if and only if

$$325\tau^2 - 427\tau + 126 > 0 \quad \text{(respectively } <0\text{)},$$

which holds if and only if $\tau > \tau_1 = .86634\ldots$ (respectively $\tau < \tau_1$). We let $\varepsilon_1 = \theta(\tau_1) = .002473\ldots$. We have thus proved

Proposition 3.2. *Let* $0 < \varepsilon < \varepsilon_0$. *Then* P_1 *is a sink and* P_2 *is a saddle point.*

Since $\dfrac{dz}{dt} = q - z$ the flow in the region $0 \leqq q < z$ is to the left, that in the region $0 < z < q$ to the right. We also make frequent use of the direction of flow on the diagonal $q = z$, where $\dfrac{dz}{dt} = 0$, given in the following lemma.

Lemma 3.3.

(i) *Let* $0 < \varepsilon < \varepsilon_0$. *Then* $G(z, z) > 0$ *for* $0 < z < \bar{k}_1(\varepsilon)^{\frac{3}{2}}$ *and for* $\bar{k}_2(\varepsilon)^{\frac{3}{2}} < z < \infty$, *while* $G(z, z) < 0$ *for* $\bar{k}_1(\varepsilon)^{\frac{3}{2}} < z < \bar{k}_2(\varepsilon)^{\frac{3}{2}}$.

(ii) *Let* $\varepsilon = \varepsilon_0$. *Then* $G(z, z) > 0$ *for all* $z > 0$, $z \neq (\tau^*)^{\frac{1}{2}}$.

(iii) *Let* $\varepsilon > \varepsilon_0$. *Then* $G(z, z) > 0$ *for all* $z > 0$.

For the purpose of studying the existence of periodic orbits it is convenient to introduce the new variable $r = F_q(z, q)$. It is easily verified, using the fact that $F_{qq} > 0$, that $(z, q) \to (z, r)$ maps $z > 0$, $q > 0$ onto $z > 0$, $r > 0$ and has a smooth inverse. Thus (3.11) is equivalent to

$$\begin{aligned} \frac{dz}{dt} &= q(z, r) - z \overset{\text{def}}{=} Z(z, r), \\ \frac{dr}{dt} &= F_z(z, q(z, r)) + \tfrac{2}{3} r \overset{\text{def}}{=} R(z, r). \end{aligned} \tag{3.18}$$

An easy computation shows that

$$\frac{\partial Z}{\partial z} + \frac{\partial R}{\partial r} = -\tfrac{1}{3}. \tag{3.19}$$

Integration of (3.19) over the region enclosed by a nontrivial periodic or homoclinic orbit gives a contradiction. We have thus proved

Proposition 3.4. *The system* (3.15) *has no nontrivial periodic orbit and no homoclinic orbit in* $z > 0$, $q > 0$.

We next study the continuation and asymptotic properties of solutions.

Proposition 3.5. *Let* $z_0 > 0$, $q_0 > 0$, *and let* $(z(t), q(t))$ *denote the unique solution of* (3.15) *with* $z(0) = z_0$, $q(0) = q_0$. *Then* $(z(t), q(t))$ *exists and remains in* $z > 0$, $q > 0$ *on a maximal interval* $(t_{\min}, \infty)$, *where* $-\infty \leqq t_{\min} < 0$. *As* $t \to \infty$, *either* $z(t) \to \infty$ *and* $q(t) \to \infty$ *or* $(z(t), q(t)) \to (k^{\frac{-3}{2}}, k^{\frac{-3}{2}})$, *a rest point. As* $t \to t_{\min}+$ *either* $(z(t), q(t)) \to (0, 0)$ *or* $z(t) \to \infty$ *and* $q(t) \to c = c(z_0, q_0) \in [0, \infty)$† *or* $(z(t), q(t)) \to (k^{\frac{-3}{2}}, k^{\frac{-3}{2}})$, *a rest point.*

Proof. Let the maximal interval in which the solution $(z(t), q(t))$ exists and remains in $z > 0$, $q > 0$ be $(t_{\min}, t_{\max})$, where $-\infty \leqq t_{\min} < 0 < t_{\max} \leqq \infty$. Observe first that if $(z(t), q(t))$ remains in a compact subset of $z > 0$, $q \geqq 0$ for all $t \in [0, t_{\max})$ (respectively $t \in (t_{\min}, 0]$) then $t_{\max} = \infty$ (respectively $t_{\min} = -\infty$), and we can apply the Poincaré-Bendixson theory (*cf.* HARTMAN [19, p. 151 ff.]). By Proposition 3.4 the only possibilities are that $(z(t), q(t))$ tends to a rest point as $t \to \infty$ (respectively $t \to -\infty$), or that the ω-limit set (respectively α-limit set) of $(z(\cdot), q(\cdot))$ contains more than one rest point (and thus $0 < \varepsilon < \varepsilon_0$). The latter case cannot occur since P_1 is asymptotically stable.

Next we note that on any open t-interval where $q(t) \neq z(t)$ we have $\dfrac{dz}{dt} \neq 0$, and thus the orbit has the representation $q = q(z)$, where by (3.15)

$$\frac{dq}{dz} = \frac{q^2}{3z(q-z)}\left[\frac{(\frac{2}{3})^{12}(1-z^2)[13q(7-z^2)-84z]\,q^{11}+\varepsilon z^4}{91(\frac{2}{3})^{12}(1-z^2)^2\,q^{12}+\varepsilon z^4}\right] \stackrel{\text{def}}{=} H(z, q, \varepsilon). \tag{3.20}$$

We first eliminate the possibility that $q(z)$ becomes unbounded as $z \to \bar{z} \in (0, \infty)$ either from above or below. By general results on ordinary differential equations we would then have $q(z) \to +\infty$ as $z \to \bar{z}+$ or $q(z) \to +\infty$ as $z \to \bar{z}-$. If $\bar{z} \neq 1$, then for q large and for z near $\bar{z}$ we have

$$\left|\frac{dq}{dz}\right| = \left|\frac{q}{3z\left(1-\dfrac{z}{q}\right)}\left[\frac{(\frac{2}{3})^{12}(1-z^2)\left[13(7-z^2)-84\dfrac{z}{q}\right]+\varepsilon\dfrac{z^4}{q^{12}}}{91(\frac{2}{3})^{12}(1-z^2)^2+\dfrac{z^4}{q^{12}}}\right]\right| \leqq Cq,$$

where here and below C denotes a generic constant. Thus q is bounded near $\bar{z}$, a contradiction. If $\bar{z} = 1$, we observe that $q(z)$ satisfies

$$\frac{d}{dz}((q-z)F_q - F) = -\tfrac{1}{3}F_q, \tag{3.21}$$

† It will be shown in Proposition 3.6 that in this case $t_{\min} = -\infty$ and $c(z_0, q_0) > 0$.

where F is given by (3.14). (This is essentially the DuBois-Reymond equation for $\hat{I}$.) Now

$$F_q = \tfrac{28}{3}\left(\frac{2q}{3}\right)^{13} z^{-\frac{14}{3}}(1-z^2)^2 + \frac{4\varepsilon}{3}\cdot\frac{2q}{3} z^{-\frac{2}{3}},$$

and

$$\psi(z, q) \overset{\text{def}}{=} (q-z)\, F_q - F$$

$$= \left(\frac{2q}{3}\right)^{14} z^{-\frac{14}{3}}(1-z^2)^2\left(13 - \frac{14z}{q}\right) + \varepsilon\left(\frac{2q}{3}\right)^2 z^{-\frac{2}{3}}\left(1 - \frac{2z}{q}\right).$$

Thus, for z near 1 and q large,

$$|F_q| \leqq C(q^{13}(1-z^2)^{\frac{13}{7}} + q)$$

$$\leqq C(q^{14}(1-z^2)^2 + q^{\frac{14}{13}})^{\frac{13}{14}}$$

$$\leqq C(q^{14}(1-z^2)^2 + q^2)^{\frac{13}{14}},$$

and so by (3.21)

$$\left|\frac{d\psi}{dz}(z, q(z))\right| \leqq C\,|\psi(z, q(z))|^{\frac{13}{14}}.$$

Thus $\psi(z, q(z))$ is bounded near $z = 1$, which is a contradiction.

The case when (z_0, q_0) is a rest point being trivial, we now consider the remaining cases. First suppose that $q_0 < z_0$. Note that $q = 0$, $0 < z < \infty$ is an orbit of (3.15), and that $G(z, q) > 0$ if $z > 0$, $q > 0$ and $z + q$ is sufficiently small. Since $\frac{dz}{dt} < 0$ for $q < z$ it now follows that either $(z(t), q(t))$ remains below the line $q = z$ on $[0, t_{\max})$, and hence by the first part of the proof tends to a rest point, or that $z(t_0) = q(t_0)$ for some $t_0 > 0$. In the latter case it may happen that $z(t_1) = q(t_1)$ for some $t_1 > t_0$, with $q(t) > z(t)$ for $t_0 < t < t_1$. If so, then by Lemma 3.3, $0 < \varepsilon < \varepsilon_0$ and $z(t_0) < \bar{k}_1(\varepsilon)^{\frac{3}{2}} < z(t_1) < \bar{k}_2(\varepsilon)^{\frac{3}{2}}$, so that, unless $(z(t), q(t)) \to P_1$ as $t \to \infty$ without a further crossing of $q = z$, $z(t_2) = q(t_2)$ for some $t_2 > t_1$. If $z(t_2) < z(t_0)$ the orbit $(z(t), q(t))$ would remain in a compact subset of $z > 0$, $q \geqq 0$ for $t_{\min} < t \leqq 0$ and hence tend to P_1 as $t \to -\infty$; this is impossible as P_1 is a sink. Thus by Proposition 3.4, $z(t_2) > z(t_0)$, which implies that $(z(t), q(t))$ remains in a compact subset of $z > 0$, $q \geqq 0$ for $0 \leqq t < t_{\max}$, and thus tends to P_1 as $t \to \infty$.

The above considerations show that, as regards the behavior for $t \geqq 0$, it suffices to examine the case when $q(t) > z(t)$ for all $t \in [0, t_{\max})$ and the corresponding solution curve $q(z)$ is defined for all $z \geqq z_0$. To show that $t_{\max} = \infty$ we examine the slope of the vector field on the line $q = \mu z$, where $\mu > 1$. On this line, as $z \to \infty$,

$$\frac{dq}{dz} = \frac{\mu^2}{3(\mu-1)}\left[\frac{(\frac{2}{3})^{12}\left(\frac{1}{z^2}-1\right)\left(\frac{91}{z^2}-13-\frac{84}{z^2\mu}\right)+\frac{\varepsilon}{\mu^{12}z^{12}}}{91(\frac{2}{3})^{12}\left(\frac{1}{z^2}-1\right)^2+\frac{\varepsilon}{\mu^{12}z^{12}}}\right]$$

$$= \frac{\mu^2}{21(\mu-1)}\left[1+o\left(\frac{1}{z}\right)\right],$$

where the $o\left(\frac{1}{z}\right)$ term is independent of μ. Hence, provided $\mu_0 > \frac{21}{20}$, there exists $\hat{z} > 0$ such that if $z \geqq \hat{z}$ and $\mu \geqq \mu_0$ then $\frac{dq}{dz}(z) < \mu$ on $q = \mu z$. Choosing $\mu > \frac{q(\hat{z})}{\hat{z}}$ we deduce that

$$\dot{z}(t) \leqq (\mu - 1)\, z(t)$$

whenever $z(t) \geqq \hat{z}$, and hence that $t_{\max} = \infty$.

We consider now the behavior of $(z(t), q(t))$ for $t \in (t_{\min}, 0]$. Suppose first that $q_0 > z_0$. If $q(t) > z(t)$ for all $t \in (t_{\min}, 0]$, then either $\inf_{t\in(t_{\min},0]} z(t) > 0$ or $z(t) \to 0$ as $t \to t_{\min}+$. In the former case, since $q(t)$ cannot become unbounded as $t \to t_{\min}+$, the curve lies in a compact set of $z > 0$, $q \geqq 0$ and we must have that $t_{\min} = -\infty$ and $(z(t), q(t))$ tends to a rest point as $t \to -\infty$. If $z(t) \to 0$ as $t \to t_{\min}+$ then by (3.20) the corresponding curve $q(z)$ satisfies $\frac{dq}{dz} > 0$ for sufficiently small $z > 0$, so that $q(t_{\min}) \overset{\text{def}}{=} \lim_{t\to t_{\min}+} q(t)$ exists. If $q(t_{\min}) > 0$ then by (3.20) $\frac{dq}{dz} \geqq \frac{C}{z}$ for sufficiently small $z > 0$, where $C > 0$ is a constant, and integration of this inequality gives a contradiction. Thus $(z(t), q(t)) \to (0, 0)$ as $t \to t_{\min}+$. On the other hand, if $q(t) = z(t)$ for some $t \in (t_{\min}, 0]$ then $q(t_1) < z(t_1)$ for some earlier time.

It only remains, therefore, to consider the case when $q_0 < z_0$. First, if $q(t) < z(t)$ for all $t \in (t_{\min}, 0]$ then either $z(t)$ remains bounded as $t \to t_{\min}+$, in which case $t_{\min} = -\infty$ and $(z(t), q(t))$ tends to a rest point as $t \to -\infty$, or $\lim_{t\to t_{\min}+} z(t) = \infty$. In the latter case, by (3.20), $\frac{dq}{dz} < 0$, for $z^2 > 7$, $q < z$, and so as $t \to t_{\min}+$ $q(t)$ tends to a nonnegative limit, which we denote by $c(z_0, q_0)$. Next, if $q(t_0) = z(t_0)$ for some $t_0 \in (t_{\min}, 0]$ then $q(\bar{t}) > z(\bar{t})$ for some $\bar{t} \in (t_{\min}, t_0)$. We have already treated the case when $q(t) > z(t)$ for all $t \in (t_{\min}, \bar{t}]$ and thus it remains to eliminate the possibility that $q(t_j) = z(t_j)$ for an infinite sequence $t_j \to t_{\min}+$, and of course this can only occur for $0 < \varepsilon < \varepsilon_0$. The corresponding orbit would spiral either inwards or outwards as $t \to t_{\min}+$. If it spiralled inwards then clearly we would have $t_{\min} = -\infty$ and $(z(t), q(t)) \to P_1$ as $t \to -\infty$, which is impossible since P_1 is a sink. It must thus spiral outwards, and of course it cannot remain in a compact subset of $z > 0$, $q \geqq 0$, since otherwise it would have to tend to P_2 as $t \to t_{\min}+$, which is clearly impossible. Furthermore the orbit must remain under that part of the stable manifold of P_2 lying in $q > z$, and so $z(t_j) \to 0$ as $t_j \to t_{\min}+$. But the solution curve $(z_r(t), q_r(t))$ of (3.15) satisfying $z_r(0) = 1$, $q_r(0) = \frac{1}{r}$ approaches the z-axis as $r \to \infty$, crossing $q = z$ arbitrarily close to the origin, which implies that $z(t_j)$ is bounded away from zero. □

Note that Propositions 3.2, 3.5 together imply that when $\varepsilon = \varepsilon_0$ the unique fixed point $q = z = k^{-\frac{3}{2}}$ is unstable.

It is possible to specify more precisely the asymptotic behavior of those solutions of (3.15) satisfying $z(t) \to \infty$, $q(t) \to \infty$ as $t \to \infty$. For such a solution we have seen in the proof of Proposition 3.5 that $\frac{q(z)}{z(t)}$ is bounded for large t. Setting $\zeta(t) = \frac{1}{z(t)}$, we see that (3.15) becomes

$$\dot{\zeta} = \zeta(1 - \varphi),$$
$$\dot{\varphi} = (1 - \varphi) + \frac{\varphi^2}{3}\left[\frac{(\frac{2}{3})^{12}(\zeta^2 - 1)\,[13\varphi(7\zeta^2 - 1) - 84\zeta^2]\,\varphi^{11} + \varepsilon\zeta^{12}}{91(\frac{2}{3})^{12}\,(\zeta^2 - 1)^2\,\varphi^{12} + \varepsilon\zeta^{12}}\right], \tag{3.22}$$

where $\varphi(t) \stackrel{\text{def}}{=} \frac{q(t)}{z(t)}$, and hence as $t \to \infty$,

$$\dot{\varphi} = \varphi(1 - \tfrac{20}{21}\varphi) + o(1).$$

Hence $\varphi(t) \to \frac{21}{20}$ as $t \to \infty$. Linearizing about the rest point $\zeta = 0$, $\varphi = \frac{21}{20}$ of (3.22) we obtain

$$\zeta(t) \geqq C_1 e^{-\frac{1}{20}t},$$
$$|\varphi(t) - \tfrac{21}{20}| \leqq C_2 e^{-t},$$

for sufficiently large t, where C_1 and C_2 are positive constants. It follows that

$$|q(t) - \tfrac{21}{20}z(t)| \leqq \frac{C_2}{C_1} e^{-\frac{19}{20}t}$$

for sufficiently large t, so that the solution curve rapidly approaches the line $q = \frac{21}{20}z$. Since $\dot{z} = q - z$ we deduce that

$$z(t) = Ae^{\frac{t}{20}} + O(e^{-\frac{19}{20}t})$$

as $t \to \infty$, where $A = A(z_0, q_0)$ is a constant, and hence that the corresponding solution u of (3.5) satisfies

$$u(x) = Ax^{\frac{7}{10}} + O(x^{\frac{1}{30}}) \quad \text{as } x \to \infty.$$

We now study the behavior of solutions in a neighborhood of the q and z axes, and in particular near the origin.

Proposition 3.6. *Every smooth solution u of* (3.5) *with* $u(0)=0,\ u'(0)>0$ *corresponds to a single orbit of* (3.15) *in* $z>0,\ q>0$ *that leaves the origin* $z=q=0$ *with slope* $\frac{3}{2}$. *The only other orbits of* (3.15) *leaving the origin correspond to solutions* u *of* (3.5) *with* $u(x_0)=0$ *for some* $x_0>0$; *these orbits satisfy*

$$\lim_{t\to\log x_0+} z(t)=\lim_{t\to\log x_0+} q(t)=0,\qquad \lim_{t\to\log x_0+}\frac{q(t)}{z(t)}=\infty. \tag{3.23}$$

Solutions $(z(\cdot),q(\cdot))$ *of* (3.15) *whose orbits have an unbounded intersection with* $0<q<z$ *correspond precisely to solutions* u *of* (3.5) *with* $u(0)>0,\ u'(0)>0$, *and thus satisfy* $\lim_{t\to-\infty} z(t)=\infty,\ \lim_{t\to-\infty} q(t)=c>0$, *where* $c=c(z(0),q(0))$ *is a constant.*

Proof. Let u be a smooth solution of (3.5) on some interval $[0,a]$, $a>0$, satisfying $u(0)=0$, $u'(0)=\alpha>0$. Then $u(x)=\alpha x+o(x)$, $u'(x)=\alpha+o(1)$, as $x\to 0+$, and hence $z=\alpha^{\frac{3}{2}}x^{\frac{1}{2}}+o(x^{\frac{1}{2}})$, $q=\frac{3}{2}(\alpha^{\frac{3}{2}}x^{\frac{1}{2}}+o(x^{\frac{1}{2}}))$. Thus the corresponding solution $(z(t),q(t))$ satisfies

$$\lim_{t\to-\infty} z(t)=\lim_{t\to-\infty} q(t)=0,\qquad \lim_{t\to-\infty}\frac{q(t)}{z(t)}=\frac{3}{2}.$$

That this solution is the same for any $\alpha>0$ (up to adding a constant to t) follows from the similarity transformation (3.13) and the uniqueness of solutions to the initial value problem for (3.5).

Let $u_\beta(x)$ denote the unique solution to (3.5) satisfying $u_\beta(1)=0$, $u_\beta'(1)=\beta>0$; this corresponds to a solution $(z_\beta(\cdot),q_\beta(\cdot))$ satisfying

$$z_\beta(t)=\frac{[\beta(e^t-1)+o(e^t-1)]^{\frac{3}{2}}}{e^t}=o(1),$$

and

$$q_\beta(t)=\tfrac{3}{2}[\beta(e^t-1)+o(e^t-1)]^{\frac{1}{2}}(\beta+o(1))=o(1),$$

as $t\to 0+$. Also

$$\lim_{t\to 0+}\frac{q_\beta(t)}{z_\beta(t)}=\lim_{t\to 0+}\frac{3}{2(e^t-1)}=\infty.$$

Let $\delta>0$ be sufficiently small. It follows from Proposition 3.5 that $z_\beta(t_\beta)=\delta$ for some minimal $t_\beta>0$. Also, since $q_\beta(t_\beta)>z_\beta(t_\beta)$, the corresponding intersection at $x=e^{t_\beta}$ of the graph of u_β with $\delta^{\frac{2}{3}}x^{\frac{2}{3}}$ is transversal, and thus by the implicit function theorem t_β depends continuously on β. Hence also $q_\beta(t_\beta)$ depends continuously on β. We examine the behavior of $q_\beta(t_\beta)$ as β varies from 0 to ∞. We first show that

$$\lim_{\beta\to\infty} q_\beta(t_\beta)=\infty. \tag{3.24}$$

Since $u'_\beta(x) > 0$ for all $x \geqq 0$, u_β is invertible; denote the inverse function by $x_\beta(u)$. By (3.5) $x_\beta(\cdot)$ satisfies the transformed equation

$$[19(x^2 - u^3)^2 + \varepsilon x_u^{12}]\, x_{uu} = x_u(x^2 - u^3)\,(28xx_u - 39u^2), \qquad (3.25)$$

where the subscripts denote derivatives with respect to u. This equation has the solution $\bar{x}(u) \equiv 1$, $u \in [0, \frac{1}{2}]$, in the neighborhood of which (3.25) can be written in the form $x_{uu} = h(u, x, x_u)$ with h continuously differentiable. Since $\bar{x}(0) = x_\beta(0) = 1$, $\bar{x}_u(0) = 0$, $(x_\beta)_u\,(0) = \dfrac{1}{\beta}$, it follows that $x_\beta \to 1$ in $C^1([0, \frac{1}{2}])$ as $\beta \to \infty$. In particular, $t_\beta \to 0$ as $\beta \to \infty$. Since $q_\beta(t_\beta) = \frac{3}{2}\, u_\beta(e^{t_\beta})^{\frac{1}{2}}\, u'_\beta(e^{t_\beta}) = \frac{3}{2}\,\dfrac{\delta^{\frac{1}{3}}\, e^{\frac{t}{3}\beta}}{(x_\beta)_u\,(\delta^{\frac{2}{3}}\, e^{\frac{2}{3}t_\beta})}$ this gives (3.24).

Next, let $\tilde{u}_\beta(x) = \beta^2 u_\beta(\beta^{-3}\, x)$, which also solves (3.5) and satisfies $\tilde{u}_\beta(\beta^3) = 0$, $\tilde{u}'_\beta(\beta^3) = 1$. Clearly $\tilde{u}_\beta \to \tilde{u}$ in $C^1([0, 1])$ as $\beta \to 0+$, where $\tilde{u}$ is the unique solution of (3.5) satisfying $\tilde{u}(0) = 0$, $\tilde{u}'(0) = 1$. But $\beta^3 e^{t_\beta}$ is the least value of $x > \beta^3$ such that $\tilde{u}_\beta(x) = \delta^{\frac{2}{3}}\, x^{\frac{2}{3}}$, and thus tends to the least positive root $\tilde{x}$ of $\tilde{u}(x) = \delta^{\frac{2}{3}} x^{\frac{2}{3}}$ as $\beta \to 0+$. Thus

$$\lim_{\beta \to 0+} q_\beta(t_\beta) = \tfrac{3}{2}(\delta\tilde{x})^{\frac{1}{3}}\, \tilde{u}'(\tilde{x}),$$

which is the value of q at the intersection of $z = \delta$ with the smooth solution orbit leaving $q = z = 0$ with slope $\frac{3}{2}$. We have thus shown that the region above this orbit in the strip $0 < z \leqq \delta$ is completely filled by the orbits $(z_\beta(\cdot), q_\beta(\cdot))$. If $x_0 > 0$ is given then $(z(t), q(t)) = (z_\beta(t - \log x_0),\ q_\beta(t - \log x_0))$ corresponds by (3.13)ff to the solution u of (3.5) satisfying $u(x_0) = 0$, $u'(x_0) = \beta x_0^{-\frac{1}{3}}$, and thus (3.23) holds.

Let $u_{\gamma,\nu}$ be the unique solution of (3.5) satisfying $u(0) = \gamma > 0$, $u'(0) = \nu > 0$. Then the corresponding solution $(z_{\gamma,\nu}(\cdot), q_{\gamma,\nu}(\cdot))$ of (3.15) satisfies $\lim\limits_{t \to -\infty} z_{\gamma,\nu}(t) = \infty$, $\lim\limits_{t \to -\infty} q_{\gamma,\nu}(t) = \frac{3}{2}\gamma^{\frac{1}{2}}\nu$. As $\gamma \to 0+$, $u_{\gamma,1} \to \tilde{u}$ in $C^1([0, 1])$ and hence, for each *fixed* t, $z_{\gamma,1}(t) \to \tilde{z}(t) \overset{\text{def}}{=} \dfrac{\tilde{u}^{\frac{3}{2}}(e^t)}{e^t}$ and $q_{\gamma,1}(t) \to \tilde{q}(t) \overset{\text{def}}{=} \frac{3}{2}\,\tilde{u}^{\frac{1}{2}}(e^t)\,\tilde{u}'(e^t)$.

Conversely, suppose that $(z(\cdot), q(\cdot))$ is a solution of (3.15) whose orbit has an unbounded intersection with $0 < q < z$. By Proposition 3.5, $\lim\limits_{t \to t_{\min}+} q(t) = c \geqq 0$. Let $x_0 = e^{t_{\min}}$. Suppose $t_{\min} > -\infty$, so that $x_0 > 0$. Then the corresponding solution u of (3.5) would satisfy

$$\lim_{x \to x_0+} v(x) = \lim_{t \to t_0} z(t) = \infty, \qquad \lim_{x \to x_0+} v'(x) = c,$$

where $v = u^{\frac{3}{2}}$, which is impossible. Thus $t_{\min} = -\infty$, $x_0 = 0$, and since $\lim\limits_{x \to 0+} v'(x) = c$ we have $v(x) \to d$ as $x \to x_0+$, where $d \geqq 0$ is a constant. But if d were zero then we would have

$$\infty = \lim_{x \to 0+} \frac{v(x)}{x} = \lim_{x \to 0+} \frac{v'(x)}{1} = c,$$

a contradiction. Hence $u(0) > 0$, $u'(0) = \dfrac{2c}{3u(0)^{\frac{1}{2}}} \geqq 0$. Now if $c = 0$, $u(x) \equiv u(0)$ by uniqueness of solutions to (3.5), and hence $q(t) \equiv 0$. Hence $u'(0) > 0$.

It follows immediately from the above that for $\delta > 0$ sufficiently small the region in $0 < z < \delta$, $q > 0$ below the smooth solution orbit is completely filled with orbits corresponding to solutions of (3.5) with $u(0) > 0$, $u'(0) > 0$. In particular there are no other orbits leaving the origin. □

We next apply the results of Section 2.

Theorem 3.7. *I attains an absolute minimum on the set* $\mathscr{A} = \{u \in W^{1,1}(0, 1)$: $u(0) = 0, u(1) = k\}$. *Let u be any minimizer. If* $\varepsilon > \varepsilon_0$ *then u is a* C^∞ *solution of* (3.5) *on* [0, 1]. *If* $0 < \varepsilon \leqq \varepsilon_0$ *then either u is a* C^∞ *solution of* (3.5) *on* [0, 1] *or u is a* C^∞ *solution of* (3.5) *on* (0, 1] *with* $u(x) \sim \bar{k}x^{\frac{2}{3}}$, $u'(x) \sim \frac{2}{3}\bar{k}x^{-\frac{1}{3}}$ *as* $x \to 0+$, *where* $\bar{k}$ *satisfies* (3.6). *In all cases u corresponds to a single semi-orbit* $(z(t), q(t))$, $t \in (-\infty, 0]$, *of* (3.15), *with* $z(t) > 0$, $q(t) > 0$ *for all* $t \in (-\infty, 0]$.

Proof. That I attains a minimum on $\mathscr{A}$ follows immediately from Theorem 2.1. Let u be any minimizer. By Theorem 2.7 and the subsequent discussion there is a closed set E of measure zero on the complement of which u is a C^3, and hence smooth, solution of (3.5). Let D_1 be a maximal relatively open interval in $[0, 1] \setminus E$, and denote by x_0, x_1 the left and right hand endpoints of D_1 respectively. We have already noted that $u'(x) \geqq 0$ a.e., and it thus follows from Theorem 2.7 that if $x_0 \neq 0$ (respectively $x_1 \neq 1$) then $\lim_{x \to x_0+} u'(x) = +\infty$ (respectively $\lim_{x \to x_1-} u'(x) = +\infty$). If $u'(x)$ were zero for some $x \in (x_0, x_1)$ we would have, by uniqueness of solutions to (3.5), that $u = \text{const.}$ in (x_0, x_1) and thus in $D_1 = [0, 1]$, contradicting $k > 0$. Thus $u'(x) > 0$ for all $x \in (x_0, x_1)$ and u generates a solution $(z(t), q(t))$, $t \in (\log x_0, \log x_1)$, to (3.15) with $z(t) > 0$, $q(t) > 0$ for all $t \in (\log x_0, \log x_1)$. But by Proposition 3.5 the solution $(z(t), q(t))$ exists for all $t > \log x_0$, and therefore

$$\lim_{x \to x_1-} u'(x) = \lim_{t \to \log x_1-} \tfrac{2}{3}\, q(t)\, z(t)^{-\frac{1}{3}}\, x_1^{-\frac{1}{3}} < \infty.$$

Hence $x_1 = 1$. Suppose that $-\infty \leqq t_{\min} < \log x_0$. Then

$$\lim_{x \to x_0+} u'(x) = \lim_{t \to \log x_0+} \tfrac{2}{3}\, q(t)\, z(t)^{-\frac{1}{3}}\, x_0^{-\frac{1}{3}} < \infty,$$

since $x_0 > 0$, yielding a contradiction. Therefore $t_{\min} = \log x_0$. By Proposition 3.5 there are three cases to consider. First, we may have $(z(t), q(t)) \to (0, 0)$ as $t \to \log x_0+$. If $x_0 > 0$ this is impossible since we would then have $u(x_0) = 0$ and hence $u(x) = 0$ for all $x \in [0, 1]$. If $x_0 = 0$ then by Proposition 3.6 u is C^∞ on [0, 1]. Second, we may have $z(t) \to \infty$ and $q(t) \to c \geqq 0$ as $t \to \log x_0+$. In this case, by Proposition 3.6 $x_0 = 0$ and $u(0) > 0$, which is impossible. Third, we may have $x_0 = 0$ and $\lim_{t \to -\infty} (z(t), q(t)) = (\bar{k}^{\frac{3}{2}}, \bar{k}^{\frac{3}{2}})$, a rest point. In this case u is C^∞ on (0, 1] with $u(x) \sim \bar{k}x^{\frac{2}{3}}$, $u'(x) \sim \frac{2}{3}\bar{k}x^{-\frac{1}{3}}$ as $x \to 0+$. □

As a preliminary result showing that every minimizer must in certain cases be singular we prove

Lemma 3.8. *Let u minimize I on* $\mathcal{A}$, *and suppose that* $0<\alpha<\beta<\min(1,k)$ *and*

$$\frac{4\varepsilon}{3}\beta^{\frac{1}{2}}<\left(\tfrac{9}{13}\right)^{13}(1-\beta^3)^2(\beta-\alpha)^{14}. \tag{3.26}$$

Then $u(x)>\alpha x^{\frac{2}{3}}$ *for all* $x\in(0,1]$.

Proof. We modify an argument of Manià [25] (see also Cesari [11] and Section 4). If the conclusion of the lemma were false then there would exist a subinterval (x_1, x_2) of $[0,1]$ such that

$$\alpha x^{\frac{2}{3}}\leqq u(x)\leqq\beta x^{\frac{2}{3}}\quad\text{for all } x_1\leqq x\leqq x_2$$

and $u(x_1)=\alpha x_1^{\frac{2}{3}}$, $u(x_2)=\beta x_2^{\frac{2}{3}}$. Thus

$$\begin{aligned}\int_0^{x_2} f(x,u,u')\,dx &\geqq \int_{x_1}^{x_2} x^4\left(1-\frac{u^3}{x^2}\right)^2(u')^{14}\,dx\\ &\geqq(1-\beta^3)^2\int_{x_1}^{x_2} x^4(u')^{14}\,dx.\end{aligned}$$

Let $x=y^{\frac{13}{9}}$. Then

$$\int_{x_1}^{x_2} x^4(u')^{14}\,dx=\left(\tfrac{9}{13}\right)^{13}\int_{x_1^{\frac{9}{13}}}^{x_2^{\frac{9}{13}}}\left(\frac{du}{dy}\right)^{14}dy,$$

and by Jensen's inequality the minimizer of this integral subject to $u\Big|_{y=x_1^{\frac{9}{13}}}=\alpha x_1^{\frac{2}{3}}$, $u\Big|_{y=x_2^{\frac{9}{13}}}=\beta x_2^{\frac{2}{3}}$ is given by the linear function $u=\alpha x_1^{\frac{2}{3}}+\left(\dfrac{\beta x_2^{\frac{2}{3}}-\alpha x_1^{\frac{2}{3}}}{x_2^{\frac{9}{13}}-x_1^{\frac{9}{13}}}\right)\left(y-x_1^{\frac{9}{13}}\right)$.

Therefore

$$\begin{aligned}\int_0^{x_2} f(x,u,u')\,dx &\geqq\left(\tfrac{9}{13}\right)^{13}(1-\beta^3)^2\frac{\left(\beta x_2^{\frac{2}{3}}-\alpha x_1^{\frac{2}{3}}\right)^{14}}{\left(x_2^{\frac{9}{13}}-x_1^{\frac{9}{13}}\right)^{13}}\\ &=\left(\tfrac{9}{13}\right)^{13}(1-\beta^3)^2x_2^{\frac{1}{3}}\frac{\left(\beta-\alpha\left(\dfrac{x_1}{x_2}\right)^{\frac{2}{3}}\right)^{14}}{\left(1-\left(\dfrac{x_1}{x_2}\right)^{\frac{9}{13}}\right)^{13}} \qquad (3.27)\\ &\geqq\left(\tfrac{9}{13}\right)^{13}(1-\beta^3)^2x_2^{\frac{1}{3}}(\beta-\alpha)^{14}.\end{aligned}$$

Define $v \in \mathscr{A}$ by

$$v(x) = \begin{cases} x^{\frac{2}{3}}, & 0 \leqq x \leqq \beta^{\frac{3}{2}} x_2 \\ \beta x_2^{\frac{2}{3}}, & \beta^{\frac{3}{2}} x_2 \leqq x \leqq x_2 \\ u(x), & x_2 \leqq x \leqq 1. \end{cases}$$

Then

$$\begin{aligned} I(v) &= \int_0^{\beta^{\frac{3}{2}} x_2} \varepsilon(\tfrac{2}{3} x^{-\frac{1}{3}})^2 \, dx + \int_{x_2}^{1} f(x, u, u') \, dx \\ &= \frac{4\varepsilon}{3} \beta^{\frac{1}{2}} x_2^{\frac{1}{3}} + I(u) - \int_0^{x_2} f(x, u, u') \, dx. \end{aligned}$$

Hence if (3.26), (3.27) hold then $I(v) < I(u)$, a contradiction. □

Remark. Although MANIÀ's device, which he developed in connection with the Lavrentiev phenomenon, is used in the proof of Lemma 3.8, our minimization problem does not exhibit this phenomenon. In fact if u is a minimizer then by Theorem 3.7 we have $|u(x)| \leqq C x^{\frac{2}{3}}$ for x near zero. Thus if

$$u_\delta(x) = \begin{cases} \dfrac{u(\delta)\, x}{\delta}, & 0 \leqq x \leqq \delta \\ u(x), & \delta \leqq x \leqq 1, \end{cases}$$

then $\lim_{\delta \to 0+} \int_0^{\delta} f(x, u_\delta, u_\delta') \, dx = 0$ and so

$$\inf_{v \in W^{1,\infty}(0,1) \cap \mathscr{A}} I(v) = I(u).$$

In order to identify the minimizer from among the various geometrically possible trajectories in the phase-plane we make use of the following lemma.

Lemma 3.9. *Let* $u \in \mathscr{A}$ *be a smooth solution of* (3.5) *on* (0, 1] *with* $u(x) > 0$, $u'(x) > 0$ *for all* $x \in (0, 1]$. *Let* $(z(\cdot), q(\cdot))$ *be the corresponding solution of* (3.15). *Then*

$$I(u) = -3\psi(k^{\frac{3}{2}}, q(0)),$$

where $\psi(z, q) = (q - z) F_q - F$.

Proof. From (3.11), (3.21) we have that

$$x^{-\frac{2}{3}} F = -3 \frac{d}{dx} [x^{\frac{1}{3}} \psi], \quad x \in (0, 1].$$

By Proposition 3.6 (see the formula for ψ in the proof of Theorem 3.7) $\lim_{x\to 0+} \psi(x)$ exists and is finite. Therefore

$$I(u) = \int_0^1 x^{-\frac{2}{3}} F\,dx = -3\psi(z(0), q(0))$$

$$= -3\psi(k^{\frac{3}{2}}, q(0)). \quad \square$$

Since $\psi_q(z, q) = (q - z) F_{qq}$ and $F_{qq} > 0$, it follows from Lemma 3.9 that, of all trajectories $(z(\cdot), q(\cdot))$ of (3.15) satisfying $z(0) = k$ and $\lim_{t\to -\infty} (z(t), q(t)) = (0, 0)$ or $(k^{\frac{-3}{2}}, k^{\frac{-3}{2}})$ (a rest point), that corresponding to an absolute minimum of I has either the greatest value of $q(0) \geqq k^{\frac{3}{2}}$ or the least value of $q(0) \leqq k^{\frac{3}{2}}$. So as to decide between these two possibilities it is convenient to restate Lemma 3.9 in the following way. Define

$$\Gamma(z, q) = \psi(z, q) + \tfrac{1}{3} \int_0^z F_q(\zeta, \zeta)\, d\zeta. \tag{3.28}$$

Then if u_1, u_2 satisfy the hypotheses of Lemma 3.9 with corresponding solutions $(z_1(\cdot), q_1(\cdot))$, $(z_2(\cdot), q_2(\cdot))$ of (3.15),

$$I(u_1) - I(u_2) = -3[\Gamma(k^{\frac{3}{2}}, q_1(0)) - \Gamma(k^{\frac{3}{2}}, q_2(0))]. \tag{3.29}$$

Note that by (3.21) we have that along solutions of (3.20)

$$\frac{d}{dz}\Gamma(z, q) = -\tfrac{1}{3}(F_q(z, q) - F_q(z, z))$$

$$= -(q - z) M(z, q, \varepsilon), \tag{3.30}$$

where $M(z, q, \varepsilon) > 0$ for $z, q > 0$. As an application of this idea we prove the following proposition.

We denote by $(z_{sm}(\cdot), q_{sm}(\cdot))$ the smooth solution orbit, which by Proposition 3.6 leaves the origin with slope $\frac{3}{2}$; this orbit is unique modulo adding an arbitrary constant to t, and we choose for convenience the normalization corresponding to the smooth solution u of (3.5) satisfying $u(0) = 0$, $u'(0) = 1$.

Proposition 3.10. *If* $z_{sm}(t) \to \infty$, $q_{sm}(t) \to \infty$ *as* $t \to \infty$ *then for any* $k > 0$ *there exists precisely one solution u of* (3.5) *belonging to* $C^\infty([0, 1])$ *and satisfying the boundary conditions* (3.2), *and u is the unique minimizer of I in* $\mathcal{A}$.

Proof. If u is a smooth solution of (3.5) on [0, 1] satisfying (3.2) then $u(x) > 0$, $u'(x) > 0$ for all $x \in (0, 1]$. Otherwise there would exist some $x_0 \in (0, 1)$ with $u'(x_0) = 0$, and hence $u(x) \equiv u(x_0)$ by uniqueness, a contradiction. Thus any such solution is represented by an appropriate portion of the smooth solution orbit $(z_{sm}(\cdot), q_{sm}(\cdot))$, and since this orbit cuts the line $z = k^{\frac{3}{2}}$ exactly, once the existence and uniqueness of u is assured.

It remains to prove that $I(u) < I(v)$ for every other $v \in \mathscr{A}$. If this were false there would exist by Theorem 3.7 an absolute minimizer u_1 of I in $\mathscr{A}$ with $u_1 \neq u$, $I(u_1) \leqq I(u)$. Let $(z_1(\cdot), q_1(\cdot))$ be the corresponding solution of (3.15); thus $z_1(0) = k^{\frac{3}{2}}$. We know by Theorem 3.7 that we must have $0 < \varepsilon \leqq \varepsilon_0$ and $\lim_{t\to-\infty} (z_1(t), q_1(t)) = (\bar{k}^{-\frac{3}{2}}, \bar{k}^{-\frac{3}{2}})$, and since P_1 is a sink we also have $\bar{k} = \bar{k}_2$ if $0 < \varepsilon < \varepsilon_0$. Since the smooth solution orbit lies entirely above any such solution, by our preceding discussion we know that $q_1(0)$ has the least value of $q(0) \leqq k^{\frac{3}{2}}$ of all solutions $(z(\cdot), q(\cdot))$ of (3.15) with $z(0) = k^{\frac{3}{2}}$, $\lim_{t\to-\infty} (z(t), q(t)) = (\bar{k}^{-\frac{3}{2}}, \bar{k}^{-\frac{3}{2}})$. It follows that $q_1(t) \leqq z_1(t) \leqq \bar{k}^{-\frac{3}{2}}$ for all $t \in (-\infty, 0]$. Let the smooth solution orbit have graph $q = q_{\mathrm{sm}}(z)$, $z > 0$. Then by (3.28)–(3.30) and the fact that $\psi_q(z, q) > 0$ for $q > z$,

$$\Gamma(k^{\frac{3}{2}}, q_1(0)) \leqq \Gamma(\bar{k}^{-\frac{3}{2}}, \bar{k}^{-\frac{3}{2}}) < \Gamma(\bar{k}^{-\frac{3}{2}}, q_{\mathrm{sm}}(\bar{k}^{-\frac{3}{2}})) \leqq \Gamma(k^{\frac{3}{2}}, q_{\mathrm{sm}}(k^{\frac{3}{2}})),$$

and thus $I(u) < I(u_1)$, a contradiction. □

We give now an alternative proof of the assertion in Proposition 3.10 that the unique smooth solution u of (3.5) minimizes I, since it illustrates the various connections between the phase-plane diagram and the field theory of the calculus of variations. We note that $u_A(x) \stackrel{\mathrm{def}}{=} A^{-\frac{2}{3}} u(Ax)$ is a smooth solution of (3.5) for any $A > 0$, and that $\dfrac{\partial u_A(x)}{\partial A} = A^{\frac{1}{3}} u'(Ax) > 0$. Also, for any $x > 0$ we have

$$\lim_{A\to 0+} u_A(x) = \lim_{A\to 0+} A^{\frac{1}{3}} x \frac{u(Ax)}{Ax} = 0 \qquad \text{and} \qquad \lim_{A\to\infty} u_A(x) = \lim_{A\to\infty} x^{\frac{2}{3}} \left[\frac{u^{\frac{3}{2}}(Ax)}{Ax}\right]^{\frac{2}{3}}$$

$= x^{\frac{2}{3}} \lim_{t\to\infty} z_{\mathrm{sm}}(t)^{\frac{2}{3}} = \infty$. Define $u_0(x) \equiv 0$. Then $\{u_A\}_{0\leqq A<\infty}$ is a field of extremals that simply covers the region $x > 0$, $u \geqq 0$. Let $v \in \mathscr{A}$, $v \neq u$, with $v(x) > 0$ for all $x \in (0, 1]$ and $v(x) \leqq Cx^{\frac{2}{3}}$ as $x \to 0+$ (we have already seen that any minimizer of I has these properties). In order to handle the singularity of the field at the origin define for $\delta > 0$.

$$v_\delta(x) = \begin{cases} u(x), & 0 \leqq x \leqq \delta, \\ u(\delta) + \dfrac{x-\delta}{\delta}(v(2\delta) - u(\delta)), & \delta \leqq x \leqq 2\delta, \\ v(x), & 2\delta \leqq x \leqq 1. \end{cases}$$

Then

$$I(v_\delta) - I(u) = \int_0^1 [f(x, v_\delta(x), v_\delta'(x)) - f(x, v_\delta(x), p(x, v_\delta(x)))$$
$$- (v_\delta'(x) - p(x, v_\delta(x))) f_p(x, v_\delta(x), p(x, v_\delta(x)))]\, dx,$$

where $p(x, v)$ denotes the slope function of the field. Since the integrand on the right-hand side is positive by convexity, and since it can be verified that

$\lim_{\delta\to 0+} \int_\delta^{2\delta} f(x, v_\delta, v'_\delta)\, dx = 0$, it follows that $I(v_\delta) \to I(v)$ as $\delta \to 0+$, and we obtain by Fatou's Lemma that

$$I(v) - I(u) \geqq \int_0^1 [f(x, v(x), v'(x)) - f(x, v(x), p(x, v(x)))$$
$$- (v'(x) - p(x, v(x)), f_p(x, v(x), p(x, v(x))))]\, dx > 0,$$

as required.

Theorem 3.11. *There exists a number* ε^* *satisfying* $0 < \varepsilon^* < \varepsilon_1 < \varepsilon_0$ *such that*

(i) *if* $0 < \varepsilon < \varepsilon^*$ *then* $(z_{sm}(t), q_{sm}(t)) \to (\bar{k}_1^{-\frac{3}{2}}, \bar{k}_1^{-\frac{3}{2}})$ as $t \to \infty$,

(ii) *if* $\varepsilon = \varepsilon^*$ *then* $(z_{sm}(t), q_{sm}(t)) \to (\bar{k}_2^{-\frac{3}{2}}, \bar{k}_2^{-\frac{3}{2}})$ *as* $t \to \infty$, *and*

(iii) *if* $\varepsilon > \varepsilon^*$ *then* $z_{sm}(t) \to \infty$, $q_{sm}(t) \to \infty$ *as* $t \to \infty$.

Proof. We first show that there exists a minimal number ε^* with $0 \leqq \varepsilon^* < \varepsilon_1$ such that (iii) holds. If $\varepsilon > \varepsilon_0$ then $z_{sm}(t) \to \infty$, $q_{sm}(t) \to \infty$ as $t \to \infty$ by Proposition 3.5. Thus suppose $0 < \varepsilon \leqq \varepsilon_0$, and let $\bar{k} = \bar{k}_2(\varepsilon)$ if $0 < \varepsilon < \varepsilon_0$, $\bar{k} = (\tau^*)^{\frac{1}{3}}$ if $\varepsilon = \varepsilon_0$ (for τ^* as in (3.6)), and set $\tau = \bar{k}^3$. For $\gamma > \frac{39}{40}$ define $v_\gamma(x) = \bar{k}x^{\frac{2\gamma}{3}}$. Then by direct calculation

$$J(\gamma) \stackrel{\text{def}}{=} \tfrac{1}{3}(\tfrac{3}{2})^{14}\, \bar{k}^{-14}(I(v_\gamma) - I(v_1))$$
$$= \tau^2 \left[\frac{\gamma^{14}}{40\gamma - 39} - \frac{13\gamma^2}{4\gamma - 3} + 12\right] + \tau\left[-\frac{2\gamma^{14}}{34\gamma - 33} + \frac{20\gamma^2}{4\gamma - 3} - 18\right]$$
$$+ \frac{\gamma^{14}}{28\gamma - 27} - \frac{7\gamma^2}{4\gamma - 3} + 6.$$

Therefore

$$J(1.1) = a\tau^2 + b\tau + c,$$

where $a = 1.52378\ldots$, $b = -2.44042\ldots$, $c = .94934\ldots$. It now follows that $J(1.1)$ is negative if $\tau_- < \tau < \tau_+$, where $\tau_- = .66576\ldots$, $\tau_+ = .93578\ldots$. Since $\tau^* > \tau_-$ it follows that $\bar{k}x^{\frac{2}{3}}$ does not minimize I if $\varepsilon > \theta(\tau_+) = .0019603\ldots$. Therefore if $\varepsilon > \theta(\tau_+)$, there is some solution $(z(\cdot), q(\cdot))$ of (3.15) with $z(0) = \bar{k}^{-\frac{3}{2}}$, $q(0) \neq \bar{k}^{-\frac{3}{2}}$ and $\lim_{t\to -\infty} (z(t), q(t)) = (0, 0)$ or a rest point, and this clearly implies that $z_{sm}(t) \to \infty$, $q_{sm}(t) \to \infty$ as $t \to \infty$. Define ε^* to be the least nonnegative number such that (iii) holds. Since $\varepsilon_1 = \theta(\tau_1) = .0024735\ldots$ it follows that $0 \leqq \varepsilon^* \leqq \varepsilon_1$, as claimed.

We next prove that $\varepsilon^* > 0$. If not we would have $z_{sm}(t) \to \infty$, $q_{sm}(t) \to \infty$ as $t \to \infty$ for every $\varepsilon > 0$. By Proposition 3.10 all minimizers of I in $\mathscr{A}$ would then be smooth for any $k > 0$. But by Lemma 3.8 this is false for $\varepsilon > 0$ sufficiently small.

For the remainder of the proof it is convenient to make the dependence on ε explicit by writing $z_{sm}(t) = z_{sm}(t, \varepsilon)$, $q_{sm}(t) = q_{sm}(t, \varepsilon)$, and where appropriate $q_{sm}(z) = q_{sm}(z, \varepsilon)$. Using the implicit function theorem it is easily shown that if $z_{sm}(\bar{t}, \bar{\varepsilon}) = \bar{z} > 0$, $q_{sm}(\bar{t}, \bar{\varepsilon}) \neq \bar{z}$ then there exists a smooth function $t(\varepsilon)$ defined for ε near $\bar{\varepsilon}$ such that $z_{sm}(t(\varepsilon), \varepsilon) = \bar{z}$. Thus if $z_{sm}(t, \varepsilon^*) \to \infty$ as $t \to \infty$ we also have $z_{sm}(t, \varepsilon) \to \infty$ as $t \to \infty$ for ε near ε^*, contradicting the minimality of ε^*. Likewise, if $(z_{sm}(t, \varepsilon^*), q_{sm}(t, \varepsilon^*)) \to (\bar{k}_1(\varepsilon^*)^{\frac{3}{2}}, \bar{k}_1(\varepsilon^*)^{\frac{3}{2}})$ as $t \to \infty$ then since $\varepsilon^* < \varepsilon_1$ we have $q_{sm}(t, \varepsilon^*) < z_{sm}(t, \varepsilon^*)$ for some t; thus $q_{sm}(t', \varepsilon) < z_{sm}(t', \varepsilon)$ for some $\varepsilon > \varepsilon^*$ and some t', a contradiction. There remains only one possibility, that $(z_{sm}(t, \varepsilon^*), q_{sm}(t, \varepsilon^*)) \to (\bar{k}_2(\varepsilon^*)^{\frac{3}{2}}, \bar{k}_2(\varepsilon^*)^{\frac{3}{2}})$ as $t \to \infty$, which proves (ii).

We next remark that for any $\varepsilon > 0$ the slope of the vector field on the curve $q = \frac{14}{13z}$ equals, by (3.20),

$$H\left(z, \frac{14}{13z}, \varepsilon\right) = \frac{14^2}{39z^2(14 - 13z^2)},$$

which is positive if $0 < z \leqq 1$. In particular $q_{sm}(z, \varepsilon^*) < \frac{14}{13z}$ for all $z \in (0, \bar{k}_2(\varepsilon^*)^{\frac{3}{2}})$. An easy computation also shows that $\frac{\partial H}{\partial \varepsilon}(z, q, \varepsilon) > 0$ for $0 < z < 1$, $z < q < \frac{14}{13z}$, $\varepsilon > 0$. Suppose that $0 < \varepsilon < \varepsilon^*$ but that $(z_{sm}(t, \varepsilon), q_{sm}(t, \varepsilon)) \nrightarrow (\bar{k}_1(\varepsilon)^{\frac{3}{2}}, \bar{k}_1(\varepsilon)^{\frac{3}{2}})$ as $t \to \infty$. Since $\bar{k}_2(\varepsilon)$ is decreasing in ε we must then have $q_{sm}(\bar{k}_2(\varepsilon^*)^{\frac{3}{2}}, \varepsilon) > \bar{k}_2(\varepsilon^*)^{\frac{3}{2}}$. Choose any q_0 with $q_{sm}(\bar{k}_2(\varepsilon^*)^{\frac{3}{2}}, \varepsilon) > q_0 > \bar{k}_2(\varepsilon^*)^{\frac{3}{2}}$ and consider the solution $(z(t, \varepsilon), q(t, \varepsilon))$ of (3.15) satisfying $z(0, \varepsilon) = \bar{k}_2(\varepsilon^*)^{\frac{3}{2}}$, $q(0, \varepsilon) = q_0$. For $t < 0$ this solution curve cannot cross the $(z_{sm}(\cdot, \varepsilon), q_{sm}(\cdot, \varepsilon))$ orbit and hence by Proposition 3.6 it crosses $q = z$. Therefore there exists $\hat{z} \in (0, \bar{k}_2(\varepsilon^*)^{\frac{3}{2}})$ with $q(\hat{z}, \varepsilon) = q_{sm}(\hat{z}, \varepsilon^*)$, $\frac{dq}{dz}(\hat{z}, \varepsilon) \geqq \frac{dq_{sm}}{dz}(\hat{z}, \varepsilon^*)$, where $q = q(z, \varepsilon)$ denotes the graph of $(z(\cdot, \varepsilon), q(\cdot, \varepsilon))$. But this contradicts the monotonicity of $H(\hat{z}, q_{sm}(\hat{z}, \varepsilon), \cdot)$. Therefore (i) holds. □

Remarks. The numerical evidence is that $\varepsilon^* = .00173\ldots$. That $(z_{sm}(t), q_{sm}(t)) \to (\bar{k}_1^{\frac{3}{2}}(\varepsilon), \bar{k}_1^{\frac{3}{2}}(\varepsilon))$ as $t \to \infty$ for $\varepsilon > 0$ sufficiently small can also be proved by trapping the smooth solution orbit in an appropriate triangular invariant region, but the calculations are rather tedious.

For $0 < \varepsilon < \varepsilon^*$ we denote by $\zeta_1(\varepsilon)$ the maximum value of $z_{sm}(t)$, $t \in \mathbb{R}$, which is achieved when the smooth solution orbit cuts $q = z$, $z > 0$, for the first time. It follows immediately from Theorem 3.11 that if $0 < \varepsilon < \varepsilon^*$, $k^{\frac{3}{2}} > \zeta_1(\varepsilon)$ (or if $\varepsilon = \varepsilon^*$ and $k \geqq \bar{k}_2(\varepsilon)$) then *there is no smooth solution to the Dirichlet problem consisting in the Euler-Lagrange equation* (3.5) *and the boundary conditions* (3.2).

In the following theorem we identify the absolute minimizer of I in $\mathscr{A}$ for every $k > 0$, $\varepsilon > 0$. If $0 < \varepsilon < \varepsilon_1$ we denote by $\zeta_0(\varepsilon)$ the minimum value of z on the unstable manifold of P_2, which is achieved when that part of the unstable manifold in $q \leqq z$ cuts $q = z$ for the first time.

Theorem 3.12

(a) *Let* $0 < \varepsilon < \varepsilon^*$. *There exists a number* $\zeta(\varepsilon)$ *with* $\zeta_0(\varepsilon) < \zeta(\varepsilon) < \zeta_1(\varepsilon)$ *such that*

(i) *if* $0 < k < \zeta(\varepsilon)^{\frac{2}{3}}$ *there is exactly one u that minimizes I in* $\mathscr{A}$ *and u is the unique smooth solution of* (3.5) *on* [0, 1] *satisfying* (3.2),

(ii) *if* $k = \zeta(\varepsilon)^{\frac{2}{3}}$ *there are exactly two functions* u_1, u_2 *that minimize I in* $\mathscr{A}$; u_1 *is the unique smooth solution of* (3.5) *on* [0, 1] *satisfying* (3.2), *and* $u_2(x) \sim \bar{k}_2(\varepsilon)\, x^{\frac{2}{3}}$ *as* $x \to 0+$ *and corresponds to that connected part of the unstable manifold of* P_2 *defined by* $q \leqq z$, $\zeta(\varepsilon) \leqq z \leqq \bar{k}_2^{\frac{3}{2}}(\varepsilon)$,

(iii) *if* $k > \zeta(\varepsilon)^{\frac{2}{3}}$ *there is exactly one u that minimizes I in* $\mathscr{A}$; $u(x) \sim \bar{k}_2(\varepsilon)\, x^{\frac{2}{3}}$ *as* $x \to 0+$ *and corresponds to that part of the unstable manifold of* P_2 *defined by* $q \leqq z$, $k^{\frac{3}{2}} \leqq z \leqq \bar{k}_2^{\frac{3}{2}}(\varepsilon)$ *if* $k \leqq \bar{k}_2(\varepsilon)$ *and by* $q \geqq z$, $\bar{k}_2^{\frac{3}{2}}(\varepsilon) \leqq z \leqq k^{\frac{3}{2}}$ *if* $k \geqq \bar{k}_2(\varepsilon)$, *so that in particular if* $k = \bar{k}_2(\varepsilon)$ *then* $u(x) = \bar{k}_2(\varepsilon)\, x^{\frac{2}{3}}$.

(b) *Let* $\varepsilon = \varepsilon^*$. *Then there is exactly one u that minimizes I in* $\mathscr{A}$. *If* $k < \bar{k}_2(\varepsilon^*)$ *then u is the unique smooth solution of* (3.5) *on* [0, 1] *satisfying* (3.2), *and, if* $k \geqq \bar{k}_2(\varepsilon^*)$, $u(x) \sim \bar{k}_2(\varepsilon^*)\, x^{\frac{2}{3}}$ *as* $x \to 0+$ *and corresponds to that connected part of the unstable manifold of* P_2 *defined by* $q \geqq z$, $\bar{k}_2^{\frac{3}{2}}(\varepsilon) \leqq z \leqq k^{\frac{3}{2}}$. *In particular if* $k = \bar{k}_2(\varepsilon^*)$ *then* $u(x) = \bar{k}_2(\varepsilon^*)\, x^{\frac{2}{3}}$.

(c) *Let* $\varepsilon > \varepsilon^*$. *Then there is exactly one u that minimizes I in* $\mathscr{A}$ *and u is the unique smooth solution of* (3.5) *on* [0, 1] *satisfying* (3.2).

Proof. Part (c) follows immediately from Theorem 3.11(iii) and Proposition 3.10. If $0 < \varepsilon < \varepsilon^*$ and $k \in (0, \zeta_0(\varepsilon)^{\frac{2}{3}}) \cup (\zeta_1(\varepsilon)^{\frac{2}{3}}, \infty)$ or if $\varepsilon = \varepsilon^*$ and $k \in (0, \zeta_0(\varepsilon^*)^{\frac{2}{3}}) \cup [\bar{k}_2(\varepsilon^*), \infty)$ then the solution specified in the theorem is the only geometrically possible one, and perforce by Theorem 3.7 is the unique minimizer. Suppose $0 < \varepsilon < \varepsilon^*$ and $k \in [\zeta_0(\varepsilon)^{\frac{2}{3}}, \zeta_1(\varepsilon)^{\frac{2}{3}}]$. By Lemma 3.9 and the subsequent discussion there are only two possibilities for a minimizer, a smooth solution $u = u_1(x, k)$ represented by part of the smooth solution orbit $q = q_{\text{sm}}(z)$, $0 \leqq z \leqq k^{\frac{3}{2}}$, $q \geqq z$ and a singular solution $u = u_2(x, k)$ represented by a part of the unstable manifold of P_2 which we denote by $q = q_{\text{un}}(z)$, $k^{\frac{3}{2}} \leqq z \leqq \bar{k}_2(\varepsilon)^{\frac{3}{2}}$,

$q \leqq z$. Define

$$R(k) = I(u_1(\cdot, k)) - I(u_2(\cdot, k)).$$

By Lemma 3.9 we have

$$R(\zeta_0(\varepsilon)^{\frac{2}{3}}) = -3(\psi(\zeta_0(\varepsilon), q_{\mathrm{sm}}(\zeta_0(\varepsilon))) - \psi(\zeta_0(\varepsilon), \zeta_0(\varepsilon))) < 0,$$

and

$$R(\zeta_1(\varepsilon)^{\frac{2}{3}}) = -3(\psi(\zeta_1(\varepsilon), \zeta_1(\varepsilon)) - \psi(\zeta_1(\varepsilon), q_{\mathrm{un}}(\zeta_1(\varepsilon)))) > 0.$$

Also, by (3.29), (3.30),

$$\frac{dR}{dk}(k) > 0 \quad \text{for } \zeta_0(\varepsilon)^{\frac{2}{3}} \leqq k \leqq \zeta_1(\varepsilon)^{\frac{2}{3}}.$$

Hence $R(\zeta(\varepsilon)^{\frac{2}{3}}) = 0$ for a unique $\zeta(\varepsilon) \in (\zeta_0(\varepsilon), \zeta_1(\varepsilon))$ and part (a) follows.

In the case $\varepsilon = \varepsilon^*$, $k \in [\zeta_0(\varepsilon^*)^{\frac{2}{3}}, \bar{k}_2(\varepsilon^*)]$ we define $R(k)$ as above and note that $R(\zeta_0(\varepsilon^*)^{\frac{2}{3}}) < 0$, $\lim_{k \to \bar{k}_2(\varepsilon^*)-} R(k) = 0$, $\frac{dR}{dk}(k) > 0$ for $k \in [\zeta_0(\varepsilon^*)^{\frac{2}{3}}, \bar{k}_2(\varepsilon^*))$. Hence part (b) holds. □

The results of Theorems 3.11, 3.22 are summarized pictorially in Figure 3.1. Note that whenever the minimizer is singular at the origin neither (IEL) nor (IDBR) holds, since then both $\lim_{x \to 0+} f_p(x, u(x), u'(x))$ and $\lim_{x \to 0+} [u'(x) f_p(x, u(x), u'(x)) - f(x, u(x), u'(x))]$ are $+\infty$. However, in all cases (WEL) and (WDBR) are satisfied. Note also that if $0 < \varepsilon < \varepsilon_0$ then $u(x) = \bar{k}_1(\varepsilon)\, x^{\frac{2}{3}}$ is never a minimizer. It is interesting to observe from the figure how for fixed ε the number of solutions $u \in C^\infty((0, 1])$ of (3.5) satisfying $u(0) = 0$, $u(1) = k$ varies with k. For example, if $0 < \varepsilon < \varepsilon_1$ then as k approaches k_1 the number of such solutions tends to infinity. An alternative proof that for $k = \bar{k}_2(\varepsilon)$ and $\varepsilon > 0$ sufficiently small $u(x) = \bar{k}_2(\varepsilon)\, x^{\frac{2}{3}}$ minimizes I in $\mathcal{A}$ has been given by CLARKE & VINTER [14].

We conclude our discussion of the phase portrait with a few remarks concerning the behavior of the branch of the unstable manifold of P_2 that near P_2 lies in $q < z$. Since, by Lemma 3.8, if $k > 0$ is arbitrary but fixed then any minimizer of I in $\mathcal{A}$ is singular provided $\varepsilon > 0$ is sufficiently small, it follows from Theorem 3.12 that $\zeta_1(\varepsilon) \to 0$ as $\varepsilon \to 0$. Hence also $\zeta_0(\varepsilon) \to 0$ as $\varepsilon \to 0$; this is consistent with the fact that the slope of the unstable manifold at P_2 tends to infinity as $\varepsilon \to 0$. As $t \to \infty$ the above branch of the unstable manifold tends to $(\bar{k}_1(\varepsilon)^{\frac{3}{2}}, \bar{k}_1(\varepsilon)^{\frac{3}{2}})$; in fact it cannot tend to $(\bar{k}_2(\varepsilon)^{\frac{3}{2}}, \bar{k}_2(\varepsilon)^{\frac{3}{2}})$ by Proposition 3.4, and it cannot tend to infinity because the upper branch of the stable manifold at P_2 would then have nowhere to go as $t \to -\infty$. For $0 < \varepsilon < \varepsilon_0$ and $\varepsilon_0 - \varepsilon$ very small an application of center manifold theory (see, for example, CARR [10]) shows that the connecting orbit from P_2 to P_1 is almost parallel to the line $q = z$.

§ 4. The case with no x-dependence

In this section we consider the problem of minimizing

$$I(u) = \int_{-1}^{1} f(u(x), u'(x))\, dx \tag{4.1}$$

in

$$\mathscr{A} = \{u \in W^{1,1}(-1, 1) : u(-1) = k_1,\, u(1) = k_2\}, \tag{4.2}$$

where $k_1, k_2 \in \mathbb{R}$. Concerning the integrand $f = f(u, p)$ we will require that

$$\begin{gathered} f \in C^\infty(\mathbb{R}^2), \quad f_{pp} > 0, \\ |p| \leqq f(u, p) \leqq \text{const.}\,(1 + p^2), \quad (u, p) \in \mathbb{R}^2. \end{gathered} \tag{4.3}$$

We will show that an absolute minimizer u_0 of I over $\mathscr{A}$ need not satisfy (WEL) or (IEL), although by Corollary 2.5 u_0 must satisfy (IDBR). In our examples u_0 is constructed directly, though as we remarked in Section 2, for the functions f appearing below the existence of u_0 also follows from known extensions of Theorem 2.1.

We first give an example where the Tonelli set $E = \{x_0\}$ is a singleton.

Theorem 4.1. *There exist an f satisfying* (4.3) *and*

$$\frac{f(u, p)}{|p|} \to \infty \quad \text{as } |p| \to \infty \text{ for each } u \neq 0 \tag{4.4}$$

and a number $k_0 > 0$ such that whenever $-k_1, k_2 > k_0$ then (4.1), (4.2) *has a unique global minimizer u_0, but $E = \{x_0\}$ for some $x_0 = x_0(k_1, k_2) \in (-1, 1)$, and*

$$f_u(u_0, u_0') \notin L^1_{\text{loc}}(-1, 1),$$

so that neither (WEL) *nor* (IEL) *is satisfied.*

Remark. The theorem shows that if (2.13) fails for just *one* value of u then the conclusion of Corollary 2.12 need not hold.

Proof of Theorem 4.1. The proof splits naturally into two parts. Part I is devoted to the construction of a strictly monotone function $g \in C^1(\mathbb{R})$ satisfying

(g1) $g \in C^1(\mathbb{R}) \cap C^\infty(\mathbb{R} \setminus \{0\})$,

(g2) $g'' \notin L^1(-\delta, \delta)$ for any $\delta > 0$,

and to the solution of the minimization problem on $\mathscr{A}$ for a certain functional J involving g. Part II then presents the construction of an integrand f satisfying (4.3), (4.4) such that the corresponding functional I has the same global minimizer as J over $\mathscr{A}$.

Part I. Select an even function $h \in C(\mathbb{R}) \cap C^\infty(\mathbb{R} \setminus \{0\})$ such that

$$\left.\begin{aligned} &h(0) = 0, \quad 0 \leqq h \leqq 1, \quad h(s) > 0 \quad \text{for } s \neq 0, \\ &h(s) = 1 \quad \text{for } |s| \geqq 1/2, \quad h' \notin L^1(-\delta, \delta) \quad \text{for any } \delta > 0. \end{aligned}\right\} \tag{4.5}$$

For instance,

$$h(s) = s^2(2 + \sin(s^{-2}))\,\eta(s) + (1 - \eta(s)), \quad s \in \mathbb{R},$$

with $\eta \in C^\infty$ an even function satisfying

$$0 \leqq \eta \leqq 1, \quad \eta(s) = 0 \quad \text{for} \quad |s| \geqq 1/2, \quad \eta(s) = 1 \quad \text{for} \quad |s| \leqq 1/4,$$

defines such a function.

Now specify $g \in C^1(\mathbb{R})$ by

$$g' = h, \quad g(0) = 0, \tag{4.6}$$

and note that g is odd, strictly monotone, and satisfies (g1), (g2). Put

$$J(u) = \int_{-1}^{1} [g'(u(x))\,u'(x)]^2\,dx, \quad u \in \mathscr{A}. \tag{4.7}$$

Since g is C^1 it readily follows that

$$g \circ u \in W^{1,1}(-1, 1) \quad \text{for all } u \in W^{1,1}(-1, 1). \tag{4.8}$$

Hence given $l \in \mathbb{R}$ one can decompose J as follows:

$$\begin{aligned} J(u) &= \int_{-1}^{1} [(g'(u(x))\,u'(x) - l)^2 + 2lg'(u(x))\,u'(x) - l^2]\,dx \\ &= \int_{-1}^{1} (g'(u(x))\,u'(x) - l)^2\,dx + 2l(g(k_2) - g(k_1)) - 2l^2, \end{aligned} \tag{4.9}$$

for all $u \in \mathscr{A}$. Thus it is clear that if $u \in \mathscr{A}$ satisfies for some l

$$g'(u(x))\,u'(x) = l, \quad \text{a.e. } x \in [-1, 1], \tag{4.10}$$

then u is a global minimizer of J in $\mathscr{A}$. By (4.8) this last condition requires that

$$g(u(x)) = lx + m, \quad x \in [-1, 1], \tag{4.11}$$

and the end conditions on u imply that

$$l = \tfrac{1}{2}(g(k_2) - g(k_1)), \quad m = \tfrac{1}{2}(g(k_1) + g(k_2)). \tag{4.12}$$

Now since g is strictly increasing and has range $\mathbb{R}$, (4.11), (4.12) determine a unique strictly increasing function $u_0 \in C([-1, 1])$. Moreover, by the inverse function theorem $\left(\text{applied to } \frac{1}{l}(g - m)\right)$ it follows that

$$u_0 \in C([-1, 1]) \cap C^\infty([-1, 1] \setminus \{x_0\}), \tag{4.13}$$

where x_0 is the unique point such that $u_0(x_0) = 0$. Finally, by (4.12) it follows that

$$u_0(-1) = k_1, \quad u_0(1) = k_2.$$

Therefore if the function u_0 defined by (4.11), (4.12) is absolutely continuous, then u_0 belongs to $\mathcal{A}$ and provides a (unique) global minimizer for J. The absolute continuity of u_0 is now verified by making use of the monotonicity of u_0 and (4.13):

$$\begin{aligned}\int_{-1}^{1} |u_0'(x)|\,dx &= \int_{-1}^{x_0} u_0'(x)\,dx + \int_{x_0}^{1} u_0'(x)\,dx = (u_0(x_0) - u_0(-1)) + (u_0(1) - u_0(x_0)) \\ &= k_2 - k_1 < \infty.\end{aligned}$$

Part II. Write

$$f^0(u, p) = (g'(u)\,p)^2, \quad (u, p) \in \mathbb{R}^2,$$

so that

$$f^0_{pp}(u, p) > 0 \quad \text{if and only if} \quad u \neq 0.$$

Using (4.11)–(4.13) and (g1), we have

$$\begin{aligned} f^0_u(u_0, u_0') &= 2(g'(u_0)\,u_0')\,g''(u_0)\,u_0' \\ &= (g(k_2) - g(k_1))\,g''(u_0)\,u_0' \quad \in C^\infty([-1, 1] \setminus \{x_0\}).\end{aligned}$$

Therefore, by (g2), if $-1 < a < x_0 < b < 1$ then

$$\begin{aligned}\int_a^b |f^0_u(u_0, u_0')|\,dx &= \lim_{h\to 0+}\left[\int_a^{x_0-h} |f^0_u(u_0, u_0')|\,dx + \int_{x_0+h}^{b} |f^0_u(u_0, u_0')|\,dx\right] \\ &= (g(k_2) - g(k_1)) \lim_{h\to 0+}\left[\int_{u_0(a)}^{u_0(x_0-h)} |g''(u)|\,du + \int_{u_0(x_0+h)}^{u_0(b)} |g''(u)|\,du\right] \\ &= \infty,\end{aligned}$$

so that

$$f^0_u(u_0, u_0') \notin L^1_{\text{loc}}(-1, 1). \tag{4.14}$$

A function $f \in C^\infty(\mathbb{R}^2)$ satisfying (4.3) as well as

$$\left.\begin{aligned} f(u, p) &\geqq f^0(u, p) + p, \quad (u, p) \in \mathbb{R}^2, \\ f(u, p) &= f^0(u, p) + p \quad \text{when} \quad g'(u)\,p \geqq l - \delta, \end{aligned}\right\} \tag{4.15}$$

where l is given by (4.12) and $\delta > 0$, is constructed below. Obviously for f satisfying (4.15) and for u_0 as in (4.11), (4.12),

$$I(u_0) \stackrel{\text{def}}{=} \int_{-1}^{1} f(u_0, u_0')\,dx = \int_{-1}^{1} [f^0(u_0, u_0') + u_0']\,dx = J(u_0) + k_2 - k_1, \tag{4.16}$$

so that u_0 is also the unique global minimizer for I over $\mathcal{A}$. Moreover, by (4.14), (4.15),

$$f_u(u_0, u_0') = f^0_u(u_0, u_0') \notin L^1_{\text{loc}}(-1, 1). \tag{4.17}$$

Also, since (4.10) implies that

$$u_0'(x) = \frac{l}{g'(u_0(x))} \to \infty \quad \text{as} \quad x \to x_0,$$

it follows by TONELLI's partial regularity theorem (Theorem 2.7) that

$$u_0'(x_0) = \infty,$$

so that the Tonelli set of u_0 is the singleton $E = \{x_0\}$, completing the conclusions of the theorem.

To construct f satisfying (4.3) and (4.15) we first construct an appropriate function $e \in C^\infty([0, 1] \times \mathbb{R})$ such that the formula

$$f(u, p) \overset{\text{def}}{=} e((g'(u))^2, p) + p \tag{4.18}$$

yields a function f with the desired properties. Let $\varrho \in C^\infty(\mathbb{R})$ be a nonnegative even function with $\operatorname{supp} \varrho \subset (-1, 1)$, $\int_{-\infty}^{\infty} \varrho(p)\, dp = 1$, and put

$$\alpha = \int_{-\infty}^{\infty} p^2 \varrho(p)\, dp.$$

Thus $\varrho_\varepsilon(p) = \varepsilon^{-1} \varrho(p/\varepsilon)$ satisfies

$$\operatorname{supp} \varrho_\varepsilon \subset (-\varepsilon, \varepsilon), \quad \int_{-\infty}^{\infty} \varrho_\varepsilon(p)\, dp = 1, \quad \int_{-\infty}^{\infty} p^2 \varrho_\varepsilon(p)\, dp = \varepsilon^2 \alpha. \tag{4.19}$$

Now let $\theta \in C^1(\mathbb{R}) \cap C^\infty(\mathbb{R} \setminus \{0\})$ be given by

$$\theta(p) = \begin{cases} 2p^2 - p + 1, & p \leqq 0 \\ (p+1)^{-1}, & p > 0. \end{cases} \tag{4.20}$$

Note that θ is strictly convex, with $\theta''(p) > 0$ for $p \neq 0$. We claim that for $\varepsilon > 0$ small and $b \in (0, 1]$ the graphs of θ and of $\nu(p) = b(p^2 - \alpha\varepsilon^2)$, $p \in \mathbb{R}$, intersect at a unique point $p_b \in [\frac{1}{2}, \infty)$. The existence and uniqueness of the intersection follows from the strict monotonicity, in opposing senses, of θ and ν on $0 \leqq p < \infty$. The condition for intersection:

$$(p+1)(p^2 - \alpha\varepsilon^2) = b^{-1}, \tag{4.21}$$

implies when $b = 1$ that

$$(p_1 + 1)\, p_1^2 > 1,$$

so that $p_1 > \frac{1}{2}$. Since the left-hand side of (4.21) is strictly increasing on $\frac{1}{2} \leqq p < \infty$ when $\alpha\varepsilon^2 < 1$, it follows that then $p_b \in [1/2, \infty)$ as required. Note also that (4.21) yields the asymptotic estimate

$$p_b \sim b^{-\frac{1}{3}} \quad \text{for} \quad b \sim 0. \tag{4.22}$$

By the inverse function theorem p_b is C^∞ on (0, 1]. Therefore on defining

$$\begin{aligned} e(b,p) &= (\varrho_\varepsilon * \max\{\theta, v\})(p) \\ &= \int_{-\infty}^{p_b} \varrho_\varepsilon(p-q)\,\theta(q)\,dq + \int_{p_b}^{\infty} \varrho_\varepsilon(p-q)\,v(q)\,dq, \end{aligned} \tag{4.23}$$

one obtains e as the sum of two functions in $C^\infty((0,1]\times\mathbb{R})$, so that $e \in C^\infty((0,1]\times\mathbb{R})$. Moreover, by (4.19), if $p > p_b + \varepsilon$ then

$$\begin{aligned} e(b,p) &= b\int_{-\infty}^{\infty} \varrho_\varepsilon(q)\,[(p-q)^2 - \alpha\varepsilon^2]\,dq \\ &= b\int_{-\infty}^{\infty} \varrho_\varepsilon(q)\,[q^2 - 2pq + p^2 - \alpha\varepsilon^2]\,dq = bp^2, \end{aligned} \tag{4.24}$$

while if $p < p_b - \varepsilon$ then

$$e(b,p) = \int_{-\infty}^{\infty} \varrho_\varepsilon(p-q)\,\theta(q)\,dq \overset{\text{def}}{=} \varphi(p). \tag{4.25}$$

It is easily verified that φ has the following properties:

$$\left.\begin{aligned} \varphi > 0; \quad \varphi(p) &= 2p^2 - p + 1 + 2\alpha\varepsilon^2 && \text{for } p \leqq -\varepsilon, \\ \varphi(p) \geqq \theta(p+\varepsilon) &= (p+\varepsilon+1)^{-1} && \text{for } p \geqq -\varepsilon. \end{aligned}\right\} \tag{4.26}$$

Thus, since by (4.23)

$$e(b,p) \geqq \max\{(\varrho_\varepsilon * \theta)(p), (\varrho_\varepsilon * v)(p)\} = \max\{\varphi(p), bp^2\}, \quad p \in \mathbb{R}, \tag{4.27}$$

it follows from (4.26) that for $\varepsilon > 0$ sufficiently small

$$e(b,p) + p \geqq \varphi(p) + p \geqq |p|, \quad (b,p) \in (0,1]\times\mathbb{R}. \tag{4.28}$$

Now set

$$e(0,p) = \varphi(p), \quad p \in \mathbb{R}, \tag{4.29}$$

and define f by (4.18). It is immediate that $f \in C^\infty((\mathbb{R}\setminus\{0\})\times\mathbb{R})$. Furthermore, since for any interval $-A \leqq p \leqq A$ there exists by (4.22) a number $\delta_A > 0$ such that

$$e(g'(u)^2, p) + p = \varphi(p) + p, \quad |g'(u)| < \delta_A, \quad p \in [-A, A],$$

it is seen that actually $f \in C^\infty$ as required. The proof that f satisfies (4.3) is straightforward: the property $f_{pp} > 0$ follows from the facts that $\theta''(p) > 0$ for all $p \neq 0$, $v''(p) > 0$ if $b \in (0,1]$, $p \in \mathbb{R}$, while the growth condition follows from (4.23), (4.26) and (4.28). The growth condition (4.4) is a consequence of (4.24). The inequality in (4.15) results directly from (4.27), (4.29). To establish the equation in (4.15) we choose $k_0 = 3/2$, so that by (4.5), (4.6) there exists $\delta > 0$ such that

$$l = \tfrac{1}{2}(g(k_2) - g(k_1)) > g(k_0) > 1 + 2\delta$$

whenever $-k_1, k_2 > k_0$. Next, we note that by (4.21)

$$p_b^3 \leqq p_b^3 + \left(p_b - \frac{\alpha\varepsilon^2}{2}\right)^2 = b^{-1} + \alpha\varepsilon^2 + \frac{\alpha^2\varepsilon^4}{4},$$

so that for sufficiently small $\varepsilon > 0$ we have

$$l - \delta > b^{\frac{1}{2}} p_b + \varepsilon b^{\frac{1}{2}} \quad \text{for all } b \in (0, 1]. \tag{4.30}$$

Now suppose that $b^{\frac{1}{2}} p \geqq l - \delta$ for some $b \in (0, 1]$, $p \in \mathbb{R}$. By (4.30) we have

$$p \geqq b^{-\frac{1}{2}}(l - \delta) > p_b + \varepsilon,$$

so that $e(b, p) = bp^2$ by (4.24). This completes the proof. □

Remarks. 1. If the construction in Theorem 4.1 is repeated with a function $h' \in L^2(-1, 1)$ then it is easily verified that the minimizer u_0 does satisfy (IEL), (WEL) even though $u'(x_0) = \infty$.

2. Let $\varphi \in C_0^\infty(-1, 1)$ be nonzero in a neighborhood of x_0. Then for any $t \neq 0$ there exist constants $c(t) > 0$, $\alpha(t) > 0$, such that

$$I(u_0 + t\varphi) \geqq c(t) \int_{x_0 - \alpha(t)}^{x_0 + \alpha(t)} (u_0'(x))^2 \, dx.$$

Since

$$\int_{x_0 - \alpha(t)}^{x_0 + \alpha(t)} (u_0'(x))^2 \, dx = \int_{u_0(x_0 - \alpha(t))}^{u_0(x_0 + \alpha(t))} l \frac{du}{h(u)},$$

it follows that if $\frac{1}{h} \notin L^1(-\delta, \delta)$, for any $\delta < 0$, which is clearly consistent with (4.5), then $I(u_0 + t\varphi) = +\infty$.

We now give an example where the Tonelli set E is any prescribed closed Lebesgue null set; this shows that the Tonelli partial regularity theorem (Theorem 2.7) is in a certain sense optimal.

Theorem 4.2. *Given any closed subset* $E \subset [-1, 1]$ *of measure zero, there exists a function* $f = f^E$ *satisfying* (4.3) *and*

$$\frac{f(u, p)}{|p|} \to \infty \quad \text{as } |p| \to \infty \quad \text{for all } u \notin F, \tag{4.4$'$}$$

with F a Lebesgue null set, such that for certain scalars $k_1, k_2 \in \mathbb{R}$, *there variational problem* (4.1), (4.2) *has a unique global minimizer* u_0, *and* u_0 *is strictly increasing with*

$$u_0'(x) = +\infty \quad \text{if and only if } x \in E.$$

Furthermore

$$f_u(u_0, u_0') \notin L^1_{\text{loc}}(-1, 1),$$

so that neither (WEL) *nor* (IEL) *is satisfied.*

Proof. Again the proof splits naturally into two parts, with Part II identical with the argument for Part II in Theorem 4.1. Hence only Part I is given here.

Part I. The construction *begins* with the global minimizer u_0 and then yields a function $g \in C^1(\mathbb{R})$ satisfying

(g1)′ $g \in C^1(\mathbb{R}) \cap C^\infty(\mathbb{R} \setminus F)$, with $F \subset \mathbb{R}$ a compact Lebesgue null set,

(g2)′ $g'' \notin L^1(a, b)$ for any (a, b) such that $F \cap (a, b) \neq \emptyset$.

Let $k \in C(\mathbb{R}) \cap C^\infty(0, 1)$ satisfy

$$\begin{gathered} k(t) = 0 \quad \text{for } t \in (-\infty, 0], \qquad k(t) = 2 \quad \text{for } t \in [1, \infty), \\ k'(t) > 1 \quad \text{for } t \in (0, 1), \qquad \lim_{t \to 0+} k'(t) = \lim_{t \to 1-} k'(t) = +\infty. \end{gathered} \tag{4.31}$$

We take the harder case when E is an infinite set such that neither -1 nor 1 belongs to E. The modifications necessary when E is finite and/or one or both endpoints belong to E are easily made. Let $x_- = \min_{x \in E} x$, $x_+ = \max_{x \in E} x$, so that $-1 < x_- < x_+ < 1$. Pick $c < -1$, $d > 1$. Then

$$(c, d) \setminus E = \bigcup_{j=1}^{\infty} \mathcal{O}_j,$$

where the $\mathcal{O}_j = (a_j, b_j)$, $j \geqq 1$, are disjoint and open, with

$$(a_1, b_1) = (c, x_-), \quad (a_2, b_2) = (x_+, d).$$

Clearly

$$\sum_{j=1}^{\infty} |\mathcal{O}_j| = d - c, \tag{4.32}$$

where $|\mathcal{O}_j| = b_j - a_j$. It follows that

$$\alpha \stackrel{\text{def}}{=} \sum_{j=1}^{\infty} \varphi(|\mathcal{O}_j|) < \infty \tag{4.33}$$

for some increasing continuous function $\varphi : (0, \infty) \to (0, \infty)$ satisfying

$$\frac{\varphi(t)}{t} \geqq 1, \quad t > 0; \qquad \lim_{t \to 0+} \frac{\varphi(t)}{t} = \infty. \tag{4.34}$$

Define $\bar{u}_0 : [c, d] \to \mathbb{R}$ by

$$\bar{u}_0(x) = \sum_{j=1}^{\infty} \varphi(|\mathcal{O}_j|)\, k\left(\frac{x - a_j}{b_j - a_j}\right). \tag{4.35}$$

By (4.31), (4.33) it follows that this series is uniformly convergent on $\mathbb{R}$. Moreover, for $x \in \mathcal{O}_j$,

$$\begin{aligned} \bar{u}_0(x) &= \sum_{i \neq j} \varphi(|\mathcal{O}_i|)\, k\left(\frac{x - a_i}{b_i - a_i}\right) + \varphi(|\mathcal{O}_j|)\, k\left(\frac{x - a_j}{b_j - a_j}\right) \\ &= \bar{u}_0(a_j) + \varphi(|\mathcal{O}_j|)\, k\left(\frac{x - a_j}{b_j - a_j}\right), \end{aligned} \tag{4.36}$$

so that

$$|\bar{u}_0(\mathcal{O}_j)| = \lim_{x\to b_j^-} \bar{u}_0(x) - \bar{u}_0(a_j) = 2\varphi(|\mathcal{O}_j|). \tag{4.37}$$

It follows from (4.36), (4.37) that $\bar{u}_0$ is strictly increasing on $[c, d]$, and $\bar{u}_0 \in C^\infty([c, d] \setminus E)$ with

$$\bar{u}_0(c) = 0, \quad \bar{u}_0(d) = 2\alpha. \tag{4.38}$$

Furthermore,

$$\begin{aligned}\int_c^d |\bar{u}_0'(x)|\, dx &= \sum_{j=1}^\infty \int_{a_j}^{b_j} \frac{\varphi(|\mathcal{O}_j|)}{b_j - a_j} k'\left(\frac{x - a_j}{b_j - a_j}\right) dx \\ &= \sum_{j=1}^\infty 2\varphi(|\mathcal{O}_j|) = 2\alpha = \bar{u}_0(d) - \bar{u}_0(c) < \infty,\end{aligned}$$

so $\bar{u}_0 \in W^{1,1}((c, d) \setminus E)$, and since $\bar{u}_0 \in C([c, d])$ it follows ([29, p. 224]) that

$$\bar{u}_0 \in W^{1,1}(c, d). \tag{4.39}$$

Now define u_0 to be the restriction of $\bar{u}_0$ to $[-1, 1]$ and let

$$k_1 = u_0(-1), \quad k_2 = u_0(1),$$

so that $0 < k_1 < k_2 < 2\alpha$. Define $g: [k_1, k_2] \to [-1, 1]$ by $g = u_0^{-1}$. It follows from (4.36) that

$$g(u) = (b_j - a_j)\, k^{-1}\left(\frac{u - u_0(a_j)}{\varphi(|\mathcal{O}_j|)}\right) + a_j \quad \text{for} \quad u \in \bar{u}_0(\mathcal{O}_j) \cap [k_1, k_2], \tag{4.40}$$

where $\bar{u}_0(\mathcal{O}_j) = (\bar{u}_0(a_j), \bar{u}_0(b_j))$, $j \geqq 1$. Consequently $g \in C^\infty(\bar{u}_0(\mathcal{O}_j) \cap [k_1, k_2])$ and

$$g'(u) = \frac{b_j - a_j}{\varphi(b_j - a_j)} (k^{-1})'\left(\frac{u - \bar{u}_0(a_j)}{\varphi(b_j - a_j)}\right), \quad u \in \bar{u}_0(\mathcal{O}_j) \cap [k_1, k_2]. \tag{4.41}$$

By (4.31), (4.40)

$$\begin{aligned}&\lim_{u\to u_0(a_j)+} g'(u) = 0, \quad j \neq 1, \\ &\lim_{u\to u_0(b_j)-} g'(u) = 0, \quad j \neq 2, \\ 0 < g'(u) < &\frac{b_j - a_j}{\varphi(b_j - a_j)} \quad \text{for} \quad u \in \bar{u}_0(\mathcal{O}_j) \cap [k_1, k_2], \quad j \geqq 1.\end{aligned} \tag{4.42}$$

By (4.34), (4.42),

$$0 < \sup_{u\in\bar{u}_0(\mathcal{O}_j)\cap[k_1,k_2]} g'(u) \to 0 \quad \text{as} \quad j \to \infty, \tag{4.43}$$

so that g' can be extended to a function $g^* \in C([k_1, k_2])$ by setting

$$g^*(u) \stackrel{\text{def}}{=} \begin{cases} 0, & u \in F, \\ g'(u), & u \in [k_1, k_2] \setminus F, \end{cases} \tag{4.44}$$

where $F \overset{\text{def}}{=} [k_1, k_2] - \bigcup_{j=1}^{\infty} \bar{u}_0(\mathcal{O}_j) = u_0(E)$. To show that $g \in C([k_1, k_2]) \cap C^\infty([k_1, k_2] \setminus F)$ is in $C^1([k_1, k_2])$ note that

$$\int_{k_1}^{k_2} |g'(u)|\, du = \sum_{j=1}^{\infty} \int_{\mathcal{O}_j \cap [k_1,k_2]} g'(u)\, du$$

$$= \sum_{j=3}^{\infty} [g(u_0(b_j)) - g(u_0(a_j))] + g(u_0(b_1)) - g(u_0(-1)) + g(u_0(1)) - g(u_0(a_2))$$

$$= \sum_{j=1}^{\infty} (b_j - a_j) - (d - c) + 2$$

$$= 2 = g(k_2) - g(k_1) < \infty,$$

where we have used the fact that, by (4.36)–(4.38), F has measure zero. Hence $g \in W^{1,1}((k_1, k_2) \setminus F)$ and thus by the continuity of g on $[k_1, k_2]$ ([29, p. 224]) $g \in W^{1,1}(k_1, k_2)$. Therefore, for each $u \in [k_1, k_2]$,

$$g(u) - g(k_1) = \int_{k_1}^{u} g'(y)\, dy = \int_{k_1}^{u} g^*(y)\, dy.$$

Since $g^* \in C([k_1, k_2])$ we deduce that $g \in C^1([k_1, k_2])$. Moreover, by (4.34), (4.42) g can be extended to a function in $C^1(\mathbb{R}) \cap C^\infty(\mathbb{R} \setminus F)$ satisfying

$$0 < g'(u) \leqq 1, \quad u \in \mathbb{R} \setminus [k_1, k_2].$$

Thus

$$0 \leqq g'(u) \leqq 1, \quad u \in \mathbb{R},$$

$$g'(u) = 0 \quad \text{if and only if } u \in F.$$

Now $g = u_0^{-1}$ on $[k_1, k_2]$ implies that

$$g'(u_0(x))\, u_0'(x) = 1 \quad \text{a.e. } x \in [-1, 1].$$

It follows as in (4.7)–(4.9) that u_0 is a (unique) global minimizer for J in $\mathcal{A}$.

It remains only to repeat the proof given in Part II of the argument of Theorem 4.1 in order to construct an f satisfying (4.3), (4.15) relative to $f^0(u, p) \overset{\text{def}}{=} (g'(u)\, p)^2$, $(u, p) \in \mathbb{R}^2$. □

§ 5. A case exhibiting the Lavrentiev phenomenon

In this section we consider the problem of minimizing

$$I(u) = \int_{-1}^{1} [(x^4 - u^6)^2\, |u'|^s + \varepsilon (u')^2]\, dx \tag{5.1}$$

over

$$\mathcal{A} = \{u \in W^{1,1}(-1, 1) : u(-1) = k_1, \quad u(1) = k_2\},$$

where $s > 3$, $\varepsilon > 0$ and $k_1, k_2 \in \mathbb{R}$. Note that the integrand

$$f(x, u, p) = (x^4 - u^6)^2 |p|^s + \varepsilon p^2 \tag{5.2}$$

is C^3, nonnegative, and satisfies $f_{pp} \geqq 2\varepsilon > 0$; furthermore, in the case when $s = 2m$ is an even integer f is a polynomial.

By Theorem 2.1 there exists at least one absolute minimizer u_0 of I in $\mathscr{A}$. Any minimizer u_0 of I in $\mathscr{A}$ is either nondecreasing or nonincreasing, since otherwise the value of I could be reduced by making u_0 constant on some interval. (In fact, if $k_1 \neq k_2$ then u_0 is *strictly* increasing or decreasing, since if u_0 were constant on some interval then, as constants satisfy (EL), by Theorem 2.7 we would have u_0 constant everywhere.)

Our first aim is to prove the following theorem. (In the statement of the theorem and below we abbreviate $W^{1,p}(-1, 1)$ by $W^{1,p}$ where convenient.)

Theorem 5.1. *Let* $s \geqq 27$. *Let* $-1 \leqq k_1 < 0 < k_2 \leqq 1$ *and* $0 < \alpha < 1$. *Then there is an* $\varepsilon_1 = \varepsilon_1(\alpha, k_1, k_2, s) > 0$ *such that when* $0 < \varepsilon < \varepsilon_1$ *each minimizer* u_0 *of* I *in* $\mathscr{A}$ *satisfies*

(i) *the Tonelli set for* u_0 *is* $E = \{0\}$,

(ii) $u_0 \in W^{1,p}$ *for all* p *such that* $1 \leqq p < 3$, *and*

$$u_0(x) \sim |x|^{\frac{2}{3}} \operatorname{sign} x \quad \text{as } x \to 0, \tag{5.3}$$

$$-|x|^{\frac{2}{3}} < u_0(x) < \alpha k_1 |x|^{\frac{2}{3}} \quad \text{for } -1 < x < 0, \tag{5.4}$$

and

$$\alpha k_2 x^{\frac{2}{3}} < u_0(x) < x^{\frac{2}{3}} \quad \text{for } 0 < x < 1, \tag{5.5}$$

(iii) u_0 *satisfies none of* (WEL), (WDBR), (IEL), (IDBR),

(iv) *for any* q, $3 \leqq q \leqq \infty$,

$$\inf_{v \in W^{1,q} \cap \mathscr{A}} I(v) > \inf_{v \in \mathscr{A}} I(v) = I(u_0) \quad \textit{(the Lavrentiev phenomenon)}.$$

The proof of Theorem 5.1 depends on some lemmas.

Lemma 5.2. *Let* $0 < \alpha < \beta < 1$, $0 < k \leqq 1$, $\gamma \geqq \frac{2}{3}$ *and* $s > 9$. *Let* $v \in W^{1,1}(0, 1)$ *satisfy*

$$\alpha k x^\gamma \leqq v(x) \leqq \beta k x^\gamma \quad \text{for } x_1 \leqq x \leqq x_2,$$

$$v(x_1) = \alpha k x_1^\gamma, \quad v(x_2) = \beta k x_2^\gamma,$$

where $0 \leqq x_1 < x_2 \leqq 1$. *Then*

$$\int_{x_1}^{x_2} (x^4 - v^6)^2 |v'|^s \, dx \geqq (1 - (\beta k)^6)^2 \theta^{s-1} k^s (\beta - \alpha)^s x_2^{(\gamma-1)s+9}$$

where $\theta = \dfrac{s-9}{s-1}$.

Proof (*cf.* MANIÀ [25]). We have

$$\int_{x_1}^{x_2} (x^4 - v^6)^2 |v'|^s \, dx = \int_{x_1}^{x_2} \left(1 - \frac{v^6}{x^4}\right)^2 x^8 |v'|^s \, dx$$
$$\geqq (1 - (\beta k)^6)^2 \int_{x_1}^{x_2} x^8 |v'|^s \, dx.$$

Setting $y = x^\theta$, $\tilde{v}(x^\theta) = v(x)$, we obtain by Jensen's inequality

$$\int_{x_1}^{x_2} x^8 |v'|^s \, dx = \theta^{s-1} \int_{x_1^\theta}^{x_2^\theta} \left|\frac{d\tilde{v}}{dy}\right|^s dy$$
$$\geqq \theta^{s-1} \frac{[v(x_2) - v(x_1)]^s}{[x_2^\theta - x_1^\theta]^{s-1}}$$
$$= \theta^{s-1} k^s x_2^{(\gamma-1)s+9} \frac{\left[\beta - \alpha\left(\frac{x_1}{x_2}\right)^\gamma\right]^s}{\left[1 - \left(\frac{x_1}{x_2}\right)^\theta\right]^{s-1}}$$
$$\geqq \theta^{s-1} k^s (\beta - \alpha)^s \, x_2^{(\gamma-1)s+9},$$

and the result follows. □

Lemma 5.3. *Let k_1, k_2 be arbitrary, $s > 3$, and let u_0 minimize I in $\mathcal{A}$. Then either the Tonelli set E of u_0 is empty, or $E = \{0\}$ and $u_0(0) = 0$.*

Proof. Suppose first that $x_0 \in [-1, 1]$ with $u_0(x_0)^6 \neq x_0^4$. Then there is a non-trivial interval $[c, d] \subset [-1, 1]$ containing x_0 and such that

$$u_0(x)^6 \neq x^4, \quad x \in [c, d]. \tag{5.6}$$

Now u_0 minimizer the integral

$$J(v) = \int_c^d [(x^4 - v^6)^2 |v'|^s + \varepsilon(v')^2] \, dx$$

in

$$\mathcal{B} = \{v \in W^{1,1}(c, d) : v(c) = u_0(c), v(d) = u_0(d)\}.$$

But by (5.6),

$$|f_u(x, u_0(x), u_0'(x))| \leqq \text{const.} \, f(x, u_0(x), u_0'(x)), \quad x \in [c, d],$$

and therefore, since $J(u_0) < \infty$, $f_u(\cdot, u_0(\cdot), u_0'(\cdot)) \in L^1(c, d)$. By Theorem 2.10, u_0 is smooth in $[c, d]$, and in particular $x_0 \notin E$.

It remains to consider the possibility that $x_0 \in E$ with $u(x_0)^6 = x_0^4 \neq 0$. Suppose $k_2 \geqq k_1$; the case $k_2 \leqq k_1$ is treated similarly. Then u is nondecreasing and so we must have $u_0'(x_0) = +\infty$. Suppose that $x_0 > 0$ and $u(x_0) = x_0^{\frac{2}{3}}$; the other three cases $x_0 > 0$, $u(x_0) = -x_0^{\frac{2}{3}}$ and $x_0 < 0$, $u(x_0) = \pm |x_0|^{\frac{2}{3}}$

are treated similarly. Then there exists $x_1 \in (0, x_0)$ such that $0 < u(x) < x^{\frac{2}{3}}$ for all $x \in (x_1, x_0)$. By the preceding argument u_0 is a C^3 (in fact C^∞) solution of the Euler-Lagrange equation

$$\frac{d}{dx}[s(x^4 - u^6)^2 |u'|^{s-1} \operatorname{sign} u' + 2\varepsilon u'] = -12u^5(x^4 - u^6)|u'|^s \tag{5.7}$$

in (x_1, x_0). Since the right-hand side of (5.7) is negative in (x_1, x_0) this contradicts $u_0'(x_0) = +\infty$. □

Lemma 5.4. *Let k_1, k_2 be arbitrary, $s > 3$ and let u_0 minimize I in $\mathscr{A}$. Suppose that $0 \leqq x_1 < x_2 \leqq 1$ and that $u_0(x_1) = x_1^{\frac{2}{3}}$, $u_0(x_2) = x_2^{\frac{2}{3}}$. Then $u_0(x) \leqq x^{\frac{2}{3}}$ for all $x \in [x_1, x_2]$.*

Proof. Suppose first that $u_0(x) > x^{\frac{2}{3}}$ for all $x \in (x_1, x_2)$. Define

$$\bar{u}(x) = \begin{cases} u_0(x) & x \notin [x_1, x_2] \\ x^{\frac{2}{3}} & x \in [x_1, x_2]. \end{cases}$$

Then

$$\begin{aligned} I(u_0) - I(\bar{u}) &= \int_{x_1}^{x_2} [(x^4 - u_0^6)^2 |u_0'|^s + \varepsilon(u_0')^2 - \varepsilon(\bar{u}')^2]\, dx \\ &= \int_{x_1}^{x_2} (x^4 - u_0^6)^2 |u_0'|^s\, dx + \varepsilon \int_{x_1}^{x_2} [(\tfrac{2}{3}x^{-\frac{1}{3}} + v'(x))^2 - (\tfrac{2}{3}x^{-\frac{1}{3}})^2]\, dx, \end{aligned}$$

where $v = u_0 - \bar{u}$. Note that $v(x_1) = v(x_2) = 0$, $v(x) > 0$ for all $x \in (x_1, x_2)$. The first integral on the right-hand side is positive, and for the second integral we obtain, using integration by parts,

$$\begin{aligned} &\int_{x_1}^{x_2} \tfrac{4}{3}x^{-\frac{1}{3}} v'\, dx + \int_{x_1}^{x_2} (v')^2\, dx \\ &\qquad = (\tfrac{4}{3}x^{-\frac{1}{3}} v)|_{x_1+}^{x_2} + \int_{x_1}^{x_2} \tfrac{4}{9}x^{-\frac{4}{3}} v\, dx + \int_{x_1}^{x_2} (v')^2\, dx > 0. \end{aligned}$$

This contradicts the minimum property for u_0. (When $x_1 = 0$ the validity of the integration by parts stems from the fact that finiteness of $I(u_0)$ ensures that $v' \in L^2(-1, 1)$, so that $v(x) = \int_0^x v'(y)\, dy$ is $o(x^{\frac{1}{2}})$ as $x \to 0+$.)

More generally, if $u_0(\bar{x}) > \bar{x}^{\frac{2}{3}}$ for any $\bar{x} \in (x_1, x_2)$ then there is an interval $(\bar{x}_1, \bar{x}_2) \subset (x_1, x_2)$ such that $u_0(\bar{x}_1) = \bar{x}_1^{\frac{2}{3}}$, $u_0(\bar{x}_2) = \bar{x}_2^{\frac{2}{3}}$ and $u_0(x) > x^{\frac{2}{3}}$ for all $x \in (\bar{x}_1, \bar{x}_2)$. Applying the preceding argument to $(\bar{x}_1, \bar{x}_2)$ gives a contradiction. □

Proof of Theorem 5.1. Fix α, β with $0 < \alpha < \beta < 1$. Let $v \in \mathscr{A}$, and suppose for the moment that

$$v(\bar{x}) \leqq \alpha k_2 \bar{x}^{\frac{2}{3}} \quad \text{for some } \bar{x} \in (0, 1]. \tag{5.8}$$

Then there exists an interval $[x_1, x_2] \subset (0, 1)$ such that

$$\left.\begin{aligned} \alpha k_2 x^{\frac{2}{3}} < v(x) < \beta k_2 x^{\frac{2}{3}}, \quad x \in (x_1, x_2) \\ v(x_1) = \alpha k_2 x_1^{\frac{2}{3}}, \quad v(x_2) = \beta k_2 x_2^{\frac{2}{3}}. \end{aligned}\right\} \tag{5.9}$$

By Lemma 5.2, with $k = k_2$ and $\gamma = 2/3$,

$$I(v) > (1 - (\beta k_2)^6)^2 \, \theta^{s-1} k_2^{\frac{s}{2}} (\beta - \alpha)^s \, x_2^{-\frac{1}{3}(s-27)}. \tag{5.10}$$

Since $s \geqq 27$, (5.10) implies that

$$I(v) > (1 - (\beta k_2)^6)^2 \, \theta^{s-1} k_2^{\frac{s}{2}} (\beta - \alpha)^s. \tag{5.11}$$

Similarly, if in place of (5.8) we assume that

$$v(\bar{x}) \geqq \alpha k_1 \, |\bar{x}|^{\frac{2}{3}} \qquad \text{for some } \bar{x} \in [-1, 0), \tag{5.12}$$

then by applying Lemma 5.2 to the function $-v(-x)$ we obtain

$$I(v) > (1 - (\beta k_1)^6)^2 \, \theta^{s-1} (-k_1)^s \, (\beta - \alpha)^s. \tag{5.13}$$

We now note that one of (5.8), (5.12) holds if either $v(0) \neq 0$ or $v \in W^{1,q} \cap \mathcal{A}$ for some q with $3 \leqq q \leqq \infty$, since if $v(0) = 0$ and $v \in W^{1,q} \cap \mathcal{A}$, $q \in [3, \infty)$, then for all $x \in [-1, 1]$

$$|v(x)| = \left| \int_0^x v'(y) \, dy \right| \leqq \left| \int_0^x |v'|^q \, dy \right|^{1/q} \left| \int_0^x 1^{q'} \, dy \right|^{1/q'} = o(1) \, |x|^{1-1/q},$$

while if $q = \infty$,

$$|v(x)| \leqq \text{const.} \, |x|.$$

In either case we therefore have

$$I(v) > \min \{h_\beta(k_1), h_\beta(k_2)\} \, \theta^{s-1} (\beta - \alpha)^s, \tag{5.14}$$

where $h_\beta(k) \stackrel{\text{def}}{=} (1 - (\beta k)^6) \, |k|^s$, this estimate being independent of $\varepsilon > 0$ and of q.

Now consider the following function $\hat{u} \in \mathcal{A}$:

$$\hat{u}(x) = \begin{cases} k_1, & -1 \leqq x \leqq -|k_1|^{\frac{3}{2}} \\ -|x|^{\frac{2}{3}}, & -|k_1|^{\frac{3}{2}} \leqq x \leqq 0 \\ x^{\frac{2}{3}}, & 0 \leqq x \leqq k_2^{\frac{3}{2}} \\ k_2, & k_2^{\frac{3}{2}} \leqq x \leqq 1. \end{cases}$$

A direct computation yields

$$I(\hat{u}) = 0 + \int_{-|k_1|^{\frac{3}{2}}}^{k_2^{\frac{3}{2}}} \varepsilon (\hat{u}')^2 \, dx = \frac{4\varepsilon}{9} \int_{-|k_1|^{\frac{3}{2}}}^{k_2^{\frac{3}{2}}} |x|^{-\frac{2}{3}} \, dx = \frac{4\varepsilon}{3} (|k_1|^{\frac{1}{2}} + k_2^{\frac{1}{2}}). \tag{5.15}$$

Together (5.14), (5.15) ensure that for $\beta = \beta(\alpha, k_1, k_2, s)$ chosen to maximize the right side of (5.14) and for $\varepsilon > 0$ sufficiently small, *i.e.* $\varepsilon < \varepsilon_0(\alpha, k_1, k_2, s)$,

$$\inf_{v \in W^{1,q} \cap \mathscr{A}} I(v) > I(\hat{u}) \geqq \inf_{v \in \mathscr{A}} I(v)$$

for all q with $3 \leqq q \leqq \infty$; furthermore any minimizer u_0 of I in $\mathscr{A}$ satisfies $u_0(0) = 0$, and

$$\left.\begin{aligned} u_0(x) &< \alpha k_1 |x|^{\frac{2}{3}}, \quad -1 \leqq x < 0 \\ u_0(x) &> \alpha k_2 x^{\frac{2}{3}}, \quad 0 < x \leqq 1. \end{aligned}\right\} \tag{5.16}$$

It follows from (5.16) that $u_0'(0) = +\infty$, so that by Lemma 5.3 we have $E = \{0\}$. Also, since $|k_i| \leqq 1$, $i = 1, 2$, it follows from Lemma 5.4 (applied to $u_0(x)$ and $-u_0(-x)$) that

$$\left.\begin{aligned} u_0(x) &\geqq -|x|^{\frac{2}{3}}, \quad -1 \leqq x \leqq 0 \\ u_0(x) &\leqq x^{\frac{2}{3}}, \quad 0 \leqq x \leqq 1. \end{aligned}\right\} \tag{5.17}$$

Since by Theorem 2.7 $u_0' : [-1, 1] \to \overline{\mathbb{R}}$ is continuous, we have

$$\begin{aligned} \lim_{x \to 0+} f_p(x, u_0(x), u_0'(x)) &= \lim_{x \to 0+} [u_0'(x) f_p(x, u_0(x), u_0'(x)) - f(x, u_0(x), u_0'(x))] \\ &= +\infty, \end{aligned}$$

and it follows immediately that none of (WEL), (WDBR), (IEL), (IDBR) hold.

We next show that

$$|u_0'(x)| \leqq \text{const.}\, |x|^{-\frac{1}{3}}, \quad x \in [-1, 1], \tag{5.18}$$

which ensures that $u_0 \in W^{1,p}$ for all p such that $1 \leqq p < 3$. For $\zeta \in [0, 1]$ define $u_\zeta \in \mathscr{A}$ by

$$u_\zeta(x) = \begin{cases} u_0(x), & -1 \leqq x \leqq -\zeta \\ u_0(-\zeta), & -\zeta \leqq x \leqq -|u_0(-\zeta)|^{\frac{3}{2}} \\ |x|^{\frac{2}{3}} \operatorname{sign} x, & -|u_0(-\zeta)|^{\frac{3}{2}} \leqq x \leqq u_0(\zeta)^{\frac{3}{2}} \\ u_0(\zeta), & u_0(\zeta)^{\frac{3}{2}} \leqq x \leqq \zeta \\ u_0(x), & \zeta \leqq x \leqq 1. \end{cases}$$

Then

$$\begin{aligned} 0 \geqq I(u_0) - I(u_\zeta) = &\int_{-\zeta}^{\zeta} [(x^4 - u_0^6)^2 (u_0')^s + \varepsilon (u_0')^2]\, dx \\ &- \int_{-|u_0(-\zeta)|^{\frac{3}{2}}}^{u_0(\zeta)^{\frac{3}{2}}} \varepsilon (\tfrac{2}{3} |x|^{-\frac{1}{3}})^2\, dx, \end{aligned}$$

and hence by (5.17)

$$\int_{-\zeta}^{\zeta} [(x^4 - u_0^6)^2 (u_0')^s + \varepsilon(u_0')^2]\, dx \leqq \text{const.}\, \{u_0(\zeta)^{\frac{1}{2}} + |u_0(-\zeta)|^{\frac{1}{2}}\}$$
$$\leqq C\zeta^{\frac{1}{3}}, \quad \zeta \in [0, 1], \tag{5.19}$$

for some constant $C > 0$. Now define

$$g(x) = u_0'(x) f_p(x, u_0(x), u_0'(x)) - f(x, u_0(x), u_0'(x)),$$

so that

$$g(x) = (s - 1)(x^4 - u_0^6)^2 (u_0')^s + \varepsilon(u_0')^2.$$

Since $u_0(x)$ is smooth for $x \neq 0$, by (DBR),

$$g'(x) = -8x^3(x^4 - u_0^6)(u_0')^s, \quad x \in [-1, 0) \cup (0, 1],$$

and so by (5.17) g is increasing on $[-1, 0)$, decreasing on $(0, 1]$. Thus for $\zeta \in (0, 1]$,

$$g(\zeta)\,\zeta = \int_0^{\zeta} g(\zeta)\, dx \leqq \int_0^{\zeta} g(x)\, dx \leqq (s - 1) \int_0^{\zeta} [(x^4 - u_0^6)^2 (u_0')^s + \varepsilon(u_0')^2]\, dx, \tag{5.20}$$

and it follows from (5.19), (5.20) that

$$g(\zeta) \leqq \text{const.}\, |\zeta|^{-\frac{2}{3}}. \tag{5.21}$$

The same argument applied on $[-1, 0)$ shows that (5.21) holds also for $\zeta \in [-1, 0)$, and (5.18) follows by the formula for g. Clearly (5.18) implies that $u_0 \in W^{1,p}$ for $1 \leqq p < 3$.

It now only remains to prove (5.3) and the strictness of the inequality in (5.17). For this we make the same substitutions as in Section 3, namely

$$z = \frac{u^{\frac{3}{2}}}{x}, \quad q = \tfrac{3}{2} u^{\frac{1}{2}} u', \quad x = e^t,$$

which for $u > 0$, $x > 0$ reduce the Euler-Lagrange equation (5.7) to the system

$$\frac{dz}{dt} = q - z,$$

$$\frac{dq}{dt} = \frac{q^2}{3z} \tag{5.22}$$

$$\times \frac{\left(\frac{2q}{3}\right)^{s-3} (1 - z^4) \left[(s - 1) q \left(\frac{s}{3}(1 - z^4) + 8z^4\right) - 8sz\right] + \varepsilon z^{\frac{s-2}{3}} e^{\left(\frac{s-26}{3}\right)t}}{\frac{s}{2}(s - 1)(1 - z^4)^2 \left(\frac{2q}{3}\right)^{s-2} + \varepsilon z^{\frac{s-2}{3}} e^{\left(\frac{s-26}{3}\right)t}},$$

which, of course, is not autonomous since $s \neq 26$. We compute the sign of $\frac{dq}{dt}$ on the diagonal $q = z$; this is the same as the sign of

$$\left(\frac{2z}{3}\right)^{s-3} (1 - z^4) \left[(s-1)\, z \left(\frac{s}{3}(1 - z^4) + 8z^4\right) - 8sz\right] + \varepsilon z^{\frac{s-2}{3}} e^{\left(\frac{s-26}{3}\right)t}$$

$$= \left(\frac{2z}{3}\right)^{s-3} (1 - z^4)\, z(s - 24) \left(\frac{s-1}{3}\right)(\sigma_s - z^4) + \varepsilon z^{\frac{s-2}{3}} e^{\left(\frac{s-26}{3}\right)t},$$

where $\sigma_s = \dfrac{s(s-25)}{(s-1)(s-24)} < 1$. Fix $\varrho \in (\sigma_s^{\frac{1}{4}}, 1)$. Then if $\varrho < z_0 < 1$ we have that, for t sufficiently large and negative, $\frac{dq}{dt} < 0$ whenever $q = z$, $\varrho < z < z_0$.

Now let $(z(\cdot), q(\cdot))$ be the solution of (5.22) corresponding to u_0 on $(0, 1]$; the behavior of u_0 on $[-1, 0)$ is handled similarly. By our results so far (taking $\alpha_\varrho = \varrho^{\frac{2}{3}}$) it follows that if $\varepsilon > 0$ is sufficiently small, *i.e.* $\varepsilon < \varepsilon_0(\alpha_\varrho, k_1, k_2, s)$, then $\varrho < z(t) \leqq 1$, $0 < q(t) \leqq$ const. for all $t \in (-\infty, 0]$. Note that $\frac{dz}{dt} < 0$ for $q < z$, $\frac{dz}{dt} > 0$ for $q > z$. In view of the bound on z, this implies that we cannot have $z(t) = 1 \neq q(t)$ for any $t \in (-\infty, 0)$. If $z(t) = q(t) = 1$ for some $t \in (-\infty, 0)$ then by (5.22), $\frac{dq}{dt}(t) = \frac{1}{3}$, $\frac{dz}{dt}(t) = 0$, $\frac{d^2z}{dt^2}(t) = \frac{1}{3}$, which by the same reasoning is impossible. Thus strict inequality holds in (5.17) for $x \neq 0$, ± 1. If $z(t) \leqq q(t)$ for all sufficiently large and negative t, or if $z(t) \geqq q(t)$ for all sufficiently large and negative t, then for some $z^* \in [0, 1]$, $z(t) \to z^*$ as $t \to -\infty$; in these cases we must have $z^* = 1$, since for $z^* < 1$ the relations $u_0(x) \sim (z^* x)^{\frac{2}{3}}$ as $x \to 0+$ and $I(u_0) < \infty$ imply that

$$\int_0^1 x^8 (u_0')^s \, dx < \infty,$$

and hence by Hölder's inequality

$$u_0(x) = \int_0^x u_0'(y)\, dy \leqq \text{const.} \left(\int_0^x y^8 u_0'(y)^s \, dy\right) x^{1-\frac{9}{s}} = o(1)\, x^{1-\frac{9}{s}},$$

which contradicts $z(t) > \varrho$ since $s \geqq 27$. We therefore need only consider the case when there exists a sequence $t_j \to -\infty$ with $z(t_j) = q(t_j) < 1$, $\frac{dq}{dt}(t_j) \geqq 0$. By our analysis of the sign of $\frac{dq}{dt}$ on the diagonal, it follows that for any such sequence $z(t_j) \to 1$ as $j \to \infty$, and the sign of $\frac{dz}{dt}$ then implies that $z(t) \to 1$ as $t \to -\infty$. This proves (5.3). $\square$

We remark that the relations (5.3)–(5.5) imply that the inverse function $x_0(u)$ of u_0 is not C^2, and in particular that the graph of u_0 is not a smooth curve in the plane. We mention this fact because of its relevance to attempts to elucidate the phenomenon of singular minimizers by consideration of some parametric problem of the calculus of variations.

We remark also that if $\varphi \in C_0^\infty(-1, 1)$ is given with $\varphi(0) \neq 0$, and if $t \neq 0$, then $I(u_0 + t\varphi) = \infty$. In fact, since $u_0'(x) \to \infty$ as $x \to 0$ by Theorem 2.7, there exist constants $c(t) > 0$, $\alpha(t) > 0$ such that

$$I(u_0 + t\varphi) \geqq c(t) \int_{-\alpha(t)}^{\alpha(t)} |u_0'(x)|^s \, dx.$$

Since by (i) and (iv) $u_0 \notin W^{1,s}(-\alpha(t), \alpha(t))$ the assertion follows.

The reader can easily verify that an appropriate version of Theorem 5.1 holds when the signs of k_1, k_2 are reversed. Note also that the comparison argument used in the proof requires that the value of ε approach zero as $k_1, k_2 \to 0$. In fact if $\varepsilon > 0$ is fixed then for sufficiently small $|k_1|, |k_2|$ the minimizer of I in $\mathscr{A}$ is unique and *smooth*. This can be proved by noting that constants satisfy (EL), and thus, by an argument similar to that used in the proof of Lemma 2.8, for $|k_1|, |k_2|$ sufficiently small there is a unique smooth solution u_1 of (EL) in $\mathscr{A}$ and u_1 can be embedded in a field of extremals simply covering the region $S = \{(x, v): |x| \leqq 1, k_1 \leqq v \leqq k_2\}$. But any minimizer $\bar{u}_0$ of I in $\mathscr{A}$ is monotone and thus has graph lying in S. By the field theory of the calculus of variations $I(u_1) \leqq I(\bar{u}_0)$ with equality if and only if $\bar{u}_0 = u_1$. Hence $\bar{u}_0 = u_1$, as claimed.

It is important to note the significance of the Lavrentiev phenomenon for numerical schemes designed to approximate minimizers of variational problems such as (5.1). Such schemes, for instance those using finite elements, are often associated with the use of approximating functions that are Lipschitz. Hence the existence of Lavrentiev's gap ensures that *no such scheme can yield a minimizing sequence* for I. On the other hand, one might suppose that a sequence $\{v_m\} \subset W^{1,\infty} \cap \mathscr{A}$ could be found satisfying the pseudo-minimizing condition

$$I(v_m) \to \inf_{v \in W^{1,\infty} \cap \mathscr{A}} I(v),$$

and such that v_m converges to the actual minimizer u_0 in some mild sense. Our next result shows that even this cannot happen.

Theorem 5.5. *Let* $s > 27$, $-1 \leqq k_1 < 0 < k_2 \leqq 1$, $0 < \alpha < 1$, $0 < \varepsilon < \varepsilon_1(\alpha, k_1, k_2, s)$ *and* $3 \leqq q \leqq \infty$. *Let* u_0 *be an absolute minimizer of* I *in* $\mathscr{A}$. *For any sequence* $\{v_m\} \subset W^{1,q} \cap \mathscr{A}$ *such that* $v_m(x) \to u_0(x)$ *for each* x *in some set containing arbitrarily small positive and negative numbers one necessarily has*

$$I(v_m) \to \infty \quad \text{as} \quad m \to \infty.$$

Proof. Let $l_j \in [-1, 0)$, $r_j \in (0, 1]$ satisfy $l_j \to 0$, $r_j \to 0$ as $j \to \infty$, and suppose that $v_m(l_j) \to u_0(l_j)$, $v_m(r_j) \to u_0(r_j)$ as $m \to \infty$ for each $j = 1, 2, \ldots$. Fix $\bar{\alpha}$ with $0 < \bar{\alpha} < \alpha$. Given j we have by (5.5) that for all sufficiently large m

$$v_m(l_j) < \alpha k_1 |l_j|^{\frac{2}{3}}, \quad v_m(r_j) > \alpha k_2 r_j^{\frac{2}{3}}.$$

Since v_m changes sign in $[l_j, r_j]$, $v_m(x_0) = 0$ for some $x_0 \in (l_j, r_j)$. If $x_0 \geqq 0$, say, then by the argument preceding (5.14) there is an interval $[y_1, y_2] \subset (0, r_j)$ such that

$$\bar{\alpha} k_2 x^{\frac{2}{3}} < v_m(x) < \alpha k_2 x^{\frac{2}{3}}, \quad x \in (y_1, y_2),$$

$$v_m(y_1) = \bar{\alpha} k_2 y_1^{\frac{2}{3}}, \quad v_m(y_2) = \alpha k_2 y_2^{\frac{2}{3}}.$$

By Lemma 5.2,

$$\begin{aligned} I(v_m) &> (1 - (\alpha k_2)^6)^2\, \theta^{s-1} k_2^s (\alpha - \bar{\alpha})^s\, y_2^{-\frac{1}{3}(s-27)} \\ &> C r_j^{-\frac{1}{3}(s-27)}, \end{aligned}$$

where $C > 0$ is a constant. If $x_0 \leqq 0$ we obtain similarly

$$I(v_m) > C\, |l_j|^{-\frac{1}{3}(s-27)}.$$

Letting $j \to \infty$ we obtain that $I(v_m) \to \infty$ as $m \to \infty$, as required. □

A more quantitative version of Theorem 5.5 may be proved. If $s, k_1, k_2, \alpha, \varepsilon, q$ and u_0 are as in the theorem and if $1 \leqq \sigma \leqq \infty$ then

$$\inf_{v \in \mathcal{A} \cap W^{1,q}} (I(v) - I(u_0))\, \|v - u_0\|_{L^\sigma(-1,1)}^{\frac{s-27}{2+3/\sigma}} > 0.$$

This has some of the features of an uncertainty principle. By Theorem 5.1(iv) it suffices for the proof to show that if $\delta = \|v - u_0\|_{L^\sigma(-1,1)}$ then

$$I(v)\, \delta^{\frac{s-27}{2+3/\sigma}} \geqq \text{const.} > 0$$

for $\delta > 0$ sufficiently small. But it is easily shown using (5.3) that for $\delta > 0$ sufficiently small there exist points $-l, r \in (0, \text{const.}\ \delta^{\frac{3}{2+3/\sigma}})$ such that

$$v(l) < \alpha k_1 |l|^{\frac{2}{3}}, \quad v(r) > \alpha k_2 r^{\frac{2}{3}},$$

and the result follows using the same proof as for Theorem 5.5.

Theorem 5.5 contrasts strongly with a claim of Lewy [23]. There it was asserted that the sequence $\{u_M\}$ constructed through the following constrained minimization procedure:

$$I(u_M) = \inf_{v \in \mathcal{A}_M} I(v),$$

where $\mathcal{A}_M = \{v \in W^{1,\infty} \cap \mathcal{A} : \|v\|_{W^{1,\infty}} \leqq M\}$, would yield a sequence $\{u_M\}$ converging to the global minimizer u_0 as $M \to \infty$. Note that the existence of a constrained minimizer u_M for M sufficiently large follows from the precompactness of $\mathcal{A}_M$ in $C([-1, 1])$. Theorem 5.5 reveals that in our example no subsequence of $\{u_M\}$ can converge to u_0 pointwise, even on a two-sided sequence of points $x_j \to 0$. A similar comment applies to any "penalty method" which involves adding to the integrand a term such as $\eta\, |u'|^{3+\gamma}$ or $\eta\, |u''|^{1+\gamma}$, $\gamma > 0$, and examining the limiting behavior of the corresponding minimizers as $\eta \to 0+$.

The above predictions for numerical methods have been confirmed experimentally in BALL & KNOWLES [6], where numerical methods are described and developed that are capable of detecting the absolute minimizer u_0. Their numerical experiments also indicate that in an example due to MANIÀ [25] (which is not regular, but to which the ideas below also apply) minimizing sequences in $W^{1,\infty}$ converge to a "pseudominimizer" $u_1 \neq u_0$. We now examine the existence of pseudominimizers corresponding to the regular integrand (5.2).

We first discuss the problem of minimizing

$$J(u) = \int_0^1 [(x^4 - u^6)^2 |u'|^s + \varepsilon(u')^2]\, dx \tag{5.23}$$

in various subsets of

$$\mathscr{A}_0 = \{u \in W^{1,1}(0,1) : u(0) = 0,\ u(1) = k\}.$$

We begin by stating an analogue of Theorem 5.1 for this problem. Note that there is no condition on the size of k.

Theorem 5.6. *Let $s \geqq 27$, $k > 0$, and $0 < \alpha < 1$. Then there is an $\varepsilon_1 = \varepsilon_1(\alpha, k, s) > 0$ such that if $0 < \varepsilon < \varepsilon_1$ each minimizer u_0 of J in $\mathscr{A}_0$ (at least one such existing by Theorem 2.1 for any $\varepsilon > 0$) satisfies*

(i) *the Tonelli set for u_0 is $E = \{0\}$,*

(ii) *$u_0 \in W^{1,p}(0,1)$ for $1 \leqq p < 3$ and satisfies*

$$u_0(x) \sim x^{\frac{2}{3}} \quad \text{as } x \to 0+; \tag{5.24}$$

if $k \leqq 1$ then

$$\alpha k x^{\frac{2}{3}} < u_0(x) < x^{\frac{2}{3}} \quad \text{for } 0 < x < 1, \tag{5.25}$$

while if $k > 1$ then there exists exactly one $\bar{x} \in (0,1)$ with $u_0(\bar{x}) = \bar{x}^{\frac{2}{3}}$ and

$$\alpha x^{\frac{2}{3}} < u_0(x) < x^{\frac{2}{3}} \quad \text{for } 0 < x < \bar{x}, \tag{5.26}$$

$$x^{\frac{2}{3}} < u_0(x) < k \quad \text{for } \bar{x} < x < 1. \tag{5.27}$$

(iii) *u_0 does not satisfy* (IEL) *or* (IDBR),

(iv) *for any q, $3 \leqq q \leqq \infty$,*

$$\inf_{v \in W^{1,q}(0,1) \cap \mathscr{A}_0} J(v) > \inf_{v \in \mathscr{A}_0} J(v) = J(u_0).$$

Proof. If $k \leqq 1$ then the proof follows the same lines as that of Theorem 5.1. We therefore suppose that $k > 1$. Choose β with $\alpha < \beta < 1$. Let $v \in \mathscr{A}_0$ with $v(x_0) \leqq \alpha x_0^{\frac{2}{3}}$ for some $x_0 \in (0,1)$; we have seen that such an x_0 exists if $v \in W^{1,q}(0,1)$ with $3 \leqq q \leqq \infty$. As in the proof of Theorem 5.1 there exists an interval $[x_1, x_2] \subset (0,1)$ such that

$$\alpha x^{\frac{2}{3}} < v(x) < \beta x^{\frac{2}{3}}, \quad x \in (x_1, x_2),$$

$$v(x_1) = \alpha x_1^{\frac{2}{3}}, \quad v(x_2) = \beta x_2^{\frac{2}{3}}$$

and thus by Lemma 5.2

$$\int_0^{x_2} [(x^4 - v^6)^2 |v'|^s + \varepsilon(v')^2]\, dx > (1-\beta^6)^2\, \theta^{s-1}(\beta-\alpha)^s\, x_2^{-\frac{1}{3}(s-27)}. \qquad (5.28)$$

Now define $\hat{v} \in \mathscr{A}_0$ by

$$\hat{v}(x) = \begin{cases} x^{\frac{2}{3}}, & 0 \leqq x \leqq v(x_2)^{\frac{3}{2}}, \\ v(x_2), & v(x_2)^{\frac{3}{2}} \leqq x \leqq x_2, \\ v(x), & x_2 \leqq x \leqq 1. \end{cases}$$

Then

$$J(v) - J(\hat{v}) = \int_0^{x_2} [(x^4 - v^6)^2 |v'|^s + \varepsilon(v')^2]\, dx - \int_0^{\beta^{\frac{3}{2}} x_2} \varepsilon(\tfrac{2}{3}x^{-\frac{1}{3}})^2\, dx,$$

and so by (5.28), with $\beta = \beta(\alpha, s)$ chosen to maximize $(1-\beta^6)^2 (\beta-\alpha)^s$, one has

$$J(v) - J(\hat{v}) > \tfrac{1}{2}(1-\beta^6)^2\, \theta^{s-1}(\beta-\alpha)^s$$

for ε sufficiently small (independently of v). In particular, (iv) holds. Also, any minimizer u_0 of J in $\mathscr{A}_0$ satisfies

$$u_0(x) > \alpha x^{\frac{2}{3}} \quad \text{for } 0 < x \leqq 1,$$

and so $E \supset \{0\}$ with $u_0'(0) = +\infty$. By inspection of (5.7) it is seen to be impossible that $u_0(x) \geqq x^{\frac{2}{3}}$ for all sufficiently small $x \in [0, 1]$; hence there exists some $\bar{x} \in (0, 1)$ with $u_0(\bar{x}) = \bar{x}^{\frac{2}{3}}$ and we may assume that $\bar{x}$ is maximal. By Lemma 5.4, $u_0(x) > x^{\frac{2}{3}}$ for $\bar{x} < x \leqq 1$ and $u_0(x) \leqq x^{\frac{2}{3}}$ for $0 \leqq x \leqq \bar{x}$. It follows as in the proof of Theorem 5.5 that $u_0(x) < x^{\frac{2}{3}}$ for $0 < x < \bar{x}$, so that $\bar{x}$ is unique. The remaining assertions in the theorem follow as before. □

We now prove the existence of a pseudominimizer for (5.23).

Theorem 5.7. *Let* $s > 27$, $k > 0$ *and* $3 \leqq q \leqq \infty$. *Then* $J(u)$ *attains an absolute minimum on* $W^{1,q}(0, 1) \cap \mathscr{A}_0$, *and any such minimizer* u_1 *belongs to* $C^\infty([0, 1])$ *and satisfies* (EL) *on* $[0, 1]$.

Proof. We first note that it suffices to prove the theorem for $q = 3$, since any minimizer for this q value is by the theorem smooth and thus a minimizer for all $q > 3$.

Let $\{v_j\}$ be a minimizing sequence for J in $W^{1,3}(0, 1) \cap \mathscr{A}_0$. Since $v_j \in W^{1,3}(0, 1)$ we have by Hölder's inequality that, as $x \to 0$,

$$v_j(x) = o(1)\, x^{\frac{2}{3}}, \quad \text{all } j \geqq 1.$$

We claim that there exists a number $\delta > 0$ such that

$$v_j(x) \leqq \tfrac{1}{2} x^{\frac{2}{3}} \quad \text{for all } x \in [0, \delta], \text{ all } j \geqq 1. \qquad (5.29)$$

If not there would exist a subsequence $\{v_\mu\}$ of $\{v_j\}$ and a sequence $x_\mu \to 0+$ with $v_\mu(x_\mu) > \frac{1}{2}x_\mu^{\frac{2}{3}}$. Therefore there would exist numbers $x_{1\mu}, x_{2\mu} \in (0, x_\mu)$ such that

$$\tfrac{1}{4}x^{\frac{2}{3}} \leqq v_\mu(x) \leqq \tfrac{1}{2}x^{\frac{2}{3}}, \quad x_{1\mu} \leqq x \leqq x_{2\mu}$$

$$v_\mu(x_{1\mu}) = \tfrac{1}{4}x_{1\mu}^{\frac{2}{3}}, \quad v_\mu(x_{2\mu}) = \tfrac{1}{2}x_{2\mu}^{\frac{2}{3}}.$$

Applying Lemma 5.2 we deduce that

$$J(v_\mu) > (1 - 2^{-6})^2\, \theta^{s-1}\, 4^{-s} x_{2\mu}^{\frac{1}{3}(27-s)}.$$

Since $s > 27$ and $x_{2\mu} \to 0+$, it follows that $J(v_\mu) \to \infty$ as $\mu \to \infty$. This contradiction establishes (5.29). By (5.29),

$$J(v_j) \geqq \int_0^\delta x^8 \left(1 - \frac{v_j^6}{x^4}\right)^2 |v_j'|^s\, dx \geqq (1 - 2^{-6})^2 \int_0^\delta x^8 |v_j'|^s\, dx. \tag{5.30}$$

But

$$\int_0^\delta |v_j'|^3\, dx \leqq \left(\int_0^\delta x^{-\frac{24}{s-3}}\, dx\right)^{1-\frac{3}{s}} \left(\int_0^\delta x^8 |v_j'|^s\, dx\right)^{\frac{3}{s}}$$

$$\leqq \text{const.} \left(\int_0^\delta x^8 |v_j'|^s\, dx\right)^{\frac{3}{s}},$$

and therefore by (5.30)

$$\int_0^\delta |v_j'|^3\, dx \leqq M < \infty, \quad j \geqq 1.$$

Since $v_j(0) = 0$ it follows that $\{v_j\}$ is bounded in $W^{1,3}(0, \delta)$. Moreover, it is obvious from the form of J that $\{v_j\}$ is also bounded in $W^{1,2}(0, 1)$. Therefore there exist a subsequence $\{v_\varrho\}$ of $\{v_j\}$ and a function $u \in W^{1,3}(0, \delta) \cap W^{1,2}(0, 1) \cap \mathscr{A}_0$ such that in the sense of weak convergence,

$$v_\varrho \rightharpoonup u \quad \text{in } W^{1,3}(0, \delta) \text{ and in } W^{1,2}(0, 1);$$

in particular, $v_\varrho(x) \to u(x)$, for all x, $0 \leqq x \leqq 1$. Since J is sequentially weakly lower semicontinuous in $W^{1,1}(0, 1)$ (see, for example, CESARI [11, p. 104]) it follows that

$$J(u) \leqq \liminf_{\varrho \to \infty} J(v_\varrho) = \inf_{v \in W^{1,3}(0,1) \cap \mathscr{A}_0} J(v). \tag{5.31}$$

For given $\bar{x} \in (0, 1)$, however small, the integral

$$J_{\bar{x}}(v) = \int_{\bar{x}}^1 [(x^4 - v^6)^2 |v'|^s + \varepsilon(v')^2]\, dx$$

attains a minimum on the set

$$\mathscr{A}_{\bar{x}} = \{v \in W^{1,1}(\bar{x}, 1) : v(\bar{x}) = u(\bar{x}),\ v(1) = k\}$$

and, by the proof of Lemma 5.3 (reformulated for the interval $(\bar{x}, 1)$), any minimizer $\bar{u}$ belongs to $C^\infty([\bar{x}, 1])$. Given any such minimizer $\bar{u}$, define

$$\bar{v}_\varrho(x) = \begin{cases} v_\varrho(x), & 0 \leqq x \leqq \bar{x} \\ v_\varrho(\bar{x}) + \eta_\varrho^{-1}[\bar{u}(\bar{x} + \eta_\varrho) - v_\varrho(\bar{x})]\,(x - \bar{x}), & \bar{x} \leqq x \leqq \bar{x} + \eta_\varrho \\ \bar{u}(x), & \bar{x} + \eta_\varrho \leqq x \leqq 1, \end{cases}$$

where $\eta_\varrho = |v_\varrho(\bar{x}) - \bar{u}(\bar{x})|$. For sufficiently large ϱ and small $\bar{x}$, $\bar{v}_\varrho$ is well defined and belongs to $W^{1,3}(0, 1) \cap \mathscr{A}_0$. Notice that $|\bar{v}_\varrho'(x)|$ is uniformly bounded in $[\bar{x}, \bar{x} + \eta_\varrho]$, independently of ϱ. Therefore, since $\eta_\varrho \to 0$,

$$\lim_{\varrho\to\infty} J_{\bar{x}}(\bar{v}_\varrho) = J_{\bar{x}}(\bar{u}).$$

By lower semicontinuity,

$$\liminf_{\varrho\to\infty} J_{\bar{x}}(v_\varrho) \geqq J_{\bar{x}}(u),$$

and hence

$$\begin{aligned} 0 \leqq \limsup_{\varrho\to\infty} [J(\bar{v}_\varrho) - J(v_\varrho)] &= \limsup_{\varrho\to\infty} [J_{\bar{x}}(\bar{v}_\varrho) - J_{\bar{x}}(v_\varrho)] \\ &\leqq J_{\bar{x}}(\bar{u}) - J_{\bar{x}}(u). \end{aligned}$$

Therefore $J_{\bar{x}}(u) = J_{\bar{x}}(\bar{u})$ and thus u minimizes $J_{\bar{x}}$ in $\mathscr{A}_{\bar{x}}$. In particular $u \in C^\infty([\bar{x}, 1])$ and satisfies (EL) in $[\bar{x}, 1]$. Since $\bar{x}$ was arbitrary it follows that $u \in C^\infty((0, 1])$ and satisfies (EL) in $(0, 1]$. Since $u \in W^{1,3}(0, \delta)$ we also have $u \in W^{1,3}(0, 1)$ and therefore by (5.31) u minimizes J in $W^{1,3}(0, 1) \cap \mathscr{A}_0$. Clearly $u'(x) > 0$ for $x \in (0, 1]$.

Our final task is to show that $u \in C^\infty([0, 1])$, and by (EL) it suffices for this to show that $u'(0)$ is finite. Passing to the limit $\varrho \to \infty$ in (5.29) we obtain

$$u(x) \leqq \tfrac{1}{2} x^{\frac{2}{3}} \quad \text{for all } x \in [0, \delta], \tag{5.32}$$

and since $J(u) < \infty$ it follows that

$$\int_0^1 x^8\, |u'|^s\, dx < \infty. \tag{5.33}$$

Since u is a smooth solution of (DBR) on $(0, 1]$ we have

$$\begin{aligned} \frac{d}{dx}(u' f_p - f) &= \frac{d}{dx}((s-1)\,(x^4 - u^6)^2\,(u')^s + (u')^2) \\ &= -8x^3(x^4 - u^6)\,(u')^s \end{aligned}$$

for $0 < x \leqq 1$, and therefore by (5.32), (5.33)

$$x \frac{d}{dx}(u' f_p - f) \in L^1(0, 1).$$

But $J(u) < \infty$ implies that

$$u' f_p - f = (s-1)\,(x^4 - u^6)^2\,(u')^s + \varepsilon(u')^2 \quad \in L^1(0, 1)$$

and thus

$$\frac{d}{dx}(x(u'f_p - f)) = x\frac{d}{dx}(u'f_p - f) + (u'f_p - f)$$

belongs to $L^1(0, 1)$. Hence

$$x(u'f_p - f) = (s-1)\left(1 - \frac{u^6}{x^4}\right)^2 x^9(u')^s + \varepsilon x(u')^2$$

is uniformly bounded, and by (5.32) this implies that $x^9(u')^s$ is bounded. Hence

$$u'(x) \leqq \text{const.}\, x^{-\frac{9}{s}}, \quad u(x) \leqq \text{const.}\, x^{1-\frac{9}{s}}, \quad x \in (0, 1]. \tag{5.34}$$

Note that since $s > 27$, $1 - 9/s > \frac{2}{3}$. Pick $\sigma_0 \in (\frac{2}{3}, 1 - 9/s)$ such that if

$$\sigma_n \overset{\text{def}}{=} (\sigma_0 - \tfrac{2}{3})\left(\frac{s+5}{s-1}\right)^n + \tfrac{2}{3} \tag{5.35}$$

then $\sigma_n \neq \dfrac{s-5}{s+5}$ for any $n = 0, 1, 2, \ldots$. This is clearly possible. We prove by induction that for any $n = 0, 1, 2, \ldots$ there is a constant $c_n > 0$ such that

$$u'(x) \leqq c_n(1 + x^{\sigma_n - 1}), \quad u(x) \leqq c_n(x + x^{\sigma_n}), \quad x \in (0, 1]. \tag{5.36}$$

This is true for $n = 0$ by (5.34). Suppose the assertion is true for n. We prove that it holds for $n + 1$. This is obvious if $\sigma_n \geqq 1$, so we consider the case $\sigma_n < 1$. Now by (EL)

$$\frac{d}{dx}f_p = -12u^5(x^4 - u^6)(u')^s, \quad x \in (0, 1],$$

and so by (5.32), (5.36)

$$\left|\frac{d}{dx}f_p\right| \leqq \text{const.}\, x^{5\sigma_n + 4 + s(\sigma_n - 1)}, \quad x \in (0, 1].$$

Since $\sigma_n \neq \dfrac{s-5}{s+5}$ it follows that

$$f_p = s(x^4 - u^6)^2(u')^{s-1} + 2\varepsilon u' \leqq \text{const.}\,(1 + x^{5-s+(s+5)\sigma_n}), \quad x \in (0, 1]. \tag{5.37}$$

If $\sigma_n > \dfrac{s-5}{s+5}$ then (5.37) implies that u' is bounded on $(0, 1]$ and thus that (5.36) holds for $n + 1$. Otherwise, $\sigma_n < \dfrac{s-5}{s+5}$ and we deduce from (5.37) that

$$x^8(u')^{s-1} \leqq \text{const.}\, x^{5-s+(s+5)\sigma_n}, \quad x \in (0, 1].$$

Therefore

$$u'(x) \leqq \text{const.}\, x^{\left(\frac{s+5}{s-1}\right)\sigma_n - \left(\frac{s+3}{s-1}\right)} = \text{const.}\, x^{\sigma_{n+1} - 1},$$

so that (5.36) holds for $n + 1$. This proves our assertion.

Since $\sigma_0 > \frac{2}{3}$, $\sigma_n \geqq 1$ for large enough n, and hence by (5.36) $u'(x)$ is bounded in $(0, 1]$. Therefore $u'(0)$ is finite and $u \in C^\infty([0, 1])$. Finally, if $\tilde{u}$ is *any* minimizer of J in $W^{1,q}(0,1) \cap \mathscr{A}_0$, $q \geqq 3$, then the above arguments applied to the minimizing sequence in $W^{1,3}(0,1) \cap \mathscr{A}_0$ given by $v_j \equiv \tilde{u}$ show that $\tilde{u} \in C^\infty([0,1])$ and satisfies (EL) on $[0,1]$. □

Note that the proof of Theorem 5.7 shows that any minimizing sequence for J in $W^{1,q}(0,1) \cap \mathscr{A}_0$, $q \geqq 3$, has a subsequence converging weakly in $W^{1,q}(0,\delta)$ and $W^{1,2}(0,1)$ to a minimizer.

Remark. Theorems 5.6 and 5.7 apply equally to the problem of minimizing

$$J_-(u) = \int_{-1}^{0} [(x^4 - u^6)^2 |u'|^s + \varepsilon(u')^2]\, dx$$

over various subsets of

$$\mathscr{A}_- = \{u \in W^{1,1}(-1,0) : u(-1) = k,\ u(0) = 0\}, \quad \text{with } k < 0,$$

as can be seen by noting that $v(\cdot) \in \mathscr{A}_-$ if and only if $\hat{v} = -v(-\,\cdot) \in \mathscr{A}_0$, and $J_-(v) = J(\hat{v})$.

We next prove the existence of pseudominimizers for our original functional $I(u)$ given by (5.1).

Theorem 5.8. *Let* $s > 27$, $3 \leqq q \leqq \infty$, *and let* k_1, k_2 *be arbitrary. Then* $I(u)$ *attains an absolute minimum in* $\mathscr{A} \cap W^{1,q}$, *and each such minimizer* u_1 *belongs to* $C^\infty([-1,1])$ *and satisfies* (EL) *on* $[-1,1]$.

Proof. If $k_1 = k_2$ then the unique minimizer of I in $\mathscr{A} \cap W^{1,q}$ is $u_1 \equiv k_1$ and there is nothing to prove. If k_1, k_2 are not equal and have the same sign then any minimizer u_0 of I in $\mathscr{A}$ is strictly monotone and by Lemma 5.3 is a smooth solution of (EL) in $[-1,1]$, and again we have finished. We therefore suppose that $k_1 < 0 < k_2$; the case $k_1 > 0 > k_2$ is treated similarly. Let $\{v_j\}$ be a minimizing sequence for I in $\mathscr{A} \cap W^{1,q}$. By extracting an appropriate subsequence, again denoted by $\{v_j\}$, we may suppose that $v_j \rightharpoonup u_1$, say, in $W^{1,2}(-1,1)$ and that either (a) $v_j(0) = 0$ for all j, or (b) $v_j(0) < 0$, for all j, or (c) $v_j(0) > 0$ for all j. If (a) holds then clearly $\{v_j\}$ (restricted to $[0,1]$) is a minimizing sequence for J (given by (5.23)) in $\mathscr{A}_0 \cap W^{1,q}(0,1)$, where $\mathscr{A}_0 = \{u \in W^{1,1}(0,1) : u(0) = 0,\ u(1) = k_2\}$, and therefore by the proof of Theorem 5.7 u_1 minimizes J in $W^{1,q}(0,1) \cap \mathscr{A}_0$. A similar argument holds on $[-1,0]$, and so by Theorem 5.7 and lower semicontinuity u_1 is smooth on $[-1,0]$ and $[0,1]$ and minimizes I in $\mathscr{A} \cap W^{1,q}$. Standard arguments then show that u_1 satisfies (EL) and is smooth in $[-1,1]$.

Suppose (b) holds; case (c) is treated similarly. Suppose first that $\lim_{j\to\infty} v_j(0) = u_1(0) < 0$. Let u_2 be any minimizer of I in $\bar{\mathscr{A}} \stackrel{\text{def}}{=} \{u \in \mathscr{A} : u(0) = u_1(0)\}$. Then by Lemma 5.3 u_2 is smooth in $[-1,0]$ and $[0,1]$, and so

$$\inf_{\mathscr{A} \cap W^{1,q}} I \leqq I(u_2) \leqq I(u_1).$$

But by lower semicontinuity $I(u_1) \leqq \inf_{\mathscr{A} \cap W^{1,q}} I$, and it follows that u_1 minimizes

I in $\bar{\mathscr{A}}$ also. Hence u_1 is smooth in $[-1, 0]$ and $[0, 1]$, minimizes I in $\mathscr{A} \cap W^{1,q}$, and by standard arguments is a smooth solution of (EL) in $[-1, 1]$.

It remains to consider the case when (b) holds and $\lim_{j\to\infty} v_j(0) = u_1(0) = 0$. Let u_3 be any minimizer of

$$J_-(u) = \int_{-1}^{0} [(x^4 - u^6)^2 \, |u'|^s + \varepsilon(u')^2] \, dx$$

in $\mathscr{A}_-$, where

$$\mathscr{A}_- = \{u \in W^{1,1}(-1, 0) : u(-1) = k_1, u(0) = 0\}.$$

On the other hand let u_4 be any minimizer of $J(u)$ in $\mathscr{A}_0 \cap W^{1,q}(0, 1)$; the existence and smoothness of u_4 is guaranteed by Theorem 5.7 and the remark following it. Define $\bar{u} \in \mathscr{A}$ by

$$\bar{u}(x) = \begin{cases} u_3(x), & -1 \leqq x \leqq 0 \\ u_4(x), & 0 \leqq x \leqq 1. \end{cases}$$

We first show that

$$I(\bar{u}) \geqq \inf_{\mathscr{A} \cap W^{1,q}} I. \tag{5.38}$$

To this end consider the sequence

$$w_j(x) = \begin{cases} u_3(x), & -1 \leqq x \leqq -\dfrac{1}{j}, \\[2mm] u_3\left(-\dfrac{1}{j}\right), & -\dfrac{1}{j} \leqq x \leqq 0, \\[2mm] u_3\left(-\dfrac{1}{j}\right) + Mx, & 0 \leqq x \leqq \beta_j, \\[2mm] u_4(x), & \beta_j \leqq x \leqq 1. \end{cases} \tag{5.39}$$

In (5.39), M is chosen greater than $\max\{u_4'(0), k_2 + |k_1|\}$ so $\beta_j \to 0+$ satisfies $u_4(\beta_j) = u_3\left(-\dfrac{1}{j}\right) + M\beta_j$. The existence of β_j follows from the intermediate value theorem. Note that by a version of Lemma 5.3 which applies to J_-, $u_3 \in C^\infty([-1, 0))$ and so $w_j \in \mathscr{A} \cap W^{1,q}$. Since, as is easily checked,

$$\lim_{j\to\infty} I(w_j) = I(\bar{u}),$$

(5.38) follows. Next, let $\delta_j > 0$ be the largest root of $v_j(x) = 0$ in $(0, 1)$, and define

$$\bar{v}_j(x) = \begin{cases} 0, & 0 \leqq x \leqq \delta_j, \\ v_j(x), & \delta_j \leqq x \leqq 1. \end{cases}$$

Then $\bar{v}_j \in \mathscr{A}_0 \cap W^{1,q}(0, 1)$ and so $J(v_j) \geqq J(\bar{v}_j) \geqq J(u_4)$. Also, by the sequential lower semicontinuity of J_-,

$$J_-(u_3) \leqq J_-(u_1) \leqq \liminf_{j\to\infty} J_-(v_j).$$

Therefore

$$\inf_{\mathscr{A}\cap W^{1,q}} I = \lim_{j\to\infty} I(v_j) = \lim_{j\to\infty} (J_-(v_j) + J(v_j))$$
$$\geqq \liminf_{j\to\infty} J_-(v_j) + \liminf_{j\to\infty} J(v_j) \tag{5.40}$$
$$\geqq J_-(u_3) + J(u_4) = I(\bar{u}).$$

Combining (5.38), (5.40) we obtain

$$I(\bar{u}) = \inf_{\mathscr{A}\cap W^{1,q}} I. \tag{5.41}$$

Suppose first that $u_3'(0)$ is finite so that u_3 is smooth on $[-1, 0]$. Then $\bar{u}$ minimizes I in $\mathscr{A} \cap W^{1,q}$, is smooth on $[-1, 0]$ and $[0, 1]$, and so is a smooth solution of (EL) on $[-1, 1]$. On the other hand by Theorem 5.6 and the Remark, we know that for $\varepsilon > 0$ sufficiently small $u_3'(0) = +\infty$. For such ε, $\lim_{j\to\infty} v_j(0) = u_1(0) = 0$ cannot occur. Indeed, solving (EL) with initial data

$$u(0) = \delta, \quad u'(0) = M, \tag{5.42}$$

for $|\delta|$ small generates by Lemma 2.8 a field of extremals covering a neighborhood of the origin. For δ sufficiently small and negative the solution u_δ of (EL) satisfying (5.42) intersects the graphs of both u_3 and u_4 at points $r_\delta < 0$ and $s_\delta > 0$ respectively, where $r_\delta, s_\delta \to 0$ as $\delta \to 0-$. It then follows from the field theory that

$$v_\delta(x) = \begin{cases} u_3(x), & -1 \leqq x \leqq r_\delta, \\ u_\delta(x), & r_\delta \leqq x \leqq s_\delta, \\ u_4(x), & s_\delta \leqq x \leqq 1, \end{cases}$$

satisfies $I(v_\delta) < I(\bar{u})$. But $v_\delta \in \mathscr{A} \cap W^{1,q}$, contradicting (5.41).

Summarizing, we have shown that in all cases I attains a minimum on $\mathscr{A} \cap W^{1,q}$ at some smooth solution u_1 of (EL). If u_1 is *any* minimizer in $\mathscr{A} \cap W^{1,q}$ then applying the proof to $v_j \equiv u_1$ shows that u_1 is smooth (the case when (b) holds and $\lim_{j\to\infty} v_j(0) = 0$ does not occur). □

We now examine what happens if $s < 27$. If $s = 26$ the integrand f given by (5.2) satisfies the scale invariance property (3.7) with $\gamma = \frac{2}{3}$ and $\varrho = -\frac{2}{3}$ and the phase-plane techniques of Section 3 are applicable to the one-sided problem of minimizing J in $\mathscr{A}_0$. We confine attention here to the observation that the same argument as in Lemma 3.8 shows that if $s \geqq 26$, $0 < \alpha < \min(1, k)$ and $\varepsilon > 0$ is sufficiently small then any minimizer u_0 of J in $\mathscr{A}_0$ satisfies $u_0(x) > \alpha x^{\frac{2}{3}}$ for all $x \in (0, 1]$ and is thus singular. If, further, $s < 27$ then the Lavrentiev phenomenon does not occur; this follows by noting that, by the proof of Theorem 5.6, $u_0(x) \leqq x^{\frac{2}{3}}$ for x sufficiently small, and by using the argument in the remark following Lemma 3.8. It remains, therefore, to consider the case $s < 26$.

Theorem 5.9. *Let* $3 < s < 26$, $\varepsilon > 0$.

(i) *Let* $k_1, k_2 \in \mathbb{R}$ *and let* u_0 *minimize* I *in* $\mathscr{A}$. *Then* $u_0 \in C^\infty([-1, 1])$ *and satisfies* (EL) *on* $[-1, 1]$.

(ii) *Let* $k \in \mathbb{R}$ *and let* u_0 *minimize* J *in* $\mathscr{A}_0$. *Then* $u_0 \in C^\infty([0, 1])$ *and satisfies* (EL) *on* $[0, 1]$.

Proof. It suffices to prove (ii), since in case (i) if $u_0(0) \neq 0$ then u_0 is smooth by Lemma 5.3.

To prove (ii) we may as before assume that $k > 0$. We note that by the same arguments as in the proofs of Theorems 5.1, 5.6 any minimizer u_0 satisfies $0 \leqq u_0(x) \leqq x^{\frac{2}{3}}$ for x sufficiently small and $|u_0'(x)| \leqq \text{const.}\, x^{-\frac{1}{3}}$, $x \in (0, 1]$. It follows from (5.7) that

$$\left|\frac{d}{dx}(s(x^4 - u_0^6)^2 (u_0')^{s-1} + 2\varepsilon u_0')\right| \leqq \text{const.}\, x^{\frac{22-s}{3}}, \quad x \in (0, 1]. \tag{5.43}$$

Hence, by integration, $u_0'(x)$ is bounded for $x \in (0, 1]$ if $s < 25$. If $25 \leqq s < 26$ we deduce by integrating (5.43) that

$$u_0'(x) \leqq \text{const.}\, x^{\tau_0 - 1}, \quad u_0(x) \leqq \text{const.}\, x^{\tau_0}, \quad x \in (0, 1], \tag{5.44}$$

for some $\tau_0 \in (\frac{2}{3}, 1)$, and we may clearly choose τ_0 such that if

$$\tau_n \stackrel{\text{def}}{=} \left(\tau_0 - \frac{s-6}{s+4}\right)(s+5)^n + \frac{s-6}{s+4}, \tag{5.45}$$

then $\tau_n \neq \dfrac{s-5}{s+5}$ for any $n = 0, 1, 2, \ldots$. We prove by induction that for any $n = 0, 1, 2, \ldots$ there is a constant $d_n > 0$ such that

$$u_0'(x) \leqq d_n(1 + x^{\tau_n - 1}), \quad u_0(x) \leqq d_n(x + x^{\tau_n}), \quad x \in (0, 1]. \tag{5.46}$$

This is true for $n = 0$ by (5.44). Suppose it is true for n. We prove that it holds for $n + 1$. This is obvious if $\tau_n \geqq 1$, so we consider the case $\tau_n < 1$. By (5.7)

$$2\varepsilon u_0'(x) \leqq \text{const.}\, (1 + x^{5-s+(s+5)\tau_n}), \quad x \in (0, 1]. \tag{5.47}$$

But $5 - s + (s + 5)\tau_n = \tau_{n+1} - 1$, so that (5.46) holds for $n + 1$.

Since $s < 26$ it follows that $\tau_0 > \dfrac{s-6}{s+4}$, and thus $\tau_n \geqq 1$ for large enough n. Hence in all cases u_0' is bounded in $(0, 1]$ and thus u_0 is a smooth solution of (EL) in $[0, 1]$. □

We end by remarking that the methods of this section apply also to the problem of minimizing

$$I(u) = \int_0^1 [(x^2 - u^3)^2 |u'|^s + \varepsilon(u')^2]\, dx$$

in

$$\mathscr{A} = \{u \in W^{1,1}(0, 1) : u(0) = 0, u(1) = k\},$$

the special case $s = 14$ having been exhaustively discussed in Section 3. For this problem any absolute minimizer is smooth for $3 < s < 14$, singular minimizers can exist without the Lavrentiev phenomenon for $14 \leqq s < 15$, singular minimizers and the Lavrentiev phenomenon can exist for $s \geqq 15$, and smooth pseudominimizers exist if $s > 15$.

Acknowledgment. We are grateful to J. C. EILBECK, W. HRUSA, G. KNOWLES, H. LEWY, F. MURAT, P. J. OLVER, and J. SIVALOGANATHAN for their comments and interest in this work. The research of J.M.B. was supported by an S.E.R.C. Senior Fellowship, that of V.J.M. partially supported by the Air Force Office of Scientific Research under AFOSR-82-0259. Parts of the research were carried out when J.M.B. visited the Department of Mathematics, University of Maryland, in 1984, when V.J.M. was an S.E.R.C. visiting fellow at the Department of Mathematics, Heriot-Watt University, in 1981, 1982, and 1984, and when both authors made brief joint visits to the Institute of Mathematics and its Applications, University of Minnesota, and to the Mathematics Research Center, University of Wisconsin, in 1983.

Bibliography

1. T. ANGELL, A note on approximation of optimal solutions of free problems of the calculus of variations, Rend. Circ. Mat. Palermo **28** (1979), 258–272.
2. J. BALL, Constitutive inequalities and existence theorems in nonlinear elastostatics, in *Nonlinear Analysis and Mechanics: Heriot-Watt Symposium*, vol. 1, pp. 187–241, Ed. R. J. KNOPS. Pitman, London, 1977.
3. J. BALL, Discontinuous equilibrium solutions and cavitation in nonlinear elasticity, Phil. Trans. R. Soc. Lond. A **306** (1982), 557–611.
4. J. BALL, Editor, *Systems of Nonlinear Partial Differential Equations*, Reidel, Dordrecht, 1983.
5. J. BALL, Minimizers and the Euler-Lagrange equations, in *Trends and Applications of Pure Mathematics to Mechanics*, ed. P. G. CIARLET & M. ROSEAU, Lecture Notes in Physics, vol. 195, Springer-Verlag, Berlin Heidelberg New York Tokyo 1984.
6. J. BALL & G. KNOWLES, (to appear).
7. J. BALL & V. MIZEL, Singular minimizers for regular one-dimensional problems in the calculus of variations, Bull. Amer. Math. Soc. **11** (1984), 143–146.
8. J. BALL & F. MURAT, $W^{1,p}$-quasiconvexity and variational problems for multiple integrals, J. Functional Analysis **58** (1984), 225–253.
9. O. BOLZA, *Lectures on the Calculus of Variations*, Dover Publications Inc., New York, 1961.
10. J. CARR, *Applications of Centre Manifold Theory* (Appl. Math. Sci. Series Vol. 35) Springer-Verlag, Berlin Heidelberg New York, 1981.
11. L. CESARI, *Optimization—Theory and Applications*, Springer-Verlag, Berlin Heidelberg New York Tokyo 1983.
12. S. CINQUINI, Sopra la continuità di una classe di integrali del calcolo delle variazioni, Riv. Mat. Univ. Parma **3** (1974), 139–161.
13. F. CLARKE & R. VINTER, Regularity properties of solutions to the basic problem in the calculus of variations (preprint).
14. F. CLARKE & R. VINTER, On the conditions under which the Euler Equation or the maximum principle hold, Appl. Math. and Opt. **12** (1984), 73–79.
15. F. CLARKE & R. VINTER, Regularity of solutions to variational problems with polynomial Lagrangians (preprint).
16. I. E. EKELAND & R. TÉMAM, *Convex Analysis and Variational Inequalities*, North-Holland Publishing, Amsterdam, 1976.

17. M. GIAQUINTA, *Multiple Integrals in the Calculus of Variations and Nonlinear Elliptic Systems*, Annals of Math. Studies # 105, Princeton University Press, Princeton, N.J., 1983.
18. M. GIAQUINTA & E. GIUSTI, Differentiability of minima of nondifferentiable functionals, Inventiones Math. **72**, 285–298.
19. P. HARTMAN, *Ordinary Differential Equations*, Wiley, New York, 1964.
20. M. HESTENES, *Calculus of Variations and Optimal Control Theory*, Krieger, Huntington, New York, 1980.
21. E. L. INCE, *Ordinary Differential Equations*, Dover, New York, 1956.
22. M. LAVRENTIEV, Sur quelques problèmes du calcul des variations, Ann. Mat. Pura Appl. **4** (1926), 7–28.
23. H. LEWY, Über die Methode der Differenzengleichungen zur Lösung von Variations- und Randwertproblemen, Math. Annalen **98** (1927), 107–124.
Math. J. **4** (1938), 132–156.
24. E. MCSHANE, Existence theorems for problems in the calculus of variations, Duke Math. J. **4** (1938), 132–156.
25. B. MANIÀ, Sopra un esempio di Lavrentieff, Bull. Un. Mat. Ital. **13** (1934), 147–153.
26. B. MANIÀ, Sull'approssimazione delle curve e degli integrali, Bull. Unione Mat. **13** (1934), 36–41.
27. M. MARCUS & V. MIZEL, Absolute continuity on tracks and mappings of Sobolev spaces, Arch. Rational Mech. Anal. **45** (1972), 294–320.
28. M. MARCUS & V. MIZEL, Nemitsky operators on Sobolev spaces, Arch. Rational Mech. Anal. **51** (1973), 347–370.
29. S. SAKS, *Theory of the Integral*, Second Edition, Dover, New York, 1964.
30. J. SERRIN & D. VARBERG, A general chain rule for derivatives and the change of variable formula for the Lebesgue integral, Amer. Math. Monthly **76** (1969), 514–520.
31. L. TONELLI, *Fondamenti di Calcolo delle Variazioni*, Zanichelli 2 vols. I, II, 1921–23.
32. L. TONELLI, Sulle estremaloidi del calcolo delle variazioni, Rend. R. Accad. Lincei **21** (1935), 289–293 (in L. TONELLI, *Opere Scelte* vol. III, # 107 Edizioni Cremonese, Roma, 1961).
33. L. TONELLI, Sugli integrali del calcolo delle variazioni in forma ordinaria, Ann. R. Scuola Norm. Sup. Pisa **3** (1934), 401–450 (in L. Tonelli *Opere Scelte* vol. III # 105, Edizioni Cremonese, Roma, 1961).
34. CH. DE LA VALLÉE POUSSIN, Sur l'intégrale de Lebesgue, Trans. Amer. Math. Soc. **16** (1915), 435–501.

Department of Mathematics
Heriot-Watt University
Edinburgh

and

Department of Mathematics
Carnegie-Mellon University
Pittsburgh

(Received January 18, 1985)

On the Finite Element–Least Squares Approximation to Higher Order Elliptic Systems

GEORGE J. FIX & ERNST STEPHAN

Dedicated to Walter Noll on his sixtieth birthday

§ 1. Introduction

The use of variational methods of the least squares type has been studied by many authors. They were first introduced for scalar elliptic equations by BRAMBLE & SCHATZ [1]. This work was extended by BAKER [2], who also treated the scalar case. Recent applications have stressed the importance of these ideas for systems. These include problems in acoustics involving the Helmholtz equation [3]–[4], transonic flow problems of the mixed elliptic-hyperbolic type [5]–[7] and elasto-plastic problems [8]. The analysis of these methods for the special case of second order systems is given in [9]–[10], and the goal of this paper is to extend these results to higher order systems.

To simplify the exposition attention will initially be confined to the following fourth order case for the region Ω in $\mathbb{R}^n$ having Γ as its boundary and $\boldsymbol{\nu}$ as the outer normal:

$$\boldsymbol{u}_0 - \nabla\varphi_0 = 0 \quad \text{in } \Omega, \tag{1.1}$$

$$\boldsymbol{u}_1 - \nabla\varphi_1 = 0 \quad \text{in } \Omega, \tag{1.2}$$

$$\operatorname{div} \boldsymbol{u}_0 - \varphi_1 = 0 \quad \text{in } \Omega, \tag{1.3}$$

$$\operatorname{div} \boldsymbol{u}_1 = f \quad \text{in } \Omega, \tag{1.4}$$

$$\varphi_0 = 0, \boldsymbol{u}_0 \cdot \boldsymbol{\nu} = 0 \quad \text{on } \Gamma. \tag{1.5}$$

This is the first order system associated with the biharmonic equation

$$\Delta^2\varphi_0 = f \quad \text{in } \Omega, \tag{1.6}$$

$$\varphi_0 = \frac{\partial\varphi_0}{\partial\nu} = 0 \quad \text{on } \Gamma, \tag{1.7}$$

where

$$\boldsymbol{u}_0 = \nabla\varphi_0, \quad \varphi_1 = \operatorname{div}\cdot\operatorname{grad}\varphi_0 = \Delta\varphi_0, \quad \boldsymbol{u}_1 = \operatorname{grad}\Delta\varphi_0. \tag{1.8}$$

The analysis for this special case will be done in such a manner that generalization to the $2m^{\text{th}}$ order case is routine. Details of how this extension is carried out is given in the final section.

One crucial assumptions that will be used throughout this paper deserves special mention here. This is the requirement that solutions to (1.1)–(1.5) (or equivalently (1.6)–(1.7)) have the standard regularity properties associated with elliptic systems. In particular, not only must the data f be sufficiently smooth but also the region Ω. (See Section 2 for details.) Our analysis and results do not apply for example to regions with re-entrant corners. In fact, it is known from the second order case [8], [11] that in such situations the entire variational formulation must be changed to get optimal approximation from finite element schemes based on least squares principles. Instead of working in standard Sobolev spaces (as is done in this paper) one works in appropriately weighted spaces. Analysis of this case will be given in a forthcoming paper.

The variational formulation and the assumptions needed to define the overall context is given in Section 2. The next section contains the basic mathematical content of the paper, and in particular the analysis of the approximations. The structure of the analysis is perhaps worth noting since it can be used not only for finite element schemes of the least squares type, but also other types of so-called mixed finite element methods [12]–[15]. Indeed, the first step is to estimate the divergence of the errors in the vector fields associated with the highest derivatives (for example $\boldsymbol{u}_1$) in dual or negative norms. These results lead to estimates for errors in the high order scalars (for example ε_1) in appropriate dual norms. At this point specific structural properties of the finite element spaces must be used (*e.g.* the discrete analog of the Helmholtz decomposition), and these plus earlier estimates give bounds for the errors in L_2 for all the variables (*i.e.*, φ_0, $\boldsymbol{u}_0$, φ_1, $\boldsymbol{u}_1$).

Implications of the error analysis for applications of the method are discussed in the last section. These will be related to the results obtained in numerical simulations reported elsewhere in the literature. Also in the last section the mathematical theory will be generalized to cover the case of $2m^{\text{th}}$ order systems.

§ 2. Variational Formulation

In the least squares approach one simply minimizes the residuals

$$\sum_{j=0}^{1} \int_{\Omega} |\boldsymbol{u}_j - \nabla\varphi_j|^2 + \int_{\Omega} \{|\operatorname{div} \boldsymbol{u}_0 - \varphi_1|^2 + |\operatorname{div} \boldsymbol{u}_1 - f|^2\} \tag{2.1}$$

as $\Phi = \{\varphi_0, \varphi_1, u_0, u_1\}$ varies over an appropriate finite dimensional subspace. For the analysis in the next section it is convenient recast the problem in terms of bilinear forms, a context familiar from Lax-Milgram theory [16].

To this end we use standard notation from Sobolev space theory. In particular let $H^r(\Omega)$ denote the rth order Sobolev space with its norm denoted by $\|\cdot\|_r$. We let $H_0^r(\Omega)$ be the subspace of $H^r(\Omega)$ obtained by taking the closure of C^∞ functions with compact support in Ω with respect to the norm $\|\cdot\|_r$. Finally,

$$H(\Omega; \operatorname{div}) = \{\boldsymbol{v} \in L^2(\Omega) \colon \operatorname{div} \boldsymbol{v} \in L^2(\Omega)\}, \tag{2.2}$$

and $H_0(\Omega;\text{div})$ is the subspace consisting in the vector fields $\boldsymbol{v} \in H(\Omega;\text{div})$ whose normal trace $\boldsymbol{v} \cdot \boldsymbol{\nu}$ vanishes on Γ.

Let

$$H := H_0^1(\Omega) \times H^1(\Omega) \times H(\Omega_0, \text{div}) \times H(\Omega, \text{div}). \tag{2.3}$$

Then for

$$\Phi = (\varphi_0, \varphi_1, \boldsymbol{u}_0, \boldsymbol{u}_1), \qquad \Psi = (\psi_0, \psi_1, \boldsymbol{v}_0, \boldsymbol{v}_1) \tag{2.4}$$

in H define the bilinear form $B(\cdot,\cdot)$ on $H \times H$ by

$$\begin{aligned} B(\Phi, \Psi) := \sum_{j=0}^{1} \int_\Omega (\boldsymbol{u}_j - \nabla\varphi_j) \cdot (\boldsymbol{v}_j - \Delta\psi_j) \\ + \int_\Omega \{(\text{div}\, \boldsymbol{u}_0 - \varphi_1)(\text{div}\, \boldsymbol{v}_0 - \psi_1) + \text{div}\, \boldsymbol{u}_1 \,\text{div}\, \boldsymbol{v}_1)\}. \end{aligned} \tag{2.5}$$

Observe that $B(\cdot,\cdot)$ is bounded on $H \times H$, and that (2.1) can be written

$$B(\Phi, \Phi) - 2 \int_\Omega f \,\text{div}\, \boldsymbol{u}_1 + \int_\Omega f^2. \tag{2.6}$$

Thus if we define the bounded linear functional $F(\cdot)$ on H by

$$F(\Psi) := \int_\Omega f \,\text{div}\, \boldsymbol{v}_1, \tag{2.7}$$

where Ψ is given by (2.4), then (1.1)–(1.5) is equivalent to finding a Φ in H such that

$$B(\Phi, \Psi) = F(\Psi) \tag{2.8}$$

holds for all Ψ in H.

To approximate we introduce finite dimensional finite element spaces to be used in place of the Sobolev spaces. In particular, let

$$\mathscr{S}_0(\boldsymbol{h}_0) \subseteq H_0^1(\Omega) \tag{2.9}$$

be the space for φ_0. Here $\boldsymbol{h}_0$ represents the vector whose components are the diameter of the simplicial subdivisions in the finite element grid. The magnitude of the largest such diameter is denoted by h_0. Similarly we introduce

$$\mathscr{S}_1(\boldsymbol{h}_1) \subseteq H^1(\Omega) \tag{2.10}$$

for the other scalar, and

$$\mathscr{V}_0(\boldsymbol{\delta}_0) \subset H_0(\Omega, \text{div}), \qquad \mathscr{V}_1(\boldsymbol{\delta}_1) \subset H(\Omega, \text{div}) \tag{2.11}$$

for $\boldsymbol{u}_0$ and $\boldsymbol{u}_1$. For $\boldsymbol{h} = (\boldsymbol{h}_0, \boldsymbol{h}_1, \boldsymbol{\delta}_0, \boldsymbol{\delta}_1)$ we define

$$H_h = \mathscr{S}_0(\boldsymbol{h}_0) \times \mathscr{S}_1(\boldsymbol{h}_1) \times \mathscr{V}_0(\boldsymbol{\delta}_0) \times \mathscr{V}_1(\boldsymbol{\delta}_1). \tag{2.12}$$

The approximation is to find a

$$\Phi_h \in H_h \tag{2.13}$$

such that

$$B(\Phi_h, \Psi_h) = F(\Psi_h) \tag{2.14}$$

for all $\Psi_h \in H_h$. The fact that both (2.8) and (2.14) have unique solutions (in H and H_h, respectively) is easy an consequence of the Riesz Representation Theorem [16] and the fact that

$$|||\Phi||| = B(\Phi, \Phi)^{\frac{1}{2}} \tag{2.15}$$

is a norm on H.

To complete this section we state the assumptions that will be needed about the system (1.1)–(1.5) as well as the finite element space H_h. As noted in the introduction the most important assumption concerns regularity. Given

$$f \in H^1(\Omega) \tag{2.16}$$

we assume that the solution

$$\Phi = \{\varphi_0, \varphi_1, \boldsymbol{u}_0, \boldsymbol{u}_1\} \tag{2.17}$$

satisfies

$$\|\varphi_0\|_5 \leqq C_R \|f\|_1 \tag{2.18}$$

for some constant C_R depending only on Ω. It follows from the theory of elliptic equations that (2.18) will hold, and more generally

$$\|\varphi_0\|_{5+r} \leqq C_R \|f\|_{r+1} \tag{2.19}$$

for $r \geqq 0$, provided Ω is sufficiently smooth [20]. Without this property not only is the analysis given in this paper invalid, but the method itself needs to be reformulated.

Two properties are needed for the finite element space H_h. The first is the standard approximation properties. In particular, we assume $\mathscr{S}_0(\boldsymbol{h}_0)$, $\mathscr{S}_1(\boldsymbol{h}_1)$, $\mathscr{V}_0(\boldsymbol{\delta}_0)$, $\mathscr{V}_1(\boldsymbol{\delta}_1)$ consist of piecewise polynomials of degree $k_0 - 1, k_1 - 1, l_0 - 1, l_1 - 1$, respectively. Thus from known results from approximation theory [21] it follows that for

$$\Psi = (\psi_0, \psi_1, \boldsymbol{v}_0, \boldsymbol{v}_1) \in H \cap \{H^{k_0}(\Omega) \times H^{k_1}(\Omega) \times H^{l_0}(\Omega) \times H^{l_1}(\Omega)\} \tag{2.20}$$

we have a constant $0 < C_A < \infty$ depending only on Ω for which the following holds:

$$\inf \|\psi_0 - \psi_{0,h}\|_r \leqq C_A h_0^{k_0 - r} \|\psi_0\|_{k_0} \quad \text{for } 0 \leqq r < k_0, \tag{2.21}$$

$$\inf \|\psi_1 - \psi_{1,h}\|_r \leqq C_A h_1^{k_1 - r} \|\psi_1\|_{k_1} \quad \text{for } 0 \leqq r < k_1, \tag{2.22}$$

$$\inf \|\boldsymbol{v}_0 - \boldsymbol{v}_{0,h}\|_r \leqq C_A \delta_0^{l_0 - r} \|\boldsymbol{v}_0\|_{l_0} \quad \text{for } 0 \leqq r < l_0, \tag{2.23}$$

$$\inf \|\boldsymbol{v}_1 - \boldsymbol{v}_{1,h}\|_r \leqq C_A \delta_1^{l_1 - r} \|\boldsymbol{v}_1\|_{l_1} \quad \text{for } 0 \leqq r < l_1. \tag{2.24}$$

The inf in the above is taken over $\mathscr{S}_0$, $\mathscr{S}_1$. $\mathscr{V}_0$, $\mathscr{V}_1$, respectively. Moreover, the estimates remain valid if in the right hand side k_0, k_1, l_0, l_1 are replaced with smaller integers greater than r.

The final property needed is the discrete analog of the Helmholtz decomposition, frequently called the discrete decomposition property (D.D.P.) for the spaces $\mathscr{V}_0(\boldsymbol{\delta}_0)$ and $\mathscr{V}_1(\boldsymbol{\delta}_1)$. Only selected finite element spaces have this property (see [17]–[19]), and numerical experience indicates that it necessary in least squares formulations for optimal performance (see [8], [11]). In particular we assume there is a number $0 < C_H < \infty$ such that the following holds for $\mathscr{V}$ being $\mathscr{V}_0(\boldsymbol{\delta}_0)$ or $\mathscr{V}_1(\boldsymbol{\delta}_1)$:

$$\text{(2.25)} \quad \begin{cases} \text{Given } \boldsymbol{v}_h \in \mathscr{V} \text{ there are } \boldsymbol{w}_h, \boldsymbol{z}_h \in \mathscr{V} \text{ such that} \\ \qquad \boldsymbol{v}_h = \boldsymbol{w}_h + \boldsymbol{z}_h \\ \text{and} \\ \qquad \operatorname{div} \boldsymbol{z}_h = 0 \text{ in } \Omega, \ \boldsymbol{z}_h \cdot \boldsymbol{\nu} = 0 \text{ on } \Gamma \\ \qquad \|\boldsymbol{w}_h\|_0 \leqq C_H \|\operatorname{div} \boldsymbol{v}_h\|_{-1}^*. \end{cases}$$

In the above the dual norm $\|\cdot\|_{-1}^*$ is defined by

$$\text{(2.26)} \qquad \|\psi\|_{-1}^* = \sup_{\theta \in H^1(\Omega)} \left\{ \int_\Omega \theta\psi / \|\theta\|_1 \right\}.$$

The superscript is used to distinguish this norm from the standard negative norm defined by

$$\text{(2.27)} \qquad \|\psi\|_{-1} = \sup_{\theta \in H_0^1(\Omega)} \left\{ \int_\Omega \psi\theta / \|\theta\|_1 \right\}.$$

For the infinite dimensional spaces $\mathscr{V} = H_0(\Omega, \operatorname{div})$ or $H(\Omega, \operatorname{div})$, the decomposition (2.25) is in fact the Helmholtz decomposition [18], and it is valid provided Ω is sufficiently smooth.

§ 3. Error Analysis

Let

$$\text{(3.1)} \qquad \Phi = \{\varphi_0, \varphi_1, \boldsymbol{u}_0, \boldsymbol{u}_1\}$$

denote the solution to (1.1)–(1.5) with

$$\text{(3.2)} \qquad \Phi_h = \{\varphi_{0,h}, \varphi_{1,h}, \boldsymbol{u}_{0,h}, \boldsymbol{u}_{1,h}\}$$

being the approximation defined by (2.14). Our task is to derive L_2 estimates for the errors

$$\text{(3.3)} \qquad E = \{\varepsilon_0, \varepsilon_1, \boldsymbol{e}_0, \boldsymbol{e}_1\} = \{\varphi_0 - \varphi_{0,h}, \varphi_1 - \varphi_{1,h}, \boldsymbol{u}_0 - \boldsymbol{u}_{0,h}, \boldsymbol{u}_1 - \boldsymbol{u}_{1,h}\}.$$

The first is an estimate for E in the norm $|||\cdot|||$ defined by (2.15).

Lemma 1. *For any Ψ_h in H_h we have*

$$\text{(3.4)} \qquad |||E||| \leqq |||\Phi - \Psi_h|||.$$

Proof. Equations (2.8) and (2.4) give

$$B(E, \Psi_h) = 0 \quad \text{for all } \Psi_h \text{ in } H_h. \tag{3.5}$$

The inequality (3.4) is an immediate consequence of this.

Corollary. *There is a number* $0 < C < \infty$ *depending only on the number* C_A *in* (2.11)–(2.24) *such that*

$$|||E||| \leqq C(h_0^{k_0-1}\|\varphi_0\|_{k_0} + h_1^{k_1-1}\|\varphi_1\|_{k_1} + \delta_0^{l_0-1}\|\boldsymbol{u}_0\|_{l_0} + \delta_1^{l_1-1}\|\boldsymbol{u}_1\|_{l_1}\}. \tag{3.6}$$

Proof. This follows directly from (3.4) and the results (2.21)–(2.24) from approximation theory.

Remark. The inequality (3.6) remains valid with $\|\varphi_1\|_{k_1}$ replaced with $\|\varphi_0\|_{k_1+2}$, $\|\boldsymbol{u}_0\|_{l_0}$ replaced with $\|\varphi_0\|_{l_0+1}$, and $\|\boldsymbol{u}_1\|_{l_1}$ replaced with $\|\varphi_0\|_{l_1+3}$. Moreover, since we are dealing only with smooth Ω, $\|\varphi_0\|_t$ can be replaced by $\|f\|_{t-4}$ for any $t \geqq 4$.

Estimates like (3.6) do not provide useful information by themselves; however, it is a starting point for a more refined analysis leading to L_2 estimates.

Lemma 2. *Let*

$$\sigma = h_0^{\min(k_0-1,4)} + \delta_0^{\min(l_0-1,3)} + h_1^{\min(k_1-1,2)}. \tag{3.7}$$

Then there is a number $0 < C < \infty$ *depending only on* C_A *and* C_R *for which*

$$\|\operatorname{div} \boldsymbol{e}_1\|_{-1}^* \leqq C(\delta_1 + \sigma)|||E|||. \tag{3.8}$$

Proof. Let

$$\theta \in H^1(\Omega) \tag{3.9}$$

be given. We solve the following system for $\Psi = \{\psi_0, \psi_1, \boldsymbol{v}_0, \boldsymbol{v}_1\}$:

$$\boldsymbol{v}_0 - \nabla\psi_0 = 0 \quad \text{in } \Omega, \tag{3.10}$$

$$\boldsymbol{v}_1 - \nabla\psi_1 = 0 \quad \text{in } \Omega, \tag{3.11}$$

$$\operatorname{div} \boldsymbol{v}_0 - \psi_1 = 0 \quad \text{in } \Omega, \tag{3.12}$$

$$\operatorname{div} \boldsymbol{v}_1 = \theta \quad \text{in } \Omega, \tag{3.13}$$

$$\psi_0 = \boldsymbol{v}_0 \cdot \boldsymbol{\nu} = 0 \quad \text{on } \Gamma. \tag{3.14}$$

Let

$$\Psi = \{\psi_0, \psi_1, \boldsymbol{v}_0, \boldsymbol{v}_1\}. \tag{3.15}$$

Then

$$B(E, \Psi) = \int_\Omega \theta \operatorname{div} \boldsymbol{e}_1$$

follows directly from (3.11)–(3.14) and the definition (2.5) of $B(\cdot,\cdot)$. On the other hand,

$$B(E, \Psi) = B(E, \Psi - \Psi_h) \tag{3.17}$$

for any Ψ_h in H_h by (3.5). Thus

$$\int_\Omega \theta \operatorname{div} \boldsymbol{e}_1 = B(E, \Psi - \Psi_h). \tag{3.18}$$

Taking the sup over $\theta \in H^1(\Omega)$ we obtain

$$\|\operatorname{div} \boldsymbol{e}_1\|_{-1}^* \leqq |||E||| \sup_{\theta \in H^1(\Omega)} \left\{\frac{|||\Psi - \Psi_h|||}{\|\theta\|_1}\right\}. \tag{3.19}$$

We now use both regularity (2.18) and approximation theory as expressed in (2.21)–(2.24) to conclude that Ψ_h in H_h can be chosen such that

$$\frac{|||\Psi - \Psi_h|||}{\|\theta\|_1} \leqq C(\delta_1 + \sigma), \tag{3.20}$$

where C depends only on C_A and C_R.

Lemma 3. *There is a number C depending only on C_A and C_R for which*

$$\|\operatorname{div} \boldsymbol{e}_0 - \varepsilon_1\|_{-1}^* \leqq C\, |||E|||\, (\delta_0 + h_0^{\min(k_0 - 1, 2)}). \tag{3.21}$$

Proof. Let $\theta \in H^1(\Omega)$ be given. The proof is similar to that for Lemma 2 except in that now we solve the second order problem

$$\Delta \psi_0 = \theta \quad \text{in } \Omega, \tag{3.22}$$

$$\psi_0 = 0 \quad \text{on } \Gamma, \tag{3.23}$$

putting $\boldsymbol{v}_0 = \nabla \psi_0$, $\psi_1 = 0$, and $\boldsymbol{v}_1 = \boldsymbol{0}$. Regularity results for this problem are like (1.6)–(1.7) controlled by the smoothness of Ω [20]. Hence, we have

$$\|\psi_0\|_3 = \|\boldsymbol{v}_0\|_2 \leqq C_R \|\theta\|_1. \tag{3.24}$$

Observe that for $\Psi = \{\psi_0, 0, \boldsymbol{v}_0, \boldsymbol{0}\}$ we also have

$$B(E, \Psi) = \int (\operatorname{div} \boldsymbol{e}_0 - \varepsilon_1) \operatorname{div} \boldsymbol{v}_0 = \int (\operatorname{div} \boldsymbol{e}_0 - \varepsilon_1)\, \theta. \tag{3.25}$$

But as before

$$B(E, \Psi) = B(E, \Psi - \Psi_h). \tag{3.26}$$

Thus

$$\|\operatorname{div} \boldsymbol{e}_0 - \varepsilon_1\|_{-1}^* \leqq |||E||| \sup_{\theta \in H^1(\Omega)} \frac{|||\Psi - \Psi_h|||}{\|\theta\|_1}. \tag{3.27}$$

Using (2.21)–(2.24), we select $\Psi_h = \{\psi_{0,h}, 0, \boldsymbol{v}_{0,h}, \boldsymbol{0}\}$ in H_h such that

$$\frac{|||\Psi - \Psi_h|||}{\|\theta\|_1} \leqq C(\delta_0 + h_0^{\min(k_0 - 1, 2)}), \tag{3.28}$$

which completes the estimate.

We are now prepared to state and prove our first L_2 estimate.

Lemma 4. *There is a number C depending on C_A, C_R and the constant C_H in the D.D.P.* (2.25) *such that*

$$\|e_1\|_0 \leqq C(\inf \|\operatorname{div}(u_1 - v_{1,h})\|^*_{-1} + \|\operatorname{div} e_1\|^*_{-1}), \tag{3.29}$$

where the inf *is taken over* $\mathscr{V}_1(\delta_1)$.

Proof. Let $v_{1,h}$ in $\mathscr{V}_1(\delta_1)$ be given, and put

$$e_1^h = u_{1,h} - v_{1,h}. \tag{3.30}$$

Use (2.25) to decompose e_1^h as follows:

$$e_1^h = w_h + z_h \tag{3.31}$$

where

$$\|w_h\|_0 \leqq C_H \|\operatorname{div} e_1^h\|^*_{-1}, \tag{3.32}$$

$$\operatorname{div} z_h = 0 \text{ in } \Omega, \quad z_h \cdot \nu = 0 \text{ on } \Gamma. \tag{3.33}$$

Since

$$\|\operatorname{div} e_1^h\|^*_{-1} \leqq \|\operatorname{div} e_1\|^*_{-1} + \|\operatorname{div}(u_1 - v_{1,h})\|^*_{-1}, \tag{3.34}$$

it is sufficient to estimate $\|z_h\|_0$. To do this we put $\Psi_h = \{0, 0, \mathbf{0}, z_h\}$ into

$$B(\Phi, \Psi_h) = B(\Phi_h, \Psi_h) \tag{3.35}$$

and use (3.34) to conclude that

$$\int \nabla\varphi_1 \, z_h = 0, \quad \int \nabla\varphi_{1,h} \, z_h = 0. \tag{3.36}$$

Then (3.33) reduces to

$$\int e_1^h \cdot z_h = \int (u_1 - v_{1,h}) \, z_h. \tag{3.37}$$

Thus (3.31) gives

$$\|z_h\|_0 \leqq \|u_1 - v_{1,h}\|_0 + \|w_h\|_0. \tag{3.38}$$

To complete the estimate we take the inf over $v_{1,h}$ in $\mathscr{V}_1(\delta_1)$ using (3.32) and (3.34).

The next result will be used in the proof of Lemma 5.

Lemma 5. *We have*

$$\int_\Omega \varepsilon_1 = 0. \tag{3.39}$$

Proof. Since the finite element space $\mathscr{S}_1(h_1)$ has no boundary conditions associated with it, it follows that the constant function is in this space:

$$1 \in \mathscr{S}_1(h_1). \tag{3.40}$$

Thus putting $\psi_{0,h} = 0$, $\boldsymbol{v}_{0,h} = \boldsymbol{v}_{1,h} = \mathbf{0}$, and $\psi_{1,h} = 1$ in

$$B(E, \Psi_h) = 0 \tag{3.41}$$

($\Psi_h = (0, 1, \mathbf{0}, \mathbf{0})$), we obtain

$$\int_\Omega (\operatorname{div} \boldsymbol{e}_0 - \varepsilon_1) = 0. \tag{3.42}$$

But $\boldsymbol{e}_0 \cdot \boldsymbol{\nu} = 0$ on Γ; hence

$$\int_\Omega \operatorname{div} \boldsymbol{e}_0 = 0, \tag{3.43}$$

which gives (3.39).

Lemma 6. *There is a number C depending only on C_A and C_R for which*

$$\|\varepsilon_1\|^*_{-1} \leqq C\{\delta_1 \, |||E||| + \|\boldsymbol{e}_1\|_0 + \|\operatorname{div} \boldsymbol{e}_1\|^*_{-1}\}. \tag{3.44}$$

Proof. Since ε_1 has zero mean it follows that $\|\varepsilon_1\|^*_{-1}$ is the sup of

$$\int_\Omega \theta\varepsilon_1 / \|\theta\|_1 \tag{3.45}$$

where θ varies on $H^1(\Omega)$ with zero mean, *i.e.*,

$$\int_\Omega \theta = 0. \tag{3.46}$$

Given such a function we can solve

$$\operatorname{div} \boldsymbol{v}_1 = \theta \quad \text{in } \Omega, \tag{3.47}$$

$$\boldsymbol{v}_1 \cdot \boldsymbol{\nu} = 0 \quad \text{on } \Gamma \tag{3.48}$$

with

$$\|\boldsymbol{v}_1\|_2 \leqq C_R \|\theta\|_1. \tag{3.49}$$

Putting $\boldsymbol{v}_0 = \mathbf{0}$, $\psi_0 = \psi_1 = 0$, we obtain (as before)

$$B(E, \Psi - \Psi_h) = B(E, \Psi) = \int \{[\boldsymbol{e}_1 - \nabla\varepsilon_1] \, \boldsymbol{v}_1 + \operatorname{div} \boldsymbol{e}_1 \operatorname{div} \boldsymbol{v}_1\} \tag{3.50}$$

for $\Psi_h = \{0, 0, \mathbf{0}, \boldsymbol{v}_{1,h}\} \in H_h$. Thus integrating by parts and using (3.48), we obtain

$$\int_\Omega \varepsilon_1\theta = -\int_\Omega \nabla\varepsilon_1 \cdot \boldsymbol{v}_1 = B(E, \Psi - \Psi_h) - \int_\Omega \{\boldsymbol{e}_1\boldsymbol{v}_1 + \operatorname{div} \boldsymbol{e}_1\theta\}. \tag{3.51}$$

Taking the sup over $\theta \in H^1(\Omega)$ satisfying (3.46), we obtain (3.44).

Remark. Using the triangle inequality we have

$$\|\operatorname{div} \boldsymbol{e}_0\|^*_{-1} \leqq \|\operatorname{div} \boldsymbol{e}_0 - \varepsilon_1\|^*_{-1} + \|\varepsilon_1\|^*_{-1}. \tag{3.52}$$

Thus Lemma 3 and 6 give a bound for $\|\operatorname{div} \boldsymbol{e}_0\|^*_{-1}$.

We now are in a position to make our final estimates.

Lemma 7. *There is a constant C depending only on C_H such that*

$$\|e_0\|_0 \leqq C\{\inf \|u_0 - v_{0,h}\|_0 + \|\operatorname{div} e_0\|^*_{-1}\}, \tag{3.53}$$

where the inf *is taken over* $\mathscr{V}_0(\delta_0)$.

Proof. Let $v_{0,h}$ in $\mathscr{V}_0(\delta_0)$ be given. Use the D.D.P. (2.25) to write

$$e_0^h := u_{0,h} - v_{0,h} = w_h + z_h, \tag{3.54}$$

where

$$\|w_h\|_0 \leqq C_H \|\operatorname{div}(e_0)\|^*_{-1}, \tag{3.55}$$

$$\operatorname{div} z_h = 0 \text{ in } \Omega, \quad z_h \cdot \nu = 0 \text{ on } \Gamma. \tag{3.56}$$

Putting $\Psi_h = \{0, 0, z_h, 0\}$ in

$$B(\Phi, \Psi_h) = B(\Phi_h, \Psi_h), \tag{3.57}$$

we obtain

$$\int_\Omega e_0^h z_h = \int_\Omega (u_0 - v_{0,h})\, z_h. \tag{3.58}$$

Thus

$$\|z_h\|_0 \leqq \|w_h\|_0 + \|u_0 - v_{0,h}\|_0. \tag{3.59}$$

Taking the inf over $v_{0,h}$ in $\mathscr{V}_0(\delta_0)$ gives (3.53).

Lemma 8. *There is a number C depending only on C_R and C_A such that*

$$\|\varepsilon_0\|_0 \leqq C(\|\operatorname{div} e_0\|_{-1} + h_0\, |||E|||). \tag{3.60}$$

Proof. Let $\psi_1 = 0$, $v_1 = v_0 = 0$, and solve

$$-\Delta\psi_0 = \varepsilon_0 \quad \text{in } \Omega, \tag{3.61}$$

$$\psi_0 = 0 \quad \text{on } \Gamma \tag{3.62}$$

for $\psi_0 \in H^2(\Omega)$. We have

$$\|\psi_0\|_2 \leqq C_R \|\varepsilon_0\|_0. \tag{3.63}$$

Moreover

$$\begin{aligned} B(E, \Psi - \Psi_h) = B(E, \Psi) &= \int (e_0 - \nabla\varepsilon_0)(-\nabla\psi_0) \\ &= \int_\Omega \psi_0 \operatorname{div} e_0 + \int \nabla\varepsilon_0 \cdot \nabla\psi_0 \\ &= \int_\Omega \psi_0 \operatorname{div} e_0 + \int_\Omega \varepsilon_0^2 \end{aligned} \tag{3.64}$$

for any $\Psi_h = \{\psi_{0,h}, 0, 0, 0\}$ in H_h. The estimate (3.60) follows from this as in preceding lemmas.

Lemma 9. *There is a constant C depending only on C_A and C_R for which*

$$\|\varepsilon_1\|_0 \leqq C(\|\operatorname{div} e_1\|_{-1}^* + \|\operatorname{div} e_0\|_{-1}^* + h_1 \,|||E|||). \tag{3.65}$$

Proof. Let $\psi_0 = 0$, $v_0 = v_1 = 0$ and solve for ψ_1 from

$$-\Delta\psi_1 + \psi_1 = \varepsilon_1 \text{ in } \Omega, \quad \frac{\partial\psi_1}{\partial\nu} = 0 \text{ on } \Gamma. \tag{3.66}$$

Then as above

$$B(E, \Psi - \Psi_h) = \int_\Omega B(E, \Psi] = (e_1 - \nabla\varepsilon_1)(-\nabla\psi_1) + (\operatorname{div} e_0 - \varepsilon_1)(-\psi_1) \tag{3.67}$$
$$= \int_\Omega \operatorname{div} e_1\psi_1 + \int_\Omega [\nabla\varepsilon_1 \nabla\psi_1 + \varepsilon_1\psi_1] - \int \operatorname{div} e_0\psi_1$$

for any $\Psi_h - = (0, \psi_{1,h}, \mathbf{0}, \mathbf{0})$ in H_h. After use of (3.67) this gives

$$\int_\Omega \varepsilon_1^2 = B(E, \Psi - \Psi_h) - \int_\Omega \operatorname{div} e_1\psi_1 + \int_\Omega \operatorname{div} e_0\psi_1. \tag{3.68}$$

Since regularity for (3.66) gives

$$\|\psi_1\|_2 \leqq C_R \|\varepsilon_1\|_0, \tag{3.69}$$

we have

$$\|\varepsilon_1\|_0^2 \leqq C[|||E|||\, h_1 \|\psi_1\|_2 + \|\operatorname{div} e_1\|_{-1} \|\psi_1\|_1 + \|\operatorname{div} e_0\|_{-1} \|\psi_1\|_1], \tag{3.70}$$

and hence (3.66).

§ 4. Discussion and Generalizations

The analysis in the previous section gives L_2 bounds in terms of $|||E|||$, where we recall that

$$|||E||| \leqq C(h_0^{k_0-1} \|f\|_{k_0-4} + \delta_0^{l_0-1} \|f\|_{l_0-3} + h_1^{k_1-1} \|f\|_{k_1-2} + \delta_1^{l_1-1} \|f\|_{l_1-1}), \tag{4.1}$$

with C depending only on the constants C_A and C_R. To complete the analysis we start with the scalar errors $\varepsilon_0 = \varphi_0 - \varphi_{0h}$, $\varepsilon_1 = \varphi_1 - \varphi_{1h}$. An important point first established by Jesperson [10] for the second order case and seen in numerical studies [8], [11] is the fact that the scalar error bounds do not depend on the number C_H, and hence are valid for arbitrary finite element spaces.

Theorem 1. *There is a number C depending only on C_A and C_R for which*

$$\|\varepsilon_0\|_0 \leqq C(h_0 + \delta_0 + h_1 + \delta_1)\, |||E||| \tag{4.2}$$

and

$$\|\varepsilon_1\|_0 \leqq C(\delta_0 + h_1 + \delta_1 + \sigma)\, |||E|||, \tag{4.3}$$

where

$$\sigma = h_0^{\min(k_0-1,2)} + \delta_0^{\min(l_0-1,3)} + h_1^{\min(k_1-1,2)}. \tag{4.4}$$

Proof. This is an immediate consequence of Lemma 2, Lemma 3, Lemma 8 and Lemma 9.

Observe that the error ε_0 in φ_0 is strongly coupled to the grids for the other variables. To reduce this error all grids must be refined. On the other hand, to improve the approximation for φ_1 it may be necessary to reduce only δ_0, h_1, δ_1 provided the other terms in the right hand side of (4.1) dominate the $O(h_0^{k_0-1})$ term. This explains conjectures that were presented on the basis of numerical experiments [8].

The errors for the vectors $\boldsymbol{e}_0 = \boldsymbol{u}_0 - \boldsymbol{u}_{0,h}$, $\boldsymbol{e}_1 = \boldsymbol{u}_1 - \boldsymbol{u}_{1,h}$ do depend on C_H, and hence are valid only for those finite element spaces for which this number is uniformly bounded independent of the grids (see Section 2).

Theorem 2. *This is a number C depending only on C_A, C_R, and C_H such that*

$$\|\boldsymbol{e}_0\|_0 \leqq C(\delta_0^{l_0-1}\|f\|_{l_0-3} + (\delta_0 + \delta_1 + h_1 + \sigma)\,|||E|||), \tag{4.5}$$

$$\|\boldsymbol{e}_1\|_0 \leqq C\delta_1^{l_1-1}\|f\|_{l_1-1} + (\delta_1 + \sigma)\,|||E|||) \tag{4.6}$$

where σ is given by (4.4).

Proof. This is a direct consequence of Lemmas 4 and 7 and the preceding Lemmas.

Observe that the error $\boldsymbol{e}_0$ in $\boldsymbol{u}_0$ is strongly coupled to all grids while the error $\boldsymbol{e}_1$ in $\boldsymbol{u}_1$ may or may not be coupled to the grids for φ_0 and $\boldsymbol{u}_0$.

There are a number of special cases of interest that have been used in practice. The first gives an analog of a standard finite element approximation [21] to the scalar equation (1.6)–(1.7). Here one chooses

$$h := h_0 = h_1 = \delta_0 = \delta_1 \tag{4.7}$$

with

$$k_0 = k, \quad l_0 = k - 1, \quad k_1 = k - 2, \quad l_1 = k - 3. \tag{4.8}$$

In this case $\varepsilon_0 = \varphi_0 - \varphi_{0,h}$ is of optimal order $O(h^k)$ in L_2; however this rate is reduced by 1 for each derivative; *i.e.*,

$$\|\varepsilon_0\|_0 = O(h^k), \quad \|\boldsymbol{e}_0\|_0 = O(h^{k-1}), \quad \|\varepsilon_1\| = O(h^{k-2}), \quad \|\boldsymbol{e}_1\|_0 = O(h^{k-3}). \tag{4.9}$$

Normally one would use a quintic polynomial ($k = 6$) in this case giving rates from $O(h^6)$ down to $O(h^3)$.

In problems like those mentioned in [6] where all of the variables φ_0, $\boldsymbol{u}_0$, φ_1, $\boldsymbol{u}_1$ are important, one typically chooses (4.7) with

$$k = k_0 = k_1 = l_0 = l_1. \tag{4.10}$$

In this case all L_2 erorrs are of order $O(h^k)$, and linear elements ($k = 2$) are commonly used in this context. The same rates can be achieved with (4.8) and with different grids satisfying

$$h_0 = h, \quad \delta_0 = O\left(h^{\frac{k}{k-1}}\right), \quad h_1 = O\left(h^{\frac{k}{k-2}}\right), \quad \delta_1 = O\left(h^{\frac{1}{k-3}}\right). \tag{4.11}$$

In elasticity the scalar φ_0 is typically not very interesting, it is related to a potential function. On the other hand, $\boldsymbol{u}_0$ (which is related to the stresses and strain and φ_1 (which is related to the bending moments) are often of great interest. As noted above the error $\boldsymbol{e}_0$ for the former is strongly coupled to all grids, hence in general one cannot expect reductions in $\|\boldsymbol{e}_0\|_0$ simply by reducing δ_0. All grids must be reduced to achieve this effect. Similarly to improve the error ε_1 in the latter, one must reduce δ_0 and δ_1 along with h_1 (and also possibly h_0).

Exactly the same trends hold for the $2m^{\text{th}}$ order case, which can be written as

$$\begin{bmatrix} I & G \\ D & C \end{bmatrix} \begin{bmatrix} \boldsymbol{u} \\ \varphi \end{bmatrix} = \boldsymbol{F}. \tag{4.12}$$

Here

$$\boldsymbol{u} = (\boldsymbol{u}_0, \ldots, \boldsymbol{u}_{m-1}), \quad \varphi = (\varphi_0, \ldots, \varphi_{m-1}) \tag{4.13}$$

with G being a matrix of gradients D being a matrix of divergences, and C being a bounded (in L_2) multiplication operator. Observe that the associated bilinear form $B(\cdot, \cdot)$ in this case is

$$B(\Phi, \Psi) = \int_\Omega \{|\boldsymbol{u} - G\varphi|^2 + |D\boldsymbol{u} + C\varphi|^2\}. \tag{4.14}$$

The error analysis precedes exactly as above. A bound for the error

$$E = \{\varphi - \varphi_h, \boldsymbol{u} - \boldsymbol{u}_h\} = \{\boldsymbol{e}, \varepsilon\} \tag{4.15}$$

in the norm

$$|||E||| = B(E, E)^{\frac{1}{2}} \tag{4.16}$$

is an immediate consequence of the variational statements (2.8) and (2.14). The next step is to solve appropriate adjoint problems to bound

$$\|De + C\varepsilon\|_{-1}^* \tag{4.17}$$

(as was done in Lemmas 2 and 3). Following this, bounds for

$$\|\varepsilon\|_0 \tag{4.18}$$

can be established which depend only on regularity and approximation but not the D.D.P. (2.25). The estimate are then complete by bounding

$$\|\boldsymbol{e}\|_0, \tag{4.19}$$

a result which will depend on the constant C_H in (2.25).

This research was supported by Army Research Office Grant DAAG29-83-K-0084.

References

1. J. H. BRAMBLE & A. H. SCHATZ, "Least Squares Method for $2m^{th}$-order Elliptic Boundary-value Problems," *Math. Comp.* **25** (1971) 1–32.
2. G. BAKER, "Simplified Proofs of Error Estimates for the Least Squares Method for Dirichlet's Problem," *Math. Comp.* **27** (1973) 229–235.
3. G. J. FIX & S. MARIN, "Variational Methods for Underwater Acoustic Problems," *Journal of Comp. Phys.* **28** (1978) 1–18.
4. G. J. FIX & R. A. NICOLAIDES, "An Analysis of Mixed Finite Element Approximations for Periodic Acoustic Wave Propagation," *SIAM J. Num. Anal.* **17**, No. 6 (1980) 779–786.
5. G. J. FIX & M. GURTIN, "On Patched Variational Methods," *Numerische Mathematik* **28** (1977) 259–271.
6. G. J. FIX & M. D. GUNZBURGER, "On Least Squares Approximation to Indefinite Problems of the Mixed Types," *Int'l. Journal for Numer. Mtd. in Eng.* **12** (1978) 453–470.
7. C. L. COX, G. J. FIX & M. D. GUNZBURGER, "A Least Squares Finite Element Scheme for Transonic Flow around Harmonically Oscillating Wings," *Journal of Comp. Physics* **51**, No. 3 (Sept. 1983) 387–403.
8. Y. LEE, *Shear Bands in Elastic-Perfectly Plastic Media*, Ph. D. thesis, Carnegie-Mellon University (1981).
9. G. J. FIX, M. D. GUNZBURGER, & R. A. NICOLAIDES, "Least Squares Finite Element Methods," NASA-ICASE Report 77-18, revised version published in *Math. and Comp. with Appls.* **5** (1979) 87–98.
10. D. C. JESPERSON, "On Least Squares Decomposition of Elliptic Boundary Value Problems," *Math. Comp.* **31** (1977).
11. C. COX & G. J. FIX, "On the Accuracy of Least Squares Methods in the Presence of Corner Singularities," *Computers and Mathematics with Applications*, to appear.
12. G. J. FIX, M. D. GUNZBURGER, & R. A. NICOLAIDES, "On Mixed Finite Element Methods for First Order Elliptic Systems, *Numerische Mathematik* **37** (1981) 29–48.
13. G. J. FIX, "On the Structure of Errors in Mixed Finite Element Methods," Nonlinear Partial Differential Equations in Engineering and Applied Science, Proceedings of a Conference Sponsored by Office of Naval Research, University of Rhode Island, R. STERNBERG, ed., M. Dekker, Inc. (1980) 53–86.
14. G. J. FIX, M. D. GUNZBURGER, & R. A. NICOLAIDES, "Theory and Applications of Mixed Finite Element Methods, paper in *Constructive Approaches to Mathematical Models* (by C. COFFMAN and G. J. FIX) Academic Press (1979) 375–393.
15. G. J. FIX, M. D. GUNZBURGER, & R. A. NICOLAIDES, "On Mixed Finite Element Methods for a Class of Nonlinear Boundary Value Problems," *Computational Methods in Nonlinear Methods*, T. ODEN, ed., North Holland (1980) 245–260.
16. K. YOSIDA, *Functional Analysis*, Academic Press (1965).
17. G. J. FIX, "A Survey of Numerical Methods for Selected Problems in Continuum Mechanics," Proceedings of Numerical Methods for Ocean-Circulations, National Academy of Sciences, October (1974) 268–283.
18. V. GIRAULT & P.-A. RAVIART, "Finite Element Approximation of the Navier-Stokes Equations," in Lecture Notes in Mathematics (A. DOLD & B. ECKMANN, eds.), Springer-Verlag, New York (1979).
19. J. M. BOLAND, *Finite Elements and the Divergence Constraint for Viscous Flows*, Ph. D. Thesis, Dept. of Mathematics Carnegie-Mellon University (1983).

20. J. L. Lions & E. Magenes, *Nonhomogeneous Boundary Value Problems*, Springer-Verlag, New York (1973).
21. G. Strang & G. J. Fix, *An Analysis of the Finite Element Method*, Prentice-Hall, Englewood Cliffs, New Jersey (1973).

Department of Mathematics
Carnegie-Mellon University
Pittsburgh

&

School of Mathematics
Georgia Institute of Technology
Atlanta

(Received December 1, 1984)

Development of Singularities in the Motion of Materials with Fading Memory

C. M. DAFERMOS

Dedicated to Walter Noll on his sixtieth birthday

1. Introduction

Consider a one-dimensional body with homogeneous reference configuration the axis $(-\infty, \infty)$ and reference density $\varrho_0 = 1$. The typical material *particle* is denoted by x. The relevant kinematic variables are *deformation gradient* u and *velocity* v. The Piola-Kirchhoff *stress* is denoted by σ and there is no body force.

When the body is *elastic*, with constitutive relation

$$\sigma(x, t) = p(u(x, t)), \tag{1.1}$$

the equations of motion read

$$\begin{aligned} \partial_t u - \partial_x v &= 0 \\ \partial_t v - p_u(u)\, \partial_x u &= 0. \end{aligned} \tag{1.2}$$

We postulate that

$$p_u(u) > 0, \tag{1.3}$$

so that the system (1.2) is strictly hyperbolic.

The amplitude of an acceleration wave propagating into the elastic body in a state of homogeneous strain, say $u = \bar{u} = \text{constant}$, $v = 0$, satisfies the ordinary differential equation

$$\dot{a} + \frac{p_{uu}(\bar{u})}{2p_u^{\frac{3}{2}}(\bar{u})}\, a^2 = 0. \tag{1.4}$$

Therefore, if

$$p_{uu}(\bar{u}) > 0 \tag{1.5}$$

and $a(0) < 0$ then the amplitude eventually explodes and, presumably, the acceleration wave breaks into a shock wave.

The above discussion suggests that motions of nonlinear elastic bodies, starting out from smooth initial states, generally develop singularities in a finite time. That this is indeed the case can be demonstrated by the following elegant argument of Lax [9]: We introduce *Riemann Invariants* of (1.2) by

$$r := v + \int_{\bar{u}}^{u} p_u^{\frac{1}{2}}(\xi)\, d\xi, \qquad s := v - \int_{\bar{u}}^{u} p_u^{\frac{1}{2}}(\xi)\, d\xi \tag{1.6}$$

and apply on them the operators of differentiation along the characteristic directions,

$$\backslash := \partial_t - p_u^{\frac{1}{2}}(u)\, \partial_x, \qquad / := \partial_t + p_u^{\frac{1}{2}}(u)\, \partial_x, \tag{1.7}$$

to get

$$\overset{\backslash}{r} = 0, \qquad \overset{/}{s} = 0. \tag{1.8}$$

Thus r stays constant along backward characteristics and s stays constant along forward characteristics. Consequently, so long as a classical solution of (1.2) exists, the functions r, s, and thereby also u, v, will stay bounded by constants that depend solely on the L^∞ norm of the initial values of deformation gradient and velocity. On the other hand, an interesting calculation shows that the partials $\partial_x r$, $\partial_x s$ of the Riemann Invariants evolve along characteristics according to the ordinary differential equations

$$\begin{aligned} \left[p_u^{\frac{1}{4}}(u)\, \partial_x r\right]^{\backslash} - \tfrac{1}{4} p_u^{-\frac{5}{4}}(u)\, p_{uu}(u) \left[p_u^{\frac{1}{4}}(u)\, \partial_x r\right]^2 &= 0 \\ \left[p_u^{\frac{1}{4}}(u)\, \partial_x s\right]^{/} - \tfrac{1}{4} p_u^{-\frac{5}{4}}(u)\, p_{uu}(u) \left[p_u^{\frac{1}{4}}(u)\, \partial_x s\right]^2 &= 0 \end{aligned} \tag{1.9}$$

and so, if p_{uu} does not change sign on the range of the solution, then $\partial_x r$, $\partial_x s$, and thereby also the first partials of u, v, will generally blow up in finite time. It is known that solutions break down even when p_{uu} changes sign [11, 8].

Our aim here is to compare the behavior of elastic materials with the behavior of materials with fading memory, in which the destabilizing influence of nonlinear elastic response coexists and competes with the damping action of memory response. Let us assume that the stress σ at the particle x and time t is determined by the history of deformation gradient u at x, relative to t, i.e., the material is *simple* in the sense of Noll [13]. For our purposes it is expedient to adopt a somewhat unconventional formulation, visualizing the history of u at x, relative to t, as the pair $\{u(x, t), u^{(t)}(x, \cdot)\}$ of the *present value* $u(x, t)$ and the *past history* $u^{(t)}(x, \cdot)$, defined by

$$u^{(t)}(x, \tau) := u(x, t - \tau), \qquad 0 < \tau < \infty. \tag{1.10}$$

Therefore, the constitutive equation reads

$$\sigma(x, t) = \mathscr{S}\{u(x, t), u^{(t)}(x, \cdot)\}. \tag{1.11}$$

In the spirit of the characterization of materials with *fading memory* by Coleman & Noll [2], we assume that $\mathscr{S}$ is defined as a continuously Fréchet differentiable functional on the Banach space $\mathbb{R} \times L_h^1(0, \infty)$*. Here h stands for a fixed positive-valued, decreasing, integrable *influence* function on $[0, \infty)$ and L_h^1 denotes the L^1 space weighted by h.

The Fréchet derivative of $\mathcal{S}$ at $\{u, w(\cdot)\} \in \mathbb{R} \times L_h^1(0, \infty)$ admits the Riesz representation

$$D\mathcal{S}\{u, w(\cdot); f, g(\cdot)\} = \mathcal{S}_u\{u, w(\cdot)\} f + \int_0^\infty \mathcal{K}\{u, w(\cdot), \tau\} g(\tau) \, d\tau \tag{1.12}$$

with $\mathcal{K}\{u, w(\cdot), \cdot\} \in L_{h^{-1}}^\infty(0, \infty)$, and so the equations of motion can be written in the form

$$\begin{gathered} \partial_t u - \partial_x v = 0 \\ \partial_t v - \mathcal{S}_u\{u, u^{(t)}\} \partial_x u = \int_{-\infty}^t \mathcal{K}\{u, u^{(t)}, t - \tau\} \partial_x u \, d\tau. \end{gathered} \tag{1.13}$$

We further require that $\mathcal{S}_u$ be Fréchet differentiable on $\mathbb{R} \times L_h^1(0, \infty)$ so its Fréchet derivative at each $\{u, w(\cdot)\}$ admits a Riesz representation $\{\mathcal{S}_{uu}\{u, w(\cdot)\}, \mathcal{M}\{u, w(\cdot), \cdot\}\}$ in $\mathbb{R} \times L_{h^{-1}}^\infty(0, \infty)$. We assume that $\mathcal{K}$ and $\mathcal{M}$ are Fréchet differentiable on the subset $\mathbb{R} \times L_h^1(0, \infty) \times \mathbb{R}^+$ of the Banach space $\mathbb{R} \times L_h^1(0, \infty) \times \mathbb{R}$. We require that $\mathcal{S}_u, \mathcal{S}_{uu}, \mathcal{K}, \mathcal{M}$, and the Fréchet derivatives of $\mathcal{K}, \mathcal{M}$ be uniformly bounded on every closed bounded subset of their domain. Moreover, in order to ensure that the modulus of instantaneous elastic response is positive, we postulate

$$\mathcal{S}_u\{u, w(\cdot)\} > 0, \qquad \{u, w(\cdot)\} \in \mathbb{R} \times L_h^1(0, \infty). \tag{1.14}$$

A simple example of a material with memory is provided by the constitutive equation

$$\sigma(x, t) = p(u(x, t)) + \int_0^\infty K(\tau) \, q(u(x, t - \tau)) \, d\tau, \tag{1.15}$$

where K is an $L^1(0, \infty)$ kernel which is bounded and differentiable on $[0, \infty)$, p and q are smooth functions and p satisfies (1.3). It conforms with the conditions of smoothness imposed above when $q(u) \equiv u$. The equations of motion for (1.15) take the form

$$\begin{gathered} \partial_t u - \partial_x v = 0 \\ \partial_t v - p_u(u) \, \partial_x u = \int_{-\infty}^t K(t - \tau) \, q_u(u) \, \partial_x u \, d\tau. \end{gathered} \tag{1.16}$$

In particular, when $p \equiv q$ (1.16) reduces to the special system

$$\begin{gathered} \partial_t u - \partial_x v = 0 \\ \partial_t v - p_u(u) \, \partial_x u = \int_{-\infty}^t K(t - \tau) \, p_u(u) \, \partial_x u \, d\tau, \end{gathered} \tag{1.17}$$

which was first considered by MACCAMY [10].

---

* Since the deformation gradient assumes positive values, the natural domain of $\mathcal{S}$ is the cone of positive-valued histories in this Banach space. Here we will be dealing with histories with range lying in a small neighborhood of a fixed strain $\bar{u} > 0$. Therefore, to avoid technical complications we tacitly assume that $\mathcal{S}$ admits a smooth extension on the entire space.

The theory of propagation of acceleration waves in materials with fading memory is discussed by COLEMAN & GURTIN [1], in great detail and generality. In the present setting, on the assumption that the material ahead of the wave front has been in a state of homogeneous strain, say $u = \bar{u} =$ constant, $v = 0$, at all times prior to the arrival of the wave, the amplitude satisfies the Bernoulli differential equation

$$(1.18)\qquad \dot{a} + \frac{\mathscr{S}_{uu}\{\bar{u}, ''\bar{u}''\}}{2\mathscr{S}_u^{\frac{3}{2}}\{\bar{u}, ''\bar{u}''\}} a^2 - \frac{\mathscr{K}\{\bar{u}, ''\bar{u}'', 0\}}{2\mathscr{S}_u\{\bar{u}, ''\bar{u}''\}} a = 0,$$

where $''\bar{u}''$ denotes the constant past history equal to $\bar{u}$. It follows from (1.18) that if the instantaneous elastic response is nonlinear, say,

$$(1.19)\qquad \mathscr{S}_{uu}\{\bar{u}, ''\bar{u}''\} > 0,$$

but the memory response is dissipative, in that $\mathscr{K}\{\bar{u}, ''\bar{u}'', 0\} < 0$, then the amplitude of the acceleration wave does not explode unless its initial strength exceeds a threshold value.

The above discussion suggests that motions of materials with fading memory, which start out from an initial history near a state of homogeneous strain, stay smooth for all time while motions that originate from initial histories with substantially nonhomogeneous strain may develop singularities in finite time. The first part of this conjecture, namely the existence of globally defined smooth motions with initial history in a small neighborhood of a state of homogeneous strain, was established by HRUSA [7], following preparatory work by several authors [10, 5]. Here we take up the question of breakdown of smooth motions. For preliminary results in this direction, see [6, 12, 3].

For simplicity, let us assume the body has been in a state of homogeneous strain

$$(1.20)\qquad u(x, t) = \bar{u} = \text{constant}, \qquad -\infty < x < \infty, -\infty < t \leqq 0,$$

and the motion is generated by an impulse at $t = 0$. This induces initial conditions

$$(1.21)\qquad u(x, 0) = \bar{u}, \quad v(x, 0) = v_0(x), \qquad -\infty < x < \infty.$$

Our program is to show that when v_0 is suitably selected, classical solutions of (1.13), (1.20), (1.21) explode in a finite time. The time of explosion gets shorter as $\max \partial_x v_0$ increases. (When the direction of the inequality (1.19) is reversed, it is $\min \partial_x v_0$ that controls the time of explosion.) The precise statement of the result is

Theorem 1.1. *Assume $\mathscr{S}$ satisfies the hypotheses of smoothness stated above as well as the inequalities* (1.14), (1.19). *Given any positive numbers N and T, there is a positive number ε, depending only on $\bar{u}$, and a positive number M, depending on $\bar{u}$, ε, N, T, such that when v_0 is a C^2 function with compact support in* $(-\infty, \infty)$

which satisfies

(1.22) $$|v_0(x)| < \varepsilon, \quad -\infty < x < \infty,$$

(1.23) $$-\partial_x v_0(x) < N, \quad -\infty < x < \infty,$$

(1.24) $$\max_{(-\infty,\infty)} \partial_x v_0(\cdot) > M,$$

then the length of the maximal time interval of existence of any classical solution of (1.13), (1.20), (1.21) *cannot exceed* T.

The general strategy in the proof of the above theorem is to show that the term on the right-hand side of $(1.13)_2$ is subordinate to the other terms and hence that the behavior of solutions of system (1.13) parallels the behavior of solutions of the hyperbolic system (1.2). To this end, we have to demonstrate that in the term

(1.25) $$\int_{-\infty}^{t} \mathcal{K}\{u, u^{(t)}, t-\tau\} \, \partial_x u \, d\tau$$

ntegration with respect to τ offsets differentiation with respect to x. Roughly, his is so because, in first approximation, all waves travel along characteristics and hence x-derivatives and t-derivatives are approximately proportional. The compensating of x-differentiation with t-integration becomes clear in the context of the special system (1.17). Indeed, $(1.17)_2$ may be viewed as a linear Volterra integral equation operating on $p_u(u)\,\partial_x u$. Upon inverting this integral operator, by means of the resolvent kernel L, we may rewrite (1.17) in the equivalent form (*cf.* [10]):

(1.26)
$$\begin{aligned} &\partial_t u - \partial_x v = 0 \\ &\partial_t v - p_u(u)\,\partial_x u = K(0)\, v - \int_0^t L'(t-\tau)\, v \, d\tau + L(t)\, v_0 . \end{aligned}$$

It is now clear that the right-hand side of $(1.26)_2$ is of lower order than the other terms and so it is plausible that solutions of (1.26) with large initial data break down in finite time, just as solutions of (1.2) do. This was indeed verified by HATTORI [6].

In the absence of the special structure of (1.17), the proof of the theorem becomes considerably harder. It is still possible to write down equations, analogous to (1.8), (1.9), that govern the evolution of Riemann Invariants and their derivatives along characteristics but these are now strongly coupled, and so they can no longer be treated as independent ordinary differential equations. We overcome this difficulty by a procedure developed in [3] which aims at monitoring the time-evolution of the functions $\max|r(\cdot, t)|$, $\max|s(\cdot, t)|$, $\max \partial_t r(\cdot, t)$, $\max \partial_t s(\cdot, t)$, $\min \partial_t r(\cdot, t)$ and $\min \partial_t r(\cdot, t)$. A parallel approach is followed in [4] to show that singularities also develop in the motion of thermoelastic bodies.

In order to avoid cumbersome notation, stemming from the need to write down representations for the Fréchet derivative of several functionals, I will present the proof of Theorem 1.1 in the context of the model system (1.16)*. The

* A seemingly similar result for this system was announced recently by NOHEL & RENARDY (paper in preparation).

reader will discern without difficulty that the approach is general and, guided by a rough sketch at the end of Section 2, will be able, if he so wishes, to carry out the proof in detail for the general system (1.13) under the stated assumptions of smoothness. In fact it will become clear in the course of the proof that the assertion of the theorem is valid under substantially weaker hypotheses which, however, are too complicated to state explicitly in advance. In particular, the result holds for the material models used in practice such as the BKZ fluid.

2. Proof of Theorem 1.1

As stated in the Introduction, we will prove Theorem 1.1 in the context of system (1.16). Hypotheses (1.14), (1.19) now reduce to (1.3), (1.5).

Assume (u, v) is a C^1-solution of (1.16), (1.20), (1.21) on a strip $(-\infty, \infty) \times [0, T]$. Insofar as the initial data are in C^2, it follows by standard analysis that (u, v) is C^2 on $(-\infty, \infty) \times [0, T]$. Moreover, for any $t \in [0, T]$ the functions $u(\cdot, t) - \bar{u}$ and $v(\cdot, t)$ have compact support in $(-\infty, \infty)$.

Let us consider the Riemann Invariants r, s, defined by (1.6), and apply on them the differentiation operators (1.7). After use of (1.16), the calculation yields

$$
\begin{aligned}
\overset{\backslash}{r} &= \int_0^t K(t-\tau)\, q_u(u)\, \partial_x u \, d\tau \\
\overset{/}{s} &= \int_0^t K(t-\tau)\, q_u(u)\, \partial_x u \, d\tau .
\end{aligned}
\tag{2.1}
$$

Since the high-order term $\partial_x u$ appears on the right-hand sides of (2.1), the only hope for success of our program lies in offsetting x-differentiation with t-integration.

On $(-\infty, \infty) \times [0, T]$ we define functions

$$
\phi := p_u^{-\frac{1}{4}}(u)\, \partial_t r, \qquad \psi := -p_u^{-\frac{1}{4}}(u)\, \partial_t s . \tag{2.2}
$$

From (2.2), (1.6), it follows that

$$
\partial_t u = \tfrac{1}{2} p_u^{-\frac{1}{4}}(u)\, (\phi + \psi), \qquad \partial_t v = \tfrac{1}{2} p_u^{\frac{1}{4}}(u)\, (\phi - \psi). \tag{2.3}
$$

Combining (2.3) with (1.16) we obtain

$$
p_u^{\frac{1}{2}}(u)\, \partial_x u + p_u^{-\frac{1}{2}}(u) \int_0^t K(t-\tau)\, \frac{q_u(u)}{p_u^{\frac{1}{2}}(u)}\, p_u^{\frac{1}{2}}(u)\, \partial_x u \, d\tau = \tfrac{1}{2} p_u^{-\frac{1}{4}}(u)\, (\phi - \psi). \tag{2.4}
$$

We realize the left-hand side of (2.4) as a linear Volterra integral operator operating on $p_u^{\frac{1}{2}}(u)\, \partial_x u$. Inverting this operator yields

$$
p_u^{\frac{1}{2}}(u)\, \partial_x u = \tfrac{1}{2} p_u^{-\frac{1}{4}}(u)\, (\phi - \psi) + \int_0^t \mathscr{L}\{u, u^{(t)}, t-\tau\}\, (\phi - \psi)\, d\tau , \tag{2.5}
$$

where $\mathscr{L}$ is a functional defined on $\mathbb{R} \times L^\infty(0, \infty) \times \mathbb{R}^+$ and continuous in the norm of the Banach space $\mathbb{R} \times L^\infty(0, \infty) \times \mathbb{R}$. From (1.7), (2.3) and (2.5) we deduce

$$
\begin{aligned}
\overset{\backslash}{u} &= p_u^{-\frac{1}{4}}(u)\, \psi - \int_0^t \mathscr{L}\{u, u^{(t)}, t-\tau\}\, (\phi - \psi)\, d\tau \\
\overset{/}{u} &= p_u^{-\frac{1}{4}}(u)\, \phi + \int_0^t \mathscr{L}\{u, u^{(t)}, t-\tau\}\, (\phi - \psi)\, d\tau .
\end{aligned}
\tag{2.6}
$$

Substituting $\partial_x u$ from (2.5) into (2.1) yields

$$
\begin{aligned}
(2.7)_{\mathrm{r}} \qquad \grave{r} &= \tfrac{1}{2} \int_0^t K(t-\tau)\, p_u^{-\frac{3}{4}}(u)\, q_u(u)\, (\phi - \psi)\, d\tau \\
&\quad + \int_0^t \left\{ \int_\tau^t K(t-\xi)\, \mathscr{L}\{u, u^{(\xi)}, \xi - \tau\}\, p_u^{-\frac{1}{2}}(u)\, q_u(u)\, d\xi \right\} (\phi - \psi)\, d\tau,
\end{aligned}
$$

$$
\begin{aligned}
(2.7)_{\mathrm{s}} \qquad \acute{s} &= -\tfrac{1}{2} \int_0^t K(t-\tau)\, p_u^{-\frac{3}{4}}(u)\, q_u(u)\, (\psi - \phi)\, d\tau \\
&\quad - \int_0^t \left\{ \int_\tau^t K(t-\xi)\, \mathscr{L}\{u, u^{(\xi)}, \xi - \tau\}\, p_u^{-\frac{1}{2}}(u)\, q_u(u)\, d\xi \right\} (\psi - \phi)\, d\tau
\end{aligned}
$$

whence

$$
\begin{aligned}
(2.8)_{\mathrm{r}} \qquad p_u^{\frac{1}{2}}(u)\, \partial_x r &= p_u^{\frac{1}{4}}(u)\, \phi - \tfrac{1}{2} \int_0^t K(t-\tau)\, p_u^{-\frac{3}{4}}(u)\, q_u(u)\, (\phi - \psi)\, d\tau \\
&\quad - \int_0^t \left\{ \int_\tau^t K(t-\xi)\, \mathscr{L}\{u, u^{(\xi)}, \xi - \tau\}\, p_u^{-\frac{1}{2}}(u)\, q_u(u)\, d\xi \right\} (\phi - \psi)\, d\tau,
\end{aligned}
$$

$$
\begin{aligned}
(2.8)_{\mathrm{s}} \qquad p_u^{\frac{1}{2}}(u)\, \partial_x s &= p_u^{\frac{1}{4}}(u)\, \psi - \tfrac{1}{2} \int_0^t K(t-\tau)\, p_u^{-\frac{3}{4}}(u)\, q_u(u)\, (\psi - \phi)\, d\tau \\
&\quad - \int_0^t \left\{ \int_\tau^t K(t-\xi)\, \mathscr{L}\{u, u^{(\xi)}, \xi - \tau\}\, p_u^{-\frac{1}{2}}(u)\, q_u(u)\, d\xi \right\} (\psi - \phi)\, d\tau.
\end{aligned}
$$

We proceed to compute $\grave{\phi}$ and $\acute{\psi}$. Combining (2.2), (1.7), (2.3), (2.6), (2.7), (2.8) and after a lengthy but straightforward calculation we get

$$
\begin{aligned}
(2.9)_{\phi} \qquad \grave{\phi} &= -\tfrac{1}{4} p_u^{-1}(u)\, p_{uu}(u)\, \grave{u} \phi + p_u^{-\frac{1}{4}}(u)\, \partial_t \grave{r} + \tfrac{1}{2} p_u^{-\frac{3}{4}}(u)\, p_{uu}(u)\, \partial_t u\, \partial_x r \\
&= \tfrac{1}{4} p_u^{-\frac{5}{4}}(u)\, p_{uu}(u)\, \phi^2 + \tfrac{1}{2} K(0)\, p_u^{-1}(u)\, q_u(u)\, (\phi - \psi) \\
&\quad + K(0)\, p_u^{-\frac{3}{4}}(u)\, q_u(u) \int_0^t \mathscr{L}\{u, u^{(t)}, t - \tau\}\, (\phi - \psi)\, d\tau \\
&\quad + \tfrac{1}{4} p_u^{-1}(u)\, p_{uu}(u) \left\{ \int_0^t \mathscr{L}\{u, u^{(t)}, t - \tau\}\, (\phi - \psi)\, d\tau \right\} \phi \\
&\quad + \tfrac{1}{2} p_u^{-\frac{1}{4}}(u) \int_0^t K'(t-\tau)\, p_u^{-\frac{3}{4}}(u)\, q_u(u)\, (\phi - \psi)\, d\tau \\
&\quad + p_u^{-\frac{1}{4}}(u) \int_0^t \left\{ \int_\tau^t K'(t-\xi)\, \mathscr{L}\{u, u^{(\xi)}, \xi - \tau\}\, p_u^{-\frac{1}{2}}(u)\, q_u(u)\, d\xi \right\} (\phi - \psi)\, d\tau \\
&\quad - \tfrac{1}{8} p_u^{-\frac{3}{2}}(u)\, p_{uu}(u) \left\{ \int_0^t K(t-\tau)\, p_u^{-\frac{3}{4}}(u)\, q_u(u)\, (\phi - \psi)\, d\tau \right\} (\phi + \psi) \\
&\quad - \tfrac{1}{4} p_u^{-\frac{3}{2}}(u)\, p_{uu}(u) \left\{ \int_0^t \left\{ \int_\tau^t K(t-\xi)\, \mathscr{L}\{u, u^{(\xi)}, \xi - \tau\} \right.\right. \\
&\quad \left.\left. \times\, p_u^{-\frac{1}{2}}(u)\, q_u(u)\, d\xi \right\} (\phi - \psi)\, d\tau \right\} (\phi + \psi),
\end{aligned}
$$

$$(2.9)_\psi \qquad \acute{\psi} = -\tfrac{1}{4}p_u^{-1}(u)\,p_{uu}(u)\,\acute{u}\psi - p_u^{-\frac{1}{4}}(u)\,\partial_t \acute{s} + \tfrac{1}{2}p_u^{-\frac{3}{4}}(u)\,p_{uu}(u)\,\partial_t u\,\partial_x s$$

$$= \tfrac{1}{4}p_u^{-\frac{5}{4}}(u)\,p_{uu}(u)\,\psi^2 + \tfrac{1}{2}K(0)\,p_u^{-1}(u)\,q_u(u)\,(\psi - \phi)$$

$$+ K(0)\,p_u^{-\frac{3}{4}}(u)\,q_u(u)\int_0^t \mathscr{L}\{u, u^{(t)}, t-\tau\}\,(\psi - \phi)\,d\tau$$

$$+ \tfrac{1}{4}p_u^{-1}(u)\,p_{uu}(u)\left\{\int_0^t \mathscr{L}\{u, u^{(t)}, t-\tau\}\,(\psi - \phi)\,d\tau\right\}\psi$$

$$+ \tfrac{1}{2}p_u^{-\frac{1}{4}}(u)\int_0^t K'(t-\tau)\,p_u^{-\frac{3}{4}}(u)\,q_u(u)\,(\psi - \phi)\,d\tau$$

$$+ p_u^{-\frac{1}{4}}(u)\int_0^t \left\{\int_\tau^t K'(t-\xi)\,\mathscr{L}\{u, u^{(\xi)}, \xi-\tau\}\,p_u^{-\frac{1}{2}}(u)\,q_u(u)\,d\xi\right\}(\psi - \phi)\,d\tau$$

$$- \tfrac{1}{8}p_u^{-\frac{3}{2}}(u)\,p_{uu}(u)\left\{\int_0^t K(t-\tau)\,p_u^{-\frac{3}{4}}(u)\,q_u(u)\,(\psi - \phi)\,d\tau\right\}(\psi + \phi)$$

$$- \tfrac{1}{4}p_u^{-\frac{3}{2}}(u)\,p_{uu}(u)\left\{\int_0^t \left\{\int_\tau^t K(t-\xi)\,\mathscr{L}\{u, u^{(\xi)}, \xi-\tau\}\right.\right.$$

$$\left.\left.\times\, p_u^{-\frac{1}{2}}(u)\,q_u(u)\,d\xi\right\}(\psi - \phi)\,d\tau\right\}(\psi + \phi).$$

In order to estimate the solution, we now introduce the following positive Lipschitz continuous functions, defined on the interval $[0, T]$:

$$(2.10) \qquad R(t) := \max_{(-\infty,\infty)} |r(\cdot, t)|, \qquad S(t) := \max_{(-\infty,\infty)} |s(\cdot, t)|,$$

$$(2.11) \qquad \Phi(t) := \max_{(-\infty,\infty)} |\phi(\cdot, t)|, \qquad \Psi(t) := \max_{(-\infty,\infty)} |\psi(\cdot, t)|,$$

$$(2.12) \qquad \Phi^+(t) := \max_{(-\infty,\infty)} \phi(\cdot, t), \qquad \Psi^+(t) := \max_{(-\infty,\infty)} \psi(\cdot, t),$$

$$(2.13) \qquad \Phi^-(t) := -\min_{(-\infty,\infty)} \phi(\cdot, t), \qquad \Psi^-(t) := -\min_{(-\infty,\infty)} \psi(\cdot, t),$$

$$(2.14) \qquad I(t) := \int_0^t \{\Phi(\tau) + \Psi(\tau)\}\,d\tau.$$

Taking account of (1.5), we fix $\delta \in (0, \bar{u})$ such that, for some $\alpha > 0$,

$$(2.15) \qquad p_u^{-\frac{5}{4}}(u)\,p_{uu}(u) \geqq 8\alpha > 0, \qquad |u - \bar{u}| < \delta,$$

then we redefine T, by making it smaller if necessary, so that, at the same time, $T \leqq 1$,

$$(2.16) \qquad |u(x, t) - \bar{u}| \leqq \delta, \qquad -\infty < x < \infty, \qquad 0 \leqq t \leqq T,$$

$$(2.17) \qquad \int_0^T I^3(t)\,dt \leqq 1.$$

A crucial step in our proof of Theorem 1.1 is to show that $T > 0$ may be selected *a priori* in such a way that (2.16), (2.17) hold for any solution on $(-\infty, \infty) \times [0, T]$ of (1.16), (1.20), (1.21), provided only v_0 satisfies (1.22), (1.23). Towards that end we proceed to derive estimates for the functions R, S, Φ, Ψ, Φ^+, Ψ^+, Φ^-, Ψ^-, defined above. In what follows, μ will denote unspecified constants that depend solely upon $\bar{u}$, δ (and so, ultimately, on $\bar{u}$ alone) while ν will denote an unspecified constant that depends at most upon $\bar{u}$, δ and the constant N in (1.23).

We fix any $t \in (0, T]$ and identify points $\hat{x}$ and $\check{x}$ in $(-\infty, \infty)$ with the property

$$R(t) = |r(\hat{x}, t)|, \qquad S(t) = |s(\check{x}, t)|. \tag{2.18}$$

For any $\Delta t \in (0, t]$, it is

$$\begin{aligned} R(t - \Delta t) &\geqq |r(\hat{x} + \Delta t\, p_u^{\frac{1}{2}}(u(\hat{x}, t)), t - \Delta t)| \\ S(t - \Delta t) &\geqq |s(\check{x} - \Delta t\, p_u^{\frac{1}{2}}(u(\check{x}, t)), t - \Delta t)|. \end{aligned} \tag{2.19}$$

Subtracting (2.19) from (2.18), dividing through by Δt and passing to the limit, as $\Delta t \downarrow 0$, we deduce

$$D^- R(t) \leqq \left|\overset{\backslash}{r}(\hat{x}, t)\right|, \qquad D^- S(t) \leqq \left|\overset{/}{s}(\check{x}, t)\right|. \tag{2.20}$$

Combining (2.20) with (2.7), (2.11), (2.16) and (2.14), we conclude that, for almost all t in $[0, T]$,

$$\frac{d}{dt}\{R(t) + S(t)\} \leqq \mu I(t) \tag{2.21}$$

whence it follows, upon use of (2.10), (1.6), (1.21), (1.22), that

$$R(t) + S(t) \leqq 2\varepsilon + \mu \int_0^t I(\tau)\, d\tau, \qquad 0 \leqq t \leqq T. \tag{2.22}$$

Next, we fix again $t \in (0, T]$ and identify points $\hat{y}$ and $\check{y}$ in $(-\infty, \infty)$ with

$$\Phi^-(t) = -\phi(\hat{y}, t), \qquad \Psi^-(t) = -\psi(\check{y}, t). \tag{2.23}$$

For any $\Delta t \in (0, t]$,

$$\begin{aligned} \Phi^-(t - \Delta t) &\geqq -\phi(\hat{y} + \Delta t\, p_u^{\frac{1}{2}}(u(\hat{y}, t)), t - \Delta t) \\ \Psi^-(t - \Delta t) &\geqq -\psi(\check{y} - \Delta t\, p_u^{\frac{1}{2}}(u(\check{y}, t)), t - \Delta t). \end{aligned} \tag{2.24}$$

We subtract (2.24) from (2.23); then we divide through by Δt and pass to the limit as $\Delta t \downarrow 0$, thus obtaining

$$D^- \Phi^-(t) \leqq -\overset{\backslash}{\phi}(\hat{y}, t), \qquad D^- \Psi^-(t) \leqq -\overset{/}{\psi}(\check{y}, t). \tag{2.25}$$

Combining (2.25) with (2.9), (2.11), (2.16), (2.14) and (2.15) we get an estimate of the form

$$\frac{d}{dt}\{\Phi^-(t) + \Psi^-(t)\} \leqq \mu I(t) + \mu\{\Phi(t) + \Psi(t)\} + \mu I(t)\{\Phi(t) + \Psi(t)\}, \tag{2.26}$$

for almost all t in $[0, T]$. Integrating (2.26) and using (2.13), (2.2), (2.1), (1.7), (1.6), (1.21), (1.23), (2.14) and (2.17) yields

$$\Phi^-(t) + \Psi^-(t) \leqq 2p_u^{\frac{1}{4}}(\bar{u})\, N + \mu \int_0^t I(\tau)\, d\tau + \mu I(t) + \mu I^2(t) \tag{2.27}$$

$$\leqq \nu + \mu I^2(t), \quad 0 \leqq t \leqq T.$$

Finally, we fix $t \in [0, T)$ and identify points $\hat{z}$, $\check{z}$ in $(-\infty, \infty)$ such that

$$\Phi^+(t) = \phi(\hat{z}, t), \quad \Psi^+(t) = \psi(\check{z}, t). \tag{2.28}$$

For any $\Delta t \in (0, T-t)$, we have

$$\begin{aligned} \Phi^+(t+\Delta t) &\geqq \phi(\hat{z} - \Delta t\, p_u^{\frac{1}{2}}(u(\hat{z}, t)), t + \Delta t) \\ \Psi^+(t+\Delta t) &\geqq \psi(\check{z} + \Delta t\, p_u^{\frac{1}{2}}(u(\check{z}, t)), t + \Delta t). \end{aligned} \tag{2.29}$$

Subtracting (2.29) from (2.28), dividing through by Δt and passing to the limit as $\Delta t \downarrow 0$, we get

$$D^+\Phi^+(t) \geqq \grave{\phi}(\hat{z}, t), \quad D^+\Psi^+(t) \geqq \acute{\psi}(\check{z}, t). \tag{2.30}$$

By combining (2.30), (2.9), (2.16), (2.15), (2.12), (2.13), (2.11), (2.14), (2.27) and the obvious inequality

$$\Phi(t) + \Psi(t) \leqq \{\Phi^+(t) + \Psi^+(t)\} + \{\Phi^-(t) + \Psi^-(t)\}, \tag{2.31}$$

we derive the differential inequality

$$\frac{d}{dt}\{\Phi^+(t) + \Psi^+(t)\} \geqq \alpha\{\Phi^+(t) + \Psi^+(t)\}^2 - \beta(t)\{\Phi^+(t) + \Psi^+(t)\} - \gamma(t), \tag{2.32}$$

in which

$$\beta(t) := \mu + \mu I(t), \tag{2.33}$$

$$\gamma(t) := \nu + \nu I^3(t), \tag{2.34}$$

which is to hold for almost all t in $[0, T]$.

We now define the function X on $[0, T]$ by

$$X(t) := e^{-\int_t^T \beta(\tau) d\tau}\{\Phi^+(t) + \Psi^+(t)\} - \int_t^T \gamma(\tau)\, d\tau \tag{2.35}$$

and use (2.32) to show that

$$\frac{dX}{dt} \geqq \alpha \left\{X(t) + \int_t^T \gamma(\tau)\, d\tau\right\}^2, \tag{2.36}$$

for almost all t in $[0, T]$. Integrating (2.36) backwards in time, starting from $t = T$, yields

$$X(t) \leqq \frac{X(T)}{1 + \alpha X(T)(T-t)} \leqq \frac{1}{\alpha(T-t)}, \quad 0 \leqq t < T, \tag{2.37}$$

and so, in view of (2.35),

$$\Phi^+(t) + \Psi^+(t) \leqq e^{\int_t^T \beta(\tau)d\tau} \left\{ \frac{1}{\alpha(T-t)} + \int_t^T \gamma(\tau)\, d\tau \right\}. \tag{2.38}$$

By virtue of (2.33), (2.34) and (2.17),

$$\int_0^T \beta(\tau)\, d\tau \leqq \mu, \quad \int_0^T \gamma(\tau)\, d\tau \leqq \nu. \tag{2.39}$$

Hence, (2.38), (2.39) imply

$$\int_0^t \{\Phi^+(\tau) + \Psi^+(\tau)\}\, d\tau \leqq -\mu \log (T-t) + \nu, \quad 0 \leqq t < T. \tag{2.40}$$

At the same time, (2.27), (2.17) yield

$$\int_0^t \{\Phi^-(\tau) + \Psi^-(\tau)\}\, d\tau \leqq \nu, \quad 0 \leqq t < T. \tag{2.41}$$

Combining (2.14), (2.31), (2.40) and (2.41), we obtain

$$I(t) \leqq -\mu \log (T-t) + \nu, \quad 0 \leqq t < T, \tag{2.42}$$

whence, upon integrating,

$$\int_0^T I(t)\, dt \leqq (-\mu \log T + \nu)\, T, \tag{2.43}$$

$$\int_0^T I^3(t)\, dt \leqq (-\mu \log^3 T + \nu)\, T. \tag{2.44}$$

From (2.22), (2.10), (1.6) and (2.43) we deduce

$$\left| \int_{\bar{u}}^{u(x,t)} p_u^{\frac{1}{2}}(\xi)\, d\xi \right| \leqq \varepsilon + (-\mu \log T + \nu)\, T, \quad -\infty < x < \infty, 0 \leqq t \leqq T. \tag{2.45}$$

At this point, holding δ fixed, we select ε and T so small that the right-hand side of (2.44) is less than 1 and, at the same time, the right-hand side of (2.45) does not exceed $\delta p_u^{\frac{1}{2}}(\bar{u} - \delta)$. Once ε and T have been thus fixed, it follows by our previous analysis and a standard continuation argument that if v_0 conforms with (1.22), (1.23) then any classical solution of (1.16), (1.20), (1.21), defined on $(-\infty, \infty) \times [0, T]$, must satisfy the differential inequality (2.36), for almost all t in $[0, T]$, as well as the bounds (2.39). It remains to show that no solution may exist on $(-\infty, \infty) \times [0, T]$ when (1.24) holds with M sufficiently large.

Equations (2.12), (2.8), (1.6), (1.21) and (1.24) together imply

$$\Phi^+(0) + \Psi^+(0) \geqq 2p^{\frac{1}{4}}(\bar{u})\, M \tag{2.46}$$

and so it follows from (2.35), (2.39) that if M is sufficiently large then

$$X(0) \geqq \frac{1}{\alpha T}. \tag{2.47}$$

On the other hand, on account of (2.36),

$$X(t) \geqq \frac{X(0)}{1 - \alpha X(0)\, t}, \qquad 0 \leqq t \leqq T, \tag{2.48}$$

and hence, when $X(0)$ satisfies (2.47), X will have to explode at some t in the interval $[0, T]$. This completes the proof of Theorem 1.1 for the case of the model material (1.15).

The proof of Theorem 1.1 for the general material (1.11) follows the same pattern. The analog of the differentiation operators (1.7) is

$$\begin{aligned} \backslash &:= \partial_t - \mathscr{S}_u^{\frac{1}{2}}\{u, u^{(t)}\}\, \partial_x \\ / &:= \partial_t + \mathscr{S}_u^{\frac{1}{2}}\{u, u^{(t)}\}\, \partial_x . \end{aligned} \tag{2.49}$$

To introduce the analog of the Riemann Invariants (1.6), we define on $\mathbb{R} \times L_h^1(0, \infty)$ the functional

$$\mathscr{G}\{u, w(\cdot)\} := \int_{\bar{u}}^{u} \mathscr{S}_u^{\frac{1}{2}}\{\xi, w(\cdot)\}\, d\xi \tag{2.50}$$

and then set

$$r := v + \mathscr{G}\{u, u^{(t)}\}, \qquad s := v - \mathscr{G}\{u, u^{(t)}\}. \tag{2.51}$$

In the place of (2.2) we have

$$\phi := \mathscr{S}_u^{-\frac{1}{4}}\{u, u^{(t)}\}\, \partial_t r, \qquad \psi := -\mathscr{S}_u^{-\frac{1}{4}}\{u, u^{(t)}\}\, \partial_t s. \tag{2.52}$$

It is now easy to derive the analogs of (2.7) and (2.9). The notation gets cumbersome because it involves representations for the Fréchet derivative of functionals like $\mathscr{S}_u$, $\mathscr{K}$ (appearing in (1.12)), $\mathscr{G}$ (defined by (2.50)) *etc.* as well as a number of resolvent kernels analogous to $\mathscr{L}$ (*cf.* (2.5)). Otherwise, the computation is straightforward. The remainder of the argument is virtually the same as our discussion for the special case; it relies upon definitions like (2.10), (2.11), (2.12), (2.13), (2.14) and differential inequalities like (2.21), (2.26) and (2.32).

Acknowledgement. I thank WILLIAM HRUSA & CLIFFORD TRUESDELL for their helpful remarks. This work was supported in part by the National Science Foundation through Grant No. DMS-8205355, in part by the U.S. Army under Contract No. DAAG-29-83-K-0029 and in part by the Office of Naval Research under Contract No. N00014-83-K-0542.

References

1. COLEMAN, B. D., & M. E. GURTIN, Waves in materials with memory, II. On the growth and decay of one-dimensional acceleration waves. *Arch. Rational Mech. Anal.* **19** (1965), 266–298.
2. COLEMAN, B. D., & W. NOLL, An approximation theorem for functionals with applications in continuum mechanics. *Arch. Rational Mech. Anal.* **6** (1960), 355–370.

3. DAFERMOS, C. M., Dissipation in materials with memory. Proc. Symp. on Viscoelasticity and Rheology held at the MRC, Univ. of Wisconsin, 1984. To be published by Academic Press.
4. DAFERMOS, C. M., & L. HSIAO, Development of singularities in solutions of the equations of nonlinear thermoelasticity. *Quart. Appl. Math.* (to appear).
5. DAFERMOS, C. M., & J. A. NOHEL, A nonlinear hyperbolic Volterra equation in viscoelasticity. *Am. J. Math. Suppl. dedicated to* P. HARTMAN (1981), 87–116.
6. HATTORI, H., Breakdown of smooth solutions in dissipative nonlinear hyperbolic equations. *Quart. Appl. Math.* **40** (1982), 113–127.
7. HRUSA, W. J., A nonlinear functional differential equation in Banach space with applications to materials with fading memory. *Arch. Rational Mech. Anal.* **84** (1983), 99–137.
8. KLAINERMAN, S., & A. MAJDA, Formation of singularities for wave equations including the nonlinear vibrating string. *Comm. Pure Appl. Math.* **33** (1980), 241–263.
9. LAX, P. D., Development of singularities of solutions of nonlinear hyperbolic partial differential equations. *J. Math. Physics* **5** (1964), 611–613.
10. MACCAMY, R. C., A model for one-dimensional, nonlinear viscoelasticity. *Quart. Appl. Math.* **35** (1977), 21–33.
11. MACCAMY, R. C., & V. J. MIZEL, Existence and nonexistence in the large of solutions of quasilinear wave equations. *Arch. Rational Mech. Anal.* **25** (1967), 299–320.
12. MALEK-MADANI, R., & J. A. NOHEL, Formation of singularities for a conservation law with memory. *SIAM J. Math. Anal.* (to appear).
13. NOLL, W., A mathematical theory for the mechanical behavior of continuous media. *Arch. Rational Mech. Anal.* **2** (1958/59), 197–226.

Lefschetz Center for Dynamical Systems
Division of Applied Mathematics
Brown University
Providence

(Received February 21, 1985)

Geometric Measure Theory and the Axioms of Continuum Thermodynamics

MORTON E. GURTIN, WILLIAM O. WILLIAMS, & WILLIAM P. ZIEMER

Dedicated to Walter Noll, whose ideas underlie much of what is presented herein

1. Introduction

Basic to all of continuum physics is the notion of a *subbody* of a given *body* B. Indeed, the starting point for most theories is the assumption that the underlying laws hold not only for B itself but also for all subbodies of B. A crucial question that arises when examining the foundations of such theories is:

What is the appropriate class of sets for B and the family of subbodies?

Indeed, standard operations require, among other things, that the union of two subbodies be a subbody and that the boundary of a subbody have a well defined normal vector field. The usual choices for subbodies are the standard domains of WHITNEY [1] and the regular regions of KELLOGG [2]. Unfortunately, such choices lead to inconsistencies in the axiomatic structure: one needs a much broader class of sets with a more general notion of normal vector field. Our main purpose here is to show that a class of sets with the desired properties is generated by the *sets of finite perimeter.*[1]

We shall demonstrate the viability of this class by using it to develop, in detail, an axiomatic structure for continuum thermodynamics.[2] We choose this particular physical theory because it entails a balance law (energy) and an inequality (entropy), as well as a certain measure-theoretic connection between the two (temperature). It is clear that all considerations pertaining to balance of energy apply trivially to all other balance laws of continuum physics; for example, to balance of momentum.

For convenience, we restrict our attention to rigid bodies. Balance of energy for an arbitrary subbody A is then the assertion that $\dot{E}(A)$, the rate of change of

---

[1] *Cf.* FEDERER [3], GIUSTI [4]. Such sets are sometimes called CACCIAPPOLI sets.

[2] Our treatment generalizes that of [5, 6, 7].

internal energy of A, be equal to $H(A, A^e)$, the heat transferred into A from its exterior, A^e:

$$\dot{E}(A) = H(A, A^e).$$

The dot over the E is meant to suggest the connotation of a time-derivative; we shall, however, work at a single *fixed* time t, and for that reason will not find it necessary to mention explicitly the underlying dependence on t.

We shall assume that $H(A, C)$, the heat transferred into A from C, is defined whenever:

(i) A is a subbody;

(ii) C is a subbody or the exterior of one;

(iii) A and C are separate (*i.e.*, touch at most along their boundaries).

We assume further that:

(A1) $\dot{E}(A)$ is bounded in absolute value by a constant times the volume of A;

(A2) $H(A, C)$ is bounded in absolute value by a constant times the contact area of A and C plus a scalar K_C times the volume of A, where K_C vanishes with the volume of C;

(A3) $H(A, C)$ is additive in A and additive in C.

We show that these assumptions (made precise) lead to the existence of an energy density $\dot{e}(x)$, a heat flux vector $q(x)$, a heat supply $r^e(x)$, and a long-range heat transfer $r(x, y)$, with more or less standard properties. In particular, these fields obey a local balance law of the form

$$\dot{e}(x) = \operatorname{div} q(x) + r^e(x) + \int_B r(x, y)\, dv(y)$$

almost everywhere in B.

Our treatment of entropy parallels the above. The primitives are $\dot{S}(A)$, the rate of change of internal entropy of A, and $M(A, C)$, the entropy transferred into A from C. These set functions are assumed to obey (A1)–(A3) with $\dot{E}$ and H replaced by $\dot{S}$ and M, respectively. We assume further that the entropy production

$$\Gamma(A) = \dot{S}(A) - M(A, A^e)$$

is bounded in absolute value by a constant times the volume of A.

Our statement of the second law is that Γ be non-negative and superadditive, and that there be no entropy transferred into A from C whenever there is no heat transferred from any part of C to any part of A. We show that these assumptions lead to the existence of an entropy density $\dot{s}(x)$ and temperatures[1] $\varphi(x)$, $\theta^e(x)$, and $\theta(x, y)$, and to the local entropy inequality

$$\dot{s}(x) \geqq \operatorname{div}(\varphi q)(x) + \theta^e(x)\, r^e(x) + \int_B \theta(x, y)\, r(x, y)\, dv(y)$$

almost everywhere in B.

[1] Actually, reciprocals of temperatures.

2. Notations and Preliminaries

Let $\mathscr{E}$ denote n-dimensional Euclidean space. We will denote the Lebesgue measure of a set $A \subset \mathscr{E}$ by $|A|$ or, occasionally, $v(A)$, and the $(n-1)$-dimensional Hausdorff measure of A by $a(A)$. We reserve the term measurable to mean Lebesgue measurable; the term a-measurable may be used when needed. By $B_r(x)$ we mean the closed ball of radius r centered at $x \in \mathscr{E}$, and we write ω_n, α_n for the constants defined by

$$|B_r(x)| = \omega_n r^n,$$
$$a\,(\text{bdry } B_r(x)) = \alpha_n r^{n-1}.$$

For $A \subset \mathscr{E}$, $x \in \mathscr{E}$ we define the **upper density** of A at x to be

$$d^-(A, x) = \limsup_{r \downarrow 0} \frac{|A \cap B_r(x)|}{\omega_n r^n},$$

and the **lower density** $d_-(A, x)$ to be the corresponding lim inf. If these two coincide we write $d(A, x)$ for the common value. If $d(A, x) = 1$ we call x a **point of density** for A; if $d(A, x) = 0$ it is a **point of rarefaction** for A. Points x which are neither points of density nor points of rarefaction for A constitute the **measuretheoretic boundary** ∂A of A. If A is measurable then points of rarefaction of A coincide with the points of density of the complement A'. Thus if we introduce the notation

$$A_* = \{x \in \mathscr{E} \mid d(A, x) = 1\}$$

then for measurable A we have the disjoint union

$$\mathscr{E} = A_* \cup \partial A \cup (A')_*,$$

and we note that $\partial(A') = \partial A$. Finally, we let

$$A^* = A_* \cup \partial A.$$

A classical result from measure theory (*cf.* [8]) is that for any measurable A

$$|A \,\Delta\, A_*| = 0,$$

where "Δ" denotes the symmetric difference. From this it follows that

$$|\partial A| = 0.$$

If, in addition, it is true that

$$a(\partial A) < \infty$$

then A is said to be of **finite perimeter**. From these observations it follows immediately that for measurable sets A and E

$$A_{**} = A_*, \qquad A^{**} = A^*,$$
$$A_*{}^* = A^*, \qquad A^*{}_* = A_*,$$
$$A_* \cup E_* \subset (A \cup E)_*,$$

and

$$(A \cap E)_* = A_* \cap E_*.$$

A measurable set E is homogeneous if $E \subset E_*$; the collection of all homogeneous subsets of $\mathscr{E}$ forms a topology, the **d-topology** of GOFFMAN & WATERMAN [9] Thus the d-interior of a set A is $A \cap A_*$ and its d-closure is $A \cup A^*$. From this we obtain

Lemma 2.1. *A set A is a regular open set in the d-topology if and only if* $A = A_*$.

Proof. If $A = d\text{-int}\,(d\text{-cl}\,(A))$ then, since $A \subset A_*$,

$$\begin{aligned} A &= (A \cup A^*)_* \cap (A \cup A^*) \\ &= A_* \cap (A \cup A^*) = A_*, \end{aligned}$$

and the steps are reversible. □

Given a unit vector $\boldsymbol{\nu}$, we write

$$P(x, \boldsymbol{\nu}) = \{y \in \mathscr{E} \mid (y - x) \cdot \boldsymbol{\nu} \geqq 0\}$$

and

$$B_r(x, \boldsymbol{\nu}) = B_r(x) \cap P(x, \boldsymbol{\nu}),$$

so that $P(x, \boldsymbol{\nu})$ is a half-space and $B_r(x, \boldsymbol{\nu})$ a half-ball. We say that $\boldsymbol{\nu}$ is a (measure-theoretic) **exterior normal** to a measurable set A at $x \in \mathscr{E}$ if both

$$\lim_{r \downarrow 0} \frac{|B_r(x, \boldsymbol{\nu}) \cap A|}{|B_r(x, \boldsymbol{\nu})|} = 0$$

and

$$\lim_{r \downarrow 0} \frac{|B_r(x, -\boldsymbol{\nu}) \cap A|}{|B_r(x, -\boldsymbol{\nu})|} = 1.$$

If $\boldsymbol{\nu}$ exists at x, it is unique; we denote it by $\boldsymbol{\nu}(A, x)$. Clearly the exterior normal to A' is defined whenever $\boldsymbol{\nu}(A, x)$ is and $\boldsymbol{\nu}(A', x) = -\boldsymbol{\nu}(A, x)$. We denote by $\partial^* A$ the collection of all points at which A has an exterior normal and note that

$$\partial^* A \subset \partial A.$$

A fundamental result of modern geometric measure theory is FEDERER'S proof (*cf.* [3], Sect. 4.5.6) that for a set of finite perimeter

$$a(\partial A \setminus \partial^* A) = 0.$$

In addition, one can show that for Lipschitzian vector fields $\boldsymbol{v}$,

$$\int_{\partial^* A} \boldsymbol{v}(x) \cdot \boldsymbol{\nu}(A, x)\, da(x) = \int_A \operatorname{div} \boldsymbol{v}(x)\, dv(x)$$

(*cf.* [3], Sect. 4.5.6).

Lemma 2.2. *Let* $A \cap C = \emptyset$. *Then, for* $x \in \partial^* A \cap \partial^* C$,

$$\boldsymbol{\nu}(A, x) = -\boldsymbol{\nu}(C, x),$$

Proof. Let $a = \nu(A, x)$, $c = \nu(C, x)$, and, for any unit vector m, write

$$P(m) = P(x, m), \qquad B_r(m) = B_r(x, m).$$

Then

$$|C \cap B_r(-c)| = |C \cap B_r(-c) \cap P(a)| + |C \cap B_r(-c) \cap P(-a)|. \tag{1}$$

Since $C = C \cap A'$, it follows that

$$C \cap B_r(-c) \cap P(-a) \subset B_r(-a) \cap A'.$$

Further,

$$\lim_{r \downarrow 0} \frac{|B_r(-a) \cap A'|}{|B_r(-a)|} = 0,$$

because $\nu(A', x) = -\nu(A, x) = -a$. Thus if we divide (1) by $|B_r(-a)| = |B_r(-c)|$ and let $r \downarrow 0$, we find that

$$1 = \lim_{r \downarrow 0} \frac{|C \cap B_r(-c) \cap P(a)|}{|B_r(-c)|} \leq \lim_{r \downarrow 0} \frac{|B_r(-c) \cap P(a)|}{|B_r(-c)|}$$

which is possible only if $-c = a$. □

3. The Material Universe. Material Surfaces

We will now use the notions of the previous section to construct the collection of sets with which we shall operate. Because a set A may be modified to within a set of measure zero without affecting A_*, ∂A, or $\partial^* A$, it is convenient to normalize our choices of sets, and we can at the same time exploit the *d*-topology to deduce properties of the collection. Thus, if we consider only *d*-regular open sets (sets A with $A = A_*$), we find that the definitions

$$A \vee B = (A \cup B)_*$$
$$A \wedge B = A \cap B$$

give the collection the structure of a Boolean algebra (*cf.* [10]).

Proposition 3.1. *With the operations* $\vee$ *and* $\wedge$ *the collection*

$$\mathcal{G} = \{A \subset \mathcal{E} \mid A = A_*\}$$

forms a complete Boolean algebra.

We shall use A^e and $A \sim B$ to denote complementation and difference in this algebra. Clearly,

$$A^e = (\mathcal{E} \setminus A)_*, \qquad A \sim B = (A \setminus B)^*; \tag{3.1}$$

we will refer to A^e as the **exterior** of A.

Our next step will be to establish some useful relations concerning the boundaries of sets in $\mathcal{G}$.

Lemma 3.2. *Let* $A, B \in \mathcal{G}$. *Then*:

(i) $(\partial A \cap B) \cup (\partial B \cap A) \subseteq (\partial A \wedge B) \subseteq \partial(A \cap B) \cup (\partial B \cap A) \cup (\partial A \cap \partial B)$,

(ii) $\partial(A \vee B) \subset (\partial A \setminus B) \cup (\partial B \setminus A)$,

(ii)i $(\partial A \setminus \partial B) \cup (\partial B \setminus \partial A) \subset \partial(A \vee B) \subset \partial A \cup \partial B$ *if* $A \wedge B = \emptyset$.

Proof. (i) The inclusion of $\partial(A \wedge B)$ is immediate:

$$\begin{aligned}\partial(A \wedge B) &= (A \cap B)^* \setminus (A \cap B)_* = (A \cap B)^* \setminus (A_* \cap B_*)\\ &\subset (A^* \cap B^*) \setminus (A_* \cap B_*) = (\partial A \cap B^*) \cap (\partial B \cap A^*)\\ &= (\partial A \cap B) \cup (\partial B \cap A) \cup (\partial A \cap \partial B).\end{aligned}$$

The inclusion on the other side requires examination of the definition of points of density: we refer the reader to [11], p. 193.

(ii) Using the definition of ∂E and earlier observations, we have

$$\begin{aligned}\partial(A \vee B) &= (A \cup B)^* \setminus (A \cup B)_* = (A^* \cup B^*) \setminus (A \cup B)_*\\ &\subset (A^* \cup B^*) \cap (A')^* \cap (B')^*\\ &= (\partial A \cap (B')^*) \cup (\partial B \cap (A')^*)\\ &= (\partial A \setminus B) \cup (\partial B \setminus A).\end{aligned}$$

This suffices to prove (ii) and one inclusion in (iii).

(iii) Suppose $A \wedge B = \emptyset$, and suppose that $x \in (A \cup B)_*$. Then since $A \cap B = \emptyset$,

$$\lim_{r \downarrow 0} \frac{|B_r(x) \cap A| + |B_r(x) \cap B|}{\omega_n r^n} = 1.$$

Thus either $d(A, x) = 1$ and $d(B, x) = 0$ or $d(A, x) = 0$ and $d(B, x) = 1$ or x is in the boundary of both. Thus

$$(A \cup B)_* \subset A \cup B \cup (\partial A \cap \partial B),$$

so

$$\partial(A \vee B) \supset (A^* \cup B^*) \setminus (A \cup B \cup (\partial B \cap \partial A)).$$

But (by an analogous argument) $A^* \cap B = \emptyset$ and $B^* \cap A = \emptyset$, so the latter is

$$(\partial A \setminus \partial B) \cup (\partial B \setminus \partial A),$$

as claimed. □

The collection of sets in $\mathcal{G}$ which are of finite perimeter is a subalgebra of $\mathcal{G}$. Firstly, $A^e = (A')_*$ implies that $\partial A^e = \partial A$ and hence that A^e is of finite perimeter whenever A is. Secondly, by Lemma 3.2(i),

$$a(\partial(A \cap B)) \leqq a(\partial A \cap B) + a(\partial B \cap A) + a(\partial A \cap \partial B)$$

so that $A \cap B$ is of finite perimeter whenever A and B both are.

Finally, we introduce the collections needed in our further work. Let $B \in \mathcal{G}$ be bounded and of finite perimeter: we shall call B the **body** and let

$$\mathbb{M} = \{A \in \mathcal{G} \mid A \subset B \text{ and } A \text{ is of finite perimeter}\}$$

be the set of **subbodies** of B. $\mathbb{M}$ is a Boolean algebra with the inherited operations and with complement

$$A^b = A^e \wedge B.$$

To discuss the interaction between B and its environment we introduce the collection $\mathbb{N}$ of subbodies and exteriors of subbodies:

$$\mathbb{N} = \{A \in \mathcal{G} \mid A \in \mathbb{M} \text{ or } A^e \in \mathbb{M}\}.$$

Clearly, $\mathbb{N}$ is a subalgebra of $\mathcal{G}$. Sets A, C in $\mathbb{N}$ are said to be **separate** if $A \wedge C = \emptyset$, and we shall have frequent occasion to deal with the collections of separate elements

$$\mathbb{M} \varnothing \mathbb{M} = \{(A, C) \in \mathbb{M} \times \mathbb{M} \mid A \wedge C = \emptyset\}$$

and

$$\mathbb{M} \varnothing \mathbb{N} = \{(A, C) \in \mathbb{M} \times \mathbb{N} \mid A \wedge C = \emptyset\}.$$

The final collection we consider is the family $\mathbb{S}$ of **material surfaces**. A material surface is a pair $(S, \boldsymbol{\nu})$, where S is a set of the form

$$S = \partial A \cap \partial C$$

for some $(A, C) \in \mathbb{M} \varnothing \mathbb{N}$ and $\boldsymbol{\nu}$ is the exterior normal to C, restricted to S.

Proposition 3.3. *If* $(A, C) \in \mathbb{M} \varnothing \mathbb{N}$, *then*

$$\partial(A \wedge C) = (\partial A \setminus \partial C) \cup (\partial C \setminus \partial A) \cup N$$

with $a(N) = 0$.

Proof. The existence of $N \subset (\partial A \cap \partial C)$ follows from Lemma 3.2(ii). In fact we can say that

$$N = (\partial A \cap \partial C) \setminus (A \cup C)_* .$$

Now if $x \in \partial^* A \cap \partial^* C$ and $B_r^{\pm} = B_r(x, \pm \boldsymbol{\nu}(A, x))$, we have

$$|B_r(x) \cap (A \cup C)| = |B_r^+ \cap (A \cup C)| + |B_r^- \cap (A \cup C)|,$$

so that, since $A \cap C = \emptyset$,

$$\frac{|B_r(x) \cap (A \cup C)|}{\omega_n r^n} = \frac{|B_r^+ \cap A|}{2\,|B_r^+|} + \frac{|B_r^+ \cap C|}{2\,|B_r^+|} + \frac{|B_r^- \cap A|}{2\,|B_r^-|} + \frac{|B_r^- \cap C|}{2\,|B_r^-|}.$$

Using Lemma 2.2, in the limit as $r \downarrow 0$ this becomes $\frac{1}{2} + \frac{1}{2}$. Thus $\partial^* A \cap \partial^* C \subset (A \cup C)_*$; since $\partial^* A \cap \partial^* C$ is a-dense in $\partial A \cap \partial C$, we have

$$a(N) = a((\partial A \cap \partial C) \setminus (A \cup C)_*) = 0. \quad \square$$

Proposition 3.4. *If A, C, D in $\mathbb{N}$ are mutually separate, then*

$$a(\partial A \cap \partial C \cap \partial D) = 0.$$

Proof. At a point x in $\partial^* A \cap \partial^* C \cap \partial^* D$ we would have

$$\boldsymbol{\nu}(A, x) = -\boldsymbol{\nu}(C, x) = +\boldsymbol{\nu}(D, x) = -\boldsymbol{\nu}(A, x).$$

Thus $\partial^* A \cap \partial^* C \cap \partial^* D$ is empty. □

Finally, we prove an approximation result central to the analysis of the next section.

Theorem 3.5. *If $(A, C) \in \mathrm{M\o N}$ and $S = \partial A \cap \partial C$, then for each $\varepsilon > 0$ there exist $\hat{A} \subset A$, $\hat{C} \subset C$ in $\mathbb{N}$ such that*

$$|\hat{A}| < \varepsilon, \quad |\hat{C}| < \varepsilon$$

and

$$a(S \setminus (\partial\hat{A} \cap \partial\hat{C})) = 0.$$

Proof. It is clear that our construction will be complete if we can find, for any $A \in \mathbb{N}$, a set $\hat{A} \in \mathbb{N}$ with $\hat{A} \subset A$, $|\hat{A}| < \varepsilon$, and $a(\partial A \setminus \partial\hat{A}) = 0$.

Now for each $x \in \partial^* A$ and for any $\eta > 0$ we may find $\varrho(x) > 0$ with the property that $r < \varrho(x)$ ensures both

$$\left|\tfrac{1}{2} - \frac{|B_r(x) \cap A|}{\omega_n r^n}\right| < \eta \tag{3.2}$$

and

$$1 - \frac{a(B_r(x) \cap \partial^* A)}{\omega_{n-1} r^{n-1}} < \eta. \tag{3.3}$$

(For the latter *cf.* [4], p. 63). Let $R > 0$ be fixed and set

$$\mathscr{H} = \left\{B_r(x) \cap \partial^* A \mid x \in \partial^* A, r < \frac{\varrho(x)}{3}, r < R\right\}.$$

We remove an (at most) countable family from $\mathscr{H}$ (*viz.* the argument in [11], p. 194) to ensure that $a(\partial B_r(x) \cap \partial A) = 0$ for all sets in $\mathscr{H}$. Even then, the family $\mathscr{H}$ forms an a-Vitali relation on $\partial^* A$ in the sense used by FEDERER. To establish this (via Theorem 2.8.17 of [3]) we need only observe that: (i) for each $x \in \partial^* A$ there are sets in $\mathscr{H}$ of arbitrarily small diameter which include x; and (ii) we can uniformly bound the measure of larger elements of $\mathscr{H}$ in terms of smaller ones, *i.e*, from (3.3)

$$\frac{a(B_{3r}(x) \cap \partial^* A)}{a(B_r(x) \cap \partial^* A)} \leq \frac{3^{n-1}}{1-\eta},$$

so that the hypothesis of the above referenced theorem, namely

$$\limsup_{r\downarrow 0}\left\{r + \frac{a(B_{3r}\cap \partial^* A)}{a(B_r \cap \partial^* A)}\right\} < \infty,$$

is satisfied. From this now follows the existence of a disjoint countable family from $\mathscr{H}$ which covers $\partial^* A$ up to an a-null set. We write

$$\partial^* A = \bigcup_{m=1}^{\infty} (B_{r_m}(x_m) \cap \partial^* A) \cup N,$$

where $a(N) = 0$.

Next we need two estimates on the measure of these sets. From (3.2) and (3.3), for any set in $\mathscr{H}$,

$$|B_r(x) \cap A| \leqq (\tfrac{1}{2} + \eta)\frac{\omega_n}{\omega_{n-1}}\frac{R}{1-\eta} a(B_r(x) \cap \partial^* A),$$

so that

$$\begin{aligned}\sum_{m=1}^{\infty} |B_{r_m}(x_m) \cap A| &\leqq R\frac{1+2\eta}{2(1-\eta)}\frac{\omega_n}{\omega_{n-1}}\sum_{m=1}^{\infty} a(B_{r_m}(x_m) \cap \partial^* A) \\ &= R\frac{1+2\eta}{2(1-\eta)}\frac{\omega_n}{\omega_{n-1}} a(\partial^* A).\end{aligned} \tag{3.4}$$

Similarly, for sets in $\mathscr{H}$,

$$a(\partial B_r(x) \cap A) \leqq \frac{\alpha_n}{\omega_n}\frac{1}{1-\eta} a(B_r(x) \cap \partial^* A).$$

Let us now set

$$B_m = B_{r_m}(x_m)_* \cap A \in \mathrm{M},$$

$$\hat{A} = \bigcap_{m=1}^{\infty} B_m.$$

We wish to show that $\hat{A}$ is in M. First, since $\mathscr{G}$ is complete, it is clear that $\hat{A} \in \mathscr{G}$. To show that $\hat{A}$ is of finite perimeter, we first define

$$\hat{A}_k = \bigcup_{m=1}^{k} B_m.$$

Now each B_m is of finite perimeter and therefore so is $\hat{A}_k$ for each k. Moreover, we know that

$$\partial B_m \subset (\partial B_{r_m}(x_m) \cap A^*) \cup (B_{r_m}(x_m) \cap \partial A)$$

and

$$a(\partial B_{r_m}(x_m) \cap A^*) \leqq \frac{\alpha_n}{\omega_n}\frac{1}{1-\eta} a(B_{r_m}(x_m) \cap \partial^* A).$$

Because the $B_{r_m}(x_m)$ are mutually disjoint, it follows that

$$a(\partial \hat{A}_k) \leqq \frac{\alpha_n}{\omega_n} \frac{1}{1-\eta} a(\partial^* A) + a(\partial A).$$

If we now identify each $\hat{A}_k$ with its characteristic function, we have that $\{\hat{A}_k\}$ is a sequence of BV functions whose BV norms are uniformly bounded. Therefore, by Rellich's Theorem, there exist $\tilde{A} \in BV$ and a subsequence such that $\hat{A}_k \to \tilde{A}$ weakly in BV. Moreover, $\hat{A}_k \to \tilde{A}$ in L^1 and therefore (by passing to another subsequence) $\hat{A}_k \to \tilde{A}$ pointwise a.e. Thus $\tilde{A} = \hat{A}$ and $\hat{A} \in BV$. □

All of these results will be used in the next section to reduce certain functions on $\mathbb{M}$ to functions defined on the family $\mathbb{S}$ of material surfaces. For these purposes it is useful to define a Boolean algebra structure on subcollections from $\mathbb{S}$. Let $C \in \mathbb{N}$, let $\boldsymbol{\nu}$ be the exterior normal to C, and define

$$\mathbb{S}_C = \{S \subset \partial C \mid (S, \boldsymbol{\nu}|_S) \in \mathbb{S}\},$$

where $\boldsymbol{\nu}|_S$ is $\boldsymbol{\nu}$ restricted to $\mathbb{S}$. $\mathbb{S}_C$ has a natural structure inherited from $\mathbb{N}$: most concisely, we note that $\mathbb{S}_C$ is isomorphic to

$$\mathbb{N}/\mathrm{A},$$

where A is the filter

$$\mathrm{A} = \{A \in \mathbb{N} \mid A \supset \partial C\}.$$

Then it follows immediately from the definition of the operations in $\mathscr{G}$ and from Lemma 2.2 that

$$S \wedge T = S \cap T,$$

while the symmetric difference of $S \wedge T$ and $S \cup T$ obeys

$$a(S \vee T \,\Delta\, S \cup T) = 0,$$

so that—as a measure space—we can regard $\mathbb{S}_C$ as having the usual set operations.

We close this section with some terminology. Let $\mathscr{W} \subset \mathbb{N}$ be closed under $\vee$. Then $\psi: \mathscr{W} \to \mathbb{R}$ is **additive** if

$$\psi(A \vee C) = \psi(A) + \psi(C)$$

for all separate $A, C \in \mathscr{W}$, and an analogous meaning applies to the term **superadditive**. Similarly, for $\mathscr{W}$ equal to $\mathbb{M}$ or $\mathbb{N}$, $\psi: \mathbb{M} \o \mathscr{W} \to \mathbb{R}$ is **biadditive** if $\psi(\cdot, C)$ is additive for each $C \in \mathscr{W}$ and $\psi(A, \cdot)$ is additive for each $A \in \mathbb{M}$.

4. Interactions

A **Cauchy interaction** for a body B is a map

$$I: \mathbb{M} \o \mathbb{N} \to \mathbb{R}$$

with the following properties:

a) I is biadditive;

b) there exists a scalar K and, for each $C \in \mathbb{N}$ a scalar K_C, with $K_C \leqq K$ and $K_C \to 0$ as $|C| \to 0$, such that for all $(A, C) \in \mathrm{M} \varnothing \mathbb{N}$,

$$|I(A, C)| \leqq K a(\partial A \cap \partial C) + K_C v(A).$$

We assume, for the remainder of this section, that I is a Cauchy interaction.

The first fundamental property of interactions is that they divide naturally into "volumetric" and "areal" parts.

Proposition 4.1. *There exist Cauchy interactions R and Q such that*

$$I = R + Q$$

and, for all $(A, C) \in \mathrm{M} \varnothing \mathbb{N}$,

$$|R(A, C)| \leqq K_C v(A),$$

$$|Q(A, C)| \leqq K a(\partial A \cap \partial C).$$

Proof. Choose $(A, C) \in \mathrm{M} \varnothing \mathbb{N}$. Then if

$$\mathcal{P}_A = \{D \in \mathrm{M} \mid D \subset A \text{ and } a(\partial A \cap \partial C \setminus \partial D \cap \partial C) = 0\},$$

we have for all $D, E \in \mathcal{P}_A$ (*cf.* (3.1))

$$\begin{aligned} &|I(D, C) - I(E, C)| \\ &= |I(D \sim E, C) - I(E \sim D, C)| \\ &\leqq K_C \max(|D|, |E|). \end{aligned} \tag{4.1}$$

By Theorem 3.5 there are elements in $\mathcal{P}_A$ of arbitrarily small volume. Thus, by (4.1), we may define

$$Q(A, C) = \lim_{\substack{|D| \to 0 \\ D \in \mathcal{P}_A}} I(D, C).$$

Clearly $|Q(A, C)| \leqq K a(\partial A \cap \partial C)$. If A_1, A_2 and C are mutually separate, and if $D_1 \in \mathcal{P}_{A_1}$, $D_2 \in \mathcal{P}_{A_2}$, then, by Proposition 3.4, $D_1 \vee D_2 \in \mathcal{P}_{A_1 \vee A_2}$ and $|D_1 \vee D_2| = |D_1| + |D_2|$. Hence

$$Q(A_1 \vee A_2, C) = \lim_{\substack{|D_i| \to 0 \\ D_i \in \mathcal{P}_{A_i}}} I(D_1 \vee D_2, C).$$

But then

$$Q(A_1 \vee A_2, C) = Q(A_1, C) + Q(A_2, C)$$

follows from the additivity of $I(\cdot, C)$. Likewise if A, C_1, C_2 are mutually separate one can use Proposition 3.4 to show that

$$Q(A, C_1 \vee C_2) = Q(A, C_1) + Q(A, C_2).$$

Finally, if we define R by

$$R(A, C) = I(A, C) - Q(A, C)$$

then R is clearly a Cauchy interaction and

$$R(A,C) = \lim_{\substack{|D| \to 0 \\ D \in \mathscr{P}_A}} I(A \sim D, C),$$

so that $|R(A, C)| \leqq K_C\, v(A)$. □

Proposition 4.2. *There exists a function*

$$\mathscr{Q} : \mathbb{S} \to \mathbb{R}$$

such that

$$Q(A, C) = \mathscr{Q}(S, \boldsymbol{\nu})$$

whenever $S = \partial A \cap \partial C$ *and* $\boldsymbol{\nu}$ *is exterior to* C.

Proof. We may define $\mathscr{Q}$ by this relation once we prove consistency. If $(A, C), (\hat{A}, \hat{C}) \in \mathbb{M} \varnothing \mathbb{N}$ are such that

$$\partial A \cap \partial C = \partial \hat{A} \cap \partial \hat{C} = S$$

and C and $\hat{C}$ share an exterior normal on S, we may define

$$D = A \wedge \hat{A}, \qquad E = C \wedge \hat{C}$$

and note that

$$Q(A, C) - Q(D, E) = Q(A \sim D, C) + Q(D, C \sim E).$$

But by Proposition 3.4,

$$a(\partial(A \sim D) \cap \partial C) = a(\partial D \cap \partial(C \sim E)) = 0$$

and thus $Q(A, C) = Q(D, E)$. A similar argument applies to $Q(\hat{A}, \hat{C})$. □

Corollary 4.3. *For any* $C \in \mathbb{N}$

$$\mathscr{Q}(S \cup T, \boldsymbol{\nu}) = \mathscr{Q}(S, \boldsymbol{\nu}|_S) + \mathscr{Q}(T, \boldsymbol{\nu}|_T)$$

for all $S, T \in \mathbb{S}_C$ *with* $a(S \cap T) = \emptyset$.

We will say that surface elements $(S, \boldsymbol{\nu}), (T, \boldsymbol{\mu})$ are **compatible** if there is a $C \in \mathbb{N}$ such that $S \subset \partial C$, $T \subset \partial C$, $a(S \cap T) = 0$, and both $\boldsymbol{\nu}$ and $\boldsymbol{\mu}$ are exterior to C. This corollary then says that $\mathscr{Q}$ is additive on compatible surface elements.

Since $R(\cdot, C)$ is additive and v-continuous, it is straight-forward to argue that it is the restriction of a unique v-continuous measure and hence represented as an integral (*cf.* [6]).

Proposition 4.4. *For each* $C \in \mathbb{N}$ *there is a v-integrable function*

$$r_C : B \to \mathbb{R}$$

for which

$$R(A, C) = \int_A r_C \, dv$$

whenever $A \wedge C = \emptyset$.

In fact, we shall use this result only in the case $C = B^e$, defining

$$r^e = r_{B^e}.$$

For $C \neq B^e$ we can detail the dependence of r_C on C. To begin, let $C \in \mathrm{M}$ and set

$$\varrho(C) = \sup\left\{\frac{|R(A, C)|}{|A|} \,\middle|\, (A, C) \in \mathrm{M} \varnothing \mathbb{N}, |A| \neq 0\right\}.$$

Lemma 4.5. ϱ *is subadditive and v-continuous.*

Proof. Since $R(A, \cdot)$ is additive,

$$\varrho(C_1 \vee C_2) \leqq \varrho(C_1) + \varrho(C_2);$$

since $|r(C)| \leqq K_C$, r is v-continuous. □

We may now apply a classical result of BURKHILL ([12], *cf.* the discussion in [6]) to deduce the existence of an integrable function (the upper derivate of ϱ)

$$g : B \to \mathbb{R}$$

with

$$\varrho(C) \leqq \int_C g \, dv$$

for all C. But this means that for any $(A, C) \in \mathrm{M} \varnothing \mathbb{N}$,

$$|R(A, C)| \leqq \varrho(C)\, v(A) \leqq \int_{A \times C} g(y)\, dv(x)\, dv(y).$$

This suffices to ensure that condition ***I*** of [7] is satisfied, and we deduce the existence of an integrable function

$$r : B \times B \to \mathbb{R}$$

such that

$$R(A, C) = \int_{A \times C} r \, dv^2.$$

We summarize our results:

Proposition 4.6. *For any* $(A, C) \in \mathrm{M} \varnothing \mathrm{M}$,

$$R(A, C) = \int_{A \times C} r(x, y)\, dv(x)\, dv(y),$$

$$R(A, B^e) = \int_A r^e(x)\, dv(x).$$

5. Balance Equations. Energy

We suppose given a map

$$\dot{E}: \mathbb{M} \to \mathbb{R},$$

which we call the **rate of change of internal energy**, and a Cauchy interaction

$$H: \mathbb{M} \varnothing \mathbb{N} \to \mathbb{R},$$

which we call the **heat transfer.** We assume there exists a constant K such that

$$|\dot{E}(A)| \leqq K|A| \tag{5.1}$$

for all $A \in \mathbb{M}$. Finally we suppose that $\dot{E}$ and H satisfy the **first law of thermodynamics,** which takes the form of the *balance equation*

$$\dot{E}(A) = H(A, A^e) \quad \text{for all } A \in \mathbb{M}.$$

Since H is an interaction, we use the results of the last section to write

$$H(A, C) = R(A, C) + Q(A, C),$$

and we introduce r, r^e and $\mathcal{Q}$ as before.

Proposition 5.1. *For each* $(S, \boldsymbol{v}) \in \mathbb{S}$

$$\mathcal{Q}(S, -\boldsymbol{v}) = -\mathcal{Q}(S, \boldsymbol{v}).$$

Proof. Let $(A, C) \in \mathbb{M} \varnothing \mathbb{N}$, $S = \partial A \cap \partial C$, with $\boldsymbol{v}$ exterior to C and, hence, $-\boldsymbol{v}$ exterior to A. Since

$$A^e = (A \vee C)^e \vee C,$$

we can apply the balance law to $A \vee C, A, C,$ and subtract, to find

$$\dot{E}(A \vee C) - \dot{E}(A) - \dot{E}(C) = -H(C, A) - H(A, C).$$

Thus

$$\begin{aligned} \mathcal{Q}(S, \boldsymbol{v}) + Q(S, -\boldsymbol{v}) &= Q(A, C) + Q(C, A) \\ &= -\dot{E}(A \vee C) + \dot{E}(A) + \dot{E}(C) - R(A, C) - R(C, A). \end{aligned}$$

But by (5.1) and Proposition 4.6 we may choose A and C of arbitrarily small measure and thus make the right-hand side arbitrarily small. Therefore

$$Q(S, \boldsymbol{v}) + Q(S, -\boldsymbol{v}) = 0. \quad \square$$

Proposition 5.2. *The limit*

$$\dot{e}(x) = \lim_{\varrho \downarrow 0} \frac{\dot{E}(B_\varrho(x))}{V(B_\varrho(x))}$$

exists for v-a.e. $x \in B$. *Moreover* $\dot{e}$ *is* v-*integrable and for all* $A \in \mathbb{M}$,

$$\dot{E}(A) = \int_A \dot{e}\, dv - \int_{A \times A} r\, dv^2.$$

Proof. From Propositions 5.1 and 4.6, for all separate A, C,

$$\dot{E}(A \vee C) - \dot{E}(A) - \dot{E}(C) = - \int_{A\times C} r\, dv^2 - \int_{C\times A} r\, dv^2 .$$

Thus the function μ defined on $\mathbb{M}$ by

$$\mu(A) = \dot{E}(A) + \int_{A\times A} r\, dv^2$$

is additive and v-continuous. Standard arguments (which rely on the observation that $\mathbb{M}$ generates the σ-ring of Borel sets in $\mathscr{E}$ and that $|\partial A| = 0$ for all such sets A, *cf.* [6]) then establish that μ is the restriction to $\mathbb{M}$ of a v-continuous measure. We let $\dot{e}$ be the density of this measure. By Proposition 4.1 the v-density of $A \mapsto R(A, A)$ is zero and hence $\dot{e}$ is the v-density of $\dot{E}$. □

Finally, we observe that $\mathscr{Q}$ is a Cauchy flux in the sense of [13]:

$$|\mathscr{Q}(S, \boldsymbol{\nu})| \leqq Ka(S),$$

$$\mathscr{Q}(S \cup T, \boldsymbol{\nu}) = \mathscr{Q}(S, \boldsymbol{\nu}|_S) + \mathscr{Q}(T, \boldsymbol{\nu}|_T),$$

$$|\mathscr{Q}(\partial C, \boldsymbol{\nu})| \leqq K|C|.$$

We may therefore appeal to [11] for the proof of what is usually referred to as NOLL's theorem (*cf.* [14]):

Proposition 5.3. *There exists a field* $q : B \times S_{n-1} \to \mathbb{R}$ *such that, for all* $(S, \boldsymbol{\nu}) \in \mathbb{S}$,

$$\mathscr{Q}(S, \boldsymbol{\nu}) = \int_S q((x, \boldsymbol{\nu}(x))\, da(x).$$

Moreover

$$q(x, -\boldsymbol{\nu}) = -q(x, \boldsymbol{\nu})$$

for all $(x, \boldsymbol{\nu}) \in B \times S_{n-1}$.

Further properties of the field q have been established in [11] and in [13]. In particular, it can be shown that there exists a vector field $\boldsymbol{q}$, defined on B, with the property

$$q(x, \boldsymbol{\nu}) = \boldsymbol{q}(x) \cdot \boldsymbol{\nu}$$

for v-a.e. $x \in B$, and that $\boldsymbol{q}$ has a weak divergence (in the sense of distributions). These results yield

Theorem 5.4. *There exists a vector field* $\boldsymbol{q}$ *on* B *such that for* v-a.e. $x \in B$,

$$\dot{e}(x) = (\operatorname{div} \boldsymbol{q})(x) + r^e(x) + \int_B r(x, y)\, dv(y),$$

where div $\boldsymbol{q}$ *is the weak divergence of* $\boldsymbol{q}$.

6. Inequalities. Entropy

Exactly as for energy we suppose given a map

$$\dot{S}: \mathbb{M} \to \mathbb{R},$$

the **rate of change of internal entropy,** and a Cauchy interaction

$$M: \mathbb{M}\text{ø}\mathbb{N} \to \mathbb{R},$$

which we call the **entropy transfer.** Further, we introduce the term

$$\Gamma(A) = \dot{S}(A) - M(A, A^e), \quad A \in \mathbb{M},$$

which we call the **entropy production.**

Crucial to many of the results below is the presumption that not only $\dot{S}$ but also the entropy production is Lipschitz continuous with respect to volume: we assume that there exists a constant K such that

$$|\dot{S}(A)| \leqq Kv(A),$$
$$|\Gamma(A)| \leqq Kv(A),$$

for all $A \in \mathbb{M}$.

Finally we suppose that $\dot{S}$, and M satisfy the **second law of thermodynamics,** which for us consists in two statements:

(i) $\Gamma(A) \geqq 0$ for all $A \in \mathbb{M}$;

(ii) Γ is super-additive.

Assertion (ii) is the requirement that no entropy be lost in interactions[1], as can be seen from its expression in terms of $\dot{S}$ and M:

$$\dot{S}(A \vee C) - \dot{S}(A) - \dot{S}(C) \geqq -M(A, C) - M(C, A)$$

for all $(A, C) \in \mathbb{M}\text{ø}\mathbb{N}$.

Since M is a Cauchy interaction, we have a decomposition

$$M(A, C) = K(A, C) + J(A, C)$$

where, exactly as above,

$$K(A, C) = \int_{A \times C} k \, dv^2,$$

$$K(A, B^e) = \int_A k^e \, dv,$$

$$J(A, C) = \mathscr{J}(S, \boldsymbol{\nu}),$$

with k integrable, k^e bounded and $\mathscr{J}$ defined on $\mathbb{S}$. Here $S = \partial A \cap \partial C$, and $\boldsymbol{\nu}$ is the outer normal to ∂C.

---

[1] For energy balance the corresponding statement, itself a consequence of the general balance, is

$$\dot{E}(A \vee C) - \dot{E}(A) - \dot{E}(C) = -H(A, C) - H(C, A).$$

The requirement (ii) was apparently first introduced in [5]; previously it had always been assumed that Γ was additive (*cf.* the discussion of the *optimal* entropy function below).

Proposition 6.1. *For each* $(S, \boldsymbol{\nu}) \in \mathbb{S}$

$$\mathcal{J}(S, -\boldsymbol{\nu}) = -\mathcal{J}(S, \boldsymbol{\nu}).$$

Proof. If $(A, C) \in \mathbb{M} ø \mathbb{N}$, $S = \partial A \cap \partial C$, with $\boldsymbol{\nu}$ the outer normal to ∂C, we find (*cf.* Proposition 5.1)

$$\begin{aligned} &\Gamma(A \vee C) - \Gamma(A) - \Gamma(C) \\ &\quad = \dot{S}(A \vee C) - \dot{S}(A) - \dot{S}(C) - M(A, C) - M(C, A). \end{aligned}$$

From this and the v-bounds on Γ and $\dot{S}$,

$$|J(A, C) + J(C, A)| \leqq 8K(V(A) + V(C)).$$

Thus we use Proposition 5.1 to establish our claim. □

Moreover, the assumed volume continuity ensures that $\mathcal{J}$, like $\mathcal{Q}$, is a Cauchy flux:

$$|\mathcal{J}(S, \nu)| \leqq ka(S),$$

$$\mathcal{J}(S \cup T, \boldsymbol{\nu}) = \mathcal{J}(S, \boldsymbol{\nu}|_S) + \mathcal{J}(T, \boldsymbol{\nu}|_T),$$

$$|\mathcal{J}(\partial C, \boldsymbol{\nu})| \leqq kv(C).$$

Thus, paralleling the results for q (*cf.* [11]), we have

Proposition 6.2. *There exists a field* $j: B \times S_{n-1} \to \mathbb{R}$ *such that, for all* $(S, \boldsymbol{\nu}) \in \mathbb{S}$,

$$\mathcal{J}(S, \boldsymbol{\nu}) = \int_S j(x, \boldsymbol{\nu}(x))\, da(x).$$

Moreover $j(\cdot, -\boldsymbol{\nu}) = -j(\cdot, \boldsymbol{\nu})$ *for all* $\boldsymbol{\nu}$ *and there exists a measurable vector field* $\boldsymbol{j}$ *on* B *such that for* v-a.e. $x \in B_j$,

$$j(x, \boldsymbol{\nu}) = \boldsymbol{j}(x) \cdot \boldsymbol{\nu}$$

for all $\boldsymbol{\nu}$.

We find that, lacking additivity, the construction of a density function for $\dot{S}$ does not follow as simply as the corresponding construction for $\dot{E}$. First, Proposition 6.1 allows a convenient extension of K, J and, hence, M to $\mathbb{M} \times \mathbb{N}$: we define, for all $A \in \mathbb{M}$,

$$K(A, A) = \int_{A \times A} k\, dv^2,$$

$$J(A, A) = 0.$$

The resulting extensions continue to be biadditive. Then, letting

$$U = B \vee B^e,$$

we define

$$\begin{aligned} \bar{S}(A) &= \dot{S}(A) + K(A, A) \\ &= \Gamma(A) + M(A, U). \end{aligned}$$

$\overline{S}$ is super-additive and Lipschitz with respect to v. At each $x \in B$ we define the **lower derivate** of $\dot{S}$:

$$\dot{s}(x) = \liminf_{V(A)\to 0} \frac{\dot{S}(A)}{V(A)},$$

where the limit is taken over regular sequences of intervals A (*cf.* [15]). It then follows that $\dot{s}$ is the lower derivate of $\overline{S}$, that it is integrable, and that

$$\overline{S}(A) \geqq \int_A \dot{s}\, dv \tag{6.1}$$

for all elements A in M (*cf.* [12]). Moreover $\dot{s}$ exceeds all integrable functions bounded above by $\overline{S}$ in the sense of (6.1). Since $M(\cdot, U)$ is a v-continuous measure bounded by $\overline{S}$, it follows that, for all $A \in$ M,

$$\int_A \dot{s}\, dv \geqq M(A, U).$$

Thus

$$\int_A \dot{s}\, dv - \int_{A\times A} k\, dv^2 \geqq M(A, A^e).$$

We introduce the notation

$$\dot{S}^*(A) = \int_A \dot{s}\, dv - \int_{A\times A} k\, dv^2$$

and note that for all $A \in$ M,

$$\dot{S}(A) \geqq \dot{S}^*(A) \geqq M(A, A^c).$$

Moreover,

$$\begin{aligned} &\dot{S}^*(A \vee C) - \dot{S}^*(A) - \dot{S}^*(C) \\ &= -K(A, C) - K(C, A) \\ &= -K(A, C) - K(C, A) - J(A, C) - J(C, A) \\ &= -M(A, C) - M(C, A). \end{aligned}$$

Thus $\dot{S}^*$ has the same properties as the entropy rate previously introduced; in fact, $\dot{S}^*$ is **optimal**, as it is a lower bound for the entropy rate $\dot{S}$ and has a corresponding entropy-production function which is *additive*. We shall henceforth replace $\dot{S}$ by $\dot{S}^*$ as "the" entropy function.

Proposition 6.3. *There exist integrable fields* s, k^c. *and* k, *and a measurable vector field* $\boldsymbol{j}$ *with weak divergence* $\operatorname{div} \boldsymbol{j}$ *such that*

$$\dot{s}(x) \geqq (\operatorname{div} \boldsymbol{j})(x) + k^e(x) + \int_B k(x, y)\, dv(y)$$

for v-a.e. $x \in B$.

The final assertion of the second law is fundamental: it asserts an intimate relation between heat transfer and entropy transfer.[1] We assume, for any $(A, C) \in \mathbb{M}ø\mathbb{N}$, that

$$\left.\begin{array}{l} M(A, C) = 0 \text{ whenever } H(\hat{A}, \hat{C}) = 0 \\ \text{for all } \hat{A} \subset A, \hat{C} \subset C \text{ with } \hat{A}, \hat{C} \in \mathbb{N}. \end{array}\right\} \tag{6.2}$$

If $C = B^e$, then this has an immediate consequence. Let us assume, then, that $C \in \mathbb{M}$. We will show that this requirement leads to corresponding relations between Q and J and R and K. Because of the nature of the condition (6.2) it is convenient to regard the extensions of H and M as measures on subsets of $A \times C$. Thus, we first define Radon measures λ_1, μ_1 by

$$\lambda_1(\Gamma) = \int_\Gamma r \, dv^2,$$

$$\mu_1(\Gamma) = \int_\Gamma k \, dv^2.$$

For the surface-dependent part of the transfers, we note, for fixed $(A, C) \in \mathbb{M}ø\mathbb{N}$, that if

$$\mathscr{S} = \partial A \cap \partial C,$$

then

$$\mathscr{S}^2 := \mathscr{S} \times \mathscr{S} \subset \{(x, y) \in B \times B \mid x = y\}$$

and in particular one can easily calculate that

$$\alpha := H_{n-1} \llcorner \mathscr{S}^2 = \frac{1}{\sqrt{2}} f_\sharp(H_{n-1} \llcorner \mathscr{S}),$$

where $f \colon B \to B \times B$ is the map $f(x) = (x, x)$ and H_{n-1} is $(n-1)$-dimensional Hausdorff measure. Here we let $f_\sharp(H_{n-1} \llcorner \mathscr{S})$ denote the f-transport of $H_{n-1} \llcorner \mathscr{S}$; the latter is defined to be the measure $A \mapsto H_{n-1}(\mathscr{S} \cap A)$. Then since $\partial\hat{A} \cap \partial\hat{C} \subset \mathscr{S}$ whenever $\hat{A} \subset A$, $\hat{C} \subset C$, if we write q for the H^{n-1}-integrable function

$$q(x) = q(x, \boldsymbol{v}(x)), \qquad x \in \mathscr{S},$$

and $\boldsymbol{v}$ the exterior normal to A, we have

$$Q(\hat{A}, \hat{C}) = \int_{\partial\hat{A} \cap \partial\hat{C}} q \, da = \sqrt{2} \int_{\hat{A} \times \hat{C}} (q \circ f^{-1}) \, d\alpha,$$

and thus we may also regard Q as a restriction of a Radon measure on $A \times C$. If we call this measure λ_2, and the measure corresponding to J we call μ_2, then

$$\lambda := \lambda_1 + \lambda_2$$

$$\mu := \mu_1 + \mu_2$$

---

[1] In the absence of other modes of energy transfer the relation loses some of its physical significance. We also omit, as irrelevant for our current purposes, the further significant requirement that the entropy transfer be of the same sign as the heat transfer (*cf.* [5]).

are the natural extensions of H and M respectively to subsets of $A \times C$. Since each may be regarded as a Lebesgue decomposition with respect to v^2, and since the sets $\hat{A} \times \hat{C}$, with $\hat{A}, \hat{C} \in \mathrm{M}$ generate in the obvious way the Borel subsets of $A \times C$, it follows from (6.2) that

$$\lambda \text{ is absolutely continuous with respect to } \mu,$$

$$\lambda_1 \text{ is absolutely continuous with respect to } \mu_1,$$

$$\lambda_2 \text{ is absolutely continuous with respect to } \mu_2.$$

This observation together with the previous representations and with the corresponding observation for $H(\cdot, B^e)$, $M(\cdot, B^e)$ lead to the following result:

Proposition 6.4. (Existence of "temperatures"). *There exist measurable functions*

$$\theta^e : B \to \mathbb{R},$$

$$\theta : B^2 \to \mathbb{R},$$

and, for each $(\mathcal{S}, \boldsymbol{v}) \in \mathbb{S}$,

$$\varphi_{\mathcal{S}} : \mathcal{S} \to \mathbb{R}$$

such that, for $(A, C) \in \mathrm{M} \varnothing \mathrm{M}$ *with* $\partial A \cap \partial C \subset \mathcal{S}$,

$$J(A, C) = \int_{\partial A \cap \partial C} \varphi_{\mathcal{S}}(x)\, q(x, \boldsymbol{v}(x))\, da(x),$$

$$K(A, C) = \int_{A \times C} \theta r \, dv^2,$$

and

$$H(A, B^e) = \int_A \theta^e r^e \, dv + \int_{\partial A \cap \partial B} \varphi_{\partial B}(x)\, q(x, \boldsymbol{v}(x))\, da(x).$$

Proposition 6.5. *There exists a function* $\varphi : B \times S_{n-1} \to \mathbb{R}$ *such that, for each* $(\mathcal{S}, \boldsymbol{v}) \in \mathbb{S}$,

$$\mathcal{J}(\mathcal{S}, \boldsymbol{v}) = \int_{\mathcal{S}} \varphi(x, \boldsymbol{v}(x))\, q(x, \boldsymbol{v}(x))\, da(x).$$

Proof. At each $(x, \boldsymbol{n}) \in B \times S_{n-1}$ for which $q(x, \boldsymbol{n}) \neq 0$, set

$$\varphi(x, \boldsymbol{n}) = \frac{j(x, \boldsymbol{n})}{q(x, \boldsymbol{n})};$$

otherwise let $\varphi(x, \boldsymbol{n}) = 1$. Choose $(\mathcal{S}, \boldsymbol{v}) \in S$. From Propositions 5.3, 6.2, and 6.5 there is a set $\mathcal{T} \subset \mathcal{S}$ with $a(\mathcal{S} \setminus \mathcal{T}) = 0$ and the property that

$$j(x, \boldsymbol{v}(x)) = \varphi_{\mathcal{S}}(x)\, q(x, \boldsymbol{v}(x))$$

for all $x \in \mathcal{T}$. But if $q(x, \boldsymbol{v}(x)) = 0$, then $j(x, \boldsymbol{v}(x)) = 0$, so that

$$j(x, \boldsymbol{v}(x)) = \varphi(x, \boldsymbol{v}(x))\, q(x, \boldsymbol{v}(x)),$$

while if $q(x, \boldsymbol{v}(x)) \neq 0$ by definition this still holds. □

Remark 6.5. This is the final reduction in form for φ (*cf.* [6], in which stronger assumptions are seen to lead to a slightly stronger result). We cannot further reduce the form of $\theta(\cdot, \cdot)$. In particular, it would be unreasonable to require that $\theta(x, y) = \Theta(x)$. The only constitutive model generally used for such distant interactions is that found in the theory of radiative transfer; there it appears reasonable (*cf.* [16]) to assume that entropy transfer into a point x should depend not only upon the amount of heat radiated into x but also upon the wave-length of the radiation from the source-point y. Thus $\int_B |r(x, y)|\, dv(y) = 0$ should *not* ensure that $\int_B k(x, y)\, dv(y) = 0$.

A final result now follows from Proposition 6.3.

Proposition 6.6. *There exists a function* $\varphi : \mathscr{B} \to \mathbb{R}$ *such that*

$$\boldsymbol{j} = \varphi \boldsymbol{q}$$

and hence such that for v-a.e. $x \in \mathscr{B}$,

$$\dot{s}(x) \geqq \operatorname{div}(\varphi \boldsymbol{q})(x) + \theta^e(x)\, r^e(x) + \int_B \theta(x, y)\, r(x, y)\, dv(y).$$

References

1. Whitney, H., *Geometric Integration Theory*, Princeton University Press, Princeton, 1957.
2. Kellogg, O. D., *Foundations of Potential Theory*, Dover, New York, 1953.
3. Federer, H., *Geometric Measure Theory*, Springer-Verlag, Berlin Heidelberg New York, 1969.
4. Giusti, E., Minimal surfaces and functions of bounded variation, (notes by G. H. Williams) *Notes in Pure Mathematics* **10**, Austral. Nat. Univ., Canberra, 1977.
5. Gurtin, M. E., & W. O. Williams, An axiomatic foundation for continuum thermodynamics, *Arch. Rational Mech. Anal.* **26**, 83–177, 1967.
6. Williams, W. O., On internal interactions and the concept of thermal isolation, *Arch. Rational Mech. Anal.* **34**, 245–258, 1969.
7. Williams, W. O., Thermodynamics of rigid continua, *Arch. Rational Mech. Anal.* **36**, 270–284, 1970.
8. Munroe, M. E., *Introduction to Measure and Integration*, Addison-Wesley, Reading, Mass., 1953.
9. Goffman, C., & D. Waterman, Approximately continuous transformations, *Proc. Am. Math. Soc.* **12**, 116–121, 1961.
10. Sikorsky, R., *Boolean Algebras*, Springer-Verlag, Berlin Heidelberg New York, 1969.
11. Ziemer, W. P., Cauchy flux and sets of finite perimeter, *Arch. Rational Mech. Anal.* **84**, 189–201, 1983.
12. Burkhill, J. C., Functions of intervals, *Proc. London Math. Soc.* (2) **22**, 275–310, 1923.

13. GURTIN, M. E., & L. C. MARTINS, Cauchy's theorem in classical physics, *Arch. Rational Mech. Anal.* **26**, 83–117, 1976.
14. NOLL, W., The foundations of classical mechanics in the light of recent advances in continuum mechanics, in *The Axiomatic Method with special Reference to Geometry and Physics*, North Holland Company, Amsterdam, 1959.
15. SAKS, S., *Theory of the Integral*, 2nd rev. ed., Monografie Mat., vol. 7, PWN, Warsaw, 1937.
16. PLANCK, M., *The Theory of Heat Radiation*, trans. M. MASIUS, Dover, New York, 1959.

Department of Mathematics
Carnegie-Mellon University
Pittsburgh
and
Department of Mathematics
Indiana University
Bloomington

(Received April 9, 1985)

Cavitation and Phase Transition of Hyperelastic Fluids

P. Podio-Guidugli, G. Vergara Caffarelli, & E. G. Virga

Dedicated to Walter Noll on his sixtieth birthday

Contents

1. Introduction

A hyperelastic material is described by a mapping $F \mapsto \sigma(F)$ delivering the stored energy, per unit volume of a fixed local reference placement, for any admissible deformation gradient F. As is physically reasonable, the mechanical response of a hyperelastic material is invariant under a certain group $\mathcal{G}$ of linear transformations interpreted as deformation gradients defining other local placements, all of which are undetectable from the one originally selected as reference. More precisely, according to Noll, $\mathcal{G}$ consists in all linear transformations H such that

$$\sigma(FH) = \sigma(F), \tag{1.1}$$

for all F in the domain of σ.

Consideration of $\mathcal{G}$ allows a distinction between *fluids*, for which $\mathcal{G} = \mathcal{U} := \{H \mid \det H = 1\}$, and *solids*, for which $\mathcal{G} \subset \mathcal{R} := \{H \in \mathcal{U} \mid H^T = H^{-1}\}$. However, one is left with the natural desire to find a rationale such as to distinguish liquid fluids from gaseous fluids.

In this paper, for a fairly broad class of hyperelastic fluids, we offer such a rationale by showing how certain expected phenomena, characteristic of liquids and gases, respectively, are reflected by the solutions of certain boundary-value problems of place and traction under suitable qualitative hypotheses on the stored energy function σ.

We assume that

$$\sigma(F) = \varphi(\det F), \tag{1.2}$$

with φ a smooth, strictly convex function which becomes infinite when $\det F$ tends to zero. For such fluids, either φ has one isolated minimum or it is monotone decreasing. We then show, among other things, that for critical boundary data:
(a) under uniform radial outwards displacement of the boundary, a fluid ball may cavitate, as is typical of liquids, if and only if φ has an isolated minimum;
(b) under uniform radial Piola-Kirchhoff pressure, a fluid ball may undergo a phase transition (*i.e.*, admit two equi-energetic equilibrium states with different radii), if and only if the mapping $\delta \mapsto \delta^{2/3}\,\varphi'(\delta)$ is not monotone.

We remark that Ball [B1, B2] and Dacorogna [D1, D2, D3] considered classes of hyperelastic materials which include the fluids (2) as a special case. Moreover, in [B3] and [PVV], certain hyperelastic solids are studied whose stored energy differs from (2) by a term that by itself defines the neo-hookean stored energy:

$$\bar{\sigma}(F) = \tfrac{1}{2}\,(F \cdot F - 3) + \varphi\,(\det F).$$

For a ball made of such a solid and subject to boundary conditions of either place or traction, discontinuous energy minimizers, possibly to be associated with cavitation, have been shown to be expected under assumptions on φ analogous to those prevailing here.

In Section 2 of this paper we derive the differential equations for the equilibrium of the fluids described by (2). We wish to point out two side results that are established in that section: (i) σ satisfies the Legendre-Hadamard condition if and only if φ is convex; (ii) the equilibrium equations cannot possibly be an elliptic system in any of the usual senses, no matter how the function φ be chosen.

In Section 3 we introduce the non-convex energy functional which governs the equilibrium problems to be considered later, and make repeated use of the classical inequality of Jensen to prove in an elementary and direct way that: (iii) minimizers must induce uniform change of volume; (iv) the mapping $F \mapsto \sigma(F)$ defined by (2) is quasiconvex in the sense of Morrey.

The developments of the remaining sections require a premiss which may also help to put our present work into proper perspective. We do not aim here to furnish full mathematical description of the phenomenology of fluid cavitation: for this, a theory based simply on a volume energy density might conceivably be too poor *ab initio*. Rather, we are content to exhibit one explicit mathematical instance which, at least when surface effects are negligible, suggests the possibility of cavitation, thereby singling out a constitutive property which allows easy conceptual distinction between liquid and gaseous elastic fluids.

Accordingly, we restrict attention to a fluid body making up a ball in a (possibly stressed) reference equilibrium placement, and, from Section 4 on, to uniformly radial further deformations of this ball which may be discontinuous only at the center (so that, in particular, in this paper the term cavitation refers to an equilibrium solution with a centered spherical hole). After briefly describing radial deformations in Section 4, in Section 5 we consider the problem of place obtained by imposing a uniform radial displacement on the boundary; finally, in Section 6, we examine the traction problems which result by prescribing either the Cauchy pressure or the Piola-Kirchhoff pressure at the boundary. Our main results in these last two sections are stated above under (a) and (b).

* * *

On reading a former version of this paper, M. E. GURTIN and W. NOLL, whom we thank for many valuable comments. remarked separately that, for a liquid, outwards displacements of the boundary large enough to annihilate the internal pressure, as well as null surface tractions, should correspond to an infinite variety of "deformations with cavitation", each giving one and the same minimal value to the energy integral and entailing a subdivision of the fluid body into possibly multi-connected parts of abitrary shape.

Prompted by this remark, we have included an Appendix in which we exemplify how some of our methods can be adapted to situations not restricted to uniformly radial deformations which may be discontinuous at the center only.

At a cursory glance, the analysis in the Appendix is in rather fair agreement with the physical expectations outlined above; moreover, an easy increase in mathematical sophistication of the setting would lead to an apparently even better agreement. However, no matter how much improved, such an agreement would be somewhat illusory. Indeed, as we have already remarked, in order to construct the "right" mathematics of fluid cavitation one should introduce physically significant modifications of the basic energy functional (perhaps calling upon constitutive assumptions extrinsic to elasticity) rather than allow a more general class possible minimizers.

2. Equilibrium Equation

Consider a continuous body which, in a reference placement, occupies the unit ball $\mathcal{B}$ of $\mathbb{R}^3$. Let f denote a deformation of $\mathcal{B}$, *i.e.*, a smooth mapping of $\mathcal{B}$ into $\mathbb{R}^3$ with deformation gradient $F := \mathrm{D}f$ consistent with

$$\det F(x) > 0 \quad \text{for } x \in \mathcal{B}. \tag{2.1}$$

We assume that the material of which the body consists is a hyperelastic fluid with stored energy

$$\sigma(F) = \varphi(\det F) \tag{2.2}$$

per unit reference volume. Furthermore, we assume that

$$\varphi \in C^2(0, \infty) \quad \text{with } \varphi'' > 0. \tag{H1}$$

In view of (2), the Piola-Kirchhoff stress S and the Cauchy stress T depend on the deformation gradient F in the following way:

$$S(F) = \mathrm{D}\sigma(F) = \varphi'(\det F)\, F^*; \tag{2.3}$$

$$T(F) = (\det F)^{-1}\, S(F)\, F^T = \varphi'(\det F)\, I. \tag{2.4}$$

Here $F^* = \mathrm{D}(\det F) = (\det F)\, F^{-T}$, F^{-T} is the inverse of the transpose F^T of F, and I denotes the identity tensor. We may write $(4)_2$ in the form

$$T = -\pi I, \tag{2.5}$$

with π, the pressure function, given by

$$\pi = -\varphi'. \tag{2.6}$$

Customarily, one requires that

$$\pi \geqq 0, \tag{2.7}$$

and this requirement, in the present context, would lead one to assume, along with (H1), that φ is strictly decreasing. However, we choose not to regard (7) as a constitutive constraint, but rather as a property, to be verified *a posteriori*, of solutions of certain equilibrium problems.

In the absence of body forces, such problems are based on the equilibrium equation

$$\operatorname{Div} S = 0, \tag{2.8}$$

which by $(3)_2$ we may write as

$$\operatorname{Div}(\varphi'(\det F)\, F^*) = 0 \tag{2.9}$$

and view as a quasilinear second-order system of partial differential equations for the unknown f. A natural expectation in elasticity theory would be that this system is (in some suitable sense) elliptic, provided that the stored energy function σ is (in some suitable sense) convex. We show hereafter that, although σ is indeed rank-one convex, the system (9) is neither strongly elliptic nor elliptic in the sense of Petrowskii.

Recall that σ is *rank-one convex* if it satisfies the inequality

$$\sigma(\tau F + (1 - \tau)\, G) \leqq \tau\sigma(F) + (1 - \tau)\, \sigma(G), \tag{2.10}$$

where $F - G = a \otimes b$, for all $\tau \in (0, 1)$ and for every pair of vectors a, b; it is known (*cf. e.g.* [B1], p. 352) that (10) holds if and only if σ obeys the *Legendre-Hadamard* condition, namely,

$$(a \otimes b) \cdot \mathrm{D}^2\sigma(F)\, [a \otimes b] \geqq 0 \tag{2.11}$$

for all vectors a, b. Recall further that a system is called *strongly elliptic* if the inequality (11) is strict for all non-null a and b; *elliptic in the sense of Petrowskii*, if

$$\det\left((\mathrm{D}^2\sigma(F))_{ijhk}\, b_j b_k\right) \neq 0 \tag{2.12}$$

for all vectors $b \neq 0$.

Now, simple computations based on $(3)_2$ yield

$$(a \otimes b) \cdot D^2\sigma(F)\,[a \otimes b] = \varphi''\,(\det F)\,(a \cdot F^*b)^2, \tag{2.13}$$

and

$$\det\,((D^2\sigma(F))_{ijhk}\, b_j b_k) = (\varphi''(\det F))^3 \det\,(F^*b \otimes F^*b). \tag{2.14}$$

Thus, for the stored energy function (2), we see from (13) that rank-one convexity of σ is equivalent to convexity of φ; moreover, as vectors $a, b \neq 0$ can be found which annihilate the right-hand side of (13), whereas the right-hand side of (14) vanishes for any b, the system (9) is not elliptic in any of the usual senses, no matter what be the constitutive function φ.

Such state of affairs indicates that troubles are to be expected in solving boundary-value problems for $\mathscr{B}$. In particular, one might be unable to predict whether or not solutions exist*. Anyway, it is not difficult to show that solutions of (9), whenever they exist, must induce uniform change of volume.

Indeed, as

$$\text{Div}\, F^* = 0, \tag{2.15}$$

(*cf.*, *e.g.*, equation (18.2) of [TT]), the equilibrium equation (9), in view also of $(\text{H1})_2$, reduces to

$$\text{Grad}\,(\det F) = 0. \tag{2.16}$$

3. Variational Formulation

The equilibrium of $\mathscr{B}$ can also be formulated as a variational problem, with the system (2.9) viewed as the Euler-Lagrange equation corresponding to the energy functional

$$\mathscr{E}\{f\} := \int_{\mathscr{B}} \varphi(\det Df). \tag{3.1}$$

Again, deformations which are minimizers of (1) may or may not exist, but if they do, they obey (2.16).

To prove the last assertion, which reflects a well known fact in the mechanics of elastic fluids, we now wish to use a classical inequality due to JENSEN (*cf. e.g.* [HPL], p. 150).

In any function class rendering both the integrals $\int_{\mathscr{B}} \det(Df)$ and (1) finite, Jensen's inequality yields

$$\int_{\mathscr{B}} \varphi(\det F) \geqq (\text{vol}\,\mathscr{B})\,\varphi\left(\frac{\int_{\mathscr{B}} \det F}{\text{vol}\,\mathscr{B}}\right), \tag{3.2}$$

---

* In this connection, see Pbm 5.11, p. 409 of [MH].

so that, on setting

$$\delta = \frac{\int_{\mathscr{B}} \det F}{\operatorname{vol} \mathscr{B}}, \tag{3.3}$$

one has

$$\int_{\mathscr{B}} \varphi(\det F) \geqq \int_{\mathscr{B}} \varphi(\delta). \tag{3.4}$$

As φ is strictly convex by (H1), equality obtains in (4) only if $\det F(x) \equiv \delta$. Therefore, all the minimizers of (1) obey (2.16). In other words, given $\delta \in (0, \infty)$,

$$\mathscr{E}\{f\} \geqq (\operatorname{vol} \mathscr{B})\, \varphi(\delta)$$

for any f in the domain of $\mathscr{E}$ such that

$$\int_{\mathscr{B}} \det(Df) = (\operatorname{vol} \mathscr{B})\, \delta.$$

We call the mapping $\delta \mapsto (\operatorname{vol} \mathscr{B})\, \varphi(\delta)$ a *lower envelope* for the functional $\mathscr{E}$.

Another straightforward use of Jensen's inequality allows us to show that the mapping $F \mapsto \varphi(\det F)$ is *quasiconvex* in the sense of [M], [B1], *i.e.*,

$$\int_{\mathscr{B}} \varphi(\det(F_0 + Dg)) \geqq \int_{\mathscr{B}} \varphi(\det F_0) = (\operatorname{vol} \mathscr{B})\, \varphi(\det F_0) \tag{3.5}$$

for any constant tensor field F_0, and for any vector field $g \in C_0^\infty(\mathscr{B})$ such that $\det(F_0 + Dg) > 0$. Although this fact follows from more general results (*cf.* [D1], p. 46), we give a direct proof.

In view of Jensen's inequality,

$$\int_{\mathscr{B}} \varphi(\det(F_0 + Dg)) \geqq (\operatorname{vol} \mathscr{B})\, \varphi\left(\frac{\int_{\mathscr{B}} \det(F_0 + Dg)}{\operatorname{vol} \mathscr{B}}\right)$$

which is (5), provided that

$$\int_{\mathscr{B}} \det(F_0 + Dg) = \int_{\mathscr{B}} \det F_0. \tag{3.6}$$

To establish (6), let $t \mapsto F(t) = F_0 + t\, Dg$. Then, firstly,

$$\int_{\mathscr{B}} \det(F_0 + Dg) - \int_{\mathscr{B}} \det F_0 = \int_0^1 \frac{d}{dt}\left(\int_{\mathscr{B}} \det F(t)\right);$$

secondly,

$$\frac{d}{dt}(\det F(t)) = F^*(t) \cdot Dg;$$

thirdly, as F^* is solenoidal by (2.15), and g has compact support,

$$\int_{\mathscr{B}} F^*(t) \cdot Dg = 0$$

by the divergence theorem. Hence the desired conclusion.

4. Radial Deformations

The difficulties we have alluded to with solving equation (2.9), or finding minimizers for the integral (3.1), disappear when we restrict ourselves to the class of radial deformations of $\mathcal{B}$, where explicit equilibrium solutions are easily found. We collect hereafter the necessary material on radial deformations, taking it from [PVV] (*vid.* also [B3]).

A deformation f of $\mathcal{B}$ is (uniformly) *radial* if there is a smooth, strictly increasing function ϱ mapping $(0, 1]$ into $\mathbb{R}^+$, such that

$$f(x) = \varrho(r)\, p(x), \tag{4.1}$$

with $r := |x|$, and $p(x) := r^{-1} x$ the radial unit vector at point x. Clearly, f is well defined in $\mathcal{B} \setminus \{0\}$. For the function ϱ, the limit

$$\lim_{r \to 0^+} \varrho(r) = \varrho(0)$$

exists, and we assume that

$$\varrho(0) \geqq 0.$$

If $\varrho(0) = 0$, we set $f(0) = 0$, and call f a *regular* deformation. If $\varrho(0) > 0$, we leave f undefined at the origin, and call f *irregular*; an irregular deformation produces a spherical hole centered at 0, of radius $\varrho(0)$. Regular deformations of special interest are the homogeneous deformations

$$f(x) = \alpha x, \qquad \alpha > 0;$$

in a homogeneous deformation

$$\varrho(r) = \alpha r.$$

It follows from (1) that

$$F(x) = \frac{\varrho(r)}{r}(I - P(x)) + \varrho'(r)\, P(x), \tag{4.2}$$

with $P(x)$ the radial projection tensor at x:

$$P(x) := p(x) \otimes p(x).$$

Consequently,

$$\det F(x) = \varrho'(r)\, \frac{\varrho^2(r)}{r^2} > 0, \tag{4.3}$$

$$F^*(x) = \varrho'(r) \frac{\varrho(r)}{r}(I - P(x)) + \frac{\varrho^2(r)}{r^2} P(x), \tag{4.4}$$

and, by (2), (4) and $(2.3)_2$, we conclude that both the deformation gradient F and the Piola-Kirchhoff stress S have as proper spaces at x the line spanned by $p(x)$ and the plane orthogonal to this line, whereas the corresponding proper numbers are, respectively,

$$f_1(x) = \varrho'(r), \quad f_2(x) = f_3(x) = \frac{\varrho(r)}{r}; \tag{4.5}$$

$$\begin{aligned} s_1(x) &= \frac{\varrho^2(r)}{r^2}\,\varphi'\left(\varrho'(r)\,\frac{\varrho^2(r)}{r^2}\right), \\ s_2(x) = s_3(x) &= \varrho'(r)\,\frac{\varrho(r)}{r}\,\varphi'\left(\varrho'(r)\,\frac{\varrho^2(r)}{r^2}\right). \end{aligned} \tag{4.6}$$

5. Problem of Place and Cavitation of Liquids

For $\varrho_1 > 0$ given, we look for a radial deformation f of $\mathscr{B}$ obeying the uniform displacement condition at the boundary $\partial\mathscr{B}$:

$$\varrho(1) = \varrho_1, \quad \text{or, equivalently,} \quad f(x) = \varrho_1 x \quad \text{for } x \in \partial\mathscr{B}, \tag{5.1}$$

and minimizing the energy integral (3.1). Prospective radial minimizers, like any minimizer, must induce a constant change of volume, a condition that, in view of (4.3), can be written as

$$\varrho'(r)\,\frac{\varrho^2(r)}{r^2} = \delta_0, \tag{5.2}$$

with δ_0 a positive constant to be determined in such a way as to minimize $\varphi(\delta_0)$. An estimate from above for δ_0 follows from the fact that the volume of $f(\mathscr{B})$, the deformed placement of $\mathscr{B}$, cannot exceed the volume of $\mathscr{B}_1$, the ball of center 0 and radius ϱ_1:

$$\operatorname{vol} f(\mathscr{B}) \leqq \operatorname{vol} \mathscr{B}_1; \tag{5.3}$$

for radial deformations, this implies that

$$\delta_0 \leqq \varrho_1^3. \tag{5.4}$$

We now stipulate that φ satisfies a further restriction of physical plausibility, namely,

$$\lim_{\delta \to 0^+} \varphi(\delta) = +\infty. \tag{H2}$$

The interpretation of (H2), as is well known, is that local implosions of matter would require infinite energy. Hypotheses (H1) and (H2) together imply that φ either has an isolated minimum or is strictly decreasing over $(0, \infty)$. We consider these two instances separately.

Firstly, suppose that φ has an isolated minimum at δ^0. Then, we have the following alternative: either $\varrho_1^3 \leqq \delta^0$, and then, in view of (3) and (4), respectively,

$$\operatorname{vol} f(\mathscr{B}) = \operatorname{vol} \mathscr{B}_1, \quad \delta_0 = \varrho_1^3; \tag{5.5}$$

or $\varrho_1^3 > \delta^0$, and then

$$\operatorname{vol} f(\mathscr{B}) < \operatorname{vol} \mathscr{B}_1, \quad \delta_0 = \delta^0. \tag{5.6}$$

Moreover, when $(5)_2$ prevails, equation (2) with the boundary condition $(1)_1$ is solved by the homogeneous deformation

$$\varrho(r) = \varrho_1 r; \tag{5.7}$$

otherwise, $(6)_2$ holds, and the solution is the irregular deformation

$$\varrho(r) = (\delta^0 r^3 + (\varrho_1^3 - \delta^0))^{1/3},$$

which is accompanied by the formation of a centered hole of radius

$$\varrho(0) = (\varrho_1^3 - \delta^0)^{1/3},$$

and by a null stress field throughout.*

Secondly, suppose that φ is strictly decreasing over $(0, \infty)$. Then, the restriction of φ to $(0, \varrho_1^3)$ attains its minimum value at ϱ_1^3, *i.e.*, $(5)_2$ holds and the homogeneous deformation (7) solves our minimization problem.

As cavitation is a phenomenon which may be exhibited by a liquid, but never is by a gas in statics, our present findings suggest that hyperelastic fluids whose stored energy function φ obeys (H1) and (H2) should be classified as liquids if φ has an isolated minimum, as gases when φ is strictly decreasing.** We adhere to this criterion in the rest of this paper.

6. Traction Problems and Phase Transition of Gases

In this section we deal with equilibrium problems where either the Cauchy traction or the Piola-Kirchhoff traction vector at the boundary is specified, and demonstrate the diverse behavior of liquids and gases of the class considered here. In closure, for the second problem above, we work out the example of a Van der Waals gas exhibiting phase transition at a critical value of the boundary pressure.

6.1 The Cauchy Traction Problem

Let π_0 be the amount of the uniform radial Cauchy traction applied throughout $f(\partial\mathcal{B})$. Within the class of radial deformations, the total energy functional has the form

$$\mathcal{E}_C\{f\} := \int_{\mathcal{B}} \varphi\,(\det Df) + \pi_0\,(\operatorname{vol}\mathcal{B}_1 - \operatorname{vol}\mathcal{B}), \tag{6.1}$$

where ϱ_1, the radius of $\mathcal{B}_1 = f(\mathcal{B})$, is now a parameter, not a datum.

---

* As $\varphi'(\delta^0) = 0$, (4.6) imply that the stress vanishes identically in $\mathcal{B}$.

** Notice that the usual assignments of φ for gases (ideal, polytropic, Van der Waals, ...) are such that φ, at a fixed temperature (higher than the critical temperature in the Van der Waals case), satisfies (H1), (H2) and is indeed strictly decreasing over the positive half-line.

The argument leading to (3.4) allows us to conclude that the lower envelope of the functional $\mathscr{E}_{\mathrm{C}}$ is the mapping $(\delta, \varrho_1) \mapsto (\mathrm{vol}\,\mathscr{B})\,(\psi_{\mathrm{C}}(\delta, \varrho_1) - \pi_0)$, where

$$\psi_{\mathrm{C}}(\delta, \varrho_1) := \varphi(\delta) + \pi_0 \varrho_1^3. \tag{6.2}$$

Modulo some inessential constants, we then have to minimize ψ_C over the set $D := \{(\delta, \varrho_1) \mid 0 \leqq \delta \leqq \varrho_1^3\}$. Three situations may occur:

(i) $\pi_0 < 0$ (uniform outward traction)

In this case, ψ_{C} is not bounded below on D, so that our equilibrium problem has no solution.

(ii) $\pi_0 = 0$ (null surface traction)

ψ_{C} is now independent of ϱ_1. If the material is a liquid, ψ_{C} has constant minimum value on the half-line $\{(\delta^0, \varrho_1) \mid \delta^0 \leqq \varrho_1^3\}$. Consequently, all minimizers give the energy integral the same value and entail cavitation, except for the one corresponding to the initial point $(\delta^0, (\delta^0)^{1/3})$ of the half-line, which has the same energy as the others, but no cavitation: we may single out this peculiar state at ease of our liquid body, and interpret the situation as an instability of this state with respect to radial deformations; when we do so, such a loose statement as "liquids have no preferred configuration" acquires a definite meaning.

If the material is a gas, for any ϱ_1^0 fixed, ψ_{C} has minimum value at the boundary point $((\varrho_1^0)^3, \varrho_1^0)$ of D. Thus, as is physically expected, a gaseous body under null surface traction expands freely so as to occupy the whole space.

(iii) $\pi_0 > 0$ (uniform pressure)

In this case, it is easily seen that ψ_{C} attains its minimum at any point $(\delta, \delta^{1/3})$ of the boundary curve of D such that δ solves the equation

$$\varphi'(\delta) + \pi_0 = 0. \tag{6.3}$$

By (H1), equation (3) has exactly one root $\bar{\delta}$, which always corresponds to a regular deformation (and, if the material is a liquid, is such that

$$\bar{\delta} < \delta^0). \tag{6.4}$$

6.2 The Piola-Kirchhoff Traction Problem

Let τ_0 be the amount of the (uniform, radial) Piola-Kirchhoff traction assigned per unit area of $\partial\mathscr{B}$. Again, restrict attention to radial deformations. Then the total energy functional has the form

$$\mathscr{E}_{\mathrm{PK}}\{f\} := \int_{\mathscr{B}} \varphi\,(\det \mathrm{D}f) + 3\tau_0 \left(\frac{\mathrm{area}\,(\partial\mathscr{B})}{\mathrm{area}\,(\partial\mathscr{B}_1)} \mathrm{vol}\,\mathscr{B}_1 - \mathrm{vol}\,\mathscr{B}\right) \tag{6.5}$$

(compare with (1)), and the lower envelope of $\mathscr{E}_{\mathrm{PK}}$ is the mapping $(\delta, \varrho_1) \mapsto$ $(\mathrm{vol}\,\mathscr{B}) \left(\psi_{\mathrm{PK}}(\delta, \varrho_1) - \dfrac{\mathrm{area}\,(\partial\mathscr{B})}{\mathrm{vol}\,\mathscr{B}} \tau_0\right)$, with

$$\psi_{\mathrm{PK}}(\delta, \varrho_1) := \varphi(\delta) + 3\tau_0 \varrho_1 \tag{6.6}$$

an ordinary function to be minimized over D in order to solve our original problem of minimization.

Now, for $\tau_0 \leqq 0$, the qualitative properties of ψ_{PK} are the same as those of ψ_C for $\pi_0 \leqq 0$, so that the discussion under (i) and (ii) of 6.1 could be paraphrased, the mechanical interpretation included. However, the case $\tau_0 > 0$ is different, and far more interesting than the corresponding case (iii) of subsection 6.1.

Indeed, if $\tau_0 > 0$, ψ_{PK} attains its minimum on the curve $\delta = \varrho_1^3$ for any δ such that

$$\delta^{2/3}\varphi'(\delta) + \tau_0 = 0. \tag{6.7}$$

On the other hand, hypotheses (H1) and (H2) together imply that

$$\lim_{\delta \to 0+} \delta^{\alpha}\varphi'(\delta) = -\infty \quad \text{for any } \alpha < 1. \tag{6.8}$$

Thus, for any given $\tau_0 > 0$, equation (7) is solvable. Again, solutions $\bar{\delta}$ of (7) correspond to regular deformations, and satisfy (4) if the material is a liquid. However, since there are functions φ which agree with (H1), (H2) but give rise to non-monotone mappings $\delta \mapsto \delta^{2/3}\varphi'(\delta)$, uniqueness may fail for (7). When this is the case, a phenomenon of *instability by phase transition* is in order, analogous to the one originally modelled by ERICKSEN in [E] for one-dimensional elastic bars in tension with non-convex stored energy. The interpretation is now that a three-dimensional fluid ball under critical Piola-Kirchhoff pressure may admit of various equilibrium placements (phases), all of these being balls of different radii. We exemplify this situation in the next subsection.

6.3 Phase Transition of Van der Waals Gases

The pressure function for a Van der Waals gas reads

$$\hat{\pi}(\xi) = \frac{1}{\xi - 1} - \varkappa \frac{1}{\xi^2},$$

with

$$\xi := \frac{v}{v_C} > 1, \quad v := v_0\,\delta, \quad \hat{\pi}(\xi) := \frac{v_C}{R\vartheta}\pi(\delta)$$

and

$$\pi(\delta) = -\varphi'(\delta)$$

(*cf.* (2.6)): moreover,

$$\varkappa := \frac{27}{8}\frac{\vartheta_C}{\vartheta}.$$

Here v is the specific volume and ϑ is the absolute temperature, with v_C and ϑ_C their critical values, respectively; R is the gas constant; v_0 is the specific volume in the reference placement.

Modulo some appropriate rescaling, the mapping $\delta \mapsto \delta^{2/3}\varphi'(\delta)$ is the same as the mapping $\xi \mapsto \gamma(\xi) = -\xi^{2/3}\,\hat{\pi}(\xi)$. It is easy to verify that if

$$\tfrac{8}{3} < \varkappa < \tfrac{27}{8},$$

then $\hat{\pi}'(\xi)$ is negative and $\gamma(\xi)$ is not monotone. The graphs of $\hat{\pi}(\xi)$ and $\gamma(\xi)$ *vs.* $\log \xi$ are sketched in figures 1 and 2, respectively, for $\varkappa = 2.9$.

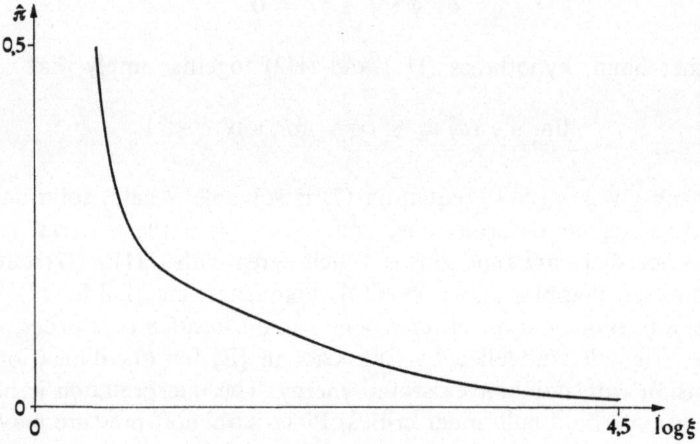

Fig. 1

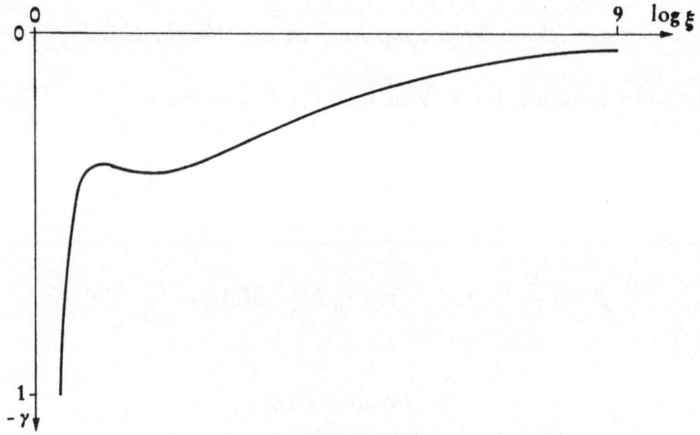

Fig. 2

The multiplicity and stability analysis of ERICKSEN, based on the humped curve of fig. 2 (interpreted as a non-monotone stress-strain law), has been repeated later, with slight variants, by many authors; we follow hereafter the neat version recently offered by CARR, GURTIN, & SLEMROD in [CGS].

Define ξ^* and ξ_* to be such that γ' is first positive on $(0, \xi^*)$, then negative on (ξ^*, ξ_*), and again positive on $(\xi_*, +\infty)$. Also, set

$$\gamma_* = \gamma(\xi_*), \quad \gamma^* = \gamma(\xi^*).$$

For $\gamma_0 \in (\gamma_*, \gamma^*)$ there are exactly three values of ξ solving the equation

$$\gamma(\xi) + \gamma_0 = 0; \tag{6.9}$$

we denote these by

$$\alpha(\gamma_0) < \lambda(\gamma_0) < \beta(\gamma_0).$$

For $\gamma_0 < \gamma_*$, we continue to denote by $\alpha(\gamma_0)$ the unique root of (9), and likewise for $\gamma_0 > \gamma^*$ and $\beta(\gamma_0)$. With slight abuse of terminology, the *maxwellian stress* γ_M is the unique value in (γ_*, γ^*) such that

$$\int_{\alpha(\gamma_M)}^{\beta(\gamma_M)} \gamma(\xi)\, d\xi = \gamma_M(\beta(\gamma_M) - \alpha(\gamma_M)).$$

For equation (9), the root of minimal energy is $\alpha(\gamma_0)$ for $\gamma_0 < \gamma_M$, and $\beta(\gamma_0)$ for $\gamma_0 > \gamma_M$; for $\gamma_0 = \gamma_M$, the energy of $\alpha(\gamma_0)$ is the same as $\beta(\gamma_0)$, and lower than the energy of $\lambda(\gamma_0)$.

Imagine now a loading process during which γ_0 is varied continuously starting from zero. As long as γ_0 is greater than the maxwellian stress γ_M, the radius ϱ_1 of the deformed ball $\mathscr{B}_1$ decreases continuously according to the law $\varrho_1 = \mu\beta^{1/3}(\gamma_0)$, with $\mu := (v_C/v_0)^{1/3}$; at $\gamma_0 = \gamma_M$, it suffers a discontinuity (of the kind called first-order phase transitions by LANDAU), jumping from $\mu\beta^{1/3}(\gamma_M)$ down to $\mu\alpha^{1/3}(\gamma_M)$; finally, for $\gamma_0 < \gamma_M$, $\varrho_1 = \mu\alpha^{1/3}(\gamma_0)$. Notice that for no value of the loading parameter γ_0 can the ball $\mathscr{B}$ be maintained at rest in the form of a ball $\mathscr{B}_1$ with radius $\varrho_1 \in (\mu\alpha^{1/3}(\gamma_M), \mu\beta^{1/3}(\gamma_M))$.

Notice also that, at variance with the one-dimensional bars of ERICKSEN, at the maxwellian stress either one or the other of the two equi-energetic phases prevails, but of course not both.

References

[B1] J. M. BALL, Convexity conditions and existence theorems in nonlinear elasticity, *Arch. Rational Mech. Anal.* **63** (1977), pp. 337–403.

[B2] J. M. BALL, Strict convexity, strong ellipticity, and regularity in the calculus of variations, *Math. Proc. Camb. Phil. Soc.* **87** (1980), pp. 501–513.

[B3] J. M. BALL, Discontinuous equilibrium solutions and cavitation in nonlinear elasticity, *Phil. Trans. R. Soc.* (London) A **306** (1982), pp. 557–611.

[CGS] J. CARR, M. E. GURTIN & M. SLEMROD, *One-dimensional structured phase transformations under prescribed loads*, MRC Report no. 2559, Mathematics Research Center, University of Wisconsin-Madison, 1983.

[D1] B. DACOROGNA, *Weak continuity and weak semicontinuity of non-linear functionals*, Lecture Notes in Mathematics no. 922, Springer-Verlag, Berlin, 1982.

[D2] B. DACOROGNA, A relaxation theorem and its application to the equilibrium of gases, *Arch. Rational Mech. Anal.* **77** (1981), pp. 359–386.

[D3] B. DACOROGNA, Quasiconvexity and relaxation of nonconvex problems in the calculus of variations, *J. Funct. Anal.* **46** (1982), pp. 102–118.

[E] J. L. ERICKSEN, Equilibrium of bars, *J. Elasticity* **5** (1975), pp. 191–201.

[G] M. E. GURTIN, *An introduction to continuum mechanics*, Academic Press, New York, 1981.

[HPL] G. H. HARDY, J. E. LITTLEWOOD & G. POLYA, *Inequalities*, Cambridge University Press, Cambridge, 1952.

[M] C. B. MORREY, Quasi-convexity and the lower semicontinuity of multiple integrals, *Pacific J. Math.* **2** (1952), pp. 25–53.

[MH] J. E. MARSDEN & T. J. R. HUGHES, *Mathematical foundations of elasticity*, Prentice-Hall, Englewood Cliffs, 1983.

[PVV] P. PODIO-GUIDUGLI, G. VERGARA CAFFARELLI & E. G. VIRGA, Discontinuous energy minimizers in nonlinear elastostatics: an example of J. Ball revisited, to appear in *J. Elasticity*, 1984.

[TT] C. TRUESDELL & R. A. TOUPIN, The classical field theories, *Handbuch der Physik* III/1, S. Flügge Ed., Springer-Verlag, Berlin, 1960.

Appendix

As is well known, in continuum mechanics a deformation is an (orientation-preserving) everywhere smooth bijection of regions in Euclidean space. Thus, strictly speaking, the irregular radial mappings introduced in Section 4 should not have been termed deformations. On the other hand, in a variational approach to equilibrium problems, any mapping which keeps the basic functional well defined is in principle an admissible minimizer; moreover, both for solids and liquids, consideration of classes of "deformations" larger than the canonical one may conceivably yield non-smooth minimizers, which one would be tempted to interpret as accounting for cavitation. As BALL has shown first, the possibility of such "cavitations" is most conveniently demonstrated by minimizing the energy integral in the radial class described in Section 4. It is the purpose of this appendix to show that, for fluids, the same can be done, with no more effort and the same methods as in the preceding sections, for not necessarily radial classes of mappings, also to be regarded as deformations in a sense weaker than the usual one, wich fit physical intuition better than the radial class.

Consider the collection of all partitions P of $\mathscr{B}$, a partition being a finite family of open disjoint parts $\mathscr{P}_i$ of $\mathscr{B}$, such that

$$\bigcup_i \overline{\mathscr{P}}_i = \overline{\mathscr{B}} \tag{A.1}$$

(here $\overline{\mathscr{P}}_i$ denotes the closure of $\mathscr{P}_i$, a measurable set in the sense of Peano-Jordan). Given a partition P, the *partitioning deformations* associated with P are all the mappings f_{P} of $\bigcup_i \mathscr{P}_i$ into $\mathbb{R}^3$ such that: (i) for all indices i, the restriction of f_{P} to $\mathscr{P}_i$ is a deformation in the usual sense: (ii) for $i \neq j$, the images under f_{P} of $\mathscr{P}_i$ and $\mathscr{P}_j$ are disjoint. When the stored energy functional (3.1) is evaluated at a partitioning deformation, it takes the form

$$\mathscr{E}\{f_{\mathrm{P}}\} = \sum_i \int_{\mathscr{P}_i} \varphi\,(\det \mathrm{D}f_{\mathrm{R}}); \tag{A.2}$$

with no substantial loss of generality we assume that the constitutive function φ attains its minimum at $\delta^0 = 1$.

Now, it is easily seen that the stored energy functional has constant minimal value

$$\mathscr{E}_{\min} = (\text{vol}\,\mathscr{B})\,\varphi(1) \tag{A.3}$$

over the set of the partitioning deformations f_{P} which are, so to speak, isochoric by the part, in the sense that the restriction of f_{P} ro $\mathscr{P}_i$ is an isochoric deformation for all indices i. Notice that, for a isochoric partitioning deformation,

$$\sum_i \text{vol}\, f_{\mathrm{P}}(\mathscr{P}_i) = \text{vol}\,\mathscr{B}. \tag{A.4}$$

For the problem of null surface traction (*cf.* (ii) of Subsection 6.1), we then have an infinite variety of minimizers, each of which corresponds to a state of ease of our liquid body, with or without cavitation.

For the place problem (*cf.* Section 5), we take as the energy functional the stored energy functional (2) restricted to the class of partitioning deformations which obey the following *confinement condition*

$$f_{\mathrm{P}}(\mathscr{P}_i) \subset \mathscr{B}_1 \quad \text{for all indices } i, \tag{A.5}$$

(with $\mathscr{B}_1$, as before, the ball of center 0 and radius ϱ_1). Now, for $\varrho_1 > 1$, any isochoric partitioning deformation satisfying (5) gives the functional the value $\mathscr{E}_{\min}$ and, due to the fact that

$$\text{vol}\,\mathscr{B} < \text{vol}\,\mathscr{B}_1, \tag{A.6}$$

induces cavitation, as is intuitively expected. Finally, consider the case $\varrho_1 \leq 1$. By using Jensen's inequality as in Section 3, one can show that, if a partitioning deformation f_{P} is a minimizer, then a list of positive numbers δ_i exists such that both

$$\det\,(\mathrm{D}f_{\mathrm{P}}(x)) = \delta_i \quad \text{for } x \in \mathscr{P}_i \tag{A.7}$$

and

$$\sum_i (\text{vol}\,\mathscr{P}_i)\,\delta_i \leq \text{vol}\,\mathscr{B}_1. \tag{A.8}$$

Moreover,

$$\mathscr{E}\{f_{\mathrm{P}}\} \geq \sum_i (\text{vol}\,\mathscr{P}_i)\,\varphi(\delta_i) \tag{A.9}$$

for any f_{P} in the domain of $\mathscr{E}$ such that

$$\int_{\mathscr{P}_i} \det\,(\mathrm{D}f_{\mathrm{P}}) = (\text{vol}\,\mathscr{P}_i)\,\delta_i \tag{A.10}$$

for all indices i. On the other hand, a direct application of the *discrete* Jensen's inequality (*v.* [HPL], p. 77) yields

$$\sum_i (\text{vol}\,\mathscr{P}_i)\,\varphi(\delta_i) \geq (\text{vol}\,\mathscr{B})\,\varphi\left(\frac{\sum_i (\text{vol}\,\mathscr{P}_i)\,\delta_i}{\text{vol}\,\mathscr{B}}\right), \tag{A.11}$$

with the equality sign holding only if $\delta_i = \delta_0$ for all indices i. As φ is strictly decreasing in $(0, \varrho_1^3]$, and as $\delta_0 \leq \varrho_1^3$ in view of (8), we conclude that

$$\mathscr{E}_{\min} = (\text{vol}\,\mathscr{B})\,\varphi(\delta_0), \quad \text{with } \delta_0 = \varrho_1^3. \tag{A.12}$$

Thus, a constant inwards displacement of the boundary does not correspond only to one homogeneous contraction, as was the case in the restricted setting of Section 5, but rather to a variety of partitioning deformations, all of which give one and the same value to the volume ratio throughout $\mathscr{B}$.

Dipartimento di Ingegneria Civile Edile
Università di Roma-Tor Vergata

&

Dipartimento di Matematica
Università di Pisa

&

Dipartimento di Ingegneria Aerospaziale
Università di Pisa

(Received April 26, 1985)

The Main Elastic Capacities of a Spheroid

PIERO VILLAGGIO

To Walter Noll for his 60th birthday

Introduction

In 1869 KIRCHHOFF [1869] gave a mathematically precise definition of elastic capacity, that is of the total electrical charge to be applied to a conductor in order to generate the constant potential one on it, while the potential at infinity is kept at level zero. According to Kirchhoff's definition, the capacity of a given solid body D and surface S can be determined by a function $u(x, y, z)$, harmonic in the infinite space outside D, and such that $u = 1$ on S, and $u = 0$ at infinite distance. Once $u(x, y, z)$ is known, the capacity is defined by

$$C = \frac{1}{4\pi} \iiint (u_x^2 + u_y^2 + u_z^2)\, dx\, dy\, dz, \tag{1.1}$$

where the triple integral is extended over the whole space outside D.

A problem of the same type was studied by THOMSON & TAIT [1867, Art. 320] and by KIRCHHOFF [1869] to describe the rigid motion of a solid through a frictionless liquid filling all space. Here, again, the motion of the fluid, if irrotational, is characterized by a single-valued function $\varphi(x, y, z)$, harmonic in the infinite space occupied by the liquid, with its normal derivative $-\dfrac{\partial \varphi}{\partial n}$ equal to the normal component of the velocity at any point of the surface of the body, while its partial derivatives φ_x, φ_y, φ_z vanish at ∞ in every direction. If the motion is defined by three angular velocities $\dot{p}, \dot{q}, \dot{r}$ about three perpendicular axes fixed in the body, and three translational velocities $\dot{u}$, $\dot{v}$, $\dot{w}$ along these axes, it is possible to write, as KIRCHHOFF did,

$$\varphi = \dot{u}\varphi_1 + \dot{v}\varphi_2 + \dot{w}\varphi_3 + \dot{p}\chi_1 + \dot{q}\chi_2 + \dot{r}\chi_3, \tag{1.2}$$

where $\varphi_1, \varphi_2, \varphi_3, \chi_1, \chi_2, \chi_3$ are harmonic functions satisfying, on the surface of the body, the boundary conditions

$$-\frac{\partial \varphi_1}{\partial n} = l, \quad -\frac{\partial \varphi_2}{\partial n} = m, \quad -\frac{\partial \varphi_3}{\partial n} = n,$$
$$-\frac{\partial \chi_1}{\partial n} = ny - mz, \quad -\frac{\partial \chi_2}{\partial n} = zl - xn, \quad -\frac{\partial \chi_3}{\partial n} = xm - yl, \tag{1.3}$$

where l, m, n denote the direction-cosines of the normal, drawn towards the fluid; in addition, the derivatives of these functions tend to zero at infinity.

When the functions $\varphi_1, \varphi_2, \varphi_2, \chi_1, \chi_2, \chi_2$ are determined, the kinetic energy of the fluid is given by

$$2T = \varrho \iiint (\varphi_x^2 + \varphi_y^2 + \varphi_z^2)\, dx\, dy\, dz, \tag{1.4}$$

where ϱ is the density and the integration extends over the space outside the body. Substituting into (1.4) the expression (1.2), one gets

$$\begin{aligned} 2T = {} & A\dot{u}^2 + B\dot{v}^2 + C\dot{w}^2 + 2A'\dot{v}\dot{w} + 2B'\dot{w}\dot{u} + 2C'\dot{u}\dot{v} \\ & + P\dot{p}^2 + Q\dot{q}^2 + R\dot{r}^2 + 2P'\dot{q}\dot{r} + 2Q'\dot{r}\dot{p} + 2R'\dot{p}\dot{q} \\ & + 2\dot{p}(F\dot{u} + G\dot{v} + H\dot{w}) + 2\dot{q}(F'\dot{u} + G'\dot{v} + H'\dot{w}) + 2\dot{r}(F''\dot{u} + G''\dot{v} + H''\dot{w}), \end{aligned} \tag{1.5}$$

where the twenty-one coefficients $A, B \ldots$, are determined by the form of the body and the position of the coordinate axes. A simple calculation shows that, for example,

$$A = \varrho \iiint (\varphi_{1,x}^2 + \varphi_{1,y}^2 + \varphi_{1,z}^2)\, dx\, dy\, dz,$$
$$A' = \varrho \iiint (\varphi_{2,x}\varphi_{3,x} + \varphi_{2,y}\varphi_{3,y} + \varphi_{2,z}\varphi_{3,z})\, dx\, dy\, dz,$$
$$P = \varrho \iiint (\chi_{1,x}^2 + \chi_{1,y}^2 + \chi_{1,z}^2)\, dx\, dy\, dz.$$

If the solid moves inside the liquid, its kinetic energy T_1 must be increased by T in order to obtain the total kinetic energy of system. Hence T represents the system of impulsive pressures exerted on the fluid by the surface of the solid, when this is instantaneously put in a general rigid motion from rest. For example, the x-component of the system of impulsive pressures is

$$\varrho \iint_S \varphi l\, dS = -\varrho \iint_S \varphi \frac{\partial \varphi_1}{\partial n}\, dS = A\dot{u} + C'\dot{v} + B'\dot{w} + F\dot{p} + F'\dot{q} + F''\dot{r} = \frac{\partial T}{\partial \dot{u}}. \tag{1.6}$$

The generai expression for the kinetic energy contains twenty-one coefficients, but, by the choice of special directions of the coordinate axes and of the origin, they can be reduced to fifteen, since the axes can be chosen so that $A' = B' = C' = 0$ and $F' = G' = H' = 0$. In addition, if the body has planes of symmetry, other coefficients of T must vanish. For a solid of revolution about the z-axis, one can show (*cf.* LAMB [1932, Art. 126]) that the expression (1.5) takes the form

$$2T = A(\dot{u}^2 + \dot{v}^2) + C\dot{w}^2 + P(\dot{p}^2 + \dot{q}^2) + R\dot{r}^2. \tag{1.7}$$

In this paper I study a problem in linear elasticity that has a strict analogy with KIRCHHOFF's problem in hydrodynamics. When a rigid body D, called the "inclusion", is immersed inside a homogeneous and isotropic elastic space, called the "matrix", it is natural to ask which is the distribution of strains and stresses in the elastic space corresponding to a rigid displacement of the body, characterized by three components of translation u, v, w along three arbitrary axes, and three rotations p, q, r about these axes. Here, of course, the problem is much harder than that considered by KIRCHHOFF, but, at least in principle, it is still possible to write the strains in the elastic space in the form

$$\begin{aligned} \varepsilon_{xx} &= u\varepsilon_{xx}^{(u)} + v\varepsilon_{xx}^{(v)} + w\varepsilon_{xx}^{(w)} + p\varepsilon_{xx}^{(p)} + q\varepsilon_{xx}^{(q)} + r\varepsilon_{xx}^{(r)}, \\ \varepsilon_{yy} &= u\varepsilon_{yy}^{(u)} + v\varepsilon_{yy}^{(v)} + w\varepsilon_{yy}^{(w)} + p\varepsilon_{yy}^{(p)} + p\varepsilon_{yy}^{(q)} + r\varepsilon_{yy}^{(r)}, \\ &\dots \\ \varepsilon_{yz} &= u\varepsilon_{yz}^{(u)} + v\varepsilon_{yz}^{(v)} + w\varepsilon_{yz}^{(w)} + p\varepsilon_{yz}^{(p)} + p\varepsilon_{yz}^{(q)} + r\varepsilon_{yz}^{(r)}, \\ &\dots \end{aligned} \tag{1.8}$$

where $\varepsilon_{xx}^{(u)}, \varepsilon_{yy}^{(u)}, \dots, \varepsilon_{yz}^{(u)}, \dots$ are the strains for $u = 1, v = w = 0, p = q = r = 0$, and so on for $\varepsilon_{xx}^{(v)}, \varepsilon_{yy}^{(v)}, \dots, \varepsilon_{yz}^{(v)}, \dots$ The stresses $\sigma_{xx}, \sigma_{yy}, \dots, \sigma_{yz} \dots$, can be similarly decomposed by a formula analogous to (1.8).

The strain energy of the elastic space is thus given by

$$2W = \iiint (\sigma_{xx}\varepsilon_{xx} + \sigma_{yy}\varepsilon_{yy} + \sigma_{zz}\varepsilon_{zz} + 2\sigma_{yz}\varepsilon_{yz} + 2\sigma_{zx}\varepsilon_{zx} + 2\sigma_{xy}\varepsilon_{xy})\, dx\, dy\, dz,$$

where the integration is extended over the space outside the body. On substituting (1.8) into the expression of the strain energy one gets

$$\begin{aligned} 2W = {} & Au^2 + Bv^2 + Cw^2 + 2A'vw + 2B'wu + 2C'uv \\ & + Pp^2 + Qq^2 + Rr^2 + 2P'qr + 2Q'rp + 2R'pq \\ & + 2p(Fu + Gv + Hw) + 2q(F'u + G'v + H'w) + 2r(F''u + G''v + H''w), \end{aligned} \tag{1.9}$$

where $A, B, C, \dots$ are twenty-one coefficients depending on the form and position of the body relative to the coordinate axes. Thus, for example,

$$\begin{aligned} A &= \iiint (\sigma_{xx}^{(u)}\varepsilon_{xx}^{(u)} + \dots + 2\sigma_{yz}^{(u)}\varepsilon_{yz}^{(u)} + \dots)\, dx\, dy\, dz, \\ A' &= \iiint (\sigma_{xx}^{(v)}\varepsilon_{xx}^{(w)} + \dots + 2\sigma_{yz}^{(v)}\varepsilon_{yz}^{(w)} + \dots)\, dx\, dy\, dz \\ &= \iiint (\sigma_{xx}^{(w)}\varepsilon_{xx}^{(v)} + \dots + 2\sigma_{yz}^{(w)}\varepsilon_{yz}^{(v)} + \dots)\, dx\, dy\, dz, \\ P &= \iiint (\sigma_{xx}^{(p)}\varepsilon_{xx}^{(p)} + \dots + 2\sigma_{yz}^{(p)}\varepsilon_{yz}^{(p)} + \dots)\, dx\, dy\, dz. \end{aligned} \tag{1.10}$$

The coefficients A, B, C, ... are called the *elastic capacities* of the body D. They also represent the forces exerted by the surrounding elastic medium on the body, when this undergoes a rigid motion. For instance, the total component of the force parallel to the x-axis is

$$\frac{\partial W}{\partial u} = Au + B'w + C'v + Fp + F'q + F''r, \tag{1.11}$$

and the moment about x is

$$\frac{\partial W}{\partial p} = Pp + Q'r + R'q + Fu + Gv + Hw. \tag{1.12}$$

The general expression for the strain energy can be simplified by a suitable choice of the coordinate axes. In fact, the directions of the axes can be so chosen that $A' = B' = C' = 0$, and the origin can be displaced so as to make $F' = G' = H' = 0$. Henceforward these simplifications will be supposed to have been made.

In particular, the case of an ellipsoid of revolution about the z-axis (a spheroid) will be considered. In this case the coefficients of the mixed terms, like $A', P', F, \ldots$, must vanish since the strain energy must be unaltered by reversing the signs of the displacements, and $A = B$, $P = Q$ as the strain energy must not change by writing $v, q, -u, -p$ for u, p, v, q, respectively, since this is equivalent to rotating the x, y-axes through a right angle. Therefore,

$$2W = A(u^2 + v^2) + Cw^2 + P(p^2 + q^2) + Rr^2. \tag{1.13}$$

Explicit determination of the capacities A, C, P, R is rather laborious, even in the special case, which is treated here, of a spheroid. For this reason, only the coefficients C and R, corresponding to a translation along z and a rotation about it, are calculated. These quantities, called the *main elastic capacities* of a spheroid, can be exactly determined by combining the classes of elastic solutions in spheroidal coordinates found by SADOWSKY & STERNBERG [1947] and extended by EDWARDS [1951].

Since the elastic capacity depends only on the shape and size, but not on the location, of the solid, one may think of fixing the volume and seeking those spheroids for which the translational or rotational capacity is a minimum. The equivalent problem in electrostatics, that is of finding the solid of lowest capacity with given volume, is classically known as the "isoperimetric problem" for capacity, considered first by POINCARÉ [1902] and completely solved by SZEGÖ [1930].

Only to anticipate the results, from the discussion of the formulae giving explicitly the main capacities C and R of a spheroid, it turns out that C becomes a minimum for a sharp prolate spheroid and R is a minimum for a needle-shaped prolate spheroid. In contrast with what happens in electrostatics, the sphere is no longer the solid that enjoys the isoperimetric property of lowest capacity.

2. Elastic Solutions in Prolate Spheroidal Coordinates

Let D be a rigid body having the shape of a spheroid, entirely surrounded by a homogeneous, isotropic elastic medium characterized by a shear modulus G and Poisson's ratio σ. The spheroid is referred to a Cartesian system of coordinates with the origin at the center, the z-axis along the axis of revolution, and the x, y-axes on the equatorial plane (Fig. 1). First the spheroid D will be supposed prolate. It is thus convenient to introduce the spheroidal coordinates α, β, γ by means of the transformation

$$x = \sinh\alpha \sin\beta \cos\gamma,$$
$$y = \sinh\alpha \sin\beta \sin\gamma, \tag{2.1}$$
$$z = \cosh\alpha \cos\beta,$$

where α, β, γ are defined in the following intervals

$$0 \leqq \alpha < \infty, \quad 0 \leqq \beta \leqq \pi, \quad 0 \leqq \gamma < 2\pi. \tag{2.2}$$

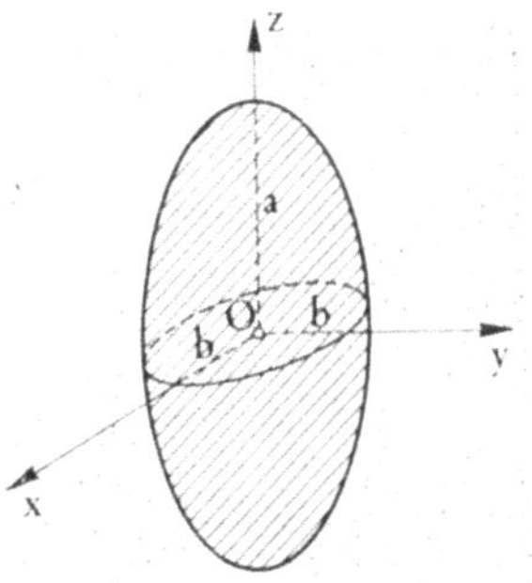

Fig. 1

The coordinate surfaces α = constant, β = constant, γ = constant form three families of orthogonal surfaces, constituted by prolate spheroids, hyperboloids of two sheets, and meridional half-planes.

In the curvilinear coordinates α, β, γ the distance between two neighboring points can be expressed in the form

$$(ds)^2 = \left(\frac{d\alpha}{h_1}\right)^2 + \left(\frac{d\beta}{h_2}\right)^2 + \left(\frac{d\gamma}{h_3}\right)^2, \tag{2.3}$$

where

$$\frac{1}{h_1^2} = \left(\frac{\partial x}{\partial\alpha}\right)^2 + \left(\frac{\partial y}{\partial\alpha}\right)^2 + \left(\frac{\partial z}{\partial\alpha}\right)^2 = \cosh^2\alpha - \cos^2\beta,$$
$$\frac{1}{h_2^2} = \left(\frac{\partial x}{\partial\beta}\right)^2 + \left(\frac{\partial y}{\partial\beta}\right)^2 + \left(\frac{\partial z}{\partial\beta}\right)^2 = \cosh^2\alpha - \cos^2\beta, \tag{2.4}$$
$$\frac{1}{h_3^2} = \left(\frac{\partial x}{\partial\gamma}\right)^2 + \left(\frac{\partial y}{\partial\gamma}\right)^2 + \left(\frac{\partial z}{\partial\gamma}\right)^2 = \sinh^2\alpha \sin^2\beta.$$

The direction cosines of the normal to the surface α = constant with respect the Cartesian coordinates are

$$\cos(\alpha, x) = h_1 \frac{\partial x}{\partial\alpha}, \quad \cos(\alpha, y) = h_1 \frac{\partial y}{\partial\alpha}, \quad \cos(\alpha, z) = h_1 \frac{\partial z}{\partial\alpha}, \tag{2.5}$$

and analogous formulae hold for the direction cosines of the normals to the surfaces $\beta =$ constant, $\gamma =$ constant.

After SADOWSKY & STERNBERG [1947], it is convenient to introduce the auxiliary variables

$$\left.\begin{aligned} q &= \cosh\alpha, & \bar{q} &= \sinh\alpha = \sqrt{q^2 - 1}, \\ p &= \cos\beta, & \bar{p} &= \sin\beta = \sqrt{1 - p^2}, \end{aligned}\right\} \tag{2.6}$$

where, as a consequence of equations (2.2),

$$1 \leqq q < \infty, \quad 0 \leqq \bar{q} < \infty, \quad -1 \leqq p \leqq 1, \quad 0 \leqq \bar{p} \leqq 1. \tag{2.7}$$

In terms of the variables p and q, the coefficients $h_1 = h_2$, h_3 are given by

$$\frac{1}{h_1^2} = \frac{1}{h_2^2} = \frac{1}{h^2} = q^2 - p^2, \quad \frac{1}{h_3} = \bar{q}\bar{p}, \tag{2.8}$$

and the complete expression of the direction cosines is

	x	y	z
α	$hq\bar{p}\cos\gamma$	$hq\bar{p}\sin\gamma$	$h\bar{q}p$
β	$h\bar{q}p\cos\gamma$	$h\bar{q}p\sin\gamma$	$-hq\bar{p}$
γ	$-\sin\gamma$	$\cos\gamma$	0

(2.9)

As the spheroid D is rigidly displaced from its initial position, stresses are produced in the surrounding elastic medium. In the absence of body forces, these stresses may be derived from four classes of elastic solutions, which, in spheroidal coordinates, appear as follows:

$$u_\alpha = \frac{h}{2G}\bar{q}\varphi_q, \quad u_\beta = -\frac{h}{2G}\bar{p}\varphi_p, \quad u_\gamma = \frac{1}{2G}\frac{\varphi_\gamma}{\bar{p}\bar{q}}, \tag{2.10}$$

$$\begin{aligned} u_\alpha &= \frac{h}{2G}[q\lambda_q - (3 - 4\sigma)\lambda]\, p\bar{q}, \\ u_\beta &= \frac{h}{2G}[-p\lambda_q - (3 - 4\sigma)\lambda]\, \bar{p}q, \\ u_\gamma &= \frac{1}{2G}\frac{pq}{\bar{p}\bar{q}}\lambda_\alpha, \end{aligned} \tag{2.11}$$

and

$$\begin{aligned} u_\alpha &= \frac{h}{2G}[\bar{q}^2\mu_q - (3 - 4\sigma)\, q\mu]\, \bar{p}\cos\gamma, \\ u_\beta &= \frac{h}{2G}[-\bar{p}^2\mu_p - (3 - 4\sigma)\, p\mu]\, \bar{q}\cos\gamma, \\ u_\gamma &= \frac{1}{2G}\frac{1}{\bar{p}\bar{q}}[\mu_\gamma\cos\gamma + (3 - 4\sigma)\,\mu\sin\gamma)\,\bar{q}\bar{p}, \end{aligned} \tag{2.12}$$

and

$$u_\alpha = \frac{h}{2G}\,[\bar{q}^2 \nu_q - (3-4\sigma)\, q\nu]\, \bar{p}\sin\gamma,$$

$$u_\beta = \frac{h}{2G}\,[-\bar{p}^2 \nu_p - (3-4\sigma)\, p\nu]\, \bar{q}\sin\gamma, \tag{2.13}$$

$$u_\gamma = \frac{1}{2G}\,\frac{1}{\bar{p}\bar{q}}\,[\nu_\gamma \sin\gamma - (3-4\sigma)\,\nu\cos\gamma]\,\bar{q}\bar{p},$$

where φ, λ, μ, ν are four harmonic functions playing the role of displacement potentials. The elastic state can be represented by superposition of the three basic solutions. Formulae (2.10), (2.11), (2.12), (2.13) are known as BOUSSINESQ [1885] or NEUBER [1937] solutions; SADOWSKY & STERNBERG [1947] gave them in the form above, which is more advantageous for problems in spheroidal coordinates.

From (2.10), (2.11), (2.12), (2.13) the curvilinear components of stress are obtained by writing the curvilinear components of strain and making use of the stress-strain relations. However their detailed expression can be omitted here since only some of the components of stresses enter into the evaluations of the capacities.

The harmonic functions φ, λ, μ, ν must be represented as products of single functions of the variables p, q, γ. For example,

$$\varphi(q, p, \gamma) = [F_n^{(m)}(q) \text{ or } S_n^{(m)}(q)]\,[F_n^{(m)}(p) \text{ or } S_n^{(m)}(p)]\,[\cos m\gamma \text{ or } \sin m\gamma], \tag{2.14}$$

with $m = 0, 1, 2, \ldots$; $n = 0, 1, 2, \ldots$; and where $F_n^{(m)}$ and $S_n^{(m)}$ are Legendre functions of the first and second kinds. For $m = 0$, φ is independent of γ and has the form

$$\varphi(q, p) = [F_n(q) \text{ or } S_n(q)]\,[F_n(p) \text{ or } S_n(p)], \tag{2.15}$$

with F_n and S_n ordinary Legendre functions of the first and second kinds.

3. Solutions for the Prolate Spheroid

When a prolate spheroid is rigidly translated by w along the z-axis, the boundary conditions for the elastic problem are

$$u_\alpha = w\cos(\alpha, z) = wh\bar{q}p, \qquad u_\beta = w\cos(\beta, z) = -whq\bar{p}, \qquad u_\gamma = 0 \tag{3.1}$$

at $\alpha = \alpha_0$, where $\alpha = \alpha_0$ is the equation of the surface of the spheroid; in addition, stresses and strains must vanish at infinity.

Since the solution is symmetric about z, it is clear that the displacement potentials are independent of γ and must be represented by formulae like (2.15). But the only stress functions that vanish at infinity and behave like (3.1) on the surfaces α = constant are

$$\begin{aligned}\varphi &= S_1(q)\,F_1(p),\\ \lambda &= S_0(q)\,F_0(p),\end{aligned} \tag{3.2}$$

where the Legendre functions S_0, F_0, S_1, F_1, of grade zero and one are defined as follows (*cf.* JAHNKE-EMDE-LÖSCH [1966, VIII. 2]):

$$S_0(q) = \frac{Q-1}{q}, \qquad F_0(p) = 1, \qquad S_1(q) = Q, \qquad F_1(p) = p,$$

where the auxiliary harmonic function $Q(q)$ is defined by

$$Q = 1 + \tfrac{1}{2} q \ln \frac{q-1}{q+1}. \tag{3.3}$$

Since $\varphi = Qp$, by applying (2.10) it is easy to derive the first class of solutions

$$\begin{aligned} u_\alpha &= \frac{h}{2G} \bar{q} p \left(\frac{Q}{q} + \frac{1}{q\bar{q}^2} \right), \\ u_\beta &= -\frac{h}{2G} \bar{p} Q, \\ u_\gamma &= 0. \end{aligned} \tag{3.4}$$

In the same way, the second class of solutions is obtained from (2.12) by putting $\lambda = \frac{Q-1}{q}$:

$$u_\alpha = \frac{h}{2G} \left[\frac{q}{\bar{q}^2} - (3-4\sigma) \frac{Q-1}{q} \right] p\bar{q}, \quad u_\beta = \frac{h}{2G} \left[(3-4\sigma) \frac{Q-1}{q} \right] \bar{p} q, \quad u_\gamma = 0. \tag{3.5}$$

In order to satisfy the boundary conditions (3.1), it is sufficient to consider a linear combination of the solutions (3.4) and (3.5) of the type $a_1 \varphi + b_1 \lambda$ and determine a_1, b_1 to satisfy

$$\begin{aligned} a_1 \frac{h_0}{2G} \bar{q}_0 p \left(\frac{Q_0}{q_0} + \frac{1}{q_0 \bar{q}_0^2} \right) + b_1 \frac{h_0}{2G} \left[\frac{q_0}{\bar{q}_0^2} - (3-4\sigma) \frac{Q_0 - 1}{q_0} \right] p\bar{q}_0 &= w h_0 \bar{q}_0 p, \\ -a_1 \frac{h_0}{2G} \bar{p} \frac{Q_0}{q_0} + b_1 \frac{h_0}{2G} (3-4\sigma)(Q_0 - 1) \bar{p} &= -w h_0 q_0 \bar{p}, \end{aligned} \tag{3.6}$$

where all quantities with the subscript zero are to be evaluated for $\alpha = \alpha_0$. The solution of the system yields

$$a_1 = \frac{2G q_0 w}{(3-4\sigma) \frac{Q_0 - 1}{q_0^2} + Q_0}, \qquad b_1 = -\frac{(2G/q_0)\, w}{(3-4\sigma) \frac{Q_0 - 1}{q_0^2} + Q_0}. \tag{3.7}$$

The corresponding curvilinear components of stress on the surface $\alpha = \alpha_0$ are (*cf.* SADOWSKY & STERNBERG [1947, eq. (19), (21)])

$$\begin{aligned}
\sigma_\alpha &= a_1[h_0^2 \bar{q}_0^2 \varphi_{qq} + h_0^4 \bar{p}^2 (q_0 \varphi_q - p\varphi_p)] + b_1[h_0^2(q_0 \lambda_{qq} - 2\lambda_q)\, p \bar{q}_0^2 \\
&\quad -2\sigma h_0^2 (q_0 \bar{p}^2 \lambda_p - p \bar{q}_0^2 \lambda_q) + h_0^4 (q_0 \lambda_q - p\lambda_p)\, p q_0 \bar{p}^2], \\
\tau_{\alpha\beta} &= a_1[-h_0^2 \varphi_{pq} + h_0^4 (q_0 \varphi_p - p\varphi_q)]\, \bar{p}\bar{q}_0 + b_1\{h_0^2[(1-2\sigma)(p\lambda_p + q_0 \lambda_q) \\
&\quad - p q_0 \lambda_{pq}]\, \bar{p}\bar{q}_0 + h_0^4 p \bar{p} q_0 \bar{q}_0 (q_0 \lambda_p - p\lambda_q)\}, \\
\tau_{\alpha\gamma} &= 0.
\end{aligned} \tag{3.8}$$

If the particular expressions (3.2) for φ and λ are used, these stresses assume the forms

$$\begin{aligned}
\sigma_\alpha &= a_1 \left(-h_0^2 \frac{2p}{\bar{q}_0^2} + h_0^4 \frac{\bar{p}^2 p}{\bar{q}_0^2}\right) + b_1 \left[-2h_0^2 \left(\frac{q_0^2}{\bar{q}_0^2} + 1\right) p + 2\sigma h_0^2 p + h_0^4 \frac{q_0^2}{\bar{q}_0^2} p \bar{p}^2\right], \\
\tau_{\alpha\beta} &= a_1 \left(-h_0^2 \frac{1}{q_0 \bar{q}_0^2} - h_0^4 \frac{p^2}{q_0 \bar{q}_0^2}\right) \bar{p}\bar{q}_0 + b_1 \left[(1-2\sigma)\, h_0^2 \frac{q_0}{\bar{q}_0} \bar{p} - h_0^4 \frac{q_0}{\bar{q}_0} p^2 \bar{p}\right], \\
\tau_{\alpha\gamma} &= 0.
\end{aligned} \tag{3.9}$$

The stress vector parallel to the z-axis is thus given by

$$\begin{aligned}
S_z &= \sigma_\alpha \cos(\alpha, z) + \tau_{\alpha\beta} \cos(\beta, z) \\
&= a_1 h_0 \left[(1-3p^2)\, h_0^2 \frac{1}{\bar{q}_0} + 2h_0^4 \frac{1}{\bar{q}_0} p^2 \bar{p}^2\right] \\
&\quad + b_1 h_0 \left[-3h_0^2 \frac{q_0^2}{\bar{q}_0} p^2 + h_0^2 \frac{1}{\bar{q}_0} p^2 - \frac{1}{\bar{q}_0} + 2\sigma \frac{1}{\bar{q}_0} + 2h_0^4 \frac{q_0^2}{\bar{q}_0} p^2 \bar{p}^2\right].
\end{aligned} \tag{3.10}$$

The resultant of these stress vectors represents the total force that the elastic medium exerts on the spheroid. This force is equal and opposite to the one that must be applied to the spheroid so as to displace it by w along the z-axis. It is given by

$$Cw = -\iint [\sigma_\alpha \cos(\alpha, z) + \tau_{\alpha\beta} \cos(\beta, z)] \frac{1}{hh_3}\, d\beta\, d\gamma,$$

where the integration extends over the boundary of the spheroid. If α and β are expressed in terms of q and p, the surface integral becomes

$$\begin{aligned}
Cw = -2\pi \int_{-1}^{1} &\{a_1[(1-3p^2)\, h_0^2 + 2h_0^4 p^2 \bar{p}^2] \\
&+ b_1[-3h_0^2 q_0^2 p^2 + h_0^2 p^2 - 1 + 2\sigma + 2h_0^4 q_0^2 p^2 \bar{p}^2]\}\, dp.
\end{aligned} \tag{3.11}$$

When a_1 and b_1 are replaced by their values (3.7), the capacity C is given by

$$C = -\frac{4\pi G(1-\sigma)}{(3-4\sigma)\dfrac{Q_0 - 1}{q_0^2} + Q_0} \int_{-1}^{1} \frac{2}{q_0}\, dp = -\frac{16\pi G(1-\sigma)}{(3-4\sigma)\dfrac{Q_0 - 1}{q_0} + Q\, q_0}. \tag{3.12}$$

An analogous procedure must be followed to evaluate the capacity R. In this case the spheroid is rotated by r about its axis of revolution and the curvilinear components of the displacements at the surface $\alpha = \alpha_0$ assume the form

$$u_\alpha = u_\beta = 0, \qquad u_\gamma = r\bar{p}\bar{q}. \tag{3.13}$$

By considering the displacement potentials vanishing at infinity, of the type

$$\begin{aligned} \mu &= S_1^{(1)}(q)\, F_1^{(1)}(p) \sin\gamma = \frac{\bar{q}}{q}\left(Q + \frac{1}{\bar{q}^2}\right) \bar{p} \sin\gamma, \\ \nu &= S_1^{(1)}(q)\, F_1^{(1)}(p) \cos\gamma = \frac{\bar{q}}{q}\left(Q + \frac{1}{\bar{q}^2}\right) \bar{p} \cos\gamma, \end{aligned} \tag{3.14}$$

it is easy to construct a linear combination of these potentials such as to satisfy (3.13). In fact, the combination $c_1(-\mu + \nu)$, where c_1 is a constant, generates, through (2.12), (2.13), a displacement such that $u_\alpha = u_\beta = 0$; the third component u_γ is

$$u_\gamma = -\frac{2}{G} c_1(1-\sigma)\, S_1^{(1)}(q)\, F_1^{(1)}(p) = -\frac{2}{G} c_1(1-\sigma) \frac{\bar{q}}{q}\left(Q + \frac{1}{\bar{q}^2}\right) p.$$

The last of (3.13) yields

$$c_1 = -\frac{G q_0}{2(1-\sigma)\left(Q_0 + \dfrac{1}{\bar{q}_0^2}\right)}. \tag{3.15}$$

Once the displacements in spheroidal coordinates are known, the strains can be obtained by the strain-displacement relations (*cf.* LOVE [1927, Art. 20]), and the stresses by the stress-strain relation. It is not necessary to find all components of stress, but only those of the type $\tau_{\alpha\gamma}$ on the surface $\alpha = \alpha_0$. The result of this computation is simply

$$\tau_{\alpha\beta} = -4c_1(1-\sigma)\, h_0 \frac{1}{\bar{q}_0^2} \bar{p}. \tag{3.16}$$

These stresses generate, around the axis of revolution, a couple which is equal and opposite to the rotational capacity of the spheroid about the z-axis:

$$Rr = -\iint \tau_{\alpha\gamma} \sqrt{x^2+y^2} \frac{1}{hh_3}\, d\beta\, d\gamma \qquad \left(\sqrt{x^2+y^2} = \bar{p}\bar{q}\right), \tag{3.17}$$

where the integration extends over the surface. By substituting the expression (3.16) into the last integral and recalling (3.15), it follows that R is given by

$$R = -\frac{4\pi G q_0}{Q_0 + \dfrac{1}{\bar{q}_0^2}} \int_{-1}^{1} \bar{p}^2\, dp = -\frac{16\pi G q_0}{3\left(Q_0 + \dfrac{1}{\bar{q}_0^2}\right)}. \tag{3.18}$$

It may be useful observe that, for large values of q_0, Q_0 can be expanded in the series

$$Q_0 = 1 - q_0 \left(\frac{1}{q_0} + \frac{1}{3q_0^3} + \dots \right),$$

so that, when terms of higher order in $\frac{1}{q_0}$ are neglected, C and R become

$$C \cong \frac{24\pi(1-\sigma)\, Gq_0}{5-6\sigma}, \qquad R \cong 8\pi G q_0^3,$$

that is, the corresponding capacities for a sphere.

4. Solutions for the Oblate Spheroid

The foregoing solutions have been established on the assumption that the spheroid is prolate, that is that $a > b$. If the spheroid is oblate (Fig. 2), it is useful to introduce the curvilinear coordinates

$$\begin{aligned} x &= q\bar{p} \cos \gamma, \\ y &= q\bar{p} \sin \gamma, \\ z &= \bar{q}p, \end{aligned} \tag{4.1}$$

so that the coefficients $h_1 = h_2, h_3$ have the forms

$$\frac{1}{h_1^2} = \frac{1}{h_2^2} = \frac{1}{h^2} = \bar{q}^2 + p^2, \qquad \frac{1}{h_3} = q\bar{p}. \tag{4.2}$$

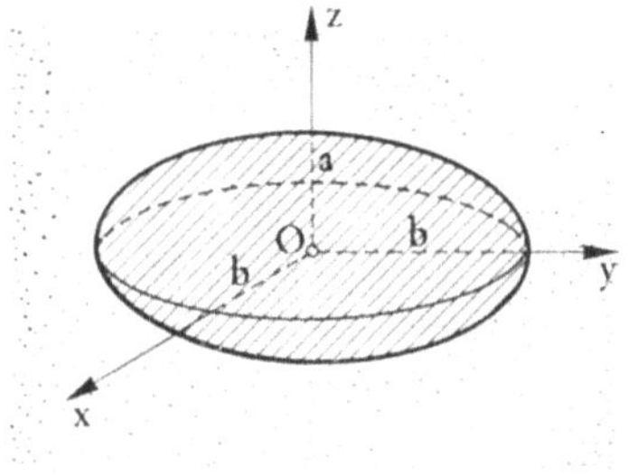

Fig. 2

The direction cosines of the normals to the coordinate surfaces are again determined by the formulae (2.5):

	x	y	z
α	$h\bar{q}\bar{p} \cos \gamma$	$h\bar{q}\bar{p} \sin \gamma$	hqp
β	$-hqp \cos \gamma$	$-hqp \sin \gamma$	$-h\bar{q}\bar{p}$
γ	$-\sin \gamma$	$\cos \gamma$	0

(4.3)

The basic solutions can be expressed by four harmonic functions, φ, λ, μ, ν, as follows:

$$u_\alpha = \frac{\bar{h}}{2G}\bar{q}\varphi_q, \quad u_\beta = -\frac{\bar{h}}{2G}\bar{p}\varphi_p, \quad u_\gamma = \frac{1}{2G}\frac{\varphi_\gamma}{q\bar{p}}, \tag{4.4}$$

and so

$$\begin{aligned} u_\alpha &= \frac{\bar{h}}{2G}[\lambda_q\bar{q}^2 - (3-4\sigma)\,q\lambda]\,p, \\ u_\beta &= \frac{\bar{h}}{2G}[-p\lambda_p + (3-4\sigma)\,\lambda]\,\bar{q}\bar{p}, \\ u_\gamma &= \frac{1}{2G}\frac{\bar{q}p}{q\bar{p}}\lambda_\gamma, \end{aligned} \tag{4.5}$$

$$\begin{aligned} u_\alpha &= \frac{\bar{h}}{2G}[q\mu_q - (3-4\sigma)\,\mu]\,\bar{q}\bar{p}\cos\gamma, \\ u_\beta &= \frac{\bar{h}}{2G}[-\bar{p}^2\mu_p - (3-4\sigma)\,p\mu]\,q\cos\gamma, \\ u_\gamma &= \frac{1}{2G}[\mu_\gamma\cos\gamma + (3-4\sigma)\,\mu\sin\gamma], \end{aligned} \tag{4.6}$$

and

$$\begin{aligned} u_\alpha &= \frac{\bar{h}}{2G}[q\nu_q - (3-4\sigma)\,\nu]\,\bar{q}\bar{p}\sin\gamma, \\ u_\beta &= \frac{\bar{h}}{2G}[-\bar{p}^2\nu_p - (3-4\sigma)\,p\nu]\,q\sin\gamma, \\ u_\gamma &= \frac{1}{2G}[\nu_\gamma\sin\gamma - (3-4\sigma)\,\nu\cos\gamma]. \end{aligned} \tag{4.7}$$

The general formulae to represent harmonic functions, like (2.14) and (2.15), remain valid for the oblate coordinates, provided that the real quantities q and $\bar{q}$ are replaced by the imaginary quantities $i\bar{q}$ and iq, respectively. In particular, the auxiliary harmonic function Q must be replaced by

$$1 + \tfrac{1}{2}i\bar{q}\ln\frac{i\bar{q}-1}{i\bar{q}+1} = 1 - \bar{q}\operatorname{arc\,cot}\bar{q} = 1 - \bar{q}T. \tag{4.8}$$

If the oblate spheroid is displaced by w along the z-axis, the displacements at the boundary are

$$u_\alpha = w\cos(\alpha, z) = w\bar{h}qp, \quad u_\beta = w\cos(\beta, z) = -w\bar{h}\bar{q}\bar{p}, \quad u_\gamma = 0, \tag{4.9}$$

while, at infinity, the stresses and strains must vanish.

If one puts $\varphi = (1 - \bar{q}T)\,p$, $\lambda = T$, the displacements, calculated through (4.4) and (4.5), assume the forms

$$
\begin{aligned}
u_\alpha &= \frac{\bar{h}}{2G}\left(-qT + \frac{\bar{q}}{q}\right) p, \\
u_\beta &= -\frac{\bar{h}}{2G}(1 - \bar{q}T)\,\bar{p}, \\
u_\gamma &= 0,
\end{aligned}
\tag{4.10}
$$

and

$$
\begin{aligned}
u_\alpha &= \frac{\bar{h}}{2G}\left[-\frac{\bar{q}}{q} - (3 - 4\sigma)\,qT\right] p, \\
u_\beta &= \frac{\bar{h}}{2G}(3 - 4\sigma)\,\bar{q}\bar{p}T, \\
u_\gamma &= 0.
\end{aligned}
\tag{4.11}
$$

The condition that a linear combination $\bar{a}_1\varphi + \bar{b}_1\lambda$ satisfies (4.9) yields

$$
\begin{aligned}
\bar{a}_1\frac{\bar{h}_0}{2G}\left(-q_0T_0 + \frac{\bar{q}_0}{q}\right) p + \bar{b}_1\frac{\bar{h}_0}{2G}\left[-\frac{\bar{q}_0}{q_0} - (3 - 4\sigma)\,q_0T_0\right] p &= w\bar{h}_0q_0p, \\
-\bar{a}_1\frac{\bar{h}_0}{2G}(1 - \bar{q}_0T_0)\,\bar{p} + \bar{b}_1\frac{\bar{h}_0}{2G}(3 - 4\sigma)\,\bar{q}_0\bar{p}T_0 &= -w\bar{h}_0\bar{q}_0\bar{p},
\end{aligned}
\tag{4.12}
$$

where the subscript zero indicates evaluation at $\alpha = \alpha_0$. The values of $\bar{a}_1$, $\bar{b}_1$, solutions of (4.12), thus are

$$
\bar{a}_1 = \frac{2G\bar{q}_0w}{(3 - 4\sigma)\dfrac{T_0}{\bar{q}_0} + (1 - \bar{q}_0T_0)}, \qquad \bar{b}_1 = -\frac{(2G/\bar{q}_0)\,w}{(3 - 4\sigma)\dfrac{T_0}{\bar{q}_0} + (1 - \bar{q}_0T_0)}. \tag{4.13}
$$

The components of stress on the surface $\alpha = \alpha_0$ are found by using the strain-displacement relations and the stress-strain relation. The result of the (lengthy) computation is

$$
\begin{aligned}
\sigma_\alpha = {} & \bar{a}_1[\bar{h}_0^2\bar{q}_0^2\varphi_{qq} + \bar{h}_0^4(q_0p^2\varphi_q + p\bar{p}^2\varphi_p)] + \bar{b}_1[\bar{h}_0^2(\bar{q}_0^2\lambda_{qq} - 2q_0\lambda_q)\,p\bar{q}_0 \\
& + 2\sigma\bar{h}_0^2(pq_0\lambda_q + \bar{p}^2\lambda_p)\,\bar{q}_0 - \bar{h}_0^4(pq_0\bar{q}_0\lambda_q + \bar{p}^2\lambda_p)\,p^2],
\end{aligned}
\tag{4.14}
$$

$$
\begin{aligned}
\tau_{\alpha\beta} = {} & \bar{a}_1[-\bar{h}_0^2\varphi_{pq} + \bar{h}_0^4(q_0\varphi_p + p\varphi_q)]\,\bar{p}\bar{q}_0 + \bar{b}_1 \\
& \times\{\bar{h}_0^2[(1 - 2\sigma)\,(\bar{q}_0^2\lambda_q + pq_0\lambda_p) - p\bar{q}_0^2\lambda_{pq}]\,\bar{p} + \bar{h}_0^4p\bar{p}\bar{q}_0^2(q_0\lambda_p + p\lambda_q)\}, \\
\tau_{\alpha\gamma} = {} & 0,
\end{aligned}
$$

which, when it is assumed that $\varphi = (1 - \bar{q}T)p$, $\lambda = T$, become

$$\sigma_\alpha = \bar{a}_1\left(-\bar{h}_0^2\frac{p}{q_0^2} - \bar{h}_0^4 p\right) + \bar{b}_1\left[-\bar{h}_0^2\left(\frac{1}{q_0^2} - 4\right)p - 2\sigma\bar{h}_0^2 p - \bar{h}_0^4 p^3\right],$$
$$\tau_{\alpha\beta} = -\bar{a}_1\bar{h}_0^4\frac{\bar{q}_0}{q_0}\bar{p} - \bar{b}_1\left[(1-2\sigma)\,\bar{h}_0^2\,\bar{p}\frac{\bar{q}_0}{q_0} + \bar{h}_0^4 p^2\bar{p}\frac{\bar{q}_0}{q_0}\right]. \tag{4.15}$$

The stress vector in the z-direction is given by an expression like (3.10), where the direction cosines must be derived from the table (4.3):

$$S_z = \bar{a}_1\bar{h}_0\left(\bar{h}_0^2\frac{\bar{p}^2}{q_0} - 2\bar{h}_0^4 p^2 q\right) + \bar{b}_1\bar{h}_0\left[(1-2\sigma)\frac{1}{q_0} + 3\bar{h}_0^2 q_0 p^2 - 2\bar{h}_0^4 p^4 q_0\right]. \tag{4.16}$$

The capacity C, which is the resultant of these surface tractions with reversed sign, is thus given by

$$Cw = -2\pi\int_{-1}^{1}\{\bar{a}_1(\bar{h}_0^2\bar{p}^2 - 2\bar{h}_0^4 p^2 q_0^2) + \bar{b}_1[(1-2\sigma) + 3\bar{h}_0^2 q_0^2 p^2 - 2\bar{h}_0^4 p^4 q_0^2]\}\,dp.$$

Once the values of $\bar{a}_1$ and $\bar{b}_1$, given by (4.13), are replaced in (4.15), it is possible to express the capacity by

$$C = \frac{4\pi G}{(3-4\sigma)\,T_0 + (1-\bar{q}_0 T_0)\,\bar{q}_0}\int_{-1}^{1}\left[(1-2\sigma) + 6q_0^2\frac{p^2}{q_0^2+p^2} + 4q_0^2\frac{p^4}{(q_0^2+p^2)^2}\right]dp,$$

which, after evaluation of the integral (*cf.* GRADSHTEYN & RYZIK [1965, 2.17]), yields

$$C = \frac{16\pi G(1-\sigma)}{(3-4\sigma)\,T_0 + (1-\bar{q}_0 T_0)\,\bar{q}_0}. \tag{4.17}$$

In order to find the rotational capacity R, it is necessary to recall that the displacements at the surface are now

$$u_\alpha = u_\beta = 0, \quad u_\gamma = q\bar{p}. \tag{4.18}$$

The appropriate displacement potentials must be of the form $\bar{c}_1(-\mu + \nu)$, where

$$\mu = \left(qT - \frac{\bar{q}}{q}\right)\bar{p}\sin\gamma,$$
$$\nu = \left(qT - \frac{\bar{q}}{q}\right)\bar{p}\cos\gamma, \tag{4.19}$$

and where $\bar{c}_1$ is a constant. The first two displacement components u_α, u_β vanish, while u_γ has the form

$$u_\gamma = -\frac{2}{G}\bar{c}_1(1-\sigma)\left(qT - \frac{\bar{q}}{q}\right)\bar{p}.$$

From the last of (4.18) the value of the constant $\bar{c}_1$ can be derived:

$$\bar{c}_1 = -\frac{Gq_0}{2(1-\sigma)\left(q_0 T_0 - \frac{\bar{q}_0}{q_0}\right)}. \tag{4.20}$$

It follows that the component of stress $\tau_{\alpha\gamma}$ on the surface $\alpha = \alpha_0$ can be written

$$\tau_{\alpha\gamma} = -4\bar{c}_1(1-\sigma)\,\bar{h}_0 \frac{1}{q_0^2}\bar{p}, \tag{4.21}$$

and the capacity R, about the axis of revolution, is the resultant moment of these stresses with reversed sign:

$$Rr = -\iint \tau_{\alpha\gamma} \sqrt{x^2+y^2}\,\frac{1}{\bar{h}h_3}\,d\beta\,d\gamma \qquad (\sqrt{x^2+y^2} = q\bar{p}),$$

and hence

$$R = -\frac{4\pi G q_0}{q_0 T_0 - \frac{\bar{q}_0}{q_0}} \int_{-1}^{1} \bar{p}^2\,dp = -\frac{16\pi G q_0}{3\left(q_0 T_0 - \frac{\bar{q}_0}{q_0}\right)}. \tag{4.22}$$

When q_0 is large, the function T_0 can be expanded in the series

$$T_0 = \frac{1}{\bar{q}_0} - \frac{1}{3\bar{q}_0^3} + \dots, \tag{4.23}$$

and, therefore, C and R are expressed approximately by

$$C \cong \frac{24\pi(1-\sigma)\,G q_0}{5-6\sigma}, \qquad R \cong 8\pi G q_0^3,$$

that is, again, the capacities of a sphere.

5. Isoperimetric Properties

The capacities C and R depend on shape and size of the spheroid, but not on its location. If the spheroid is a sphere, the capacities C and R have been evaluated at the end of Sections 3 and 4 as limiting cases. These formulae show that the translational capacity C is proportional to the radius q_0, and the rotational capacity R is proportional to q_0^3.

If the spheroid is not a sphere, the elastic capacities do not have such simple expressions. Nevertheless it is convenient to define the *volume radius* of the solid (introduced by PÓLYA & SZEGÖ [1951, 1.4]) in the form

$$\bar{V} = (q_0\bar{q}_0^2)^{\frac{1}{3}} \tag{5.1}$$

for the prolate spheroid, and

$$\overline{V} = (\overline{q}_0 q_0^2)^{\frac{1}{3}} \tag{5.2}$$

for the oblate spheroid, and, next, to consider the behavior of the ratios $C/\overline{V}$ and $R/\overline{V}^3$ as $1 \leqq q_0 < \infty$.

For the prolate spheroid, the curve for $C/\overline{V}$, sketched in Fig. 3, attains a minimum at a value q_0^* close to $q_0 = 1$. The precise values of q_0^* are, in fact, $q_0^* = 2{,}05$; $1{,}50$; $1{,}17$, respectively for $\sigma = -1, 0, \frac{1}{2}$.

For the oblate spheroid, the graph of the function $C/\overline{V}$, again represented in Fig. 3, shows that the minimum is reached at infinity, that is to say for the sphere.

To study the dependence of the rotational capacity on the shape it is necessary to consider the ration $R/\overline{V}^3$.

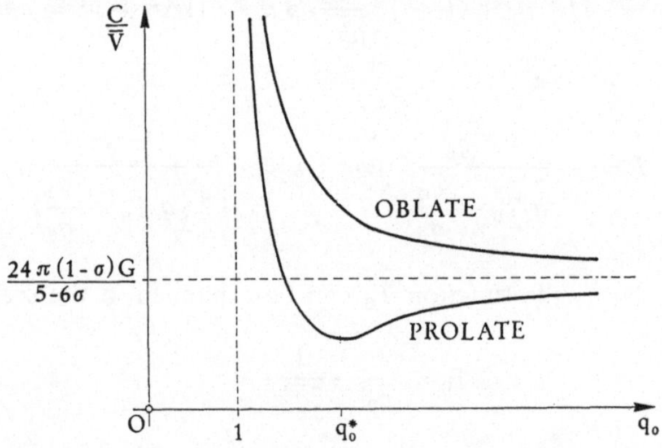

Fig. 3

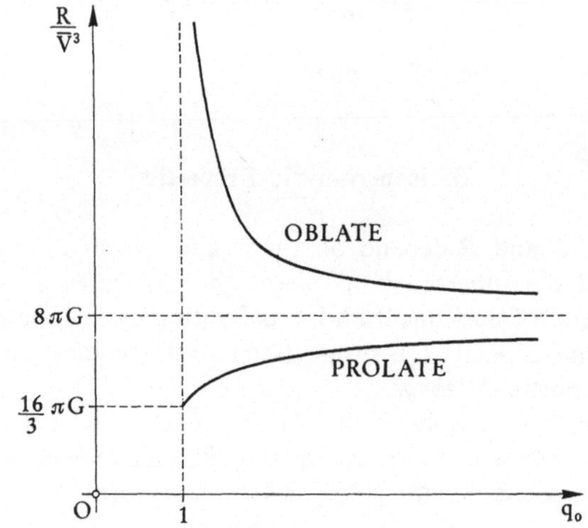

Fig. 4

For the prolate spheroid, R is given by (3.18) and $\bar{V}$ by (5.1). The curve $R/\bar{V}^3$ represented in Fig. 4, shows that the minimal capacity is reached for $q_0 = 1$.

When the spheroid is oblate, R must be evaluated according to (4.22) and $\bar{V}$ from (5.2). This second curve attains its minimum at infinity, that is to say again for the sphere.

References

[1867] THOMSON, W., & TAIT, P. G.: *Natural Philosophy*. First Ed. Oxford.

[1869] KIRCHHOFF, G.: "Über die Bewegung eines Rotationskörpers in einer Flüssigkeit." J. f. Math. (Crelle), Bd. 71, p. 237.

[1885] BOUSSINESQ, J.: *Applications des Potentiels à l'étude de l'équilibre et du mouvement des solides élastiques*. Paris: Gauthier-Villars.

[1902] POINCARÉ, H.: *Figures d'Équilibre d'une Masse Fluide*. Paris.

[1927] LOVE, A. E. H.: *A Treatise on the Mathematical Theory of Elasticity*. Fourth Ed., New York: Dover.

[1930] SZEGÖ, G.: "Über einige Extremalaufgaben der Potentialtheorie." Math. Zeit. Bd. 31, pp. 583–593.

[1932] LAMB, H. *Hydrodynamics*. Sixth Ed., New York: Dover.

[1937] NEUBER, H.: *Kerbspannungslehre*. Berlin: Springer.

[1947] SADOWSKY, M. A., & STERNBERG, E.: "Stress Concentration Around an Ellipsoidal Cavity in an Infinite Body Under Plane Stress Perpendicular to the Axis of Revolution of Cavity." Journ. Appl. Mech. Vol. 69, pp. A-191, A-201.

[1951] EDWARDS, R. H.: "Stress Concentrations Around Spherical Inclusions and Cavities." Journ. Appl. Mech. Vol. 73, pp. A-19, A-30.

[1951] PÓLYA, G., & SZEGÖ, G.: *Isoperimetric Inequalities in Mathematical Physics*. Princeton: Princeton Univ. Press.

[1965] GRADSHTEYN, I. S., & RYZIK, I. M.: *Table of Integral Series and Products*. New York-London: Acad. Press.

[1966] JAHNKE-EMDE-LÖSCH: *Tafeln Höherer Funktionen*. Stuttgart: Teubner.

Istituto di Scienza delle Costruzioni
Universitá di Pisa

(Received June 26, 1985)

On the Gibbsian Thermostatics of Mixtures

ROGER FOSDICK & JORGE PATIÑO

We dedicate this paper to Walter Noll on the occasion of his sixtieth birthday and in appreciation of his thoughtful contributions to the foundations of continuum mechanics

1. Introduction

In 1970, NOLL [N] developed a modern version of the elements of Gibbsian thermostatics for describing the equilibrium of mixtures. An important aspect of NOLL's work is that he considered the old fundamental basis of thermostatics within a modern variational setting and, with mild hypotheses, gave proofs to some central theorems in the subject. NOLL considered the stable states of a mixture to be those that maximize an entropy functional over a certain convex set of measures, and for the analysis of this variational problem he employed an elegant, albeit not standard, mathematical theory. While his work could be considered a precursor to much of the present-day studies on coexistent phases in continuum thermomechanics, nevertheless it has remained relatively, and undeservedly, obscure, perhaps because of its unusual mathematical setting.

In this paper we shall reconsider the classical problem of determining those configurations of an N-component ideal mixture, of fixed total energy, total volume, and N total masses, that maximize the total entropy. We formulate the problem in Section 2 and introduce the notion of a distributional measure in order to identify, at the outset, a main connection between NOLL's study and the more conventional calculus of variations approach, which we shall adopt. Then, in Section 3, we apply variational arguments and some elements from convex analysis to investigate the existence, uniqueness, and structure of solutions of this maximum problem.† Whenever possible, correlations are made between the conclusions reached here and those reached earlier by NOLL [N].

The results of Section 3 are summarized in five theorems. In Theorem 1 we show that the only states that a maximizer can possibly assume are those which correspond to global concave parts of the specific entropy graph. Theorems 2 and 3 then give, respectively, necessary and sufficient conditions on the prescribed total energy and N total masses in order to ensure the existence of a

† DUNN & FOSDICK [DF] and FOSDICK [F] have considered the single component system in a similar manner.

maximizer. In the proof of Theorem 3 a maximizer in the form of a step function is constructed. Theorem 4 and the remarks immediately following its proof define and characterize the temperature, the N chemical potentials, and the pressure which is associated with a maximizing state. In Theorem 5 we are concerned about uniqueness, and we record a sufficient condition for maximizing states to be unique up to a rearrangement. In this case, a maximizer can coexist in an N-component mixture in at most $N+2$ phases.

In Section 4, the final section of this paper, we consider some elementary comparison principles in thermostatics. For these, the general idea is to suppose that the total energy, the total volume, and the N total masses of the mixture are to take on a prescribed sequence of values, and we seek restrictions on the manner in which the associated pressure, temperature, and N chemical potentials of a corresponding sequence of maximizers, for the mixture, can change. In any sequence for which the prescribed total energy, total volume, and N total masses depend smoothly on a single parameter, we find that global forms of the Gibbs and Gibbs-Duhem equations must hold. In addition, whenever such a sequence supports multiple coexistent phases as the parameter is varied, we find that a generalized Clausius-Clapeyron equation must hold between any two of these phases. Finally, for any sequence, smooth or not, we obtain a generalized Van't Hoff-LeChatelier inequality.

The Appendix of this paper contains a brief summary of some elements from convex analysis that we have found helpful in this work.

2. Notation and Setting

For the purpose of this paper a *configuration* $\mathscr{C}$ of a mixture of N fluid bodies (herein called *components*) is a prescribed bounded open region $R \subset \mathbb{R}^3$, together with an ordered set of functions

$$(\eta, \varepsilon, \underset{1}{\varrho}, \ldots, \underset{N}{\varrho}) : R \to \mathbb{R}^{N+2}.$$

Here, $\eta(x)$, $\varepsilon(x)$, and $\underset{\mathrm{a}}{\varrho}(x)$, $\mathrm{a} = 1, 2, \ldots, \mathrm{N}$, are, respectively, the *specific entropy of the mixture*, *specific energy of the mixture*, and the *specific mass densities of the components* at the point $x \in R$; all measured per unit volume of R. Thus, if v represents the *volume measure* of R,

$$(H, E, V, \underset{\mathrm{a}}{M})(\mathscr{C}) \equiv \int_R (\eta, \varepsilon, 1, \underset{\mathrm{a}}{\varrho})(x)\, dv(x), \qquad a = 1, 2, \ldots, \mathrm{N},$$

denotes, respectively, the *total entropy*, *total energy*, *total volume*, and the *total masses of the components* associated with the configuration $\mathscr{C}$. The *mass density of the mixture* at the point $x \in R$, measured per unit volume of R, is given by

$$\varrho(x) = \sum_{\mathrm{a}=1}^{N} \underset{\mathrm{a}}{\varrho}(x),$$

so that $\nu(x) \equiv 1/\varrho(x)$ is the corresponding *specific volume*, measured per unit mass of the mixture.

We shall assume throughout this work that the constitutive relation†,

$$\eta = \hat{\eta}(\varepsilon, \underset{1}{\varrho}, \underset{2}{\varrho}, \ldots, \underset{N}{\varrho}),$$

is continuous in all arguments, and refer to the ordered $(N+1)$-tuple

$$\vec{s}(x) \equiv (\varepsilon, \underset{1}{\varrho}, \underset{2}{\varrho}, \ldots, \underset{N}{\varrho})\,(x)$$

as a *state* at $x \in R$. Thus, the domain of $\hat{\eta}$, denoted by $D(\hat{\eta})$, is contained in $\mathbb{R}^{N+1}$ and we shall assume that this domain is open in $\mathbb{R}^{N+1}$. For latter reference, we note that in classical thermostatics when the constitutive function $\hat{\eta}$ is smooth, the *temperature* θ, *chemical potentials* $\underset{a}{\mu}$, a $= 1, 2, \ldots$, N, and *pressure* p corresponding to a state $\vec{s}$ are related to η through

$$\operatorname{grad}_{\vec{s}}\, \hat{\eta}(\vec{s}) = (1/\theta, -\underset{a}{\mu}/\theta),$$

and

$$\hat{\eta}(\vec{s}) - \vec{s} \cdot \operatorname{grad}_{\vec{s}}\, \hat{\eta}(\vec{s}) = p/\theta.$$

We wish to consider the classical thermostatics variational problem of determining those configurations of the mixture that maximize the total entropy H for given values of the total energy E, total volume V, and total masses $\underset{a}{M}$, a $= 1, 2, \ldots$, N. More precisely, the problem is to

$$\underset{\vec{s} \in \mathscr{A}(\vec{c})}{\text{maximize}}\; H[\vec{s}], \tag{2.1$_1$}$$

where

$$H[\vec{s}] \equiv \int_R \hat{\eta}(\vec{s}(x))\, dv(x), \tag{2.1$_2$}$$

and

$$\mathscr{A}(\vec{c}) \equiv \left\{ \vec{s} : R \to D(\hat{\eta}) \subset \mathbb{R}^{N+1} \mid \vec{s} \quad \text{is v-integrable on } R,\ \int_R \vec{s}(x)\, dv(x) = \vec{c}\, V \right\}. \tag{2.1$_3$}$$

The $(\mathrm{N}+1)$-tuple $\vec{c}$ is given, the volume V is given, and the class $\mathscr{A}(\vec{c})$ is called the *class of admissible states under the constraint* $\vec{c}$.

Since the constitutive function $\hat{\eta}$ is a function of $\vec{s}$, but not of the spatial gradients of $\vec{s}$, it is reasonable to expect that the value $H[\vec{s}]$ would be unaffected by a v-measure preserving rearrangement of a given field $\vec{s}(x)$ for $x \in R$. A precise interpretation of this observation will be helpful in this work and so we record the following two definitions and proposition.

† Dunn & Fosdick [DF] consider the Gibbsian thermostatics of a single body without assuming that the constitutive relation can be written as a function. In their work $(\eta, \varepsilon, \varrho)$ was required to lie on a continuous two-dimensional manifold.

Definition. *Let μ be a measure on a set $E \subset \mathbb{R}^m$ and let $y: E \to \mathbb{R}^n$ be μ-measurable. Then, for any Borel set $B \subset \mathbb{R}^n$, the measure*

$$\omega_y(B) \equiv \mu\{x \in E \mid y(x) \in B\} \tag{2.2}$$

is the ***distribution*** *of y.*

This notion of distribution is natural for characterizing the idea of rearrangement, as we note in the following

Definition. *Let $E \subset \mathbb{R}^m$ and suppose $y: E \to \mathbb{R}^n$ and $z: E \to \mathbb{R}^n$ are μ-measurable on E. Then $\begin{Bmatrix} y \\ z \end{Bmatrix}$ is a* ***rearrangement*** *of $\begin{Bmatrix} z \\ y \end{Bmatrix}$ on E if $\omega_y = \omega_z$.*

Clearly, *rearrangement* is a relation which divides the set of μ-measurable functions on E into equivalence classes, each of which we shall denote as

$$\mathscr{E}(\omega) \equiv \{y, \mu\text{-measurable on } E \mid \omega_y = \omega\}. \tag{2.3}$$

In this case, each equivalence class is said to consist of functions that are equi-distributed.

We now have

Proposition 1. *Let $g: \mathbb{R}^n \to \mathbb{R}^p$ be continuous and let $y: E \to \mathbb{R}^n$, $E \subset \mathbb{R}^m$, be μ-measurable on E. Then,*

$$\int_E g \circ y(x)\, d\mu(x) = \int_{\mathbb{R}^n} g\, d\omega_y = \int_{\operatorname{supp} \omega_y} g\, d\omega_y, \tag{2.4}$$

where $\operatorname{supp} \omega_y$ *denotes the support of the measure ω_y.*

Proof. This is a standard result which may be found in HALMOS [H, p. 163]. □

Suppose we replace E in (2.4) by R and, correspondingly, interpret μ as the volume measure v. Then, by taking g to be the identity map and y to be any v-measurable state $\vec{s}$ on R in Proposition 1, we see that *for two such states $\vec{s}_1$ and $\vec{s}_2$, if $\vec{s}_1 \in \mathscr{A}(\vec{c})$, then $\vec{s}_2 \in \mathscr{A}(\vec{c})$ provided $\omega_{\vec{s}_1} = \omega_{\vec{s}_2}$, i.e.*, provided $\vec{s}_1$ and $\vec{s}_2$ are rearrangements of one another on R. Moreover, by identifying g with the constitutive function $\hat{\eta}$, we see that if the two states $\vec{s}_1$ and $\vec{s}_2$ in $D(\hat{\eta})$, are, indeed, rearrangements of one another then $H[\vec{s}_1] = H[\vec{s}_2]$. Thus, if $\vec{s}_*$ maximizes H in $\mathscr{A}(\vec{c})$ it follows that every v-measurable state $\vec{s}$ on R which is in the equivalence class

$$\mathscr{E}_* \equiv \{\vec{s}: R \to D(\hat{\eta}) \subset \mathbb{R}^{N+1}, v\text{-measurable on } R \mid \omega_{\vec{s}} = \omega_{\vec{s}_*}\}$$

also maximizes H in $\mathscr{A}(\vec{c})$. The set of all such maximizing states is called a maximizing class. We then borrow a phrase from DUNN & FOSDICK [DF] and say that a solution $\vec{s}_* \in \mathscr{A}(\vec{c})$ of problem (2.1) is *unique up to a rearrangement* if $H[\vec{s}_*] \geq H[\vec{s}]$ for all $\vec{s} \in \mathscr{A}(\vec{c})$, with equality holding if and only if $\omega_{\vec{s}} = \omega_{\vec{s}_*}$. In this case, problem (2.1) is said to have a unique maximizing class of equi-distributed states.

In 1970 NOLL [N] considered problem (2.1) from the novel and elegant point of view of maximizing over a set of measures. In the present paper we shall consider the more classical approach in the calculus of variations of maximizing over a set of admissible functions. Our aim is to illustrate similarities between these two approaches by emphasizing the above natural relationship between the distribution of a state in R and a measure. We hope that our considerations will lend additional perspective to NOLL's contribution beyond that already contained in his early study.

For latter purposes, we record without proof the following companion result to Proposition 1, which represents a mean value property:

Proposition 2 (NOLL [N, p. 3]).† *Under the same hypotheses as in Proposition* 1, *it follows that*

$$\frac{1}{\omega_y(\mathbb{R}^n)} \int_{\mathbb{R}^n} g \, d\omega_y \in \mathscr{H}(g\,(\operatorname{supp} \omega_y)) \,.^{\dagger\dagger}$$

3. Existence and Uniqueness of Maximizers

We now turn to the question of necessary and sufficient conditions for the existence of a maximizer of problem (2.1). In Theorems 1 and 4 we shall consider, respectively, the analogs of the necessary conditions of Weierstrass and Euler-Lagrange in the calculus of variations. Theorems 2 and 3 show how the choice of the constraint $(N+1)$-tuple $\vec{c}$ is related to both necessary and sufficient conditions for existence.

Although here we have a somewhat different set-up and we consider a larger class of admissible states than was considered by DUNN & FOSDICK [DF], the following theorem has, essentially, the structure of their Theorem 1:

Theorem 1. *If $\vec{s}_*$ is a maximizer of H in $\mathscr{A}(\vec{c})$, then*

$$\operatorname{supp} \omega_{\vec{s}_*} \subset \operatorname{Dcon}(\hat{\eta}) \,.^{\dagger\dagger\dagger}$$

Proof. Let $\vec{q}$ be any point in $\operatorname{supp} \omega_{\vec{s}_*}$ and suppose that $\vec{q}_1$, $\vec{q}_2$ and α are such that

$$\vec{q} = \alpha \vec{q}_1 + (1-\alpha)\,\vec{q}_2, \qquad 0 \leqq \alpha \leqq 1,$$

with $\vec{q}_1$ and $\vec{q}_2$ in $D(\hat{\eta})$. Further, let n be a positive integer and consider the state $\vec{s}_n : R \to \mathbb{R}^{N+1}$ defined by

$$\vec{s}_n(x) \equiv \begin{cases} \vec{s}_*(x), & x \in R \setminus B_n, \\ \vec{q}_1, & x \in B_n', \\ \vec{q}_{2n}, & x \in B_n'', \end{cases}$$

† See also DUNN & FOSDICK [DF, p. 90] for a related theorem.

†† $\mathscr{H}(E)$ denotes the *convex hull* of a set E, as defined in the Appendix.

††† Dcon $(\hat{\eta})$ denotes the *domain of concavity* of the function $\hat{\eta}$ as defined in the Appendix.

where

$$B_n \equiv \left\{ x \in R \mid d(\vec{s}_*(x), \vec{q}) < \frac{1}{n} \right\} = B'_n \cup B''_n .$$

Here, $d(\cdot, \cdot)$ is the usual Euclidean distance in $\mathbb{R}^{N+1}$, the v-measures of the sets B'_n and B''_n are given by

$$v(B'_n) = \alpha v(B_n), \quad v(B''_n) = (1 - \alpha)\, v(B_n),$$

and $\vec{q}_{2n}$ is such that

$$\int_R \vec{s}_n(x)\, dv(x) = \vec{c}\, V .$$

Note that since

$$\int_R \vec{s}_n(x)\, dv(x) = \int_R \vec{s}_*(x)\, dv(x) = \vec{c}\, V,$$

we have

$$\frac{1}{v(B_n)} \int_{B_n} \vec{s}_*(x)\, dv(x) = \alpha \vec{q}_1 + (1 - \alpha)\, \vec{q}_{2n} .$$

Then, since $\vec{s}_* : R \to \mathbb{R}^{N+1}$ is v-integrable on R and $\vec{q} \in \operatorname{supp} \omega_{\vec{s}_*}$, it follows that

$$\lim_{n \to \infty} \frac{1}{v(B_n)} \int_{B_n} \vec{s}_*(x)\, dv(x) = \vec{q} ,$$

which shows that

$$\lim_{n \to \infty} \vec{q}_{2n} = \vec{q}_2 .$$

Thus, $\vec{s}_n \in \mathscr{A}(\vec{c})$ for sufficiently large n.

Now, since, by hypothesis $H[\vec{s}_*] \geqq H[\vec{s}_n]$ for all sufficiently large n, it follows that

$$\int_{B_n} \hat{\eta}(\vec{s}_*(x))\, dv(x) \geqq \int_{B'_n} \hat{\eta}(\vec{q}_1)\, dv(x) + \int_{B''_n} \hat{\eta}(\vec{q}_{2n})\, dv(x),$$

which yields

$$\frac{1}{v(B_n)} \int_{B_n} \hat{\eta}(\vec{s}_*(x))\, dv(x) \geqq \alpha \hat{\eta}(\vec{q}_1) + (1 - \alpha)\, \hat{\eta}(\vec{q}_{2n}) .$$

Finally, since $\hat{\eta} : D(\hat{\eta}) \to \mathbb{R}$ is continuous, $\vec{s}_* : R \to D(\hat{\eta})$ is v-integrable on R, and $\vec{q} \in \operatorname{supp} \omega_{\vec{s}_*}$, we see that in the limit $n \to \infty$,

$$\hat{\eta}(\vec{q}) \geqq \alpha \hat{\eta}(\vec{q}_1) + (1 - \alpha)\, \hat{\eta}(\vec{q}_2) .$$

Then, since $\vec{q}$ is any convex combination of $\vec{q}_1$ and $\vec{q}_2$, we conclude that $\vec{q}$ is a point of concavity of $\hat{\eta}$, and this, together with the fact that $\vec{q} \in \sup \omega_{\vec{s}_*}$, completes this proof. □

In order that a maximizer of H in $\mathscr{A}(\vec{c})$ exist, the constraint $(N+1)$-tuple $\vec{c}$ must be suitably restricted. In fact, we have

Theorem 2. *It is necessary for the existence of a maximizer of H in $\mathscr{A}(\vec{c})$ that*

$$(\vec{c}, {}_u\hat{\eta}(\vec{c})) \in \mathscr{H}(\Gamma(\hat{\eta})).^\dagger$$

Proof. Suppose, in Propositions 1 and 2, we take $n = N+1$, $p = N+2$, $m = 3$, $E = R$, let μ be the volume measure v, identify y as any v-measurable state $\vec{s}: R \to D(\hat{\eta})$, and let $g: \mathbb{R}^{N+1} \to \mathbb{R}^{N+2}$ be the continuous map

$$g(\vec{q}) \equiv (\vec{q}, \hat{\eta}(\vec{q})) \ \forall \vec{q} \in D(\hat{\eta}).$$

Then, since $\omega_{\vec{s}}(\mathbb{R}^{N+1}) = v(R) = V$, it readily follows that

$$\frac{1}{V} \int_R [\vec{s}(x), \hat{\eta}(\vec{s}(x))] \, dv(x) \in \mathscr{H}(\operatorname{supp} \omega_{\vec{s}}, \hat{\eta}(\operatorname{supp} \omega_{\vec{s}})),$$

and we then see that for any $\vec{s} \in \mathscr{A}(\vec{c})$

$$\left(\vec{c}, \frac{1}{V} \int_R \hat{\eta}(\vec{s}(x)) \, dv(x)\right) \in \mathscr{H}(\Gamma(\hat{\eta})).$$

Now, since the upper convex envelope ${}_u\hat{\eta}$ has the property that

$${}_u\hat{\eta}(\vec{c}) = \sup \{\gamma \in \mathbb{R} \mid (\vec{c}, \gamma) \in \mathscr{H}(\Gamma(\hat{\eta}))\},$$

if we can show that for each $\gamma \in \mathbb{R}$ with $(\vec{c}, \gamma) \in \mathscr{H}(\Gamma(\hat{\eta}))$ there is a state $\vec{s}^\# \in \mathscr{A}(\vec{c})$ such that

$$\frac{1}{V} \int_R \hat{\eta}(\vec{s}^\#(x)) \, dv(x) = \gamma,$$

then

$${}_u\hat{\eta}(\vec{c}) = \sup_{\vec{s} \in \mathscr{A}(\vec{c})} \frac{1}{V} \int_R \hat{\eta}(\vec{s}(x)) \, dv(x), \tag{3.1}$$

and accordingly, it would be necessary that

$$(\vec{c}, {}_u\hat{\eta}(\vec{c})) \in \mathscr{H}(\Gamma(\hat{\eta})).$$

Thus, to complete this proof, we first observe that Theorem A1 (see Appendix), as applied to the case where the set E is the graph $\Gamma(\hat{\eta})$ in $\mathbb{R}^{N+2}$, guarantees the existence of $N+3$ points $g(\vec{q}_i) \equiv (\vec{q}_i, \hat{\eta}(\vec{q}_i))$ in $\Gamma(\hat{\eta})$, $\vec{q}_i \in D(\hat{\eta})$ for each $i = 1, 2, \dots, N+3$, such that

$$\sum_{i=1}^{N+3} \alpha_i(\vec{q}_i, \hat{\eta}(\vec{q}_i)) = (\vec{c}, \gamma)$$

---

† ${}_u\hat{\eta}$ denotes the *upper convex envelope* of $\hat{\eta}$, and $\Gamma(\hat{\eta})$ represents the *graph* of $\hat{\eta}$. See the Appendix for the formal definitions.

for $\alpha_i \geqq 0$ and $\sum_{i=1}^{N+3} \alpha_i = 1$. Then, by defining

$$\vec{s}^{\#}(x) \equiv \vec{q}_i, \quad x \in R_i,$$

where the v-measures of the disjoint sets R_i in R are given by

$$v(R_i) = \alpha_i V,$$

we see that $\vec{s}^{\#} \in \mathscr{A}(\vec{c})$ and

$$\frac{1}{V} \int_R \hat{\eta}(\vec{s}^{\#}(x))\, dv(x) = \sum_{i=1}^{N+3} \alpha_i \hat{\eta}(\vec{q}_i) = \gamma. \quad \square$$

The construction of the state $\vec{s}^{\#} \in \mathscr{A}(\vec{c})$ in the preceding proof suggests the following sufficient condition for the existence of a maximizer. We have

Theorem 3. *It is sufficient for the existence of a maximizer of H in $\mathscr{A}(\vec{c})$ that*

$$(\vec{c}, u\hat{\eta}(\vec{c})) \in \mathscr{H}(\Gamma(\hat{\eta})).$$

Proof. Since $u\hat{\eta}$ is a concave function, it follows that

$$u\hat{\eta}(\vec{q}) \leqq u\hat{\eta}(\vec{c}) + \vec{g} \cdot (\vec{q} - \vec{c}) \tag{3.2}$$

for any $\vec{q} \in D(u\hat{\eta}) \equiv \mathscr{H}(D(\hat{\eta}))$ and for any supergradient $\vec{g} \in \partial u\hat{\eta}(\vec{c})$.[†] Thus, by setting $\vec{q} = \vec{s}(x)$, where $\vec{s} \in \mathscr{A}(\vec{c})$, and integrating, we get

$$\int_R \hat{\eta}(\vec{s}(x))\, dv(x) \leqq V u\hat{\eta}(\vec{c}),$$

where we have used the fact that $\hat{\eta}(\vec{q}) \leqq u\hat{\eta}(\vec{q})$ for any $\vec{q} \in D(\hat{\eta})$.

Now, because of the hypothesis of this theorem, we may replace γ by $u\hat{\eta}(\vec{c})$ in the construction given in the proof of Theorem 2 and conclude that there exists a state $\vec{s}^{\#} \in \mathscr{A}(\vec{c})$ such that

$$\frac{1}{V} \int_R \hat{\eta}(\vec{s}^{\#}(x))\, dv(x) = u\hat{\eta}(\vec{c}).$$

Returning to the previous inequality, we then see that $H[\vec{s}] \leqq H[\vec{s}^{\#}]$ for all $\vec{s} \in \mathscr{A}(\vec{c})$ to complete this proof. $\square$

It should be noted that the maximizer $\vec{s}^{\#}$ constructed in Theorem 3 has the structure of a step function with at most $N+3$ distinct steps. This, together with Theorems 2 and 3 may be summarized as follows:[††] *A necessary and sufficient condition for the existence of a maximizer of H in $\mathscr{A}(\vec{c})$ is $(\vec{c}, u\hat{\eta}(\vec{c})) \in \mathscr{H}(\Gamma(\hat{\eta}))$. In this case, one maximizer is a step function $\vec{s}^{\#}$ for which* supp $\omega_{\vec{s}^{\#}}$ *has at most $N+3$ points.* In general, not all maximizers of the problem (2.1) are step functions.

[†] $\partial u\hat{\eta}(\vec{c})$ denotes the *superdifferential* of $u\hat{\eta}$ at $\vec{c}$ as defined in the Appendix.

[††] See Theorem 3 of NOLL [N] and Theorem 7 of DUNN & FOSDICK [DF].

The following theorem may be interpreted as a counterpart of the classical necessary condition of Euler-Lagrange in the calculus of variations. The proof given here emphasizes global arguments.

Theorem 4.† *If* $\vec{s}_*$ *is a maximizer of* H *in* $\mathscr{A}(\vec{c})$, *then*

$$\bigcap_{\vec{q} \in \operatorname{supp} \omega_{\vec{s}_*}} \partial \hat{\eta}(\vec{q}) \neq \emptyset .$$

In particular, if $\hat{\eta}: D(\hat{\eta}) \to \mathbb{R}$ *is differentiable, then*

$$\operatorname{grad}_{\vec{q}} \hat{\eta}(\vec{q})\big|_{\vec{q} \in \operatorname{supp} \omega_{\vec{s}_*}} = \vec{g}_*,$$

where $\vec{g}_* \in \mathbb{R}^{N+1}$ *is constant. Also, in this case,*

$$[\hat{\eta}(\vec{q}) - \vec{q} \cdot \operatorname{grad}_{\vec{q}} \hat{\eta}(\vec{q})]\big|_{\vec{q} \in \operatorname{supp} \omega_{\vec{s}_*}} = c_*,$$

where $c_* \in \mathbb{R}$ *is constant. The constants* $\vec{q}_*$ *and* c_* *determine the temperature* θ_*, *the chemical potentials* $\underset{a}{\mu_*}$, *and the pressure* p_* *of a maximizer via,*††

$$\vec{g}_* = (1/\theta_*, -\underset{a}{\mu_*}/\theta_*), \qquad c_* = p_*/\theta_* .$$

Proof. As in the first step of the proof of Theorem 3, if we take $\vec{q} = \vec{s}_*(x)$ in the initial inequality, integrate over R, and recall from Theorem 1 that

$$\hat{\eta}(\vec{q}) = \mathscr{u}\hat{\eta}(\vec{q}) \quad \textit{for} \quad \vec{q} \in \operatorname{supp} \omega_{\vec{s}_*}, \tag{3.3}$$

we get

$$\int_R \hat{\eta}(\vec{s}_*(x))\, dv(x) \leqq V \mathscr{u}\hat{\eta}(\vec{c}),$$

where, in this case, equality holds if and only if

$$\mathscr{u}\hat{\eta}(\vec{q}) = \mathscr{u}\hat{\eta}(\vec{c}) + \vec{g} \cdot (\vec{q} - \vec{c})$$

for every $\vec{q} \in \operatorname{supp} \omega_{\vec{s}_*}$ and any $\vec{g} \in \partial \mathscr{u}\hat{\eta}(\vec{c})$. But, because $\vec{s}_*$ is a maximizer of H in $\mathscr{A}(\vec{c})$, we see from (3.1) that equality must hold, above. Thus, by recalling (3.2) and (3.3) we may write

$$\mathscr{u}\hat{\eta}(\vec{q}\,') \leqq \hat{\eta}(\vec{q}) + \vec{g} \cdot (\vec{q}\,' - \vec{q})$$

for any $\vec{q}\,' \in D(\mathscr{u}\hat{\eta}) = \mathscr{H}(D(\hat{\eta}))$, $\vec{q} \in \operatorname{supp} \omega_{\vec{s}_*}$, and any supergradient $\vec{g} \in \partial \mathscr{u}\hat{\eta}(\vec{c})$. Therefore, since $\hat{\eta}(\vec{q}\,') \leqq \mathscr{u}\hat{\eta}(\vec{q}\,')$ for any $\vec{q}\,' \in D(\hat{\eta})$, we may replace the left

† This theorem should be compared to (5.12) of Noll [N] and Theorem 7 of Dunn & Fosdick [DF]. Noll uses the idea of affine bounds to support his proof while Dunn & Fosdick employ an argument using convex envelopes and touching planes.

†† Recall the classical relations from thermostatics for the temperature, chemical potentials and the pressure that were introduced early in Section 2.

hand side of the above inequality by $\hat{\eta}(\vec{q}\,')$ and conclude, by the definition[†] of supergradient and superdifferential, that

$$\bigcap_{\vec{q} \in \operatorname{supp} \omega_{\vec{s}_*}} \partial\hat{\eta}(\vec{q}) = \partial_{\mathscr{U}}\hat{\eta}(\vec{c}) \neq \emptyset, \tag{3.4}$$

to complete the essential part of this proof.

While the first special conclusion of this theorem, for the case when $\hat{\eta}\colon D(\hat{\eta}) \to \mathbb{R}$ is differentiable, follows immediately, to substantiate the second claim will require the additional use of Theorem 1. Because of the differentiability of $\hat{\eta}$ we may state the result of that theorem as

$$\hat{\eta}(\vec{s}) \leqq [\hat{\eta}(\vec{q}) + (\vec{s} - \vec{q}) \cdot \operatorname{grad}_{\vec{q}} \hat{\eta}(\vec{q})]\big|_{\vec{q} \in \operatorname{supp} \omega_{\vec{s}_*}}$$

for all $\vec{s} \in D(\hat{\eta})$. Thus, by identifying $\vec{s}$ and $\vec{q}$ in this inequality with any two points in supp $\omega_{\vec{s}_*}$, say $\vec{q}_1$ and $\vec{q}_2$, and then reversing the order of this identification, we finally find that

$$\hat{\eta}(\vec{q}_1) - \vec{q}_1 \cdot \operatorname{grad}_{\vec{q}} \hat{\eta}(\vec{q}_1) = \hat{\eta}(\vec{q}_2) - \vec{q}_2 \cdot \operatorname{grad}_{\vec{q}} \eta(\vec{q}_2). \ \square$$

It may be of interest to note the following immediate consequence of Theorem 3 and the arguments given in support of Theorem 4: *Suppose* $\vec{c}$ *is such that* $(\vec{c}, \mathscr{U}\hat{\eta}(\vec{c})) \in \mathscr{H}(\Gamma(\hat{\eta}))$. *Then, for any maximizer* $\vec{s}_* \in \mathscr{A}(\vec{c})$, $\vec{c}$ *determines the unique set of associated supergradients given by* (3.4). When $\mathscr{U}\hat{\eta}$ is differentiable at $\vec{c}$, the elements, *i.e.*, supergradients, of the set $\partial_{\mathscr{U}}\hat{\eta}(\vec{c})$ are all parallel to and have the same orientation as a single vector in $\mathbb{R}^{N+1}$. When $\hat{\eta}$, itself, is differentiable, it is natural to take this vector to be the constant $\vec{g}_* \equiv (1/\theta_*, -\underset{a}{\mu}_*/\theta_*)$ of this theorem. In this case, if an $(N+1)$-dimensional plane with normal direction equal to $(-1, \vec{g}_*)$ is constructed through any point $\vec{q} \in \operatorname{supp} \omega_{\vec{s}_*}$ on the graph $\Gamma(\hat{\eta})$, it will support the graph (*i.e.* surface) all to one side, and it is straightforward to show that the plane so constructed will intersect the η-axis in $\mathbb{R}^{N+2}$ at a height which is equal to the constant $c_* = p_*/\theta_*$ of this theorem. Thus, in the more general situation, when $\hat{\eta}$ is not differentiable, it seems reasonable to structure each of the possible supergradients $\vec{g}$ in the unique set $\partial_{\mathscr{U}}\hat{\eta}(\vec{c})$ in the form $\vec{g} \equiv (1/\theta, -\underset{a}{\mu}/\theta)$, and refer to each of the corresponding numbers θ and $\underset{a}{\mu}$ as an admissible temperature and chemical potential N-tuple for the maximizer $\vec{s}_*$. Then, corresponding to this $\vec{g}$, an admissible pressure is obtained by use of $[\hat{\eta}(\vec{q}) - \vec{q} \cdot \vec{g}]\big|_{\vec{q} \in \operatorname{supp} \omega_{\vec{s}_*}} \equiv c \equiv p/\theta$.

In many applications, it is not possible to assume, as we have, that the domain $D(\hat{\eta})$ is open in $\mathbb{R}^{N+1}$. One may wish to restrict the specific mass densities so that $\underset{a}{\varrho} \geqq 0$, $a = 1, 2, \ldots, N$, and allow for the possibility that any or all but one of them may vanish. In this case, it is possible that the support of the distribution of a maximizer $\vec{s}_*$ of problem (2.1), *i.e.* supp $\omega_{\vec{s}_*}$, may intersect the boundary

[†] See the Appendix.

$\partial D(\hat{\eta})$ at points where some of the mass densities vanish. In such situations, the first part of Theorem 4 still remains valid. In addition, when $\hat{\eta}: D(\hat{\eta}) \to \mathbb{R}$ is differentiable, the second part holds as stated at any $\vec{q} \in \operatorname{supp} \omega_{\vec{s}_*} \cap$ Interior $D(\hat{\eta})$. However, at those $\vec{q} \in \operatorname{supp} \omega_{\vec{s}_*} \cap \partial D(\hat{\eta})$ for which $\underset{a}{\varrho} = 0$, it can be argued that $\partial\hat{\eta}/\partial \underset{a}{\varrho} \leqq \underset{a}{g}$, where $\underset{a}{g}$ represents the corresponding component of the constant vector $\vec{g}$. NOLL [N, p. 11] arrived at this result and concluded that "*In a stable state* [of positive temperature], *the temperature and the pressure are the same in all phases. The chemical potential of a compound is the same in all phases in which that compound is present. In those phases in which the compound is absent, the potential is not less.*"

We noted earlier that the total entropy H has the same fixed value for every v-measurable state within a set which is equally distributed. Thus, a maximizer of problem (2.1) can be, at best, unique up to a rearrangement. In fact, we have

Theorem 5. *Let $\vec{c} \in \mathbb{R}^{N+1}$ be such that $(\vec{c}, \mathscr{u}\hat{\eta}(\vec{c})) \in \mathscr{H}(\Gamma(\hat{\eta}))$, let $\pi_{\vec{c}}$ denote the set of support hyperplanes to* epi $\mathscr{u}\hat{\eta}$† *at the point $(\vec{c}, \mathscr{u}\hat{\eta}(\vec{c}))$, and consider the set of points*

$$\sigma_{\vec{c}} \equiv \bigcap_{\pi \in \pi_{\vec{c}}} (\pi \cap \Gamma(\hat{\eta})) \subset \mathbb{R}^{N+2}.$$

If $\sigma_{\vec{c}}$ contains k points with $k \leqq N+2$, then every maximizer of problem (2.1) *belongs to a single equivalence class of equidistributed v-measurable states on R.*

Proof. Suppose $\vec{s}_* \in \mathscr{A}(\vec{c})$ is a maximizer of H in problem 2.1. Then, because of Theorem 4, the set of all points $(\vec{q}, \hat{\eta}(\vec{q})) \in \Gamma(\hat{\eta})$ for which $\vec{q} \in \operatorname{supp} \omega_{\vec{s}_*}$ must lie on at least one single supporting hyperplane to $\mathscr{H}(\Gamma(\hat{\eta}))$ at the point $(\vec{c}, \mathscr{u}\hat{\eta}(\vec{c}))$. In addition, because of the integral constraint which $\vec{s}_*$ satisfies in $\mathscr{A}(\vec{c})$, the point $(\vec{c}, \mathscr{u}\hat{\eta}(\vec{c}))$ must be in the convex hull of this set of coplanar points, and, therefore, be representable as a convex combination of such points. Since this support hyperplane is contained in the set $\pi_{\vec{c}}$, as defined in the hypothesis of this theorem, we see from Corollary 2 of Theorem A2 in the Appendix that

$$\{(\vec{q}, \hat{\eta}(\vec{q})) \in \Gamma(\hat{\eta}) \mid \vec{q} \in \operatorname{supp} \omega_{\vec{s}_*}\} \subset \sigma_{\vec{c}}.$$

Now, since $\sigma_{\vec{c}}$ contains $k \leqq N+2$ points, there is a *unique* convex combination of no more than k points $(\vec{q}_i, \hat{\eta}(\vec{q}_i))$ in $\sigma_{\vec{c}}$ such that

$$(\vec{c}, \mathscr{u}\hat{\eta}(\vec{c})) = \sum_{i=1}^{k} \alpha_i (\vec{q}_i, \hat{\eta}(\vec{q}_i))$$

for $\alpha_i \geqq 0$, $\sum_{i=1}^{k} \alpha_i = 1$.

Finally, to complete this proof, we note that the measure (*i.e.*, distribution)

$$\omega \equiv \sum_{i=1}^{k} \alpha_i \, \delta_{\vec{q}_i},$$

† The *epigraph* of a function f, denoted epi f, is defined in the Appendix.

where $\delta_{\vec{q}_i}$ is the Dirac measure at $\vec{q}_i \in \mathbb{R}^{N+1}$, determines the unique equivalence class of v-measurable states on R that solve problem 2.1. □

The preceding theorem is analogous to Corollary 1 of NOLL [N, p. 6][†] which shows, in NOLL's terminology, that a measure is an extreme point, in a certain prescribed set, if and only if it is a convex combination of at most $n+1$ Dirac measures. NOLL used n to denote the dimension of what is equivalent to our state space here, *i.e.* $N+1$.

According to Theorem 5, we may conclude that a maximizer of problem (2.1) is unique up to a rearrangement if the smallest face[††] containing $(\vec{c}, \mathscr{u}\hat{\eta}(\vec{c}))$ intersects the graph $\Gamma(\hat{\eta})$ at no more than $N+2$ points. MAN [M] recently has investigated the question of why thermostatic surfaces such as that described by the graph $\Gamma(\hat{\eta})$, are expected to meet this condition. When such is the case, the GIBBS [G, p. 97] phase rule is said to hold.

4. Comparison Principles

In the remainder of this paper we shall assume, for convenience, that $\hat{\eta}$ is differentiable and show how the pressure, the temperature, and the chemical potentials of a maximizer must change when the constraint $(N+1)$-tuple $\vec{c}$ and the given total volume V are changed within a prescribed manner. Within elementary classical thermostatics, this kind of study leads to such things as, for example, a Van't Hoff or LeChatelier inequality, the Gibbs equation, the Clausius-Clapeyron equation, and the Gibbs-Duhem equation.

To begin with, let us suppose that the given constraint data $\vec{c}_1$ and $\vec{c}_2$ are such that both $(\vec{c}_1, \mathscr{u}\hat{\eta}(\vec{c}_1))$ and $(\vec{c}_2, \mathscr{u}\hat{\eta}(\vec{c}_2))$ are in the convex hull $\mathscr{H}(\Gamma(\hat{\eta}))$. Further, let us suppose that $\vec{s}_1 \in \mathscr{A}(\vec{c}_1)$ and $\vec{s}_2 \in \mathscr{A}(\vec{c}_2)$ are two maximizers of problem (2.1) which correspond to the given total volumes V_1 and V_2, respectively, and the associated regions R_1 and R_2. Then, by applying the results of Theorem 1 and Theorem 4 to the case when $x \in R_1$ and $y \in R_2$ are any two points for which $\vec{s}_1(x) \in \operatorname{supp} \omega_{\vec{s}_1}$ and $\vec{s}_2(y) \in \operatorname{supp} \omega_{\vec{s}_2}$, we may write the following:

$$\hat{\eta}(\vec{s}_2(y)) \leqq \hat{\eta}(\vec{s}_1(x)) + \vec{g}_1 \cdot [\vec{s}_2(y) - \vec{s}_1(x)],$$
$$\hat{\eta}(\vec{s}_1(x)) \leqq \hat{\eta}(\vec{s}_2(y)) + \vec{g}_2 \cdot [\vec{s}_1(x) - \vec{s}_2(y)],$$

where

$$\operatorname{grad}_{\vec{s}} \hat{\eta}(\vec{s})\big|_{\vec{s}=\vec{s}_1(x)} = \vec{g}_1 \equiv (1/\theta_1, -\underset{a}{\mu_1}/\theta_1),$$

$$\operatorname{grad}_{\vec{s}} \hat{\eta}(\vec{s})\big|_{\vec{s}=\vec{s}_2(y)} = \vec{g}_2 \equiv (1/\theta_2, -\underset{a}{\mu_2}/\theta_2),$$

$$\hat{\eta}(\vec{s}_1(x)) - \vec{g}_1 \cdot \vec{s}_1(x) = p_1/\theta_1,$$

$$\hat{\eta}(\vec{s}_2(y)) - \vec{g}_2 \cdot \vec{s}_2(y) = p_2/\theta_2,$$

[†] See, also, Theorem 8 and its Corollary in DUNN & FOSDICK [DF] for a related result.

[††] The smallest face containing $(\vec{c}, \hat{\eta}(\vec{c}))$ is the intersection of all supporting hyperplanes to $(\Gamma(\hat{\eta}))$ at $(\vec{c}, \hat{\eta}(\vec{c}))$.

and where $p_1, p_2, \theta_1, \theta_2, \underset{a}{\mu_1}, \underset{a}{\mu_2}$, for $a = 1, 2, \ldots, N$, are constant pressures, temperatures, and chemical potentials associated with the respective maximizers. Thus, we see that

$$\hat{\eta}(\vec{s}_2(y)) - \vec{g}_1 \cdot \vec{s}_2(y) \leqq p_1/\theta_1,$$

and

$$\hat{\eta}(\vec{s}_1(x)) - \vec{g}_2 \cdot \vec{s}_1(x) \leqq p_2/\theta_2,$$

and by integrating the last four results over the appropriate regions R_1 and R_2, and using $(2.1)_2$ and $(2.1)_3$, we get, respectively,

$$\begin{aligned} H_1 - \vec{g}_1 \cdot \vec{c}_1 V_1 &= p_1 V_1/\theta_1, \\ H_2 - \vec{g}_2 \cdot \vec{c}_2 V_2 &= p_2 V_2/\theta_2, \\ H_2 - \vec{g}_1 \cdot \vec{c}_2 V_2 &\leqq p_1 V_2/\theta_1, \\ H_1 - \vec{g}_2 \cdot \vec{c}_1 V_1 &\leqq p_2 V_1/\theta_2. \end{aligned}$$

It is now straightforward to see from these conclusions that *if θ_1 and θ_2 are positive, then*

$$(p_1 - p_2)(V_1 - V_2) \leqq (\theta_1 - \theta_2)(H_1 - H_2) - (\theta_1 \vec{g}_1 - \theta_2 \vec{g}_2) \cdot (\vec{c}_1 V_1 - \vec{c}_2 V_2).$$

More explicitly, since

$$\theta_1 \vec{g}_1 = (1, -\underset{a}{\mu_1}), \qquad \theta_2 \vec{g}_2 = (1, -\underset{a}{\mu_2}),$$

and

$$\vec{c}_1 V_1 = (E_1, \underset{a}{M_1}), \qquad \vec{c}_2 V_2 = (E_2, \underset{a}{M_2}),$$

we have, equivalently, the following generalized Van't Hoff-LeChatelier inequality:

$$(p_1 - p_2)(V_1 - V_2) \leqq (\theta_1 - \theta_2)(H_1 - H_2) + \sum_{a=1}^{N} (\underset{a}{\mu_1} - \underset{a}{\mu_2})(\underset{a}{M_1} - \underset{a}{M_2}). \tag{4.1}$$

When the system has only a single component the terms in the summation do not appear in (4.1). This case was considered by DUNN & FOSDICK [DF] in their Theorem 12 and a full interpretation of the resulting Van't Hoff-LeChatelier inequality was given in that work. In the present more general situation we may draw from (4.1) the additional conclusion that *if in a sequence of maximizers of problem* (2.1) *either $H_1 = H_2$ or $\theta_1 = \theta_2$, either $V_1 = V_2$ or $p_1 = p_2$, and either $\underset{a}{M_1} = \underset{a}{M_2}$ or $\underset{a}{\mu_1} = \underset{a}{\mu_2}$ for all $a = 1, 2, \ldots, N$ except $a = b$, then*

$$(\underset{b}{\mu_1} - \underset{b}{\mu_2})(\underset{b}{M_1} - \underset{b}{M_2}) \geqq 0,$$

and we see that *an increase in the chemical potential of the bth component in such a sequence*, i.e., at fixed temperature, pressure, and chemical potential of all other components, *will shift the "equilibrium" in a direction such that mass of that component will be absorbed* (*or, at best, conserved*).

Now, to obtain a generalized Clausius-Clapeyron equation, suppose $\vec{c}(\tau)$, $0 \leqq \tau \leqq 1$, is a smooth path in the domain $D(\hat{\eta})$ and let R_τ be a sequence of bounded open regions in $\mathbb{R}^3$ with corresponding smoothly varying total volume $V(\tau)$. Suppose that for each value of τ the state $\vec{s}_\tau \in \mathscr{A}(\vec{c}(\tau))$ is a maximizer of problem (2.1) for the region R_τ. In addition, suppose that for all τ in an open interval $\mathscr{I} \subset (0, 1)$ the support supp $\omega_{\vec{s}_\tau}$ contains at least two separate smooth paths of states:

$$\vec{q}_1(\tau) \equiv [\varepsilon_1(\tau), \underset{a}{\varrho_1}(\tau)], \qquad \vec{q}_2(\tau) \equiv [\varepsilon_2(\tau), \underset{a}{\varrho_2}(\tau)].$$

In this case the maximizer $\vec{s}_\tau$ consists of at least two coexistent states (*i.e.*, phases) in R_τ corresponding to $\vec{q}_1(\tau)$ and $\vec{q}_2(\tau)$ for each $\tau \in \mathscr{I}$. And, because of Theorems 1 and 4, we may write

$$\hat{\eta}(\vec{q}_2(\tau)) \leqq \hat{\eta}(\vec{q}_1(\tau)) + \vec{g}_1(\tau) \cdot [\vec{q}_2(\tau) - \vec{q}_1(\tau)],$$

and

$$\hat{\eta}(\vec{q}_1(\tau)) \leqq \hat{\eta}(\vec{q}_2(\tau)) + \vec{g}_2(\tau) \cdot [\vec{q}_1(\tau) - \vec{q}_2(\tau)],$$

where

$$\vec{g}_1(\tau) = \vec{g}_2(\tau) = \operatorname{grad}_{\vec{s}} \eta(\vec{s})\big|_{\vec{s} = \vec{q}_1(\tau) \text{ or } \vec{q}_2(\tau)}$$

$$\equiv \vec{g}(\tau) \equiv ([1/\theta(\tau), -\underset{a}{\mu}(\tau)/\theta(\tau)]. \tag{4.2}$$

The numbers $\theta(\tau)$ and $\underset{a}{\mu}(\tau)$, $a = 1, 2, \ldots, N$, correspond to the temperature and the chemical potentials of the components in both of the two coexistent states. Clearly we have

$$\hat{\eta}(\vec{q}_2(\tau)) = \hat{\eta}(\vec{q}_1(\tau)) + \vec{g}(\tau) \cdot [\vec{q}_2(\tau) - \vec{q}_1(\tau)].$$

Theorem 4 implies also that

$$p(\tau) = \theta(\tau)\, \hat{\eta}(\vec{q}_1(\tau)) - \theta(\tau)\, \vec{g}(\tau) \cdot \vec{q}_1(\tau), \tag{4.3}$$

where $p(\tau)$ is the pressure associated with the maximizer $\vec{s}_\tau$. By differentiation of (4.3) and use of (4.2) we readily obtain

$$\dot{p}(\tau) = \dot{\theta}(\tau)\, \hat{\eta}(\vec{q}_1(\tau)) + \sum_{a=1}^{N} \underset{a}{\dot{\mu}}(\tau)\, \underset{a}{\varrho_1}(\tau),$$

and, similarly, we also have the analogous formula

$$\dot{p}(\tau) = \dot{\theta}(\tau)\, \hat{\eta}(\vec{q}_2(\tau)) + \sum_{a=1}^{N} \underset{a}{\dot{\mu}}(\tau)\, \underset{a}{\varrho_2}(\tau).$$

It is now convenient to divide the last two equations through by the respective mass densities of each of the phases of the mixture, $\varrho_1(\tau) = \sum_{a=1}^{N} \underset{a}{\varrho_1}(\tau)$ and

$\varrho_2(\tau) = \sum_{a=1}^{N} \underset{a}{\varrho_2}(\tau)$, and write

$$\dot{p}(\tau)\, v_1(\tau) = \dot{\theta}(\tau)\, n_1(\tau) + \sum_{a=1}^{N} \underset{a}{\dot{\mu}}(\tau)\, \underset{a}{m_1}(\tau),$$

and

$$\dot{p}(\tau)\, v_2(\tau) = \dot{\theta}(\tau)\, n_2(\tau) + \sum_{a=1}^{N} \underset{a}{\dot{\mu}}(\tau)\, \underset{a}{m_2}(\tau). \tag{4.4}$$

Here, for $\alpha = 1$ or 2, $v_\alpha(\tau) \equiv 1/\varrho_\alpha(\tau)$ is the specific volume of each of the phases, $n_\alpha(\tau) \equiv \hat{\eta}(\vec{q}_\alpha(\tau))/\varrho_\alpha(\tau)$ *is the entropy per unit mass of each of the phases*, and $\underset{a}{m_\alpha}(\tau) > \underset{a}{\varrho_\alpha}(\tau)/\varrho_\alpha(\tau)$ *is the mass concentration of the* a^{th} *component in each of the phases for* $a = 1, 2, \ldots, N$. Thus, whenever $\dot{\theta}(\tau) \neq 0$ and $v_2(\tau) - v_1(\tau) \neq 0$ we may subtract the two equations of (4.4) and write

$$\frac{\dot{p}(\tau)}{\dot{\theta}(\tau)} = \frac{[n_2(\tau) - n_1(\tau)]}{[v_2(\tau) - v_1(\tau)]} + \sum_{a=1}^{N} \left[\frac{\underset{a}{\dot{\mu}}(\tau)}{\dot{\theta}(\tau)}\right] \frac{[\underset{a}{m_2}(\tau) - \underset{a}{m_1}(\tau)]}{[v_2(\tau) - v_1(\tau)]},$$

which is the generalized Clausius-Clapeyron equation that was sought. This equation relates the changes that must take place in the temperature, pressure and chemical potentials of the components of a mixture in a multi-phase sequence of maximizing states of problem (2.1).

In principle, the support supp $\omega_{\vec{s}_\tau}$ can contain any number of separate smooth paths of states for $\tau \in \mathscr{I}$. As a rule, however, just as the doctrine of the phase rule in classical thermostatics suggests that at most $N + 2$ coexistent phases are possible, here the same doctrine allows the support supp $\omega_{\vec{s}_\tau}$ to contain at most $N + 1$ paths of states for $\tau \in \mathscr{I}$. In this case there are $N + 1$ equations like (4.4) from which $\underset{a}{\dot{\mu}}(\tau)$, $a = 1, 2, \ldots, N$, can be eliminated so as to obtain a differential equation for the ratio $\dot{p}(\tau)/\dot{\theta}(\tau)$ alone in terms of all of the specific volumes, entropies, and mass concentrations.†

Finally, to complete this section, let us again suppose that $\vec{s}_\tau \in \mathscr{A}(\vec{c}(\tau))$, $\tau \in \mathscr{I}$, is a sequence of maximizers of problem (2.1) corresponding to the smooth data $\vec{c}(\tau)$. Because of Theorem 4, we may then write

$$p(\tau) = \theta(\tau)\, \hat{\eta}(\vec{s}_\tau(x)) - \theta(\tau)\, \vec{g}(\tau) \cdot \vec{s}_\tau(x)$$

for all $x \in R_\tau$ for which $\vec{s}_\tau(x) \in \operatorname{supp} \omega_{\vec{s}_\tau}$, where $\operatorname{grad}_{\vec{s}}\, \hat{\eta}(\vec{s}_\tau(x)) \equiv \vec{g}(\tau) \equiv [1/\theta(\tau), -\underset{a}{\mu}(\tau)/\theta(\tau)]$, and where the numbers $p(\tau)$, $\theta(\tau)$ and $\underset{a}{\mu}(\tau)$, $a = 1, 2, \ldots, N$, represent the pressure, temperature, and chemical potentials, respectively, associated with the maximizer. Thus, by differentiating we obtain

$$\dot{p}(\tau) = \dot{\theta}(\tau)\, \hat{\eta}(\vec{s}_\tau(x)) - [0, -\underset{a}{\dot{\mu}}(\tau)] \cdot \vec{s}_\tau(x),$$

† This equation is like that of Gibbs [G, p. 98, eq. (129)].

and by integrating the last two equations over R_τ we reach

$$p(\tau)\,V(\tau) = \theta(\tau)\,H(\tau) - E(\tau) + \sum_{a=1}^{N} \underset{a}{\mu}(\tau)\,\underset{a}{M}(\tau), \tag{4.5}$$

and

$$\dot{p}(\tau)\,V(\tau) = \dot{\theta}(\tau)\,H(\tau) + \sum_{a=1}^{N} \underset{a}{\dot{\mu}}(\tau)\,\underset{a}{M}(\tau). \tag{4.6}$$

To obtain these two results we also used $(2.1)_2$ and $(2.1)_3$.

Equation (4.6) is recognized as a global form of the Gibbs-Duhem equation. By differentiating (4.5) and using (4.6) it follows immediately that

$$\dot{E}(\tau) = -p(\tau)\,\dot{V}(\tau) + \theta(\tau)\,\dot{H}(\tau) - \sum_{a=1}^{N} \underset{a}{\mu}(\tau)\,\underset{a}{\dot{M}}(\tau), \tag{4.7}$$

which is a global form of the classical Gibbs equation. All three of (4.5)–(4.7) must hold during a sequence of maximizing states of problem (2.1).

Appendix: Elements of Convex Analysis

We present here some of the elements of convex analysis that are used in this work. The terminology and all of the results except the two corollaries contained herein are taken with minor modification from ROCKAFELLAR [R].

A set E in $\mathbb{R}^m$ is *convex* (*affine*) if for any two points x_1 and x_2 in E, the whole line $\alpha x_1 + (1-\alpha)\,x_2$, $\alpha \in [0, 1]$ $(\alpha \in (-\infty, \infty))$, is in E. The *convex hull* $\mathscr{H}(E)$ (*affine hull* $\mathscr{A}(E)$) of a set E is the intersection of all convex (affine) sets containing E. The sum $\sum_{i=1}^{k} \alpha_i x_i$, $\alpha_i \geqq 0$, $\sum_{i=1}^{k} \alpha_i = 1$, is a *convex combination* of the points $x_1, x_2, \ldots, x_k$. The convex hull of E consists of all of the convex combinations of the points of E. The set E is *affinely dependent* if there is a finite set of points $x_1, x_2, \ldots, x_k$ in E and numbers $\alpha_1, \alpha_2, \ldots, \alpha_k$, not all zero, such that $\sum_{i=1}^{k} \alpha_i = 0$ and $\sum_{i=1}^{k} \alpha_i x_i = 0$. If E is not affinely dependent it is *affinely independent*.

Every non-empty affine set E is parallel to a unique subspace L in the sense that $E = L \oplus e$ for some $e \in \mathbb{R}^m$. This allows one to define the *dimension of an affine set* as the dimension of the subspace to which it is parallel. The *dimension of a set* is the dimension of its affine hull. An $m-1$ dimensional affine set in $\mathbb{R}^m$ is a *hyperplane*, π, and has the representation $\pi = \{x \in \mathbb{R}^m \mid x \cdot a = \lambda\}$ for some non-zero $a \in \mathbb{R}^m$ and real number λ. a and λ are unique up to a common scalar multiplier, and we use "dot" to denote the usual inner product. A *supporting hyperplane* of a set E in $\mathbb{R}^m$ is a hyperplane π as above with $x \cdot a \leqq \lambda$ for every $x \in E$, and $x \cdot a = \lambda$ for at least one x in the closure $\bar{E}$. A point $x \in E$ is a *support point* of E if it is a member of a support hyperplane of $\mathscr{H}(E)$. Thus, open sets do not have support points.

The following is a major representation theorem of Carathéodory:

Theorem A1. *Let E be a set in* $\mathbb{R}^m$. *Then,* $x \in \mathbb{R}^m$ *is in the convex hull* $\mathscr{H}(E)$ *if and only if it can be written as a convex combination of* $m+1$ *(not necessarily distinct) points in E.*

A *face* of a convex set C is a convex subset $C' \subset C$ with the property that every closed line segment in C which has a relative interior point in C' has both end points in C'. An *exposed face* of a convex set C is a face of the form $C \cap \pi$ where π is a supporting hyperplane to C. We then have

Theorem A2. *Let* C' *be a face of the convex hull* $\mathscr{H}(E)$ *and let* E' *be the set of points in E which belong to* C'. *Then,* $C' = \mathscr{H}(E')$.

Corollary 1. *Let E be a set in* $\mathbb{R}^m$, *and let p be a support point of* $\mathscr{H}(E)$. *Then, p is the convex combination of m (not necessarily distinct) points in E. Further, p and these points share a common support hyperplane to E.*

Proof. Let π be a supporting hyperplane to $\mathscr{H}(E)$ at the point p and consider the face $C' \equiv \pi \cap \mathscr{H}(E)$, which is, in fact, an exposed face of $\mathscr{H}(E)$. Theorem A2 implies that C' is, then, the convex hull of the set of points in E that belong to C'. Since $C' \subset \pi$, the dimension of C' is $\leqq m-1$, and because of Theorem A1 we are assured of the existence of m (not necessarily distinct) points $x_1, x_2, \ldots, x_m$ in $E \cap C'$ such that $p = \sum_{i=1}^{m} \alpha_i x_i$, $\alpha_i \geqq 0$, $\sum_{i=1}^{m} \alpha_i = 1$. Since π is, in fact, a supporting hyperplane to C', then the points p, and $x_1, x_2, \ldots, x_m$ share a common supporting hyperplane. □

Corollary 2. *Let E be a set in* $\mathbb{R}^m$, *let p be a point in* $\mathscr{H}(E)$, *and let* π_p, *the collection of all supporting hyperplanes to* $\mathscr{H}(E)$ *at p, be non-empty. Then, if p is a convex combination of a set of points* $\{x_i\}$ *in* $\pi \cap E$ *for some* π *in* π_p, *it follows that*

$$\{x_i\} \subset \sigma_p \equiv \bigcap_{\pi \in \pi_p} (\pi \cap E).$$

Proof. By Corollary 1, above, we see that p is in the convex hull $\mathscr{H}(\pi \cap E)$ for any π in the collection π_p. Thus, given any two supporting hyperplanes to $\mathscr{H}(E)$ at p, *i.e.*, $\pi_1 \equiv \{x \in \mathbb{R}^m \mid x \cdot a_1 = \lambda_1\}$ and $\pi_2 \equiv \{x \in \mathbb{R}^m \mid x \cdot a_2 = \lambda_2\}$, we may write both of the convex combinations

$$p = \sum_{i=1}^{m+1} \alpha_i x_i, \qquad \alpha_i \geqq 0, \qquad \sum_{i=1}^{m+1} \alpha_i = 1,$$

and

$$p = \sum_{i=1}^{m+1} \beta_i y_i, \qquad \beta_i \geqq 0, \qquad \sum_{i=1}^{m+1} \beta_i = 1,$$

where $x_i \in \pi_1 \cap E$ and $y_i \in \pi_2 \cap E$. Further, since p belongs to both π_1 and π_2, we see that

$$\lambda_2 = p \cdot a_2 = \sum_{i=1}^{m+1} \alpha_i (x_i \cdot a_2).$$

Now, because π_2 is a supporting hyperplane to $\mathscr{H}(E)$ at p, it follows, in particular, that $(x_i \cdot a_2) \leqq \lambda_2$ for all $x_i \in \pi_1 \cap E$, and since $\alpha_i \geqq 0$, $\sum_{i=1}^{m+1} \alpha_i = 1$ we conclude that $(x_i \cdot a_2) = \lambda_2$ so that $x_i \in \pi_2 \cap E$. Finally, by reversing the roles of π_1 and π_2 in the above argument we find $y_i \in \pi_1 \cap E$. □

Let f be a real valued function whose *domain* $D(f)$ is open in $\mathbb{R}^m$. The sets epi f and $\Gamma(f)$ in $\mathbb{R}^{m+1}$, defined by

$$\text{epi}\, f \equiv \{(x, y),\ y \in \mathbb{R},\ x \in D(f) \mid y \leqq f(x)\},$$

$$\Gamma(f) \equiv \{(x, y),\ y \in \mathbb{R},\ x \in D(f) \mid y = f(x)\},$$

are called the *epigraph* of f and *graph* of f, respectively. The function is *concave* at a point $x \in D(f)$ if there is a supporting hyperplane to epi f at $(x, f(x))$. The set of points in $D(f)$ for which f is concave is the *domain of concavity* of f, denoted Dcon (f), and a function is concave if it is concave at every point in its domain. A concave function f whose domain is convex has the property that epi f is a convex set.

The *upper convex envelope* of a function f, denoted $\mathscr{U}f$, is defined for every $x \in \mathscr{H}(D(f))$ by

$$\mathscr{U}f(x) \equiv \sup \{y \in \mathbb{R} \mid (x, y) \in \mathscr{H}(\Gamma(f))\}.$$

Since epi $\mathscr{U}f = \mathscr{H}(\text{epi}\, f)$, it follows that $\mathscr{U}f$ is a concave function.

A vector g in $\mathbb{R}^m$ is a *supergradient* of the function f at a point $y \in D(f)$ if

$$f(x) \leqq f(y) + g \cdot (x - y)$$

for all $x \in D(f)$. If we rewrite this as

$$(x, f(x)) \cdot (-g, 1) \leqq (y, f(y)) \cdot (-g, 1) \equiv \lambda,$$

it follows that

$$(x, y) \cdot (-g, 1) \leqq \lambda$$

for all points (x, y) in epi f, where equality holds at the point $(y, f(y))$. Thus, *whenever f has a supergradient at $y \in D(f)$, it is necessary and sufficient that* epi f *have a supporting hyperplane at the point $(y, f(y))$ and f must be concave at y.* The set of all supergradients of f at $y \in D(f)$ is called the *superdifferential* set of f at y and is denoted by $\partial f(y)$. This set is equivalent to the set of all supporting hyperplanes to epi f at $(y, f(y))$. Whenever f has a supergradient (equivalently, is concave) at a point $y \in D(f)$ at which f is differentiable, then epi f has a single supporting hyperplane at $(y, f(y))$, and the superdifferential set $\partial f(y)$ consists in all positive multiples of the usual gradient at y, *i.e.* $\text{grad}_x f(x)|_{x=y}$. In this case we may write

$$f(x) \leqq f(y) + \operatorname{grad}_x f(x)|_{x=y} \cdot (x - y)$$

for all $x \in D(f)$. In general, if f is a concave function it follows that the super-differential set $\partial f(x)$ is a monotone function of x in the sense that $(g - h) \cdot (x - y) \leqq 0$ for every $g \in \partial f(x)$ and $h \in \partial f(y)$, and all points x and y in $D(f)$.

Acknowledgement. The support of the U.S. National Science Foundation and the Brasilian CNPq is greatfully acknowledged.

References

[N] NOLL, WALTER, On certain convex sets of measures and on phases of reacting mixtures. *Arch. Rational Mech. Anal.* **38**, 1–12 (1970).

[DF] DUNN, J. E. & R. L. FOSDICK, The morphology and stability of material phases. *Arch. Rational Mech. Anal.* **74**, 1–99 (1980).

[F] FOSDICK, R. L., Structure and dynamical stability of Gibbsian states. *Proceedings of the Laws and Structure of Continuum Thermomechanics, Minneapolis, June* 1983. *New Perspectives in Thermodynamics*, ed. by J. SERRIN, Ch 8, 125–155, Heidelberg: Springer-Verlag, 1986.

[R] ROCKAFELLAR, R. T., *Convex Analysis*, Princeton University Press, Princeton, 1970.

[G] GIBBS, J. W., On the equilibrium of heterogeneous substances. (Trans. Conn. Acad. 1875–1878). *The Scientific Papers*, Vol I, 55–353. New Haven: Yale University Press, 1907.

[M] MAN, C.-S., Material stability, the Gibbs conjecture and the first phase rule for substances. *Arch. Rational Mech. Anal.* **91**, 1–53 (1985).

[H] HALMOS, P., Measure Theory. D. Van Nostrand, New York, 1950.

Department of Aerospace
Engineering and Mechanics
University of Minnesota
Minneapolis

and

Departmento de Matematica
Pontificia Universidade Catolica
do Rio de Janeiro

(Received July 24, 1985)

Stability of Equilibrium for a Nonlinear Hyperbolic System Describing Heat Propagation by Second Sound in Solids

BERNARD D. COLEMAN, WILLIAM J. HRUSA, & DAVID R. OWEN

Dedicated to our friend and colleague Walter Noll on his sixtieth birthday

1. Introduction

We consider here the propagation of heat in a unidimensional case in which the only nonzero component of the heat flux is its x-component, q, and not only q, but also the absolute temperature, θ, and the internal energy density, e, are functions of x and the time, t. When, as we assume here, such a unidimensional flow of heat is not accompanied by appreciable deformation, the law of balance of energy reduces to

$$e_t + q_x = 0 \tag{1.1}$$

(with the subscripts t and x indicating partial derivatives), and the constitutive assumptions behind Fourier's theory of heat conduction reduce to

$$q = -\varkappa(\theta)\,\theta_x \tag{1.2}$$

and

$$e = e_0(\theta), \quad \text{i.e.,} \quad e_t = c_0(\theta)\,\theta_t \quad \text{with} \quad c_0(\theta) = e_0'(\theta), \tag{1.3}$$

where $\varkappa$ and c_0 are positive-valued functions characterizing the material under consideration. Thus, in FOURIER's theory, the evolution of q and θ is governed by the following system of equations:

$$\begin{aligned} q + \varkappa(\theta)\,\theta_x &= 0, \\ q_x + c_0(\theta)\,\theta_t &= 0. \end{aligned} \tag{1.4}$$

When, as is common practice, $\varkappa$ and c_0 are taken to be independent of θ, the system (1.4) yields the familiar "linear heat equation", $\theta_t = (\varkappa/c_0)\,\theta_{xx}$, for $\theta = \theta(x, t)$. Although this *parabolic* equation provides a description of heat conduction that is useful under a very broad range of conditions, it predicts that a thermal disturbance at any one point has an instantaneous effect at every other point, no matter how distant.

In 1948, CATTANEO [1948, 1] suggested that the constitutive relation (1.2) be replaced by

$$\tau(\theta)\, q_t + q = -\varkappa(\theta)\, \theta_x, \tag{1.5}$$

with $\tau(\theta)$, like $\varkappa(\theta)$, positive; he observed that when the relations (1.5) and (1.3) are placed in the balance law (1.1), and τ, $\varkappa$, and c_0 are taken to be constant, the resulting equations for q and θ are hyperbolic (in fact, they reduce to KELVIN's "equation of telegraphy") and hence are free from the "paradox of instantaneous propagation". Subsequently, several physicists, among whom are CHESTER [1963, 1], GUYER & KRUMHANSL [1966, 1], and ACKERMAN & GUYER [1968, 1] (see also the references in [1982, 1], [1983, 1]) observed that the phonon theory of heat conduction in pure dielectric crystals at low temperatures indicates that there can be a range of temperatures in which (1.5) holds approximately with $\varkappa/\tau \sim \frac{1}{3} c_0 V^2$, where V^2 is the mean-square phonon speed. (In such a range, both τ and $\varkappa$ vary rapidly with temperature, but remain (strictly) positive.) Heat-pulse propagation methods have been used to show that disturbances which are almost entirely thermal can propagate with finite speed, in apparent accord with equation (1.5), in dielectric crystals of high purity in appropriate temperature ranges. This phenomenon, in crystals, is called *second sound*; it is related to a similar (but not identical) phenomenon of the same name occurring in liquid helium. Experimental demonstrations of second sound have been achieved for solid helium-4 [1966, 2], solid helium-3 [1969, 1], sodium fluoride crystals of very high purity [1970, 2], [1971, 1], and for pure bismuth [1972, 1] in a temperature range (1.2°–4.0° K), in which it is an electrical insulator.

COLEMAN, FABRIZIO, & OWEN [1982, 1] recently have shown that, when the constitutive assumption (1.5) (with $\tau > 0$ and $\varkappa > 0$) is adopted, the second law of thermodynamics requires that e be given by a constitutive relation of the form[#]

$$e = \tilde{e}(q, \theta) = e_0(\theta) + a(\theta)\, q^2 \tag{1.6}$$

with

$$a(\theta) = \frac{1}{\theta} Z(\theta) - \tfrac{1}{2} Z'(\theta), \qquad Z = \tau/\varkappa. \tag{1.7}$$

[#] COLEMAN, FABRIZIO, & OWEN simplify their original derivation [1982, 1] of (1.6) in [1982, 2]. Both those papers are concerned with three-dimensional forms of (1.5) and (1.6). Some implications of (1.6) with further background material are given in [1983, 1]. Years ago, CHEN & GURTIN [1970, 1] treated unidimensional examples in the theory of GURTIN & PIPKIN [1968, 2] and discussed extensions of that theory to deformable media. It is clear from the example in § 6 of [1970, 1] that, despite differences in language, methods, and initial assumptions, in the special case in which τ and $\varkappa$ are independent of θ, CHEN & GURTIN's theory intersects that of COLEMAN, FABRIZIO, & OWEN and agrees with it in the conclusion that an equation equivalent to (1.6) can hold. CHEN & GURTIN do not observe that (1.6) is implied by (1.5), *i.e.*, that (1.6) with $a(\theta)$ as in (1.7), is the only internal energy function compatible with (1.5). (Of course, the term $-\frac{1}{2} Z'(\theta)$ in (1.7) does not occur in the example treated in detail by CHEN & GURTIN.)

It is important to note that the function a in (1.6) is determined by τ and $\varkappa$. Both molecular theory and experiments on second sound in solids give values for $\tau(\theta)/\varkappa(\theta)$ which, in conjunction with (1.7), yield

$$a(\theta) > 0; \tag{1.8}$$

in particular, a is not zero. Thus, the classical equation (1.3) for e is not appropriate for materials obeying (1.5).

The new constitutive relation (1.6) implies that e_t is given by

$$e_t = [c_0(\theta) + a'(\theta)\, q^2]\, \theta_t + 2a(\theta)\, qq_t, \qquad c_0(\theta) = e_0'(\theta); \tag{1.9}$$

in the present context c_0 is called the *equilibrium heat capacity*. Thus, for materials obeying (1.5), the evolution of q and θ is governed by the following system of partial differential equations:

$$\begin{aligned} &\tau(\theta)\, q_t + q + \varkappa(\theta)\, \theta_x = 0, \\ &q_x + c_0(\theta)\, \theta_t + a'(\theta)\, q^2\theta_t + 2a(\theta)\, qq_t = 0, \qquad x \in J, \qquad t \geqq 0; \end{aligned} \tag{1.10}$$

the interval J characterizes the region occupied by the (materially homogeneous) body in question.

This paper is about the global existence and asymptotic behavior of classical (*i.e.*, continuously differentiable) solutions of (1.10) for an appropriate class of smooth initial data. We discuss pure initial value problems for infinite bodies with $J = \mathbb{R}$, and some related problems for finite bodies with appropriate boundary conditions. We assume throughout that τ, $\varkappa$, c_0, and a are smooth functions defined on $(0, \infty)$ with τ, $\varkappa$, and c_0 positive. [Observe that $(0, \infty)$ is the natural domain of these functions because θ represents absolute temperature.] The proofs of our theorems do not require either that a be positive or that (1.7) hold. However, positivity of a does have an effect on the asymptotic behavior of θ in the case of a finite body with insulated boundary; the magnitude of this effect is such that it might be used to provide experimental verification of the dependence of e on q.

Although (1.7) is not employed as an hypothesis for our mathematical results, because that relation between a and $\tau/\varkappa$ must hold in order for the system (1.10) to be physically meaningful, we have avoided assumptions conflicting with (1.7) in any way.

It is easily seen that the *steady-state solutions* (*i.e.*, the solutions with $q_t \equiv \theta_t \equiv 0$) of (1.10) are precisely the same as those of the classical system (1.4). In particular, each pair (q, θ) with $q \equiv 0$ and $\theta \equiv \bar{\theta} > 0$, is a steady state solution; such a solution is called an *equilibrium solution* and the corresponding pair of numbers $(0, \bar{\theta})$ is called an *equilibrium state*. It is not difficult to show that if $J = \mathbb{R}$ (and $\varkappa > 0$), the only steady-state solutions of (1.10) with θ bounded are the equilibrium solutions.

Each equilibrium state has a neighborhood in which the system (1.10) is hyperbolic. If a' is nonnegative, then (1.10) is everywhere hyperbolic. In practice it appears that a' is actually negative, and this leads to changes of type for large values of q. (The article of COLEMAN & OWEN [1983, 1][#] contains a detailed discussion of these matters.) In the present paper, no assumption is made regarding the sign of a'. However, the solutions whose existence is established here remain close to an equilibrium state, and hence stay in the hyperbolic region.

Linearization of (1.10) about an equilibrium state $(0, \bar{\theta})$ yields a hyperbolic system for which the characteristic speeds are $\pm U(\bar{\theta})$ with $U(\bar{\theta}) = [\varkappa(\bar{\theta})/\tau(\bar{\theta})\, c_0(\bar{\theta})]^{\frac{1}{2}}$. In the physics literature $U(\bar{\theta})$ is referred to as the *speed of second sound* at the temperature $\bar{\theta}$. COLEMAN, FABRIZIO, & OWEN [1982, 1] have discussed the propagation of certain types of thermal disturbances, called *temperature-rate waves*[##], for the *nonlinear* system (1.10). If such a wave is moving into a region in which $q = 0$ and $\theta = \bar{\theta}$, its speed must be $U(\bar{\theta})$. The heat-pulse propagation experiments have been performed under conditions such that the observed pulse speeds may be interpreted as experimental values of $U(\bar{\theta})$. For two solids, sodium fluoride (in the range 10°–18°K) [1971, 1] and bismuth (1.0°–4.0°K) [1972, 1], there are published values of the speed of second sound as a function of temperature. For these materials accurate values of $c_0(\theta)$ are known, and consequently experimental values of the ratio $Z(\theta) = \tau(\theta)/\varkappa(\theta)$ can be determined through the formula for U given above; once Z is known as a function of θ, (1.6) can be calculated from (1.7) (such calculations have been performed by COLEMAN & OWEN [1983, 1].)

GURTIN & PIPKIN [1968, 2] have introduced a theory of heat conduction that leads to partial-functional differential equations for the temperature and heat flux in a rigid heat conductor. Their equations include the instance of our equations that occurs when the functions τ and $\varkappa$ are constants. MACCAMY [1977, 1], DAFERMOS & NOHEL [1979, 1], and STAFFANS [1980, 1] have studied equations that include some of the non-linearities and memory effects that arise in the theory of GURTIN & PIPKIN, and they obtain theorems of existence and stability similar in spirit to those obtained here.

To indicate the nature of our results, let us assume temporarily that $J = \mathbb{R}$. For smooth initial data in the hyperbolic region, the existence and uniqueness of local (in time) solutions to (1.10) follows from standard theory for hyperbolic systems. However, it is well known that smooth solutions of quasilinear hyperbolic systems may develop singularities in finite time, even for very regular initial data. (See, for example, [1964, 1], [1967, 1].) The constitutive relation (1.5) includes a damping mechanism which, as we shall show, prevents the formation of singularities in solutions of (1.10) for initial data that are close to an equi-

[#] COLEMAN & OWEN also discuss the propagation of periodic travelling waves in rigid heat conductors obeying (1.10), and they show that smooth functions q and θ cannot be used to describe waves of this type. Their discussion of periodic, discontinuous, travelling waves is speculative, because they do not address the issue of assigning a meaning to discontinuous solutions of (1.10). When $(1.10)_1$ is written in the form $q_t + \tau(\theta)^{-1} q + \tau(\theta)^{-1} \varkappa(\theta)\, \theta_x = 0$, the system (1.10) becomes meaningful, in the sense of distributions, for discontinuous solutions, and it can be shown that the discontinuous, travelling waves discussed in [1983, 1] obey one, but not both, of the jump conditions required of distributional solutions.

[##] This term was introduced by GURTIN & PIPKIN [1968, 2] in their study of waves in materials for which the history of θ and θ_x influence q. See also the related work of CHEN & GURTIN [1970, 1].

librium state $(0, \bar{\theta})$ in an appropriate Sobolev norm: for such initial data, globally defined classical solutions (q, θ) exist and approach $(0, \bar{\theta})$ as $t \to \infty$. We also show that the situation is similar for certain problems involving finite bodies, although the precise asymptotic behavior depends on the type of boundary conditions. In Section 2 we state our main results, Theorems 1, 2, and 3, and we discuss several extensions and variants of these theorems. We prove Theorem 1 in Section 3 by deriving estimates of energy type that also provide, without significant changes, proofs of Theorems 2 and 3. The basic strategy which we employ in the proof of Theorem 1 is due to MATSUMURA [1977, 2].

2. Statement of Results

We consider first the following initial value problem for which J in (1.10) is the entire real line:

$$\left.\begin{aligned} &\tau(\theta)\, q_t + q + \varkappa(\theta)\, \theta_x = 0, && (2.1)\\ &q_x + c_0(\theta)\, \theta_t + a'(\theta)\, q^2\theta_t + 2a(\theta)\, qq_t = 0, && (2.2)\end{aligned}\right\} \quad x \in \mathbb{R}, \quad t \geqq 0,$$

$$q(x, 0) = q_0(x), \qquad \theta(x, 0) = \theta_0(x), \qquad x \in \mathbb{R}. \tag{2.3}$$

The functions q_0 and θ_0 give the initial heat flux and temperature distribution in the body. Because of the physical background for this problem, we assume that τ, $\varkappa$, and c_0 are positive-valued functions; in addition, we employ hypotheses of smoothness on these functions and on the function a:

$$\begin{gathered} \tau, \varkappa, c_0 > 0, \\ \tau, \varkappa, c_0 \in C^2((0, \infty)), \qquad a \in C^3((0, \infty)). \end{gathered} \tag{H}$$

Throughout our discussion, all derivatives are to be interpreted in the distributional sense.

Theorem 1. *If* (H) *is satisfied, then for each* $\bar{\theta} > 0$ *there is a* $\delta > 0$ *such that when* q_0 *and* θ_0 *obey*

$$q_0, \quad \theta_0 - \bar{\theta} \in H^2(\mathbb{R}) \tag{2.4}$$

and[#]

$$\int_{-\infty}^{\infty} \{q_0(x)^2 + q_0'(x)^2 + q_0''(x)^2 + (\theta_0(x) - \bar{\theta})^2 + \theta_0'(x)^2 + \theta_0''(x)^2\}\, dx \leqq \delta^2, \tag{2.5}$$

the initial value problem (2.1), (2.2), (2.3) *has a unique solution* (q, θ) *with*

$$q, q_x, q_t, q_{xx}, q_{xt}, q_{tt}, (\theta - \bar{\theta}), \theta_x, \theta_t, \theta_{xx}, \theta_{xt}, \theta_{tt} \in C([0, \infty); L^2(\mathbb{R})), \tag{2.6}$$

[#] Observe that the Sobolev embedding theorem and (2.5) with δ sufficiently small relative to $\bar{\theta}$ imply that θ_0 has only positive values.

and

$$\theta(x, t) > 0, \quad \text{for all } x \in \mathbb{R} \quad \text{and} \quad t \geqq 0; \tag{2.7}$$

the solution is not only globally defined in time, but approaches equilibrium in the sense that, as $t \to \infty$,

$$\theta(\cdot, t) \to \bar{\theta} \quad \text{uniformly on } \mathbb{R}, \tag{2.8}$$

and

$$q(\cdot, t), q_x(\cdot, t), q_t(\cdot, t), \theta_x(\cdot, t), \theta_t(\cdot, t) \to 0 \quad \text{in } L^2(\mathbb{R}) \quad \text{and uniformly on } \mathbb{R}. \tag{2.9}$$

The solution (q, θ) given by this theorem has, in addition to (2.6) through (2.9), the properties stated below.

Remark 1. The Sobolev embedding theorem and (2.6) imply that the solution (q, θ) is classical in the sense that

$$q, \theta \in C^1(\mathbb{R} \times [0, \infty)). \tag{2.10}$$

Remark 2. It will be evident from the proof of Theorem 1 that the equilibrium state is stable with respect to the H^2-norm; *i.e.*, for each $\varepsilon > 0$ the number δ in (2.5) can be chosen so that

$$\|q(\cdot, t)\|_2^2 + \|\theta(\cdot, t) - \bar{\theta}\|_2^2 \leqq \varepsilon^2, \quad \text{for all } t \geqq 0, \tag{2.11}$$

with $\|\cdot\|_2$ the standard norm on $H^2(\mathbb{R})$.

Remark 3. The quantities τ, $\varkappa$, and c_0 need not be positive on all of $(0, \infty)$. In fact the theorem remains valid if τ, $\varkappa$, and c_0 are positive at the temperature $\bar{\theta}$. It follows from Remark 2 and the Sobolev embedding theorem that $|\theta(x, t) - \bar{\theta}|$ can be held small for all $x \in \mathbb{R}$ and $t \geqq 0$ by choosing δ sufficiently small.

Let us now consider the following problem for which $J = [0, 1]$ in (1.10), and both initial data and boundary data are given:

$$\tau(\theta)\, q_t + q + \varkappa(\theta)\, \theta_x = 0, \tag{2.12}$$

$$q_x + c_0(\theta)\, \theta_t + a'(\theta)\, q^2 \theta_t + 2a(\theta)\, q q_t = 0, \tag{2.13}$$

$$\left.\right\} \quad 0 \leqq x \leqq 1, \quad t \geqq 0,$$

$$q(0, t) = q(1, t) = 0, \qquad t \geqq 0, \tag{2.14}$$

$$q(x, 0) = q_0(x), \quad \theta(x, 0) = \theta_0(x), \qquad 0 \leqq x \leqq 1. \tag{2.15}$$

Our principal results for this problem are given in Theorem 2, stated below. Of course, normalization to [0, 1] of the interval of points occupied by a finite body can be achieved by scaling.

It is important that initial data be compatible with boundary conditions.

If (q, θ) is a solution of (2.12), (2.13), (2.14), (2.15) with $q, \theta \in C^1([0,1]\times[0,T])$ for some $T>0$, differentiation of (2.14) with respect to t yields

$$q_t(0,t) = q_t(1,t) = 0, \qquad t \in [0,T]. \tag{2.16}$$

If (2.12), (2.14), (2.15), and (2.16) are to hold at $x=0$, $x=1$, and $t=0$, then we must have

$$q_0(0) = q_0(1) = \theta_0'(0) = \theta_0'(1) = 0, \tag{2.17}$$

and this is assumed in Theorem 2. A violation of this compatibility condition would play the role of a singularity in the initial data at the boundary, and, when the data are in the hyperbolic region, singularities tend to propagate from the boundary to the interior of the interval.

In Theorem 3, (2.14) is replaced by

$$\theta(0,t) = \theta(1,t) = \bar{\theta}, \qquad \text{for } t \geqq 0, \tag{2.18}$$

with $\bar{\theta}$ a positive constant; the corresponding compatibility conditions are

$$\theta_0(0) = \theta_0(1) = \bar{\theta}, \tag{2.19}$$

and

$$\begin{aligned} &\tau(\bar{\theta})\, q_0'(0) - 2a(\bar{\theta})\, q_0(0)\, [q_0(0) + \varkappa(\bar{\theta})\, \theta_0'(0)] \\ &\qquad = \tau(\bar{\theta})\, q_0'(1) - 2a(\bar{\theta})\, q_0(1)\, [q_0(1) + \varkappa(\bar{\theta})\, \theta_0'(1)] = 0. \end{aligned} \tag{2.20}$$

We may note that (2.20) is satisfied in the special case in which $q_0(0) = q_0(1) = q_0'(0) = q_0'(1) = 0$.

The equations (2.12) and (2.13) have precisely one steady-state solution obeying (2.18), namely, that corresponding to the equilibrium state $(0, \bar{\theta})$. On the other hand, there are infinitely many steady-state solutions obeying (2.14). For (q, θ) to be a steady-state solution of (2.12)–(2.14), it is necessary and sufficient that q vanish identically and θ be constant, *i.e.*, that it describe some equilibrium state $(0, \tilde{\theta})$, $\tilde{\theta}$ a positive constant. This fact makes the question of asymptotic behavior more interesting when the boundary is insulated rather than held at fixed temperature.

Suppose that (q, θ) is a solution of (2.12)–(2.15) and that (q, θ) converges to an equilibrium state $(0, \bar{\theta}_*)$ as $t \to \infty$. By virtue of (2.14), on this solution the energy $\int_0^1 \{e_0(\theta) + a(\theta)\, q^2\}(x,t)\, dx$ is independent of time. Therefore, the equilibrium state $(0, \bar{\theta}_*)$ should have the same energy as the initial state (q_0, θ_0), *i.e.*, $\bar{\theta}_*$ should satisfy

$$e_0(\bar{\theta}_*) = \int_0^1 \{e_0(\theta_0(x)) + a(\theta_0(x))\, q_0(x)^2\}\, dx. \tag{2.21}$$

As $c_0 (= e_0')$ is positive, e_0 is strictly monotone, and hence for each (q_0, θ_0) there is at most one solution $\bar{\theta}_*$ of (2.21). For initial data in an appropriate class, there is precisely one positive number $\bar{\theta}_*$ obeying (2.21); moreover (2.12)–(2.15) has a unique solution (q, θ) and that solution converges to $(0, \bar{\theta}_*)$ as $t \to \infty$. In view of this, it appears to us possible that, by measuring the final equilibrium temperature $\bar{\theta}_*$ corresponding to initial data (q_0, θ_0) in an experiment in which (2.14) holds, an experimenter would be able to verify directly the non-classical dependence of e on q seen in (1.6), and perhaps even obtain reliable values of $a(\theta)$.

Theorem 2. *Under the hypothesis* (H), *for each* $\bar{\theta} > 0$ *there is a* $\delta > 0$ *such that when* $q_0, \theta_0 \in H^2(0, 1)$ *obey* (2.17) *and*

$$\int_0^1 \{q_0(x)^2 + q_0'(x)^2 + q_0''(x)^2 + (\theta_0(x) - \bar{\theta})^2 + \theta_0'(x)^2 + \theta_0''(x)^2\}\, dx \leqq \delta^2, \tag{2.22}$$

there is a unique number $\bar{\theta}_* > 0$ *obeying* (2.21), *and the initial-boundary value problem* (2.12)–(2.15) *has a unique solution* (q, θ) *with*

$$q, q_x, q_t, q_{xx}, q_{xt}, q_{tt}, \theta, \theta_x, \theta_t, \theta_{xx}, \theta_{xt}, \theta_{tt} \in C([0, \infty); L^2(0, 1)), \tag{2.23}$$

$$\theta(x, t) > 0, \quad \text{for all } x \in [0, 1] \quad \text{and} \quad t \geqq 0; \tag{2.24}$$

the solution approaches equilibrium, i.e.,

$$\theta(\cdot, t) \to \bar{\theta}_* \quad \text{uniformly on } [0, 1], \tag{2.25}$$

and

$$q(\cdot, t), q_x(\cdot, t), q_t(\cdot, t), \theta_x(\cdot, t), \theta_t(\cdot, t) \to 0 \quad \text{uniformly on } [0, 1]. \tag{2.26}$$

Theorem 3. *If* τ, $\varkappa$, c_0 *and* a *satisfy* (H), *then for each* $\bar{\theta} > 0$ *there is a* $\delta > 0$ *such that when* $q_0, \theta_0 \in H^2(0, 1)$ *obey* (2.19), (2.20), *and* (2.22), *the initial-boundary value problem* (2.12), (2.13), (2.18), (2.15) *has a unique, globally defined solution* (q, θ) *obeying* (2.23) *and* (2.24); *this solution also approaches equilibrium as* $t \to \infty$, i.e., (2.26) *holds and*

$$\theta(\cdot, t) \to \bar{\theta} \quad \text{uniformly on } [0, 1]. \tag{2.27}$$

Remark 4. There are superfluous terms in (2.22) that can be eliminated by using the compatibility conditions on the initial data. It follows from (2.17), (2.19), and the inequalities of Poincaré, that (2.22) can be replaced by

$$\int_0^1 \{q_0''(x)^2 + (\theta_0(x) - \bar{\theta})^2 + \theta_0''(x)^2\}\, dx \leqq \delta^2 \tag{2.28}$$

in Theorem 2 and by

$$\int_0^1 \{q_0(x)^2 + q_0'(x)^2 + q_0''(x)^2 + \theta_0''(x)^2\}\, dx \leqq \delta^2 \tag{2.29}$$

in Theorem 3.

Remark 5. Analogues of Theorems 2 and 3 hold when the boundary condition (2.14) in Theorem 2 or (2.18) in Theorem 3 is replaced by

$$q(0, t) = 0 \quad \text{and} \quad \theta(1, t) = \bar{\theta}, \quad \text{for } t \geqq 0 \tag{2.30}$$

(or $\theta(0, t) = \bar{\theta}$, $q(1, t) = 0$, $t \geqq 0$). In these analogues, (2.26) and (2.27) hold again as $t \to \infty$.

Remark 6. Statements parallel to those of Remarks 1 through 3 hold for the initial-boundary value problems of Theorems 2 and 3 and Remark 5. If it were assumed only that τ, $\varkappa$, and c_0 are positive at $\bar{\theta}$ (see Remark 3), then the claim made in Theorem 2 about the uniqueness of the number $\bar{\theta}_*$ obeying (2.21) would require modification.

The proofs of Theorems 1, 2, and 3 are very similar. We give the details for Theorem 1 in the next section and discuss there the modifications required for Theorems 2 and 3.

The results presented here can be modified and extended in various ways, a few of which are mentioned below.

(i) *Heat Sources.* If a smooth supply of heat, $r = r(x, t)$, that is small and tends to zero appropriately as $t \to \infty$, be included in the energy balance equation (2.2), our Theorems 1, 2, and 3 remain valid with suitable modifications of the relations (2.5), (2.17), (2.20), (2.21), and (2.22).

(ii) *Regularity.* In each of our theorems, an assumption of additional smoothness for the coefficients and initial data implies additional smoothness for the solution for all $t > 0$. In the case of Theorem 1, *if*, for an integer $k \geqq 2$, the second line of the hypothesis (H) is replaced by

$$\tau, \varkappa, c_0 \in C^k((0, \infty)), \qquad a \in C^{k+1}((0, \infty)),$$

and if (2.4) is replaced by

$$q_0, (\theta_0 - \bar{\theta}) \in H^k(\mathbb{R}), \tag*{(2.4)$_k$}$$

then

$$q, \theta \in C^{k-1}(\mathbb{R} \times [0, \infty)); \tag*{(2.10)$_k$}$$

moreover, the derivatives of q and θ of orders 2 through k then belong to $C([0, \infty); L^2(\mathbb{R}))$, and derivatives through order $k - 1$ converge to zero uniformly and in $L^2(\mathbb{R})$ as $t \to \infty$. These conclusions do not require that derivatives of order higher than two of the initial data be small. Analogous results hold for initial-boundary value problems when the derivatives of the initial data satisfy appropriate compatibility conditions at the boundary.

(iii) *Other Boundary Conditions.* Several problems involving other types of boundary conditions, including some that depend on time and some appropriate to a half-line, can be treated by the methods employed here. We note, however, that when such boundary conditions lead to steady states that are not spatially homogeneous [*e.g.*, $\theta(0,t)=\bar{\theta}_0$, $\theta(1,t)=\bar{\theta}_1 \neq \bar{\theta}_0$] the proofs increase in complexity.

(iv) *Multidimensional Problems.* The methods we employ can be used to establish certain multidimensional analogues of the present results. If the spatial dimension were $n>1$, our arguments would require that the coefficients in the differential equations have additional smoothness and that the data be small in the sense that their derivatives of order through $[n/2]+2$ be small in L^2.

3. Proof of Theorem 1

We begin by introducing a modified initial value problem that is more convenient to analyze and is equivalent to (2.1)–(2.3) for an appropriate class of solutions. The new system is constructed as follows: (i) We solve (2.1) for q_t and use the result to rewrite (2.2) as

$$q_x + c_0(\theta)\,\theta_t + a'(\theta)\,q^2\theta_t - \frac{2a(\theta)}{\tau(\theta)}\,q[q+\varkappa(\theta)\,\theta_x] = 0. \tag{3.1}$$

(ii) We divide (2.1) by $\varkappa(\theta)\,c_0(\theta)$ and (3.1) by $c_0(\theta)$. (iii) We make the change of variable

$$\varphi = \theta - \bar{\theta}. \tag{3.2}$$

(iv) We introduce new coefficients that have desirable global properties and agree with the original coefficients for (q,φ) near $(0,0)$.

These steps are designed to facilitate the derivation of our global estimates. Step (iv) will be justified when we show (*a posteriori*) that the pair (q,φ) remains in a neighborhood of $(0,0)$ on which the new and original coefficients agree.

To carry out the procedure described above, we choose a sufficiently small positive number $\varepsilon\in(0,\bar{\theta})$ and construct smooth functions $A, B, C, E, F:\mathbb{R}\to\mathbb{R}$ and $D:\mathbb{R}^2\to\mathbb{R}$ such that (3.3)–(3.9) hold:

$$A, B, C, E, F \in C_b^2(\mathbb{R}); \tag{3.3}$$

$$A(\xi) = \frac{\tau(\xi+\bar{\theta})}{\varkappa(\xi+\bar{\theta})\,c_0(\xi+\bar{\theta})}, \qquad B(\xi) = \frac{1}{\varkappa(\xi+\bar{\theta})\,c_0(\xi+\bar{\theta})}$$

$$C(\xi) = \frac{1}{c_0(\xi+\bar{\theta})}, \qquad E(\xi) = \frac{-2a(\xi+\bar{\theta})\,\varkappa(\xi+\bar{\theta})}{\tau(\xi+\bar{\theta})\,c_0(\xi+\bar{\theta})} \tag{3.4}$$

$$F(\xi) = \frac{-2a(\xi+\bar{\theta})}{\tau(\xi+\bar{\theta})\,c_0(\xi+\bar{\theta})}, \qquad \xi\in(-\varepsilon,\varepsilon),$$

and for some constants $\underline{A}$, $\underline{B}$, $\underline{C}$, there holds

$$A(\xi) \geqq \underline{A} > 0, \quad B(\xi) \geqq \underline{B} > 0, \quad C(\xi) \geqq \underline{C} > 0, \quad \xi \in \mathbb{R}; \tag{3.5}$$

$$D \in C_b^2(\mathbb{R}^2); \tag{3.6}$$

$$D(\eta, \xi) = \frac{a'(\xi + \bar{\theta})\,\eta^2}{c_0(\xi + \bar{\theta})}, \quad \eta, \xi \in (-\varepsilon, \varepsilon) \tag{3.7}$$

$$|D(\eta, \xi)| \leqq \tfrac{1}{2}, \quad (\eta, \xi) \in \mathbb{R}^2, \tag{3.8}$$

and for some constant K,

$$|D(\eta, \xi)| \leqq K\eta^2, \quad (\eta, \xi) \in \mathbb{R}^2. \tag{3.9}$$

Here, C_b^2 denotes the space of functions which together with their derivatives of first and second order are bounded and continuous. It is clear that such a construction always is possible by virtue of (H).

In place of (2.1), (2.2), (2.3), we consider the initial value problem

$$\left.\begin{aligned} &A(\varphi)\, q_t + B(\varphi)\, q + C(\varphi)\, \varphi_x = 0, &(3.10)\\ &C(\varphi)\, q_x + \varphi_t + D(q, \varphi)\, \varphi_t + E(\varphi)\, q\varphi_x + F(\varphi)\, q^2 = 0, &(3.11) \end{aligned}\right\} \quad x \in \mathbb{R}, \quad t \geqq 0,$$

$$q(x, 0) = q_0(x), \quad \varphi(x, 0) = \varphi_0(x), \quad x \in \mathbb{R}, \tag{3.12}$$

where

$$\varphi_0(x) = \theta_0(x) - \bar{\theta}, \quad x \in \mathbb{R}, \tag{3.13}$$

and consequently

$$q_0, \varphi_0 \in H^2(\mathbb{R}), \tag{3.14}$$

by (2.4). If (q, φ) is a solution of (3.10), (3.11), (3.12) satisfying

$$|q(x, t)| < \varepsilon, \quad |\varphi(x, t)| < \varepsilon, \quad x \in \mathbb{R},\ t \geqq 0, \tag{3.15}$$

then a straightforward computation shows that if we put

$$\theta(x, t) = \varphi(x, t) + \bar{\theta}, \quad x \in \mathbb{R},\ t \geqq 0, \tag{3.16}$$

then (q, θ) is a solution of (2.1), (2.2), (2.3). Moreover, (2.7) holds, since $\varepsilon \in (0, \bar{\theta})$.

We shall prove first the existence (for small q_0 and φ_0) of a global solution to (3.10)–(3.12) that satisfies (3.15) and then show that the corresponding pair (q, θ) has the properties listed in Theorem 1.

Observe that the coefficient of φ_t in (3.11) is $[1 + D(q, \varphi)] \geqq \frac{1}{2}$. It is convenient for our purposes to continue to write this coefficient as the sum of a positive constant and a term that is small when q is small. It follows from (3.5) and (3.8) that the system (3.10), (3.11) is always hyperbolic. Moreover, it is symmetric (in the sense of the article [1975, 1]). Thus the initial value problem is well-posed locally in time. The relevant local existence result is recorded below.

Proposition. Assume that (3.3), (3.5), (3.6), (3.8) hold, and let (q_0, φ_0) satisfying (3.14) be given. The initial value problem (3.10)–(3.12) then has a unique local solution (q, φ) that can be continued to a maximal time interval $[0, T_0)$ with

$$(3.17) \qquad q, q_x, q_t, q_{xx}, q_{xt}, q_{tt}, \varphi, \varphi_x, \varphi_t, \varphi_{xx}, \varphi_{xt}, \varphi_{tt} \in C([0, T_0); L^2(\mathbb{R})).$$

Moreover, if

$$(3.18) \qquad \sup_{t \in [0, T_0]} \int_{-\infty}^{\infty} \{q^2 + q_x^2 + q_t^2 + q_{xx}^2 + q_{xt}^2 + q_{tt}^2 + \varphi^2 + \varphi_x^2 + \varphi_t^2 + \varphi_{xx}^2 + \varphi_{xt}^2 + \varphi_{tt}^2\}(x, t)\, dx < \infty,$$

then $T_0 = \infty$.

The existence and uniqueness of a local solution (q, φ) with the desired regularity follows, for example, from results of KATO [1975, 1]. That this solution can be continued to a maximal interval $[0, T_0)$ such that (3.18) implies $T_0 = \infty$ can be established in the usual way. (See, for example, [1976, 1].)[#]

We want to show that if q_0 and φ_0 are sufficiently small, then (3.18) holds. For this purpose, we set

$$(3.19) \qquad \begin{aligned} \mathscr{E}(t) = \max_{s \in [0, t]} \int_{-\infty}^{\infty} &\{q^2 + q_x^2 + q_t^2 + q_{xx}^2 + q_{xt}^2 + q_{tt}^2 \\ &+ \varphi^2 + \varphi_x^2 + \varphi_t^2 + \varphi_{xx}^2 + \varphi_{xt}^2 + \varphi_{tt}^2\}(x, s)\, dx \\ &+ \int_0^t \int_{-\infty}^{\infty} \{q^2 + q_x^2 + q_t^2 + q_{xx}^2 + q_{xt}^2 + q_{tt}^2 \\ &+ \varphi_x^2 + \varphi_t^2 + \varphi_{xx}^2 + \varphi_{xt}^2 + \varphi_{tt}^2\}(x, s)\, dx\, ds, \end{aligned}$$

$$t \in [0, T_0),$$

[#] KATO's [1975, 1] arguments draw on the theory of semigroups. Energy methods will also yield a proof of our proposition. See, for example, a recent paper [1985, 1] in which initial value problems and initial-boundary value problems for hyperbolic systems are discussed; an analogue of the proposition for initial-boundary value problems also can be obtained by means of arguments similar to those given there.

and

$$(3.20)\qquad Q_0 = \int_{-\infty}^{\infty} \{q_0(x)^2 + q_0'(x)^2 + q_0''(x)^2 + \varphi_0(x)^2 + \varphi_0'(x)^2 + \varphi_0''(x)^2\}\,dx.$$

Our basic objective is to show that if Q_0 is sufficiently small, then $\mathscr{E}(t)$ is suitably small for all $t \in [0, T_0)$. Boundedness of $\mathscr{E}$ on $[0, T_0)$ will imply that $T_0 = \infty$ (by the proposition). Moreover, standard embedding inequalities and boundedness of $\mathscr{E}$ on $[0, \infty)$ will imply the desired decay of q and φ as $t \to \infty$. Finally, it follows from Sobolev's embedding theorem that if $\mathscr{E}$ is suitably small on $[0, \infty)$ then (3.15) holds.

A time-independent bound for $\int_0^t \int_{-\infty}^{\infty} \varphi^2(x, s)\,dx\,ds$ is not to be expected. Therefore this term has been omitted in the definition of $\mathscr{E}(t)$.

The appropriate bound for $\mathscr{E}$ will be obtained from a chain of estimates of the energy type. To express these estimates in a concise form, it is convenient to define

$$(3.21)\qquad \nu(t) = \sup_{\substack{x\in\mathbb{R}\\ s\in[0,t]}} \{q^2 + q_x^2 + q_t^2 + \varphi^2 + \varphi_x^2 + \varphi_t^2\}^{\frac{1}{2}}(x, s), \qquad t \in [0, T_0).$$

We shall make repeated use of the elementary inequalities

$$(3.22)\qquad |\alpha\beta| \leqq \tfrac{1}{2}(\alpha^2 + \beta^2), \qquad \alpha, \beta \in \mathbb{R}$$

and

$$(3.23)\qquad \left(\sum_{i=1}^{N} \alpha_i\right)^2 \leqq N \sum_{i=1}^{N} \alpha_i^2, \qquad \alpha_1, \ldots, \alpha_N \in \mathbb{R},$$

and we shall use the symbol Γ to denote a (possibly large) generic positive constant which can be chosen to be independent of q_0, φ_0, and T_0.

The aim of the computations that follow is that of establishing the inequality (3.62) below. Three basic types of estimates will be used:

(i) those obtained directly from energy integrals;
(ii) inequalities obtained by using equations (3.10) and (3.11) to express derivatives of q and φ in terms of quantities for which we already have estimates; and
(iii) interpolation inequalities.

A reader unfamiliar with energy methods who seeks further motivation for our computation may wish to look at (3.62) *et seq.* before examining our derivation of (3.62).

There are numerous combinations of estimates of types (i), (ii), and (iii) that will lead to an inequality of the form (3.62). The particular argument we use here has been chosen so that it can be used also for initial-boundary value problems. In particular, the boundary terms (at $x = 0, 1$) for the integrations by parts are annihilated under the boundary conditions (2.14), (2.18), or (2.30).

To obtain the basic energy integral, we multiply (3.10) by q and (3.11) by φ, add the resulting equations, and integrate over $\mathbb{R} \times [0, t]$, $t \in [0, T_0)$. After several integrations by parts and use of (3.12) we obtain the equation

$$\begin{aligned} \text{(3.24)} \quad & \tfrac{1}{2} \int_{-\infty}^{\infty} \{A(\varphi)\, q^2 + \varphi^2\}(x, t)\, dx + \int_0^t \int_{-\infty}^{\infty} B(\varphi)\, q^2(x, s)\, dx\, ds \\ &= \tfrac{1}{2} \int_{-\infty}^{\infty} \{A(\varphi_0)\, q_0^2 + \varphi_0^2\}(x)\, dx \\ &\quad + \int_0^t \int_{-\infty}^{\infty} \{\tfrac{1}{2} A'(\varphi)\, q^2 \varphi_t + [C'(\varphi) - E(\varphi)]\, q\varphi\varphi_x \\ &\quad - D(q, \varphi)\, \varphi\varphi_t - F(\varphi)\, q^2 \varphi\}(x, s)\, dx\, ds, \qquad t \in [0, T_0). \end{aligned}$$

The relation

$$\begin{aligned} \text{(3.25)} \quad & \int_{-\infty}^{\infty} \{q^2 + \varphi^2\}(x, t)\, dx + \int_0^t \int_{-\infty}^{\infty} q^2(x, s)\, dx\, ds \\ &\leqq \Gamma Q_0 + \Gamma\{\nu(t) + \nu(t)^2\}\, \mathscr{E}(t), \qquad t \in [0, T_0), \end{aligned}$$

then follows because, by (3.5), the left-hand side of (3.24) is bounded from below by

$$\tfrac{1}{2} \int_{-\infty}^{\infty} \{\underline{A} q^2 + \varphi^2\}(x, t)\, dx + \underline{B} \int_0^t \int_{-\infty}^{\infty} q^2(x, s)\, dx\, ds$$

for all $t \in [0, T_0)$, and (3.3), (3.9), (3.22) give the required upper bound for the right-hand side of (3.24). We shall give the details of our derivation of the upper bound for one term on the right-hand side of (3.24); a similar procedure can be used to bound the remaining terms. Putting $\Gamma_0 := \sup_{\xi \in \mathbb{R}} \{|C'(\xi)| + |E(\xi)|\}$, we have

$$\begin{aligned} & \left| \int_0^t \int_{-\infty}^{\infty} [C'(\varphi) - E(\varphi)]\, q\varphi\varphi_x(x, s)\, dx\, ds \right| \\ &\leqq \Gamma_0 \int_0^t \int_{-\infty}^{\infty} |q\varphi\varphi_x|\, (x, s)\, dx\, ds \\ \text{(3.26)} \quad &\leqq \tfrac{1}{2} \Gamma_0 \int_0^t \int_{-\infty}^{\infty} |\varphi|\, \{q^2 + \varphi_x^2\}(x, s)\, dx\, ds \\ &\leqq \tfrac{1}{2} \Gamma_0 \sup_{\substack{x \in \mathbb{R} \\ s \in [0, t]}} \{|\varphi(x, s)|\} \cdot \int_0^t \int_{-\infty}^{\infty} \{q^2 + \varphi_x^2\}(x, s)\, dx\, ds \\ &\leqq \tfrac{1}{2} \Gamma_0 \nu(t)\, \mathscr{E}(t), \qquad t \in [0, T_0). \end{aligned}$$

Differentiation of (3.10), (3.11) with respect to t yields

$$A(\varphi)\, q_{tt} + B(\varphi)\, q_t + C(\varphi)\, \varphi_{xt} + A'(\varphi)\, q_t\varphi_t + B'(\varphi)\, q\varphi_t + C'(\varphi)\, \varphi_x\varphi_t = 0 \tag{3.27}$$

$$\begin{aligned} &\varphi_{tt} + C(\varphi)\, q_{xt} + C'(\varphi)\, q_x\varphi_t + D(q, \varphi)\, \varphi_{tt} \\ &\quad + D_q(q, \varphi)\, q_t\varphi_t + D_\varphi(q, \varphi)\, \varphi_t^2 + E(\varphi)\, q\varphi_{xt} \\ &\quad + E(\varphi)\, q_t\varphi_x + E'(\varphi)\, q\varphi_x\varphi_t + 2F(\varphi)\, qq_t + F'(\varphi)\, q^2\varphi_t = 0. \end{aligned} \tag{3.28}$$

We multiply (3.27) by q_t and (3.28) by φ_t, add the resulting equations, and integrate over $\mathbb{R}\times[0, t]$ to obtain

$$\begin{aligned} &\tfrac{1}{2} \int_{-\infty}^{\infty} \{A(\varphi)\, q_t^2 + \varphi_t^2\}\,(x, t)\, dx + \int_0^t \int_{-\infty}^{\infty} B(\varphi)\, q_t^2(x, s)\, dx\, ds \\ &\quad = \tfrac{1}{2} \int_{-\infty}^{\infty} \{A(\varphi)\, q_t^2 + \varphi_t^2\}\,(x, 0)\, dx \\ &\qquad - \int_0^t \int_{-\infty}^{\infty} \{\tfrac{1}{2} A'(\varphi)\, q_t^2\varphi_t + [B'(\varphi) + 2F(\varphi)]\, qq_t\varphi_t \\ &\qquad + C'(\varphi)\, q_x\varphi_t^2 + D(q, \varphi)\, \varphi_t\varphi_{tt} + D_q(q, \varphi)\, q_t\varphi_t^2 + D_\varphi(q, \varphi)\, \varphi_t^3 \\ &\qquad + E(\varphi)\, q\varphi_t\varphi_{xt} + E(\varphi)\, q_t\varphi_x\varphi_t + E'(\varphi)\, q\varphi_x\varphi_t^2 \\ &\qquad + F'(\varphi)\, q_t^2\varphi_t^2\}\,(x, s)\, dx\, ds, \qquad t\in[0, T_0). \end{aligned} \tag{3.29}$$

It is easy to express $q_t(\cdot, 0)$ and $\varphi_t(\cdot, 0)$ in terms of q_0 and φ_0. Indeed, using (3.10) and (3.11), we find that

$$q_t(\cdot, 0) = \frac{-1}{A(\varphi_0)}\,[B(\varphi_0)\, q_0 + C(\varphi_0)\, \varphi_0'], \tag{3.30}$$

from which follows the inequality

$$\int_{-\infty}^{\infty} A(\varphi_0)\, q_t^2(x, 0)\, dx \leqq \Gamma Q_0. \tag{3.31}$$

We use a slightly different type of estimate for $\int_{-\infty}^{\infty} \varphi_t^2(x, 0)\, dx$. The relations (3.11), (3.8), and (3.23) imply that

$$\varphi_t^2(x, 0) \leqq 6[C(\varphi)^2 q_x^2 + E(\varphi)^2 q^2\varphi_x^2 + F(\varphi)^2 q^4]\,(x, 0), \qquad x\in\mathbb{R}, \tag{3.32}$$

and consequently

$$\begin{aligned} \int_{-\infty}^{\infty} \varphi_t^2(x, 0)\, dx &\leqq 6 \int_{-\infty}^{\infty} C(\varphi_0(x))^2 q_0'(x)^2\, dx \\ &\quad + 6 \int_{-\infty}^{\infty} \{E(\varphi)^2 q^2\varphi_x^2 + F(\varphi)^2 q^4\}\,(x, 0)\, dx \\ &\leqq \Gamma Q_0 + \Gamma \nu(t)^2\, \mathscr{E}(t), \quad t\in[0, T_0). \end{aligned} \tag{3.33}$$

By appeal to Sobolev's embedding theorem, one can replace $\Gamma \nu(t)^2 \mathscr{E}(t)$ by ΓQ_0^2 in the last line of (3.33). However, as we have a term of the form $\Gamma \nu(t)^2 \mathscr{E}(t)$ in (3.25), there is no harm in having another here, and it is desirable to avoid terms nonlinear in Q_0.

It follows from (3.29), (3.31), (3.33), and a routine computation that

$$(3.34) \qquad \int_{-\infty}^{\infty} \{q_t^2 + \varphi_t^2\}(x,t)\,dx + \int_0^t \int_{-\infty}^{\infty} q_t^2(x,s)\,dx\,ds$$
$$\leqq \Gamma Q_0 + \Gamma\{\nu(t) + \nu(t)^2\}\,\mathscr{E}(t), \qquad t \in [0, T_0).$$

By combining (3.25) and (3.34), we obtain the estimate

$$(3.35) \qquad \int_{-\infty}^{\infty} \{q^2 + q_t^2 + \varphi^2 + \varphi_t^2\}(x,t)\,dx + \int_0^t \int_{-\infty}^{\infty} \{q^2 + q_t^2\}(x,s)\,dx\,ds$$
$$\leqq \Gamma Q_0 + \Gamma\{\nu(t) + \nu(t)^2\}\,\mathscr{E}(t), \qquad t \in [0, T_0).$$

We can obtain additional information directly from equations (3.10) and (3.11). It follows from (3.10), (3.5), and (3.23) that

$$(3.36) \qquad \varphi_x^2 \leqq 2(\underline{C})^{-2}\,[A(\varphi)^2 q_t^2 + B(\varphi)^2 q^2]$$

which, in conjunction with (3.35), yields

$$(3.37) \qquad \int_{-\infty}^{\infty} \varphi_x^2(x,t)\,dx + \int_0^t \int_{-\infty}^{\infty} \varphi_x^2(x,s)\,dx\,ds$$
$$\leqq \Gamma Q_0 + \Gamma\{\nu(t) + \nu(t)^2\}\,\mathscr{E}(t), \qquad t \in [0, T_0).$$

By applying the same kind of argument to (3.11), we obtain

$$(3.38) \qquad \int_{-\infty}^{\infty} q_x^2(x,t)\,dx \leqq \Gamma Q_0 + \Gamma\{\nu(t) + \nu(t)^2\}\,\mathscr{E}(t), \qquad t \in [0, T_0).$$

The relations (3.35), (3.37), and (3.38), yield

$$\int_{-\infty}^{\infty} \{q^2 + q_x^2 + q_t^2 + \varphi^2 + \varphi_x^2 + \varphi_t^2\}(x,t)\,dx + \int_0^t \int_{-\infty}^{\infty} \{q^2 + q_t^2 + \varphi_x^2\}(x,s)\,dx\,ds$$
$$(3.39) \qquad \leqq \Gamma Q_0 + \Gamma\{\nu(t) + \nu(t)^2\}\,\mathscr{E}(t), \qquad t \in [0, T_0).$$

It remains to obtain similar bounds for[#]

$$(3.40) \qquad \int_0^t \int_{-\infty}^{\infty} \{q_x^2 + \varphi_t^2\}(x,s)\,dx\,ds$$

[#] One can bound the integral (3.40) with an energy integral obtained by differentiating (3.10) and (3.11) with respect to x, multiplying the respective results by q_x and φ_x, adding the resulting equations, and integrating over $\mathbb{R} \times [0, t]$. However, we choose to bound (3.40) by using (3.39), (3.10), (3.11), our subsequent estimates for second order derivatives, and a simple interpolation inequality.

and for the terms in the definition (3.19) in $\mathscr{E}(t)$ that contain derivatives of order two.

Our final energy integral (3.42) can be obtained formally by differentiating (3.27) and (3.28) with respect to t, using q_{tt} and φ_{tt} as multipliers, and proceeding as in the derivation of (3.29). Because third derivatives of q and φ do not necessarily exist as functions, we shall employ an approximation argument based on difference operators in order to obtain a rigorous derivation of (3.42).

For $h \in (0, T_0)$ and functions $w: \mathbb{R} \times [0, T_0) \to \mathbb{R}$, we define the forward difference operator Δ_h of stepsize h by

$$[\Delta_h w](x, t) := w(x, t+h) - w(x, t), \quad x \in \mathbb{R}, \quad t \in [0, T_0 - h). \tag{3.41}$$

We fix $h \in (0, T_0)$, apply Δ_h to (3.27) and (3.28), multiply the results by $\Delta_h q_t$ and $\Delta_h \varphi_t$, respectively, add the equations so obtained, and integrate over $\mathbb{R} \times [0, t]$, $t \in [0, T_0 - h)$. After a number of integrations by parts, we divide by h^2 and let h tend to zero. The outcome of this straightforward but tedious computation is

$$\begin{aligned}
&\tfrac{1}{2} \int_{-\infty}^{\infty} \{A(\varphi)\, q_{tt}^2 + \varphi_{tt}^2\}(x, t)\, dx + \int_0^t \int_{-\infty}^{\infty} B(\varphi)\, q_{tt}^2(x, s)\, dx\, ds \\
&\quad = \tfrac{1}{2} \int_{-\infty}^{\infty} \{A(\varphi)\, q_{tt}^2 + [1 + D(q, \varphi)]\, \varphi_{tt}^2\}(x, 0)\, dx - \tfrac{1}{2} \int_{-\infty}^{\infty} D(q, \varphi)\, \varphi_{tt}^2(x, t)\, dx \\
&\quad - \int_0^t \int_{-\infty}^{\infty} \{\tfrac{3}{2} A'(\varphi)\, \varphi_t q_{tt}^2 + A'(\varphi)\, q_t q_{tt} \varphi_{tt} + A''(\varphi)\, q_t \varphi_t^2 q_{tt} + 2B'(\varphi)\, \varphi_t q_t q_{tt} \\
&\quad + B'(\varphi)\, q q_{tt} \varphi_{tt} + B''(\varphi)\, q \varphi_t^2 q_{tt} + 2C'(\varphi)\, \varphi_t \varphi_{xt} q_{tt} + 2C'(\varphi)\, \varphi_t q_{xt} \varphi_{tt} \\
&\quad + C'(\varphi)\, q_x \varphi_{tt}^2 + C''(\varphi)\, \varphi_x \varphi_t^2 q_{tt} + C''(\varphi)\, q_x \varphi_t^2 \varphi_{tt} + \tfrac{3}{2} D_q(q, \varphi)\, q_t \varphi_{tt}^2 \\
&\quad + \tfrac{5}{2} D_\varphi(q, \varphi)\, \varphi_t \varphi_{tt}^2 + D_q(q, \varphi)\, \varphi_t q_{tt} \varphi_{tt} + D_{qq}(q, \varphi)\, q_t^2 \varphi_t \varphi_{tt} + 2D_{q\varphi}(q, \varphi)\, q_t \varphi_t^2 \varphi_{tt} \\
&\quad + D_{\varphi\varphi}(q, \varphi)\, \varphi_t^3 \varphi_{tt} + E(\varphi)\, \varphi_x q_{tt} \varphi_{tt} + 2E(\varphi)\, q_t \varphi_{xt} \varphi_{tt} - \tfrac{1}{2} E(\varphi)\, q_x \varphi_{tt}^2 \\
&\quad + 2E'(\varphi)\, q \varphi_t \varphi_{xt} \varphi_{tt} + 2E'(\varphi)\, q_t \varphi_x \varphi_t \varphi_{tt} + \tfrac{1}{2} E'(\varphi)\, q \varphi_x \varphi_{tt}^2 + E''(\varphi)\, q \varphi_x \varphi_t^2 \varphi_{tt} \\
&\quad + 2F(\varphi)\, q q_{tt} \varphi_{tt} + 2F(\varphi)\, q_t^2 \varphi_{tt} + 4F'(\varphi)\, q q_t \varphi_t \varphi_{tt} + F'(\varphi)\, q^2 \varphi_{tt}^2 \\
&\quad + F''(\varphi)\, q^2 \varphi_t^2 \varphi_{tt}\}(x, s)\, dx\, ds, \qquad t \in [0, T_0).
\end{aligned} \tag{3.42}$$

We note that the pattern of integrations by parts used to derive (3.42) is slightly different from that used to obtain previous energy integrals, because we wish to avoid terms involving third derivatives of q or φ.

Application to (3.42) of the procedure used to derive (3.34) from (3.29) yields

$$\begin{aligned}
&\int_{-\infty}^{\infty} \{q_{tt}^2 + \varphi_{tt}^2\}(x, t)\, dx + \int_0^t \int_{-\infty}^{\infty} q_{tt}^2(x, s)\, dx\, ds \\
&\qquad \leq \Gamma Q_0 + \Gamma\{\nu(t) + \nu(t)^3\}\, \mathscr{E}(t), \qquad t \in [0, T_0).
\end{aligned} \tag{3.43}$$

Here we have made use of the inequality $\alpha^2 \leqq \alpha + \alpha^3$ (holding for $\alpha \geqq 0$) to eliminate the need for a term of the form $\Gamma\nu(t)^2\,\mathscr{E}(t)$ on the right-hand side of (3.43). Observe that the term $-\frac{1}{2}\int_{-\infty}^{\infty} D(q,\varphi)\,\varphi_{tt}^2(x,t)\,dx$ in (3.42), which does not have an analogue in (3.29), causes no difficulties; it can either be absorbed by $\frac{1}{2}\int_{-\infty}^{\infty}\varphi_{tt}^2(x,t)\,dx$ (in view of (3.8)) or it can be bounded by $\Gamma\nu(t)^2\,\mathscr{E}(t)$ (by virtue of (3.9)).

It follows from (3.27), (3.5), and (3.23) that

$$\text{(3.44)}\qquad \int_{-\infty}^{\infty}\varphi_{xt}^2(x,t)\,dx \leqq 5(\underline{C})^{-2}\int_{-\infty}^{\infty}\{A(\varphi)^2\,q_{tt}^2 + B(\varphi)^2 q_t^2 + A'(\varphi)^2\,q_t^2\varphi_t^2 + B'(\varphi)^2\,q^2\varphi_t^2 + C'(\varphi)^2\,\varphi_x^2\varphi_t^2\}(x,t)\,dx, \qquad t\in[0,T_0).$$

Using (3.3), (3.39), and (3.43), we conclude from (3.44) that

$$\text{(3.45)}\qquad \int_{-\infty}^{\infty}\varphi_{xt}^2(x,t)\,dx + \int_0^t\int_{-\infty}^{\infty}\varphi_{xt}^2(x,s)\,dx\,ds \leqq \Gamma Q_0 + \Gamma\{\nu(t)+\nu(t)^3\}\,\mathscr{E}(t), \qquad t\in[0,T_0).$$

A similar argument applied to (3.28) yields, from the estimates (3.39), (3.43), and (3.45),

$$\text{(3.46)}\qquad \int_{-\infty}^{\infty} q_{xt}^2(x,t)\,dx \leqq \Gamma Q_0 + \Gamma\{\nu(t)+\nu(t)^4\}\,\mathscr{E}(t), \qquad t\in[0,T_0).$$

We have used the inequality $\alpha^2+\alpha^3 \leqq 2\alpha + \alpha^4$, $\alpha \geqq 0$, to eliminate the need for terms of the form $\Gamma\nu(t)^2\,\mathscr{E}(t)$ and $\Gamma\nu(t)^3\,\mathscr{E}(t)$ in (3.46).

Differentiation of (3.10) and (3.11) with respect to x yields

$$\text{(3.47)}\qquad -C(\varphi)\,\varphi_{xx} = A(\varphi)\,q_{xt} + B(\varphi)\,q_x + A'(\varphi)\,q_t\varphi_x + B'(\varphi)\,q\varphi_x + C'(\varphi)\,\varphi_x^2,$$

$$\text{(3.48)}\qquad -C(\varphi)\,q_{xx} = [1+D(q,\varphi)]\,\varphi_{xt} + C'(\varphi)\,q_x\varphi_x + D_q(q,\varphi)\,q_x\varphi_t + D_\varphi(q,\varphi)\,\varphi_x\varphi_t + E(\varphi)\,q\varphi_{xx} + E(\varphi)\,q_x\varphi_x + E'(\varphi)\,q_x\varphi_x^2 + 2F(\varphi)\,qq_x + F'(\varphi)\,q^2\varphi_x.$$

It follows from (3.47), (3.3), (3.5), (3.23), (3.39), and (3.46) that

$$\text{(3.49)}\qquad \int_{-\infty}^{\infty}\varphi_{xx}^2(x,t)\,dx \leqq \Gamma Q_0 + \Gamma\{\nu(t)+\nu(t)^4\}\,\mathscr{E}(t), \qquad t\in[0,T_0).$$

Similarly, we deduce from (3.48), (3.3), (3.5), (3.8), (3.23), and (3.45) that

$$\text{(3.50)}\qquad \int_{-\infty}^{\infty} q_{xx}^2(x,t)\,dx + \int_0^t\int_{-\infty}^{\infty} q_{xx}^2(x,s)\,dx\,ds \leqq \Gamma Q_0 + \Gamma\{\nu(t)+\nu(t)^4\}\,\mathscr{E}(t), \qquad t\in[0,T_0).$$

By combining (3.39), (3.43), (3.45), (3.46), (3.49), and (3.50), we obtain

$$\begin{aligned}\int_{-\infty}^{\infty} \{q^2 + q_x^2 + q_t^2 + q_{xx}^2 + q_{xt}^2 + q_{tt}^2 + \varphi^2 + \varphi_x^2 \\ + \varphi_t^2 + \varphi_{xx}^2 + \varphi_{xt}^2 + \varphi_{tt}^2\}(x, t)\,dx \\ + \int_0^t \int_{-\infty}^{\infty} \{q^2 + q_t^2 + q_{xx}^2 + q_{tt}^2 + \varphi_x^2 + \varphi_{xt}^2\}(x, s)\,dx\,ds \\ \leqq \Gamma Q_0 + \Gamma\{\nu(t) + \nu(t)^4\}\,\mathscr{E}(t), \qquad t \in [0, T_0).\end{aligned} \tag{3.51}$$

The next group of estimates will be obtained by a simple interpolation argument. Integration by parts yields

$$\int_{-\infty}^{\infty} q_x^2(x, t)\,dx = -\int_{-\infty}^{\infty} q q_{xx}(x, t)\,dx, \qquad t \in [0, T_0), \tag{3.52}$$

and consequently, by virtue of (3.22),

$$\int_0^t \int_{-\infty}^{\infty} q_x^2(x, s)\,dx\,ds \leqq \tfrac{1}{2} \int_0^t \int_{-\infty}^{\infty} \{q^2 + q_{xx}^2\}(x, s)\,dx\,ds, \qquad t \in [0, T_0). \tag{3.53}$$

From (3.53) and (3.51) we get

$$\int_0^t \int_{-\infty}^{\infty} q_x^2(x, s)\,dx\,ds \leqq \Gamma Q_0 + \Gamma\{\nu(t) + \nu(t)^4\}\,\mathscr{E}(t), \qquad t \in [0, T_0). \tag{3.54}$$

The identity

$$\begin{aligned}\int_0^t \int_{-\infty}^{\infty} q_{xt}^2(x, s)\,dx\,ds = \int_0^t \int_{-\infty}^{\infty} q_{xx} q_{tt}(x, s)\,dx\,ds - \int_{-\infty}^{\infty} q_t q_{xx}(x, t)\,dx \\ + \int_{-\infty}^{\infty} q_t q_{xx}(x, 0)\,dx, \qquad t \in [0, T_0)\end{aligned} \tag{3.55}$$

can be derived formally via integration by parts with respect to space and time. It is easy to see that this derivation can be made rigorous and that (3.55) indeed holds for our local solution, by virtue of (3.17). Using (3.22), we conclude from (3.55) that

$$\begin{aligned}\int_0^t \int_{-\infty}^{\infty} q_{xt}^2(x, s)\,dx\,ds \leqq \tfrac{1}{2} \int_{-\infty}^{\infty} \{q_t^2 + q_{xx}^2\}(x, 0)\,dx + \tfrac{1}{2} \int_{-\infty}^{\infty} \{q_t^2 + q_{xx}^2\}(x, t)\,dx \\ + \tfrac{1}{2} \int_0^t \int_{-\infty}^{\infty} \{q_{xx}^2 + q_{tt}^2\}(x, s)\,dx\,ds, \qquad t \in [0, T_0).\end{aligned} \tag{3.56}$$

In view of (3.30), (3.3), (3.5), and (3.22), it follows from (3.56) and (3.51) that

$$\int_0^t \int_{-\infty}^{\infty} q_{xt}^2(x, s)\,dx\,ds \leqq \Gamma Q_0 + \Gamma\{\nu(t) + \nu(t)^4\}\,\mathscr{E}(t), \qquad t \in [0, T_0). \tag{3.57}$$

To obtain our final set of estimates, we go back to equations (3.11), (3.28), and (3.47). It follows from (3.11), (3.8), and (3.23) that

$$(3.58)\quad \int_0^t \int_{-\infty}^{\infty} \varphi_t^2(x, s)\, dx\, ds$$
$$\leqq 6 \int_0^t \int_{-\infty}^{\infty} \{C(\varphi)^2\, q_x^2 + E(\varphi)^2 q^2 \varphi_x^2 + F(\varphi)^2 q^4\}\,(x, s)\, dx\, ds, \qquad t \in [0, T_0).$$

Using (3.3) and (3.54), we deduce from (3.58) that

$$(3.59)\quad \int_0^t \int_{-\infty}^{\infty} \varphi_t^2(x, s)\, dx\, ds \leqq \Gamma Q_0 + \Gamma\{\nu(t) + \nu(t)^4\}\, \mathscr{E}(t), \qquad t \in [0, T_0).$$

In a similar way we conclude from (3.28), with the aid of (3.3), (3.8), (3.23), and (3.56) that

$$(3.60)\quad \int_0^t \int_{-\infty}^{\infty} \varphi_{tt}^2(x, s)\, dx\, ds \leqq \Gamma Q_0 + \Gamma\{\nu(t) + \nu(t)^4\}\, \mathscr{E}(t), \qquad t \in [0, T_0).$$

Finally, it follows from (3.47), (3.3), (3.5), (3.22), and (3.57) that

$$(3.61)\quad \int_0^t \int_{-\infty}^{\infty} \varphi_{xx}^2(x, s)\, dx\, ds \leqq \Gamma Q_0 + \Gamma\{\nu(t) + \nu(t)^4\}\, \mathscr{E}(t), \qquad t \in [0, T_0).$$

The main estimate in our proof of Theorem 1 can now be obtained by combining (3.51), (3.54), (3.57), (3.59), (3.60), and (3.61):

$$(3.62)\quad \mathscr{E}(t) \leqq \bar{\Gamma} Q_0 + \bar{\Gamma}\{\nu(t) + \nu(t)^4\}\, \mathscr{E}(t), \qquad t \in [0, T_0),$$

where $\bar{\Gamma}$ denotes a (fixed) positive constant that can be chosen independently of q_0, φ_0, and T_0. We choose $\bar{\mathscr{E}}$, $\delta > 0$ such that

$$(3.63)\quad \bar{\mathscr{E}} \leqq \tfrac{1}{2}\varepsilon^2, \qquad \bar{\Gamma}\{(2\bar{\mathscr{E}})^{\frac{1}{2}} + (2\bar{\mathscr{E}})^2\} \leqq \tfrac{1}{4}, \qquad \bar{\Gamma}\delta^2 \leqq \tfrac{1}{4}\bar{\mathscr{E}}.$$

Here ε is the constant which was used in constructing the modified initial value problem.

Suppose now that (2.5) holds with the above choice of δ. Then, we have $\bar{\Gamma} Q_0 \leqq \frac{1}{4}\bar{\mathscr{E}}$. Moreover, it follows from Sobolev's embedding theorem that

$$(3.64)\quad \nu(t)^2 \leqq 2\mathscr{E}(t), \qquad t \in [0, T_0).$$

We therefore conclude from (3.62) that for any $t \in [0, T_0)$ with $\mathscr{E}(t) \leqq \bar{\mathscr{E}}$, we actually have $\mathscr{E}(t) \leqq \frac{1}{2}\bar{\mathscr{E}}$. Consequently, by continuity,

$$(3.65)\quad \mathscr{E}(t) \leqq \tfrac{1}{2}\bar{\mathscr{E}}, \qquad t \in [0, T_0),$$

provided that $\mathscr{E}(0) \leqq \frac{1}{2}\bar{\mathscr{E}}$.

We can always choose a smaller $\delta > 0$ (if necessary) such that $Q_0 \leqq \delta^2$ implies $\mathscr{E}(0) \leqq \frac{1}{2}\bar{\mathscr{E}}$. (See, *e.g.* (3.30), (3.32).) Observe that (3.63) still holds if the size of δ is reduced. Thus, if (2.5) is satisfied with our revised choice of δ, then (3.65) holds. This implies that $T_0 = \infty$. Moreover, by virtue of Sobolev's embedding theorem,

$$\varphi(x, t)^2 + q(x, t)^2 \leqq \mathscr{E}(t) \leqq \tfrac{1}{2}\bar{\mathscr{E}} \leqq \tfrac{1}{4}\varepsilon^2, \qquad x \in \mathbb{R}, \quad t \geqq 0, \tag{3.66}$$

and consequently

$$|\varphi(x, t)| \leqq \tfrac{1}{2}\varepsilon, \qquad |q(x, t)| \leqq \tfrac{1}{2}\varepsilon, \qquad x \in \mathbb{R}, \quad t \geqq 0. \tag{3.67}$$

Therefore the corresponding pair (q, θ), with θ as in (3.16), is a solution of (2.1), (2.2), (2.3) obeying (2.7). The uniqueness claim in Theorem 1 follows easily from (3.67) and local uniqueness of solutions to (3.10), (3.11), (3.12).

Finally, we note that (2.8) and (2.9) follow from (3.65), (3.16), and standard embedding inequalities. For example, we see from (3.65) that

$$\varphi \in L^\infty([0, \infty); L^2(\mathbb{R})), \tag{3.68}$$

$$\varphi_x, \varphi_{xt} \in L^2([0, \infty); L^2(\mathbb{R})). \tag{3.69}$$

It follows from (3.69) that

$$\varphi_x(\cdot, t) \to 0 \quad \text{in } L^2(\mathbb{R}) \quad \text{as } t \to \infty. \tag{3.70}$$

Then, we observe that

$$\varphi^2(x, t) = 2 \int_{-\infty}^{x} \varphi\varphi_x(\bar{x}, t)\, d\bar{x} \leqq 2 \int_{-\infty}^{\infty} |\varphi| \cdot |\varphi_x|\, (\bar{x}, t)\, d\bar{x}, \qquad x \in \mathbb{R}, \quad t \geqq 0. \tag{3.71}$$

Using the Cauchy-Schwarz inequality, (3.68), and (3.70), we conclude from (3.71) that

$$\varphi(\cdot, t) \to 0 \quad \text{uniformly on } \mathbb{R} \quad \text{as } t \to \infty. \tag{3.72}$$

We have thus shown that $\theta(\cdot, t) \to \bar{\theta}$ uniformly on $\mathbb{R}$ and $\theta_x(\cdot, t) \to 0$ in $L^2(\mathbb{R})$ as $t \to \infty$. The remaining parts of (2.9) can be established in a similar fashion. This completes the proof of Theorem 1.

As noted in Section 2, the proofs of Theorems 2 and 3 (and of Remark 7) and the proof of Theorem 1 given above are virtually identical. For an initial-boundary value problem, the arguments concerning asymptotic behavior are slightly different (but still very similar). To establish (2.27) in Theorem 3 and in Remark 7, one uses the fact the boundary conditions (2.30) or (2.18) hold the value of θ fixed at one (or both) of the endpoints. To establish (2.25) in Theorem 2, one first shows that the boundedness of ξ yields (2.26) as well as boundedness of θ on $[0, 1] \times [0, \infty)$. It follows that every sequence of times tending to infinity has a subsequence on which θ converges uniformly to a constant. By the argument given before equation (2.21), the constant must be the unique solution $\bar{\theta}_*$ of (2.21), and one may conclude that $\theta(\cdot, t)$ converges uniformly on [0, 1] to $\bar{\theta}_*$ as t tends to infinity.

Acknowledgments. While working on this paper, HRUSA was a visitor at the Mathematics Research Center of the University of Wisconsin-Madison, and his research was supported by the U.S. Army under Contract DAAG29-80C-0041 and by the National Science Foundation under Contract MCS-82-10950. The research of COLEMAN and OWEN was supported by the National Science Foundation under Contract MCS-82-02647.

References

1948 1. CATTANEO, C., Sulla conduzione del calore, *Atti Sem. Mat. Fis. Univ. Modena* **3**, 83–101.

1963 1. CHESTER, M., Second sound in solids, *Phys. Rev.* **131**, 2013–2015.

1964 1. LAX, P. D., Development of singularities in solutions of nonlinear hyperbolic partial differential equations, *J. Math. Phys.* **5**, 611–613.

1966 1. GUYER, R. A., & J. A. KRUMHANSL, Solution of the linearized phonon Boltzmann equation, *Phys. Rev.* **148**, 766–778.

2. ACKERMAN, C. C., B. BERTMAN, H. A. FAIRBANK, & R. A. GUYER, Second sound in solid helium, *Phys. Rev. Lett.* **16**, 789–791.

1967 1. MACCAMY, R. C., & V. J. MIZEL, Existence and nonexistence in the large of solutions of quasilinear wave equations, *Arch. Rational Mech. Anal.* **25**, 299–320.

1968 1. ACKERMAN, C. C., & R. A. GUYER, Temperature pulses in dielectric solids, *Ann. Phys.* (N.Y.) **50**, 128–185.

2. GURTIN, M. E., & A. C. PIPKIN, A general theory of heat conduction with finite wave speeds, *Arch. Rational Mech. Anal.* **31**, 113–126.

1969 1. ACKERMAN, C. C., & W. C. OVERTON, Jr., Second-sound in solid helium-3, *Phys. Rev. Lett.* **22**, 764–766.

1970 1. CHEN, P. J., & M. E. GURTIN, On second sound in materials with memory, *Z. ang. Math. Phys.* **21**, 232–241.

2. JACKSON, H. E., C. T. WALKER, & T. F. MCNELLY, Second sound in NaF, *Phys. Rev. Lett.* **25**, 26–28.

1971 1. JACKSON, H. E., & C. T. WALKER, Thermal conductivity, second sound, and phonon-phonon interactions in NaF, *Phys. Rev.* **83**, 1428–1439.

1972 1. NARAYANAMURTI, V., & R. C. DYNES, Observation of second sound in bismuth, *Phys. Rev. Lett.* **28**, 1461–1465.

1975 1. KATO, T., The Cauchy problem for quasilinear symmetric hyperbolic systems, *Arch. Rational Mech. Anal.* **58**, 181–205.

1976 1. REED, M., *Abstract Nonlinear Wave Equations*, Lecture Notes in Mathematics **507**, Springer-Verlag, Berlin, Heidelberg, New York.

1977 1. MACCAMY, R. C., An integro-differential equation with application in heat flow, *Q. Appl. Math.* **35**, 1–19.

2. MATSUMURA, A., Global existence and asymptotics of the solutions of the second order quasilinear hyperbolic equations with first order dissipation, *Publ. Res. Inst. Math. Science*, Kyoto Univ., Ser. A **13**, 349–379.

1979 1. DAFERMOS, C. M., & J. A. NOHEL, Energy methods for nonlinear hyperbolic Volterra integrodifferential equations, *Comm. PDE* **4**, 219–278.

1980 1. STAFFANS, O., On a nonlinear hyperbolic Volterra equation, *SIAM J. Math. Anal.* **11**, 793–812.
1982 1. COLEMAN, B. D., M. FABRIZIO, & D. R. OWEN, On the thermodynamics of second sound in dielectric crystals, *Arch. Rational Mech. Anal.* **80**, 135–158.
2. COLEMAN, B. D., M. FABRIZIO, & D. R. OWEN, Il secondo suono nei cristalli: termodinamica ed equazioni costitutive, *Rend. Seminario Mat. Univ. Padova* **68**, 208–277.
1983 1. COLEMAN, B. D., & D. R. OWEN, On the nonequilibrium behavior of solids that transport heat by second sound, *Comp. & Maths., with Appls.* **9**, 527–546.
1985 1. DAFERMOS, C. M., & W. J. HRUSA, Energy methods for quasilinear hyperbolic initial-boundary value problems. Applications to elastodynamics, *Arch. Rational Mech. Anal.* **87**, 267–292.

Department of Mathematics
Carnegie-Mellon University
Pittsburgh, Pennsylvania

(Received August 19, 1985)

Steady-State Problems of Nonlinear Electro-Magneto-Thermo-Elasticity

ROBERT C. ROGERS & STUART S. ANTMAN

This paper is dedicated to Walter Noll on the occasion of his sixtieth birthday

Contents

Part I. The General Theory

1. Introduction

In this paper we study a class of boundary value problems for a quasilinear system of functional-differential equations describing the steady-state behavior of solids that can sustain mechanical, electromagnetic, and thermal effects. We treat partial differential equations in Part I and ordinary differential equations in Part II. Our primary goals are to show that there is a simple way to formulate the governing equations, which illuminates the physics and promotes the analysis of the equations, to actually analyze important classes of problems, and to contribute to the development of an effective constitutive theory for such materials by showing how our physically and mathematically natural constitutive restrictions support existence and regularity theories for our problems. The problems we study are simple enough to be tractable, interesting enough to possess a very rich class of solutions, and yet complicated enough to require new approaches, both in the formulation and treatment of electromagnetism in solids and in the use of techniques of nonlinear analysis.

Our constitutive equations give the stress, heat flux, dielectric displacement, magnetic induction, and electric current as arbitrary functions of the deformation gradient, temperature, temperature gradient, electric fields, and magnetic field. These constitutive functions must of course be invariant under rigid motions. In order to reduce the governing equations to ordinary differential equations for our semi-inverse problems of Part II, we further require that the bodies under study have some material symmetry. Our basic constitutive assumption is that the constitutive equations satisfy the Strong Ellipticity Condition. Originating in the theory of partial differential equations, this condition proves to be eminently natural on physical grounds. Indeed, this condition, when precisely formulated, is equivalent to the requirement that each component of the dependent constitutive variables is a strictly increasing function of the corresponding component of the independent constitutive variables (when the other components of the independent variables are held fixed). Roughly speaking, a typical consequence of this assumption is that a change in the temperature gradient produces a far more pronounced change in the heat flux vector than it does in the stress, dielectric displacement, and magnetic induction. Thus the Strong Ellipticity Condition implies a very mild uncoupling in the constitutive equations. True uncouplings, such as the independence of stress, dielectric displacement, and magnetic induction on the temperature gradient, may be interpreted as consequences of the Clausius-Duhem version of the Second Law of Thermodynamics. But we have no need for such true uncouplings anywhere in our analysis. Indeed, wherever the Clausius-Duhem inequality is more restrictive than the Strong Ellipticity Condition, we have no need for its consequences, and wherever it is less restrictive, it is inadequate for our needs. We supplement the Strong Ellipticity Condition with compatible growth conditions.

Rather than adhering to the classical tradition, (followed at least in part by TOUPIN (1956), FANO, CHU, & ADLER (1960), PENFIELD & HAUS (1967), DE GROOT & SUTTORP (1972), and others), of deriving or motivating the constitutive equations of electromagnetism from discrete microscopic models, we employ the phenomenological approach of continuum mechanics and simply lay down general constitutive laws. We thereby gain great economy and generality in our formulation of electromagnetism. (*Cf.* TRUESDELL & TOUPIN (1960).)

For simplicity and clarity in our mathematical analysis, it is crucial not only that we give Maxwell's equations a material (Lagrangian) formulation, but also that we introduce new variables in place of the dielectric displacement, magnetic induction, *etc.* In this regard we are merely extending to the theory of electromagnetism in deformable media the methodology that has proved most natural and successful for boundary value problems of nonlinear elasticity.

In the overwhelming majority of texts on continuum mechanics the emphasis placed on the spatial formulation of the governing equations and, in particular, on the use of the Cauchy stress tensor overshadows that placed on the material formulation and on the use of the Piola-Kirchhoff stress tensors. (We define these stress tensors in Section 3.) The reason for this emphasis is largely historical: The two most highly developed branches of continuum mechanics are Newtonian fluids and linear elasticity. The constitutive equations for a homogeneous Newtonian fluid are especially simple in a spatial formulation. Moreover, in this formulation the constraint of incompressibility (valid for liquids) has an elegant characterization as the linear equation expressing the vanishing of the divergence of the velocity field defined over points in space. (In contrast, incompressibility is characterized in a material formulation by the nonlinear equation requiring the Jacobian of the deformation gradient to equal unity.) For problems involving nonhomogeneous fluids or fluids with free surfaces, there are compensating disadvantages requiring some version of a material formulation, possibly disguised, for their successful treatment. In linear elasticity there is no distinction between material and spatial formulations, although the derivation of these equations from a nonlinear spatial formulation is much more difficult to carry out than that from a nonlinear material formulation. The advantages of a material formulation are evident for the boundary value problems of nonlinear solid mechanics: (i) The prescription of constitutive equations for the first Piola-Kirchhoff stress tensor in terms of the past history of deformation is natural and does not suffer from complications due to nonhomogeneity. (ii) The governing equations are posed on a *fixed* region of space, the region occupied by the body in a reference configuration, rather than on the unknown and possibly moving region actually occupied by the body. These factors have not, however, proved to be compelling in shifting the emphasis of texts towards material formulations because there have been so few studies of nonlinear boundary value problems of solid mechanics. (*Cf.* ANTMAN (1978, 1979, 1983), BALL (1977, 1982).)

Maxwell's equations have been posed almost exclusively in spatial coordinates because the most important case of the vacuum can be posed in no other way and because in the second most important case of a rigid medium there is no essential distinction between material and spatial coordinates. Moreover, the most actively cultivated field of electromagnetism in deformable media is that of magnetohydrodynamics. For the reasons mentioned in our comments on fluid dynamics, many problems for this theory are most easily set in spatial coordinates. Only recently has the use of material coordinates begun to appear in treatments of electromagnetism in media. (*Cf.* WALKER, PIPKIN, & RIVLIN (1965), HUTTER (1975), PAO & HUTTER (1975), HUTTER & VAN DE VEN (1978), McCARTHY & TIERSTEN (1978), and MAUGIN (1981).) These authors have also introduced new fields suitable for material coordinates in place of the classical fields.

The large deformation of solids in the presence of large electromagnetic fields is a problem of growing technological importance (*cf.* MOON (1978, 1984)). Awareness of this importance is evidenced by the number of papers recently devoted to this subject. (*Cf.* PARKUS (1979), AMBARTSUMIAN (1982), MAUGIN (1983).) To our knowledge, ours is the first mathematical analysis of general nonlinear boundary value problems in this area.

Much of the previous work in the electromagnetism of deformable solids can be divided into two general areas: the development of a general theory governing such media and the solution of specific nonlinear problems. General theories of the dynamics of deformable solids have been proposed by FANO, CHU & ADLER (1960), TOUPIN (1963), DIXON & ERINGEN (1965), PAO & HUTTER (1975), and MAUGIN & ERINGEN (1977). These developments consist in the derivation of some form of Maxwell's equation and associated forms of the electric body force, body couple, and internal energy supply from some discrete model of the material. Comparisons of various theories are to be found in PENFIELD & HAUS (1967), DEGROOT & SUTTORP (1972), HUTTER & VAN DE VEN (1978), and PAO (1978). Our work is more closely related to the static theories of TOUPIN (1956) and BROWN (1966). Specific problems for general nonlinear dielectrics were solved by TOUPIN (1956), ERINGEN (1963), VERMA (1964), PIPKIN & RIVLIN (1960), and SINGH & PIPKIN (1966) via the inverse methods of modern nonlinear elasticity. (SINGH & PIPKIN also provide a review of the earlier work.) There is an extensive literature on specific nonlinear materials with polynomial constitutive equations and associated problems (*cf.* JORDAN & ERINGEN (1964) and PIPKIN & RIVLIN (1966)). MAUGIN (1981) reviews the modern work on wave motion in magnetizable deformable solids and includes both general nonlinear and specific (approximate) constitutive equations.

2. Notation

The *Euclidean* 3-*space* $\mathbb{E}^3$ is defined to be abstract 3-dimensional real inner-product space. Its inner product, the dot product, is the natural source of the geometric properties of the space. We interpret $\mathbb{E}^3$ as physical space. We distinguish $\mathbb{E}^3$ from $\mathbb{R}^3$, the space of triples of real numbers equipped with any norm (which is necessarily topologically equivalent to the Euclidean norm). But we assign no natural geometrical significance to the norm on $\mathbb{R}^3$.

Vectors, which we define to be elements of $\mathbb{E}^3$, and vector-valued functions are denoted by bold-face, lower-case Latin letters. *Second-order tensors*, which form the space Lin of linear operators from $\mathbb{E}^3$ into itself, are denoted by bold-face upper-case Latin letters. The subspace of Lin consisting of symmetric second-order tensors is denoted Sym. Its subset of positive-definite tensors is denoted Psym. The group $SL(3)$ of all members of Lin with positive determinant is denoted Lin^+. *Scalars* and scalar-valued functions are denoted with light-face letters. Elements of $\mathbb{R}^n$ for $n > 1$ and functions with values in $\mathbb{R}^n$ are denoted by bold-face sanserif lower-case Latin letters and by bold-face lower-case Greek letters.

We employ the dyadic notation of Gibbs (*cf.* GIBBS & WILSON (1901)), which we now describe. (This notation is both admirably suited for treatment of problems in curvilinear coordinates and completely compatible with modern invariant formulations of linear algebra in $\mathbb{E}^3$.) The dot product of vectors $\boldsymbol{u}$ and $\boldsymbol{v}$ is denoted $\boldsymbol{u} \cdot \boldsymbol{v}$. The cross product of two vectors $\boldsymbol{u}$ and $\boldsymbol{v}$ is denoted $\boldsymbol{u} \wedge \boldsymbol{v}$. The value of the second order tensor $\boldsymbol{A}$ at $\boldsymbol{u}$ is denoted $\boldsymbol{A} \cdot \boldsymbol{u}$. The *transpose* $\boldsymbol{A}^*$ of $\boldsymbol{A}$ is defined by $\boldsymbol{v} \cdot (\boldsymbol{A} \cdot \boldsymbol{u}) = \boldsymbol{u} \cdot (\boldsymbol{A}^* \cdot \boldsymbol{v})$ for all $\boldsymbol{u}$ and $\boldsymbol{v}$. We accordingly write $\boldsymbol{A}^* \cdot \boldsymbol{v} = \boldsymbol{v} \cdot \boldsymbol{A}$. $\boldsymbol{A}$ is *symmetric* if $\boldsymbol{A} = \boldsymbol{A}^*$ and *skew* if $\boldsymbol{A} = -\boldsymbol{A}^*$. If $\boldsymbol{A}$ is skew, there is a unique vector $\boldsymbol{a}$, called the *axial vector* of $\boldsymbol{A}$, such that $\boldsymbol{A} \cdot \boldsymbol{v} = \boldsymbol{a} \wedge \boldsymbol{v}$ for all $\boldsymbol{v} \in \mathbb{E}^3$. The *dyadic product* $\boldsymbol{uv}$ of vectors $\boldsymbol{u}$ and $\boldsymbol{v}$ is defined to be the second-order tensor satisfying $(\boldsymbol{uv}) \cdot \boldsymbol{w} = (\boldsymbol{v} \cdot \boldsymbol{w})\, \boldsymbol{u}$ for all $\boldsymbol{w}$. Thus $(\boldsymbol{uv})^* = \boldsymbol{vu}$ and $\operatorname{tr}(\boldsymbol{vu}) = \boldsymbol{u} \cdot \boldsymbol{v}$ where tr denotes trace. The *product* $\boldsymbol{A} \cdot \boldsymbol{B}$ of tensors is defined by $(\boldsymbol{A} \cdot \boldsymbol{B}) \cdot \boldsymbol{v} = \boldsymbol{A} \cdot (\boldsymbol{B} \cdot \boldsymbol{v})$ for all $\boldsymbol{v}$. Thus $\boldsymbol{A} \cdot (\boldsymbol{uv}) = (\boldsymbol{A} \cdot \boldsymbol{u})\, \boldsymbol{v}$ and $(\boldsymbol{uv}) \cdot (\boldsymbol{A}) = \boldsymbol{u}(\boldsymbol{v} \cdot \boldsymbol{A})$. We set $\boldsymbol{A} : \boldsymbol{B} = \operatorname{tr}(\boldsymbol{A} \cdot \boldsymbol{B}^*)$. Hence $\operatorname{tr} \boldsymbol{A} = \boldsymbol{I} : \boldsymbol{A} = \boldsymbol{A} : \boldsymbol{I} = \boldsymbol{A}^* : \boldsymbol{I} = \operatorname{tr} \boldsymbol{A}^*$, $(\boldsymbol{uv}) : (\boldsymbol{xy}) = (\boldsymbol{u} \cdot \boldsymbol{x})\,(\boldsymbol{v} \cdot \boldsymbol{y})$, and $\boldsymbol{A} : (\boldsymbol{uv}) = \boldsymbol{u} \cdot \boldsymbol{A} \cdot \boldsymbol{v} = (\boldsymbol{uv}) : \boldsymbol{A}$. ($\boldsymbol{I}$ denotes the identity tensor.) It is easy to see that ":" is an inner product on Lin.

We accordingly define the *norm* $|A|$ of A by $|A| = \sqrt{A : A}$. If $\boldsymbol{a}$ and $\boldsymbol{b}$ are unit vectors, then $|\boldsymbol{ab}| = 1$ so that $\boldsymbol{ab}$ is a unit tensor. In this case we can represent any tensor A by the orthogonal decomposition $A = (\boldsymbol{a} \cdot A \cdot \boldsymbol{b})\, \boldsymbol{ab} + [A - (\boldsymbol{a} \cdot A \cdot \boldsymbol{b})\, \boldsymbol{ab}]$ where $\boldsymbol{a} \cdot A \cdot \boldsymbol{b}$ is the *component of* A *along* $\boldsymbol{ab}$ and $[A - (\boldsymbol{a} \cdot A \cdot \boldsymbol{b})\, \boldsymbol{ab}]$ is the *projection of* A *onto the orthogonal complement of* $\boldsymbol{ab}$. If $\{\boldsymbol{a}_k\}$ and $\{\boldsymbol{b}_j\}$ are each bases for $\mathbb{E}^3$, then $\{\boldsymbol{a}_k \boldsymbol{b}_j\}$ is a basis for Lin. Thus we can use all of our dyadic identities to construct the familiar componental formulas for all the expressions we have introduced in invariant form. Repeated indices are summed over their obvious ranges.

The (*Gâteaux*) *differentials* of $\boldsymbol{u} \mapsto \boldsymbol{f}(\boldsymbol{u})$ at $\boldsymbol{a}$ in the direction of $\boldsymbol{b}$ and of $U \mapsto \boldsymbol{F}(U)$ at A in the direction of $\boldsymbol{B}$ are defined to be the vector $[\partial \boldsymbol{f}(\boldsymbol{a})/\partial \boldsymbol{u}] \cdot \boldsymbol{b}$ and the tensor $[\partial \boldsymbol{F}(\boldsymbol{a})/\partial U] \cdot \boldsymbol{B}$ given by

$$(2.1)\qquad [\partial \boldsymbol{f}(\boldsymbol{a})/\partial \boldsymbol{u}] \cdot \boldsymbol{b} = \frac{d}{dt} \boldsymbol{f}(\boldsymbol{a} + t\boldsymbol{b})\big|_{t=0},$$

$$(2.2)\qquad [\partial \boldsymbol{F}(A)/\partial U] \cdot \boldsymbol{B} = \frac{d}{dt} \boldsymbol{F}(A + t\boldsymbol{B})\big|_{t=0}.$$

Other differentials are defined similarly. If $U \mapsto \boldsymbol{F}(U)$ and $V \mapsto \boldsymbol{G}(V)$ are (Fréchet) differentiable, then $V \mapsto \boldsymbol{F}(\boldsymbol{G}(V)) \equiv \boldsymbol{H}(V)$ is also, and its differential satisfies the *chain rule*:

$$(2.3)\qquad [\partial \boldsymbol{H}(A)/\partial V] : \boldsymbol{B} = [\partial \boldsymbol{F}(\boldsymbol{G}(A))/\partial U] : \{[\partial \boldsymbol{G}(A)/\partial V] : \boldsymbol{B}\}.$$

(The braces can be omitted from the right side of (2.3).) Our notational scheme embodied in (2.1) and (2.2) causes the chain rule (2.3) to have a form analogous to that for scalar functions. As we shall see in the next chapter this virtue is counterbalanced by the increased complexity of defining and representing the action of the classical differential operators grad, div, curl on tensor functions.

We denote n copies of a function space $\mathscr{X}$ by $\mathscr{X}$ itself. The distinction will be clear from the context: Thus a statement of the form $\boldsymbol{w} \in L_p(\mathscr{B})$ is to imply that this $L_p(\mathscr{B})$ is the space of all measurable vector-valued functions

$$\mathbb{E}^3 \supset \mathscr{B} \ni \boldsymbol{z} \mapsto \boldsymbol{w}(\boldsymbol{z}) \in \mathbb{E}^3 \text{ such that } \int_{\mathscr{B}} [\boldsymbol{w}(\boldsymbol{z}) \cdot \boldsymbol{w}(\boldsymbol{z})]^{p/2}\, dv(\boldsymbol{z}) < \infty.$$

The norm of a Banach space $\mathscr{X}$ is denoted $\|\cdot, \mathscr{X}\|$.

3. Formulation of the Governing Equations

In this and the next section we formulate the equations for steady-state problems of electro-magneto-thermo-elasticity. There are several different theories that are at least formally equivalent in the classical nonrelativistic setting we employ. (*Cf.* HUTTER & VAN DE VEN (1978) and PAO (1978).) Comparisons of the various theories is made difficult by the fact that the same symbol used in different theories has different meanings. Fortunately, the mathematical form of the governing equations expressing the balance of linear momentum, the balance of energy, Maxwell's laws, and the conservation of charge is the same for all these theories. We shall refer to the various fields that occur in our equations by their traditional names, realizing that their precise physical significance inheres in the slots they occupy in the equations for a specific theory. We employ a purely classical interpretation of space-time.

To make our presentation as transparent as possible, we assume that all the functions and boundaries that appear are smooth enough for all the classical operations that appear to make sense. (A careful treatment of these issues without such blanket smoothness assumptions can be modelled on that of ANTMAN & OSBORN (1979).) Of course, we abandon this optimistic formalism when we afterwards analyze our specific boundary value problems.

We identify a *material body* with the closure $\mathscr{B}$ of a domain in $\mathbb{E}^3$ and we identify *material points* of the body with their positions $\boldsymbol{z}$ in $\mathscr{B}$. For each $\boldsymbol{z}$ in $\mathscr{B}$ let $\boldsymbol{y}(\boldsymbol{z})$ denote the positions of $\boldsymbol{z}$ in some deformed configuration. The (*transposed*) *deformation gradient* $\boldsymbol{F}$ and the *right Cauchy-Green deformation tensor* $\boldsymbol{C}$ for the configuration $\boldsymbol{y}$ are defined by

$$\text{(3.1)} \qquad \boldsymbol{F} = \partial \boldsymbol{y}/\partial \boldsymbol{z}, \qquad \boldsymbol{C} = \boldsymbol{F}^* \cdot \boldsymbol{F}.$$

We require that no two distinct material points simultaneously occupy the same position in a given configuration. Thus each map $\boldsymbol{y}$ must be one-to-one. Since this global condition is so difficult to treat, we ignore it and content ourselves with the local condition that the deformation $\boldsymbol{y}$ merely preserve orientation, *i.e.*, that

$$\text{(3.2)} \qquad \det \boldsymbol{F} > 0,$$

where det denotes the determinant.

Let $\tilde{\lambda}(\boldsymbol{y})$ denote the *logarithm of the absolute temperature* at position $\boldsymbol{y}$ in space. (It is finite-valued if and only if the absolute temperature is positive-valued.) We set $\tilde{\boldsymbol{g}}(\boldsymbol{y}) = \partial\tilde{\lambda}(\boldsymbol{y})/\partial\boldsymbol{y}$. Let $\tilde{\boldsymbol{e}}(\boldsymbol{y})$ and $\tilde{\boldsymbol{h}}(\boldsymbol{y})$ denote the *electric* and *magnetic fields* at $\boldsymbol{y}$. We set

$$\text{(3.3)} \qquad \begin{gathered} \lambda(\boldsymbol{z}) \equiv \tilde{\lambda}(\boldsymbol{y}(\boldsymbol{z})),\ \boldsymbol{g}(\boldsymbol{z}) \equiv \partial\lambda(\boldsymbol{z})/\partial\boldsymbol{z} \equiv \tilde{\boldsymbol{g}}(\boldsymbol{y}(\boldsymbol{z}))\cdot\boldsymbol{F}(\boldsymbol{z}), \\ \boldsymbol{e}(\boldsymbol{z}) = \tilde{\boldsymbol{e}}(\boldsymbol{y}(\boldsymbol{z}))\cdot\boldsymbol{F}(\boldsymbol{z}),\ \boldsymbol{h}(\boldsymbol{z}) = \tilde{\boldsymbol{h}}(\boldsymbol{y}(\boldsymbol{z}))\cdot\boldsymbol{F}(\boldsymbol{z}). \end{gathered}$$

$\boldsymbol{g}$, $\boldsymbol{e}$, and $\boldsymbol{h}$ are the *material logarithmic temperature gradient*, *electric field*, and *magnetic field*. (We shall soon see that $\boldsymbol{e}$ and $\boldsymbol{h}$ can be represented in terms of gradients. Consequently they transform in (3.3) just like $\boldsymbol{g}$.)

Let $\tilde{\boldsymbol{T}}(\boldsymbol{y})$ denote the *effective Cauchy stress*, *i.e.*, the sum of the mechanical Cauchy stress and the Maxwell stress. $\tilde{\boldsymbol{T}}(\boldsymbol{y})$ measures force per unit actual area at $\boldsymbol{y}$. (There are several versions of $\tilde{\boldsymbol{T}}$, depending on alternative representations and decompositions of the Lorentz force and torque.) Let $\tilde{\boldsymbol{q}}(\boldsymbol{y})$ denote the *heat flux* per unit actual area at $\boldsymbol{y}$. Let $\tilde{\boldsymbol{d}}(\boldsymbol{y})$, $\tilde{\boldsymbol{b}}(\boldsymbol{y})$, $\tilde{\boldsymbol{j}}(\boldsymbol{y})$ be the *dielectric displacement*, *magnetic induction*, and *current density* at $\boldsymbol{y}$. Then we introduce material versions of these fields by

$$\text{(3.4)} \qquad \begin{gathered} \boldsymbol{T}(\boldsymbol{z})^* = \det \boldsymbol{F}(\boldsymbol{z})\,\boldsymbol{F}^{-1}(\boldsymbol{z})\cdot\tilde{\boldsymbol{T}}(\boldsymbol{y}(\boldsymbol{z}))^*, \\ \boldsymbol{q}(\boldsymbol{z}) = \det \boldsymbol{F}(\boldsymbol{z})\,\boldsymbol{F}^{-1}(\boldsymbol{z})\cdot\tilde{\boldsymbol{q}}(\boldsymbol{y}(\boldsymbol{z})),\ \text{etc.} \end{gathered}$$

$\boldsymbol{T}$ is the *effective first Piola-Kirchhoff stress*.

For simplicity let us assume that the body force and heat source have purely electromagnetic origin. Then the local form of the balance of forces, the balance of energy, and Maxwell's equations for a steady state are

$$\operatorname{Div} \boldsymbol{T} + \boldsymbol{f} = \boldsymbol{0}, \tag{3.5}$$

$$\operatorname{Div} \boldsymbol{q} + \boldsymbol{j} \cdot \boldsymbol{e} = 0, \tag{3.6}$$

$$\operatorname{Div} \boldsymbol{d} = \sigma, \tag{3.7}$$

$$\operatorname{Div} \boldsymbol{b} = 0, \tag{3.8}$$

$$\operatorname{Curl} \boldsymbol{e} = \boldsymbol{0}, \tag{3.9}$$

$$\operatorname{Curl} \boldsymbol{h} = \boldsymbol{j}. \tag{3.10}$$

The material divergence Div of a tensor is defined by Green's Theorem

$$\int_{\partial \mathscr{P}} \boldsymbol{T} \cdot \boldsymbol{n}\, da = \int_{\mathscr{P}} \operatorname{Div} \boldsymbol{T}\, dv \tag{3.11}$$

where $\mathscr{P} \subset \mathscr{B}$ and $\boldsymbol{n}$ is the unit outer normal to $\mathscr{P}$. In (3.5) $\boldsymbol{f}$ represents body forces of electromagnetic origin not absorbed by the Maxwell stress. Since every term in the usual prescriptions of the Lorentz force is a divergence, we could absorb this force entirely into the Maxwell stress and hence into the effective stress. We accordingly take $\boldsymbol{f} = \boldsymbol{0}$. (*Cf.* Hutter & van de Ven (1978).) The term $\boldsymbol{j} \cdot \boldsymbol{e}$ in (3.6) is the *Joule heating*. In (3.7) σ represents the *free charge*. We regard it as an assigned function of $\boldsymbol{z}$.

The balance of torque has the local form

$$\boldsymbol{L} = \boldsymbol{T} \cdot \boldsymbol{F}^* - \boldsymbol{F} \cdot \boldsymbol{T}^* \tag{3.12}$$

where $\boldsymbol{L}$ is a skew tensor depending upon the electromagnetic fields and the choice of the Maxwell stress tensor. We assume that (3.12) is identically satisfied when the constitutive functions, to be introduced in the next section, are substituted into (3.12). Hutter & van de Ven (1978) show that it is permissible to take $\boldsymbol{L} = \boldsymbol{0}$.

Equations (3.9) and (3.10) imply that there exist scalar functions φ and ψ, called the *electric* and *magnetic scalar potentials* such that

$$\boldsymbol{e} = \partial\varphi/\partial\boldsymbol{z}, \tag{3.13}$$

$$\boldsymbol{h}(\boldsymbol{z}) = \partial\psi(\boldsymbol{z})/\partial\boldsymbol{z} + \int_B [\boldsymbol{j}(\boldsymbol{u}) \wedge (\boldsymbol{z} - \boldsymbol{u})]\, |\boldsymbol{z} - \boldsymbol{u}|^{-3}\, dv(\boldsymbol{u}), \tag{3.14}$$

as is shown in standard books on electromagnetism. (These formulas justify the remarks following (3.3).)

4. Constitutive Equations

Of all the variables that have appeared only σ is prescribed. The remaining variables are related by constitutive equations. As our independent constitutive variables we choose

$$\boldsymbol{\Gamma} \equiv (\boldsymbol{F}, \boldsymbol{g}, \lambda, \boldsymbol{e}, \boldsymbol{h}) \tag{4.1}$$

because they are physically reasonable and mathematically convenient. The range of $\boldsymbol{F}$ is Lin^+, the range of $\boldsymbol{g}$, $\boldsymbol{e}$, $\boldsymbol{h}$ is $\mathbb{E}^3$, and the range of λ is $\mathbb{R}$. We first suppose that $\boldsymbol{g}, \boldsymbol{d}, \boldsymbol{b}, \boldsymbol{j}$ depend on these variables and on $\boldsymbol{z}$. Thus $\boldsymbol{j} \cdot \boldsymbol{e}$, appearing in (3.6), likewise depends on (4.1). We finally prescribe $\boldsymbol{T}$ and $\boldsymbol{L}$ to depend on (4.1) and $\boldsymbol{z}$ so that they satisfy (3.12). Henceforth we shall have no need for (3.12). Thus we have constitutive functions $\hat{\boldsymbol{T}}, \hat{\boldsymbol{q}}, \hat{\boldsymbol{d}}, \hat{\boldsymbol{b}}, \hat{\boldsymbol{j}}, \hat{\boldsymbol{p}}$ such that

$$\boldsymbol{T}(\boldsymbol{z}) = \hat{\boldsymbol{T}}(\boldsymbol{\Gamma}(\boldsymbol{z}), \boldsymbol{z}), \text{ etc.} \tag{4.2}$$

The functions $\hat{\boldsymbol{T}}$, *etc.*, must be invariant under rigid motions, *i.e.*, be frame-indifferent. (*Cf.* TRUESDELL & NOLL (1965).) We do not pause to exhibit the specific representations of the constitutive functions that are necessary and sufficient for frame-indifference because we have no need for them in our analysis. (See HUTTER & VAN DE VEN (1978) for a treatment of frame-indifference for electromagnetic solids in classical space-time with nonzero velocity.)

For simplicity, we assume that our constitutive functions are continuously differentiable.

5. Potentials

It is convenient in our analysis to employ the potentials φ and ψ instead of $\boldsymbol{e}$ and $\boldsymbol{h}$ as the fundamental variables defining the electromagnetic state. $\boldsymbol{e}$ is expressed as the gradient of φ in (5.15). If the current $\boldsymbol{j} = \boldsymbol{0}$, then $\boldsymbol{h}$ is likewise expressed as the gradient of ψ by (5.16). We seek conditions ensuring that $\boldsymbol{h}$ can be expressed in terms of

$$(\boldsymbol{F}, \boldsymbol{g}, \lambda, \partial\varphi/\partial\boldsymbol{z}, \partial\psi/\partial\boldsymbol{z}) \equiv \boldsymbol{\Delta} \tag{5.1}$$

when the current is not zero. Note that each entry in $\boldsymbol{\Delta}$ except λ is a gradient. Let us substitute our constitutive equation for $\boldsymbol{j}$ into (3.14) to obtain

(5.2)

$$\boldsymbol{h}(\boldsymbol{z}) - \partial\psi(\boldsymbol{z})/\partial\boldsymbol{z} = \int_{\mathscr{B}} [\hat{\boldsymbol{j}}(\boldsymbol{h}(\boldsymbol{x}), \boldsymbol{\Sigma}(\boldsymbol{x}), \boldsymbol{x}) \wedge (\boldsymbol{z} - \boldsymbol{x})] \, |\boldsymbol{z} - \boldsymbol{x}|^{-3} \, dv(\boldsymbol{x}) \equiv \boldsymbol{k}(\boldsymbol{h}, \boldsymbol{\Sigma})(\boldsymbol{z})$$

where $\boldsymbol{\Sigma} \equiv (\boldsymbol{F}, \boldsymbol{g}, \lambda, \boldsymbol{e})$. Now in the classical form of Ohm's Law, $\hat{\boldsymbol{j}}$ depends only on the electric field. More generally, if $\hat{\boldsymbol{j}}$ is independent of $\boldsymbol{h}$, then (5.2) gives an explicit representation for $\boldsymbol{h}$ in terms of $\boldsymbol{\Delta}$. There are a variety of results available for the case in which $\hat{\boldsymbol{j}}$ depends on $\boldsymbol{h}$. Typical is the following:

5.3. Theorem. *Let $\alpha > 1$ and let $\mathscr{B}$ lie in the ball B_γ of radius γ and center $\boldsymbol{0}$. Let $\boldsymbol{\Delta}$ be fixed in $L_\alpha(\mathscr{B})$. Suppose that there are positive numbers μ, θ, ζ with $3\zeta < \alpha$ such that*

$$|\hat{\boldsymbol{j}}(\boldsymbol{\Gamma}, \boldsymbol{z})| \leqq \mu(1 + |\boldsymbol{\Gamma}|^{1+\zeta}), \tag{5.4}$$

$$|\partial\hat{\boldsymbol{j}}(\boldsymbol{\Gamma}, \boldsymbol{z})/\partial\boldsymbol{h}| \leqq \theta(1 + |\boldsymbol{\Gamma}|^{\zeta}). \tag{5.5}$$

If γ and θ are small enough, then (5.2) *has a unique solution of the form*

$$h(z) = \partial\psi(z)/\partial z + \hat{k}(\Delta)(z) \tag{5.6}$$

where $L_\alpha(\mathscr{B}) \ni \Delta \mapsto \hat{k}(\Delta)(\cdot) \in L_\alpha(\mathscr{B})$ *is continuous and compact.*

Proof. It suffices to take $\mathscr{B} = B_\gamma$. We use the following amalgamation of results of SOBOLEV and KANTOROVICH (*cf.* SOBOLEV (1950, § 6) and KANTOROVICH & AKILOV (1977, Chap. XI, § 3): *If* $f \in L_\beta(B_\gamma)$, *with* $\beta \geqq 1$, *then there is a continuous function* $\gamma \mapsto \varkappa(\beta, \gamma)$ *that strictly increases from* 0 *to* ∞ *as* γ *increases from* 0 to ∞ *such that*

$$\|Kf, L_\nu(B_\gamma)\| \leqq \varkappa(\beta, \gamma)\, \|f, L_\beta(B_\gamma)\| \tag{5.7a}$$

where

$$(Kf)(z) \equiv \int_{B_\gamma} \frac{f(x)}{|x - z|^2}\, dv(x), \tag{5.7b}$$

$\nu = \infty$ *if* $\beta > 3$, $\nu < 3\beta/(3 - \beta)$ *if* $\beta \leqq 3$. *Moreover, K is compact (and continuous) from* $L_\beta(B_\gamma)$ *to* $L_\nu(B_\gamma)$.

We wish to show that $h \mapsto k(h, \Sigma, \cdot)$ is a contraction from $L_\alpha(B_\gamma)$ to itself. Let $h \in L_\alpha(B_\gamma)$. We first identify f of (5.7) with (the components of) $\hat{j}(h, \Sigma)(\cdot)$ and chose $\beta = \alpha(1 + \zeta)^{-1}$. (Then $\beta \leqq 3$ if and only if $\alpha - 3\zeta \leqq 3$.) Then (5.4) ensures that $\hat{j} \in L_\beta(B_\gamma)$. Since $3\beta(3 - \beta)^{-1} > \alpha$ when $\beta \leqq 3$, we can take $\nu = \alpha$ in this case. Thus (5.7) implies that $h \mapsto k(h, \Sigma)(\cdot)$ maps $L_\alpha(B_\gamma)$ into itself.

We now show that $h \mapsto k(h, \Sigma, \cdot)$ is a contraction. Let $h_1, h_2 \in L_\alpha(B_\gamma)$, $\delta h \equiv h_1 - h_2$. Then

$$\begin{aligned} |\delta k(h_1, h_2, \Sigma)(z)| &\equiv |k(h_1, \Sigma)(z) - k(h_2, \Sigma)(z)| \\ &\leqq \int_{B_\gamma} \frac{\left|\dfrac{\partial \hat{j}}{\partial h}(t h_1(x) + (1 - t)\, h_2(x), \Sigma(x), x)\right| |\delta h(x)|}{|z - x|^2}\, dv(x) \end{aligned} \tag{5.8}$$

where $t \in [0, 1]$. We now identify $f(x)$ with the numerator of the integrand in the right-most term of (5.8). We henceforth suppress the arguments of the functions appearing in this numerator. Let us choose β and ν as above, noting that $\beta < \alpha$. From (5.7), (5.8), and the Hölder inequality we then obtain

$$\begin{aligned} \|\delta k, L_\alpha(B_\gamma)\| &\leqq \varkappa(\beta, \gamma) \left\{ \int_{B_\gamma} |\partial \hat{j}/\partial h|^\beta\, |\delta h|^\beta\, dv \right\}^{1/\beta} \\ &\leqq \varkappa(\beta, \gamma)\, \|\partial \hat{j}/\partial h, L_{\alpha\beta/(\alpha-\beta)}(B_\gamma)\|\, \|\delta h, L_\alpha(B_\gamma)\|. \end{aligned} \tag{5.9}$$

Since $\alpha\beta/(\alpha - \beta) = \alpha/\zeta$, condition (5.5) implies that $\partial \hat{j}/\partial h \in L_{\alpha\beta/(\alpha-\beta)}(B_\gamma)$, so that the rightmost term of (5.9) is well defined.

We now prove the compactness of $\boldsymbol{\Delta} \mapsto \hat{\boldsymbol{k}}(\boldsymbol{\Delta}, \cdot)$. The arguments showing that K of (5.76) is compact ensure also that

$$L_\beta(B_\gamma) \ni \boldsymbol{j} \mapsto \left[\boldsymbol{z} \mapsto \int_{B_\gamma} \frac{\boldsymbol{j}(\boldsymbol{x}) \wedge (\boldsymbol{z} - \boldsymbol{x})}{|\boldsymbol{z} - \boldsymbol{x}|^3} \, dv(\boldsymbol{x}) \right] \in L_\alpha(B_\gamma) \tag{5.10}$$

is compact. Condition (5.4) ensures that $(\boldsymbol{h}, \boldsymbol{\Sigma}) \mapsto \hat{\boldsymbol{j}}(\boldsymbol{h}, \boldsymbol{\Sigma}, \cdot)$ takes $L_\alpha(B_\gamma)$ to $L_\beta(B_\gamma)$. By the properties of Nemytskii operators (*cf.* KRASNOSEL'SKII (1956, Sec. I.2), this mapping is continuous. Since

$$L_\alpha(B_\gamma) \ni \boldsymbol{\Delta} \mapsto \int_{B_\gamma} \frac{\hat{\boldsymbol{j}}(\partial\psi(\boldsymbol{x})/\partial \boldsymbol{z} + \hat{\boldsymbol{k}}(\boldsymbol{\Delta}, \boldsymbol{x}), \boldsymbol{\Sigma}(\boldsymbol{x}), \boldsymbol{x}) \wedge (\boldsymbol{z} - \boldsymbol{x}) \, dv(\boldsymbol{x})}{|\boldsymbol{z} - \boldsymbol{x}|^3} \equiv \hat{\boldsymbol{k}}(\boldsymbol{\Delta}, \boldsymbol{z})$$

is the composition of a compact with continuous operators, it is compact. □

It follows from (5.5) and the properties of $\varkappa$ that we can make the coefficient of $\|\delta \boldsymbol{h}, L_\alpha\|$ in the right-most term of (5.9) less than unity by fixing θ and taking γ small enough or by fixing γ and taking θ small enough or by taking each small enough. The Contraction Mapping Principle ensures that (5.2) has a unique solution giving $\boldsymbol{h}$ as a continuous function of $\boldsymbol{\Delta}$. The composite function obtained by substituting this solution $\boldsymbol{h}$ into $\boldsymbol{k}(\boldsymbol{h}, \boldsymbol{\Sigma}, \cdot)$ has value denoted by $\hat{\boldsymbol{k}}(\boldsymbol{\Delta}, \boldsymbol{z})$.

A number of related results, including some for unbounded domains, can be based on the techniques presented by SOBOLEV (1950, §§ 6, 9), STEIN (1970, Ch. V), and KANTOROVICH & AKILOV (1977, Chap. XI). Note that Theorem 5.4 says that (5.6) is valid provided the dependence of $\hat{\boldsymbol{j}}$ on $\boldsymbol{h}$ becomes weaker as the body $\mathscr{B}$ becomes larger.

We now suppose that $\hat{\boldsymbol{j}}$ is such that $\boldsymbol{h}$ admits a representation of the form (5.6). We substitute (5.6) into the right side of (4.2) to get

$$\boldsymbol{T}(\boldsymbol{z}) = \hat{\boldsymbol{T}}(\boldsymbol{F}(\boldsymbol{z}), \boldsymbol{g}(\boldsymbol{z}), \lambda(\boldsymbol{z}), \partial\varphi(\boldsymbol{z})/\partial\boldsymbol{z}, \partial\psi(\boldsymbol{z})/\partial\boldsymbol{z} + \hat{\boldsymbol{k}}(\boldsymbol{\Delta}, \boldsymbol{z}), \boldsymbol{z}), \text{ etc.} \tag{5.11}$$

Our *governing equations* are obtained by substituting (5.11) into (3.5)–(3.8):

$$\operatorname{Div} \hat{\boldsymbol{T}} = \boldsymbol{0}, \tag{5.12}$$

$$\operatorname{Div} \hat{\boldsymbol{q}} + \hat{\boldsymbol{j}} \cdot (\partial\varphi/\partial\boldsymbol{z}) = 0, \tag{5.13}$$

$$\operatorname{Div} \hat{\boldsymbol{d}} = \sigma, \tag{5.14}$$

$$\operatorname{Div} \hat{\boldsymbol{b}} = 0, \tag{5.15}$$

where the arguments of the constitutive functions, decorated with carets, are indicated in (5.11). Equations (5.12)–(5.15), having six scalar components, form a quasilinear system of partial functional differential equations for the six unknown components of $\boldsymbol{y}, \lambda, \varphi, \psi$. All other variables we have introduced can be expressed in terms of these.

TOUPIN (1956) took the polarization $\tilde{\boldsymbol{p}}$ and magnetization $\tilde{\boldsymbol{m}}$ as independent constitutive variables. One of the goals of our paper is to exhibit the mathematical advantages of choosing $\boldsymbol{d}, \boldsymbol{b}, \boldsymbol{j}$ to be dependent constitutive variables and choosing $\boldsymbol{e}$ and $\boldsymbol{h}$ to be independent constitutive variables. In this regard we generalize formulations of PAO & HUTTER (1978), JORDAN & ERINGEN (1964), and ERSOY & KIRAL (1978) *inter alia*.

6. Ellipticity and Growth Conditions

Our basic constitutive assumptions are expressed in terms of the quadratic form

(6.1)

$$\begin{aligned}\omega(\boldsymbol{A}, \boldsymbol{t}, \boldsymbol{u}, \boldsymbol{v}) \equiv \\ &\boldsymbol{A} : (\partial \hat{\boldsymbol{T}}/\partial \boldsymbol{F}) : \boldsymbol{A} + \boldsymbol{A} : (\partial \hat{\boldsymbol{T}}/\partial \boldsymbol{g}) \cdot \boldsymbol{t} + \boldsymbol{A} : (\partial \hat{\boldsymbol{T}}/\partial \boldsymbol{e}) \cdot \boldsymbol{u} + \boldsymbol{A} : (\partial \hat{\boldsymbol{T}}/\partial \boldsymbol{h}) \cdot \boldsymbol{v} \\ &+ \boldsymbol{t} \cdot (\partial \hat{\boldsymbol{q}}/\partial \boldsymbol{F}) : \boldsymbol{A} + \boldsymbol{t} \cdot (\partial \hat{\boldsymbol{q}}/\partial \boldsymbol{g}) \cdot \boldsymbol{t} + \boldsymbol{t} \cdot (\partial \hat{\boldsymbol{q}}/\partial \boldsymbol{e}) \cdot \boldsymbol{u} + \boldsymbol{t} \cdot (\partial \hat{\boldsymbol{q}}/\partial \boldsymbol{h}) \cdot \boldsymbol{v} \\ &+ \boldsymbol{u} \cdot (\partial \hat{\boldsymbol{d}}/\partial \boldsymbol{F}) : \boldsymbol{A} + \boldsymbol{u} \cdot (\partial \hat{\boldsymbol{d}}/\partial \boldsymbol{g}) \cdot \boldsymbol{t} + \boldsymbol{u} \cdot (\partial \hat{\boldsymbol{d}}/\partial \boldsymbol{e}) \cdot \boldsymbol{u} + \boldsymbol{u} \cdot (\partial \hat{\boldsymbol{d}}/\partial \boldsymbol{h}) \cdot \boldsymbol{v} \\ &+ \boldsymbol{v} \cdot (\partial \hat{\boldsymbol{b}}/\partial \boldsymbol{F}) : \boldsymbol{A} + \boldsymbol{v} \cdot (\partial \hat{\boldsymbol{b}}/\partial \boldsymbol{g}) \cdot \boldsymbol{t} + \boldsymbol{v} \cdot (\partial \hat{\boldsymbol{b}}/\partial \boldsymbol{e}) \cdot \boldsymbol{u} + \boldsymbol{v} \cdot (\partial \hat{\boldsymbol{b}}/\partial \boldsymbol{h}) \cdot \boldsymbol{v}.\end{aligned}$$

If $\omega(\boldsymbol{A}, \boldsymbol{t}, \boldsymbol{u}, \boldsymbol{v}) > 0 \quad \forall (\boldsymbol{A}, \boldsymbol{t}, \boldsymbol{u}, \boldsymbol{v}) \neq (\boldsymbol{O}, \boldsymbol{0}, \boldsymbol{0}, \boldsymbol{0})$, then $(\hat{\boldsymbol{T}}, \hat{\boldsymbol{q}}, \hat{\boldsymbol{d}}, \hat{\boldsymbol{b}})$ is said to be *strictly monotone*. The use of this attractive mathematical restriction would deprive the theory of much of its physical versatility. Among its adverse consequences (discussed in detail by ANTMAN (1983)) is that the uniqueness theorems it implies effectively prevent the buckling of a column of such a material however slender under a compressive load however large.

We can eliminate this kind of uniqueness in the mechanical response while preserving it fully in electromagnetic response and partially in the thermal response by weakening the strict monotonicity condition. If $\omega(\boldsymbol{rs}, \boldsymbol{t}, \boldsymbol{u}, \boldsymbol{v}) > 0 \ \forall (\boldsymbol{r}, \boldsymbol{s}, \boldsymbol{t}, \boldsymbol{u}, \boldsymbol{v}) \neq (\boldsymbol{0}, \boldsymbol{0}, \boldsymbol{0}, \boldsymbol{0}, \boldsymbol{0})$, then we say that $(\hat{\boldsymbol{T}}, \hat{\boldsymbol{q}}, \hat{\boldsymbol{d}}, \hat{\boldsymbol{b}})$ satisfies the (strict form of the) *restricted strong ellipticity condition*. (Note that $\boldsymbol{A}$ equals the dyadic product $\boldsymbol{rs}$ if and only if $\boldsymbol{A}$ has rank 1.) In Section 8b we discuss the physical significance of the restricted strong ellipticity condition. If $\omega(\boldsymbol{rs}, \xi_1 \boldsymbol{s}, \xi_2 \boldsymbol{s}, \xi_3 \boldsymbol{s}) > 0$ $\forall (\boldsymbol{r}, \boldsymbol{s}, \xi_1, \xi_2, \xi_3] \neq (\boldsymbol{0}, \boldsymbol{0}, 0, 0, 0)$, then $(\hat{\boldsymbol{T}}, \hat{\boldsymbol{q}}, \hat{\boldsymbol{d}}, \hat{\boldsymbol{b}})$ satisfies the (strict form of the) *strong ellipticity condition*. This condition is the generalization to elliptic systems in divergence form of the Legendre-Hadamard condition of the calculus of variations.

In this paper we shall study the strong ellipticity condition and its restricted form. Our subject is insufficiently developed to determine whether phenomena permitted by the strong ellipticity condition, but prohibited by its restricted form, are observed (*cf.* Sec. 8b). The intuitive content of these conditions is described in Section 1. Most special theories of material behavior of electro-magneto-thermo-elasticity satisfy the restricted strong ellipticity condition because many of the "off-diagonal" terms in (6.1) are zero. (But recent work on the study of plastic effects and phase changes treats theories of elastic solids for which even $\boldsymbol{rs} : (\partial \hat{\boldsymbol{T}}/\partial \boldsymbol{F}) : \boldsymbol{rs}$ need not be positive for $\boldsymbol{rs} \neq \boldsymbol{O}$. *Cf.* ERICKSEN (1980).)

We now study the behavior of the constitutive equations at extreme values of their arguments. The conditions we impose must be consistent with the strong ellipticity condition. Since our work is just a generalization of that of Antman (1983), we omit an extensive commentary. In Sections 8 and 11 we describe more specific conditions appropriate for special problems.

Recall that

$$\boldsymbol{\Gamma} = (\boldsymbol{F}, \boldsymbol{g}, \lambda, \boldsymbol{e}, \boldsymbol{h}). \tag{6.2}$$

Let $\boldsymbol{a}$ and $\boldsymbol{c}$ be unit vector fields depending on $\boldsymbol{\Gamma}, \boldsymbol{y}, \boldsymbol{z}$. The strong ellipticity condition implies that

$$\frac{\partial(\boldsymbol{a} \cdot \hat{\boldsymbol{T}} \cdot \boldsymbol{c})}{\partial(\boldsymbol{a} \cdot \boldsymbol{F} \cdot \boldsymbol{c})} > 0 \quad \text{if } \boldsymbol{a} \text{ and } \boldsymbol{c} \text{ are independent of } \boldsymbol{a} \cdot \boldsymbol{F} \cdot \boldsymbol{c}, \tag{6.3a}$$

$$\frac{\partial(\hat{\boldsymbol{q}} \cdot \boldsymbol{a})}{\partial(\boldsymbol{g} \cdot \boldsymbol{a})} > 0 \quad \text{if } \boldsymbol{a} \text{ is independent of } \boldsymbol{g} \cdot \boldsymbol{a}, \tag{6.3b}$$

$$\frac{\partial(\hat{\boldsymbol{d}} \cdot \boldsymbol{a})}{\partial(\boldsymbol{e} \cdot \boldsymbol{a})} > 0 \quad \text{if } \boldsymbol{a} \text{ is independent of } \boldsymbol{e} \cdot \boldsymbol{a}, \tag{6.3c}$$

$$\frac{\partial(\hat{\boldsymbol{b}} \cdot \boldsymbol{a})}{\partial(\boldsymbol{h} \cdot \boldsymbol{a})} > 0 \quad \text{if } \boldsymbol{a} \text{ is independent of } \boldsymbol{h} \cdot \boldsymbol{a}. \tag{6.3d}$$

Moreover, if $\boldsymbol{a}$ and $\boldsymbol{c}$ are independent of $\boldsymbol{a} \cdot \boldsymbol{F} \cdot \boldsymbol{c}$, then

$$\mathscr{D}(\boldsymbol{ac}) \equiv \{\boldsymbol{a} \cdot \boldsymbol{F} \cdot \boldsymbol{c} \in R : \det \boldsymbol{F} > 0\} \tag{6.4}$$

is either an open half-line, or the whole line, or empty and can then be written as

$$\mathscr{D}(\boldsymbol{ac}) = (l^-(\boldsymbol{ac}), l^+(\boldsymbol{ac})). \tag{6.5}$$

We suppress the dependence of $\mathscr{D}$ and $l^\pm$ on $\boldsymbol{\Gamma}, \boldsymbol{y}, \boldsymbol{z}$. The facts motivate the following

6.6. Hypothesis. *Let $\boldsymbol{a}$ and $\boldsymbol{c}$ be unit vector fields depending on $\boldsymbol{\Gamma}, \boldsymbol{y}, \boldsymbol{z}$. If $\mathscr{D}(\boldsymbol{ac})$ is an open half-line or the whole line and if $\boldsymbol{a} \cdot \boldsymbol{F} \cdot \boldsymbol{c} \mapsto \boldsymbol{a} \cdot \hat{\boldsymbol{T}} \cdot \boldsymbol{c}$ is strictly increasing, then*

(6.7a)

$$\boldsymbol{a} \cdot \hat{\boldsymbol{T}} \cdot \boldsymbol{c} \to \pm\infty \text{ as } \boldsymbol{a} \cdot \boldsymbol{F} \cdot \boldsymbol{c} \to l^\pm(\boldsymbol{ac}) \text{ for fixed } \boldsymbol{\Gamma} - ((\boldsymbol{a} \cdot \boldsymbol{F} \cdot \boldsymbol{c})\, \boldsymbol{ac}, 0, 0, 0, 0), \boldsymbol{y}, \boldsymbol{z}.$$

If $\boldsymbol{g} \cdot \boldsymbol{a} \mapsto \hat{\boldsymbol{q}} \cdot \boldsymbol{a}$ is strictly increasing, then

$$\hat{\boldsymbol{q}} \cdot \boldsymbol{a} \to \pm\infty \text{ as } \boldsymbol{g} \cdot \boldsymbol{a} \to \pm\infty \text{ for fixed } \boldsymbol{\Gamma} - (\boldsymbol{O}, (\boldsymbol{g} \cdot \boldsymbol{a})\, \boldsymbol{a}, 0, \boldsymbol{0}, \boldsymbol{0}), \boldsymbol{y}, \boldsymbol{z}. \tag{6.7b}$$

If $\boldsymbol{e} \cdot \boldsymbol{a} \mapsto \hat{\boldsymbol{d}} \cdot \boldsymbol{a}$ is strictly increasing, then

$$\hat{\boldsymbol{d}} \cdot \boldsymbol{a} \to \pm\infty \text{ as } \boldsymbol{e} \cdot \boldsymbol{a} \to \pm\infty \text{ for fixed } \boldsymbol{\Gamma} - (\boldsymbol{O}, \boldsymbol{0}, 0, (\boldsymbol{e} \cdot \boldsymbol{a})\, \boldsymbol{a}, \boldsymbol{0}), \boldsymbol{y}, \boldsymbol{z}. \tag{6.7c}$$

If $\boldsymbol{h} \cdot \boldsymbol{a} \mapsto \hat{\boldsymbol{b}} \cdot \boldsymbol{a}$ *is strictly increasing, then*

(6.7d) $\hat{\boldsymbol{b}} \cdot \boldsymbol{a} \to \pm\infty$ *as* $\boldsymbol{h} \cdot \boldsymbol{a} \to \pm\infty$ *for fixed* $\boldsymbol{\Gamma} - (\boldsymbol{O}, \boldsymbol{0}, 0, \boldsymbol{0}, (\boldsymbol{h} \cdot \boldsymbol{a})\, \boldsymbol{a}), \boldsymbol{y}, \boldsymbol{z}$.

If $\mathscr{D}(\boldsymbol{ac})$ is a half-line, which happens exactly when the cofactor of $\boldsymbol{a} \cdot \boldsymbol{F} \cdot \boldsymbol{c}$ in $\det \boldsymbol{F}$ does not vanish, then $\partial\mathscr{D}(\boldsymbol{ac})$ is a point (either $l^+(\boldsymbol{ac})$ or $l^-(\boldsymbol{ac})$). Then (6.7a) implies that we can define a function

(6.8) $$\boldsymbol{ac} \mapsto \delta(\boldsymbol{ac}) \in \{-1, 1\}$$

such that

(6.9) $$\delta(\boldsymbol{ac})\, \boldsymbol{a} \cdot \hat{\boldsymbol{T}} \cdot \boldsymbol{c} \to -\infty \text{ as } \boldsymbol{a} \cdot \boldsymbol{F} \cdot \boldsymbol{c} \to \partial\mathscr{D}(\boldsymbol{ac})$$

for fixed $\boldsymbol{\Gamma} - ((\boldsymbol{a} \cdot \boldsymbol{F} \cdot \boldsymbol{c})\, \boldsymbol{ac}, \boldsymbol{0}, 0, \boldsymbol{0}, \boldsymbol{0}), \boldsymbol{y}, \boldsymbol{z}$. Note that $\det \boldsymbol{F} \to 0$ so the local volume ratio shrinks to 0 as $\boldsymbol{a} \cdot \boldsymbol{F} \cdot \boldsymbol{c} \to \partial\mathscr{D}(\boldsymbol{ac})$.

We now complement Hypothesis 6.6 in a way that promotes the analysis of Part II by describing the behavior of the constitutive function $\hat{\boldsymbol{T}}$ as more than one component of $\boldsymbol{\Gamma}$ are allowed to vary.

6.10. Hypothesis. *Let* $\boldsymbol{z}$ *be fixed. Let* $\{\boldsymbol{E}_\tau, \tau = 1, \ldots, 9\}$ *be a basis for* Lin *consisting of dyadic products of unit vectors, which may depend on* $\boldsymbol{\Gamma}, \boldsymbol{y}, \boldsymbol{z}$, *and let* $\{\boldsymbol{E}_\tau^*\}$ *be the basis dual to* $\{\boldsymbol{E}_\tau\}$. *Let* $\boldsymbol{F} : \boldsymbol{E}_\tau \mapsto \boldsymbol{T} : \boldsymbol{E}_\tau^*$ *be strictly increasing for each* τ. *Let* $\{\boldsymbol{\Gamma}_n\}$ *be a sequence of states such that the* $l^\pm(\boldsymbol{E}_\tau)$ *formed from* $\{\boldsymbol{\Gamma}_n\}$ *are actually independent of* n. *Let the set of integers* $\{1, \ldots, 9\}$ *be written as a disjoint union* $\mathscr{a} \cup \mathscr{b} \cup \mathscr{c} \cup \mathscr{d} \cup \mathscr{e} \cup \mathscr{f}$ *with*

(6.11a) $$\partial\mathscr{D}(\boldsymbol{E}_\tau) \neq \emptyset \text{ and } \delta(\boldsymbol{E}_\tau)\, \hat{\boldsymbol{T}}(\boldsymbol{\Gamma}_n, \boldsymbol{z}) : \boldsymbol{E}_\tau^* \to -\infty \text{ for } \tau \in \mathscr{a},$$

(6.11b) $$\partial\mathscr{D}(\boldsymbol{E}_\tau) \neq \emptyset \text{ and } \boldsymbol{F}_n : \boldsymbol{E}_\tau \to \partial\mathscr{D}(\boldsymbol{E}_\tau) \text{ for } \tau \in \mathscr{b},$$

(6.11c) $$\partial\mathscr{D}(\boldsymbol{E}_\tau) \neq \emptyset \text{ and } \boldsymbol{F}_n : \boldsymbol{E}_\tau \in \text{compact subset of } (l^-(\boldsymbol{E}_\tau), l^+(\boldsymbol{E}_\tau)) \text{ for } \tau \in \mathscr{c},$$

(6.11d) $$\partial\mathscr{D}(\boldsymbol{E}_\tau) \neq \emptyset \text{ for } \tau \in \mathscr{d},$$

(6.11e) $$\partial\mathscr{D}(\boldsymbol{E}_\tau) = \emptyset \text{ and } \boldsymbol{F}_n : \boldsymbol{E}_\tau \in \text{compact subset of } (l^-(\boldsymbol{E}_\tau), l^+(\boldsymbol{E}_\tau)) \text{ for } \tau \in \mathscr{e},$$

(6.11f) $$\partial\mathscr{D}(\boldsymbol{E}_\tau) = \emptyset \text{ for } \tau \in \mathscr{f}.$$

If $\mathscr{a} \neq \emptyset$, *then*

(6.12a) $$\delta(\boldsymbol{E}_\tau)\, \hat{T}(\boldsymbol{\Gamma}_n, z) : \boldsymbol{E}_\tau^* \to -\infty \;\; \forall\, \tau \in \mathscr{a} \cup \mathscr{b} \cup \mathscr{c}; \text{ for each } \tau \text{ in } \mathscr{d} \text{ either}$$
$$|\boldsymbol{F}_n : \boldsymbol{E}_\tau| \to \infty \text{ or } \delta(\boldsymbol{E}_\tau)\, \hat{\boldsymbol{T}}(\boldsymbol{\Gamma}_n, \boldsymbol{z}) : \boldsymbol{E}_\tau^* \to -\infty.$$

If $\mathscr{b} \neq \emptyset$, *then*

(6.12b) *either* (i) $\delta(\boldsymbol{E}_\tau)\, \hat{\boldsymbol{T}}(\boldsymbol{\Gamma}_n, \boldsymbol{z}) : \boldsymbol{E}_\tau^* \to -\infty \;\; \forall\, \tau \in \mathscr{a} \cup \mathscr{b} \cup \mathscr{c} \cup \mathscr{d}$ *or else*

(ii) $\exists\, \tau \in \mathscr{a} \cup \mathscr{d}$ *such that* $|\boldsymbol{F}_n : \boldsymbol{E}_\tau| \to \infty$ *and* $\exists\, \tau \in \mathscr{a} \cup \mathscr{b} \cup \mathscr{c} \cup \mathscr{d}$ *such that* $|\boldsymbol{T}(\boldsymbol{\Gamma}_n, \boldsymbol{z}) : \boldsymbol{E}_\tau^*| \to \infty$.

Moreover, the dualization obtained by respectively replacing the statements

$$\boldsymbol{F}_n : \boldsymbol{E}_\tau \to \partial\mathscr{D}(\boldsymbol{E}_\tau), \quad |\boldsymbol{F}_n : \boldsymbol{E}_\tau| \to \infty, \quad \delta\boldsymbol{T}(\boldsymbol{\Gamma}_n, \boldsymbol{z}) : \boldsymbol{E}_\tau^* \to -\infty$$

appearing in (6.11), (6.12) *by their opposites*

$$|\boldsymbol{F}_n : \boldsymbol{E}_\tau| \to \infty, \boldsymbol{F}_n : \boldsymbol{E}_\tau \to \partial\mathscr{D}(\boldsymbol{E}_\tau), \quad \delta\hat{\boldsymbol{T}}(\boldsymbol{\Gamma}_n, \boldsymbol{z}) : \boldsymbol{E}_\tau^* \to \infty$$

is also valid.

The statement containing (6.12b) may be loosely interpreted thus: If there are fibers compressed to zero length in some directions, then either the material is squeezed out with an infinite stretch in another direction or else it is prevented from doing so by infinite compressive stresses in all other directions. The other statements have similar interpretations. The whole hypothesis effectively says that extreme behavior in one direction must be accompanied by extreme behavior in some other direction. This observation is used in Part II to establish regularity results by showing that behavior could be extreme in only one direction and therefore cannot be extreme. It would be easy to generalize Hypothesis 6.7 to account for extreme couplings between the mechanical, electromagnetic, and thermal effects, but the intuitive evidence for such a generalization is not compelling.

It is important to note that the transformations (3.3) and (3.4) ensure that the Maxwell stress contains terms with $(\det \boldsymbol{F})^{-1}$ as factors. These terms could compete with the "purely mechanical part of this stress" when $\det \boldsymbol{F}$ is small. Our constitutive hypotheses on the effective stress control this competition. They say that the material response in large compression is dominated by that for purely mechanical response.

7. Boundary Conditions. The Principle of Virtual Work

At a boundary point $\boldsymbol{z} \in \partial\mathscr{B}$ we may prescribe the position $\boldsymbol{y}(\boldsymbol{z})$ or merely subject it to certain constraints, such as the requirement that it be confined to a fixed surface. To account for the varied possibilities it is convenient to describe such boundary conditions in the language of holonomic constraints. We accordingly specify

$$\boldsymbol{y}(\boldsymbol{z}) = \bar{\boldsymbol{y}}(\boldsymbol{z}, \mathbf{r}) \quad \text{for each } \boldsymbol{z} \in \partial\mathscr{B} \tag{7.1}$$

where $\bar{\boldsymbol{y}}$ is a given function continuously differentiable in $\mathbf{r} \in \mathbb{R}^3$, which represents the set of generalized coordinates for $\boldsymbol{y}(\boldsymbol{z})$. The rank of $\partial\bar{\boldsymbol{y}}/\partial\mathbf{r}$ is the number of degrees of freedom of $\boldsymbol{z}$. Equation (7.1) restricts $\boldsymbol{y}(\boldsymbol{z})$ to a manifold. The set of vectors $\boldsymbol{y}^\#(\boldsymbol{z})$ of the form $[\partial\bar{\boldsymbol{y}}(\boldsymbol{z}, \mathbf{r})/\partial\mathbf{r}] \cdot \mathbf{r}^\#(\boldsymbol{z})$ for $\mathbf{r}^\#(\boldsymbol{z}) \in \mathbb{R}^3$ form the tangent space to this manifold at $\boldsymbol{y}(\boldsymbol{z})$. The elements $\boldsymbol{y}^\#(\boldsymbol{z})$ of this tangent space are called *virtual displacements*. We complement (7.1) by specifying the projection of the traction $\boldsymbol{T} \cdot \boldsymbol{n}$ on this tangent space:

$$(\boldsymbol{n} \cdot \boldsymbol{T}^* - \bar{\boldsymbol{t}}) \cdot (\partial\bar{\boldsymbol{y}}/\partial\mathbf{r}) = 0 \quad \text{at each } \boldsymbol{z} \in \partial\mathscr{B} \tag{7.2}$$

where $\bar{\boldsymbol{t}}$ is a given function of $\boldsymbol{y}(\boldsymbol{z})$, $\lambda(\boldsymbol{z})$, $\varphi(\boldsymbol{z})$, $\psi(\boldsymbol{z})$, $\boldsymbol{z}$ and possibly other variables.

Thus

$$(n \cdot T^* - \bar{t}) \cdot y^\# = 0 \quad \text{on } \partial\mathscr{B}. \tag{7.3}$$

At each $z \in \partial\mathscr{B}$ we also prescribe

(7.4a, b) either $\lambda(z) = \bar{\lambda}(z)$ or $q(z) \cdot n(z) = \bar{\gamma}(y(z), \lambda(z), \varphi(z), \psi(z), z)$,

(7.5a, b) either $\varphi(z) = \bar{\varphi}(z)$ or $d(z) \cdot n(z) = \bar{\delta}(y(z), \lambda(z), \varphi(z), \psi(z), z)$,

(7.6a, b) either $\psi(z) = \bar{\psi}(z)$ or $b(z) \cdot n(z) = \bar{\beta}(y(z), \lambda(z), \varphi(z), \psi(z), z)$.

Let $\lambda^\#, \varphi^\#, \psi^\#$ be arbitrary continuous functions on $\partial\mathscr{B}$ that respectively vanish where (7.4a), (7.5a), (7.6a) hold. They are *virtual* fields. Then in analogy with (7.3) we have

$$(q \cdot n - \bar{\gamma})\,\lambda^\# + (d \cdot n - \bar{\delta})\,\varphi^\# + (b \cdot n - \bar{\beta})\,\psi^\# = 0 \quad \text{on } \partial\mathscr{B}. \tag{7.7}$$

Our fundamental equations of balance are the integral versions of (3.5)-(3.7), which are to hold over "almost all" nice subbodies of B. These equations can be supplemented by appropriately weakened forms of the boundary conditions we have just listed. ANTMAN & OSBORN (1979) show (strictly speaking, for the purely mechanical problem) that when all the integrals make sense as Lebesgue integrals, then these equations and boundary conditions are equivalent to the *Principle of Virtual Work*

$$\begin{aligned} &\int_{\mathscr{B}} [\hat{T} : (\partial y^\#/\partial z) + \hat{q} \cdot (\partial\lambda^\#/\partial z) + \hat{d} \cdot (\partial\varphi^\#/\partial z) + \hat{b} \cdot (\partial\psi^\#/\partial z)]\, dv \\ &\quad - \int_{\mathscr{B}} (f \cdot y^\# + \hat{\varrho}\lambda^\# - \sigma\varphi^\#)\, dv \\ &\quad = \int_{\partial\mathscr{B}} (\bar{t} \cdot y^\# + \bar{\gamma}\lambda^\# + \bar{\delta}\varphi^\# + \bar{\beta}\psi^\#)\, da \end{aligned} \tag{7.8}$$

for all reasonably nice fields $y^\#, \lambda^\#, \varphi^\#, \psi^\#$ having the boundary behavior specified above. Equation (7.8) is just the weak formulation of our boundary value problem consisting of (5.12)–(5.15) subject to (7.1), (7.2), (7.4)–(7.6). The arguments of $\hat{T}$, *etc.*, are given in (5.11).

In many circumstances the deformation of a body subjected to the action of external electromagnetic fields changes the ambient fields. Thus there would be a complete coupling between the fields interior and exterior to the body. Since our goal is to study the role of the constitutive assumptions of Section 6, we are avoiding such coupled problem by restricting electromagnetic boundary conditions to (8.5) and (8.6). Methods for treating fully coupled problems would be similar to those of Section 5. Moreover, our results would be very useful for the treatment of such problems.

8. General Existence Theorems

In this section we obtain existence theorems for two important special classes of problems, which can be readily treated by means of recent results for elliptic systems. For the first problem we assume that there is neither thermal nor electrical conduction and that there is a stored energy function; thus this reduced problem admits a variational formulation. For the second problem we assume that the material is rigid. The restricted strong ellipticity condition then reduces to a monotonicity condition, which is capable of handling our nonlocal operators.

a) Conservative Problems

We assume that the material does not conduct electricity so that the constitutive function $\hat{\boldsymbol{j}} = \boldsymbol{0}$. Thus the Joule heating is zero (*cf.* (5.13)). Moreover, (5.2) reduces to

$$\boldsymbol{h} = \partial\psi/\partial\boldsymbol{z}. \tag{8.1}$$

We assume that one of the following conditions holds:

i) $\hat{\boldsymbol{q}}, \hat{\varrho}, \bar{\gamma}$ depend only on $\boldsymbol{g}, \lambda, \boldsymbol{z}$ and the boundary value problem (5.13), (7.4) has a weak solution λ in a suitable Sobolev space. (In part (b) below, we show how a slight strengthening of our hypotheses ensures the existence of λ.)

ii) The boundary value problem (5.13), (7.4) has a solution λ independent of the fields $\boldsymbol{F}$, $\partial\varphi/\partial\boldsymbol{z}$, $\partial\psi/\partial\boldsymbol{z}$. This situation would occur if $\hat{\boldsymbol{q}} = \boldsymbol{0}$ when $\boldsymbol{g} = \boldsymbol{0}$ and if λ is prescribed to be constant λ_0 on $\partial\mathscr{B}$, for then the boundary value problem would admit a solution $\lambda = \lambda_0$ on all of $\mathscr{B}$.

iii) The constitutive functions $\hat{\boldsymbol{T}}, \hat{\boldsymbol{d}}, \hat{\boldsymbol{b}}$ are independent of $\boldsymbol{g}$ and λ.

In cases (i) and (ii) we can substitute the solution λ and its gradient $\boldsymbol{g}$ into (5.12), (5.14), (5.15), (7.1), (7.2), (7.5), (7.6). Since λ is known, its presence in these equations merely changes the dependence of the constitutive functions on $\boldsymbol{z}$. In case (iii), these equations are unaffected by the solution λ.

We assume that σ is a prescribed function of $\boldsymbol{z}$. We finally assume that there is a stored energy function W depending on $\boldsymbol{F}, \boldsymbol{g}, \lambda, \boldsymbol{e}, \boldsymbol{h}, \boldsymbol{z}$ with W continuously differentiable in $\boldsymbol{F}, \boldsymbol{e}, \boldsymbol{h}$, continuous in $\boldsymbol{g}$ and λ, and measurable in $\boldsymbol{z}$, for all values of the remaining arguments, such that

$$\hat{\boldsymbol{T}} = \partial W/\partial\boldsymbol{F}, \quad \hat{\boldsymbol{d}} = \partial W/\partial\boldsymbol{e}, \quad \hat{\boldsymbol{b}} = \partial W/\partial\boldsymbol{h}. \tag{8.2}$$

(The Clausius-Duhem inequality would deliver a specific thermodynamic function for W and show that it would be independent of $\boldsymbol{g}$.) The discussion following assumptions (i), (ii), (iii) motivates us to suppress the dependence of W on $\boldsymbol{g}$ and λ, their effects being absorbed by the dependence of W on $\boldsymbol{z}$.

We suppose that the body force $\boldsymbol{f}$ is conservative so that there is a function $(\boldsymbol{y}, \boldsymbol{z}) \mapsto U(\boldsymbol{y}, \boldsymbol{z})$, with $U(\cdot, \boldsymbol{z})$ continuously differentiable for all $\boldsymbol{z}$ in $\mathscr{B}$ and with $U(\boldsymbol{y}, \cdot)$ measurable for all $\boldsymbol{y} \in \mathrm{E}^3$, such that

$$\boldsymbol{f} = -\partial U/\partial\boldsymbol{y}. \tag{8.3}$$

We suppose that $\bar{\boldsymbol{\tau}}, \bar{\delta}, \bar{\beta}$ of (7.2), (7.5), (7.6) are conservative so that there is a function

$$\mathrm{E}^3 \times \mathrm{E}^3 \times \mathrm{E}^3 \times \partial\mathscr{B} \ni (\boldsymbol{y}, \boldsymbol{e}, \boldsymbol{h}, \boldsymbol{z}) \mapsto V(\boldsymbol{y}, \varphi, \psi, \boldsymbol{z})$$

with $V(\cdot, \cdot, \cdot, \boldsymbol{z})$ continuously differentiable for all $\boldsymbol{z}$ in $\partial\mathscr{B}$ and with $V(\boldsymbol{y}, \varphi, \psi, \cdot)$ measurable for all $\boldsymbol{y}, \boldsymbol{e}, \boldsymbol{h}$, such that

$$(8.4) \qquad \boldsymbol{t} = \bar{\partial} V/\partial \boldsymbol{y}, \quad \bar{\delta} = \partial V/\partial \varphi, \quad \bar{\beta} = \partial V/\partial \psi.$$

(The domain of $V(\boldsymbol{y}, \varphi, \psi, \cdot)$ may be taken to be the closure of $\partial\mathscr{B} \setminus \{\boldsymbol{z} \in \partial\mathscr{B} : \boldsymbol{y}^{\#}(\boldsymbol{z}) = \boldsymbol{0}, \varphi^{\#}(\boldsymbol{z}) = 0, \psi^{\#}(\boldsymbol{z}) = 0\}$. (See Section 7.) V could conceivably depend on g and λ. We suppress any such dependence in accord with the policy we have adopted above.

We finally assume that $\partial\mathscr{B}$ is bounded and has a locally Lipschitz continuous graph. Moreover, we require that the supports of $\varphi^{\#}$, $\psi^{\#}$, and the components of $\boldsymbol{y}^{\#}$ be nice enough to ensure that the boundary conditions (7.1), (7.5a), (7.6a) are assumed in the sense of trace when $\boldsymbol{y}, \varphi, \psi$ lie in Sobolev spaces of the form $W_p^1(\mathscr{B})$ with $p > 1$. (Necessary conditions for these properties are not known. See the discussion of ANTMAN & OSBORN (1979).)

Under these conditions the weak form of the Euler-Lagrange equations for the functional

$$I(\boldsymbol{y}, \varphi, \psi) = \int_{\mathscr{B}} \left[W\left(\frac{\partial \boldsymbol{y}}{\partial \boldsymbol{z}}(\boldsymbol{z}), \frac{\partial \varphi}{\partial \boldsymbol{z}}(z), \frac{\partial \psi}{\partial \boldsymbol{z}}(z), \boldsymbol{z} \right) + U(\boldsymbol{y}(\boldsymbol{z}), \boldsymbol{z}) + \sigma(\boldsymbol{z})\, \varphi(\boldsymbol{z}) \right] dv(\boldsymbol{z})$$

$$(8.5) \qquad + \int_{\partial\mathscr{B}} V(\boldsymbol{y}(\boldsymbol{z}), \varphi(\boldsymbol{z}), \psi(\boldsymbol{z}), \boldsymbol{z})\, da(\boldsymbol{z})$$

for $\boldsymbol{y}, \varphi, \psi$ satisfying (7.1), (7.5a), (7.6a) have exactly the form of (7.8) with $\lambda^{\#} = 0$. (Of course, many authors take a variational principle, such as this, as the starting point for the derivation of the governing equation for electromechanical interactions. See, *e.g.*, TOUPIN (1956), BROWN (1966), NELSON (1979).)

Let $\boldsymbol{F}^{\times}$ denote the cofactor tensor of $\boldsymbol{F}$. W is said to be *polyconvex* (*cf.* BALL (1977)) if it can be written in the form

$$W(\boldsymbol{F}, \boldsymbol{e}, \boldsymbol{h}, \boldsymbol{z}) = \Omega(\boldsymbol{F}, \boldsymbol{F}^{\times}, \det \boldsymbol{F}, \boldsymbol{e}, \boldsymbol{h}, \boldsymbol{z})$$

with $\Omega(\cdot, \cdot, \cdot, \cdot, \cdot, \boldsymbol{z})$ convex on $\mathrm{Lin} \times \mathrm{Lin} \times (0, \infty) \times \mathbb{E}^3 \times \mathbb{E}^3$ for each $\boldsymbol{z} \in \mathscr{B}$. The work of BALL (1977) shows that if W is polyconvex, then (8.2) satisfies the restricted strong ellipticity condition of Section 6. To account for (6.7) we require that

$$(8.6) \qquad \Omega(\boldsymbol{F}, \boldsymbol{F}^{\times}, \delta, \boldsymbol{e}, \boldsymbol{h}, \boldsymbol{z}) \to \infty \quad \text{as } \delta \to 0.$$

8.7. Theorem. *Let W be polyconvex and satisfy* (8.6). *Let there be numbers $k > 0$, $p \geqq 2$, $q > p/(p-1)$, $r > 1$ and functions $\omega \in L_1(\mathscr{B})$ and $\chi \in L_1(\partial\mathscr{B})$ such that*

$$(8.8) \qquad \Omega(\boldsymbol{F}, \boldsymbol{F}^{\times}, \delta, \boldsymbol{e}, \boldsymbol{h}, \boldsymbol{z}) \geqq \omega(\boldsymbol{z}) + k(|\boldsymbol{F}|^p + |\boldsymbol{F}^{\times}|^q + \delta^r + |\boldsymbol{e}|^p + |\boldsymbol{h}|^p)$$

for all $\boldsymbol{z} \in \mathscr{B}$,

$$(8.9) \qquad U(\boldsymbol{y}, \boldsymbol{z}) \geqq \omega(\boldsymbol{z}) \quad \textit{for all } \boldsymbol{z} \in \mathscr{B},$$

$$(8.10) \qquad \sigma \in L_{p/(p-1)}(\mathscr{B}),$$

$$(8.11) \qquad V(\boldsymbol{y}, \varphi, \psi, \boldsymbol{z}) \geqq \chi(\boldsymbol{z}) \quad \textit{for all } \boldsymbol{z} \in \partial\mathscr{B}.$$

Let $\partial\mathcal{B}$ have the properties specified above. Let

$$\mathcal{W} \equiv \{(\boldsymbol{y}, \varphi, \psi) \in W_p^1(\mathcal{B}) : \boldsymbol{F}^\times \in L_q(\mathcal{B}), \det \boldsymbol{F} \in L_r(\mathcal{B}), \det \boldsymbol{F} > 0 \text{ a.e. for } \boldsymbol{F} = \partial\boldsymbol{y}/\partial\boldsymbol{z};\ (7.1),\ (7.5\text{a}),\ (7.6\text{a}) \textit{ are satisfied in the sense of trace where they are prescribed on } \partial\mathcal{B}.\} \tag{8.12}$$

If there exists an element $(\boldsymbol{y}_1, \varphi_1, \psi_1) \in \mathcal{W}$ such that $I(\boldsymbol{y}_1, \varphi_1, \psi_1) < \infty$, then there exists an element $(\bar{y}, \bar{\varphi}, \bar{\psi})$ that minimizes I on $\mathcal{W}$.

The proof of this theorem is effected by making minor adjustments to that of BALL (1977) and is accordingly omitted. (Further developments of BALL'S theory, useful for our class of problems, are given by BALL & MURAT (1984), DACOROGNA (1982), and the references cited therein.)

b) *Rigid Conductors*

We now study the effects of the conduction of heat and electricity, but confine our attention to rigid bodies, for which $\boldsymbol{F}$ is constrained to be the identity $\boldsymbol{I}$. We accordingly take the virtual displacement $\boldsymbol{y}^\#$, appearing in (7.8), to be $\boldsymbol{0}$. This choice ensures that the first Piola-Kirchhoff stress tensor, which now is the Lagrange multiplier maintaining the constraint of rigidity, does not enter into (7.8). We drop $\boldsymbol{F}$ from the list of variables constituting $\boldsymbol{\Delta}$ in (5.1) and from the arguments of the constitutive functions $\hat{\boldsymbol{q}}, \hat{\boldsymbol{d}}, \hat{\boldsymbol{b}}, \hat{\boldsymbol{j}}$ (cf. (5.11)). Our boundary value problem reduces to (5.13)-(5.15), (7.4)-(7.6), whose weak form is the suitably specialized version of (7.8).

We assume that the restricted strong ellipticity condition holds. Thus $(\boldsymbol{g}, \boldsymbol{e}, \boldsymbol{h}) \mapsto (\hat{\boldsymbol{q}}(\boldsymbol{g}, \lambda, \boldsymbol{e}, \boldsymbol{h}, \boldsymbol{z}), \hat{\boldsymbol{d}}(\boldsymbol{g}, \lambda, \boldsymbol{e}, \boldsymbol{h}, \boldsymbol{z}), \hat{\boldsymbol{b}}(\boldsymbol{g}, \lambda, \boldsymbol{e}, \boldsymbol{h}, \boldsymbol{z}))$ is strictly monotone. This condition prohibits certain kinds of nonuniqueness.

Since hysteresis frequently is associated with nonuniqueness and since hysteresis is one of the most important phenomena of ferromagnetism, it might appear that our use of the restricted strong ellipticity condition precludes us from dealing with ferromagnetic materials. But molecular theories of ferromagnetism (*cf.* TEBBLE (1969)) suggest that hysteresis is associated with constitutive equations with nonlocal effects. If we accept such theories, then to account for ferromagnetism it is necessary to generalize the form of our constitutive functions before relaxing the ellipticity conditions. We do not attempt such a generalization here: Our analysis should be regarded as merely applying to paramagnetic materials. We do, however, examine nonlocal operators that are introduced by the mathematical approach we use to handle electric currents. Some of the methods we use can be applied to more general kinds of nonlocal behavior.

We now outline an existence theory that can be applied directly to our specialized version of (7.8). We first present the theory in an abstract form in order to facilitate a comparison of it with presentations in the mathematical literature. Afterward we make the requisite identifications.

Let $\mathscr{B}$, as before, be the closure of a domain in $\mathbb{R}^3$. We assume that $\partial\mathscr{B}$ has a locally Lipschitz continuous graph. A typical point in $\mathscr{B}$ is denoted z. Let $\mathbf{u}(z) = (u_1(z), \ldots, u_m(z))$. For $p \in (1, \infty)$, let the operator

$$L_p(\mathscr{B})^m \times L_p(\mathscr{B})^{3m} \ni (\mathbf{u}, \partial\mathbf{v}/\partial z) \mapsto \hat{\mathbf{k}}(\mathbf{u}, \partial\mathbf{v}/\partial z)(\cdot) \in [L_p(\mathscr{B})]^r \tag{8.13}$$

take bounded sets into bounded sets. Let

$$\mathbb{R}^m \times \mathbb{R}^{3m} \times \mathbb{R}^r \times \mathscr{B} \ni (\xi, \boldsymbol{\eta}, \zeta, z) \mapsto \left\{\begin{matrix} \boldsymbol{a}^i(\xi, \boldsymbol{\eta}, \zeta, z) \in \mathbb{R}^3 \\ \beta^i(\xi, \boldsymbol{\eta}, \zeta, z) \in \mathbb{R} \end{matrix}\right\}, \quad i = 1, \ldots, m, \tag{8.14}$$

$$\mathbb{R}^m \times \partial\mathscr{B} \ni (\xi, z) \mapsto \gamma^i(\xi, z) \in \mathbb{R}, \quad i = 1, \ldots, m \tag{8.15}$$

satisfy

(8.16a) For almost all z in $\mathscr{B}$, the functions $\boldsymbol{a}^i(\cdot, \cdot, \cdot, z)$, $\beta^i(\cdot, \cdot, \cdot, z)$ are continuous and for all $\xi, \boldsymbol{\eta}, \zeta$, the functions $\boldsymbol{a}^i(\xi, \boldsymbol{\eta}, \zeta, \cdot)$, $\beta^i(\xi, \boldsymbol{\eta}, \zeta, \cdot)$ are measurable. (These are the Carathéodory conditions.)

(8.16b) For almost all $z \in \partial\mathscr{B}$, the functions $\gamma^i(\cdot, z)$ are continuous and for all ξ, the functions $\gamma^i(\xi, \cdot)$ are measurable on $\partial\mathscr{B}$ (with respect to two-dimensional Lebesgue measure).

(8.16c) There exist a constant $c_1 > 0$ and a function $k_1 \in L_{p^*}(\mathscr{B})$ (with $p^* = p/(p-1)$) such that

$$|\boldsymbol{a}^i(\xi, \boldsymbol{\eta}, \zeta, z)|, |\beta^i(\xi, \boldsymbol{\eta}, \zeta, z)| \leq c_1[|\xi|^{p-1} + |\eta|^{p-1} + |\zeta|^{p-1} + k_1(z)]$$

for $i = 1, \ldots, m$; $\alpha = 1, 2, 3$.

The Hölder inequality then implies that the functions

$$\boldsymbol{a}^i\left(\mathbf{u}(\cdot), \frac{\partial\mathbf{v}}{\partial z}(\cdot), \hat{k}\left(\mathbf{u}, \frac{\partial\mathbf{u}}{\partial z}\right)(\cdot), \cdot\right), \quad \beta^i\left(\mathbf{u}, \frac{\partial\mathbf{v}}{\partial z}(\cdot), \hat{k}\left(\mathbf{u}, \frac{\partial\mathbf{u}}{\partial z}\right)(\cdot), \cdot\right)$$

are in $L_{p^*}(\mathscr{B})$ for all $\mathbf{u}, \mathbf{v} \in (W_p^1)^m$. It follows that the functional

$$\begin{aligned} a(\mathbf{u}, \mathbf{w}) \equiv \int_{\mathscr{B}} \Bigg[\sum_i \boldsymbol{a}^i\left(\mathbf{u}(z), \frac{\partial\mathbf{u}}{\partial z}(z), \hat{k}\left(\mathbf{u}, \frac{\partial\mathbf{u}}{\partial z}\right)(z), z\right) \cdot \frac{\partial w_i}{\partial z}(z) \\ + \sum_i \beta^i\left(\mathbf{u}(z), \frac{\partial\mathbf{u}}{\partial z}(z), \hat{k}\left(\mathbf{u}, \frac{\partial\mathbf{u}}{\partial z}\right)(z), z\right) w_i(z) \Bigg] dv(z) \\ + \int_{\partial\mathscr{B}} \sum_i \gamma^i(\mathbf{u}(z))\, w_i(z)\, da(z) \end{aligned} \tag{8.17}$$

is well defined for all $\mathbf{u}, \mathbf{w} \in W_p^1(\mathscr{B})$.

We shall prescribe $u_1, \ldots, u_m$ respectively on subsets $\mathscr{S}_1, \ldots, \mathscr{S}_m$ of $\partial\mathscr{B}$. We assume that these subsets are measurable. Let V be the closed subspace of $[W_p^1(\mathscr{B})]^m$ containing $[\mathring{W}_p^1(\mathscr{B})]^m$ that consists of functions $(w_1, \ldots, w_m)$ for which $w_1 = 0$ on $\mathscr{S}_1, \ldots, w_m = 0$ on $\mathscr{S}_m$ in the sense of trace. Let $\bar{\mathbf{u}}$ be a given element of $[W_p^1(\mathscr{B})]^m$. We require that u_1 agree with $\bar{u}_1$ on $\mathscr{S}_1$, *etc.*, in the sense of trace by seeking solutions $\mathbf{u}$ of our equations in $[W_p^1(\mathscr{B})]^m$ for which $\mathbf{u} - \bar{\mathbf{u}} \in V$. (This prescription of boundary conditions enables us to avoid the very delicate questions of whether functions defined on $\mathcal{S}_1, \ldots, \mathcal{S}_m$ can be extended to functions in $W_p^1(\mathcal{B})$.)

Since $V \ni \mathbf{w} \mapsto a(\mathbf{u}, \mathbf{w})$ is a bounded linear functional for each $\mathbf{u}$ in $W_p^1(\mathcal{B})$, we set

$$a(\mathbf{u}, \mathbf{w}) = \langle \mathbf{A}(\mathbf{u}), \mathbf{w} \rangle, \tag{8.18}$$

where $\mathbf{w} \mapsto \langle \mathbf{v}, \mathbf{w} \rangle$ is an element of V^*. If the $\boldsymbol{a}^i$ are continuously differentiable if $\mathbf{u}$ is twice continuously differentiable on $\mathcal{B}$, and if $\mathbf{w}$ vanishes on $\partial\mathcal{B}$, then

$$\langle A(\mathbf{u}), \mathbf{w} \rangle = \int_{\mathcal{B}} \sum_i A^i(\mathbf{u})\, w_i dv, \tag{8.19}$$

$$A^i(\mathbf{u}) \equiv -\operatorname{Div} \boldsymbol{a}^i\left(\mathbf{u}, \frac{\partial \mathbf{u}}{\partial \boldsymbol{z}}, \hat{\mathbf{k}}\left(\mathbf{u}, \frac{\partial \mathbf{u}}{\partial \boldsymbol{z}}, \boldsymbol{z}\right)(\boldsymbol{z}), \boldsymbol{z}\right) + \beta^i\left(\mathbf{u}, \frac{\partial \mathbf{u}}{\partial \boldsymbol{z}}, \mathbf{k}\left(\mathbf{u}, \frac{\partial \mathbf{u}}{\partial \boldsymbol{z}}, \boldsymbol{z}\right)(\boldsymbol{z}), \boldsymbol{z}\right).$$

Let us set $\boldsymbol{\eta} = (\boldsymbol{\eta}_1, \ldots, \boldsymbol{\eta}_m)$. Our basic abstract result is the following:

8.20. Theorem. *Let $\partial\mathcal{B}$ have a locally Lipschitz continuous graph. Let $p \in (1, \infty)$. Let* (8.16) *hold. Suppose that*

$$\frac{a(\mathbf{v}, \mathbf{v})}{\|\mathbf{v}, V\|} \to \infty \quad \text{as } \|\mathbf{v}, V\| \to \infty \quad \text{for} \quad \mathbf{v} \in V, \tag{8.21}$$

(8.22)

$$\sum_i [\boldsymbol{a}^i(\boldsymbol{\xi}, \boldsymbol{\eta} + \boldsymbol{\varrho}, \boldsymbol{\zeta}, \boldsymbol{z}) - \boldsymbol{a}^i(\boldsymbol{\xi}, \boldsymbol{\eta}, \boldsymbol{\zeta}, \boldsymbol{z})] \cdot \boldsymbol{\varrho}_i > 0 \quad \forall \boldsymbol{\varrho} \equiv (\boldsymbol{\varrho}_1, \ldots, \boldsymbol{\varrho}_m) \neq \mathbf{0}.$$

If $\mathcal{B}$ is bounded, let

$$\sum_i \boldsymbol{a}^i(\boldsymbol{\xi}, \boldsymbol{\eta}, \boldsymbol{\zeta}, \boldsymbol{z}) \cdot \boldsymbol{\eta}_i\, [|\boldsymbol{\eta}| + |\boldsymbol{\eta}|^{p-1}]^{-1} \to \infty \quad \text{as } |\boldsymbol{\eta}| \to \infty \tag{8.23a}$$

for almost all $\boldsymbol{z}$ in $\mathcal{B}$ and for bounded $\boldsymbol{\xi}, \boldsymbol{\eta}$. If $\mathcal{B}$ is unbounded, let the following stronger restriction hold: There is a number $c_2 > 0$ and a function $k_2 \in L_1(\mathcal{B})$ such that

$$\sum_i \boldsymbol{a}^i(\boldsymbol{\xi}, \boldsymbol{\eta}, \boldsymbol{\zeta}, \boldsymbol{z}) \cdot \boldsymbol{\eta}_i \geqq c_2 |\boldsymbol{\eta}|^p - k_2(\boldsymbol{z}). \tag{8.24b}$$

Define $\tilde{\boldsymbol{k}}$ by

$$[W_p^1(B)]^m \ni \boldsymbol{u} \mapsto \tilde{\mathbf{k}}(\mathbf{u})(\cdot) \equiv \hat{\boldsymbol{k}}(\mathbf{u}, \partial\mathbf{u}/\partial\boldsymbol{z})(\cdot) \tag{8.25}$$

where $\hat{\boldsymbol{k}}$ is defined in (8.13). *Let $\chi_{\mathcal{C}}$ be the characteristic function of a set $\mathcal{C}$ in $\mathcal{B}$. For every subdomain $\mathcal{C}$ of $\mathcal{B}$ with compact closure in $\mathcal{B}$ let*

$$[W_p^1(B)]^m \ni \mathbf{u} \mapsto \chi_{\mathcal{C}}(\cdot)\, \tilde{\boldsymbol{k}}(\mathbf{u})(\cdot) \in [L_p(C)]^r \tag{8.26}$$

be compact. Then for every $\mathbf{f} \in V^*$ and for every $\bar{\mathbf{u}} \in [W_p^1(\mathcal{B})]^m$ there exists a $\mathbf{u} \in [W_p^1(\mathcal{B})]^m$ with $\mathbf{u} - \bar{\mathbf{u}} \in V$ such that

$$\langle \mathbf{A}(\mathbf{u}), \mathbf{v} \rangle = \langle \mathbf{f}, \mathbf{v} \rangle \quad \forall \mathbf{v} \in V. \tag{8.27}$$

The proof of this theorem is obtained by making minor adjustments to those of BREZIS (1968) (*cf.* LIONS (1969, p. 297)) and BROWDER (1977). We note the following points: In a bounded domain the operator $\mathbf{A}$ is of the "calculus of variations type" because of its monotonicity in the local values of its highest-order derivatives and because of its compactness (through $\boldsymbol{k}$) in the global values of the highest-order derivatives. Since our integral operator (7.6) for constant electric currents is not compact on unbounded domains, we have to use the theory of BROWDER (1977) based upon the compactness of (8.26) to support our intended applications.

We identify the variables appearing in Theorem 8.20 with those used in the problem outlined at the beginning of this subsection. In particular, we set

$$(8.28) \qquad \mathbf{u} = (\lambda, \varphi, \psi), \quad \tilde{\mathbf{k}}(\mathbf{u})(\cdot) = \hat{\boldsymbol{k}}\left(\frac{\partial\lambda}{\partial\boldsymbol{z}}, \lambda, \frac{\partial\varphi}{\partial\boldsymbol{z}}, \frac{\partial\psi}{\partial\boldsymbol{z}}\right)(\cdot),$$

where $\hat{\boldsymbol{k}}$ is defined in (5.6). We identify the variables appearing in (8.17) with those of (7.8):

(8.29)

$$\boldsymbol{a}^1\left(\mathbf{u}(z), \frac{\partial\mathbf{u}}{\partial\boldsymbol{z}}(\boldsymbol{z}), \tilde{\mathbf{k}}(\mathbf{u})(\boldsymbol{z}), \boldsymbol{z}\right) = \hat{\boldsymbol{q}}\left(\frac{\partial\lambda}{\partial\boldsymbol{z}}(\boldsymbol{z}), \lambda(\boldsymbol{z}), \frac{\partial\varphi}{\partial\boldsymbol{z}}(\boldsymbol{z}), \frac{\partial\psi}{\partial\boldsymbol{z}}(\boldsymbol{z}) + \hat{\boldsymbol{k}}(\boldsymbol{\Delta})(\boldsymbol{z}), \boldsymbol{z}\right),$$

$$\boldsymbol{a}^2\left(\mathbf{u}(z), \frac{\partial\mathbf{u}}{\partial\boldsymbol{z}}(\boldsymbol{z}), \tilde{\mathbf{k}}(\mathbf{u})(\boldsymbol{z}), \boldsymbol{z}\right) = \hat{\boldsymbol{d}}\left(\frac{\partial\lambda}{\partial\boldsymbol{z}}(\boldsymbol{z}), \lambda(\boldsymbol{z}), \frac{\partial\varphi}{\partial\boldsymbol{z}}(\boldsymbol{z}), \frac{\partial\psi}{\partial\boldsymbol{z}}(\boldsymbol{z}) + \hat{\boldsymbol{k}}(\boldsymbol{\Delta})(\boldsymbol{z}), \boldsymbol{z}\right),$$

$$\boldsymbol{a}^3\left(\mathbf{u}(z), \frac{\partial\mathbf{u}}{\partial\boldsymbol{z}}(\boldsymbol{z}), \tilde{\mathbf{k}}(\mathbf{u})(\boldsymbol{z}), \boldsymbol{z}\right) = \hat{\boldsymbol{b}}\left(\frac{\partial\lambda}{\partial\boldsymbol{z}}(\boldsymbol{z}), \lambda(\boldsymbol{z}), \frac{\partial\varphi}{\partial\boldsymbol{z}}(\boldsymbol{z}), \frac{\partial\psi}{\partial\boldsymbol{z}}(\boldsymbol{z}) + \hat{\boldsymbol{k}}(\boldsymbol{\Delta})(\boldsymbol{z}), \boldsymbol{z}\right),$$

$$\beta^1\left(\mathbf{u}(z), \frac{\partial\mathbf{u}}{\partial\boldsymbol{z}}(\boldsymbol{z}), \tilde{\mathbf{k}}(\mathbf{u})(\boldsymbol{z}), \boldsymbol{z}\right) = -\hat{\varrho}\left(\frac{\partial\lambda}{\partial\boldsymbol{z}}(\boldsymbol{z}), \lambda(\boldsymbol{z}), \frac{\partial\varphi}{\partial\boldsymbol{z}}(\boldsymbol{z}), \frac{\partial\psi}{\partial\boldsymbol{z}}(\boldsymbol{z}) + \hat{\boldsymbol{k}}(\boldsymbol{\Delta})(\boldsymbol{z}), \boldsymbol{z}\right),$$

$$\beta^2\left(\mathbf{u}(z), \frac{\partial\mathbf{u}}{\partial\boldsymbol{z}}(\boldsymbol{z}), \tilde{\mathbf{k}}(\mathbf{u})(\boldsymbol{z}), \boldsymbol{z}\right) = \sigma(\boldsymbol{z}),$$

$$\beta^3\left(\mathbf{u}(z), \frac{\partial\mathbf{u}}{\partial\boldsymbol{z}}(\boldsymbol{z}), \tilde{\mathbf{k}}(\mathbf{u})(\boldsymbol{z}), \boldsymbol{z}\right) = 0,$$

$$\gamma^1(\mathbf{u}, \boldsymbol{z}) = -\bar{\gamma}(\lambda(\boldsymbol{z}), \varphi(\boldsymbol{z}), \psi(\boldsymbol{z}), \boldsymbol{z}),$$

$$\gamma^2(\mathbf{u}, \boldsymbol{z}) = -\bar{\delta}(\lambda(\boldsymbol{z}), \varphi(\boldsymbol{z}), \psi(\boldsymbol{z}), \boldsymbol{z}),$$

$$\gamma^3(\mathbf{u}, \boldsymbol{z}) = -\bar{\beta}(\lambda(\boldsymbol{z}), \varphi(\boldsymbol{z}), \psi(\boldsymbol{z}), \boldsymbol{z}).$$

We identify $\mathbf{w}$ with $(\lambda^\#, \varphi^\#, \psi^\#)$. Note that hypothesis (8.22) is ensured by the restricted strong ellipticity condition. We then have

8.30. Theorem. *Let $\hat{q}, \hat{d}, \hat{b}, \hat{\varrho}, \bar{\gamma}, \bar{\delta}, \bar{\beta}$ satisfy the hypotheses of Theorem* 8.20 *with the identifications* (8.28) *and* (8.29). *Then* (7.8) *with $\boldsymbol{y}^\# = \mathbf{0}$ is satisfied for all $(\lambda^\#, \varphi^\#, x^\#)$ in V.*

The question of regularity of solutions for the types of systems described in this section remains open. GIAQUINTA (1983) gives partial regularity results for more restricted systems. However, it is by no means clear how much regularity is physically reasonable for either of the more general types of problems presented here. BALL (1982) suggests that discontinuous solutions of problems such as those treated in (8.7) can be used to model rupture of solid bodies. In addition, we suggest above that operators such as (8.13) can be used in constitutive equations to model the nonlocal behavior of ferromagnetic materials, and the physical evidence of so-called "domain structures" (*cf.* TEBBLE (1969)) suggests that highly discontinuous magnetic fields are to be expected from a good model of such materials.

Part II. The Semi-Inverse Problem

9. Formulation of the Semi-Inverse Problem

Let $\{\boldsymbol{i}_1, \boldsymbol{i}_2, \boldsymbol{i}_3\}$ be a fixed right-handed orthonormal basis for $\mathbb{E}^3$ and let $\mathbf{x} = (s, \theta, z)$ be the set of cylindrical polar coordinates for $\mathbb{E}^3$ defined by

$$\boldsymbol{z} = \tilde{\boldsymbol{z}}(\mathbf{x}) \equiv s\boldsymbol{k}_1(\theta) + z\boldsymbol{k}_3(\theta) \tag{9.1}$$

where

$$\boldsymbol{k}_1(\theta) = \cos\theta\boldsymbol{i}_1 + \sin\theta\boldsymbol{i}_2, \quad \boldsymbol{k}_2(\theta) = -\sin\theta\boldsymbol{i}_1 + \cos\theta\boldsymbol{i}_2, \quad \boldsymbol{k}_3(\theta) = \boldsymbol{i}_3. \tag{9.2}$$

Let $\tilde{\mathbf{x}}$ denote the usual inverse of $\tilde{\boldsymbol{z}}$ so that (9.1) is equivalent to $\mathbf{x} = \tilde{\mathbf{x}}(\boldsymbol{z})$. Each triple $\mathbf{x}$ also identifies a material point. We set

$$\tilde{\boldsymbol{y}}(\mathbf{x}) \equiv \boldsymbol{y}(\tilde{\boldsymbol{z}}(\mathbf{x})), \text{ etc.} \tag{9.3}$$

We consider semi-inverse problems in which $\tilde{\boldsymbol{y}}, \tilde{\lambda}, \tilde{\varphi}, \tilde{\boldsymbol{h}}$ have the form

$$\tilde{\boldsymbol{y}}(\mathbf{x}) = w_1(s)\,\boldsymbol{e}_1(\mathbf{x}) + [w_3(s) + \alpha_{32}\theta + \alpha_{33}z]\,\boldsymbol{e}_3(\mathbf{x}) \tag{9.4a}$$

with

$$\boldsymbol{e}_1(\mathbf{x}) = \cos\omega(\mathbf{x})\,\boldsymbol{i}_1 + \sin\omega(\mathbf{x})\,\boldsymbol{i}_2, \tag{9.4b}$$

$$\boldsymbol{e}_2(\mathbf{x}) = -\sin\omega(\mathbf{x})\,\boldsymbol{i}_1 + \cos\omega(\mathbf{x})\,\boldsymbol{i}_2, \quad \boldsymbol{e}_3(\mathbf{x}) = \boldsymbol{i}_3,$$

$$\omega(\mathbf{x}) = w_2(x) + \alpha_{22}\theta + \alpha_{23}z, \tag{9.4c}$$

$$\tilde{\lambda}(\mathbf{x}) = w_4(s), \tag{9.4d}$$

$$\tilde{\varphi}(\mathbf{x}) = w_5(s) + \alpha_{52}\theta + \alpha_{53}z, \tag{9.4e}$$

$$\tilde{\boldsymbol{h}}(\mathbf{x}) = h_i(s)\,\boldsymbol{k}_i(\theta). \tag{9.4f}$$

(Here i is summed from 1 to 3.) We shall make constitutive assumptions on $\hat{\boldsymbol{j}}$ to ensure that $\tilde{\psi}(\mathbf{x})$ (*cf.* Sec. 5) has the form

$$\tilde{\psi}(\mathbf{x}) = w_6(s) + \alpha_{62}\theta + \alpha_{63}z. \tag{9.4g}$$

We take the body to be

(9.5) $$\mathscr{B} = \tilde{z}([a, 1]\times[-\Theta, \Theta]\times[-Z, Z])$$

with $0 < a < 1$, $0 < \Theta \leqq \pi$, $Z > 0$. Then $\mathscr{B}$ is a cylindrical tube (possibly slit) if $\Theta = \pi$ and is a sector thereof if $\Theta < \pi$. For simplicity we do not treat the interesting and technically complicated case that $a = 0$; the methods for doing so are virtually identical to those used by ANTMAN (1983). The deformations defined by (9.4a-c) constitute "family 2" of TRUESDELL & NOLL (1965, Sec. 59). The other functions of (9.4) are so specified as to ensure that our final problem consists of ordinary functional differential equations.

The chain rule implies that

(9.6a)
$$\begin{aligned} \boldsymbol{F}(\tilde{\boldsymbol{z}}(\mathbf{x})) &= \frac{\partial \boldsymbol{y}}{\partial \boldsymbol{z}}(\tilde{\boldsymbol{z}}(\mathbf{x})) = \frac{\partial \tilde{\boldsymbol{y}}}{\partial \mathbf{x}}(\mathbf{x}) \cdot \frac{\partial \tilde{\mathbf{x}}}{\partial \boldsymbol{z}}(\tilde{\boldsymbol{z}}(\mathbf{x})) \\ &= [w_1'(s)\, \boldsymbol{e}_1(\mathbf{x}) + w_1(s)\, w_2'(s)\, \boldsymbol{e}_2(\mathbf{x}) + w_3'(s)\, \boldsymbol{e}_3]\, \boldsymbol{k}_1(\theta) \\ &\quad + s^{-1}[\alpha_{22} w_1(s)\, \boldsymbol{e}_2(\mathbf{x}) + \alpha_{32}\boldsymbol{e}_3]\, \boldsymbol{k}_2(\theta) + [\alpha_{23} w_1(s)\, \boldsymbol{e}_2(\mathbf{x}) + \alpha_{33}\boldsymbol{e}_3]\, \boldsymbol{k}_3, \end{aligned}$$

(9.6b) $$\boldsymbol{g}(\tilde{\boldsymbol{z}}(\mathbf{x})) = \frac{\partial \lambda}{\partial \boldsymbol{z}}(\tilde{\boldsymbol{z}}(\mathbf{x})) = w_4'(s)\, \boldsymbol{k}_1(\theta),$$

(9.6c) $$\boldsymbol{e}(\tilde{\boldsymbol{z}}(\mathbf{x})) = \frac{\partial \varphi}{\partial \boldsymbol{z}}(\tilde{\boldsymbol{z}}(\mathbf{x})) = w_5'(s)\, \boldsymbol{k}_1(\theta) + \alpha_{52} s^{-1}\, \boldsymbol{k}_2(\theta) + \alpha_{53}\boldsymbol{k}_3,$$

(9.6d) $$\frac{\partial \psi}{\partial \boldsymbol{z}}(\tilde{\boldsymbol{z}}(\mathbf{x})) = w_6'(s)\, \boldsymbol{k}_1(\theta) + \alpha_{62} s^{-1}\, \boldsymbol{k}_2(\theta) + \alpha_{63}\boldsymbol{k}_3.$$

The representation (9.6a) reduces (3.2) to the requirement that

(9.7a) $$(\alpha_{22}\alpha_{33} - \alpha_{23}\alpha_{32})\,(w_1/s)\, w_1' > 0 \text{ a.e.}$$

Since w_1 is a radial distance, we require that

(9.7b) $$w_1(s) > 0 \quad \text{for } s \in [a, 1],$$

whence (9.7a) reduces to

(9.7c) $$(\alpha_{22}\alpha_{33} - \alpha_{23}\alpha_{32})\, w_1' > 0 \text{ a.e.}$$

For simplicity we require that

(9.7d) $$w_1' > 0 \text{ a.e.};$$

the opposite case corresponds to an eversion (*cf.* ANTMAN (1979)) and provides no further technical difficulties.

Note that the components of (9.6), (9.4d, f) with respect to the indicated base vectors and dyads are independent of θ and z. (It is easy to show that (9.4a) and (9.4e) are the most general forms whose gradients have this property.) We denote the ordered set of the components of $\boldsymbol{\Gamma}$ corresponding to (9.4), (9.6) by the single symbol

(9.8)
$$\begin{aligned} \gamma(s) &= (\gamma_1(s), \ldots, \gamma_{15}(s)) \\ &\equiv (w_1'(s),\, w_1(s)\, w_2'(s),\, w_3'(s),\, \alpha_{22} s^{-1}\, w_1(s),\, \alpha_{32} s^{-1},\, \alpha_{23} w_1(s),\, \alpha_{33}, \\ &\qquad w_4'(s),\, w_4(s),\, w_5'(s),\, \alpha_{52} s^{-1},\, \alpha_{53},\, h_1(s),\, h_2(s),\, h_3(s)). \end{aligned}$$

We define the physical components of the dependent constitutive variables by

$$\hat{T}_{ij}(\gamma, s) \equiv \boldsymbol{e}_i(\mathbf{x}) \cdot \hat{\boldsymbol{T}}(\boldsymbol{\Gamma}, z) \cdot \boldsymbol{k}_j(\theta), \quad \hat{q}_j(\gamma, s) \equiv \hat{\boldsymbol{q}}(\boldsymbol{\Gamma}, z) \cdot \boldsymbol{k}_j(\theta), \tag{9.9a, b}$$

$$\hat{d}_j(\gamma, s) \equiv \hat{\boldsymbol{d}}(\boldsymbol{\Gamma}, z) \cdot \boldsymbol{k}_j(\theta), \quad \hat{b}_j(\gamma, s) \equiv \hat{\boldsymbol{b}}(\boldsymbol{\Gamma}, z) \cdot \boldsymbol{k}_j(\gamma, s), \tag{9.9c, d}$$

$$\hat{j}_j(\gamma, s) \equiv \hat{\boldsymbol{j}}(\boldsymbol{\Gamma}, z) \cdot \boldsymbol{k}_j(\theta) \tag{9.9e}$$

when $\boldsymbol{\Gamma}$ has the form corresponding to (9.4) and (9.6), assuming that the constitutive functions $\hat{\boldsymbol{T}}$, *etc.*, are such that these constitutive functions for the physical components depend only on γ and s. These representations are valid when the constitutive functions $\hat{\boldsymbol{T}}$, *etc.*, are hemitropic and depend on z only through s. They are also valid for special forms of aeolotropy. It then follows that $\hat{\boldsymbol{j}} \cdot \partial\varphi/\partial z$ depends only on γ and s. We also assume that σ depends only on s.

We now obtain an alternative representation for $\boldsymbol{h}$ in terms of $\partial\psi/\partial z$ directly for the semi-inverse problem; the specialization of the results of Section 5 does not yield the new representation. Substituting (9.4f) and (9.9e) into (3.10) we get

$$0 = \hat{j}_1(\gamma(s), s), \tag{9.10}$$

$$[s h_2(s)]' = s \hat{j}_3(\gamma(s), s), \quad h_3'(s) = -\hat{j}_2(\gamma(s), s). \tag{9.11}$$

Thus $\boldsymbol{h}$ must have the form

$$h_1(s) = w_6'(s), \quad h_2(s) = s^{-1}\left[\alpha_{62} - \int_s^1 t \hat{j}_3(\gamma(t), t)\, dt\right], \tag{9.12a, b}$$

$$h_3(s) = \alpha_{63} + \int_s^1 \hat{j}_2(\gamma(t), t)\, dt. \tag{9.12c}$$

Therefore $\boldsymbol{h}$ can be written as the sum of the gradient of (9.4g) and an integral operator (cf. (5.2)).

Condition (9.10) may be regarded either as a restriction on γ or else as being identically satisfied by virtue of choosing the constitutive function $\hat{j}_1$ to be the zero function, in which case the material is incapable of conducting electricity in the radial direction. In the former case we assume that (9.10) can be uniquely solved for w_5' in terms of the other elements of γ:

$$w_5'(s) = \hat{e}_1(\gamma^-(s), s), \tag{9.13}$$

where γ^- stands for all the components of γ except w_5'. A sufficient condition for (9.13) to be equivalent to (9.10) is that $w_5' \mapsto \hat{j}_1(\gamma, s)$ be strictly increasing and assume both negative and positive values. That this function be strictly increasing is ensured by the strict monotonicity of $\boldsymbol{e} \mapsto \hat{\boldsymbol{j}}(\boldsymbol{\Gamma}, z)$. Equations (9.10) and (9.13) are also equivalent in the important special case that $\hat{j}_1(\gamma, s)$ has the same sign as w_5', which occurs, *e.g.*, if $\hat{\boldsymbol{j}}(\boldsymbol{\Gamma}, z) = J(\boldsymbol{\Gamma}, z)\, \boldsymbol{e}$ where J is a positive-valued scalar function. In this case (9.13) reduces to $w_5' = 0$.

Let

$$\mathbf{w} = (w_1, \ldots, w_6), \quad \boldsymbol{\alpha} = (\alpha_{22}, \ldots, \alpha_{63}). \tag{9.14}$$

By controlling the dependence of $\hat{j}_2$ and $\hat{j}_3$ on h_2 and h_3 we can imitate the development of Section 5 to show that (9.12b, c) can be uniquely solved for h_2 and h_3 in terms of the other variables. Thus we can replace these equations with

(9.15a) $$h_2(s) = s^{-1}[\alpha_{62} + \varkappa_2(\mathbf{w}(\cdot), \alpha, s)],$$

(9.15b) $$h_3(s) = \alpha_{63} + \varkappa_3(\mathbf{w}(\cdot), \alpha, s).$$

Alternatively we may observe that (9.12b, c) is equivalent to an initial value problem for h_2, h_3. If we assume that there is a number $p > 1$ such that $\hat{j}_2(\gamma(\cdot), \cdot)$ and $\hat{j}_3(\gamma(\cdot), \cdot)$ are integrable on $[a, 1]$ when $\gamma \in L_p([a, 1])$ and that there is a number $K(\mathbf{w}, \alpha)$ such that

$$\sum_{\alpha,\beta=1} |\partial \hat{j}_\alpha(\gamma(s), s)/\partial h_\beta| \leq K(\mathbf{w}, \alpha)$$

when $\gamma \in L_p([a, 1])$, then the standard theory of ordinary differential equations (*cf.* HALE (1969, Secs. I.5, I.6)) implies that (9.12b, c) has a unique absolutely continuous solution on $[a, 1]$, which we can represent by (9.15a, b). Note that this result does not require restrictions like those of Theorem 5.3 on the size of $\partial\hat{\boldsymbol{j}}/\partial\boldsymbol{h}$ and on the size of the domain. The Arzelà-Ascoli Theorem implies that $W_p^1((a, 1)) \ni \mathbf{w}(\cdot) \mapsto \varkappa_2(\mathbf{w}(\cdot), \alpha, \cdot), \varkappa_3(\mathbf{w}(\cdot), \alpha, \cdot) \in C^0([a, 1])$ are compact (when this construction of h_2 and h_3 is used).

We henceforth assume that the representation (9.15) is valid and that $\varkappa_2$ and $\varkappa_3$ have this compactness property.

We are now ready to write down the governing equations for our semi-inverse problem when the only body force, the Lorentz force, is absorbed into the effective stress and when the only heat source is that of (3.6), due to Joule heating. Let $l_i\boldsymbol{e}_i$ be the axial vector corresponding to $\boldsymbol{L}$. Let us set

(9.16) $$\hat{\xi} = (\hat{\xi}_1, \ldots, \hat{\xi}_6), \quad \hat{\eta} = (\hat{\eta}_1, \hat{\eta}_2, 0, \hat{\eta}_4, \hat{\eta}_5, 0)$$

with

(9.17a) $$\hat{\xi}(\mathbf{w}', \mathbf{w}, \alpha, v(\cdot), s) \equiv (\hat{T}_{11}, w_1\hat{T}_{21}, \hat{T}_{31}, \hat{q}_1, \hat{d}_1, \hat{b}_1),$$

(9.17b) $$\hat{\eta}_1(\mathbf{w}', \mathbf{w}, \alpha, v(\cdot), s) \equiv \hat{T}_{21}w_2' + \alpha_{22}s^{-1}\hat{T}_{22} + \alpha_{23}\hat{T}_{23},$$

(9.17c) $$\hat{\eta}_2(\mathbf{w}', \mathbf{w}, \alpha, v(\cdot), s) \equiv \hat{l}_3,$$

(9.17d) $$\hat{\eta}_4(\mathbf{w}', \mathbf{w}, \alpha, v(\cdot), s) \equiv \alpha_{52}s^{-1}\hat{j}_2 + \alpha_{53}\hat{j}_3,$$

(9.17e) $$\hat{\eta}_5(s) = -\sigma,$$

where the arguments of the constitutive functions appearing on the right sides of (9.17a–d) are γ, s and with every h_1, h_2, h_3 appearing on these right sides replaced by w_6', $\alpha_{62}s^{-1} + \varkappa_2(v(\cdot), \alpha, s)$, $\alpha_{63} + \varkappa_3(v(\cdot), \alpha, s)$ respectively. Note the definition of $\hat{\xi}_2$. In line with the remark following (3.12) there is no loss of physical content in taking $\eta_2 = l_3 = 0$. We do so because it simplifies the ensuing analysis. Then by using the componential form of (3.12), we reduce the governing equations (5.13)-(5.16) to the following system of ordinary-functional differential equations for $\mathbf{w}, \boldsymbol{\alpha}$:

$$(s\hat{\boldsymbol{\xi}})' = s\hat{\boldsymbol{\eta}} = \mathbf{0} \tag{9.18}$$

where the arguments of $\hat{\boldsymbol{\xi}}$ and $\hat{\boldsymbol{\eta}}$ are $\mathbf{w}', \mathbf{w}, \boldsymbol{\alpha}, \mathbf{w}(\cdot), s$. (We have introduced our constitutive functions in (9.17) with the argument $\mathbf{v}(\cdot)$ so as to avoid confusion in Section 10 when we take certain partial derivatives of these functions.)

If (9.10) is equivalent to (9.13), then w_5' is completely determined by the other components of $\mathbf{w}$ and $\boldsymbol{\alpha}$, which can be found from the remaining equations and side conditions. We accordingly discard the fifth equation of (9.18), which is

$$(s\,\hat{d}_1)' = s\sigma. \tag{9.19}$$

We regard this equation as determining the σ necessary to maintain the semi-inverse state (9.4). This interpretation of (9.19) smells fishy, but is in fact quite reasonable: Consider, *e.g.*, constitutive equations of the form

$$j_1 = Je_1, \qquad d_1 = De_1$$

where J and D are positive-valued scalar functions. Then (9.13) and (9.19) require that $\sigma = 0$. Thus when (9.10) is equivalent to (9.13), we shall simply ignore (9.19), regarding (9.18) as the suitably truncated system. We shall comment on boundary conditions below.

If $\hat{j}_1$ is the zero function, then we need take no action with respect to (9.19).

We now specify boundary conditions. Our prescription is compatible with the formalism of Section 7. On the cylindrical face $s = 1$ of $\partial\mathcal{B}$ we either fix the outer radius:

$$w_1(1) = \bar{w}_1(1) \tag{9.20a}$$

where $\bar{w}_1(1)$ is a given positive number or we prescribe the traction:

$$\hat{\xi}_1(\mathbf{w}'(1), \mathbf{w}(1), \boldsymbol{\alpha}, \mathbf{w}(\cdot), 1) = \bar{\xi}_1(1) \tag{9.20b}$$

where $\bar{\xi}_1(1)$ is a given number. (More generally, we could replace $\bar{\xi}_1(1)$ with $\bar{\xi}_1(\mathbf{w}(1), \boldsymbol{\alpha})$ where the new $\bar{\xi}_1$ is a prescribed function. Since only minor technical difficulties are introduced by such a replacement in this and other such Neumann conditions, we do not bother to pursue such generality.) We fix the deformation to within a rigid displacement by setting

$$w_2(1) = 0, \tag{9.21}$$

$$w_3(1) = 0. \tag{9.22}$$

On this face we either prescribe the temperature:

$$w_4(1) = \bar{w}_4(1) \tag{9.23a}$$

where $\overline{w}_4(1)$ is a given number or we prescribe the heat flux:

(9.23b) $$\hat{\xi}_4(\mathbf{w}'(1), \mathbf{w}(1), \boldsymbol{\alpha}, \mathbf{w}(\cdot), 1) = \bar{\xi}_4(1)$$

where $\bar{\xi}_4(1)$ is a given number. Finally we fix the data of the potentials φ and ψ by taking

(9.24) $$w_5(1) = 0,$$

(9.25) $$w_6(1) = 0.$$

On the cylindrical face $s = a$ we prescribe alternative boundary conditions expressed in an analogous notation:

(9.26a, b) $$w_i(a) = \overline{w}_i(a) \text{ or } \hat{\xi}_i(\mathbf{w}'(a), \mathbf{w}(a), \boldsymbol{\alpha}, \mathbf{w}(\cdot), a) = \bar{\xi}_i(a) \text{ for } i = 1, \ldots, 6$$

where $\overline{w}_i(a), i = 1, \ldots, 6$ and $\bar{\xi}_i(a), i = 1, 3, 4, 5, 6$ are given constants and where

(9.26c) $$\bar{\xi}_2(a) \equiv w_1(a)\,\tau$$

with τ a given constant. The form of $\bar{\xi}_2(a)$ reflects its definition and the fact that it is a torque. In conformity with the condition that $w_1' > 0$, we require that $\overline{w}_1(1) > \overline{w}_1(a)$ when both these numbers are prescribed. Note that (9.13) implies that

(9.27) $$w_5(1) - w_5(a) = \int_a^1 \hat{e}_1(\gamma^-(s), s)\, ds$$

so we are not free to prescribe both $w_5(1)$ and $w_5(a)$ when (9.13) holds.

To avoid dealing with the minor technical difficulties that can arise when all boundary conditions are of the Neumann type, we assume that the temperature w_4 is prescribed on at least one of the faces $s = a$ and $s = 1$. We need make no such provision for the variable w_1 because the growth conditions we shall impose on our constitutive functions preclude any trouble with coercivity ultimately due to Neumann conditions. If $\mathscr{B}$ is an entire tube, then $\Theta = \pi$. If we require that $\mathbf{y}$, φ, and ψ be continuous, then

(9.28) $$\alpha_{22} = \pm 1, \quad \alpha_{32} = 0, \quad \alpha_{52} = 0, \quad \alpha_{62} = 0.$$

We obtain various kinds of dislocations by suspending (9.28a, b). If $\mathscr{B}$ is a sector of a tube, *i.e.*, if $\Theta < \pi$, or if $\mathscr{B}$ is a slit tube *i.e.*, if $\Theta = \pi$ but with the faces $\theta = -\pi$ and $\theta = \pi$ not identified, then we can prescribe certain degenerate boundary conditions on the faces $\theta = \pm\Theta$. We likewise prescribe such conditions on $z = \pm Z$.

We adopt the following alternative conditions for the faces $\theta = \pm\Theta$:

(9.29a, b) $$\alpha_{22} = \bar{\alpha}_{22} \quad \text{or}$$

$$\hat{A}_{22}[\mathbf{w}, \boldsymbol{\alpha}] \equiv \int_a^1 w_1(s)\, \hat{T}_{22}(\gamma(s), s)\, ds = \bar{A}_{22}[w_1] \equiv \int_a^1 w_1(s)\, \hat{T}_{22}(s)\, ds$$

where $\bar{\alpha}_{22}$ is a given number and $\bar{A}_{22}$ is a given functional of w_1 having the indicated form. In the argument $\gamma(s)$ of $\hat{T}_{22}$, $h_1(s)$ is replaced by (9.12a) and $h_2(s)$, $h_3(s)$ by (9.15). It is easy to see that $-Z\hat{A}_{22}[\mathbf{w}, \alpha]$ is the resultant effective torque about $\mathbf{e}_3$ on the material face $\theta = -\Theta$ needed to maintain the state (9.4) (*cf.* ANTMAN (1983)). Similarly we prescribe

$$(9.30\text{a, b}) \quad \alpha_{32} = \bar{\alpha}_{32} \quad \text{or} \quad \hat{A}_{32}[\mathbf{w}, \alpha] \equiv \int_a^1 \hat{T}_{32}(\gamma(s), s)\, ds = \bar{A}_{32},$$

$$(9.31\text{a, b}) \quad \alpha_{52} = \bar{\alpha}_{52} \quad \text{or} \quad \hat{A}_{52}[\mathbf{w}, \alpha] \equiv \int_a^1 d_2(\gamma(s), s)\, ds = \bar{A}_{52},$$

$$(9.32\text{a, b}) \quad \alpha_{62} = \bar{\alpha}_{62} \quad \text{or} \quad \hat{A}_{62}[\mathbf{w}, \alpha] \equiv \int_a^1 b_2(\gamma(s), s)\, ds = \bar{A}_{62}.$$

Here γ has the form just described. $-Z\hat{A}_{32}[\mathbf{w}, \alpha]$ is the effective resultant force in the $\mathbf{e}_3$-direction on the material face $\theta = -\Theta$. ${}_6\bar{A}_{32}, \bar{A}_{52}, \bar{A}_{62}$ are just numbers.

For the faces $z = \pm Z$, we likewise prescribe

$$(9.33\text{a, b}) \quad \alpha_{23} = \bar{\alpha}_{23} \quad \text{or} \quad \hat{A}_{23}[\mathbf{w}, \alpha] \equiv \int_a^1 s w_1(s)\, \hat{T}_{23}(\gamma(s), s)\, ds = \bar{A}_{23}[w_1],$$

$$(9.34\text{a, b}) \quad \alpha_{33} = \bar{\alpha}_{33} \quad \text{or} \quad \hat{A}_{33}[\mathbf{w}, \alpha] \equiv \int_a^1 s\hat{T}_{33}(\gamma(s), s)\, ds = \bar{A}_{33},$$

$$(9.35\text{a, b}) \quad \alpha_{53} = \bar{\alpha}_{53} \quad \text{or} \quad \hat{A}_{53}[\mathbf{w}, \alpha] \equiv \int_a^1 s d_3(\gamma(s), s)\, ds = \bar{A}_{53},$$

$$(9.36\text{a, b}) \quad \alpha_{63} = \bar{\alpha}_{63} \quad \text{or} \quad \hat{A}_{63}[\mathbf{w}, \alpha] \equiv \int_a^1 s b_3(\gamma(s), s)\, ds = \bar{A}_{63}.$$

$-\Theta\hat{A}_{23}[\mathbf{w}, \alpha]$ is the resultant effective torque about $\mathbf{e}_3$ and $-\Theta\hat{A}_{33}[\mathbf{w}, \alpha]$ is the resultant effective force in the $\mathbf{e}_3$-direction on the face $z = -Z$.

10. Consequences of the Strong Ellipticity Condition

In this section $\hat{\xi}$ and $\hat{\eta}$ have the arguments listed in (9.17). Thus the derivative of $\bar{\eta}_1$ with respect to w_1, say, is a pure partial derivative; no differentiation with respect to $\mathbf{v}(\cdot)$ is required.

Since $\partial\hat{\boldsymbol{T}}/\partial w_1' = (\partial\hat{\boldsymbol{T}}/\partial\boldsymbol{F}) : \boldsymbol{e}_1\boldsymbol{k}_1$, *etc.*, definition (6.1) and the strong ellipticity condition imply that

$$(10.1\text{a}) \quad \boldsymbol{\nu} \cdot (\partial\hat{\xi}/\partial\mathbf{w}') \cdot \boldsymbol{\nu} = \omega((\nu_1\boldsymbol{e}_1 + \nu_2 w_1\boldsymbol{e}_2 + \nu_3\boldsymbol{e}_3)\,\boldsymbol{k}_1, \nu_4\boldsymbol{k}_1, \nu_5\boldsymbol{k}_1, \nu_6\boldsymbol{k}_1) > 0$$

for all $\boldsymbol{\nu} \equiv (\nu_1, \ldots, \nu_6) \neq \mathbf{0}$ when $w_1 > 0$. Slightly abusing the notation, we likewise obtain

$$(10.1\text{b}) \qquad \boldsymbol{\nu} \cdot \frac{\partial\hat{\xi}}{\partial\mathbf{x}} \cdot \boldsymbol{\nu} > 0 \quad \forall \boldsymbol{\nu} \neq \mathbf{0} \text{ when } w_1 > 0 \text{ where}$$

$$\mathbf{x} \equiv (w_1', w_1 w_2', w_3', w_4', w_5', w_6').$$

Next we observe that

$$w_1 |\boldsymbol{c}|^2 = \boldsymbol{a} \cdot \boldsymbol{F} \cdot \boldsymbol{c}, \quad \hat{\eta}_1 = \boldsymbol{a} \cdot \hat{\boldsymbol{T}} \cdot \boldsymbol{c} \tag{10.2}$$

with $\boldsymbol{a} = \boldsymbol{e}_2$, $\boldsymbol{c} = w_2' \boldsymbol{k}_1 + (\alpha_{22}/s)\, \boldsymbol{k}_2 + \alpha_{23} \boldsymbol{k}_3$, so that the strong ellipticity condition implies that

$$\partial \hat{\eta}_1 / \partial w_1 \equiv \boldsymbol{ac} : (\partial \hat{\boldsymbol{T}} / \partial w_1) \equiv (\boldsymbol{ac}) : (\partial \hat{\boldsymbol{T}} / \partial \boldsymbol{F}) : \boldsymbol{ac} > 0. \tag{10.3}$$

(Note that $\boldsymbol{a}$ and $\boldsymbol{c}$ are not independent of $\boldsymbol{a} \cdot \boldsymbol{F} \cdot \boldsymbol{c}/|\boldsymbol{c}|$).)

Suppose that $\bar{\alpha}_{32}$ and $\bar{\alpha}_{33}$ are prescribed. Set

$$\mu = \bar{a}_{33} \alpha_{22} - \bar{\alpha}_{32} \alpha_{23}, \quad \nu = \bar{\alpha}_{32} \alpha_{22} + \bar{\alpha}_{33} \alpha_{23}, \tag{10.4}$$

$$\hat{M} = \bar{\alpha}_{33} \hat{T}_{22} - \bar{\alpha}_{32} s \hat{T}_{23}, \quad \hat{N} = \bar{\alpha}_{32} \hat{T}_{22} + \bar{\alpha}_{33} s \hat{T}_{23}. \tag{10.5}$$

We solve (10.4) for α_{22} and α_{23} in terms of μ and ν and substitute the resulting expressions into the arguments of $\hat{M}$ and $\hat{N}$. The strong ellipticity condition then implies that

$$\begin{pmatrix} \partial \hat{M} / \partial \mu & \partial \hat{M} / \partial \nu \\ \partial \hat{M} / \partial \mu & \partial \hat{N} / \partial \nu \end{pmatrix} \quad \text{is positive-definite.} \tag{10.6}$$

The combination of (10.1b) and Hypothesis 6.6 supports a global implicit function theorem (based on degree theory) that ensures that the function

$$\mathbf{x} \mapsto \hat{\boldsymbol{\xi}}(\mathbf{w}', \mathbf{w}, \boldsymbol{\alpha}, \mathbf{v}(\cdot), s) \tag{10.7a}$$

has a strictly monotone inverse

$$\boldsymbol{\xi} \mapsto \mathbf{f}(\boldsymbol{\xi}, \mathbf{w}, \boldsymbol{\alpha}, \mathbf{v}(\cdot), s). \tag{10.7b}$$

In particular, f_1, which delivers w_1', is positive on its domain.

11. Growth Conditions and Function Spaces

We introduce some notation to be used in the rest of this paper. Let

$$\begin{aligned} &\hat{\zeta}_{22} \equiv s^{-1} w_1 \hat{T}_{22}, \quad \hat{\zeta}_{23} \equiv w_1 \hat{T}_{23}, \quad \hat{\zeta}_{32} = s^{-1} \hat{T}_{32}, \quad \hat{\zeta}_{33} = \hat{T}_{33}, \\ &\hat{\zeta}_{52} \equiv s^{-1} \hat{d}_2, \quad \hat{\zeta}_{53} \equiv \hat{d}_3, \quad \hat{\zeta}_{62} \equiv s^{-1} \hat{b}_2, \quad \hat{\zeta}_{63} \equiv \hat{b}_3, \\ &\hat{\boldsymbol{\zeta}} \cdot \boldsymbol{\alpha}_\# \equiv \sum_{\mu,\nu} \hat{\zeta}_{\mu\nu} \alpha_{\#\mu\nu}, \end{aligned} \tag{11.1}$$

the summation being taken over $\mu = 2, 3, 5, 6$, $\nu = 2, 3$. Let

$$\boldsymbol{\omega} \equiv (\mathbf{w}, \boldsymbol{\alpha}). \tag{11.2}$$

We set

$$(11.3)\quad \langle \mathbf{m}(\boldsymbol{\omega}), \boldsymbol{\omega}_\sharp \rangle \equiv \int_a^1 (\hat{\xi}\cdot\mathbf{w}_\sharp' + \hat{\eta}\cdot\mathbf{w}_\sharp + \hat{\zeta}\cdot\boldsymbol{\alpha}_\sharp)\, s\, ds$$

$$+ a\bar{\xi}(a)\cdot\mathbf{w}_\sharp(a) - \bar{\xi}(1)\cdot\mathbf{w}_\sharp(1) - \sum_{\mu\nu} \bar{A}_{\mu\nu}[w_1]\,\alpha_{\sharp\mu\nu},$$

where the arguments of $\hat{\xi}$, $\hat{\eta}$, $\hat{\zeta}$ are $\mathbf{w}'$, $\mathbf{w}$, $\boldsymbol{\alpha}$, $\mathbf{w}(\cdot)$, s. Observe that

$$(11.4)\qquad \int_a^1 \hat{\zeta}\cdot\boldsymbol{\alpha}_\sharp\, s\, ds = \sum_{\mu\nu} \hat{A}_{\mu\nu}[\boldsymbol{\omega}]\,\alpha_{\sharp\mu\nu}.$$

The weak form of the boundary value problem of Section 9 is

$$(11.5)\qquad \langle \mathbf{m}(\boldsymbol{\omega}), \boldsymbol{\omega}_\sharp \rangle = 0 \text{ "for all" } \boldsymbol{\omega}_\sharp.$$

In the next section, we give precise interpretations to relatives of (11.5).

We pose the basic growth conditions in terms of a scalar function W, which might be interpreted as a sort of stored energy function. It allows us to replace standard L_p-spaces by related spaces better equipped to handle possible aeolotropy.

11.6. Hypothesis. *There are numbers* $q > p > 1$, $c_1 > 0$, $c_2 > 0$, $c_3 > 0$, $c_4 > 0$ *and a function* $\mathbb{R}^{15} \ni \gamma \mapsto W(\gamma, s) \in \mathbb{R}$ *having the following properties:*

$$(11.7)\qquad W(\cdot, s) \text{ is strictly convex},$$

$$(11.8)\qquad W(\mathbf{0}, s) = 0,$$

$$(11.9)\qquad W(\cdot, s) \text{ is invariant under the change of sign of any component of } \gamma,$$

$$(11.10)\qquad c_1|\gamma|^p - c_2 \leqq W(\gamma, s) \leqq c_3|\gamma|^q + c_4.$$

Let $\mathring{\mathbf{w}}$ *be an affine function satisfying whatever Dirichlet conditions from* (9.20)–(9.26) *are prescribed. Let* $\mathring{\boldsymbol{\alpha}}$ *be a vector of the form* $\boldsymbol{\alpha}$ *of* (9.14) *with its entries taken to be* $\bar{\alpha}_{22}, \dots$ *whenever these numbers are prescribed in* (9.29)-(9.36) *and otherwise taken to be arbitrary with* $\alpha_{22}\alpha_{33} - \alpha_{23}\alpha_{32} > 0$. *Let* $\mathring{\gamma}$ *be generated from* $\mathring{\mathbf{w}}$ *and* $\mathring{\boldsymbol{\alpha}}$ *by* (9.8), (9.12), (9.15). *Then*

$$(11.11)\quad \hat{\xi}\cdot(\mathbf{w}' - \mathring{\mathbf{w}}') + \hat{\eta}\cdot(\mathbf{w} - \mathring{\mathbf{w}}) + \hat{\zeta}\cdot(\boldsymbol{\alpha} - \mathring{\boldsymbol{\alpha}}) \geqq W(\gamma, s) - W(\mathring{\gamma}, s)$$

when the arguments of the constitutive functions on the left side of (11.11) *are* $\mathbf{w}'$, $\mathbf{w}$, $\boldsymbol{\alpha}$, $\mathbf{w}(\cdot)$, s *and with* γ *and* $\mathring{\gamma}$ *expressed in terms of these variables by* (9.8), (9.12a), (9.15).

Remarks. Condition (11.9) is a sort of isotropy condition. Its provenance is described by ANTMAN (1983, eq. (7.7)). The mechanical terms from (11.11) correspond to a certain stress power. This issue is likewise treated at great length by ANTMAN (1983).

We now introduce function spaces naturally associated with W. Let n be a positive integer. We set

$$\mathcal{G} \equiv \left\{ \gamma : \int_a^1 sW(\gamma(s), s)\, ds < \infty \right\}, \tag{11.12}$$

$$\mathcal{E} \equiv \{(\mathbf{w}, \alpha) : \gamma \in \mathcal{G}\}, \tag{11.13}$$

$$\mathcal{E}_n = \{(\mathbf{w}, \alpha) \in \mathcal{E} : nw_1' \geqq 1, \text{ a.e., } nw_1(s) \geqq s, n(\alpha_{22}\alpha_{33} - \alpha_{23}\alpha_{32}) \geqq 1\}, \tag{11.14}$$

$$\mathcal{A} \equiv \{(\mathbf{w}, \alpha) \in \mathcal{E}, w_1' > 0, w_1 > 0 \text{ a.e., } \alpha_{22}\alpha_{33} - \alpha_{23}\alpha_{32} > 0, \text{ a fixed subset of } \{\mathbf{w}(a), \mathbf{w}(1), \alpha\} \text{ is prescribed as in Sec. 9}\}, \tag{11.15}$$

$$\mathcal{A}_n \equiv \mathcal{A} \cap \mathcal{E}_n. \tag{11.16}$$

$\mathcal{A}$ is the set of *admissible functions*. Conditions (11.7)-(11.10) ensure that $W(\cdot, s)$ satisfies the Δ_2-condition of Orlicz space theory, whence it follows that $\mathcal{G}$ is a reflexive separable Banach space satisfying

$$L_q((a, 1)) \subset \mathcal{G} \subset L_p((a, 1)) \tag{11.17}$$

(*cf.* Krasnosel'skii & Rutitskii (1968)). Since some components of γ are products of components of $(\mathbf{w}, \alpha)$, neither $\mathcal{E}$ nor $\mathcal{E}_n$ is a Banach space. It is easy, however, to construct a suitable Banach space for $(\mathbf{w}, \alpha)$: Let $(\mathring{\mathbf{w}}, \mathring{\alpha}) \in \mathcal{E}_n$, let $\gamma^\sharp$ be defined by (9.8) with $(\mathring{w}_2, \ldots, \mathring{w}_6, \mathring{\alpha})$ replacing $(w_2, \ldots, w_6, \alpha)$, and let $\gamma^\flat$ be defined by (9.8) with $\mathring{w}_1$ replacing w_1. Define

$$\mathcal{Y}_n \equiv \left\{ (\mathbf{w}, \alpha) : \int_a^1 sW(\gamma^\sharp(s), s)\, ds < \infty, \int_a^1 sW(\gamma^\flat(s), s)\, ds < \infty \right\}. \tag{11.18}$$

11.19. Proposition. *$\mathcal{Y}_n$ is a reflexive separable Banach space. $\mathcal{E}_n$ and $\mathcal{A}_n$ are closed subsets of $\mathcal{Y}_n$. ($\mathcal{A}_n$ is not empty if n is sufficiently large.)*

The proof of this result is the same as that of Antman (1983, Prop. 7.25.) We now refine (11.11):

11.20. Hypothesis. *There are positive constants $c_5, c_6, c_7, c_8, \varepsilon$, depending on $\mathring{w}_1, \mathring{w}_2$, such that*

$$(w_1' - \mathring{w}_1')\,\hat{\xi}_1 + (w_1 - \mathring{w}_1)\,\hat{\eta}_1 \geqq c_5(|w_1'|^p + |w_1/s|^p) - c_6(1 + |\gamma|^{p-\varepsilon}), \tag{11.21}$$

where the arguments of $\hat{\xi}_1, \hat{\eta}_1$ are those listed in (9.17).

The preceding hypotheses ensure that the material is not too weak; the following hypothesis ensures that it is not too strong.

11.22. Hypothesis. *Let the constitutive function introduced in* (9.9) *depend only on γ, s. Let $\boldsymbol{a}(\mathbf{x})$ and $\tilde{\boldsymbol{a}}(\mathbf{x})$ be vectors that are linear combinations of $\boldsymbol{e}_1, \boldsymbol{e}_2, \boldsymbol{e}_3$ with coefficients depending only on s and let $\mathbf{c}(\gamma)$ and $\tilde{\mathbf{c}}(\mathbf{x})$ be vectors that are linear combinations of $\boldsymbol{k}_1, \boldsymbol{k}_2, \boldsymbol{k}_3$ with coefficients depending only on s. Let $\boldsymbol{a} \cdot \boldsymbol{F} \cdot \mathbf{c} \mapsto \tilde{\boldsymbol{a}} \cdot \hat{\boldsymbol{T}} \cdot \tilde{\mathbf{c}}, \boldsymbol{g} \cdot \boldsymbol{a} \mapsto \hat{\boldsymbol{q}} \cdot \tilde{\boldsymbol{a}}, \boldsymbol{e} \cdot \boldsymbol{a} \mapsto \hat{\boldsymbol{d}} \cdot \tilde{\boldsymbol{a}}, \boldsymbol{h} \cdot \boldsymbol{a} \mapsto \hat{\boldsymbol{b}} \cdot \tilde{\boldsymbol{a}}$ be strictly increasing and let*

$l^{\pm}(ac)$ be constant when F has the form (9.6). Then there are continuous functions $(\boldsymbol{\Gamma}, z) \mapsto \boldsymbol{T}^+(\boldsymbol{\Gamma}, z), \boldsymbol{q}^+(\boldsymbol{\Gamma}, z), \boldsymbol{d}^+(\boldsymbol{\Gamma}, z), \boldsymbol{b}^+(\boldsymbol{\Gamma}, z), \varrho^+ \boldsymbol{\Gamma}, z)$ with $\tilde{\boldsymbol{a}}(\mathbf{x}) \cdot \boldsymbol{T}^+(\boldsymbol{\Gamma}(z), z) \cdot \tilde{\boldsymbol{c}}(\mathbf{x}), \boldsymbol{q}^+(\boldsymbol{\Gamma}(z), z) \cdot \tilde{\boldsymbol{a}}(\mathbf{x})$, etc., depending only on s when $\boldsymbol{\Gamma}$ has the form corresponding to (9.4) and (9.6) such that

$$\text{(11.23)} \qquad \tilde{a} \cdot T^+(\Gamma, z) \cdot \tilde{c} \geqq \begin{cases} \delta(ac)\, \tilde{a} \cdot \hat{T}(\Gamma, z) \cdot \tilde{c} & \text{if } \partial\mathcal{D}(ac) \neq \emptyset, \\ |\tilde{a} \cdot \hat{T}(\Gamma, z) \cdot \tilde{c}| & \text{if } \partial\mathcal{D}(ac) = \emptyset, \end{cases}$$

$$q^+ \cdot \tilde{a} \geqq |\hat{q} \cdot \tilde{a}|, \quad d^+ \cdot \tilde{a} \geqq |\hat{d} \cdot \tilde{a}|, \quad b^+ \cdot \tilde{a} \geqq |\hat{b} \cdot \tilde{a}|,$$

$$\varrho^+ \geqq |\hat{j} \cdot e|$$

when $\boldsymbol{\Gamma}$ has the form corresponding to (9.4) and (9.6). Let $\boldsymbol{\xi}^+, \boldsymbol{\eta}^+, \boldsymbol{\zeta}^+$ be expressed in terms of $\boldsymbol{T}^+, \ldots$ just as $\hat{\boldsymbol{\xi}}, \hat{\boldsymbol{\eta}}, \hat{\boldsymbol{\zeta}}$ are expressed in terms of $\boldsymbol{T}, \ldots$ by (9.17) and (11.1). If $\boldsymbol{\omega}, \boldsymbol{\omega}_\# \in \mathscr{E}$, then $s \mapsto \boldsymbol{\xi}^+ \cdot \mathbf{w}'_\#(s) + \boldsymbol{\eta}^+ \cdot \mathbf{w}_\#(s) + \boldsymbol{\zeta}^+ \cdot \boldsymbol{\alpha}_\#(s)$, where the arguments of $\boldsymbol{\xi}^+, \boldsymbol{\eta}^+, \boldsymbol{\zeta}^+$ are $\mathbf{w}'(s), \mathbf{w}(s), \boldsymbol{\alpha}, \mathbf{w}(\cdot), s$, is integrable on $[a, 1]$. Moreover, if $\boldsymbol{\omega}$ is confined to a subset of $\mathscr{E}$ corresponding to a bounded subset of $\mathscr{G}$, then the corresponding $(\boldsymbol{\xi}^+, \boldsymbol{\eta}^+, \boldsymbol{\zeta}^+)$ and $\mathbf{m}(\boldsymbol{\omega})$ generate elements confined to a bounded subset of $\mathscr{G}^*$. In particular,

$$\text{(11.24)} \qquad \hat{\xi}_1 \leqq \xi_1^+, \quad \hat{\eta}_1 \leqq \eta_1^+, \quad \hat{M} \leqq M^+,$$

$$|\hat{\xi}_2| \leqq \xi_2^+, \quad |\hat{\xi}_3| \leqq \xi_3^+, \quad |\hat{N}| \leqq N^+$$

where $s \mapsto \xi_1^+(\mathbf{w}'(s), \mathbf{w}(s), \boldsymbol{\alpha}, \mathbf{w}(\cdot), s), \ldots$ are in $L_1([a, 1])$ and are confined to a bounded subset of L_1 when their arguments correspond to γ's in a bounded subset of $\mathscr{G}$.

Condition (11.24) restricts the response of $\hat{\eta}_1$ (as well as other functions) in tension. We now formulate an hypothesis to control its behavior in compression. It furnishes a quantitative statement of how $\hat{\eta}_1$ is influenced more by changes in w_1 than by changes in w'_1. Let the function with values $\eta_1^\dagger(s^{-1} w_1, w_4, \xi, \alpha, \mathbf{v}(\cdot), s)$ be the composite function obtained from $\hat{\eta}_1$ by using (10.7b) to replace its first set of arguments $\mathbf{w}'$ with those of $\mathbf{f}$ in (10.7b). Let $f_1^\dagger$ be f_1 with its arguments in the same order as those of $\eta_1^\dagger$.

11.25. Hypothesis. *Let there be a number $C > 0$, an octuple α, a scalar-valued function w_4, a function ξ with values in $\mathbb{R}^6$, and a function $\mathbf{v}$ with $(\mathbf{v}, \alpha) \in \mathscr{A}$ such that*

$$\text{(11.26)} \qquad |\alpha| \leqq C, \alpha_{22}\alpha_{33} - \alpha_{23}\alpha_{32} \geqq 1/C, \quad |w_4| \leqq C, \quad |\xi| \leqq C,$$

$$\hat{\xi}_1(\mathbf{v}', \mathbf{v}, \alpha, \mathbf{v}(\cdot), s) \leqq C, \quad |\hat{\xi}_j(\mathbf{v}', \mathbf{v}, \alpha, \mathbf{v}(\cdot), s)| \leqq C \quad \text{for } j = 2, \ldots, 6.$$

Then there is a number $m > 0$ (depending on C) such that

$$\text{(11.27a)} \qquad (0, m) \ni u \mapsto f_1^\dagger(u, w_4, \xi, \alpha, \mathbf{v}(\cdot), s) \text{ is decreasing,}$$

$$\text{(11.27b)} \qquad (0, m) \ni u \mapsto \eta_1^\dagger(u, w_4, \xi, \alpha, \mathbf{v}(\cdot), s) \text{ is increasing.}$$

Moreover,

(11.28)

$$\limsup_{\varepsilon\to 0} \int_a^{x} \eta_1^\dagger \left(\varepsilon + s^{-1} \int_a^{s} f_1^\dagger\,(\varepsilon, w_4(t), \xi(t), \mathbf{v}(\cdot), t)\,dt,\, w_4(s),\, \xi(s),\, \mathbf{v}(\cdot),\, s\right) ds = -\infty$$

for each fixed $x \in (a, 1]$.

The motivation for this hypothesis is given by ANTMAN (1983).
Our final growth condition is

11.29. Hypothesis. *There is a positive constant* c_9 *such that*

$$|\bar{A}_{22}[w_1]| + |\bar{A}_{23}[w_1]| \leqq c_9\, \|w_1, L_p\|. \tag{11.30}$$

12. Existence of Classical Solutions

In this section we prove that when the Strong Ellipticity Condition and growth conditions hold, then a certain set of the boundary value problems posed in Section 9 have regular solutions. We restrict the data prescribed in the alternatives (9.29), (9.30), (9.33), (9.34) to be one of the following nine sets

$$\begin{array}{lll} (\alpha_{22}, \alpha_{23}, \alpha_{32}, \alpha_{33}), & (\hat{A}_{22}, \alpha_{23}, \alpha_{32}, \alpha_{33}), & (\alpha_{22}, \hat{A}_{23}, \alpha_{32}, \alpha_{33}), \\ (\alpha_{22}, \alpha_{23}, \hat{A}_{32}, \alpha_{33}), & (\alpha_{22}, \alpha_{23}, \alpha_{32}, \hat{A}_{33}), & (\hat{A}_{22}, \hat{A}_{23}, \alpha_{32}, \alpha_{33}), \\ (\hat{A}_{23}, \alpha_{23}, \hat{A}_{32}, \alpha_{33}), & (\alpha_{22}, \hat{A}_{23}, \alpha_{32}, \hat{A}_{33}), & (\alpha_{22}, \alpha_{23}, \hat{A}_{32}, \hat{A}_{33}) \end{array} \tag{12.1}$$

because the unprescribed variables from $(\alpha_{22}, \alpha_{23}, \alpha_{32}, \alpha_{33})$ are then confined by (9.7) to an open half-line or open half-plane. It then follows that the corresponding $\mathscr{A}_n$ is a closed *convex* subset of $\mathscr{Y}_n$.

Since our present work generalizes that of ANTMAN (1983), we emphasize only those aspects that are novel. His work may be consulted for motivations and further discussion of such matters as the data omitted in (12.1). Our presentation also tacitly corrects some flaws in his arguments.

Our basic result is

12.2. Theorem. *Let the monotonicity conditions* (10.1), (10.3), (10.6) *hold. Let the Growth Hypotheses* 6.6, 6.10, 11.6, 11.20, 11.22, 11.25, 11.29 *hold. Let one of the sets of data of* (12.1) *be prescribed. Then the corresponding boundary value problems of Section* 9 *have classical solutions.*

Proof.

Step I. Existence of a solution to a truncated variational inequality. We can write

$$\langle \mathbf{m}(\boldsymbol{\omega}), \boldsymbol{\omega}_\# \rangle \equiv \langle \mathbf{n}(\boldsymbol{\omega}, \boldsymbol{\omega}), \boldsymbol{\omega}_\# \rangle, \tag{12.3}$$

(12.4)

$$\begin{aligned}\langle \mathbf{n}(\boldsymbol{\omega}^1, \boldsymbol{\omega}^2), \boldsymbol{\omega}_\# \rangle \equiv \int_a^1 &\{\hat{\boldsymbol{\xi}}((\mathbf{w}^1)'(s), \mathbf{w}^2(s), \boldsymbol{\alpha}^2, \mathbf{w}^2(\cdot), s) \cdot \mathbf{w}_\#'(s) \\ &+ \hat{\boldsymbol{\eta}}((\mathbf{w}^2)'(s), \mathbf{w}^2(s), \boldsymbol{\alpha}^2, \mathbf{w}^2(\cdot), s) \cdot \mathbf{w}_\#(s)\}\, s\, ds \\ &+ a\bar{\boldsymbol{\xi}}(a) \cdot \mathbf{w}_\#(a) - \bar{\boldsymbol{\xi}}(1) \cdot \mathbf{w}_\#(1) + \Sigma\{\hat{A}_{\mu\nu}[\boldsymbol{\omega}^2] - \bar{A}_{\mu\nu}[w_1^2]\}\, \alpha_{\#\mu\nu}\end{aligned}$$

where the last term is summed over $\mu = 2, 3, 5, 6$, $\nu = 2, 3$.

Since $\mathscr{A}_n$ is a cosed convex subset of $\mathscr{Y}_n$, since (10.1) ensures that $\mathbf{m}$ is semi-monotone on $\mathscr{A}_n$ in the sense that

$$\langle \mathbf{n}(\boldsymbol{\omega}^1, \boldsymbol{\omega}^2) - \mathbf{n}(\boldsymbol{\omega}^2, \boldsymbol{\omega}^2), \boldsymbol{\omega}^1 - \boldsymbol{\omega}^2 \rangle \geqq 0 \quad \forall\, \boldsymbol{\omega}^1, \boldsymbol{\omega}^2 \in \mathscr{A}_n, \tag{12.5}$$

and since $\boldsymbol{\omega} \mapsto \varkappa_2(\boldsymbol{\omega}, \cdot), \varkappa_3(\boldsymbol{\omega}, \cdot)$ are compact by assumption, we can use Hypothesis 11.22 to show that $\mathbf{m}$ is an operator of the "type of the Calculus of Variations" (*cf.* LIONS (1969)) from $\mathscr{A}_n$ to the dual space $\mathscr{Y}_n^*$ of $\mathscr{Y}_n$. Thus $\mathbf{m}$ is pseudo-monotone on $\mathscr{A}_n$. Hypothesis 11.6 ensures that $\mathbf{m}$ is coercive. A theorem of BREZIS (1968), (*cf.* LIONS (1969, p. 297)) then implies that for n sufficiently large there exists an $\boldsymbol{\omega}_n \in \mathscr{A}_n$ satisfying the variational inequality

$$0 \geqq \langle \mathbf{m}(\boldsymbol{\omega}_n), \boldsymbol{\omega}_n - \tilde{\boldsymbol{\omega}} \rangle \quad \forall\, \tilde{\boldsymbol{\omega}} \in \mathscr{A}_n. \tag{12.6}$$

Step II. Bounds on $\boldsymbol{\gamma}_n$. Let $\tilde{\boldsymbol{\omega}} = \mathring{\boldsymbol{\omega}} = (\mathring{\mathbf{w}}, \mathring{\boldsymbol{\alpha}})$ where $(\mathring{\mathbf{w}}, \mathring{\boldsymbol{\alpha}})$ is defined in Hypothesis 11.6. Let $\mathring{\boldsymbol{\gamma}}$ and $\boldsymbol{\gamma}_n$ correspond to $\mathring{\boldsymbol{\omega}}$ and $\boldsymbol{\omega}_n$. Then (12.6) and (11.11) imply that

$$\begin{aligned}0 &\geqq \int_a^1 \{\boldsymbol{\xi}_n \cdot (\mathbf{w}_n' - \mathring{\mathbf{w}}') + \boldsymbol{\eta}_n \cdot (\mathbf{w}_n - \mathring{\mathbf{w}})\}\, s\, ds \\ &\quad + a\bar{\boldsymbol{\xi}}(a) \cdot [\mathbf{w}_n(a) - \mathring{\mathbf{w}}(a)] - \bar{\boldsymbol{\xi}}(1) \cdot [\mathbf{w}_n(1) - \mathring{\mathbf{w}}(a)] \\ &\quad + \sum_{\mu,\nu} \{\hat{A}_{\mu\nu}[\boldsymbol{\omega}_n] - \bar{A}_{\mu\nu}[w_{n1}]\}\, (\alpha_{n\mu\nu} - \mathring{\alpha}_{\mu\nu}) \\ &\geqq \int_a^1 W(\boldsymbol{\gamma}_n(s), s)\, s\, ds - \int_a^1 W(\mathring{\boldsymbol{\gamma}}(s), s)\, s\, ds - \Lambda\end{aligned} \tag{12.7}$$

where $\boldsymbol{\xi}_n(s) \equiv \hat{\boldsymbol{\xi}}(\mathbf{w}_n'(s), \mathbf{w}_n(s), \boldsymbol{\alpha}_n, \mathbf{w}_n(\cdot), s)$, *etc.*, and

$$\begin{aligned}\Lambda \equiv &- \int_a^1 \sigma(s)\, [w_{n5}(s) - \mathring{w}_5(s)]\, s\, ds \\ &+ \bar{\boldsymbol{\xi}}(1) \cdot [\mathbf{w}_n(1) - \mathring{\mathbf{w}}(1)] - a\bar{\boldsymbol{\xi}}(a) \cdot [\mathbf{w}_n(a) - \hat{\mathbf{w}}(a)] \\ &+ \sum_{\mu,\nu} \bar{A}_{\mu\nu}[w_{n1}]\, (\alpha_{n\mu\nu} - \mathring{\alpha}_{\mu\nu}).\end{aligned} \tag{12.8}$$

In the following development we let C represent a positive constant independent of n, which can always be always be estimated in terms of the available data. The meaning of C can vary with each appearance. Note that (9.21) implies that $w_{n2}(1) - \mathring{w}_2(1) = 0$. Conditions (9.26c) and (9.21), the positivity of w'_{n1}, and the Hölder inequality then imply that

$$\begin{aligned}(12.9)\qquad |a\bar{\xi}_2(a)\,[w_{n2}(a) - \mathring{w}_2(a)]| &\leqq Caw_{n1}(a)\,[|w_{n2}(a) - w_{n2}(1)| + 1] \\ &\leqq C\int_a^1 w_{n1}(s)\,|w'_{n2}(s)|\,s\,ds + Cw_{n1}(1) \\ &\leqq C\,\|w_{n1}w'_{n2}, L_p\| + Cw_{n1}(1).\end{aligned}$$

In this way, by using (9.21)–(9.26), the Hölder and Poincaré inequalities, and the estimates (11.30), we find that

$$(12.10)\qquad \Lambda \leqq C\{1 + \|\gamma_n, L_p\| + \|w_{n1}, W^1_p\|\}.$$

(For certain sets of data in (12.1), $\|w_{n1}, W^1_p\| \leqq C\,\|\gamma_n, L_p\|$.) Combining (11.11) with (12.7), (12.9) we obtain

$$(12.11)\qquad \|\gamma_n, L_p\|^p \leqq C\{1 + \|\gamma_n, L_p\| + \|w_{n1}, W^1_p\|\}.$$

Next we take all the components of $\tilde{\omega}$ except $\tilde{w}_1$ to equal the components of $\boldsymbol{\omega}_n$ and we take $\tilde{w}_1 = \mathring{w}_1$. Then (12.6) and (11.21) likewise yield

$$(12.12)\qquad \|w_{n1}, W^1_p\|^p \leqq C\{1 + \|\gamma_n, L_p\|^{p-\varepsilon} + \|w_{n1}, W^1_p\|\}.$$

Inequalities (12.11) and (12.12) imply that

$$(12.13\text{a, b})\qquad \|\gamma_n, L_p\| \leqq C, \qquad \|w_{n1}, W^1_p\| \leqq C,$$

whence $\Lambda \leqq C$. It follows from (12.7) that

$$(12.14)\qquad \int_a^1 W(\gamma_n(s), s)\,s\,ds \leqq C.$$

Thus, by the definition of the norm of $\mathscr{G}$ (by duality according to the theory of Orlicz spaces), we obtain

$$(12.15)\qquad \|\gamma_n, \mathscr{G}\| \leqq C.$$

We accordingly get corresponding bounds on all the components of $\boldsymbol{\omega}_n$ except $w_{n2}, \alpha_{n22}, \alpha_{n23}$. To bound these variables we need a uniform positive lower bound for w_{n1}.

Step III. Integral inequalities. We now make judicious choices for $\tilde{\omega}$ in (12.6) in order to extract useful consequences from it.

If $\bar{\xi}_1(1)$ is prescribed in (9.20b), then we let $a < x < 1$, $0 < \varepsilon < x - a$, and set

$$\tilde{w}_1(s) \equiv \begin{cases} w_{n1}(s) & \text{for } a \leqq s \leqq x - \varepsilon, \\ w_{n1}(s) + [s - (x - \varepsilon)]/\varepsilon & \text{for } x - \varepsilon \leqq s \leqq x, \\ w_{n1}(s) + 1 & \text{for } x \leqq s \leqq 1, \end{cases} \tag{12.16}$$

$$\tilde{\mathbf{w}} = (\tilde{w}_1(s), w_{n2}, \ldots, w_{n6}), \qquad \tilde{\alpha} = \alpha_n. \tag{12.17}$$

Then $\tilde{\mathbf{w}} = (\tilde{\mathbf{w}}, \tilde{\alpha}) \in \mathscr{A}_n$, when $\tilde{\mathbf{w}}$ and $\tilde{\alpha}$ are given by (12.16) and (12.17). Substituting (12.16), (12.17) into (12.6) and letting $\varepsilon \downarrow 0$ we obtain

$$x\xi_{n1}(x) \geqq \bar{\xi}_1(1) - \int_x^1 s\eta_{n1}(s)\, ds \tag{12.18}$$

for almost all x in $(a, 1)$. Since the right-hand side of (12.18) is a continuous function of x, we can assume that (12.18) holds for all x in $(a, 1)$.

If on the other hand $w_1(1)$ is prescribed to equal $\bar{w}_1(1)$ by (9.20a), then we require a more delicate construction. Let $\mathscr{P}_n$ be the set of all $y \in (a, 1]$ for which

$$\lim_{\varepsilon \downarrow 0} \varepsilon^{-1} \int_{y-\varepsilon}^{y} w_{n1}'(s)\, ds \tag{12.19}$$

exists and exceeds $1/n$; let $\mathscr{P}_n^c$ be its complement in $[a, 1]$. (The theory of differentiation ensures that (12.19) exists a.e. On $\mathscr{P}_n^c$, $w_{n1}'(s) = 1/n$, a.e. Below we show that the Lebesgue measure of $\mathscr{P}_n^c$ approaches 0 as $n \to \infty$. For now, all we require is that $\mathscr{P}_n$ not be empty.)

Let us choose $y \in \mathscr{P}_n$, $x \in (a, y)$ and

$$0 < \lambda \leqq w_{n1}(y) - w_{n1}(x) - \frac{1}{n}(y - x), \tag{12.20}$$

the rightmost term of which is positive by the definition of $\mathscr{P}_n$. Since

$$s \mapsto w_{n1}(s) + (y - s)/n \equiv \varphi_n(s) \tag{12.21}$$

is continuous, since $\varphi_n(y) = w_{n1}(y) > w_{n1}(y) - \lambda$, and since $\varphi_n(x) \leqq w_{n1}(y) - \lambda$ by (12.20), the intermediate value theorem ensures that the equation

$$w_{n1}(t) + \lambda = w_{n1}(y) - (y - t)/n \tag{12.22}$$

has a solution $t_n(\lambda) \in [x, y)$. Since φ_n is nowhere decreasing on $[x, y]$, all solutions of (12.22) lie in an interval, which is closed because φ_n is continuous. Since $t_n(\lambda)$ satisfies (12.22) and since $y \in \mathscr{P}_n$, it follows that there is a positive number θ, depending on w_{n1} and y, such that

$$[y - t_n(\lambda)]\,[n^{-1} + \theta] < \int_{\varrho_n(\lambda)}^{\zeta} w_{n1}'(s)\, ds = [y - t_n(\lambda)]/n + \lambda, \tag{12.23}$$

which implies that

$$t_n(\lambda) \to y \quad \text{as } \lambda \downarrow 0 \quad (\text{for fixed } n). \tag{12.24}$$

When $w_1(1)$ is prescribed to equal $\overline{w}_1(1)$, we let $y \in \mathscr{P}_n$, $x \in (a, y)$, $\varepsilon \in (0, x-a)$ and take

$$\tilde{w}_1(s) \equiv \begin{cases} w_{n1}(s) & \text{for } a \leqq s \leqq x-\varepsilon, \\ w_{n1}(s) + [s-(x-\varepsilon)]\,\lambda/\varepsilon & \text{for } x-\varepsilon \leqq s \leqq x, \\ w_{n1}(s) + \lambda & \text{for } x \leqq s \leqq t_n(\lambda), \\ w_{n1}(y) - (y-s)/n & \text{for } t_n(\lambda) \leqq s \leqq y. \end{cases} \tag{12.25}$$

We define $\tilde{\boldsymbol{\omega}}$ by (12.25), (12.17), observing that $\tilde{\boldsymbol{\omega}} \in \mathscr{A}_n$. We substitute this $\tilde{\boldsymbol{\omega}}$ into (12.6), let $\varepsilon \downarrow 0$, and then let $\lambda \downarrow 0$ to obtain

$$x\xi_{n1}(x) \geqq y\xi_{n1}(y) - \int_x^y s\eta_{n1}(s)\, ds \tag{12.26}$$

for all y in $\mathscr{P}_n$ and for all $x \in (a, y)$.

By the simpler, classical version of the process leading to (12.18) or (12.26) we likewise obtain

$$x\xi_{nj}(x) = \xi_{nj}(1) - \int_x^1 s\eta_{nj}(s)\, ds, \quad j = 2, \ldots, 6, \tag{12.27}$$

for almost all x in $(a, 1)$.

To be specific in the rest of our analysis, we suppose that $\overline{\alpha}_{32}$ and $\overline{\alpha}_{33}$ are prescribed. (Thus we can exploit (10.4)–(10.6).) By substituting $\tilde{\boldsymbol{\omega}} = (\mathbf{w}_n, \tilde{\boldsymbol{\alpha}})$ with

$$\overline{\boldsymbol{\alpha}} = (\alpha_{n22} + \overline{\alpha}_{33}, \alpha_{n23} - \overline{\alpha}_{32}, \overline{\alpha}_{32}, \overline{\alpha}_{33}) \tag{12.28a}$$

and with

$$\tilde{\boldsymbol{\alpha}} = (\alpha_{n22} + \lambda\overline{\alpha}_{32}, \alpha_{n23} + \lambda\overline{\alpha}_{33}, \overline{\alpha}_{32}, \overline{\alpha}_{33}), \quad \lambda \in \mathbb{R} \tag{12.28b}$$

into (12.6) we deduce from the arbitrariness of λ that

$$\overline{\alpha}_{33}\hat{A}_{22}[\boldsymbol{\omega}_n] - \overline{\alpha}_{32}\hat{A}_{23}[\boldsymbol{\omega}_n] \geqq \overline{\alpha}_{33}\overline{A}_{22}[w_{n1}] - \overline{\alpha}_{32}\overline{A}_{23}[w_{n1}], \tag{12.29a}$$

$$\overline{\alpha}_{32}\hat{A}_{22}[\boldsymbol{\omega}_n] + \overline{\alpha}_{33}\hat{A}_{23}[\boldsymbol{\omega}_n] = \overline{\alpha}_{32}\overline{A}_{22}[w_{n1}] + \overline{\alpha}_{33}A_{23}[w_{n1}].$$

Step IV. Uniform lower bounds for w'_{n1}, w_{n1}, μ_n. Inequality (12.15) enables us to use the second part of Hypothesis 11.22 supported by (10.2), (10.3) to deduce from (12.18) that

$$x\xi_{n1}(x) \geqq -C \tag{12.30}$$

for almost all x in $(a, 1]$, and from (12.26b) that

$$x\xi_{n1}(x) \geqq y\xi_{n1}(y) - C \tag{12.31}$$

for almost all y in $\mathscr{P}_n$ and for almost all x in (a, y). Since (12.30) and (12.31) yield essential lower bounds, we regard these inequalities as holding for all such x and y. Since $\eta_2 = 0 = \eta_3$, and since (12.27) consequently implies that

$$(12.32) \qquad (1-a)\,\xi_{nj}(1) = \int_a^1 s\xi_{nj}(s)\,ds \quad \text{for } j = 2, 3,$$

we obtain from (12.15), (12.27), and Hypothesis 11.22

$$(12.33) \qquad |s\xi_{n2}(s)|, |s\xi_{n3}(s)| \leqq C \quad \forall\, s \in [a, 1].$$

Inequalities (11.26), (12.13b) enable us to deduce from (12.29) that there is a $C > 0$ such that

$$(12.34\text{a}) \qquad \bar{\alpha}_{33}\hat{A}_{22}[\omega_n] - \bar{\alpha}_{32}\hat{A}_{23}[\omega_n] \equiv \int_a^1 w_{n1}(s)\,\hat{M}(\gamma_n(s), s)\,ds \geqq -C,$$

$$(12.34\text{b}) \qquad |\bar{\alpha}_{32}\hat{A}_{22}[\omega_n] + \bar{\alpha}_{33}A_{23}[\omega_n]| \equiv \left|\int_a^1 w_{n1}(s)\,\hat{N}(\gamma_n(s), s)\,ds\right| \leqq C.$$

The analysis in the rest of this step relies critically on Hypothesis 6.10. We identify the basis $\{\boldsymbol{E}_\tau\}$ thus:

$$(12.35) \qquad \boldsymbol{E}_1 = \boldsymbol{e}_1\boldsymbol{k}_1, \quad \boldsymbol{E}_2 = \boldsymbol{e}_1\boldsymbol{k}_2, \quad \boldsymbol{E}_3 = \boldsymbol{e}_1\boldsymbol{k}_3, \quad \boldsymbol{E}_4 = \boldsymbol{e}_2\boldsymbol{k}_1,$$

$$\boldsymbol{E}_5 = \frac{\boldsymbol{e}_2(s\bar{\alpha}_{33}\boldsymbol{k}_2 - \bar{\alpha}_{32}\boldsymbol{k}_3)}{(s^2\bar{\alpha}_{33}^2 + \bar{\alpha}_{32}^2)^{1/2}}, \quad \boldsymbol{E}_6 = \frac{\boldsymbol{e}_2(s\bar{\alpha}_{32}\boldsymbol{k}_2 + \bar{\alpha}_{33}\boldsymbol{k}_3)}{(s^2\bar{\alpha}_{32}^2 + \bar{\alpha}_{33}^2)^{1/2}}$$

$$\boldsymbol{E}_7 = \boldsymbol{e}_3\boldsymbol{k}_1, \quad \boldsymbol{E}_8 = \boldsymbol{e}_3\boldsymbol{k}_2, \quad \boldsymbol{E}_9 = \boldsymbol{e}_3\boldsymbol{k}_3.$$

The dual basis is given by

$$(12.36) \qquad \boldsymbol{E}_5^* = \frac{(s^2\bar{\alpha}_{33}^2 + \bar{\alpha}_{32}^2)^{1/2}}{s(\bar{\alpha}_{32}^2 + \bar{\alpha}_{33}^2)}\,\boldsymbol{e}_2(\bar{\alpha}_{33}\boldsymbol{k}_2 - s\bar{\alpha}_{32}\boldsymbol{k}_3),$$

$$\boldsymbol{E}_6^* = \frac{(s^2\bar{\alpha}_{32}^2 + \bar{\alpha}_{33}^2)^{1/2}}{s(\bar{\alpha}_{32}^2 + \bar{\alpha}_{33}^2)}\,\boldsymbol{e}_2(\bar{\alpha}_{32}\boldsymbol{k}_2 + s\bar{\alpha}_{33}\boldsymbol{k}_3),$$

$$\boldsymbol{E}_\tau^* = \boldsymbol{E}_\tau \quad \text{for } \tau \neq 5, 6.$$

Thus

$$(12.37) \qquad \boldsymbol{F} : \boldsymbol{E}_1 = w_1', \quad \boldsymbol{F} : \boldsymbol{E}_2 = 0, \quad \boldsymbol{F} : \boldsymbol{E}_3 = 0, \quad \boldsymbol{F} : \boldsymbol{E}_4 = w_1 w_2',$$

$$\boldsymbol{F} : \boldsymbol{E}_5 = w_1\mu(s^2\bar{\alpha}_{33}^2 + \bar{\alpha}_{32}^2)^{-1/2}, \quad \boldsymbol{F} : \boldsymbol{E}_6 = w_1\nu(s^2\bar{\alpha}_{32}^2 + \bar{\alpha}_{33}^2)^{-1/2},$$

$$\boldsymbol{F} : \boldsymbol{E}_7 = w_3', \quad \boldsymbol{F} : \boldsymbol{E}_8 = s^{-1}\bar{\alpha}_{32}, \quad \boldsymbol{F} : \boldsymbol{E}_9 = \bar{\alpha}_{33},$$

$$(12.38) \quad \hat{\boldsymbol{T}} : \boldsymbol{E}_1^* = \hat{T}_{11} = \xi_1, \quad \hat{\boldsymbol{T}} : \boldsymbol{E}_2^* = \hat{T}_{12}, \quad \hat{\boldsymbol{T}} : \boldsymbol{E}_3^* = \hat{T}_{13}, \quad \hat{\boldsymbol{T}} : \boldsymbol{E}_4^* = \hat{T}_{21},$$

$$\hat{\boldsymbol{T}} : \boldsymbol{E}_5^* = \frac{(s^2\bar{\alpha}_{33}^2 + \bar{\alpha}_{32}^2)^{1/2}}{s(\bar{\alpha}_{32}^2 + \bar{\alpha}_{33}^2)}\,\hat{M}, \quad \hat{\boldsymbol{T}} : \boldsymbol{E}_6^* = \frac{(s^2\bar{\alpha}_{32}^2 + \bar{\alpha}_{33}^2)}{s(\bar{\alpha}_{32}^2 + \bar{\alpha}_{33}^2)}\,\hat{N},$$

$$\hat{\boldsymbol{T}} : \boldsymbol{E}_7^* = \hat{T}_{31}, \quad \hat{\boldsymbol{T}} : \boldsymbol{E}_8^* = \hat{T}_{32}, \quad \hat{\boldsymbol{T}} : \boldsymbol{E}_9^* = \hat{T}_{33}.$$

(We could alternatively take $\boldsymbol{E}_4 = \boldsymbol{e}_2\boldsymbol{c}/|\boldsymbol{c}|$ where $\boldsymbol{c}$ is given by (10.2).)

We now obtain a lower bound w_1^* for w_{n1} that is positive on $(a, 1]$. Suppose for the sake of contradiction that w_{n1} have no such lower bound. Then there would be an x in $(a, 1]$ and a subsequence ω_n such that $w_{n1}(x) \to 0$. Then $w_{n1} \to 0$ uniformly on $[a, x]$. The representation of $w_{n1}(x) - w_{n1}(a)$ as an integral of w'_{n1} over $[a, x]$ shows that $w'_{n1} \to 0$ in $L_1(a, x)$, whence ω_n has a further subsequence with $w'_{n1} \to 0$ pointwise a.e. on $[a, x]$. It follows from (12.15) that $\{1, 5\} \subset \mathscr{E}$, $\{2, 3, 4, 6, 7, 8, 9\} \subset e$ for almost all $y \in [a, x]$. Condition (12.34a) ensures that no alternative of (6.12b) is tenable for almost all $y \in [a, x]$. Hence there is a function w_1^* such that

$$w_{n1}(s) \geqq w_1^*(s) > 0 \quad \forall\, s \in (a, 1] \text{ and } \forall\, n. \tag{12.39}$$

Note that we can define w_1^* by

$$w_1^*(s) \equiv \inf w_{n1}(s). \tag{12.40}$$

It follows that w_1^* is nowhere decreasing, for if $x < y$, then

$$\begin{aligned} w_1^*(y) - w_1^*(x) &= \inf w_{n1}(y) - \inf w_{n1}(x) \\ &= \inf w_{n1}(y) + \sup(-w_{n1}(x)) \\ &\geqq \inf(w_{n1}(y) - w_{n1}(x)) \geqq 0 \end{aligned} \tag{12.41}$$

since w_{n1} is nowhere decreasing. If $\bar{w}_1(a)$ is prescribed, then $w_1^*(a) = \bar{w}_1(a) > 0$. Otherwise, we have yet to show that $w_1^*(a) > 0$.

A simple version of the preceding arguments shows that there is a $C > 0$ such that

$$\mu_n > 1/C. \tag{12.42}$$

We now confront the weakness of (12.26) and (12.31) inhering in the membership of y in $\mathscr{P}_n$, which conceivably could be sparsely distributed over $[a, 1]$.

12.43. Lemma. *Let* $w_{n1}(1) = \bar{w}_1(1)$. *The Lebesgue measure of* $\mathscr{P}_n^c$ *approaches* 0 *as* $n \to \infty$.

Proof. Were the conclusion false, there would be a $C > 0$ and a subsequence $\{\omega_n\}$ such that the measure of $\mathscr{P}_n^c$ exceeds C^{-1}. Condition (12.15) implies that for any $\varepsilon \in (0, \frac{1}{2}(1 - a))$ there is a subset $\mathscr{Q}_\varepsilon$ of $[a + \varepsilon, 1]$ of measure $1 - a - 2\varepsilon$ and a positive number C_ε such that the absolute value of each component of $y_n(s)$ except $w'_{n1}(s)$ is bounded by C_ε when $s \in \mathscr{Q}_\varepsilon$. Now we choose ε so small that the measure of $\mathscr{P}_n^c \cap \mathscr{Q}_\varepsilon \equiv \mathscr{R}_n^c$ has a positive lower bound (independent of n). We fix ε. Since $w_{n1} \leqq \bar{w}(1)$, inequality (11.24) implies that

$$\int_{\mathscr{R}_n} w_{n1} M_n \, ds \leqq C, \tag{12.44}$$

so that (12.34a) yields

$$\int_{\mathscr{R}_n^c} w_{n1} M_n \, ds \geqq -C. \tag{12.45}$$

The properties of $\mathcal{Q}_\varepsilon$ and the alternatives of (6.12b) imply that there is a sequence m_n of numbers with $m_n \to \infty$ as $n \to \infty$ such that $M_n \leqq -m_n$ on $\mathcal{R}_n^c$. But this inequality is incompatible with (12.45) if the measure of $\mathcal{R}_n^c$ has a positive lower bound independent of n. □

We next obtain a lower bound ξ_1^* for ξ_{n1} that is continuous on $[a, 1)$. This bound is given by (12.30) when $\bar{\xi}_1(1)$ is prescribed. We prove this by showing that for any $z \in [a, 1)$ there is a positive real number $h(z)$ such that

$$\xi_{n1}(s) \geqq -h(z) \quad \forall s \in [a, z]. \tag{12.46}$$

By choosing a sequence of such z's approaching 1 we obtain a sequence of constant lower bounds for ξ_n whose graphs are horizontal line segments in the (s, ξ_{n1})-plane. By joining parts of these segments with straight lines we readily construct a lower bound continuous on the half-open interval $[a, 1)$. Suppose that for given z there were no such $h(z)$. Then there would be a sequence $x_n \in [a, z]$ such that $\xi_{n1}(x_n) \to -\infty$ as $n \to \infty$. Then (12.31) would imply that

$$\xi_{n1}(y_n) \to -\infty \quad \forall\, y_n \in \mathcal{P}_n \cap (z, 1], \tag{12.47}$$

so that $1 \in a$ for $(\mathbf{w}_n(y_n), \alpha_n)$. Since μ_n is bounded by (12.15), (12.39), (12.42), condition (6.12a) would imply that $M_n(y_n) \to -\infty$, which is impossible by an argument like that centered on (12.45). Thus there is a function ξ_1^* continuous on $[a, 1)$ such that

$$\xi_{n1}(s) \geqq \xi_1^*(s) \quad \forall\, s \in [a, 1], \quad \forall\, n. \tag{12.48}$$

We now obtain a lower bound for $\{w_n'\}$. Let $\mathbf{f}$ be defined by (10.7). Then (12.48) implies that

$$\begin{aligned} w_{n1}'(s) &= f_1(\xi_n(s), \mathbf{w}_n(s), \alpha_n, \mathbf{w}_n(\cdot), s) \\ &\geqq f_1(\xi_1^*(s), \xi_{n2}(s), \ldots, \xi_{n6}(s), \mathbf{w}_n(s), \alpha_n, \mathbf{w}_n(\cdot), s) \\ &\geqq \min_{k \leqq n} f_1(\xi_1^*(s), \xi_{k2}(s), \ldots) \equiv \psi_n(s). \end{aligned} \tag{12.49}$$

ψ_n is continuous on $[a, 1)$. The equation (12.32), the bounds (12.33), and their analogs for $\xi_{n4}, \xi_{n5}, \xi_{n6}$, the embedding theorem, the bounds on μ_n and ν_n, and inequality (12.39) all show that on any compact subset of $(a, 1)$ the sequence $\{\psi_n\}$ is uniformly bounded, bounded below by a positive function, equicontinuous, and decreasing. The Arzelà-Ascoli Theorem implies that the whole sequence ψ_n converges uniformly on any compact subset of $(a, 1)$ to a continuous limit function ψ^*, which is positive on $(a, 1)$. (If $\bar{\xi}_1(1)$ is prescribed, then ψ^* is positive on $(a, 1]$; if $\bar{w}_1(a)$ is prescribed, then ψ^* is positive on $[a, 1)$.) We thus have

$$w_{n1}(s) \geqq \frac{a}{n} + \int_a^s \psi^*(t)\, dt. \tag{12.50}$$

Step V. Classical solutions. Let $0 < \varepsilon < \frac{1}{2}(1 - a)$. Let g be any piecewise continuously differentiable function with $g(s) = 0$ for $s \in [a, a + \varepsilon] \cup [1 - \varepsilon, 1]$ and with $|g'| \leqq \frac{1}{2}\psi^*$. Set

$$\tilde{w}_1 = w_{n1} + g, \quad \tilde{\mathbf{w}} = (\tilde{w}_1, w_{n1}, \ldots). \tag{12.51}$$

Then for n sufficiently large, $\tilde{\omega} \in \mathscr{A}_n$ (since ψ^* has a positive lower bound on $[a+\varepsilon, 1-\varepsilon]$). We substitute (12.51) into (12.6) and use the arbitrariness of g to obtain in place of (12.18) and (12.26) the equality

$$x\xi_{n1}(x) = y\xi_{n1}(y) - \int_x^y s\eta_{n1}(s)\, ds \qquad \forall x, y \in [a+\varepsilon, 1-\varepsilon]. \tag{12.52}$$

We use $\hat{\mathbf{f}}$ of (10.7) to convert (12.52), (12.27) into a form yielding an explicit representation for $\mathbf{w}_n'$. By the standard boot-strap argument it follows that $\boldsymbol{\omega}_n$ generates a twice continuously differentiable solution of (9.18) on $[a+\varepsilon, 1-\varepsilon]$ when n is sufficiently large. Indeed, if k is a positive integer, then the representations for $\mathbf{w}_n'$ supported by the estimates of Steps II and IV show that $\mathbf{w}_n'$ is uniformly bounded and equicontinuous on $[a+k^{-1}, 1-k^{-1}]$ and has a subsequence $\{\mathbf{w}_{n,k}'\}$ that converges uniformly on $[a+k^{-1}, 1-k^{-1}]$ while $\boldsymbol{\alpha}_{n,k}$ converges in $\mathbb{R}^8$. We assume without loss of generality that $\{\mathbf{w}_{n,k+1}'\}$ is a subsequence of $\{\mathbf{w}_{n,k}'\}$. It follows that the diagonal subsequence $(\mathbf{w}_{n,n}, \boldsymbol{\alpha}_{n,n})$ converges in $C^1([a+\varepsilon, 1-\varepsilon]) \times \mathbb{R}^8$ to a limit $(\mathbf{w}, \boldsymbol{\alpha})$ for every $\varepsilon \in (0, \frac{1}{2}(1-a))$. It is easily verified that $(\mathbf{w}, \boldsymbol{\alpha})$ satisfies the differential equations (9.18) on $(a, 1)$. Next we replace (12.28a) by

$$\boldsymbol{\alpha} = (\alpha_{n22} + \lambda\bar{\alpha}_{33}, \alpha_{n23} - \lambda\bar{\alpha}_{32}, \bar{\alpha}_{32}, \bar{\alpha}_{33}), \tag{12.52}$$

which is admissible for small negative λ by virtue of (12.42). Thus in place of (12.29a) we obtain the corresponding equality. By letting $n \to \infty$ through the diagonal subsequence in this modification of (12.29a) and in (12.29b) we find that $(\mathbf{w}, \boldsymbol{\alpha})$ satisfies the obvious limit form, whence we obtain

$$\hat{A}_{22}[\boldsymbol{\omega}] = \bar{A}_{22}[w_1], \qquad \hat{A}_{23}[\boldsymbol{\omega}] = \bar{A}_{23}[w_1]. \tag{12.53}$$

We verify by a similar process that other side conditions are met.

If $\bar{\xi}_1(1)$ is prescribed, we can carry out our construction of $\mathbf{w}_{n,k}'$ on intervals of the form $[a+k^{-1}, 1]$, thereby obtaining the equality corresponding to (12.18) for the limit $(\mathbf{w}, \boldsymbol{\alpha})$. Hence $\xi_1(\mathbf{w}'(s), \mathbf{w}(s), \boldsymbol{\alpha}, \mathbf{w}(\cdot), s) \to \bar{\xi}_1(1)$ as $s \to 1$.

From now on, $\{\boldsymbol{\omega}_n\}$ is understood to stand for the diagonal subsequence $\{\boldsymbol{\omega}_{n,n}\}$ or a further subsequence thereof.

We can now show that $\xi_{n1}(1)$ is bounded below. Note that (10.2), (10.5) imply that

$$\eta_{n1} = w_{n1}^{-1} w_{n2}' \xi_{n2} + s^{-1}(\bar{\alpha}_{32}^2 + \bar{\alpha}_{33}^2)^{-1}(\mu_n M_n + \nu_n N_n). \tag{12.54}$$

Hypothesis 11.22 or estimate (12.34b) together with (12.39) imply that $\{|N_n|\}$ is bounded by a fixed integrable function of s on any compact subinterval of $(a, 1]$. Hypothesis 11.22 and estimate (12.34b) imply the same for $|M_n|$. Hypothesis 11.22 alone implies the same for ξ_{n2}. Our estimates for μ_n, ν_n and our representation for w_{n2}' then show that for each $x \in (a, 1]$ there is a number $h(x)$ such that

$$\left| \int_x^1 \eta_{n1}(s)\, ds \right| < h(x). \tag{12.55}$$

Combining this estimate with the limiting equality

$$x\xi_1(x) = y\xi_1(y) - \int_x^y s\eta_1(s)\,ds \quad \text{for } a < x < y < 1 \tag{12.56}$$

corresponding to (12.26), we find that $\xi_1(1) > -\infty$, which implies that $\xi_{n1}(1)$ is bounded below.

To show that

$$\xi_1(\mathbf{w}'(s), \mathbf{w}(s), \alpha, \mathbf{w}(\cdot), s) \to \bar{\xi}_1(a) \quad \text{as } s \to a \tag{12.57}$$

when $\bar{\xi}_1(a)$ is prescribed, we require a positive lower bound for $w_{n1}(a)$ so that we can choose an $\tilde{\boldsymbol{\omega}}$ with $\tilde{w}_1(a) < w_{n1}(a)$ for large enough n. Indeed, using (12.56) n the limiting form of (12.6) we obtain

$$0 \leqq [\xi_1(a) - \bar{\xi}_1(a)]\,[w_1(a) - \tilde{w}_1(a)] \tag{12.58}$$

so that

$$\xi_1(a) \leqq \bar{\xi}_1(a). \tag{12.59}$$

We can now use (12.56) to conclude that (12.55) holds with $x = a$. Thus

$$\xi_{n1}(s) \leqq \bar{\xi}_1(a) + C \leqq C. \tag{12.60}$$

We now prove the existence of a lower bound for $w_{n1}(a)$. Since f_1 is increasing in ξ_1, we obtain from (12.59) that

$$w'_{n1}(s) \leqq f_1^\dagger(s^{-1}w_{n1}(s), w_{n4}(s), C, \xi_{n2}(s), \ldots, \xi_{n6}(s), \mathbf{w}_n(\cdot), s). \tag{12.61}$$

Now suppose for the sake of contradiction that there is a subsequence for which $w_{n1}(a) \to 0$ as $n \to \infty$. Estimate (12.39) ensures that

$$s^{-1}w_{n1}(s) \geqq s^{-1}w_1^*(s) \geqq w_1^*(s) \geqq a^{-1}w_{n1}(a) \tag{12.62}$$

for this subsequence for n sufficiently large. Property (11.27a) enables us to deduce from (12.60), (12.61) that

$$w'_{n1}(s) \leqq f_1^\dagger(a^{-1}w_{n1}(a), w_{n4}(s), \ldots) \tag{12.63}$$

for s sufficiently close to a and for n sufficiently large. Hence

$$w_{n1}(s) \leqq w_{n1}(a) + \int_a^s f_1^\dagger(a^{-1}w_{n1}(a), w_{n4}(t), \ldots)\,dt. \tag{12.64}$$

Property (11.27b) then yields

$$\eta_{n1}(s) \leqq \eta_1^\dagger\left(a^{-1}w_{n1}(a) + s^{-1}\int_a^s f_1^\dagger(a^{-1}w_{n1}(a), w_{n4}(t), \ldots)\,dt,\; w_{n4}(s), \ldots\right). \tag{12.65}$$

But then (12.55) contradicts (11.28). It follows that $\tilde{w}_1(a) \geqq 1/C$ and that we can consequently choose $\tilde{w}_1(a)$ to reverse the inequality in (12.58). Hence (12.57) holds. The demonstration that other Neumann conditions at a are satisfied is routine. These results complete the proof of Theorem 12.2.

Note added in proof. The question of injectivity, raised in the paragraph containing (3.2), has been effectively treated for variational problems of nonlinear elasticity by P. G. CIARLET & J. NEČAS in their paper "Injectivity and Self-Contact in Nonlinear Elasticity" (*Arch. Rational Mech. Anal.*, to appear). For time-dependent problems of electromagnetism it is mathematically convenient to regard $\boldsymbol{e}$ and $\boldsymbol{h}$ as depending on $\boldsymbol{d}$ and $\boldsymbol{b}$. The restricted strong ellipticity condition of Section 6 together with corresponding growth conditions permit an easy inversion of the constitutive functions from the form used in our paper to this form. But the issues raised in the fourth paragraph of section 6 still remain.

Acknowledgment. The work of ANTMAN was partially supported by National Science Foundation Grant No. DMS-8503317. ROGERS was partially sponsored by the United States Army under Contract No. DAAG29-80-C-0041, and this material is partially based upon work supported by the National Science Foundation under Grant No. DMS-8210950, Mod. 1. Some of the results reported here were developed in the doctoral dissertation of ROGERS (1984). We are indebted to R. A. TOUPIN for a number of helpful comments.

References

S. A. AMBARTSUMIAN (1982), Magneto-elasticity of thin plates and shells, *Appl. Mech. Rev.* **35**, 1–5.

S. S. ANTMAN (1978), A family of semi-inverse problems of nonlinear elasticity, in *Contemporary Developments in Continuum Mechanics and Partial Differential Equations*, ed. by G. M. DE LA PENHA & L. A. MADEIROS, North-Holland, 1–24.

S. S. ANTMAN (1979), The eversion of thick spherical shells, *Arch. Rational Mech. Anal.* **70**, 113–123.

S. S. ANTMAN (1983), Regular and singular problems for large elastic deformations of tubes, wedges, and cylinders, *Arch. Rational Mech. Anal.* **83**, 1–52. Corrigenda, *ibid.*, pp. 391–393, below.

S. S. ANTMAN & J. E. OSBORN (1979), The principle of virtual work and integral laws of motion, *Arch. Rational Mech. Anal.* **69**, 231–262.

J. M. BALL (1977), Convexity conditions and existence theorems in nonlinear elasticity, *Arch. Rational Mech. Anal.* **63**, 337–403.

J. M. BALL (1982), Discontinuous equilibrium solutions and cavitation in nonlinear elasticity, *Phil. Trans. Roy. Soc. London* **A 306**, 557–611.

J. M. BALL & F. MURAT (1984), $W^{1,p}$-quasiconvexity and variational problems for multiple integrals, *J. Functional Anal.* **58**, 225–253.

H. BREZIS (1968), Équations et inéquations non linéaires dans les éspaces vectoriels en dualité, *Ann. Inst. Fourier* **18**, 115–175.

F. E. BROWDER (1977), Pseudo-monotone operators and nonlinear elliptic boundary value problems on unbounded domains, *Proc. Natl. Acad. Sci.* **74**, 2659–2661.

W. F. BROWN, Jr. (1966), *Magnetoelastic Interactions*, Springer.

B. DACOROGNA (1982), *Weak Continuity and Weak Lower Semi-continuity of Non-Linear Functionals*, Springer Lecture Notes in Mathematics **922**.

S. R. DE GROOT & L. G. SUTTORP (1972), *Foundations of Electrodynamics*, North Holland.

R. C. DIXON & A. C. ERINGEN (1965), A dynamical theory of polar elastic dielectrics, *Int. J. Eng. Sci.* **3**, 359–398.

J. L. ERICKSEN (1980), Some phase transitions in crystals, *Arch. Rational Mech. Anal.* **73**, 99–124.

A. C. ERINGEN (1963), On the foundations of electroelastodynamics, *Int. J. Eng. Sci.* **1**, 127–153.

Y. ERSOY & E. KIRAL (1978), Dynamic theory for polarizable and magnetizable magneto-electric thermoviscoelastic electrically and thermally conductive anisotropic solids having magnetic symmetry, *Int. J. Eng. Sci.* **16**, 483–492.

R. M. FANO, L. J. CHU, & R. B. ADLER (1960), *Electromagnetic Fields, Energy, and Forces*, Wiley.

M. GIAQUINTA (1983), *Multiple Integrals in the Calculus of Variations and Nonlinear Elliptic Systems*, Princeton Univ. Press.

J. W. GIBBS & E. B. WILSON (1901), *Vector Analysis*, Yale Univ. Press.

J. K. HALE (1969), *Ordinary Differential Equations*, Wiley-Interscience.

K. HUTTER (1975), On thermodynamics and thermostatics of viscous thermoelastic solids in the electromagnetic fields. A Lagrangian formulation, *Arch. Rational Mech. Anal.* **58**, 339–368.

K. HUTTER & A. A. F. VAN DE VEN (1978), *Field Matter Interactions in Thermoelastic Solids*, Springer Lecture Notes in Physics **88**.

N. F. JORDAN & A. C. ERINGEN (1964), On the static nonlinear theory of electromagnetic thermoelastic solids I, II, *Int. J. Eng. Sci.* **2**, 59–114.

L. V. KANTOROVICH & G. P. AKILOV (1977), *Functional Analysis*, 2nd ed., (in Russian), Nauka, Moscow; Engl. transl. (1982), Pergamon.

M. A. KRASNOSEL'SKII (1956), *Topological Methods in the Theory of Nonlinear Integral Equations*, (in Russian), Gostekhteorizdat, English transl. (1964), Pergamon Press.

M. A. KRASNOSEL'SKII & YA. B. RUTITSKII (1958), *Convex Functions and Orlicz Spacses*, (in Russian), Fizmatgiz; English transl. (1961), Noordhoff.

J. L. LIONS (1969), *Quelques Méthodes de Résolution des Problèmes aux Limites non Linéaires*, Dunod, Gauthier-Villars.

G. A. MAUGIN (1981), Wave motion in magnetizable deformable solids, *Int. J. Eng. Sci.* **19**, 321–388.

G. A. MAUGIN, ed. (1984), *The Mechanical Behavior of Electromagnetic Solid Continua*, Proc. IUTAM-IUPAP Symp., 1983, North Holland.

G. A. MAUGIN & A. C. ERINGEN (1977), On the equations of the electrodynamics of deformable bodies of finite extent, *J. de Mécanique* **16**, 101–145.

M. F. MCCARTHY & H. F. TIERSTEN (1978), On integral forms of the balance laws for deformable semiconductors, *Arch. Rational Mech. Anal.* **68**, 27–36.

F. C. MOON (1978), Problems in magneto-solid mechanics, in *Mechanics Today*, Vol. **4**, edited by S. Nemat-Nasser, 307–390.

F. MOON (1984), *Magneto-solid Mechanics*, Wiley.

D. F. NELSON (1979), *Electric, Optic, and Acoustic Interactions in Dielectrics*, Wiley.

Y. H. PAO (1978), Electromagnetic forces in deformable continua. *Mechanics Today*, Vol. 4, Pergamon Press, 209–305.

Y. H. PAO & K. HUTTER (1975), Electrodynamics for moving solids and viscous fluids, *Proc. IEEE* **63**, 1011–1021.

H. PARKUS, ed. (1979), *Electromagnetic Interaction in Elastic Solids*, C.I.S.M. 257, Springer-Verlag, Vienna.

P. PENFIELD & H. A. HAUS (1967), *Electrodynamics of Moving Media*, M.I.T. Press.

R. C. ROGERS (1984), Analysis of the nonlinear equations describing the elastic thermal, and electromagnetic behavior of solids: Existence of solutions of semi-inverse problems, Doctoral dissertation, University of Maryland.

H. SINGH & A. C. PIPKIN (1966), Controllable states of elastic dielectrics, *Arch. Rational Mech. Anal.* **21**, 169–209.

S. L. SOBOLEV (1950), *Applications of Functional Analysis in Mathematical Physics*, Leningrad State Univ. Press, English transl. 1963, Amer. Math. Soc.

E. M. STEIN (1970), *Singular Integrals and Differentiability Properties of Functions*, Princeton Univ. Press.

R. S. TEBBLE (1969), *Magnetic Domains*, Methuen & Co.

R. A. TOUPIN (1956), The elastic dielectric, *J. Rational Mech. Anal.* **5**, 849–915.

R. A. TOUPIN (1963), A dynamical theory of elastic dielectrics, *Int. J. Eng. Sci.* **1**, 101–126.

C. TRUESDELL & W. NOLL (1965), *The Non-linear Field Theories of Mechanics, Handbuch der Physik III/3*, Springer-Verlag.

C. TRUESDELL & R. A. TOUPIN (1960), *The Classical Field Theories, Handbuch der Physik*, III/1, Springer-Verlag.

P. D. S. VERMA (1964), Symmetrical expansions of a hollow spherical dielectric, *Int. J. Eng. Sci.* **2**, 21–27.

J. B. WALKER, A. C. PIPKIN, & R. S. RIVLIN (1965), Maxwell's equations in a deformed body, *Rend. Acc. Naz. Lincei*, Ser. VIII **37**, 674–676.

Mathematics Research Center
University of Wisconsin
Madison
and
Department of Mathematics
University of Maryland
College Park

(Received January 6, 1986)

Existence of Weak Solutions to Linear Indefinite Systems of Differential Equations

R. J. KNOPS

Dedicated to Walter Noll on his sixtieth birthday

Introduction

It is well-known that a solution exists to a linear system of differential equations for all data in a given Banach space if and only if the solution depends continuously in a suitable norm on the data (see, for example, FICHERA [4]). Here, however, we are interested in the situation when continuous dependence may break down, so that a solution need not generally exist. Consequentially, it is natural to ask for what data, if any, a solution does exist. The question is not vacuous since eigensolutions exist, although admittedly for a highly restricted set of data.

Specific motivation for this study arises from topics such as instability, bifurcation and dynamic phase transitions in the linearised theory of elasticity, so that of concern is the Cauchy or initial-boundary value problem. More generally, our analysis has relevance in the theory of ill-posed problems where frequently attention has concentrated upon uniqueness and continuous dependence and the treatment of existence has tended to be left aside. Nevertheless, due to their technological importance, ill-posed problems are often required to be solved numerically and without a thorough understanding of existence, such computations might well be meaningless.

In fact, we prove that a solution in a certain weak sense always exists for a dense set of data. The method of proof adopted employs a decomposition into invariant subspaces, on each of which the equation becomes definite thus allowing the use of techniques familiar in the theory of elliptic and hyperbolic operators. These arguments are combined to establish that the operator, used in the definition of weak solution, is closed. An application of known results from the theory of adjoint operators then yields existence. Such an approach is equivalent to a spectral decomposition, but we do not explicitly use this method nor invest in its terminology. Of course, the various *stages* of our argument follow a standard pattern (see, *e.g.* [4], [16]), but the operators we discuss, mainly differing from those previously considered in the literature, require the development of completely new proofs to those usually employed.

Elements of the general analysis presented here were first announced in [7], [8] within the context of elasticity theory, and are subsumed into the generalised treatment of this paper in order to make the mathematical discussion reasonably self-contained. The abstract formulation also broadens the applicability of our results beyond problems encountered in elasticity.

Section 2, commencing with a brief motivation of the abstract Cauchy problem, contains the definitions of the basic subspaces and operators in terms of which the weak solution is later defined. An essential ingredient of the existence theorem proved in Section 4 demands that these operators are closed and the proof of this property occupies Section 3. Section 4 also sets down the definition of the weak solution, briefly discusses its uniqueness and then disposes of the existence proof.

Other authors who have discussed existence of solutions to ill-posed problems, notably either the Dirichlet problem or backwards parabolic equations, include Littman [14], Zaidman [21], Bates [1], Showalter [17], Vajvoda & Stedry [19] and Colton & Wimp [3]. The interested reader may also wish to refer to the books by Lions & Magenes [13] and by Tihonov & Arsenin [18] for an account of quasi-reversibility and other methods used in earlier discussions of existence.

2. Motivation. Basic definitions and assumptions

In order to motivate our analysis properly, we commence this Section with a brief description of the system of partial differential equations whose abstract formulation and associated weak solution provide the primary objects of study. The abstract version is constructed from two operators defined on invariant subspaces, and a major task in proving the existence of a weak solution is to establish that these operators are closed. Once this is accomplished the proof of existence is fairly straightforward. These considerations are deferred to Section 3 and 4. In this Section, apart from the basic motivation, our object is to define the invariant subspaces and also the operators involved in the abstract formulation.

Thus, let $\Omega \subset \mathbb{R}^3$ be a bounded region of three-dimensional euclidean space, and let the boundary, $\partial\Omega$, be smooth. Let $u: \Omega \times [0, T] \to \mathbb{R}^3$ be a smooth function. We consider the system of equations

$$(c_{ijkl}u_{i,j})_{,k} + f_i = \varrho \ddot{u}_i, \quad i, j, k, l = 1, 2, 3, \tag{2.1}$$

where $c_{ijkl}(x)$ and $\varrho(x)$ are smooth functions, the usual comma and summation conventions apply and a superposed dot indicates partial differentiation with respect to time. We also suppose that $\varrho(x) \geqq \varrho_0 > 0$, where ϱ_0 is constant, and that $f: \Omega \to \mathbb{R}^3$ is smooth. The coefficients are further assumed to possess the symmetry property

$$c_{ijkl} = c_{klij},$$

but are not assumed to satisfy any definiteness condition. Furthermore, the boundary and initial conditions respectively are given by

$$u_i = 0, \quad x \in \partial\Omega, \quad i = 1, 2, 3, \tag{2.3}$$

$$u_i(x, 0) = u_{0i}(x), \quad \dot{u}_i(x, 0) = u_{1i}(x), \quad x \in \Omega. \tag{2.4}$$

The initial-boundary value problem (2.1)–(2.4) arises, for instance, in the theory of linearised elasticity where $\varrho(x)$ is the density and it is customary to suppose that the elasticities c_{ijkl} are either positive-definite or at least strongly-elliptic. Under either condition, a solution to the problem is known to exist (see FICHERA [5], HUGHES & MARSDEN [15]). However, when both conditions fail, the problem still possesses eigensolutions and hence it becomes instructive to examine the general question of existence when the elasticities are indefinite: solutions other than eigensolutions might also exist. Indefinite elasticities imply instability [10] but unless there is an accompanying existence proof the analysis must remain purely formal. It is this aspect that in part prompts the present investigation. Other circumstances where a knowledge of existence in the absence of strong-ellipticity is desirable are mentioned in [7, 8].

In preparation for the abstract formulation, let us recall that by suitable adjustment of the body-force term, f, there is no loss in assuming that the initial (Cauchy) data is homogeneous. Thus, we may, on using a compact notation, consider the equivalent problem

$$Cu + f = \varrho\ddot{u}, \qquad (x, t) \in \Omega \times [0, T], \tag{2.5}$$

$$u = 0, \qquad (x, t) \in \partial\Omega \times [0, T], \tag{2.6}$$

$$u = 0, \qquad \dot{u} = 0, \qquad x \in \Omega, \tag{2.7}$$

where the second-order differential operator C is given by

$$Cu = \operatorname{Div} \mathbb{C}\, \nabla u, \tag{2.8}$$

in which Div denotes the usual tensor divergence operator, the gradient operator is denoted by ∇, and $\mathbb{C}$ is the elasticity tensor assumed to satisfy:

$$\mathbb{C} = \mathbb{C}^T. \tag{2.9}$$

Here, a superscript T denotes the transposed tensor.

We now construct an abstract version of the problem (2.5)–(2.9) which is later used in the proof of existence.

Let X be a Hilbert space. We denote by $\|.\|_X$, $\langle,\rangle_X$ the corresponding norm and inner product defined on X. The domain, range and null-space of an operator are represented respectively by $D(.)$, $R(.)$ and $N(.)$. Other notation will be introduced without explanation, since it has become standard (*cp.*, for example, LIONS [12]).

Let $C: D(C) \subset L^2(\Omega) \to L^2(\Omega)$ be a self-adjoint unbounded densely defined operator on $L^2(\Omega)$ whose spectrum is contained in $(-\infty, \infty) \setminus (-\varkappa, 0)$, where $\varkappa$ is a positive constant. Thus, for $u \in D(C)$, it is assumed that the operator C can satisfy either the condition

$$\langle Cu, u\rangle_{L^2(\Omega)} \geqq \varkappa \|u\|^2_{L^2(\Omega)} \tag{2.10}$$

or the condition

$$\langle Cu, u\rangle_{L^2(\Omega)} \leqq 0. \tag{2.11}$$

Motivated by these spectral properties of the operator C, we decompose $L^2(\Omega)$ into the direct sum of two orthogonal subspaces $H_\pm$:

$$L^2(\Omega) = H_+ \oplus H_- . \tag{2.12}$$

Let $P_\pm$ denote the corresponding projection operators, so that $P_\pm : L^2(\Omega) \to H_\pm$. Moreover, we also have,

$$\langle P_+ u, P_- u\rangle_{L^2(\Omega)} = 0, \quad u \in L^2(\Omega), \tag{2.13}$$

$$I = P_+ + P_-, \tag{2.14}$$

where I denotes the identity operator. We make the *assumption* that the subspaces $H_\pm$ are invariant with respect to C, in the sense that

(H1) $$P_\pm u \in H_\pm \Rightarrow CP_\pm u \in H_\pm, \quad u \in L^2(\Omega). \tag{2.15}$$

Let us observe that in order for (H1) to hold, we require that

$$CP_\pm = P_\pm C. \tag{2.16}$$

In terms of the invariant subspaces $H_\pm$ and the projections $P_\pm$, we may now describe the spectral properties of the operator C according to the following: for $u \in D(C)$, there holds either the condition

$$\langle CP_+ u, u\rangle_{L^2(\Omega)} \geqq \varkappa \, \| P_+ u\|^2_{L^2(\Omega)}, \tag{2.17}$$

or the condition

$$\langle CP_- u, u\rangle_{L^2(\Omega)} \leqq 0. \tag{2.18}$$

Our next aim is to introduce operators defined on the subspaces $H_\pm$ in terms of which the weak solution to our problem is later constructed in Section 4. For this purpose, we observe that, formally, the differential equation (2.5) may be expressed as

$$\left.\begin{aligned} Au &= f \\ \text{where} \qquad Au &= \ddot{u} - Cu, \end{aligned}\right\} \tag{2.19}$$

and without loss, we have set $\varrho \equiv 1$. We define the operator A in terms of operators $A_\pm$ according to the following definitions. In the next section, by means of the properties (2.17), (2.18), we show that $A_\pm$ are closed operators and hence that A is also closed.

Let $Q = \Omega \times [0, T)$, where $T > 0$, and let $v : Q \to L^2(Q)$.

Definition 2.1. The operators $A_\pm$ are respectively given by

$$A_\pm v = \ddot{v} - Cv, \quad v \in D(A_\pm), \tag{2.20}$$

where

$$D(A_+) = \{v \in W^{2,2}(Q) : v(t) \in H_+, \ \forall\, t \in [0, T];\ v(T) = \dot{v}(T) = 0;\ v|_{\partial\Omega} = 0\} \tag{2.21}$$

$$\begin{aligned} D(A_-) = \{v \in C^0([0, T]; X) \cap C^1([0, T]; L^2(\Omega)) : v(t) \in H_-, \ \forall\, t \in [0, T]; \\ A_- v \in L^2(Q);\ v(T) = \dot{v}(T) = 0;\ v|_{\partial\Omega} = 0\}, \end{aligned} \tag{2.22}$$

and X is the subspace of $L_2(\Omega)$ equipped with the norm

$$\|v\|_X^2 = \|v\|_{L^2(\Omega)}^2 - \langle Cv, v\rangle_{L^2(\Omega)}, \qquad v \in H_- . \tag{2.23}$$

Remark 2.1. By taking less general function spaces for $D(A_\pm)$, the existence of a (weak) solution can be established in a straightforward manner using well-known methods (see, *e.g.*, NIRENBERG [16]).

Remark 2.2. It easily follows from the definition that $v \in D(A_+) \Rightarrow A_+ v \in L^2(Q)$. In fact, it is possible to prove somewhat more.

Proposition 2.1. $v \in D(A_+) \;\Rightarrow\; A_+ v \in H_+$.

Proof. The linearity of the projection operator P_+ implies that

$$(P_+ v)^{\cdot} = P_+ \dot{v},$$

where the superposed dot indicates differentiation with respect to time. Hence we have that

$$v \in H_+ \;\Rightarrow\; \ddot{v} \in H_+ .$$

Furthermore, by the assumed invariance of subspace H_+ with respect to the operator C, it follows that $v \in H_+ \;\Rightarrow\; Cv \in H_\pm$ and the proof is complete.

Remark 2.3. It may be shown in exactly the same way that $v \in D(A_-) \;\Rightarrow\; v \in H_-$.

The operator A is next defined in terms of the operators $A_\pm$ according to the following definition.

Definition 2.2. The operator A is given by

$$Av = A_+ P_+ v + A_- P_- v, \qquad v \in D(A), \tag{2.24}$$

where

$$D(A) = \{v \in L^2(Q) : P_+ v \in D(A_+),\, P_- v \in D(A_-)\}. \tag{2.25}$$

Remark 2.4. By means of Remark 2.2 and Definition 2.1, we may easily show that $v \in D(A) \;\Rightarrow\; Av \in L^2(Q)$. This property is needed in the definition of a weak solution to the abstract version of the Cauchy problem (2.5)–(2.7), given in the next Section.

Remark 2.5. Definition 2.2 accords with the formal notion of the operator A introduced in equation (2.19), since for $v \in D(A)$ we have

$$\begin{aligned} Av &= A_+ P_+ v + A_- P_- v \\ &= (P_+ v)^{\cdot\cdot} - C P_+ v + (P_- v)^{\cdot\cdot} - C P_- v \\ &= (P_+ v + P_- v)^{\cdot\cdot} - C(P_+ v + P_- v) \\ &= \ddot{v} - Cv . \end{aligned}$$

Finally in this Section, we demonstrate the consistency of Definitions 2.1 and 2.2 by proving that the projections of the operator A onto the subspaces $H_\pm$ are the operators $A_\pm$. This result will also be required in Section 4 when we prove that the operator A is closed.

Proposition 2.2. *Let* $v \in D(A)$. *Then* $P_\pm Av = A_\pm(P_\pm v)$, *where the signs are taken in conjunction.*

Proof. It follows from Definitions (2.1) and (2.2) that when $v \in D(A)$ we have

$$P_\pm Av = P_\pm(A_+P_+v + A_-P_-v),$$

and therefore on using Proposition (2.1) together with orthogonality of the subspaces $H_\pm$, we see that, for instance,

$$P_+Av = A_+P_+v.$$

A similar equality holds for P_-Av, and hence the result is established.

The proof that a weak solution exists, contained in Section 4, depends upon standard operator theory but, however, requires that the operator A is closed. This property cannot be immediately established due to the somewhat general function spaces employed in the definition of $D(A)$. Instead, we first prove that the component operators $A_\pm$ are closed and then deduce that A is closed.

3. Proof that the operators $A_\pm$, A are closed

We prove that the operator A is closed by first establishing that each of the operators A_+ and A_- is closed. The techniques employed for the operators A_+, A_- are different and reflect the elliptic and hyperbolic character which is readily suggested from the spectral decomposition (2.10), (2.11) of the operator C onto the subspaces H_+, H_-. The proof that A_+ is a closed operator has been sketched previously in [7], but for convenience the detailed proof is presented here, with appropriate notational change.

In preparation for the proof, we state a preliminary lemma.

Lemma 3.1. *Let* $\{v_n\}$ *be a sequence such that* $v_n \in L^2(Q)$ *and* $v_n \xrightarrow{L^2(Q)} v$. *Then*

$$P_\pm v_n \xrightarrow{L^2(Q)} P_\pm v. \tag{3.1}$$

Proof. The decomposition (2.12) into the orthogonal subspaces $H_\pm$ yields the equality

$$\| P_+v_n - P_+v\|^2_{L^2(Q)} + \| P_-v_n - P_-v\|^2_{L^2(Q)} = \| v_n - v\|^2_{L^2(Q)} \tag{3.2}$$

from which the result immediately follows.

We observe that Lemma 3.1 enables us to conclude that, in particular, when

$v_n \in H_\pm$, the limit function v to which the sequence converges in $L^2(Q)$ also lies in $H_\pm$. We now present three propositions which establish that A_+, A_-, A are closed.

Proposition 3.1. *The operator A_+ is closed on $L^2(Q)$.*

Proof. We are required to prove that for a given sequence $v_n \in D(A_+)$ such that

$$v_n \xrightarrow{L^2(Q)} v, \tag{3.3}$$

$$A_+ v_n \xrightarrow{L^2(Q)} w, \tag{3.4}$$

then $v \in D(A_+)$ and $A_+ v = w$. That is, the graph of A_+ is closed.
Now, $v_n \in D(A_+) \Rightarrow v_n \in H_+$ and hence by (2.17), the operator A_+ is elliptic. Hence by standard *a priori* estimates, we have

$$\|v_n\|_{W^{2,2}(Q)} \leqq c(\|v_n\|_{L^2(Q)} + \|A_+ v_n\|_{L^2(Q)}), \tag{3.5}$$

where c is a positive constant. But by hypothesis, $v_n \in L^2(Q)$ and hence from (3.5), the sequence $\{v_n\}$ is bounded in $W^{2,2}(Q)$, which is a reflexive Banach space. Thus, $\{v_n\}$ contains a weakly convergent subsequence $\{v_{n_k}\}$ such that

$$v_{n_k} \xrightarrow{W^{2,2}(Q)} v. \tag{3.6}$$

and it therefore follows that $v \in W^{2,2}(Q)$. By Lemma 3.1, we see that also $v \in H_+$.
Furthermore, standard trace theorems (see, *e.g.*, [12]) show that

$$v_{n_k}(T) \to v(T), \quad \dot{v}_{n_k}(T) \to \dot{v}(T), \quad v_{n_k}|_{\partial\Omega} \to v|_{\partial\Omega}. \tag{3.7}$$

We may therefore conclude that $v \in D(A_+)$. Finally, from (3.6) we have

$$A_+ v_n \xrightarrow{L^2(Q)} A_+ v, \tag{3.8}$$

so that from (3.5) $A_+ v = w$ and the Proposition is proved.

Proposition 3.2. *The operator A_- is closed on $L^2(Q)$.*

Proof. The method we employ is based upon "energy" estimates of the usual kind for linear hyperbolic operators. We first establish such *a priori* inequalities for the elements of the sequence $\{v_n\}$, which satisfy $v_n \in D(A_-)$ and the convergence properties

$$v_n \xrightarrow{L^2(Q)} v, \tag{3.9}$$

$$A_- v_n \xrightarrow{L^2(Q)} w. \tag{3.10}$$

As before, we must then show that $v \in D(A_-)$ and that $A_- v = w$.
Thus, let $f_n \in L^2(Q)$ be defined by the relation

$$f_n = A_- v_n, \quad v_n \in D(A_-). \tag{3.11}$$

Then, by Definition (2.1) for A_-, we see that the quantity $E_n(t)$, given by

$$E_n(t) = \tfrac{1}{2}\langle \dot{v}_n, \dot{v}_n\rangle_{L^2(\Omega)} - \tfrac{1}{2}\langle C v_n, v_n\rangle_{L^2(\Omega)}, \tag{3.12}$$

satisfies the equation

$$E_n(t) = \int_T^t \langle f_n, \dot{v}_n\rangle_{L^2(\Omega)}\, d\eta, \quad t \in [0, T]. \tag{3.13}$$

Schwarz's inequality then gives

$$E_n(t) \leqq 2^{\frac{1}{2}} \|f_n\|^2_{L^2(Q)} \left(\int_t^T E_n(\eta)\, d\eta\right)^{\frac{1}{2}}, \quad t \in [0, T], \tag{3.14}$$

which on setting

$$E_n = \max_{t\in[0,T]} E_n(t), \tag{3.15}$$

enables us to conclude that

$$E_n^{\frac{1}{2}}(t) \leqq (2T)^{\frac{1}{2}} \|f_n\|_{L^2(Q)}, \quad t \in [0, T]. \tag{3.16}$$

The last inequality implies that $\{E_n(t)\}$ is a bounded sequence in $C^0([0, T])$, so that on using (3.12) and the Definition (2.23), we may conclude that $\{v_n\}$ forms a bounded sequence in the Banach space $C^0[0, T]; X)$, and that $\{\dot{v}_n\}$ forms a bounded sequence in the Banach space $C^0([0, T]; L^2(\Omega))$. Hence, by duality (*cp.* [13]), we have

$$v_n \xrightarrow[*]{L^\infty([0,T];X)} v, \tag{3.17}$$

$$\dot{v}_n \xrightarrow[*]{L^\infty([0,T];L^2(\Omega))} \dot{v}, \tag{3.18}$$

and therefore in the sense of distributions, we obtain

$$A_- v_n \to A_- v. \tag{3.19}$$

Also, by trace theorems, we see that

$$v_n(T) \to v(T), \quad \dot{v}_n(T) \to \dot{v}(T), \quad v_n|_{\partial\Omega} \to v|_{\partial\Omega}. \tag{3.20}$$

On recalling Proposition (3.1), we deduce that $v \in D(A_-)$. Finally, we obtain from (3.10) and (3.19) that $A_- v = w$, and the Proposition is proved.

Proposition 3.3. *The operator A is closed on $L^2(Q)$.*

Proof. We utilise the results of the previous two Propositions. Thus, let $v_n \in D(A)$ and let

$$v_n \xrightarrow{L^2(Q)} v, \tag{3.21}$$

$$A v_n \xrightarrow{L^2(Q)} w. \tag{3.22}$$

We again must show that the limit function $v \in D(A)$ and that $Av = w$.

Definition 2.2 enables us to write:

$$Av_n = A_+P_+v_n + A_-P_-v_n. \tag{3.23}$$

As already noted after Lemma 3.1, we have $P_\pm v_n \in H_\pm \Rightarrow P_\pm v \in H_\pm$. Hence, on appealing to Lemma 3.1 together with the fact that $A_\pm$ are closed operators, we obtain

$$A_\pm P_\pm v_n \xrightarrow{L^2(Q)} A_\pm P_\pm v, \qquad P_\pm v \in D(A_\pm), \tag{3.24}$$

where, as usual, the upper or lower sign is consistently taken throughout Thus, on using (3.24) in (3.23), we see that

$$\begin{aligned} Av_n &\xrightarrow{L^2(Q)} A_+P_+v + A_-P_-v \\ &= Av, \end{aligned}$$

where by Definition 2.2, $v \in D(A)$. It therefore follows that $Av = w$, and the proof is complete.

4. The weak solution and its existence

After having defined the operator A and established that it is closed, we next turn our attention to the primary objective of the paper, namely the existence of a weak solution to the Cauchy problem (2.1)–(2.4) (or equivalently, (2.5)–(2.3)). We first define the weak solution in terms of the operator A, then discuss briefly the questions of non-existence and uniqueness, before finally stating and proving the basic theorem of existence. However, in view of the results already established the proof of existence is relatively short.

Definition 4.1. A weak solution to the Cauchy problem (2.5)–(2.3) is given by the function $u: Q \to L^2(Q)$ which for prescribed $f \in L^2(Q)$ satisfies the equation

$$\langle u, Av\rangle_{L^2(Q)} = \langle f, v\rangle_{L^2(Q)}, \qquad \forall\, v \in D(A), \tag{4.1}$$

where A is the operator defined in Definition (2.2).

Since the right side of (4.1) for each $f \in L^2(Q)$, is a bounded linear functional of $v \in L^2(Q)$ we may define the adjoint operator A^* to A in the usual way and thus obtain the identity

$$\langle v, A^*u\rangle_{L^2(Q)} = \langle f, v\rangle_{L^2(Q)}, \qquad u \in D(A^*). \tag{4.2}$$

On recalling that the operator C is self-adjoint, it is easy to show that the operator A likewise enjoys this property, so that $D(A^*)$ may be defined according to the definition (2.25), except that in (2.21) and (2.23) the conditions

$$v(T) = \dot{v}(T) = 0 \tag{4.3}$$

are to be replaced by

$$u(0) = \dot{u}(0) = 0. \tag{4.4}$$

From (4.2), it immediately follows that the weak solution in a distributional sense satisfies the equation

$$A^*u = f, \qquad u \in D(A^*). \tag{4.5}$$

Remark 4.1. It has been shown by MARSDEN & HUGHES [15], (see also [20]) using estimates based on logarithmic convexity arguments (see [10]), that the non-homogeneous Cauchy problem corresponding to (2.5)–(2.8) is not well-posed, *i.e.*, the solution does not depend continuously in $L^2(\Omega)$ on initial data in $W^{1,2}(\Omega)\times L^2(\Omega)$. In view of the closed graph theorem, this result combined with the property that A is a closed operator, strongly suggests that a solution to (4.5) need not exist for arbitrary f. The theorem proved below indicates for what data $f\in L^2(Q)$ a solution is guaranteed.

Remark 4.2. The *uniqueness* of the weak solution to (4.5) has been proved by LEVINE [11] who used an argument based upon the Lagrange identity first developed for classical solutions by BRUN [2].

We now prove our main result:

Theorem 4.1. *A weak solution exists to equation* (4.5) *for a set of data f dense in* $L^2(Q)$.

Proof. We prove the theorem by establishing that $R(A^*)$ is dense in $L^2(Q)$.

Now, since the operator C is densely defined, we see from Remark 2.5 that the operator A inherits this property and hence the orthogonal complement in $L^2(Q)$ of $R(A^*)$ satisfies

$$R(A^*)^{\perp} = N(A^{**}).$$

But since A is a closed operator in $L^2(Q)$, we have

$$N(A^{**}) = N(A),$$

and by uniqueness (see Remark 4.2) it follows that $N(A) = 0$. Thus $R(A^*)^{\perp} = 0$ and the theorem is proved.

Remark 4.3. We may use the existence theorem to show that a weak solution depends Hölder continuously upon any data $\hat{f}\in L^2(Q)$. Let f be an element of the data set, dense in $L^2(Q)$, for which a weak solution to (4.5) exists. Then, on using the estimate established in [9], or using the weak solution defined by LEVINE [11] in combination with the continuous dependence results of [6] and [10], we know that for $t\in[0,T)$

$$\|u\|^2_{L^2(D)} \leqq k\,\|f\|^{2\delta}_{L^2(Q)},$$

where k is some (known) positive constant, $0<\delta\leqq 1$ and $D=\Omega\times[0,t]$. For given $\hat{f}$, we may choose f such that

$$\|f-\hat{f}\|_{L^2(Q)} \leqq \varepsilon,$$

where $\varepsilon > 0$ is arbitrary. Then

$$\|u\|^2_{L^2(D)} \leqq k(\varepsilon + \|\hat{f}\|_{L^2(Q)})^{2\delta}.$$

But ε is arbitrary and so

$$\|u\|^2_{L^2(D)} \leqq k \|\hat{f}\|^{2\delta}_{L^2(Q)}.$$

Hence the assertion is proved.

Acknowledgement. The author is grateful to Professor C. M. DAFERMOS for helpful discussions during which the method for the treatment of existence was developed. The author also wishes to express his gratitude to Professors J. M. BALL and L. E. PAYNE for further constructive advice in the preparation of the paper.

References

1. P. BATES, Solutions of nonlinear elliptic systems with meshed spectra, Nonlinear Analysis, Theory and Applications **4**, 1023–1030, 1979.
2. L. BRUN, Méthodes énergétiques dans les systèmes évolutifs linéaires. Deuxième Partie: Théorèmes d'unicité. J. de Méchanique **8**, 167–192, 1969.
3. D. COLTON & J. WIMP, The construction of solutions to the heat equation backward in time. Math. Methods in Appl. Sci. **1**, 32–39, 1979.
4. G. FICHERA, Linear elliptic differential systems and eigenvalue problems. Lecture Notes in Mathematics, No. 8, Springer-Verlag, Berlin Heidelberg New York 1965.
5. G. FICHERA, Existence theorems in elasticity. Handbuch der Physik, Vol. VIa/2, 347–389, Springer, Berlin Heidelberg New York 1974.
6. G. P. GALDI, R. J. KNOPS & S. RIONERO, Uniqueness and continuous dependence in the linear elastodynamic exterior and half-space problems, Math. Proc. Camb. Phil. Soc. **99**, 357–366, 1986.
7. R. J. KNOPS, Instability and the ill-posed Cauchy problem in elasticity, Mechanics of Solids, The Rodney Hill 60th Anniversary Volume, Edited by H. G. HOPKINS & M. J. SEWELL, Pergamon, Oxford 1981.
8. R. J. KNOPS, Existence of solutions to unstable problems in linear elastodynamics Stability in the Mechanics of Continua, (Ed. F. H. SCHROEDER), 281–289, IUTAM Symposium Nümbrecht (Germany) 1981, Springer, Berlin Heidelberg New York 1982.
9. R. J. KNOPS & L. E. PAYNE, Continuous data dependence for the equations of classical elastodynamics. Proc. Camb. Phil. Soc. **66**, 481–491, 1969.
10. R. J. KNOPS & L. E. PAYNE, Growth estimates for solutions of evolutionary equations in Hilbert space with applications in elastodynamics, Arch. Rational Mech. Anal., **41**, 363–398, 1971.
11. H. A. LEVINE, An equipartition of energy theorem for weak solutions of evolutionary equations in Hilbert space: The Lagrange identity method, J. Diff. Eqns. **24**, 197–210, 1977.
12. J.L. LIONS, Quelques méthodes de résolution des problèmes aux limites non linéaires, Dunod Gauthier-Villars, Paris 1969.
13. J.-L. LIONS & E. MAGENES, Problèmes aux limites non homogènes et applications. Dunod 1968–70.
14. W. LITTMAN, Remarks on the Dirichlet problem for general linear partial differential equations, Comm. Pur. Appl. Maths. **XI**, 145–151, 1958.

15. J. E. MARSDEN & T. J. R. HUGHES, Mathematical Foundations of Elasticity, Prentice-Hall Inc. 1983.
16. L. NIRENBERG, Linear Functional Analysis, N.Y.U. Lecture Notes 1975.
17. R. E. SCHOWALTER, Quasi-reversibility of first and second order parabolic evolution equations. Improperly Posed Boundary Value Problems. Research Notes in Mathematics, Vol. 1, 76–93. Pitman Publishing, London 1975.
18. A. H. TIHONOV & V. Y. ARSENIN, Solutions of ill-posed problems, Winston, 1977.
19. O. VEJVODA & M. ŠTĚDRY, Existence of classical periodic solutions of the wave equation. The relation of the number-theoretic character of the period and the geometric properties of solutions, Differential Equations **20**, 1243–1248, 1984.
20. N. S. WILKES, On the non-existence of semi-groups for some equations of continuum mechanics, Proc. Roy. Soc. Edin. **A 86**, 303–306, 1980.
21. S. ZAIDMAN, Some remarks concerning regularity of solutions for abstract differential equations, Rend. Sem. Math. Univ. Padova **62**, 47–64, 1982.

Department of Mathematics
Heriot-Watt University
Edinburgh

(Received July 29, 1986)

Publications of Walter Noll

Books

I. *The Non-Linear Field Theories of Mechanics*, *Encyclopedia of Physics*, Volume III/3, 602 pages, Berlin-Heidelberg-New York, Springer-Verlag, 1965 (co-author C. TRUESDELL)

II. *Viscometric Flows of Non-Newtonian Fluids*, *Theory and Experiment*, Springer Tracts in Natural Philosophy, Volume 5, 130 pages, Berlin-Heidelberg-New York, Springer-Verlag, 1966 (co-authors B. D. COLEMAN & H. MARKOVITZ)

III. *The Foundations of Mechanics and Thermodynamics*, *Selected Papers*, with a preface by C. TRUESDELL, Berlin-Heidelberg-New York, 1974. (This volume contains corrected reprints of papers and notes nos. 8, 9, 10, 13, 14, 16, 20, 21, 22, 23, 24, 25, 28, 29, 35, 36.)

IV. *Finite-Dimensional Spaces*; *Algebra, Geometry, and Analysis*, The Hague, 1986, Martinus Nijhoff/Dr. W. Junk Publishers.

Papers and Notes

1. Eine Bemerkung zur Schwarzschen Ungleichheit, *Mathematische Nachrichten*, Volume 7, pp. 55–59 (1952) (co-author E. MOHR)
2. On the Continuity of the Solid and Fluid States, *Journal of Rational Mechanics and Analysis*, Volume 4, pp. 3–81 (1955)
3. Die Herleitung der Grundgleichungen der Thermomechanik der Kontinua aus der statistischen Mechanik, *Journal of Rational Mechanics and Analysis*, Volume 4, pp. 627–646 (1955)
4. Verschiebungsfunktionen für elastische Schwingungsprobleme, *Zeitschrift für angewandte Mathematik und Mechanik*, Volume 37, pp. 81–87 (1957)
5. On the Uniqueness and Non-existence of Stokes Flow, *Archive for Rational Mechanics and Analysis*, Volume 1, pp. 97–106 (1957) (co-author R. FINN)
6. On the Rotation of an Incompressible Continuous Medium in Plane Motion, *Quarterly of Applied Mathematics*, Volume 15, pp. 317–319 (1957)
7. On Exterior Boundary Value Problems in Linear Elasticity, *Archive for Rational Mechanics and Analysis*, Volume 2, pp. 191–196 (1958) (co-author R. J. DUFFIN)
8. A Mathematical Theory of the Mechanical Behavior of Continuous Media, *Archive for Rational Mechanics and Analysis*, Volume 2, pp. 197–226 (1958)

9. The Foundations of Classical Mechanics in the Light of Recent Advances in Continuum Mechanics, pp. 266–281 of *The Axiomatic Method, with Special Reference to Geometry and Physics* (Symposium at Berkeley, 1957), Amsterdam, North-Holland Publishing Co., 1959
10. On Certain Steady Flows of General Fluids, *Archive for Rational Mechanics and Analysis*, Volume 3, pp. 289–303 (1959) (co-author B. D. COLEMAN)
11. Helical Flow of General Fluids, *Journal of Applied Physics*, Volume 30, pp. 1508–1512 (1959) (co-author B. D. COLEMAN)
12. Conditions for Equilibrium at Negative Absolute Temperatures, *The Physical Review*, Volume 151, pp. 262–265 (1959) (co-author B. D. COLEMAN)
13. On the Thermostatics of Continuous Media, *Archive for Rational Mechanics and Analysis*, Volume 4, pp. 97–128 (1959) (co-author B. D. COLEMAN)
14. An Approximation Theorem for Functionals with Applications in Continuum Mechanics, *Archive for Rational Mechanics and Analysis*, Volume 6, pp. 355–370 (1960) (co-author B. D. COLEMAN)
15. Recent Results in the Continuum Theory of Viscoelastic Fluids, *Annals of the New York Academy of Science*, Volume 89, pp. 672–714 (1961) (co-author B. D. COLEMAN)
16. Foundations of Linear Viscoelasticity, *Reviews of Modern Physics*, Volume 33, pp. 239–249 (1961) (co-author B. D. COLEMAN)
17. Normal Stresses in Second-order Viscoelasticity, *Transactions of the Society of Rheology*, Volume 5, pp. 41–46 (1961) (co-author B. D. COLEMAN)
18. Steady Extension of Incompressible Simple Fluids, *The Physics of Fluids*, Volume 5, pp. 840–843 (1962) (co-author B. D. COLEMAN)
19. Simple Fluids with Fading Memory, pp. 530–552 of *Second-Order Effects in Elasticity, Plasticity, and Fluid Dynamics* (Symposium at Haifa, 1962), Oxford etc., Pergamon Press, 1964 (co-author B. D. COLEMAN)
20. Motions with Constant Stretch History, *Archive for Rational Mechanics and Analysis*, Volume 11, pp. 97–105 (1962)
21. La Mécanique Classique, Basée sur un Axiome d'Objectivé, pp. 47–56 of *La Méthode Axiomatique dans les Mécaniques Classiques et Nouvelles* (Colloque International, Paris, 1959), Paris, Gauthier-Villars, 1963
22. The Thermodynamics of Elastic Materials with Heat Conduction and Viscosity, *Archive for Rational Mechanics and Analysis*, Volume 13, pp. 167–178 (1963) (co-author B. D. COLEMAN)
23. Material Symmetry and Thermostatic Inequalities in Finite Elastic Deformations, *Archive for Rational Mechanics and Analysis*, Volume 15, pp. 87–111 (1964) (co-author B. D. COLEMAN)
24. Euclidean Geometry and Minkowskian Chronometry, *American Mathematical Monthly*, Volume 71, pp. 129–144 (1964)
25. Proof of the Maximality of the Orthogonal Group in the Unimodular Group, *Archive for Rational Mechanics and Analysis*, Volume 18, pp. 100–102 (1965)
26. The Equations of Finite Elasticity, pp. 93–101 of *Symposium on Applications of Non-linear Partial Differential Equations in Mathematical Physics* (1964), Providence, American Mathematical Society, 1965
27. The Foundations of Mechanics, pp. 159–200 of *Non-Linear Continuum Theories* (C.I.M.E. Lectures, 1965), Rome, Cremonesi, 1966
28. Space-Time Structures in Classical Mechanics, pp. 28–34 of *Delaware Seminar in the Foundations of Physics*, Berlin-Heidelberg-New York, Springer-Verlag, 1967
29. Materially Uniform Simple Bodies with Inhomogeneities, *Archive for Rational Mechanics and Analysis*, Volume 27, pp. 1–32 (1967)

30. Inhomogeneities in Materially Uniform Simple Bodies, pp. 239–246 of *IUTAM Symposium on Mechanics of Generalized Continua* (1967), Berlin-Heidelberg-New York, Springer-Verlag, 1968
31. Quasi-invertibility in a Staircase Diagram, *Proceedings of the American Mathematical Society*, Volume 23, pp. 1–4 (1969)
32. Representations of Certain Isotropic Tensor Functions, *Archiv der Mathematik*, Volume 21, pp. 87–90 (1970)
33. On certain Convex Sets of Measures and on Phases of Reacting Mixtures, *Archive for Rational Mechanics and Analysis*, Volume 38, pp. 1–12 (1970)
34. Annihilators of Linear Differential Operators, *Journal d'Analyse*, Volume 24, pp. 205–284 (1971) (co-author H. D. DOMBROWSKI)
35. A New Mathematical Theory of Simple Materials, *Archive for Rational Mechanics and Analysis*, Volume 48, pp. 1–50 (1972)
36. Lectures on the Foundations of Continuum Mechanics and Thermodynamics, *Archive for Rational Mechanics and Analysis*, Volume 52, pp. 62–92 (1973)
37. The Role of the Second Law of Thermodynamics in Classical Continuum Physics, pp. 117–119 of *Modern Developments in Thermodynamics*, Jerusalem, 1974, Israel University Press
38. On the concept of the symmetry group of a physical system, pp. 83–99 of *Proceedings of the Symposium on Symmetry, Similarity, and Group Theoretic Methods in Mechanics*, Calgary, Canada, 1974
39. The representation of monotonous processes by exponentials, *Indiana University Mathematics Journal*, Volume 25, pp. 209–214 (1974)
40. Orders, Gauge, and Distance in Faceless Linear Cones; with Examples relevant to Continuum Mechanics and Relativity, *Archive for Rational Mechanics and Analysis*, Volume 66, pp. 345–377 (1977) (co-author J. J. SCHÄFFER)
41. A General Framework for Problems in the Statics of Finite Elasticity, pp. 363–387 of *Contemporary Developments in Continuum Mechanics and Partial Differential Equations*, North-Holland Mathematics Studies, Volume 30, 1978
42. Order-isomorphisms in Affine Spaces, *Annali di Matematica Pura ed Applicata*, Volume 117, pp. 243–262 (1978) (co-author J. J. SCHÄFFER)

Several of the above-listed papers have been reprinted in collective volumes other than NOLL's *Foundations*, as follows:

Nos. 2, 8, 10, 11, and 14 in
Continuum Mechanics II, The Rational Mechanics of Materials, International Science Review Series, Volume VIII, Part 2, New York *etc.*, Gordon & Breach, 1965.

Nos. 13 and 16 in
Continuum Mechanics III, Foundations of Elasticity Theory, Interational Science Review Series, Volume VIII, Part 3, New York *etc.*, Gordon & Breach, 1965.

Nos. 8, 29, and 30 in
Continuum Theory of Inhomogeneities in Simple Bodies, New York, Springer-Verlag, 1968.

Places of First Publication

of the articles reprinted above, all in
Archive for Rational Mechanics and Analysis